Stochastic Differential Equations and Applications

Stochastic Differential Equations and Applications

TWO VOLUMES BOUND AS ONE

Avner Friedman
THE OHIO STATE UNIVERSITY

Dover Publications, Inc.
Mineola, New York

Copyright

Bibliographical Note

This Dover edition, first published in 2006, is an unabridged republication, in one volume, of the work originally published in two volumes by Academic Press, Inc., New York, in 1975 and 1976, respectfully. The Table of Contents have been combined, and the Index originally appearing at the end of Volume 1 has been moved to page 527.

Library of Congress Cataloging-in-Publication Data

Friedman, Avner.
Stochastic differential equations and applications / Avner Friedman.
p. cm.
Originally published (in 2 v.) : New York : Academic Press, 1975–1976, in series: Probability and mathematical statistics series ; v. 28.
"Two volumes bound as one."
Includes index.
ISBN-13: 978-0-486-45359-0
ISBN-10: 0-486-45359-6
1. Stochastic differential equations. I. Title.

QA274.23.F74 2006
519.2—dc22

2006047226

Manufactured in the United States by LSC Communications
45359607 2020
www.doverpublications.com

Contents

Volume 1

5 Stochastic Differential Equations

6 Elliptic and Parabolic Partial Differential Equations and Their Relations to Stochastic Differential Equations

7 The Cameron–Martin–Girsanov Theorem

8 Asymptotic Estimates for Solutions

9 Recurrent and Transient Solutions

Volume 2

10 Auxiliary Results in Partial Differential Equations

11 Nonattainability

12 Stability and Spiraling of Solutions

13 The Dirichlet Problem for Degenerate Elliptic Equations

14 Small Random Perturbations of Dynamical Systems

15 Fundamental Solutions for Degenerate Parabolic Equations

16 Stopping Time Problems and Stochastic Games

Part I. The Stationary Case

Part II. The Nonstationary Case

To the memory of my mother,
Hanna Friedman

Stochastic Differential Equations and Applications

Volume 1

Preface–Volume 1

The object of this book is to develop the theory of systems of stochastic differential equations and then give applications in probability, partial differential equations and stochastic control problems.

In Volume 1 we develop the basic theory of stochastic differential equations and give a few selected topics. Volume 2 will be devoted entirely to applications.

Chapters 1–5 form the basic theory of stochastic differential equations. The material can be found in one form or another also in other texts. Chapter 6 gives connections between solutions of partial differential equations and stochastic differential equations. The material in partial differential equations is essentially self-contained; for some of the proofs the reader is referred to an appropriate text.

In Chapter 7 Girsanov's formula is established. This formula is becoming increasingly useful in the theory of stochastic control.

In Chapters 8 and 9 we study the behavior of sample paths of the solution of a stochastic differential system, as time increases to infinity.

The book is written in a textlike style, namely, the material is essentially self-contained and problems are given at the end of each chapter. The only prerequisite is elementary probability; more specifically, the reader is assumed to be familiar with concepts such as conditional expectation, independence, and with elementary facts such as the Borel–Cantelli lemma. This prerequisite material can be found in any probability text, such as Breiman [1; Chapters 3,4].

I would like to thank my colleague, Mark Pinsky, for some helpful conversations.

General Notation

All functions are real valued, unless otherwise explicity stated.

In Chapter n, Section m the formulas and theorems are indexed by $(m.k)$ and $m.k$ respectively. When in Chapter l, we refer to such a formula $(m.k)$ (or Theorem $m.k$), we designate it by $(n.m.k)$ (or Theorem $n.m.k$) if $l \neq n$, and by $(m.k)$ (or Theorem $m.k$) if $l = n$.

Similarly, when referring to Section m in the same chapter, we designate the section by m; when referring to Section m of another chapter, say n, we designate the section by $n.m$.

Finally, when we refer to conditions (A), (A_1), (B) etc., these conditions are usually stated earlier in the *same* chapter.

1

Stochastic Processes

1. The Kolmogorov construction of a stochastic process

We write a.s. for almost surely, a.e. for almost everywhere, and a.a. for almost all.

We shall use the following notation:

R^n is the real euclidean n-dimensional space;

$\mathfrak{B}_n$ is the Borel σ-field of R^n, i.e., the smallest σ-field generated by the open sets of R^n;

R^∞ is the space consisting of all infinite sequences $(x_1, x_2, \ldots, x_n, \ldots)$ of real numbers;

$\mathfrak{B}_\infty$ is the smallest σ-field of subsets of R^∞ containing all k-dimensional rectangles, i.e., all sets of the form

$$\{(x_1, x_2, \ldots); x_1 \in I_1, \ldots, x_k \in I_k\}, \qquad k > 0,$$

where $I_1, \ldots, I_k$ are intervals.

Clearly $\mathfrak{B}_\infty$ is also the smallest σ-field containing all sets

$$\{(x_1, x_2, \ldots); (x_1, \ldots, x_k) \in B\}, \qquad B \in \mathfrak{B}_k.$$

An *n-dimensional distribution function* $F_n(x_1, \ldots, x_n)$ is a real-valued function defined on R^n and satisfying:

(i) for any intervals $I_k = [a_k, b_k)$, $1 \leqslant k \leqslant n$,

$$\Delta_{I_1} \cdots \Delta_{I_n} F_n(x) \geqslant 0 \tag{1.1}$$

where

$$\Delta_{I_k} f(x) = f(x_1, \ldots, x_{k-1}, b_k, x_{k+1}, \ldots, x_n) - f(x_1, \ldots, x_{k-1}, a_k, x_{k+1}, \ldots, x_n);$$

(ii) if $x_j^{(k)} \uparrow x_j$ as $k \uparrow \infty$ $(1 \leqslant j \leqslant n)$, then

$$F_n(x_1^{(k)}, \ldots, x_n^{(k)}) \uparrow F_n(x_1, \ldots, x_n) \quad \text{as} \quad k \uparrow \infty;$$

We take $\Omega = R^I$, $\mathcal{F} = \mathcal{B}_I$. Given a sequence $T = \{t_j\}$ in I, denote by $\mathcal{B}_T$ the class of all sets

$$\{x(\cdot);\ (x(t_1), x(t_2), \ldots) \in D\}, \qquad D \in \mathcal{B}_\infty.$$

By Theorem 1.1 and the remark following it, there exists a unique probability measure P_T on $(\Omega, \mathcal{B}_T)$ such that

$$P_T\{x(\cdot);\ x(t_1) < x_1, \ldots, x(t_n) < x_n\} = F_{t_1, \ldots, t_n}(x_1, \ldots, x_n), \qquad n \geq 1. \tag{1.7}$$

Set $P(B) = P_T(B)$ if $B \in \mathcal{B}_T$. For this definition to make sense we must show that if $B \in \mathcal{B}_{T'}$ where $T' = \{t_j'\}$ then

$$P_T(B) = P_{T'}(B). \tag{1.8}$$

Setting $S = T \cup T'$ we have $T \subset S$, $\mathcal{B}_T \subset \mathcal{B}_S$. Since $P_T = P_S$ on all finite-dimensional rectangles with base in T, $P_T = P_S$ on any set of $\mathcal{B}_T$. Thus $P_T(B) = P_S(B)$. Similarly, $P_{T'}(B) = P_S(B)$, and (1.8) follows.

Having defined the probability space $(\Omega, \mathcal{F}, P)$, we now take

$$X(t, x(\cdot)) = x(t);$$

(1.4) then immediately follows.

We refer to the stochastic process constructed in this proof as the process obtained by the *Kolmogorov construction*.

Definition. A *ν-dimensional countable stochastic process* (or a *ν-dimensional stochastic sequence*) is a sequence $\{X_n\}$ of ν-dimensional random variables. A family $\{X(t), t \in I\}$ of ν-dimensional random variables is called a *ν-dimensional stochastic process*.

A ν-dimensional stochastic sequence $\{X_n\}$ gives rise to functions

$$F_n(\bar{x}_1, \ldots, \bar{x}_n) = P(X_1 < \bar{x}_1, \ldots, X_n < \bar{x}_n) \tag{1.9}$$

where $\bar{x}_i \in R^\nu$ and $a < b$ means that $a_j < b_j$ for $1 \leq j \leq \nu$, where $a = (a_1, \ldots, a_\nu)$, $b = (b_1, \ldots, b_\nu)$. Setting

$$G_{(n-1)\nu+k}(x_1, \ldots, x_{(n-1)\nu+k}) = F_n(\bar{x}_1, \ldots, \bar{x}_{n-1}, \bar{x}_n^*) \qquad (1 \leq k \leq \nu) \tag{1.10}$$

where

$$\bar{x}_i = (x_{(i-1)\nu+1}, x_{(i-1)\nu+2}, \ldots, x_{i\nu}) \qquad \text{if } 1 \leq i \leq n-1,$$

$$\bar{x}_n^* = (x_{(n-1)\nu+1}, \ldots, x_{(n-1)\nu+k}, \infty, \ldots, \infty),$$

it is clear that the sequence $\{G_i(x_1, \ldots, x_i)\}$ is a consistent sequence of distribution functions.

Definition. A function $F_n(\bar{x}_1, \ldots, \bar{x}_n)$ defined for all $\bar{x}_i \in R^\nu$ is called an *n-dimensional distribution function with respect to* R^ν if:

(i) condition (1.1) holds with a_k, b_k replaced by $\bar{a}_k = (a_{k1}, \ldots, a_{k\nu})$, $\bar{b}_k = (b_{k1}, \ldots, b_{k\nu})$, where $\bar{a}_k < \bar{b}_k$, and x_j replaced by $\bar{x}_j$ in the definition of $\Delta_{I_k} f$;

(ii) $F_n(\bar{x}_1^{(k)}, \ldots, \bar{x}_n^{(k)}) \uparrow F_n(\bar{x}_1, \ldots, \bar{x}_n)$ if $\bar{x}_i^{(k)} \uparrow \bar{x}_i$ for $1 \leqslant k \leqslant n$ (i.e., each component of $\bar{x}_i^{(k)}$ increases to the corresponding component of $\bar{x}_i$);

(iii) if some kth component of some $\bar{x}_i$ decreases to $-\infty$, then $F_n(\bar{x}_1, \ldots, \bar{x}_n) \downarrow 0$; and if all components of all the $\bar{x}_j$ increase to ∞, then $F_n(\bar{x}_1, \ldots, \bar{x}_n) \uparrow 1$.

Set

$$F_n(\bar{x}_1, \ldots, \bar{x}_{n-1}, \bar{x}_n^*) = \lim_{k \to \infty} F_n(\bar{x}_1, \ldots, \bar{x}_{n-1}, \bar{x}_n^k) \tag{1.11}$$

where the first k components of $\bar{x}_n^*$, $\bar{x}_n^k$ agree, whereas the last $\nu - k$ components of $\bar{x}_n^*$ and $\bar{x}_n^k$ are equal, respectively, to ∞ and k. Then, obviously, the set of distribution functions $G_{(n-1)\nu+1}, \ldots, G_{n\nu}$ defined by (1.10) is consistent.

A sequence of distributions $\{F_n\}$ with respect to R^ν is said to be *consistent* if

$$\lim_{\bar{x}_n \uparrow \infty} F_n(\bar{x}_1, \ldots, \bar{x}_n) = F_{n-1}(\bar{x}_1, \ldots, \bar{x}_{n-1}),$$

where $\bar{x}_n \uparrow \infty$ means that each component of $\bar{x}_n$ increases to ∞. This is clearly the case if and only if the corresponding sequence of distributions $G_i(x_1, \ldots, x_i)$, defined by (1.10), (1.11), is consistent.

If we apply Theorem 1.1 to the latter sequence, we conclude that for any consistent sequence of distribution functions, with respect to R^ν, there exists a countable ν-dimensional stochastic process $\{X_n\}$ whose distribution functions are the F_n, i.e., (1.9) holds. The probability space $(\Omega, \mathcal{F}, P)$ can be taken such that $\Omega = R^{\nu, \infty}$ = the space of all infinite sequences $(\bar{x}_1, \bar{x}_2, \ldots)$, $\bar{x}_n \in R^\nu$, and $\mathcal{F} = \mathcal{B}_{\nu, \infty}$ the smallest σ-field containing all finite-dimensional rectangles

$$\{(\bar{x}_1, \bar{x}_2, \ldots); \bar{x}_1 \in \bar{I}_1, \bar{x}_2 \in \bar{I}_2, \ldots\}$$

where $\bar{I}_j$ is an interval in R^ν.

A set of distribution functions $\{F_{t_1 \cdots t_n}(\bar{x}_1, \ldots, \bar{x}_n)\}$ where $t_1 < \cdots < t_n$, $t_i \in I$, $n = 1, 2, \ldots,$ is said to be *consistent* if

$$\lim_{\bar{x}_k \uparrow \infty} F_{t_1 \cdots t_n}(\bar{x}_1, \ldots, \bar{x}_n) = F_{t_1 \cdots t_{k-1} t_{k+1} \cdots t_n}(\bar{x}_1, \ldots, \bar{x}_{k-1}, \bar{x}_{k+1}, \ldots, \bar{x}_n).$$

Theorem 1.3. *Given a consistent set of distribution functions with respect to* R^ν, $\{F_{t_1 \cdots t_n}(\bar{x}_1, \ldots, \bar{x}_n);\ t_1 < \cdots < t_n,\ t_i \in I\}$, *there exists* *probability space* $(\Omega, \mathcal{F}, P)$ *and a* ν*-dimensional stochastic process* $X(t)$ $(t \in I)$ *defined on it such that*

$$P[X(t_1) < \bar{x}_1, \ldots, X(t_n) < \bar{x}_n] = F_{t_1 \cdots t_n}(\bar{x}_1, \ldots, \bar{x}_n). \tag{1.12}$$

The functions $F_{t_1 \cdots t_n}$ are called the finite-dimensional joint distribution functions (with respect to R^ν) of the process $X(t)$.

The proof is similar to the proof of Theorem 1.2; it is based on the extension of Theorem 1.1 to distribution functions with respect to R^ν, mentioned above. Here Ω is the space $R^{\nu, I}$ of all functions $x(\cdot)$ from I into R^ν and $\mathcal{F} = \mathcal{B}_{\nu, I}$ consists of all sets (1.6) with $D \in \mathcal{B}_{\nu, \infty}$. The construction of the process $X(t)$ in this proof is called the *Kolmogorov construction*.

2. Separable and continuous processes

Definition. A ν-dimensional stochastic process $\{X(t), t \in I\}$ is called *separable* if there exists a countable sequence $T = \{t_j\}$ that is a dense subset of I and a subset N of Ω with $P(N) = 0$ such that, if $\omega \notin N$,

$$\{X(t, \omega) \in F \text{ for all } t \in J\} = \{X(t_j, \omega) \in F \text{ for all } t_j \in J\}$$

for any open subset J of I and for any closed subset F of R^ν. T is called a *set of separability*, or briefly, a *separant*.

Two ν-dimensional stochastic processes $\{X(t)\}$ and $\{X'(t)\}$ $(t \in I)$ defined on the same probability space are said to be *stochastically equivalent* if

$$P[X(t) \neq X'(t)] = 0 \qquad \text{for all} \quad t \in I.$$

We then say that $\{X'(t)\}$ is a *version* of $\{X(t)\}$.

Theorem 2.1. *Any* ν*-dimensional process is stochastically equivalent to a separable stochastic process.*

The random variables of the separable process may actually take the values $\pm\infty$, even though the random variables of the original process are real valued.

For proof the reader is referred to Doob [1]. Theorem 2.1 will not be used in the text of this book.

From Theorem 2.1 it follows that when one constructs a stochastic process from a given set of distribution functions (as in the Kolmogorov construction), one may always take this process to be separable.

Let $X(t)$ be a ν-dimensional stochastic process for $t \in I$. Thus, for each t, $X(t)$ is a random variable $X(t, \omega)$. The functions $t \to X(t, \omega)$ (ω fixed) are called the *sample paths* of the process. If for a.a. ω the sample paths are continuous functions for all $t \in I$, then we say that the stochastic process is *continuous*. If for a.a. ω the sample paths are right (left) continuous, then we say that the stochastic process is *right (left) continuous*.

It is easy to see that a right (or left) continuous stochastic process is separable and any dense sequence in I is a set of separability.

A stochastic process $\{X(t), t \in I\}$ is said to be *continuous in probability* if for any $s \in I$ and $\epsilon > 0$,

$$P[|X(t) - X(s)| > \epsilon] \to 0 \quad \text{if} \quad t \in I, \quad t \to s.$$

The following theorem is due to Kolmogorov.

Theorem 2.2. *Let $X(t)$ $(t \in I)$ be a ν-dimensional stochastic process such that*

$$E|X(t) - X(s)|^\beta \leqslant C|t - s|^{1+\alpha} \tag{2.1}$$

for some positive constants C, α, β. Then there is a version of $X(t)$ that is a continuous process.

Proof. Take for simplicity $I = [0, 1]$. Let $0 < \gamma < \alpha/\beta$ and let δ be a positive number such that

$$(1 - \delta)(\alpha + 1 - \beta\gamma) > 1 + \delta. \tag{2.2}$$

Then, by (2.1),

$$\begin{aligned} &P\{|X(j2^{-n}) - X(i2^{-n})| > [(j - i)2^{-n}]^\gamma \\ &\qquad\qquad \text{for some } 0 \leqslant i \leqslant j \leqslant 2^n, j - i < 2^{n\delta}\} \\ &\leqslant \sum_{\substack{0 \leqslant i \leqslant j < 2^n \\ j - i \leqslant 2^{n\delta}}} [(j - i)2^{-n}]^{-\beta\gamma} E|X(j2^{-n}) - X(i2^{-n})|^\beta \\ &\leqslant C_1 \sum_{\substack{0 \leqslant i \leqslant j < 2^n \\ j - i \leqslant 2^{n\delta}}} [(j - i)2^{-n}]^{1+\alpha-\beta\gamma} \leqslant C_2 2^{n[1+\delta-(1-\delta)(1+\alpha-\beta\gamma)]} \\ &= C_2 2^{-n\mu} \end{aligned}$$

where $\mu > 0$ by (2.2), and C_1, C_2 are positive constants. Since $\Sigma 2^{-n\mu} < \infty$, the Borel–Cantelli lemma implies that for a.a. ω

$$\begin{aligned} |X(j2^{-n}) - X(i2^{-n})| \leqslant h((j - i)2^{-n}) \quad &\text{if} \quad 0 \leqslant i \leqslant j \leqslant 2^n, \\ &j - i \leqslant 2^{n\delta}, \quad n \geqslant m(\omega) \end{aligned} \tag{2.3}$$

where $h(t) = t^\gamma$ and $m(\omega)$ is some sufficiently large positive integer.

Let t_1, t_2 be any rational numbers in the interval $(0, 1)$ such that $t_1 < t_2$,

$t = t_2 - t_1 < 2^{-m(1-\delta)}$ where $m = m(\omega)$ is as in (2.3). Choose $n > m$ such that

$$2^{-(n+1)(1-\delta)} \leq t < 2^{-n(1-\delta)}.$$

We can expand t_1, t_2 in the form

$$t_1 = i2^{-n} - 2^{-p_1} - \cdots - 2^{-p_k},$$

$$t_2 = j2^{-n} + 2^{-q_1} + \cdots + 2^{-q_l}$$

where $n < p_1 < \cdots < p_k$, $n < q_1 < \cdots < q_l$. Then $t_1 \leq i2^{-n} \leq j2^{-n} \leq t_2$ and $j - i \leq t2^n < 2^{n\delta}$.

By (2.3),

$$|X(i2^{-n} - 2^{-p_1} - \cdots - 2^{-p_r}) - X(i2^{-n} - 2^{-p_1} - \cdots - 2^{-p_{r-1}})| \leq h(2^{-p_r}).$$

Hence

$$|X(t_1) - X(i2^{-n})| \leq \sum_{r=1}^{k} h(2^{-p_r}) \leq C_3 h(2^{-n}) \qquad (C_3 \text{ constant}).$$

Similarly,

$$|X(t_2) - X(j2^{-n})| \leq C_3 h(2^{-n}).$$

Finally, by (2.3),

$$|X(j2^{-n}) - X(i2^{-n})| \leq h((j-i)2^{-n}) \leq h(t).$$

Hence

$$|X(t_2) - X(t_1)| \leq Ch(t_2 - t_1) \qquad (C \text{ constant}). \tag{2.4}$$

Let $T = \{t_j\}$ be the sequence of all rational numbers in the interval (0, 1). From (2.4) it follows that

$$|X(t_j) - X(t_i)| \leq Ch(t_j - t_i) \quad \text{if} \quad 0 \leq t_j - t_i \leq 2^{-m(1-\delta)}, \quad m = m(\omega). \tag{2.5}$$

Thus $X(t, \omega)$, when restricted to $t \in T$ is uniformly continuous. Let $\tilde{X}(t, \omega)$ be its unique continuous extension to $0 \leq t \leq 1$. Then the process $\tilde{X}(t)$ is continuous. We shall complete the proof of Theorem 2.2 by showing that $\tilde{X}(t)$ is a version of $X(t)$.

Let $t^* \in [0, 1]$ and let $t_{j'} \in T$, $t_{j'} \to t^*$ if $j' \to \infty$. Since $\tilde{X}(t)$ is a continuous process,

$$\tilde{X}(t_{j'}, \omega) \to \tilde{X}(t^*, \omega) \quad \text{a.s.} \qquad \text{as} \quad j' \to \infty.$$

From (2.1) it also follows that

$$X(t_{j'}, \omega) \to X(t^*, \omega) \quad \text{in probability,} \qquad \text{as} \quad j' \to \infty.$$

Since $\tilde{X}(t_{j'}, \omega) = X(t_{j'}, \omega)$ a.s., we get $\tilde{X}(t^*, \omega) = X(t^*, \omega)$ a.s.

From (2.5) we see that

$$|\tilde{X}(t) - \tilde{X}(s)| \leq Ch(t-s) = C(t-s)^{\gamma}$$

if $t-s<2^{-m(1-\delta)}$, $m=m(\omega)$. Thus:

Corollary 2.3. *Let $X(t)$ $(t\in I)$ be a ν-dimensional process satisfying (2.1). Then for any bounded subinterval $J\subset I$ and for any $\epsilon>0$, a.a. sample paths of the continuous version of $X(t)$ satisfy a Hölder condition with exponent $(\alpha/\beta)-\epsilon$ on J.*

3. Martingales and stopping times

Definition. A stochastic sequence $\{X_n\}$ is called a *martingale* if $E|X_n|<\infty$ for all n, and

$$E(X_{n+1}|X_1, \ldots, X_n) = X_n \quad \text{a.s.} \qquad (n = 1, 2, \ldots).$$

It is a *submartingale* if $E|X_n| < \infty$ for all n, and

$$E(X_{n+1}|X_1, \ldots, X_n) \geq X_n \quad \text{a.s.} \qquad (n = 1, 2, \ldots).$$

Theorem 3.1. *If $\{X_n\}$ is a submartingale, then*

$$P(\max_{1\leq k\leq n} X_k > \lambda) \leq \frac{1}{\lambda} EX_n^+ \qquad \text{for any } \lambda > 0, \quad n \geq 1. \tag{3.1}$$

This inequality is called the *martingale inequality*.

Proof. Let

$$A_k = \bigcap_{j<k} (X_j < \lambda)\cap(X_k > \lambda) \qquad (1 \leq k \leq n),$$

$$A = (\max_{1\leq k\leq n} X_k > \lambda).$$

Then A is a disjoint union $\bigcup_{k=1}^{n} A_k$. Therefore

$$\lambda P(A) = \sum_{k=1}^{n} \lambda P(A_k) \leq \sum_{k=1}^{n} E(X_k\chi_{A_k}) \tag{3.2}$$

where χ_B is the indicator function of a set B. Since A_k belongs to the σ-field of $X_1, \ldots, X_k$,

$$EX_n^+ \geq \sum_{k=1}^{n} E(X_n^+ \chi_{A_k}) = \sum_{k=1}^{n} EE(X_n^+ \chi_{A_k} | X_1, \ldots, X_k)$$
$$= \sum_{k=1}^{n} E\chi_{A_k} E(X_n^+ | X_1, \ldots, X_k)$$
$$\geq \sum_{k=1}^{n} E\chi_{A_k} E(X_n | X_1, \ldots, X_k) \geq \sum_{k=1}^{n} E(\chi_{A_k} X_k).$$

Comparing this with (3.2), the assertion (3.1) follows.

Corollary 3.2. *If $\{X_n\}$ is a martingale and $E|X_n|^\alpha < \infty$ for some $\alpha \geq 1$ and all $n \geq 1$, then*

$$P\Big(\max_{1 \leq k \leq n} |X_k| > \lambda\Big) \leq \frac{1}{\lambda^\alpha} E|X_n|^\alpha \qquad \textit{for any} \quad \lambda > 0, \quad n \geq 1.$$

This follows from Theorem 3.1 and the fact that $\{|X_n|^\alpha\}$ is a submartingale (see Problem 8).

Definition. A stochastic process $\{X(t), t \in I\}$ is called a *martingale* if $E|X(t)| < \infty$ for all $t \in I$, and

$$E\{X(t)|X(\tau), \tau \leq s, \tau \in I\} = X(s) \qquad \text{for all} \quad t \in I, \quad s \in I, \quad s \leq t.$$

It is called a *submartingale* if $E|X(t)| < \infty$ and

$$E\{X(t)|X(\tau), \tau \leq s, \tau \in I\} \geq X(s) \qquad \text{for all} \quad t \in I, \quad s \in I, \quad s \leq t.$$

Corollary 3.3. (i) *If $\{X(t)\}$ $(t \in I)$ is a separable submartingale, then*

$$P\Big\{\max_{s \leq t, s \in I} X(s) > \lambda\Big\} \leq \frac{1}{\lambda} EX^+(t) \qquad \textit{for all} \quad \lambda > 0, \quad t \in I.$$

(ii) *If $\{X(t)\}$ $(t \in I)$ is a separable martingale and if $E|X(t)|^\alpha < \infty$ for some $\alpha \geq 1$ and all $t \in I$, then*

$$P\Big\{\max_{s \leq t, s \in I} X(s) > \lambda\Big\} \leq \frac{1}{\lambda^\alpha} E|X(t)|^\alpha \qquad \textit{for all} \quad \lambda > 0, \quad t \in I.$$

This follows by applying Theorem 3.1 and Corollary 3.2 with $X_j = X(s_j)$ if $1 \leq j \leq n-1$, $X_j = X(t)$ if $j \geq n$, where $s_1 < \cdots < s_{n-1} < t$, taking the set $\{s_1, \ldots, s_{n-1}\}$ to increase to the set of all t_j with $t_j < t$, where $\{t_j\}$ is a set of separability (cf. Problem 1(b)).

We denote by $\mathcal{F}(X(\lambda), \lambda \in I)$ the σ-field generated by the random variables $X(\lambda)$, $\lambda \in I$.

Definition. Let $X(t)$, $\alpha \leqslant t \leqslant \beta$, be a ν-dimensional stochastic process. (If $\beta = \infty$, we take $\alpha \leqslant t < \infty$.) A (finite-valued) random variable τ is called a *stopping time* with respect to the process $X(t)$ if $\alpha \leqslant \tau \leqslant \beta$ and if, for any $\alpha \leqslant t \leqslant \beta$, $\{\tau \leqslant t\}$ belongs to $\mathcal{F}(X(\lambda), \alpha \leqslant \lambda \leqslant t)$. If $\beta = \infty$ and τ may also assume the value $+\infty$, then we call τ an *extended stopping time.*

Notice that the sets $\{\tau > t\}$, $\{s < \tau \leqslant t\}$ (where $s < t$) also belong to $\mathcal{F}(X(\lambda), \alpha \leqslant \lambda \leqslant t)$.

Lemma 3.4. *If τ is a stopping time with respect to a ν-dimensional process $X(t)$, $t \geqslant 0$, then there exists a sequence of stopping times τ_n such that*

(i) *τ_n has a discrete range,*
(ii) *$\tau_n \geqslant \tau$ everywhere,*
(iii) *$\tau_n \downarrow \tau$ everywhere as $n \uparrow \infty$.*

Proof. Define $\tau_n = 0$ if $\tau = 0$ and

$$\tau_n = \frac{i+1}{n} \qquad \text{if} \quad \frac{i}{n} < \tau \leqslant \frac{i+1}{n} \qquad (i = 1, 2, \dots).$$

Then clearly (i)–(iii) hold. Finally, τ_n is a stopping time since

$$\{\tau_n \leqslant t\} = \bigcup_{(i+1)/n \leqslant t} \left\{\tau_n = \frac{i+1}{n}\right\}$$

$$= \bigcup_{(i+1)/n \leqslant t} \left\{\frac{i}{n} < \tau \leqslant \frac{i+1}{n}\right\} \in \mathcal{F}(X(\lambda), \lambda \leqslant t).$$

If $X(t)$ is right continuous, then

$$X(t \wedge \tau_n) \to X(t \wedge \tau) \quad \text{a.s.} \qquad \text{as} \quad n \to \infty.$$

Since $X(t \wedge \tau_n)$ is obviously a random variable, the same is true of $X(t \wedge \tau)$. We shall prove:

Theorem 3.5. *If τ is a stopping time with respect to a right continuous martingale $X(t)$, $t \geqslant 0$, then the process $Y(t) = X(t \wedge \tau)$, $t \geqslant 0$, is also a right continuous martingale.*

Before giving the proof, we need some preliminaries.

Definition. A sequence of random variables X_n is said to be *uniformly integrable* if $E|X_n| < \infty$ for any $n \geqslant 1$, and

$$\overline{\lim_{n\to\infty}} \int_{|X_n| > \lambda} |X_n| \, dP \to 0 \qquad \text{as} \quad \lambda \to \infty. \tag{3.3}$$

Lemma 3.6. *If $\{X_n\}$ is uniformly integrable, then $\overline{\lim}\, E|X_n| < \infty$. If further $X_n \to X$ a.s. or in probability, then $E|X - X_n| \to 0$ and $EX_n \to EX$.*

Proof. For any $\lambda > 0$,

$$E|X_n| \leq \lambda + \int_{|X_n|>\lambda} |X_n|\, dP.$$

From this and (3.3) it follows that $E|X_n| \leq C$, where C is a constant independent of n. To prove the last assertion, notice, by Fatou's lemma, that if $X_n \to X$ a.s., then

$$\int |X|\, dP \leq \underline{\lim} \int |X_n|\, dP < \infty.$$

Next, by Egoroff's theorem, $X_n \to X$ almost uniformly, i.e., for any $\epsilon > 0$ there is an event A with $P(A) < \epsilon$ such that $X_n \to X$ uniformly on $\Omega \backslash A$. Hence

$$\overline{\lim_{n\to\infty}} \int |X_n - X|\, dP = \overline{\lim_{n\to\infty}} \int_A |X_n - X|\, dP \leq \overline{\lim_{n\to\infty}} \int_A |X_n|\, dP + \int_A |X|\, dP.$$

Now, by (3.3),

$$\int_A |X_n|\, dP \leq \int_{|X_n|>\lambda} |X_n|\, dP + \int_{(|X_n|<\lambda)\cap A} |X_n|\, dP \leq \eta(\lambda) + \lambda\epsilon,$$

where $\eta(\lambda) \to 0$ if $\lambda \to \infty$, and $\eta(\lambda)$ does not depend on n. By Fatou's lemma, the same inequality holds for X. Taking $\epsilon \downarrow 0$, we get

$$\overline{\lim_{n\to\infty}} \int |X_n - X|\, dP \leq 2\eta(\lambda) \to 0 \qquad \text{if} \quad \lambda \to \infty.$$

If $X_n \to X$ in probability, then any subsequence $X_{n'}$ of X_n has a subsubsequence $X_{n''}$ which converges a.s. to X. By what we have proved, $E|X_{n''} - X| \to 0$ if $n'' \to \infty$. But this implies that $E|X_n - X| \to 0$ as $n \to \infty$.

Proof of Theorem 3.5. Let $0 \leq t_1 < t_2 < \cdots < t_{m-1} < t_m = s < t$,

$$A = \bigcap_{i=1}^{m} [X(t_i \wedge \tau) \in B_i], \qquad B_i \text{ Borel sets in } R^1, \tag{3.4}$$

and let τ_n be the stopping times constructed in the proof of Lemma 3.4. Assume first that there is an integer j such that $s = (j+1)/n$. On the set $\tau \leq (j+1)/n$, $\tau_n \leq (j+1)/n$ so that

$$X(s \wedge \tau_n) = X(\tau_n) = X(t \wedge \tau_n).$$

Consequently

$$\int_{A\cap[\tau\leq(j+1)/n]} X(s \wedge \tau_n)\, dP = \int_{A\cap[\tau\leq(j+1)/n]} X(t \wedge \tau_n)\, dP. \tag{3.5}$$

Next,

$$\int_{A\cap[\tau>(j+1)/n]} X(t\wedge\tau_n)\,dP = \sum_{l=j+1}^{M}\int_{A\cap[l/n<\tau\le(l+1)/n]} X\left(\frac{l+1}{n}\right)dP + \int_{A\cap[\tau>(M+1)/n]} X(t)\,dP \tag{3.6}$$

where $(M+1)/n < t \le (M+2)/n$.

Notice that

$$A\cap\left[\tau>\frac{M+1}{n}\right]=\left[\bigcap_{i=1}^{m}(X(t_i)\in B_i)\right]\cap\left[\tau>\frac{M+1}{n}\right].$$

Since each set on the right belongs to $\mathcal{F}(X(\lambda), \lambda \le (M+1)/n)$, we can use the martingale property of $X(t)$ to deduce that

$$\int_{A\cap[\tau>(M+1)/n]} X(t)\,dP = \int_{A\cap[\tau>(M+1)/n]} X\left(\frac{M+1}{n}\right)dP.$$

Hence, from (3.6),

$$\int_{A\cap[\tau>(j+1)/n]} X(t\wedge\tau_n)\,dP = \sum_{l=j+1}^{M-1}\int_{A\cap[l/n<\tau\le(l+1)/n]} X\left(\frac{l+1}{n}\right)dP + \int_{A\cap[\tau>M/n]} X\left(\frac{M+1}{n}\right)dP.$$

Proceeding in this manner step by step, we arrive at

$$\int_{A\cap[\tau>(j+1)/n]} X(t\wedge\tau_n)\,dP = \int_{A\cap[\tau>(j+1)/n]} X\left(\frac{j+2}{n}\right)dP$$
$$= \int_{A\cap[\tau>(j+1)/n]} X(s)\,dP = \int_{A\cap[\tau>(j+1)/n]} X(s\wedge\tau_n)\,dP,$$

since $s = (j+1)/n$,

$$A\cap\left[\tau>\frac{j+1}{n}\right]=\left[\bigcap_{i=1}^{m}(X(t_i)\in B_i)\right]\cap\left[\tau>\frac{j+1}{n}\right]$$

is in $\mathcal{F}(X(\lambda), \lambda \le (j+1)/n)$, and $s\wedge\tau_n = s$ on this set.

Recalling (3.5) we conclude that

$$\int_A X(s\wedge\tau_n)\,dP = \int_A X(t\wedge\tau_n)\,dP. \tag{3.7}$$

So far we have assumed that the number s is in the range of τ_n. If this is not the case, i.e., if $j/n < s < (j+1)/n$ for some j, then we modify the definition of τ_n, by taking

$$\tau_n = \begin{cases} s & \text{if } \frac{j}{n} < \tau \leqslant s, \\ \frac{j+1}{n} & \text{if } s < \tau \leqslant \frac{j+1}{n}. \end{cases}$$

The assertion (3.7) then follows as before, and the τ_n satisfy the properties asserted in Lemma 3.4.

We shall now complete the proof of the theorem in case τ is bounded, i.e., $\tau \leqslant N < \infty$ (N constant).

The range $\{s_{in}\}$ of τ_n lies in $[0, N+1)$.

Notice now that $|X(t)|$ is a submartingale. If we apply the proof of (3.7) to $|X(t)|$, instead of $X(t)$, we obtain

$$\int_A |X(s \wedge \tau_n)|\, dP \leqslant \int_A |X(t \wedge \tau_n)|\, dP$$

whenever $t > s$. Taking, in particular,

$$\tau = \tau_n, \qquad A = \{|X(s \wedge \tau_n)| > \lambda\}, \qquad t = N + 1,$$

we get

$$\int_{|X(s\wedge\tau_n)|>\lambda} |X(s \wedge \tau_n)|\, dP \leqslant \int_{|X(s\wedge\tau_n)|>\lambda} |X(\tau_n)|\, dP.$$

But since $|X(t)|$ is a submartingale,

$$\int_{\substack{|X(s\wedge\tau_n)|>\lambda \\ \tau_n = s_{in}}} |X(\tau_n)|\, dP \leqslant \int_{\substack{|X(s\wedge\tau_n)|>\lambda \\ \tau_n = s_{in}}} |X(N+1)|\, dP.$$

Summing over i,

$$\int_{|X(s\wedge\tau_n)|>\lambda} |X(s \wedge \tau_n)|\, dP \leqslant \int_{|X(s\wedge\tau_n)|>\lambda} |X(N+1)|\, dP. \tag{3.8}$$

Now, since $X(t)$ is right continuous and τ is bounded, the random variable

$$\sup_n \sup_{s \geqslant 0} |X(s \wedge \tau_n)|$$

is bounded in probability. This implies that the right-hand side of (3.8) converges to 0 if $\lambda \to \infty$, uniformly with respect to n. Thus the sequence $\{X(s \wedge \tau_n)\}$ is uniformly integrable. Similarly, one shows that the sequence $\{X(t \wedge \tau_n)\}$ is uniformly integrable. By Lemma 3.6 it follows that $E|X(t \wedge \tau)| < \infty$.

If we now take $n \uparrow \infty$ in (3.7) and apply Lemma 3.6, we get

$$\int_A X(s \wedge \tau)\, dP = \int_A X(t \wedge \tau)\, dP. \tag{3.9}$$

Since this is true for any set A of the form (3.4), $\{X(t \wedge \tau)\}$ is a martingale.

We have thus completed the proof in case τ is bounded. If τ is unbounded, then $\tau \wedge N$ is a bounded stopping time. Hence $E|X(t \wedge \tau)| < \infty$ if $N > t$. Further, (3.9) holds with τ replaced by $\tau \wedge N$, and taking $N > t$ the relation (3.9) follows. This completes the proof.

Remark. Let $\{X(t), 0 \leqslant t \leqslant T\}$ be a right continuous process and let $\mathcal{F}_t$ $(0 \leqslant t \leqslant T)$ be an increasing family of σ-fields such that $\mathcal{F}(X(\lambda), 0 \leqslant \lambda \leqslant t)$ is a subset of $\mathcal{F}_t$. If $E|X(t)| < \infty$ and

$$E(X(t)|\mathcal{F}_s) = X(s) \qquad \text{for all} \quad 0 \leqslant s \leqslant t \leqslant T,$$

then we say that $X(t)$ is a *martingale with respect to* $\mathcal{F}_t$ $(0 \leqslant t \leqslant T)$. This clearly implies that $\{X(t), 0 \leqslant t \leqslant T\}$ is a martingale. The proof of Theorem 3.5 shows that if $X(t)$ is a martingale with respect to $\mathcal{F}_t$ $(0 \leqslant t \leqslant T)$, then the process $X(t \wedge \tau)$ is also a martingale with respect to $\mathcal{F}_t$; here τ is any random variable such that $\tau \geqslant 0$ and $\{\tau \leqslant t\} \in \mathcal{F}_t$ for all $0 \leqslant t \leqslant T$.

PROBLEMS

1. Let $X(t)$, $t \geqslant 0$ be a stochastic process and let $\{t_j\}$ be a set of separability. Prove that for a.a. ω:

(a)
$$\lim_{n\to\infty} \inf_{|t_j - t| < 1/n} X(t_j, \omega) \leqslant X(t, \omega) \leqslant \lim_{n\to\infty} \sup_{|t_j - t| < 1/n} X(t_j, \omega) \qquad (t > 0).$$

(b)
$$\sup_{a<t<b} X(t, \omega) = \sup_{a<t_j<b} X(t_j, \omega), \quad \inf_{a<t<b} X(t, \omega) = \inf_{a<t_j<b} X(t_j, \omega).$$

(Notice that, as a consequence, $\sup_{a<t<b} X(t)$ and $\inf_{a<t<b} X(t)$ are finite- or infinite-valued random variables.)

(c) $\sup_{a\leqslant t\leqslant b}|X(t)|$, $\limsup_{t\to s} X(t)$, and $\liminf_{t\to s} X(t)$ are finite- or infinite-valued random variables.

(d) For any positive numbers ϵ, δ,

$$\sup_{|t_i - t_j| < \delta} |X(t_i, \omega) - X(t_j, \omega)| \leqslant \epsilon$$

if and only if

$$\sup_{|t-s| < \delta} |X(t, \omega) - X(s, \omega)| \leqslant \epsilon.$$

2. Let $X(t)$, $X'(t)$ $(t \geqslant 0)$ be ν-dimensional stochastic processes in $(\Omega, \mathcal{F}, P)$ and $(\Omega', \mathcal{F}', P')$ respectively, having the same finite-dimensional distribution functions. Then for any sequence $\{t_n\}$, $t_n \geqslant 0$, and for any set $D \in \mathcal{B}_{\nu,\infty}$,

$$P\{(X(t_1), X(t_2), \ldots) \in D\} = P'\{(X'(t_1), X'(t_2), \ldots) \in D\}.$$

3. Suppose two ν-dimensional stochastic processes $X(t)$ and $X'(t)$ $(t \geq 0)$ defined in probability spaces $(\Omega, \mathcal{F}, P)$ and $(\Omega', \mathcal{F}', P')$, respectively, are separable and have the same finite-dimensional distribution functions. If $X(t)$ is continuous, then $X'(t)$ is continuous.

[*Hint*: Let $\nu = 1$. It suffices to prove continuity for $0 \leq t \leq t^*$, $t^* < \infty$. Let $T = \{t_j\}$ be a set of separability for both processes for $0 \leq t \leq t^*$. The continuity of $X(t)$ implies that

$$P\left\{ \bigcap_{n=1}^{\infty} \bigcup_{m=1}^{\infty} \sup_{|t_i - t_j| < 1/m} |X(t_i) - X(t_j)| < \frac{1}{n} \right\} = 1.$$

By Problem 2, with

$$D = \bigcap_n \bigcup_m \bigcap_{|t_i - t_j| < 1/m} D_{mn}^{ij}, \qquad D_{mn}^{ij} = \left\{ (x_1, x_2, \ldots), |x_i - x_j| < \frac{1}{n} \right\},$$

$$P'\left\{ \bigcap_{n=1}^{\infty} \bigcup_{m=1}^{\infty} \sup_{|t_i - t_j| < 1/m} |X'(t_i) - X'(t_j)| < \frac{1}{n} \right\} = 1.$$

Now use Problem 1(d).]

4. Let $X(t)$ and $X'(t)$ $(t \geq 0)$ be ν-dimensional separable stochastic processes defined in probability spaces $(\Omega, \mathcal{F}, P)$ and $(\Omega', \mathcal{F}', P')$ respectively, and having the same finite-dimensional distribution functions. Then:

(a) $P\{|X(t)| \to \infty \text{ as } t \to \infty\} = P'\{|X'(t)| \to \infty \text{ as } t \to \infty\}$;
(b) $P\{|X(t)|/t^{\alpha} \to M \text{ as } t \to \infty\} = P'\{|X'(t)|/t^{\alpha} \to M \text{ as } t \to \infty\}$,

for any $\alpha > 0$, $M \geq 0$.

[*Hint*: $\{|X(t)| \to \infty \text{ as } t \to \infty\} = \{\bigcap_n \bigcup_m \inf_{t_j > m} |X(t_j)| > n\}$.]

5. If two stochastic processes are stochastically equivalent, then they have the same finite-dimensional distribution functions.

6. Let X_n, $n \geq 1$ be independent random variables with $E|X_n| < \infty$, $EX_n = 0$. Then the sequence of partial sums $X_1 + \cdots + X_n$ is a martingale.

7. Let X be a random variable in $(\Omega, \mathcal{F}, P)$ with $E|X| < \infty$, and let $\phi(x)$ be a convex function defined on the real line such that $E|\phi(X)| < \infty$. Prove:

(a) $\phi(EX) \leq E\phi(X)$ (Jensen's inequality);
(b) $\phi(E(X|\mathcal{F}_0)) \leq E(\phi(X)|\mathcal{F}_0)$ a.e., where $\mathcal{F}_0$ is any σ-subfield of $\mathcal{F}$.

8. If $\{X_n\}$ is a martingale and $\phi(x)$ is a convex function on the real line, and if $E|\phi(X_n)| < \infty$ for all n, then $\{\phi(X_n)\}$ is a submartingale.

9. If $\{X_n\}$ is a submartingale and $\phi(x)$ is a convex function and monotone nondecreasing on the range of the X_n, and if $E|\phi(X_n)| < \infty$ for all n, then $\{\phi(X_n)\}$ is a submartingale.

10. If τ is a bounded stopping time with respect to right continuous martingale $X(t)$, then $EX(\tau) = EX(0)$.

11. Prove that a continuous process is continuous in probability. Give an example showing that the converse is not true in general.

12. Let $X(t)$, $\tilde{X}(t)$ $(t \in I)$ be two stochastically equivalent and separable processes. Show that if one of them is continuous, then the other is also continuous, and

$$P\{X(t) = \tilde{X}(t) \text{ for all } t \in I\} = 1.$$

13. A stochastic process $\{X(t),\ t \in I\}$ is said to be *measurable* if the function $(t, \omega) \to X(t, \omega)$ (from the product measure space into R^1) is measurable. Show that a continuous stochastic process is measurable.

2

Markov Processes

1. Construction of Markov processes

If $x(t)$ is a stochastic process, we denote by $\mathcal{F}(x(\lambda), \lambda \in J)$ the smallest σ-field generated by the random variables $x(t)$, $t \in J$.

Definition. Let $p(s, x, t, A)$ be a nonnegative function defined for $0 \leq s < t < \infty$, $x \in R^\nu$, $A \in \mathcal{B}_\nu$ and satisfying:

(i) $p(s, x, t, A)$ is Borel measurable in x, for fixed s, t, A;
(ii) $p(s, x, t, A)$ is a probability measure in A, for fixed s, x, t;
(iii) p satisfies the *Chapman-Kolmogorov equation*

$$p(s, x, t, A) = \int_{R^\nu} p(s, x, \lambda, dy) p(\lambda, y, t, A) \qquad \text{for any} \quad s < \lambda < t. \tag{1.1}$$

Then we call p a *Markov transition function*, a *transition probability function*, or a *transition probability*.

Theorem 1.1. *Let p be a transition probability function. Then, for any $s \geq 0$ and for any probability distribution $\pi(dx)$ on $(R^\nu, \mathcal{B}_\nu)$, there exists a ν-dimensional stochastic process $\{x(t), s \leq t < \infty\}$ such that:*

$$P[x(s) \in A] = \pi(A), \tag{1.2}$$

$$P\{x(t) \in A | \mathcal{F}(x(\lambda), s \leq \lambda \leq \bar{s})\} = p(\bar{s}, x(\bar{s}), t, A) \quad a.s. \qquad (s \leq \bar{s} < t). \tag{1.3}$$

Proof. Let $M[[0, \infty); R^\nu]$ be the space of all functions ω from $[0, \infty)$ into R^ν. Let $\mathcal{F}_t^s$ be the smallest σ-field with respect to which all functions $\omega(u)$ are measurable, $s \leq u \leq t$, i.e., $\mathcal{F}_t^s$ is the σ-field generated by all the sets $\{\omega; \omega(u) \in A\}$ where $s \leq u \leq t$, $A \in \mathcal{B}_\nu$.

Let $\mathcal{F}^s_\infty$ be the smallest σ-field with respect to which all functions $\omega(u)$ are measurable, $s \leqslant u < \infty$. We take $\Omega = M[[0, \infty); R^\nu]$, $\mathcal{F} = \mathcal{F}^s_\infty$.

We next define a family of finite-dimensional distribution functions with respect to R^ν: if $s = t_0 < t_1 < \cdots < t_n$ and $\bar{x}_i \in R^\nu$ $(0 \leqslant i \leqslant n)$, then

$$F_{t_0 t_1 \cdots t_n}(\bar{x}_0, \bar{x}_1, \ldots, \bar{x}_n) = \int_{-\infty}^{\bar{x}_n} \cdots \int_{-\infty}^{\bar{x}_1} \int_{-\infty}^{\bar{x}_0} p(t_{n-1}, y_{n-1}, t_n, dy_n)$$
$$\cdots p(t_1, y_1, t_2, dy_2) p(t_0, y_0, t_1, dy_1) \pi(dy_0). \quad (1.4)$$

It is easy to verify that the F_n are in fact n-dimensional distribution functions with respect to R^ν. Using the Chapman–Kolmogorov equation it follows that

$$F_{t_0 \cdots t_n}(\bar{x}_0, \ldots, \bar{x}_n) \to F_{t_0 \cdots t_{k-1} t_{k+1} \cdots t_n}(\bar{x}_0, \ldots, \bar{x}_{k-1}, \bar{x}_{k+1}, \ldots, \bar{x}_n)$$
$$\text{if} \quad \bar{x}_k \uparrow \infty;$$

thus the family of distribution functions in (1.4) is consistent.

By the Kolmogorov construction, there exists a probability P_s in $(\Omega, \mathcal{F}^s_\infty)$ and a ν-dimensional stochastic process $x(t)$, $s \leqslant t < \infty$, such that the finite-dimensional distribution functions of this process coincide with the given distributions functions $F_{t_0 \cdots t_n}$.

Notice that $x(t, \omega) = \omega(t)$.

Now, if $s = t_0 < t_1 < \cdots < t_n$,

$$P_s\{x(t_0) \in B_1, \ldots, x(t_n) \in B_n\}$$
$$= \int_{B_n} \cdots \int_{B_1} \int_{B_0} p(t_{n-1}, y_{n-1}, t_n, dy_n) \cdots p(t_0, y_0, t_1, dy_1) \pi(dy_0) \quad (1.5)$$

provided each B_i is a ν-dimensional interval $(-\infty, \bar{x}_i)$. It is easily seen that both sides of (1.5) coincide also when the B_i are any ν-dimensional intervals $(\bar{z}_i, \bar{x}_i)$, or disjoint unions of such intervals. Since two probabilities that agree on a field $\mathcal{Q}$ agree also on the σ-field generated by $\mathcal{Q}$, it follows that (1.5) holds for all B_i in $\mathcal{B}_\nu$.

Taking $n = 0$ in (1.5) we get the assertion (1.2). Thus it remains to prove (1.3). Since the random variable on the right-hand side of (1.3) is measurable with respect to $\mathcal{F}(x(\lambda), s \leqslant \lambda \leqslant \bar{s})$, it remains to prove that

$$\int_D \chi_{[x(t) \in A]} \, dP_s = \int_D p(\bar{s}, x(\bar{s}), t, A) \, dP_s \qquad \text{if} \quad D \in \mathcal{F}(x(\lambda), s \leqslant \lambda \leqslant \bar{s}). \quad (1.6)$$

If we prove (1.6) for any set D of the form

$$D = \{x(t_0) \in B_0, x(t_1) \in B_1, \ldots, x(t_n) \in B_n\} \quad (1.7)$$

where $s = t_0 < t_1 < \cdots < t_n = \bar{s}$, then, by the countable additivity of both sides of (1.6), the assertion (1.6) will follow for any $D \in \mathcal{F}(x(\lambda), s \leq \lambda \leq \bar{s})$.

Set $t_{n+1} = t$, $B_{n+1} = A$. When D is given by (1.7), the left-hand side of (1.6) becomes

$$P_s[D \cap (x(t) \in A)] = P_s\{x(t_0) \in B_0, \ldots, x(t_n) \in B_n, x(t_{n+1}) \in B_{n+1}\}$$
$$= \int_{B_{n+1}} \int_{B_n} \cdots \int_{B_1} \int_{B_0} p(t_n, y_n, t_{n+1}, dy_{n+1}) p(t_{n-1}, y_{n-1}, t_n, dy_n)$$
$$\cdots p(t_0, y_0, t_1, dy_1) \pi(dy_0)$$
$$= \int_{B_n} \cdots \int_{B_1} \int_{B_0} p(t_n, y_n, t_{n+1}, A) p(t_{n-1}, y_{n-1}, t_n, dy_n)$$
$$\cdots p(t_0, y_0, t_1, dy_1) \pi(dy_0).$$

Therefore it suffices to show that

$$\int_{B_n} \cdots \int_{B_1} \int_{B_0} f(y_n) p(t_{n-1}, y_{n-1}, t_n, dy_n) \cdots p(t_0, y_0, t_1, dy_1) \pi(dy_0)$$
$$= \int_{x(t_n) \in B_n} \cdots \int_{x(t_0) \in B_0} f(x(t_n)) \, dP_s \tag{1.8}$$

where f is a Borel measurable function. It clearly suffices to prove (1.8) when f is any indicator function χ_F. Setting $B_n' = B_n \cap F$, (1.8) becomes

$$\int_{B_n'} \int_{B_{n-1}} \cdots \int_{B_1} \int_{B_0} p(t_{n-1}, y_{n-1}, t_n, dy_n) \cdots p(t_0, y_0, t_1, dy_1) \pi(dy_0)$$
$$= \int_{x(t_n) \in B_n'} \int_{x(t_{n-1}) \in B_{n-1}} \cdots \int_{x(t_0) \in B_0} dP_s. \tag{1.9}$$

Since the right-hand side is equal to

$$P_s\{x(t_0) \in B_0, \ldots, x(t_{n-1}) \in B_{n-1}, x(t_n) \in B_n'\},$$

(1.9) is a consequence of (1.5). This completes the proof of (1.3).

For any $x \in R^n$ let π_x be defined by

$$\pi_x(A) = \begin{cases} 1 & \text{if } x \in A, \\ 0 & \text{if } x \notin A. \end{cases}$$

Denote by $P_{x,s}$ the probability P_s constructed on the measure space $(\Omega, \mathcal{F}_\infty^s)$ when $\pi = \pi_x$, and denote the expectation by $E_{x,s}$. Set $\mathcal{F} = \mathcal{F}_\infty^0$. Then we have the following situation:

(a) $(\Omega, \mathcal{F})$ is a measure space, $\mathcal{F}_t^s$ are σ-subfields such that $\mathcal{F}_t^s \subset \mathcal{F}_{t'}^{s'}$ if $s' \leq s$, $t \leq t'$, and $\mathcal{F}$ is the smallest σ-field containing all the $\mathcal{F}_t^s$.

(b) There is a function $x(t, \omega)$ from $[0, \infty) \times \Omega$ into R^ν such that $x(t, \omega)$ is $\mathcal{F}_t^s$ measurable for each $t \geq s$.

(c) For each $x \in R^\nu$, $s \geq 0$ there is a probability $P_{x,s}$ defined on $(\Omega, \mathcal{F}_\infty^s)$

such that

$$P_{x,s}\{x(s,\omega) = x\} = 1, \tag{1.10}$$

$$P_{x,s}\{x(t+h,\omega)\in A|\mathcal{F}_t^s\} = p(t, x(t,\omega), t+h, A) \quad \text{a.s.} \qquad (t \geq s, h > 0). \tag{1.11}$$

Definition. A collection $\{\Omega, \mathcal{F}, \mathcal{F}_t, x(t), P_{x,s}\}$ satisfying (a), (b), (c), where p is a transition probability function, is called a (ν-dimensional) *Markov process* corresponding to p.

The property (1.11) is called the *Markov property*.

Notice that in the model constructed in Theorem 1.1 $\mathcal{F}_t^s$ is the smallest σ-field generated by $x(u)$, $s \leq u \leq t$. Notice also that $x(t,\omega) = \omega(t)$.

The model constructed in Theorem 1.1 will be called the *Kolmogorov model.*

Taking the expectation $E_{x,s}$ of both sides of (1.11), we get (when $t = s$ and $t + h$ is denoted by t),

$$E_{x,s}f(x(t)) = \int_{R^\nu} p(s,x,t,dy)f(y) \tag{1.12}$$

if $f = \chi_A$. By approximation, this relation holds also for any bounded Borel measurable function f.

We can therefore write (1.11) in the form

$$\begin{aligned} E_{x,s}\{f(x(t+h))|\mathcal{F}_t^s\} &= \int_{R^\nu} p(t,x(t),t+h,dy)f(y) \\ &= E_{x(t),t}f(x(t+h)) \quad \text{a.s.} \end{aligned} \tag{1.13}$$

if $f = \chi_A$. By approximation, this relation holds also for any bounded Borel measurable function f.

We have the following generalization of (1.5) (in case $\pi = \pi_x$):

$$\begin{aligned} &P_{x,s}\{(x(t_1),\dots,x(t_n))\in D\} \\ &\quad = \int \underset{D}{\cdots} \int p(t_{n-1},y_{n-1},t_n,dy_n)\cdots p(s,x,t_1,dy_1) \end{aligned} \tag{1.14}$$

where D is any set in $\mathcal{B}_{\nu,n}$ ($\mathcal{B}_{\nu,n}$ is the σ-field of Borel sets in $R^\nu \times R^\nu \times \cdots \times R^\nu$ (n times)). Indeed, if we denote the left-hand and right-hand sides of (1.14) by $P(D)$ and $Q(D)$ respectively, then $P(D)$ and $Q(D)$ define probability measures on $\mathcal{B}_{\nu,n}$. Since, by (1.5), these measures agree on all rectangular sets, they must agree on $\mathcal{B}_{\nu,n}$.

Theorem 1.2. *If $\{\Omega, \mathcal{F}, \mathcal{F}_t, x(t), P_{s,x}\}$ is a ν-dimensional Markov process, then*

$$P_{x,s}(B|\mathcal{F}_t^s) = P_{x,s}(B|x(t)) \quad a.s. \qquad \textit{for any} \quad B\in\mathcal{F}(x(\lambda), \lambda \geq t). \tag{1.15}$$

Proof. Let $0 < h_1 < h_2$ and let f_1, f_2 be bounded Borel measurable functions in R^ν. Using (1.13), we have

$$\begin{aligned}
&E_{x,s}\{f_1(x(t+h_1))f_2(x(t+h_2))|\mathcal{F}_t^s\}\\
&\quad= E_{x,s}\{f_1(x(t+h_1))E_{x,s}[f_2(x(t+h_2))|\mathcal{F}_{t+h_1}^s]|\mathcal{F}_t^s\}\\
&\quad= E_{x,s}\left\{\left[f_1(x(t+h_1))\int p(t+h_1, x(t+h_1), t+h_2, dy_2)f_2(y_2)\right]\Big|\mathcal{F}_t^s\right\}\\
&\quad= \int p(t, x(t), t+h_1, dy_1)f_1(y_1)\int p(t+h_1, y_1, t+h_2, dy_2)f_2(y_2).
\end{aligned}$$

Similarly one can prove by induction on n that, if $0 < h_1 < \cdots < h_n$,

$$\begin{aligned}
&E_{x,s}[f_1(x(t+h_1))\cdots f_n(x(t+h_n))|\mathcal{F}_t^s]\\
&\quad= \int\cdots\int p(t, x(t), t+h_1, dy_1)f_1(y_1)\\
&\qquad\cdots p(t+h_{n-1}, y_{n-1}, t+h_n, dy_n)f_n(y_n). \qquad (1.16)
\end{aligned}$$

For any set $D \in \mathcal{B}_{\nu,n}$ its indicator function $f(x_1, \ldots, x_n) = \chi_D(x_1, \ldots, x_n)$ can be uniformly approximated by finite linear combinations of bounded measurable functions of the form $f_1(x_1)\cdots f_n(x_n)$ (each x_i varies in R^ν). Employing (1.16) we deduce (by the Lebesgue bounded convergence theorem) that

$$\begin{aligned}
&E_{x,s}[f(x(t+h_1), \ldots, x(t+h_n))|\mathcal{F}_t^s]\\
&\quad= \int\cdots\int f(y_1, \ldots, y_n)p(t, x(t), t+h_1, dy_1)\\
&\qquad\cdots p(t+h_{n-1}, y_{n-1}, t+h_n, dy_n) \quad \text{a.s.} \qquad (1.17)
\end{aligned}$$

Setting

$$B = \{(x(t_1), \ldots, x(t_n)) \in D\}, \qquad (1.18)$$

we conclude that $P_{x,s}(B|\mathcal{F}_t^s)$ is $\mathcal{F}(x(t))$ measurable and (1.15) holds for this B; here $\mathcal{F}(x(t))$ denotes the smallest σ-subfield with respect to which $x(t)$ is measurable.

Since the class $\mathcal{C}$ of sets B for which (1.15) holds is a monotone class containing all the sets of the form (1.18), it must contain the σ-field generated by these sets, i.e., $\mathcal{C} = \mathcal{F}(x(\lambda), \lambda \geq t)$.

Theorem 1.3. *Let $x(t)$ and $y(t)$ be ν-dimensional processes defined for $t \geq 0$ and having the same set of joint distribution functions. Suppose $x(t)$ satisfies the assertions in Theorem 1.1. Then $y(t)$ also satisfies the same assertions.*

An outline of the proof is given in Problem 4.

2. The Feller and the strong Markov properties

Denote by $M(R^\nu)$ the space of all bounded Borel measurable functions f from R^ν into R^1 with the norm $\|f\| = \text{ess sup}_{x\in R^\nu}|f(x)|$. Consider the mappings $T_{s,t}$ from $M(R^\nu)$ into $M(R^\nu)$:

$$(T_{s,t}f)(x) = \int_{R^\nu} p(s, x, t, dy)f(y) \qquad (0 \leqslant s < t < \infty).$$

It is easily seen that $\|T_{s,t}\| = 1$, and

$$T_{s,t}T_{t,u} = T_{s,u} \qquad \text{if} \quad 0 \leqslant s < t < u.$$

We define $T_{t,t}$ = identity. The family $\{T_{s,t}\}$ is called the *semigroup* associated with the transition probability function p (or with the corresponding Markov process).

Definition. If for any bounded continuous function $f(x)$, the function

$$(t, z) \to (T_{t,t+\lambda}f)(z) = \int_{R^\nu} p(t, z, t + \lambda, dy)f(y)$$

is continuous, for any $\lambda > 0$, then we say that the transition probability p (or the corresponding Markov process) satisfies the *Feller property*.

We denote by $\mathcal{F}^s_{t+}$ the intersection of the σ-fields $\mathcal{F}^s_{t'}$, $t' > t$.
Similarly we denote by $\mathcal{F}^s_{t-}$ the smallest σ-field containing $\mathcal{F}^s_{t'}$, $t' < t$.

Theorem 2.1. *Let $\{\Omega, \mathcal{F}, \mathcal{F}^s_t, x(t), P_{x,s}\}$ be a ν-dimensional Markov process, right continuous and satisfying the Feller property. Then $\{\Omega, \mathcal{F}, \mathcal{F}^s_{t+}, x(t), P_{x,s}\}$ is also a Markov process with the same transition probability function.*

Proof. Let f be a bounded continuous function, and let $\epsilon > 0$, $h > 0$. By (1.13),

$$E_{x,s}\{f(x(t + h + \epsilon))|\mathcal{F}^s_{t+\epsilon}\} = \int p(t + \epsilon, x(t + \epsilon), t + h + \epsilon, dy)f(y).$$

Taking the conditional expectation with respect to $\mathcal{F}^s_{t+}$, we get

$$E_{x,s}\{f(x(t + h + \epsilon))|\mathcal{F}^s_{t+}\}$$
$$= E_{x,s}\left\{\int p(t + \epsilon, x(t + \epsilon), t + h + \epsilon, dy)f(y)|\mathcal{F}^s_{t+}\right\}. \qquad (2.1)$$

Let $\epsilon \downarrow 0$. Since $x(t + \epsilon) \to x(t)$ a.s., the Feller property implies that

$$\int p(t + \epsilon, x(t + \epsilon), t + h + \epsilon, dy) f(y) \to \int p(t, x(t), t + h, dy)f(y) \quad \text{a.s.}$$

By the Lebesgue bounded convergence theorem for conditional expectation, we then get

$$E_{x,s}\left\{ \int p(t+\epsilon, x(t+\epsilon), t+h+\epsilon, dy) f(y) | \mathcal{F}_{t+}^{s} \right\}$$

$$\to E_{x,s}\left\{ \int p(t, x(t), t+h, dy) f(y) | \mathcal{F}_{t+}^{s} \right\} = \int p(t, x(t), t+h, dy) f(y),$$

since $x(t)$ is $\mathcal{F}_{t+}^{s}$ measurable.

On the other hand

$$f(x(t+h+\epsilon)) \to f(x(t)) \qquad \text{as} \quad \epsilon \downarrow 0,$$

so that

$$E_{x,s}\{ f(x(t+h+\epsilon)) | \mathcal{F}_{t+}^{s} \} \to E_{x,s}\{ f(x(t+h)) | \mathcal{F}_{t+}^{s} \}.$$

Thus, upon taking $\epsilon \downarrow 0$ in (2.1), we get

$$E_{x,s}\{ f(x(t+h)) | \mathcal{F}_{t+}^{s} \} = \int p(t, x(t), t+h, dy) f(y).$$

This completes the proof.

From Theorem 2.1 it follows that whenever one has constructed, from a given transition probability function, a Markov process that is right continuous and satisfies the Feller property, one may assume (without loss of generality) that $\mathcal{F}_{t+}^{s} = \mathcal{F}_{t}^{s}$.

Remark. If a Markov process is left continuous, and if $\mathcal{F}_{t}^{s} = \mathcal{F}\{x(\lambda); s \leqslant \lambda \leqslant t\}$, then $\mathcal{F}_{t-}^{s} = \mathcal{F}_{t}^{s}$. This follows by noting that, for any $\alpha \in R^{\nu}$, the set $\{\alpha \cdot x(t) \leqslant \lambda\}$ coincides with the set

$$\bigcap_{m=1}^{\infty} \bigcap_{n=1}^{\infty} \bigcup_{k=n}^{\infty} \left[\alpha \cdot x\left(t - \frac{1}{k}\right) < \lambda + \frac{1}{m} \right]$$

which belongs to $\mathcal{F}_{t-}^{s}$.

Definition. An extended real-valued random variable τ defined on Ω is called an *s-Markov time* or an *s-stopping time* with respect to a given ν-dimensional Markov process $\{\Omega, \mathcal{F}, \mathcal{F}_{t}^{s}, x(t), P_{x,s}\}$ if $\tau \geqslant s$ and the sets $\{\tau \leqslant t\}$ belong to $\mathcal{F}_{t}^{s}$ for every $t \geqslant s$.

Any constant function $\tau = t$ with $t \geqslant s$ is an s-stopping time.

Since $\{\tau \leqslant t\}$ is the complement of $\{\tau > t\}$, τ is an s-stopping time if and only if $\tau \geqslant s$ and the sets $\{\tau > t\}$ belong to $\mathcal{F}_{t}^{s}$, $t \geqslant s$. Notice that the sets

$$\{\lambda < \tau \leqslant t\} = \{\tau \leqslant t\} \setminus \{\tau \leqslant \lambda\}$$

also belong to $\mathcal{F}_{t}^{s}$.

Consider the events $A \subset \mathcal{F}^s_\infty$ such that

$$A \cap \{\tau \leqslant t\} \in \mathcal{F}^s_t \qquad \text{for all} \quad t \geqslant s.$$

They form a σ-field $\mathcal{F}^s_\tau$.

Definition. A ν-dimensional Markov process $\{\Omega, \mathcal{F}, \mathcal{F}^s_t, x(t), P_{x,s}\}$ with transition probability p is said to have the *strong Markov property* if for every $x \in R^\nu$ and for every real-valued s-Markov time τ,

$$P_{x,s}\{x(t+\tau) \in A | \mathcal{F}^s_\tau\} = p(\tau, x(\tau), t+\tau, A) \quad \text{a.e.} \tag{2.2}$$

For $\tau = h$ (h constant), this coincides with the Markov property (1.11).

Using (1.12) it is clear that the strong Markov property is equivalent to

$$E_{x,s}\{f(t+\tau)|\mathcal{F}^s_\tau\} = \int p(\tau, x(\tau), t+\tau, dy) f(y) = E_{x(\tau),\tau} f(x(t+\tau)) \tag{2.3}$$

for any $f = \chi_A$; hence for any bounded Borel measurable function f.

We shall give some examples of s-stopping times.

Theorem 2.2. *Let $\{\Omega, \mathcal{F}, \mathcal{F}^s_t, x(t), P_{x,s}\}$ be a ν-dimensional continuous Markov process and let F be a nonempty closed subset of R^ν. Let $\tau(\omega) =$ first time $t \geqslant s$ such that $x(t, \omega) \in F$. Then τ is an s-stopping time, for any $s \geqslant 0$.*

If $s = 0$, τ is called the *(first) hitting time* of the set F or the *exit time* from the open set $R^\nu \setminus F$; if $s > 0$, τ is called hitting time, after s, of F (or the exit time, after s, from $R^\nu \setminus F$).

Proof. Let $\{t_j\}$ be a dense sequence in the interval $[s, \infty)$. Denote by $\rho(x, F)$ the distance from x to F. Let $\{\zeta_j\}$ be a dense sequence in F. Since

$$\{|x(t) - \zeta_j| < \alpha\} = \{\omega; x(t, \omega) \in D\} \qquad (\alpha > 0)$$

where $D = \{x \in R^\nu, |x - \zeta_j| < \alpha\}$, it follows that the set $\{|x(t) - \zeta_j| < \alpha\}$ is in $\mathcal{F}^s_t$, if $t \geqslant s$. Consequently, also the set

$$\bigcap_{n=1}^{\infty} \bigcup_{t_j < t} \left\{\rho(x(t_j), F) \leqslant \frac{1}{n}\right\} = \bigcap_{n=1}^{\infty} \bigcup_{t_j < t} \bigcup_{k} \left\{|x(t_j) - \zeta_k| < \frac{1}{n}\right\}$$

is in $\mathcal{F}^s_t$. But as is easily seen, this set coincides with the set $\{\tau \leqslant t\}$.

Let G be a nonempty open set in R^ν. Let

$$\tau(\omega) = \inf\{t; t \geqslant s, x(t, \omega) \in G\}.$$

Then τ is called the *(first) hitting time*, after s, of G. Notice that, for a continuous process,

$$\begin{aligned}\tau(\omega) &= \inf\{t; t > s, x(t, \omega) \in G\} \\ &= \inf\{t; t > s, \text{meas}\,[s < t' < t; x(t', \omega) \in G] > 0\} \quad \text{a.s.}\end{aligned}$$

For any set G, the extended random variable defined by the last expression is called the *(first) penetration time*, after s, of G.

Theorem 2.3. *Let* $\{\Omega, \mathcal{F}, \mathcal{F}_t^s, x(t), P_{x,s}\}$ *be a* ν*-dimensional continuous Markov process with* $\mathcal{F}_{t+}^s = \mathcal{F}_t^s$, *and let* τ *be the penetration time, after* s, *of an open set* G. *Then* τ *is an* s*-stopping time, for any* $s \geq 0$.

Proof. Let $\{t_j\}$ be a dense sequence in $[s, \infty)$, and let

$$A_n = \bigcup_{t_j < t + 1/n} \bigcup_{k=1}^{\infty} \left[\rho(x(t_j), R^\nu \setminus G) > \frac{1}{k} \right].$$

It is clear that if $\tau(\omega) \leq t$, then $\omega \in A_n$; and if $\omega \in A_n$, then $\tau(\omega) < t + 1/n$. Hence $(\tau \leq t) = \lim A_n$. Since $A_n \in \mathcal{F}_{t+1/n}^s$, it follows that the set $(\tau \leq t)$ belongs to $\mathcal{F}_{t+}^s$.

Remark. Theorems 2.2, 2.3 are clearly valid not only for a continuous Markov process, but also for any continuous process $x(t)$; the definition of (extended) stopping time is given in Chapter 1, Section 3.

Theorem 2.4. *Let* $\{\Omega, \mathcal{F}, \mathcal{F}_t^s, x(t), P_{x,s}\}$ *be a* ν*-dimensional Markov process, right continuous and satisfying the Feller property, and let* $\mathcal{F}_{t+}^s = \mathcal{F}_t^s$. *Then it satisfies the strong Markov property.*

We begin with a special case.

Lemma 2.5. *The strong Markov property* (2.2) *holds for any Markov process provided* τ *is an* s*-stopping time with a countable range.*

Proof. Denote by $\{t_j\}$ the range of τ. Since $\{\tau \leq t_j\}$ and

$$\{\tau < t_j\} = \bigcup_{t_l < t_j} (\tau \leq t_l)$$

belong to $\mathcal{F}_{t_j}^s$, it follows that

$$G_j \equiv \{\tau = t_j\} \in \mathcal{F}_{t_j}^s.$$

Note next that G_j belongs to $\mathcal{F}_\tau^s$, since

$$\begin{aligned} \{\tau = t_j\} \cap \{\tau \leq t\} &\in \mathcal{F}_t^s \qquad \text{if} \quad t \geq t_j, \\ &= \phi \in \mathcal{F}_t^s \qquad \text{if} \quad t < t_j. \end{aligned}$$

Hence, for any $A \in \mathcal{B}_\nu$,

$$\{x(\tau) \in A\} = \bigcup_{x(t_j) \in A} \{\tau = t_j\} \qquad \text{is in } \mathcal{F}_\tau^s.$$

This shows that $x(\tau)$ is $\mathcal{F}_\tau^s$ measurable.

Now let $E \in \mathcal{F}_\tau^s$. Then

$$E \cap G_j = E \cap (\tau \leqslant t_j) \cap G_j \in \mathcal{F}_{t_j}^s.$$

The Markov property gives

$$P_{x,s}\{x(t + t_j) \in A \mid \mathcal{F}_{t_j}^s\} = p(t_j, x(t_j), t_j + t, A) \quad \text{a.s.,}$$

and since $E \cap G_j \in \mathcal{F}_{t_j}^s$,

$$P_{x,s}\{[x(t + t_j) \in A] \cap E \cap G_j\} = \int_{E \cap G_j} P(t_j, x(t_j), t_j + t, A)\, dP_{x,s}.$$

Noting that $\tau = t_j$ on G_j and summing over j, we get, upon recalling that $\bigcup G_j = \Omega$,

$$P_{x,s}\{[x(t + \tau) \in A] \cap E\} = \int_E p(\tau, x(\tau), \tau + t, A)\, dP_{x,s}. \tag{2.4}$$

To complete the proof of (2.2) we have to show that the integrand on the right-hand side of (2.2) is $\mathcal{F}_\tau^s$ measurable. Since τ and $x(\tau)$ are $\mathcal{F}_\tau^s$ measurable, the Feller property implies that

$$\int p(\tau, x(\tau), \tau + t, dy) f(y)$$

is $\mathcal{F}_\tau^s$ measurable if $f(y)$ is a bounded continuous function. Taking a sequence $f = f_m$ of uniformly bounded and continuous functions that converge a.e. to χ_A, we conclude that

$$\lim_{m \to \infty} \int p(\tau, x(\tau), \tau + t, dy) f_m(y) = p(\tau, x(\tau), \tau + t, A)$$

is $\mathcal{F}_\tau^s$ measurable.

Proof of Theorem 2.4. Choose τ_n as in Lemma 1.3.4. We claim that $\mathcal{F}_\tau^s \subset \mathcal{F}_{\tau_n}^s$. Indeed, since $\tau_n \geqslant \tau$,

$$A \cap (\tau_n \leqslant t) = [A \cap (\tau \leqslant t)] \cap (\tau_n \leqslant t)$$

for any set A and for any $t \geqslant s$. If $A \in \mathcal{F}_\tau^s$, then $A \cap (\tau \leqslant t) \in \mathcal{F}_t^s$. Since also $(\tau_n \leqslant t) \in \mathcal{F}_t^s$,

$$A \cap (\tau_n \leqslant t) \in \mathcal{F}_t^s, \qquad \text{i.e.,} \qquad A \in \mathcal{F}_{\tau_n}.$$

If $f = \chi_B$, then the strong Markov property for τ_n gives

$$E_{x,s}[f(x(t + \tau_n)) \mid \mathcal{F}_{\tau_n}^s] = \int p(\tau_n, x(\tau_n), t + \tau_n, dy) f(y) \quad \text{a.s.} \tag{2.5}$$

But then this relation holds also for any bounded continuous function f.

Let $E \in \mathcal{F}_\tau^s$. Then (as proved above) $E \in \mathcal{F}_{\tau_n}^s$ and therefore, by (2.5),

$$\int_E f(x(t + \tau_n))\, dP_{x,s} = \int_E \left[\int p(\tau_n, x(\tau_n), t + \tau_n, dy) f(y) \right] dP_{x,s}. \tag{2.6}$$

The function $\int p(\lambda, x, t+\lambda, dy)f(y)$ is continuous in λ, x. Also, if $n \uparrow \infty$, $\tau_n \downarrow \tau$ and $x(\tau_n) \to x(\tau)$ (by the right continuity of the $x(t)$). Hence, if we let $n \uparrow \infty$ in (2.6) and employ the Lebesgue bounded convergence theorem, we get

$$\int_E f(x(t+\tau))\, dP_{x,s} = \int_E \left[\int p(\tau, x(\tau), t+\tau, dy) f(y) \right] dP_{x,s}. \qquad (2.7)$$

This relation holds also if $f = \chi_A$, $A \in \mathcal{B}_\nu$.

If we show that $x(\tau)$ is $\mathcal{F}^s_\tau$ measurable, then the proof of the theorem follows from (2.7) and the argument following (2.4).

Let $\alpha \in R^\nu$. Then, by the right continuity of $x(t)$,

$$[\alpha \cdot x(\tau) \leq \lambda] = \bigcap_{m=N}^{\infty} \bigcap_{k=1}^{\infty} \bigcup_{n=k}^{\infty} \left[\alpha \cdot x(\tau_n) < \lambda + \frac{1}{m} \right]$$

for any positive integer N. We also have

$$[\tau \leq t] = \bigcap_{p=N}^{\infty} \bigcap_{q=1}^{\infty} \bigcup_{r=q}^{\infty} \left[\tau_r < t + \frac{1}{p} \right]$$

and, more generally,

$$\begin{aligned} Q &\equiv [\alpha \cdot x(\tau) \leq \lambda] \cap [\tau \leq t] \\ &= \bigcap_{m=N}^{\infty} \bigcap_{k=1}^{\infty} \bigcup_{n=k}^{\infty} \left[\alpha \cdot x(\tau_n) \leq \lambda + \frac{1}{m} \right] \cap \left[\tau_n < t + \frac{1}{m} \right]. \end{aligned}$$

Since $x(\tau_n) \in \mathcal{F}^s_{\tau_n}$,

$$\left[\alpha \cdot x(\tau_n) \leq \lambda + \frac{1}{m} \right] \cap \left[\tau_m < t + \frac{1}{m} \right] \in \mathcal{F}^s_{t+1/m} \subset \mathcal{F}^s_{t+1/N}.$$

It follows $Q \in \mathcal{F}^s_{t+1/N}$ for any N. Hence $Q \in \mathcal{F}^s_{t+}$. Since $\mathcal{F}^s_{t+} = \mathcal{F}^s_t$, we conclude that $[\alpha \cdot x(\tau) \leq \lambda] \in \mathcal{F}^s_\tau$. But since α and λ are arbitrary, $x(\tau)$ is $\mathcal{F}^s_\tau$ measurable.

Corollary 2.6. *Let $\{\Omega, \mathcal{F}, \mathcal{F}^s_t, x(t), P_{x,s}\}$ be a ν-dimensional continuous Markov process satisfying the Feller property. Then it satisfies the strong Markov property.*

Notice that we do not assume here that $\mathcal{F}^s_{t+} = \mathcal{F}^s_t$.

Proof. The proof of Theorem 2.4 applies here except for the last step, namely, the proof that $x(\tau)$ is $\mathcal{F}^s_\tau$ measurable. To prove this, suppose for simplicity that $s = 0$. Let $\{t_l\}$ be a dense sequence in $[0, \infty)$. For any $\alpha \in R^\nu$ and λ real,

$$[\alpha \cdot x(\tau) \leq \lambda] \cap [\tau \leq t]$$
$$= \bigcap_{n=1}^{\infty} \bigcap_{m=1}^{\infty} \bigcup_{i=0}^{n-1}{}^{*} \bigcup_{(i/n)t < t_l < ((i+1)/n)t} \left[\alpha \cdot x(t_l) \leq \lambda + \frac{1}{m}\right]$$
$$\cap \left[\frac{i}{n}t < \tau \leq \frac{i+1}{n}t\right]$$

where * indicates that for $i = 0$ one takes $0 \leq \tau \leq t/n$. Since each set $[\cdots]$ on the right-hand side is in $\mathcal{F}_t^0$, it follows that $[\alpha \cdot x(\tau) \leq \lambda] \in \mathcal{F}_\tau^0$. But since α and λ are arbitrary, $x(\tau)$ is $\mathcal{F}_\tau^0$ measurable.

It can be shown (see Stroock and Varadhan [1]) that, for a continuous Markov process, $\mathcal{F}_\tau^s$ is actually generated by $x(t \wedge \tau)$, $t \geq s$.

The next result is the *Blumenthal zero–one law*.

Theorem 2.7. *Let $\{\Omega, \mathcal{F}, \mathcal{F}_t^s, x(t), P_{x,s}\}$ be a ν-dimensional Markov process, right continuous and satisfying the Feller property. If, for some $s \geq 0$, $x(s) = x$ a.s. (x a point in R^n), then*

$$P_{x,s}(B) = 0 \text{ or } 1 \qquad \text{for any} \quad B \in \mathcal{F}_{s+}^s. \tag{2.8}$$

Proof. By Theorem 2.1, formula (1.17) remains valid when $\mathcal{F}_t^s$ is replaced by $\mathcal{F}_{t+}^s$. Taking $t = s$, we get

$$P_{x,s}\{[(x(s+h_1), \ldots, x(s+h_n)) \in D] | \mathcal{F}_{s+}^s\}$$
$$= \int_D \cdots \int p(s, x, s+h_1, dy_1) \cdots p(s+h_{n-1}, y_{n-1}, s+h_n, dy_n) \quad \text{a.s.}$$

By (1.14), the right-hand side is equal to

$$P_{x,s}\{(x(s+h_1), \ldots, x(s+h_n)) \in D\}.$$

Hence,

$$P_{x,s}(B | \mathcal{F}_{s+}^s) = P_{x,s}(B) \quad \text{a.s.} \tag{2.9}$$

for any set B of the form (1.18) with $t_i > s$. But then (2.9) follows also for any set B in $\mathcal{F}_\infty^s$. Taking, in particular, $B \in \mathcal{F}_{s+}^s$ and noting that $P_{x,s}(B | \mathcal{F}_{s+}^s) = \chi_B$, we get

$$\chi_B(\omega) = P_{x,s}(B) \quad \text{a.s.}$$

Since the right-hand side is constant, $\chi_B = \text{const}$ a.s. Hence either $\chi_B = 0$ a.s. or $\chi_B = 1$ a.s. This gives (2.8).

3. Time-homogeneous Markov processes

Definition. A Markov process is said to have a *stationary transition probability function* p if

$$p(s, x, t, A) = p(0, x, t-s, A).$$

We then also say that the Markov process is *time-homogeneous*, or *temporally homogeneous*.

It is clear that, when p is stationary, $P_{x,s} = P_{x,0}$, $E_{x,s} = E_{x,0}$ for all $s \geqslant 0$. Therefore, from the probabilistic point of view it suffices to work just with $P_{x,s}$, $E_{x,s}$, $\mathcal{F}_t^s$ when $s = 0$. We set

$$P_x = P_{x,0}, \qquad E_x = E_{x,0}, \qquad p(t, x, A) = p(0, x, t, A), \qquad \mathcal{F}_t = \mathcal{F}_t^0.$$

A 0-stopping time τ will be called, simply, a *stopping time* or a *Markov time*. (Notice that τ is actually an extended stopping time according to the definition of Chapter 1, Section 3.) The Markov property now takes the form

$$E_x\{f(x(t+h))\,|\,\mathcal{F}_t\} = E_{x(t)}f(x(h)) \quad \text{a.s.}; \tag{3.1}$$

(1.12) for $s = 0$ becomes

$$E_x f(x(t)) = \int p(t, x, dy) f(y), \tag{3.2}$$

and the strong Markov property takes the form

$$E_x\{f(x(t+\tau))\,|\,\mathcal{F}_\tau\} = E_{x(\tau)}f(x(t)) \tag{3.3}$$

where τ is a stopping time and $\mathcal{F}_\tau = \mathcal{F}_\tau^0$.

We shall denote a time-homogeneous Markov process by $\{\Omega, \mathcal{F}, \mathcal{F}_t, x(t), P_x\}$.

For a time-homogeneous Markov process, the Feller property states that the function

$$z \to \int p(t, z, dy) f(y)$$

is continuous for any bounded continuous function f and for any $t > 0$.

The semigroup associated with a stationary transition probability function is given by

$$(T_t f)(x) = \int p(t, x, dy) f(y). \tag{3.4}$$

Denote by M_0 the space of all functions from $M(R^\nu)$ into R^1 for which

$$\|T_t f - f\| \to 0 \qquad \text{if} \quad t \to 0,$$

where $\|f\| = \text{ess sup}_{x \in R^\nu} |f(x)|$.

Definition. The *infinitesimal generator* $\mathcal{Q}$ of the time-homogeneous Markov process is defined by

$$\mathcal{A}f = \lim_{t \to 0} \frac{T_t f - f}{t} . \tag{3.5}$$

Its domain $D_{\mathcal{A}}$ consists of all the functions in M_0 for which the limit in (3.5) exists for a.a. x.

Theorem 3.1. *Let $\mathcal{A}$ be the infinitesimal generator of a continuous time-homogeneous Markov process satisfying the Feller property. If $f \in D_{\mathcal{A}}$ and $\mathcal{A}f$ is continuous at a point y, then*

$$(\mathcal{A}f)(y) = \lim_{U \downarrow \{y\}} \frac{E_y f(x(\tau_U)) - f(y)}{E_y \tau_U} \tag{3.6}$$

where U is an open neighborhood of y and τ_U is the exit time from U.

This is called *Dynkin's formula*. Since we shall not use this formula in the sequel, we relegate the proof to the problems.

For nonstationary transition probability function, one defines the infinitesimal generators by

$$\mathcal{A}_s f = \lim_{t \downarrow 0} \frac{T_{s,t+s} f - f}{t} . \tag{3.7}$$

The Dynkin formula extends to this case (with obvious changes in the proof).

PROBLEMS

1. Let $\{\Omega, \mathcal{F}, \mathcal{F}_t^s, x(t), P_{x,s}\}$ be a continuous Markov process. Let τ be the first time that $t \geqslant s$ and $(t, x(t))$ hits a closed set F in $R^{\nu+1}$. Prove that τ is an s-stopping time, for any $s \geqslant 0$.

2. Let $\{\Omega, \mathcal{F}, \mathcal{F}_t^s, x(t), P_{x,s}\}$ be a ν-dimensional Markov process with transition probability p. Let X denote a variable point (x_0, x) in $R^{\nu+1}$ and let $X(t, \omega) = (t, x(t, \omega))$. Prove that $\{\Omega, \mathcal{F}, \mathcal{F}_t^s, X(t), P_{X,s}\}$ is a Markov process with transition probability $\tilde{p}$,

$$\tilde{p}(s, X, t, I \times A) = \chi_I(x_0) p(s, x, t, A)$$

where I is any Borel set in $[0, \infty)$.

3. If $\{\Omega, \mathcal{F}, \mathcal{F}_t, x(t), P_x\}$ is a time-homogeneous Markov process and if

$$\int |y_1 - y_2|^\alpha p(t, x, dy_1) p(s, y_1, dy_2) \leqslant C|t - s|^{1+\epsilon}$$

for some $\alpha > 0$, $\epsilon > 0$, $C > 0$, then $x(t)$ has a continuous version.

[*Hint*: Use Theorem 1.2.2.]

4. Prove Theorem 1.3.

[*Hint*: Since $p(\bar{s}, y(\bar{s}), t, A)$ is $\mathcal{F}(y(\lambda), s \leq \lambda < \bar{s})$ measurable, it suffices to show that if $t_1 < \cdots < t_n = \bar{s} < t_{n+1} = t$,

$$P[y(t_1) \in A_1, \ldots, y(t_n) \in A_n, y(t_{n+1}) \in A]$$
$$= Ep(t_n, y(t_n), t_{n+1}, A)\chi_{A_1}(y(t_1)) \cdots \chi_{A_n}(y(t_n)).$$

Denote by Q_1 the probability distribution of $(x(t_1), \ldots, x(t_n), x(t_{n+1}))$ and by Q_2 the probability distribution of $(y(t_1), \ldots, y(t_n), y(t_{n+1}))$. Make use of the relations

$$\int \chi_{A_1}(x_1) \cdots \chi_{A_n}(x_n)\chi_A(x_{n+1})\, dQ_1(x_1, \ldots, x_n, x_{n+1})$$
$$= \int p(t_n, x_n, t_{n+1}, A)\chi_{A_1}(x_1) \cdots \chi_{A_n}(x_n)\, dQ_1(x_1, \ldots, x_n, x_{n+1}),$$
$$\int \Phi(x_1, \ldots, x_{n+1})\, dQ_1 = \int \Phi(x_1, \ldots, x_{n+1})\, dQ_2,$$

where Φ is bounded and measurable.]

5. A ν-dimensional stochastic process $x(t)$ $(t \geq 0)$ is said to satisfy the *Markov property* if

$$P[x(t) \in A \mid \mathcal{F}(x(\lambda)), \lambda \leq s)] = P[x(t) \in A \mid x(s)] \quad \text{a.s.} \qquad (\star)$$

for any $0 \leq s < t$ and for any Borel set A. It is well known (see Breiman [1]) that there exists a regular conditional probability $p(s, x, t, A)$ of $(\star)$, i.e., $p(s, x, t, A)$ is a probability in A for fixed (s, x, t) and Borel measurable in x for (s, t, A) fixed. Prove that p satisfies

$$\int p(s, x, \lambda, dy)p(\lambda, y, t, A) = p(s, x, t, A) \quad \text{a.s.} \qquad (s < \lambda < t).$$

6. Let $x(t)$ be a process satisfying the Markov property. Then, for any $0 = t_0 < t_1 < \cdots < t_n$ and for any Borel sets $B_0, \ldots, B_n$,

$$P\{x(t_n) \in B_n, \ldots, x(t_1) \in B_1, x(t_0) \in B_0\}$$
$$= \int_{B_n} \cdots \int_{B_1} \int_{B_0} p(t_{n-1}, x_{n-1}, t_n, dx_n) \cdots p(t_0, x_0, t_1, dx_1)\pi(dx_0)$$

where $\pi(dx)$ is the probability distribution of $x(0)$.

7. Let $\{\Omega, \mathcal{F}, \mathcal{F}_t^s, x(t), P_{x,s}\}$ be a ν-dimensional continuous Markov process satisfying the Feller property. Let τ be a real-valued s-stopping time. Denote by $\mathcal{F}_\infty^\tau$ the σ-field generated by $x(t + \tau)$, $t > 0$. Prove that

$$P_{x,s}(B \mid \mathcal{F}_\tau^s) = P_{x,s}(B \mid x(\tau)) \qquad \text{for any} \quad B \in \mathcal{F}_\infty^\tau. \tag{3.8}$$

[*Hint*: Prove that for any bounded measurable function $f(x_1, \ldots, x_n)$,

$$E_{x,s}[f(x(t_1 + \tau), \ldots, x(t_n + \tau)) \mid \mathcal{F}_\tau^s] = E_{x(\tau),\tau} f(x(t_1 + \tau), \ldots, x(t_n + \tau)) \tag{3.9}$$

where $s \leqslant t_1 < \cdots < t_n$; cf. the proof of Theorem 1.2.]

In Problems 8–16, $\{\Omega, \mathcal{F}, \mathcal{F}_t, x(t), P_x\}$ is a continuous time-homogeneous ν-dimesional Markov process satisfying the Feller property. By Corollary 2.6, the process also satisfies the strong Markov property.

8. Prove that for every open (closed) set G, $p(t, x, G)$ is lower (upper) semicontinuous.

9. Let A be an open set if R^ν with boundary ∂A, and let Γ be a closed subset of ∂A. Assume that there is an increasing sequence of closed subsets Γ'_m of ∂A such that $\Gamma'_m \uparrow \Gamma' = \partial A \backslash \Gamma$ as $m \uparrow \infty$. Assume that $P_x\{x(t)$ hits $\partial A\} = 1$, and let $E = \{x(t)$ hits Γ before it hits $\Gamma'\}$. Then there exists a sequence $f_n(x_1, \ldots, x_n)$ of bounded measurable functions such that

$$f_n(x(t_1), \ldots, x(t_n)) \to \chi_E \quad \text{a.s.,}$$

and the functions $f_n(x_1, \ldots, x_n)$ do not depend on the particular process $x(t)$.

[*Hint*: Let $\{\lambda_i\}$ be a dense sequence in $[0, \infty)$. Check that

$$E = \bigcup_j \bigcap_m \left\{ \bigcap_p \bigcup_{t_i < \lambda_j} \left[\rho(x(t_i), \Gamma) < \frac{1}{p} \right] \right\}$$

$$\cap \left\{ \bigcup_r \bigcap_{t_l < \lambda_j} \left[\rho(x(t_l), \Gamma'_m) \geqslant \frac{1}{r} \right] \right\}$$

where j, m, p, i, r, l vary over $1, 2, \ldots$. Show that $P(E) = \lim P(E_\alpha)$ where E_α is defined as in E but with j, m, p, i, r, l varying only over a finite set. Deduce that $\chi_E = \lim \chi_{E_\alpha}$, $\chi_{E_\alpha} = f_n(x(t_1), \ldots, x(t_n))$ for some $n = n(\alpha)$, and $f_n(x_1, \ldots, x_n)$ does not depend on the particular process $x(t)$.]

10. Let A and B be open sets with boundaries ∂A and ∂B respectively, and assume that $A \cup \partial A$ is contained in B, and that ∂A consists of a finite number of manifolds. Denote by t_A and t_B the exit times from A and B respectively. Suppose that $P_x(t_B < \infty) = 1$ for all $x \in A \cup \partial A$. Prove that for any bounded measurable function f,

$$E_x f(x(t_B)) = E_x E_{x(t_A)} f(x(t_B)) \qquad (x \in A). \tag{3.10}$$

[*Hint*: For time-homogeneous processes, (3.9) becomes

$$E_x[f(x(t_1 + \tau), \ldots, x(t_n + \tau)) | \mathcal{F}_\tau] = E_{x(\tau)} f(x(t_1), \ldots, x(t_n)). \tag{3.11}$$

Suppose $f = \chi_\Gamma$, Γ closed, and introduce the events

$$E = \{x(t) \text{ hits } \Gamma \cap \partial B \text{ before it hits } \partial B \backslash \Gamma\},$$

$$E' = \{x(t + t_A) \text{ hits } \Gamma \cap \partial B \text{ before it hits } \partial B \backslash \Gamma\}.$$

By Problem 9,

$$\chi_E = \lim f_n(x(t_1), \ldots, x(t_n)), \qquad \chi_{E'} = \lim f_n(x(t_1 + t_A), \ldots, x_n(t_n + t_A)),$$

Apply (3.11) to f_n to obtain

$$E_x f(x(t_B)) = E_x \chi_{E'} = E_x E_{x(t_A)} \chi_E = E_x E_{x(t_A)} f(x(t_B)).]$$

11. Let τ be the exit time from an open set G. Suppose $P_x(\tau > \alpha) \leqslant \beta$ for all $x \in G$. Prove that

(i) $P_x(\tau > n\alpha) \leqslant \beta^n$;
(ii) $E_x \tau \leqslant \alpha/(1 - \beta)$;
(iii) $E_x e^{\lambda\tau} \leqslant e^{\lambda\alpha}/(1 - e^{\lambda\alpha}\beta)$ if $\beta < 1$, $\lambda < (-\alpha^{-1} \log \beta)$.

[*Hint*: Write

$$\chi_{[x(t) \in G \text{ if } t < \alpha]} = \lim f_m(x(t_1), \ldots, x(t_m)) \qquad (t_i < \alpha),$$

$$\chi_{[x(t) \in G \text{ if } t < \alpha + \beta]} = \lim f_m(x(t_1 + \beta), \ldots, x(t_m + \beta))\chi_{[x(t) \in G \text{ if } t < \alpha]}$$

for any $\beta > 0$. Use (3.11) and induction to verify that

$$P_x[\tau > (n+1)\alpha] = E_x \chi_{\tau > \alpha} \chi_{\tau > (n+1)\alpha} = E_x \chi_{\tau > \alpha}(E_x \chi_{\tau > (n+1)\alpha} | \mathcal{F}_\alpha)$$
$$= E_x \chi_{\tau > \alpha} E_{x(\alpha)} \chi_{\tau > n\alpha} \leqslant \beta^{n+1}].$$

12. Suppose $\nu = 1$ and $\tau = \inf\{s, x(s) = y\}$ is finite valued, where y is a real number. Prove, for any closed set A,

$$P[x(t) \in A, \tau \leqslant t] = E_x \chi_{x(t) \in A} \chi_{\tau \leqslant t} = E_x E_x[\chi_{x(t) \in A} \chi_{\tau \leqslant t} | \mathcal{F}_\tau]$$
$$= E_x \chi_{\tau \leqslant t} E_x[\chi_{x(t) \in A} | \mathcal{F}_\tau] = E_x \chi_{\tau \leqslant t} E_{x(\tau)} \chi_{x(t - \tau) \in A}$$
$$= E_x \chi_{\tau \leqslant t} P_{x(\tau)}[x(t - \tau) \in A] = \int_0^t P_x(\tau \in ds) P_y[x(t - s) \in A].$$

[*Hint*: To prove the fourth equality, choose a sequence of functions as in Problem 10 with $E = \{x(t - \tau) \in A\}$.]

13. A point y is called *absorbing* if $p(t, y, \{y\}) = 1$ for all $t \geqslant 0$. It follows that $(T_t f)(y) = f(y)$. Hence, if $f \in D_{\mathcal{Q}}$, then $(\mathcal{Q}f)(y) = 0$. Show that if y is nonabsorbing and U is a sufficiently small neighborhood of y, τ_U = exit time from U, then $\sup_{x \in U} E_x \tau_U < \infty$.

[*Hint*: $p(t_0, y, N) < 1$ for some $t_0 > 0$ and a closed neighborhood N of y. Hence $p(t_0, x, R^\nu \backslash N) \geqslant \delta > 0$ if $|x - y|$ is sufficiently small, and then $P_x(\tau_U > t_0) \leqslant 1 - \delta$.]

14. Prove that M_0 is a closed subspace of $M(R^\nu)$, that $T_t M_0 \subset M_0$ if $t \geqslant 0$, and that $T_t f$ is a continuous function in $t \in [0, \infty)$ for any $f \in M_0$.

15. From the theory of semigroups (see, for instance, Dynkin [2]) one knows that:

(i) $D_{\mathcal{Q}}$ is dense in M_0, $\mathcal{Q}f \in M_0$ if $f \in D_{\mathcal{Q}}$;

(ii) $T_t D_{\mathcal{Q}} \subset D_{\mathcal{Q}}$;
(iii) if $f \in D_{\mathcal{Q}}$, $dT_t f/dt = T_t \mathcal{Q} f = \mathcal{Q} T_t f$ for $t \geqslant 0$;
(iv) the operator $R_\lambda f = \int_0^\infty e^{-\lambda t}(T_t f)\,dt$ ($f \in M_0$) is defined for all $\lambda > 0$ and satisfies:

(a) $R_\lambda f \in D_{\mathcal{Q}}$ if $f \in M_0$;
(b) $\|R_\lambda f\| \leqslant \lambda^{-1}\|f\|$;
(c) $(\lambda I - \mathcal{Q})R_\lambda f = f$ if $f \in M_0$, $R_\lambda(\lambda I - \mathcal{Q})f = f$ if $f \in D_{\mathcal{Q}}$.

Now let y be nonabsorbing and take a neighborhood U of y with $\sup_{x \in U} E_x \tau_U < \infty$. Let $h_\lambda = \lambda f - \mathcal{Q} f$. Check that

$$f(x) = R_\lambda h_\lambda(x)$$

$$= E_x \int_0^{\tau_U} e^{-\lambda t} h_\lambda(x(t))\,dt + E_x\left[e^{-\lambda \tau_U} \int_0^\infty e^{-\lambda t} h_\lambda(x(t + \tau_U))\,dt \right].$$

Use the strong Markov property to deduce

$$E_x f(x(\tau_U)) - f(x) = E_x\left[(e^{-\lambda \tau_U} - 1) f(x(\tau_U)) \right] - E_x\left[\int_0^{\tau_U} e^{-\lambda t} h_\lambda(x(t))\,dt \right].$$

16. Use Problems 13 and 15 to complete the proof of Dynkin's formula.

3

Brownian Motion

1. Existence of continuous Brownian motion

Definition. A *Brownian motion* or a *Wiener process* is a stochastic process $x(t)$, $t \geq 0$ satisfying;

(i) $x(0) = 0$;

(ii) for any $0 \leq t_0 < t_1 < \cdots < t_n$, the random variables $x(t_k) - x(t_{k-1})$ $(1 \leq k \leq n)$ are independent;

(iii) if $0 \leq s < t$, $x(t) - x(s)$ is normally distributed with

$$E(x(t) - x(s)) = (t - s)\mu, \qquad E(x(t) - x(s))^2 = (t - s)\sigma^2$$

where μ, σ are real constants, $\sigma \neq 0$.

μ is called the *drift* and σ^2 is called the *variance*.

Property (ii) implies that $x(t) - x(s)$ is independent of $\mathcal{F}(x(\lambda), \lambda \leq s)$ or, more generally, that $\mathcal{F}(x(\mu) - x(s), \mu \geq s)$ is independent of $\mathcal{F}(x(\lambda), \lambda \leq s)$.

If $x(t)$ is a Brownian motion with drift μ and variance σ^2, and if $0 \leq t_0 < t_1 < \cdots < t_n$, then

$$\Gamma(t_j, t_k) \equiv E(x(t_j) - \mu t_j)(x(t_k) - \mu t_k) = \sigma^2 \min(t_j, t_k). \tag{1.1}$$

Indeed, if $t_j > t_k$,

$$\Gamma(t_j, t_k) = E[x(t_j) - x(t_k) - \mu(t_j - t_k) + x(t_k) - \mu t_k][x(t_k) - \mu t_k]$$
$$= E(x(t_k) - \mu t_k)^2 = t_k\sigma^2.$$

If $\mu = 0$, $\sigma^2 = 1$, then we speak of *normalized* Brownian motion. Notice that for any Brownian motion with drift μ and variance σ^2, $(x(t) - \mu t)/\sigma$ is a normalized Brownian motion.

From now on we consider only normalized Brownian motion and refer to it briefly as Brownian motion.

Let $X_1, \ldots, X_n$ be random variables with joint normal distribution $N(0, \Gamma)$ where $\Gamma = (\Gamma_{ij})$, $\Gamma_{ij} = \min(t_i, t_j)$, $0 < t_1 < \cdots < t_n$. It is easily checked that $\det \Gamma \neq 0$. Let (Γ_{ij}^{-1}) be the inverse matrix to Γ. Let X_0 be the random variable $X_0 = 0$; its distribution function is

$$X_0(x_0) = \begin{cases} 0 & \text{if } x_0 \le 0, \\ 1 & \text{if } x_0 > 0. \end{cases}$$

Then (see Problem 1) the joint distribution function $F_{t_0 t_1 \cdots t_n}$ of X_0, $X_1, \ldots, X_n$ (where $t_0 = 0$) is given by

$$F_{t_0 t_1 \cdots t_n}(x_0, x_1, \ldots, x_n) = \frac{\chi(x_0)}{(2\pi)^{n/2}(\det \Gamma)^{1/2}} \int_{-\infty}^{x_n} \cdots \int_{-\infty}^{x_1} \exp\left[-\tfrac{1}{2} \sum_{i,j=1}^{n} \Gamma_{ij}^{-1} y_i y_j\right] dy_1 \cdots dy_n. \tag{1.2}$$

It follows that

$$F_{t_0 \cdots t_n}(x_0, \ldots, x_n) \uparrow F_{t_0 \cdots t_{k-1} t_{k+1} \cdots t_n}(x_0, \ldots, x_{k-1}, x_{k+1}, \ldots, x_n) \quad \text{if } x_k \uparrow \infty.$$

Thus, the functions $F_{t_0 t_1 \cdots t_n}$ given by (1.2) form a consistent family of distribution functions. By the Kolmogorov construction there exists a probability space $(\Omega, \mathcal{F}, P)$ and a process $x(t)$ such that the joint distribution functions of $x(t)$ are given by (1.2). But then, by Problem 2, $x(t)$ is a Brownian motion.

We have thus proved that the Kolmogorov construction produces a Brownian motion.

It is easily seen that a Brownian motion is a martingale. Thus the martingale inequality can be applied to Brownian motion. Another important inequality is given in the following theorem.

Theorem 1.1. *Let $x(t)$ be a Brownian motion and let $0 = t_0 < t_1 < \cdots < t_n$. Then, for any $\lambda > 0$,*

$$P\left[\max_{0 \le j \le n} x(t_j) > \lambda\right] \le 2P[x(t_n) > \lambda], \tag{1.3}$$

$$P\left[\max_{0 \le j \le n} |x(t_j)| > \lambda\right] \le 2P[|x(t_n)| > \lambda]. \tag{1.4}$$

Proof. Let j^* = first j such that $x(t_j) > \lambda$. Since $x(t_n) - x(t_j)$ $(0 \le j < n)$ has normal distribution, its probability distribution is symmetric about the origin, i.e., $P[x(t_n) - x(t_j) > 0] = P[x(t_n) - x(t_j) < 0]$. Using also the facts that $x(t_n) - x(t_j)$ is independent of $\mathcal{F}(x(t_0), \ldots, x(t_j))$ and that $\{j^* = j\}$

belongs to this σ-field, we can write

$$P\left[\max_{0\le j\le n} x(t_j) > \lambda, x(t_n) < \lambda\right] = \sum_{k=1}^{n-1} P[j^* = k, x(t_n) < \lambda]$$

$$\le \sum_{k=1}^{n} P[j^* = k, x(t_n) - x(t_k) < 0] = \sum_{k=1}^{n} P(j^* = k)P[x(t_n) - x(t_k) < 0]$$

$$= \sum_{k=1}^{n} P(j^* = k)P[x(t_n) - x(t_k) > 0] = \sum_{k=1}^{n} P[j^* = k, x(t_n) - x(t_k) > 0]$$

$$\le \sum_{k=1}^{n} P[j^* = k, x(t_n) > \lambda] \le P[x(t_n) > \lambda].$$

On the other hand,

$$P\left[\max_{0\le j\le n} x(t_j) > \lambda, x(t_n) > \lambda\right] = P[x(t_n) > \lambda].$$

Adding these inequalities, (1.3) follows. Noting that $-x(t)$ is also a Brownian motion and employing (1.3), we get

$$P\left[\max_{0\le j\le n} |x(t_j)| > \lambda\right] \le P\left[\max_{0\le j\le n} x(t_j) > \lambda\right] + P\left[\max_{0\le j\le n} (-x(t_j)) > \lambda\right]$$

$$\le 2P[x(t_n) > \lambda] + 2P[(-x(t_n)) > \lambda] = 2P[|x(t_n)| > \lambda].$$

If X is a random variable with normal distribution $N(0, \sigma^2)$, then

$$EX^{2n} = \frac{(2n)!}{2^n n!}\sigma^{2n}, \qquad EX^{2n+1} = 0 \qquad (n = 0, 1, 2, \ldots). \tag{1.5}$$

In particular, for any Brownian motion $x(t)$,

$$E|x(t) - x(s)|^{2n} = C_n|t - s|^n \qquad (C_n \text{ constant}) \tag{1.6}$$

for $n = 1, 2, \ldots$. Taking $n = 2$ and applying Theorem 1.2.2, we get:

Theorem 1.2. *There is a continuous version of a Brownian motion.*

Corollary 1.3. *Every separable Brownian motion is continuous.*

This follows from Theorem 1.2 and Problem 3, Chapter 1.

From now on, when we speak of a Brownian motion, it is always tacitly assumed that we speak of a continuous version.

From (1.6) we see that Corollary 1.2.3 can be applied to a Brownian motion with $\beta = 2n$, $\alpha = n - 1$, for any $n \ge 1$. Consequently:

Corollary 1.4. *For any $\alpha < \frac{1}{2}$, almost all the sample paths of a Brownian motion are Hölder continuous, with exponent α, on every bounded t-interval.*

2. Nondifferentiability of Brownian motion

We have proved the existence of a Hölder continuous Brownian motion $x(t)$, with any exponent $\alpha < \frac{1}{2}$.

Lévy [1] has proved that the sample paths of $x(t)$ satisfy:

$$\overline{\lim_{\substack{0 \leq s < t \leq T \\ t-s \downarrow 0}}} \frac{|x(t) - x(s)|}{\{2(t-s)\log 1/(t-s)\}^{1/2}} = 1 \quad \text{a.s.,} \qquad \text{for any} \quad T > 0;$$

for proof see also McKean [1].

We shall prove here only a weaker result:

Theorem 2.1. *For any $\alpha > \frac{1}{2}$, almost all simple paths of a Brownian motion are nowhere Hölder continuous with exponent α.*

Proof. Let T be any positive integer and let N be a positive integer such that $N(\alpha - \frac{1}{2}) > 1$. If $x(t, \omega)$ is Hölder continuous at s ($s \in [0, T]$) with exponent α, then

$$|x(t, \omega) - x(s, \omega)| < \beta_0 |t - s|^\alpha \qquad \text{if} \quad |t - s| \leq N/n$$

for some $\beta_0 > 0$ and some positive integer n.

Let β be a positive number and consider the event

$$A_n = \{\text{there exists an } s \in [0, T] \text{ such that } |x(t) - x(s)| < \beta |t - s|^\alpha \text{ if } |t - s| \leq N/n\}.$$

Then $A_n \uparrow A$ as $n \uparrow \infty$. If we prove that, for any β, T, $P(A) = 0$, then the assertion of the theorem follows.

Let

$$Z_k = \max_{1 \leq i \leq N} \left| x\left(\frac{k+i}{n}\right) - x\left(\frac{k+i-1}{n}\right) \right| \qquad (k = 0, 1, \ldots, nT),$$

$$B_n = \{\omega;\, Z_k(\omega) \leq 2\beta N^\alpha / n^\alpha \text{ for some } k\}.$$

Note that if $\omega \in A_n$ and $|x(t, \omega) - x(s, \omega)| < \beta |t - s|^\alpha$ when $|t - s| \leq N/n$, and if we take k to be the largest integer such that $k/n \leq s$, then

$$Z_k(\omega) \leq 2\beta (N/n)^\alpha.$$

Consequently $A_n \subset B_n$, so that

$$\begin{aligned} P(A) &= \lim_{n \to \infty} P(A_n) \\ &\leq \underline{\lim}_{n \to \infty} P(B_n). \end{aligned} \tag{2.1}$$

Next,

$$P(B_n) = P\left\{ \bigcup_{k=0}^{nT} \left[Z_k < \frac{2\beta N^\alpha}{n^\alpha} \right] \right\}$$

$$< \sum_{k=0}^{nT} P\left[Z_k < \frac{2\beta N^\alpha}{n^\alpha} \right] = nT\, P\left[Z_0 < \frac{2\beta N^\alpha}{n^\alpha} \right].$$

Since the random variables

$$x\left(\frac{i}{n}\right) - x\left(\frac{i-1}{n}\right) \qquad (1 \leq i \leq N)$$

are independent and identically distributed,

$$P(B_n) < nT\left\{ P\left[\left| x\left(\frac{1}{n}\right) \right| < \frac{2\beta N^\alpha}{n^\alpha} \right] \right\}^N$$

$$= nT\left\{ \sqrt{\frac{n}{2\pi}} \int_{-\gamma/n^\alpha}^{\gamma/n^\alpha} e^{-nx^2/2}\, dx \right\}^N \qquad (\gamma = 2\beta N^\alpha).$$

Substituting $n^\alpha x = y$, we get

$$P(B_n) = nT\left\{ \frac{1}{\sqrt{2\pi}\, n^{\alpha - 1/2}} \int_{-\gamma}^{\gamma} e^{-y^2/(2n^{2\alpha-1})}\, dy \right\}^N \to 0 \qquad \text{if} \quad n \to \infty,$$

since $N(\alpha - 1/2) > 1$. Using (2.1) we then conclude that $P(A) = 0$. This completes the proof.

Corollary 2.2. (i) *Almost all sample paths of a Brownian motion are nowhere differentiable.*

(ii) *Almost all sample paths of a Brownian motion have infinite variation on any finite interval.*

The assertion (i) follows from the fact that if a function is differentiable at a point, then it is Lipschitz continuous at that point. Since a function $f(t)$ with finite variation is almost everywhere differentiable, (ii) is a consequence of (i).

3. Limit theorems

Theorem 3.1. *For a Brownian motion $x(t)$,*

$$\overline{\lim_{t \downarrow 0}} \frac{x(t)}{\sqrt{2t \log\log(1/t)}} = 1 \quad a.s., \tag{3.1}$$

$$\overline{\lim_{t \uparrow \infty}} \frac{x(t)}{\sqrt{2t \log\log t}} = 1 \quad a.s. \tag{3.2}$$

These formulas are called the *laws of the iterated logarithm.*

Proof. We first prove that

$$\overline{\lim_{t \downarrow 0}} \frac{x(t)}{\sqrt{2t \log \log(1/t)}} \leqslant 1. \tag{3.3}$$

Let $\delta > 0$, $\phi(t) = \sqrt{2t \log \log(1/t)}$, and take a sequence $t_n \downarrow 0$. Consider the event

$$A_n = \{x(t) > (1 + \delta)\phi(t) \text{ for at least one } t \in [t_{n+1}, t_n]\}.$$

Since $\phi(t) \uparrow$ if $t \uparrow$,

$$A_n \subset \Big\{ \sup_{0 < t < t_n} x(t) > (1 + \delta)\phi(t_{n+1}) \Big\}.$$

By Theorem 1.1, if $x > 0$

$$P\Big[\sup_{0 < t < t_n} x(t) > x\sqrt{t_n} \Big] \leqslant 2P\big[x(t_n) > x\sqrt{t_n}\,\big] = 2P\bigg[\frac{x(t_n)}{\sqrt{t_n}} > x \bigg]$$

$$= \sqrt{\frac{2}{\pi}} \int_x^\infty e^{-z^2/2}\, dz \leqslant \sqrt{\frac{2}{\pi}}\, \frac{1}{x} \int_x^\infty z e^{-z^2/2}\, dz = \sqrt{\frac{2}{\pi}}\, \frac{e^{-x^2/2}}{x}.$$

Taking $x = x_n = (1 + \delta)\phi(t_{n+1})/\sqrt{t_n}$, we get

$$P(A_n) \leqslant \sqrt{\frac{2}{\pi}}\, \frac{\exp(-x_n^2/2)}{x_n}.$$

Now take $t_n = q^n$ where $0 < q < 1$ and $\lambda = q(1 + \delta)^2 > 1$. Then

$$x_n = \big\{2 \log\big[\alpha(n + 1)^\lambda\big]\big\}^{1/2}, \qquad \alpha = \log(1/q).$$

It follows that

$$P(A_n) \leqslant \frac{C}{(n + 1)^\lambda \log(n + 1)} \qquad (C \text{ constant}).$$

Hence $\Sigma\, P(A_n) < \infty$, so that, by the Borel–Cantelli lemma, $P(A_n \text{ i.o.}) = 0$ (where "i.o." stands for infinitely often). But this means that

$$\overline{\lim_{t \downarrow 0}} \frac{x(t)}{\sqrt{2t \log \log(1/t)}} \leqslant 1 + \delta.$$

Since δ is arbitrary, (3.3) follows.

We shall next prove that

$$\overline{\lim_{t \downarrow 0}} \frac{x(t)}{\sqrt{2t \log \log(1/t)}} \geqslant 1. \tag{3.4}$$

We again start with a sequence $t_n \downarrow 0$. Let $Z_n = x(t_n) - x(t_{n+1})$. For any $x > 0$, $\epsilon > 0$,

$$P(Z_n > x\sqrt{t_n - t_{n+1}}) = P\left(\frac{x(t_n) - x(t_{n+1})}{\sqrt{t_n - t_{n+1}}} > x\right) = \frac{1}{\sqrt{2\pi}} \int_x^{\infty} e^{-z^2/2}\, dz.$$

By integration by parts,

$$\frac{1}{x} e^{-x^2/2} = \int_x^{\infty}\left(1 + \frac{1}{z^2}\right)e^{-z^2/2}\, dz \leq \left(1 + \frac{1}{x^2}\right)\int_x^{\infty} e^{-z^2/2}\, dz,$$

so that

$$\int_x^{\infty} e^{-z^2/2}\, dz \geq \left(x + \frac{1}{x}\right)^{-1} e^{-x^2/2}.$$

Thus

$$P(Z_n > x\sqrt{t_n - t_{n+1}}) \geq \frac{1}{2\sqrt{2\pi}\, x} e^{-x^2/2} \qquad \text{if} \quad x > 1.$$

Taking $t_n = q^n$, $0 < q < 1$,

$$x = x_n = (1 - \epsilon)\frac{\phi(t_n)}{\sqrt{t_n - t_{n+1}}} = \frac{1 - \epsilon}{\sqrt{1 - q}} \sqrt{2 \log[n \log(1/q)]} = \sqrt{\beta \log(\alpha n)}$$

where $\alpha = \log(1/q)$, $\beta = 2(1 - \epsilon)^2/(1 - q)$, and choosing q sufficiently small so that $\beta < 2$, we get

$$P[Z_n > (1 - \epsilon)\phi(t_n)] \geq \frac{c}{n^{\beta/2}\sqrt{\log n}} \qquad (c \text{ positive constant})$$

and, consequently,

$$\sum_{n=1}^{\infty} P[Z_n > (1 - \epsilon)\phi(t_n)] = \infty.$$

Since the events $[Z_n > (1 - \epsilon)\phi(t_n)]$ are independent, the Borel–Cantelli lemma implies that

$$P[Z_n > (1 - \epsilon)\phi(t_n) \text{ i.o.}] = 1. \tag{3.5}$$

By the proof of (3.1) applied to the Brownian motion $-x(t)$,

$$P[x(t_{n+1}) < -(1 + \epsilon)\phi(t_{n+1}) \text{ i.o.}] = 0.$$

Putting it together with (3.5), we deduce that a.s.

$$\begin{aligned} x(t_n) &= Z_n + x(t_{n+1}) > (1 - \epsilon)\phi(t_n) - (1 + \epsilon)\phi(t_{n+1}) \\ &= \phi(t_n)\left[1 - \epsilon - (1 + \epsilon)\frac{\phi(t_{n+1})}{\phi(t_n)}\right] \quad \text{i.o.} \end{aligned}$$

For any given $\delta > 0$, if we choose ϵ and q so small that

$$1 - \epsilon - (1 + \epsilon)\sqrt{q} > 1 - \delta,$$

and note that $\phi(t_{n+1})/\phi(t_n) \to \sqrt{q}$ if $n \to \infty$, we get that

$$P[x(t_n) > (1 - \delta)\phi(t_n) \text{ i.o.}] = 1.$$

Since δ is arbitrary, this completes the proof of (3.4).

In order to prove (3.2), consider the process

$$\tilde{x}(t) = \begin{cases} tx\left(\dfrac{1}{t}\right) & \text{if } t > 0, \\ 0 & \text{if } t = 0. \end{cases}$$

It is easy to check that $\tilde{x}(t)$ satisfies the conditions (i)–(iii) in the definition of a Brownian motion. Since it is clearly also a separable process, it is a (continuous) Brownian motion. If we now apply (3.1) to $\tilde{x}(t)$, then we obtain (3.2).

Corollary 3.2. *For a Brownian motion $x(t)$,*

$$\varliminf_{t \downarrow 0} \frac{x(t)}{\sqrt{2t \log \log(1/t)}} = -1 \quad \text{a.s.,} \tag{3.6}$$

$$\varliminf_{t \uparrow \infty} \frac{x(t)}{\sqrt{2t \log \log t}} = -1 \quad \text{a.s.} \tag{3.7}$$

This follows by applying Theorem 3.1 to the Brownian motion $-x(t)$.

If $f(t)$ is a continuously differentiable function in an interval $[a, b]$, and if

$$\Pi_n = \{t_{n,1}, \ldots, t_{n,m_n}\} \tag{3.8}$$

is a sequence of partitions of $[a, b]$ with mesh $|\Pi_n| = \max(t_{n,j} - t_{n,j-1})$ converging to 0, then

$$\sum_{j=1}^{m_n} [f(t_{n,j}) - f(t_{n,j-1})]^2 \to 0 \quad \text{if } n \to \infty.$$

Such a conclusion does not hold for a Brownian motion. Instead we have:

Theorem 3.3. *Let $x(t)$ be a Brownian motion, and Π_n a sequence of partitions* (3.8) *of a finite closed interval $[a, b]$ with mesh $|\Pi_n| \to 0$ if $n \to \infty$. Let*

$$S_n = \sum_{j=1}^{m_n} [x(t_{n,j}) - x(t_{n,j-1})]^2.$$

Then $S_n \to b - a$ in the mean.

Proof. Write $t_j = t_{n,j}$, $m = m_n$. Then

$$S_n - (b-a) = \sum_{j=1}^{m} \left[(x(t_j) - x(t_{j-1}))^2 - (t_j - t_{j-1}) \right].$$

Since the summands are independent and of mean zero,

$$E[S_n - (b-a)]^2 = E\sum_{j=1}^{n} \left[(x(t_j) - x(t_{j-1}))^2 - (t_j - t_{j-1}) \right]^2$$
$$= \sum_{j=1}^{m} E\left[(Y_j^2 - 1)(t_j - t_{j-1}) \right]^2.$$

where

$$Y_j = \frac{x(t_j) - x(t_{j-1})}{(t_j - t_{j-1})^{1/2}}.$$

Since the Y_j are equally distributed with normal distribution,

$$E[S_n - (b-a)]^2 = E(Y_1^2 - 1)^2 \sum_{j=1}^{m} (t_j - t_{j-1})^2$$
$$\leqslant E(Y_1^2 - 1)^2 \cdot (b-a)|\Pi_n| \to 0 \qquad \text{if} \quad n \to \infty.$$

4. Brownian motion after a stopping time

Let τ be a stopping time for a Brownian motion $x(t)$. Denote by $\mathfrak{F}_\tau$ the σ-field of all events A such that

$$A \cap (\tau \leqslant t) \qquad \text{is in} \qquad \mathfrak{F}(x(\lambda), 0 \leqslant \lambda \leqslant t).$$

From the considerations of Section 1.3 we know that $x(\tau)$ is a random variable.

Theorem 4.1. *If τ is a stopping time for a Brownian motion $x(t)$, then the process*

$$y(t) = x(t+\tau) - x(\tau), \qquad t \geqslant 0$$

is a Brownian motion, and $\mathfrak{F}(y(t), t \geqslant 0)$ is independent of $\mathfrak{F}_\tau$.

Thus the assertion is that a Brownian motion starts afresh at any stopping time.

Proof. Notice that if $\tau = s$ (a constant), then the assertion is obvious. Suppose now that the range of τ is a countable set $\{s_j\}$. Let $B \in \mathfrak{F}_\tau$ and

$0 \le t_1 < t_2 < \cdots < t_n$. Then, for any Borel sets $A_1, \ldots, A_n$,

$$\begin{aligned} P[y(t_1) \in A_1, \ldots, y(t_n) \in A_n, B] &= \sum_k P[y(t_1) \in A_1, \ldots, y(t_n) \in A_n, \tau = s_k, B] \\ &= \sum_k P[(x(t_1 + s_k) - x(s_k)) \in A_1, \ldots, (x(t_n + s_k) - x(s_k)) \\ &\qquad \in A_n, \tau = s_k, B]. \end{aligned} \tag{4.1}$$

Since

$$(\tau = s_k) \cap B = [B \cap (\tau \le s_k)] \cap (\tau = s_k) \quad \text{is in} \quad \mathcal{F}(x(\lambda), 0 \le \lambda \le s_k)$$

and $\mathcal{F}(x(\lambda + s_k) - x(s_k)), \lambda \ge 0)$ is independent of $\mathcal{F}(x(\lambda), 0 \le \lambda \le s_k)$, and since the assertion of the theorem is true if $\tau \equiv s_k$, the kth term on the right-hand side of (4.1) is equal to

$$\begin{aligned} &P[(x(t_1 + s_k) - x(s_k)) \in A_1, \ldots, (x(t_n + s_k) - x(s_k)) \in A_k]P[\tau = s_k, B] \\ &\quad = P[x(t_1) \in A_1, \ldots, x(t_n) \in A_n]P[\tau = s_k, B]. \end{aligned}$$

Summing over k, we get

$$P[y(t_1) \in A_1, \ldots, y(t_n) \in A_n, B] = P[x(t_1) \in A_1, \ldots, x(t_n) \in A_n]P(B). \tag{4.2}$$

Taking $B = \Omega$, it follows that the joint distribution functions of the process $y(t)$ are the same as those for $x(t)$. Since $y(t)$ is clearly a continuous process, it is a (continuous) Brownian motion.

From (4.2) we further deduce that the σ-field $\mathcal{F}(y(t), t \ge 0)$ is independent of $\mathcal{F}_\tau$.

In order to prove Theorem 4.1 for a general stopping time, we approximate τ by a sequence of stopping times τ_n, defined in Lemma 1.3.4. We have (see the first paragraph of the proof of Theorem 2.2.4)

$$\mathcal{F}_\tau \subset \mathcal{F}_{\tau_n}.$$

Set $y_n(t) = x(t + \tau_n) - x(\tau_n)$. By what we have already proved, if $B \in \mathcal{F}_\tau$, then

$$P[y_n(t_1) < x_1, \ldots, y_n(t_k) < x_k, B] = P[x(t_1) < x_1, \ldots, x(t_k) < x_k]P(B)$$

for any $0 \le t_1 < \cdots < t_k$. Notice that $y_n(t) \to y(t)$ a.s. for all $t \ge 0$, as $n \to \infty$. Hence, if $(x_1, \ldots, x_k)$ is a point of continuity of the k-dimensional distribution function $F_k(x_1, \ldots, x_k)$ of $(x(t_1), \ldots, x(t_k))$, then

$$P[y(t_1) < x_1, \ldots, y(t_k) < x_k, B] = P[x(t_1) < x_1, \ldots, x(t_k) < x_k]P(B). \tag{4.3}$$

Since (see Problem 1) $F_k(x_1, \ldots, x_k)$ is actually an integral

$$\int_{-\infty}^{x_k} \cdots \int_{-\infty}^{x_1} \rho(z_1, \ldots, z_k)\, dz_1 \cdots dz_k,$$

it is continuous everywhere. Thus (4.3) holds for all $x_1, \ldots, x_k$. This implies that

$$P[y(t_1) \in A_1, \ldots, y(t_k) \in A_k, B] = P[x(t_1) \in A_1, \ldots, x(t_k) \in A_k] P(B)$$

for any Borel sets $A_1, \ldots, A_k$, and the proof of the theorem readily follows.

5. Martingales and Brownian motion

If $x(t)$ is a Brownian motion and $\mathcal{F}_t = \mathcal{F}(x(\lambda), 0 \leqslant \lambda \leqslant t)$, then

$$E[(x(t) - x(s)) | \mathcal{F}_s] = 0, \tag{5.1}$$

$$E\left[(x(t) - x(s))^2 | \mathcal{F}_s\right] = t - s \tag{5.2}$$

a.s. for any $0 \leqslant s < t$. Note that (5.1), (5.2) hold if and only if $x(t)$ and $x^2(t) - t$ are martingales.

We shall now prove the converse.

Theorem 5.1. *Let $x(t)$, $t \geqslant 0$ be a continuous process and let $\mathcal{F}_t$ $(t \geqslant 0)$ be an increasing family of σ-fields such that $x(t)$ is $\mathcal{F}_t$ measurable and* (5.1), (5.2) *hold a.s. for all $0 \leqslant s < t$. Then $x(t)$ is a Brownian motion.*

Proof. For any $\epsilon > 0$ and a positive integer n, let τ_0 be the first value of t such that

$$\max_{\substack{|s' - s''| \leqslant 1/n \\ 0 \leqslant s', s'' \leqslant t}} |x(s') - x(s'')| = \epsilon;$$

if no such t exists, set $\tau_0 = \infty$. Let $\tau = \tau_0 \wedge 1$. Since, for $0 < s < 1$,

$$\{\tau > s\} = \bigcup_{m=1}^{\infty} \bigcap_{\substack{|s' - s''| \leqslant 1/n \\ 0 < s', s'' \leqslant s}} \left\{ |x(s') - x(s'')| \leqslant \epsilon - \frac{1}{m} \right\} \quad \text{is in } \mathcal{F}_s,$$

τ is a stopping time.

Let $y(t) = x(t \wedge \tau)$. Since $x(t)$ and $x^2(t) - t$ are continuous martingales, Theorem 1.3.5 implies that $y(t)$ and $y^2(t) - t$ are also continuous martingales. Hence,

$$\int_A \left[y^2(s) - s \wedge \tau \right] dP = \int_A \left[y^2(t) - t \wedge \tau \right] dP \qquad \text{if } A \in \tilde{\mathcal{F}}_s, \quad s < t$$

where $\tilde{\mathfrak{F}}_s = \mathfrak{F}(y(\lambda), 0 \leq \lambda \leq s)$. It follows that

$$\int_A \{E[y^2(t) - y^2(s)] | \tilde{\mathfrak{F}}_s\} \, dP = \int_A [y^2(t) - y^2(s)] \, dP$$
$$= \int_A [t \wedge \tau - s \wedge \tau] \, dP \leq \int_A (t - s) \, dP.$$

Since the integrand on the left is $\tilde{\mathfrak{F}}_s$ measurable, we conclude that

$$E\{[y^2(t) - y^2(s)] | \tilde{\mathfrak{F}}_s\} \leq t - s \quad \text{a.s.,} \qquad s < t. \tag{5.3}$$

Our aim is to prove that if $0 \leq t_0 < t_1 < \cdots < t_k$, then

$$E \exp\left\{ i \sum_{j=1}^{k} \lambda_j [x(t_j) - x(t_{j-1})] \right\} = \exp\left[- \sum_{j=1}^{k} \tfrac{1}{2} \lambda_j^2 (t_j - t_{j-1}) \right] \tag{5.4}$$

for any real numbers λ_j. In view of the result stated in Problem 1(a), this will show that $x(t)$ is a Brownian motion.

We begin with the special case $k = 1$, $t_0 = 0$, $t_1 = 1$. Thus, we want to prove that

$$Ee^{i\lambda x(1)} = e^{-\lambda^2/2}. \tag{5.5}$$

Our approach will be to write

$$x(1) = \sum_{j=1}^{n} \left[x\left(\frac{j}{n}\right) - x\left(\frac{j-1}{n}\right) \right]$$

and make use of (5.1), (5.2) while at the same time, letting n increase to infinity. However, a technical difficulty arises due to the fact that the random variables $x(j/n) - x((j-1)/n)$ are not uniformly bounded. It is for this reason that we shall work, instead, with the process $y(t)$ and with the random variables

$$\zeta_j = y\left(\frac{j}{n}\right) - y\left(\frac{j-1}{n}\right) \qquad (1 \leq j \leq n).$$

In view of the definitions of τ and $y(t)$,

$$|\zeta_j| \leq \epsilon. \tag{5.6}$$

Since $y(t)$ is a martingale and since $y^2(t)$ satisfies (5.3),

$$E\zeta_1 = 0, \qquad E\{\zeta_j | \zeta_1, \ldots, \zeta_{j-1}\} = 0 \qquad (2 \leq j \leq n),$$
$$\sigma_1^2 = E\zeta_1^2 \leq \frac{1}{n}, \qquad \sigma_j^2 = E\{\zeta_j^2 | \zeta_1, \ldots, \zeta_{j-1}\} \leq \frac{1}{n} \qquad (2 \leq j \leq n) \tag{5.7}$$

a.s. Hence, if $j \geq 1$,

$$
\begin{aligned}
Ee^{i\lambda y(j/n)} &= E\left\{e^{i\lambda y((j-1)/n)}E\left[e^{i\lambda\zeta_j}|\zeta_1,\ldots,\zeta_{j-1}\right]\right\} \\
&= Ee^{i\lambda y((j-1)/n)}\left[1-\sigma_j^2\frac{\lambda^2}{2}(1+o(1))\right] \\
&= E\exp\left[i\lambda y\left(\frac{j-1}{n}\right)-\sigma_j^2\frac{\lambda^2}{2}(1+o(1))\right]
\end{aligned}
$$

where, in view of (5.6), $o(1)\to 0$ if $\epsilon\to 0$, uniformly with respect to λ, n provided λ varies in a bounded interval.

It follows that

$$
\begin{aligned}
&\left|E\exp\left[i\lambda y\left(\frac{j}{n}\right)\right]-E\exp\left[i\lambda y\left(\frac{j-1}{n}\right)-\frac{\lambda^2}{2n}\right]\right| \\
&\quad=\left|E\left\{\left\{\exp\left[i\lambda y\left(\frac{j-1}{n}\right)-\frac{\lambda^2}{2n}\right]\right\}\right.\right. \\
&\qquad\left.\left.\times\left\{\exp\left[-\frac{\lambda^2}{2}\left(\sigma_j^2(1+o(1))-\frac{1}{n}\right)\right]-1\right\}\right\}\right| \\
&\quad\leq E\left\{\left|\exp\left[\frac{\lambda^2}{2}\left(\frac{1}{n}-\sigma_j^2\right)+\sigma_j^2 o(1)\right]-1\right|\right\} \\
&\quad\leq C\,E\left[\left(\frac{1}{n}-\sigma_j^2\right)+\frac{o(1)}{n}\right] \\
&\quad\leq C\left[\frac{1}{n}-E(\zeta_j^2)\right]+\frac{o(1)}{n};
\end{aligned}
$$

we have used here (5.7). The symbol C denotes any one of various different positive constants which do not depend on n, ϵ, λ, provided λ is restricted to bounded intervals, $\epsilon < 1$, $n \geq 1$.

The last estimate implies that

$$
\begin{aligned}
&\left|E\exp\left[i\lambda y\left(\frac{j}{n}\right)+\frac{j}{2n}\lambda^2\right]-E\exp\left[i\lambda y\left(\frac{j-1}{n}\right)+\frac{j-1}{2n}\lambda^2\right]\right| \\
&\quad\leq C\left[\frac{1}{n}-E(\zeta_j^2)\right]+\frac{o(1)}{n}.
\end{aligned}
$$

Summing over j and noting that

$$
1 \geq \sum_{j=1}^{n} E(\zeta_j^2) = E(y^2(1)), \tag{5.8}
$$

we get

$$\left|E[e^{i\lambda y(1)+\lambda^2/2}-1]\right| \leq C\left[1-E\left(y^2(1)\right)\right]+o(1).$$

Consequently,

$$\left|E[e^{i\lambda y(1)}-e^{-\lambda^2/2}]\right| \leq C\left[1-E\left(y^2(1)\right)\right]+o(1). \tag{5.9}$$

Since $x(t)$ is a continuous process,

$$P(\tau=1)\to 1 \quad \text{and} \quad P[y(1)=x(1)]\to 1 \qquad \text{if} \quad n\to\infty.$$

Hence, by the Lebesgue bounded convergence theorem,

$$Ee^{i\lambda y(1)}\to Ee^{i\lambda x(1)} \qquad \text{if} \quad n\to\infty.$$

Therefore from (5.9) it follows that, for any $\gamma>0$, if $n \geq n_0(\epsilon, \gamma)$, then

$$\left|E[e^{i\lambda x(1)}-e^{-\lambda^2/2}]\right| \leq C\left[1-E\left(y^2(1)\right)\right]+o(1)+\gamma. \tag{5.10}$$

Note next that

$$1-E(y^2(1)) = E[x^2(1)-y^2(1)] = \int_{\tau<1}[x^2(1)-y^2(\tau)]\,dP$$
$$\leq \int_{\tau<1} x^2(1)\,dP\to 0$$

if $P(\tau=1)\to 1$. Thus, if $n \geq n_1(\epsilon,\gamma)$,

$$C\left[1-E\left(y^2(1)\right)\right] \leq \gamma.$$

Using this in (5.10) we get

$$|Ee^{i\lambda x(1)}-e^{-\lambda^2/2}| \leq o(1)+2\gamma.$$

Taking $\gamma\to 0$ (and $n \geq n_0(\epsilon,\gamma)$, $n \geq n_1(\epsilon,\gamma)$) and then $\epsilon\to 0$, the assertion (5.5) follows.

Similarly one can prove that

$$E\exp\left[i\lambda_j(x(t_j)-x(t_{j-1}))\right]=\exp\left[-\tfrac{1}{2}\lambda_j^2\right] \tag{5.11}$$

for any $0 \leq t_{j-1} < t_j$, λ_j real.

By reviewing the proof of (5.11) one finds that the same proof with minor changes actually gives a better result, namely,

$$E\left\{\exp\left[i\lambda_j(x(t_j)-x(t_{j-1}))\right]\middle|\mathcal{F}(x(\lambda), 0\leq\lambda\leq t_{j-1}\right\}=\exp\left[-\tfrac{1}{2}\lambda_j^2\right] \quad \text{a.s.} \tag{5.12}$$

(Notice that if τ_{j-1} is defined analogously to τ, with $t=0$ replaced by $t=t_{j-1}$, and if $y(t)=x(t\wedge\tau_{j-1})$, then

$$\mathcal{F}(y(\lambda), 0\leq\lambda\leq t_{j-1}) = \mathcal{F}(x(\lambda), 0\leq\lambda\leq t_{j-1})$$

since $\tau_{j-1}>t_{j-1}$.)

Now let $0 \leq t_0 < \cdots < t_{k-1} < t_k$ and let $\lambda_1, \ldots, \lambda_k$ be real numbers.

Applying (5.12) successively, we find that

$$E \exp\left[i \sum_{j=1}^{k} \lambda_j (x(t_j) - x(t_{j-1})) \right]$$
$$= E\left\{ \exp\left[i \sum_{j=1}^{k-1} \lambda_j (x(t_j) - x(t_{j-1})) \right] \right.$$
$$\left. \times E\{\exp[i\lambda_k(x(t_k) - x(t_{k-1}))] \mid \mathfrak{F}(x(\lambda), 0 \leq \lambda \leq t_{k-1})\} \right\}$$
$$= E\left\{ \exp\left[i \sum_{j=1}^{k-1} \lambda_j (x(t_j) - x(t_{j-1})) \right] \right\} \exp\left[-\tfrac{1}{2}\lambda_k^2 (t_k - t_{k-1}) \right]$$
$$= \cdots = \exp\left[-\sum_{j=1}^{k} \tfrac{1}{2}\lambda_j^2 (t_j - t_{j-1}) \right].$$

This completes the proof of (5.4).

Remark. By the remark at the end of Section 1.3 we have that $y(t)$ and $y^2(t) - t$ (introduced in the last proof) are martingales with respect to $\mathfrak{F}_t$. Hence instead of (5.3) we actually have

$$E\{ y^2(t) - y^2(s) \mid \mathfrak{F}_s \} \leq t - s \quad \text{a.s.}, \qquad s < t.$$

This implies

$$E\{\exp[i\lambda(x(t) - x(s))] \mid \mathfrak{F}_s\} = \exp\left[-\tfrac{1}{2}\lambda^2 (t - s) \right]. \tag{5.13}$$

6. Brownian motion in n dimensions

Definition. An n-dimensional process $w(t) = (w_1(t), \ldots, w_n(t))$ is called an n-dimensional *Brownian motion* (or *Wiener process*) if each process $w_i(t)$ is a Brownian motion and if the σ-fields $\mathfrak{F}(w_i(t), t \geq 0)$, $1 \leq i \leq n$, are independent.

Theorem 6.1. *Let $w(t)$ be an n-dimensional Brownian motion. Then*

$$P \varlimsup_{t \downarrow 0} \frac{|w(t)|}{\sqrt{2t \log\log(1/t)}} = 1 \quad \text{a.s.}, \tag{6.1}$$

$$P \varlimsup_{t \uparrow \infty} \frac{|w(t)|}{\sqrt{2t \log\log t}} = 1 \quad \text{a.s.} \tag{6.2}$$

Proof. It is enough to prove (6.1). If (6.1) is false, then there is a $\delta > 0$ such that

$$\overline{\lim_{t \downarrow 0}} \frac{|w(t)|}{\sqrt{2t \log\log(1/t)}} \geq 1 + \delta \quad \text{with positive probability.}$$

Given any $\epsilon > 0$, cover the unit sphere by a finite number of spherical regions K_i with opening ϵ with respect to the origin. Then, for some i,

$$P\left\{ \overline{\lim_{t \downarrow 0}} \frac{|w(t)|}{\sqrt{2t \log\log(1/t)}} \geq 1 + \delta, \frac{w(t)}{|w(t)|} \in K_i \right\} > 0.$$

But if z is the center of K_i and $(1 + \delta) \cos \epsilon > 1$, then

$$P\left\{ \overline{\lim_{t \to 0}} \frac{z \cdot w(t)}{\sqrt{2t \log\log(1/t)}} \geq (1 + \delta) \cos \epsilon > 1 \right\} > 0.$$

This is impossible since $z \cdot w(t)$ is a Brownian motion (cf. Problem 7).

Theorem 4.1 and its proof extend immediately to n-dimensional Brownian motion.

We shall now extend Theorem 5.1 to n-dimensions.

Theorem 6.2. *Let $x(t) = (x_1(t), \ldots, x_n(t))$, $t \geq 0$ be a continuous, n-dimensional process and let $\mathcal{F}_t$ $(t \geq 0)$ be an increasing family of σ-fields such that $x(t)$ is $\mathcal{F}_t$ measurable and, for all $0 \leq s < t < \infty$,*

$$E[x(t) - x(s) | \mathcal{F}_s] = 0 \quad \text{a.s.,}$$
$$E[(x_i(t) - x_i(s))(x_j(t) - x_j(s))] = \delta_{ij}(t - s) \quad \text{a.s.}$$

where $\delta_{ij} = 0$ if $i \neq j$, $\delta_{ii} = 1$. Then $x(t)$ is a Brownian motion.

Proof. For any $\gamma \in R^n$, $|\gamma| = 1$, the conditions of Theorem 5.1 are satisfied for $\gamma \cdot x(t)$. Hence $\gamma \cdot x(t)$ is a Brownian motion. Thus, in particular, each $x_i(t)$ is a Brownian motion. It remains to show that these motions are mutually independent. For this it suffices to show that $\mathcal{F}(x_i(t), t \geq 0)$ is independent of $\mathcal{F}(x_j(t), t \geq 0)$ if $i \neq j$. Take, for simplicity, $i = 1, j = 2$.

Since $\gamma_1 x_1 + \gamma_i x_2$ is a Brownian motion if $\gamma_1^2 + \gamma_2^2 = 1$,

$$E[\gamma_1 x_1(t) + \gamma_2 x_2(t)]^2 = t.$$

Since also $Ex_i^2(t) = t$,

$$Ex_1(t)x_2(t) = 0.$$

But then

$$Ex_1(t + s)x_2(t) = E\{x_2(t)E[x_1(t + s)|\mathcal{F}_t]\} = Ex_2(t)x_1(t) = 0,$$

i.e., $x_1(t+s)$ is independent of $x_2(t)$. Similarly $x_2(t+s)$ is independent of $x_1(t)$. Since t and s are arbitrary nonnegative numbers, it follows that $\mathcal{F}(x_1(\lambda), \lambda \geqslant 0)$ is independent of $\mathcal{F}(x_2(\lambda), \lambda \geqslant 0)$. This completes the proof.

One can easily check that an n-dimensional Brownian motion satisfies the Markov property with the stationary transition probability function

$$p(t, x, A) = \int_A \frac{1}{(2\pi t)^{n/2}} \exp\left[-\frac{|x-y|^2}{2t} \right] dy. \tag{6.3}$$

By the method of Section 2.1 we can construct a time-homogeneous Markov process corresponding to the transition probability (6.3). The sample space consists of all R^n-valued functions on $[0, \infty)$. But we can also construct another model $\{\Omega_0, \mathfrak{M}, \mathfrak{M}_t, \xi(t), P_x\}$ where Ω_0 is the space of all continuous functions $x(\cdot)$ from $[0, \infty)$ into R^n, $\mathfrak{M}_t$ is the smallest σ-field such that the process

$$\xi(s, x(\cdot)) = x(s) \tag{6.4}$$

is measurable for all $0 \leqslant s \leqslant t$, and

$$\begin{aligned} P_x\{x(\cdot); \xi(t_1) \in A_1, \ldots, \xi(t_k) \in A_k\} \\ = P\{\omega; x + w(t_1) \in A_1, \ldots, x + w(t_k) \in A_k\}. \end{aligned} \tag{6.5}$$

Theorem 6.3. *$\{\Omega_0, \mathfrak{M}, \mathfrak{M}_t, \xi(t), P_x\}$ is a time-homogeneous Markov process with the transition probability function given by* (6.3).

Proof. For any set B in $\mathfrak{M}$, define

$$P_x(B) = P\{\omega; x + w(\cdot, \omega) \text{ is in } B\}.$$

It is easily seen that P_x is a measure on $(\Omega_0, \mathfrak{M})$, satisfying (6.5). The rest follows from Theorem 2.1.3.

To any continuous ν-dimensional Markov process $\{\Omega^*, \mathcal{F}^*, \mathcal{F}^{*s}_t, x^*(t), P^*_{x,s}\}$ we can correspond a Markov process $\{\Omega, \mathfrak{M}, \mathfrak{M}^s_t, \xi(t), P_{x,s}\}$ having the same transition probability function p. Ω is the space of all continuous functions $x(\cdot)$ from $0 \leqslant t < \infty$ into R^ν, $\mathfrak{M}^s_t$ is the smallest σ-field such that $x(\lambda)$ is measurable for any $s \leqslant \lambda \leqslant t$, $\mathfrak{M} = \mathfrak{M}^0_\infty$, $\xi(t, x(\cdot)) = x(t)$ and, finally, if $s \leqslant t_1 < \cdots < t_k$,

$$P_{x,s}\{x(\cdot); \xi(t_1) \in A_1, \ldots, \xi(t_k) \in A_k\} = P^*_{x,s}\{\omega; x^*(t_1) \in A_1, \ldots, x^*(t_k) \in A_k\}. \tag{6.6}$$

In fact, for any set B in $\mathfrak{M}$, define

$$P_{x,s}(B) = P^*_{x,s}\{\omega; x^*(\cdot, \omega) \in B\}.$$

It is easily seen that $P_{x,s}$ is a measure on $(\Omega, \mathfrak{M})$, and it satisfies (6.6). In view of Theorem 2.1.3, $\{\Omega, \mathfrak{M}, \mathfrak{M}_t^s, \xi(t), P_{x,s}\}$ is then a Markov process with the same transition probability function as for $\{\Omega^*, \mathcal{F}^*, \mathcal{F}_t^{*s}, x^*(t), P_{x,s}^*\}$.

PROBLEMS

1. Recall that n random variables $X_1, \ldots, X_n$ are said to have *joint normal distribution* (μ, Γ), where $\mu = (\mu_1, \ldots, \mu_n)$, $\Gamma = (\Gamma_{ij})_{i,j=1}^n$ symmetric, if the characteristic function

$$f(u) = E \exp\left[i \sum_{j=1}^n u_j X_j\right] \qquad (u = (u_1, \ldots, u_n))$$

has the form

$$f(u) = \exp\left[\sum_{j=1}^n \mu_j u_j - \tfrac{1}{2} \sum_{j,k=1}^n \Gamma_{jk} u_j u_k\right].$$

Prove that if this is the case then:

(a) The random variables $X_1, \ldots, X_n$ are independent if and only if $\Gamma_{ij} = 0$ whenever $i \neq j$.

(b) If X_i, X_j are independent for any pair (i, j), $i \neq j$, then $X_1, \ldots, X_n$ are mutually independent.

(c) If $Y_i = \sum_{j=1}^n a_{ij} X_j$ $(1 \leq i \leq m)$ where a_{ij} are constants, then $Y_1, \ldots, Y_n$ have joint normal distribution.

(d) $\mu_i = EX_i$, $\Gamma_{ij} = E(X_i - \mu_i)(X_j - \mu_j)$.

(e) If $\det \Gamma \neq 0$, then $(X_1 - \mu_1, \ldots, X_n - \mu_n)$ has a distribution function $\rho(y)$ given by

$$\rho(y) = \frac{1}{(2\pi)^{n/2} (\det \Gamma)^{1/2}} \exp\left[-\tfrac{1}{2} \sum_{i,j=1}^n \Gamma_{ij}^{-1} y_i y_j\right]$$

where (Γ_{ij}^{-1}) is the inverse matrix to Γ.

2. If $x(t)$, $t \geq 0$ is a process with $x(0) = 0$ such that for any $0 \leq t_0 < t_1 < \cdots < t_n$ the random variables $x(t_0), \ldots, x(t_n)$ have joint normal distribution $N(\hat{\mu}, \Gamma)$ where $\hat{\mu} = (\mu, \ldots, \mu)$, $\Gamma = (\Gamma_{jk})$, $\Gamma_{jk} = \sigma^2 \min(t_j, t_k)$, then $x(t)$ is a Brownian motion with drift μ and variance σ^2.

3. Verify (1.5) for a random variable X with normal distribution $N(0, \sigma^2)$.

4. Give a direct proof of (3.2). [*Hint*: Apply the method of proof of (3.1) with $q > 1$.]

5. If $x(t)$ is a Brownian motion, $[x(t)/t] \to 0$ a.s. as $t \to \infty$.

6. Let $x(t)$ be a Brownian motion and let I be an interval in $[0, \infty)$. Then for any $\delta > 0$, $P\{x(n\delta) \in I \text{ i.o.}\} = 1$.

7. If $\tilde{w}_i(t) = \sum_{j=1}^n a_{ij} w_j(t)$ where (a_{ij}) is an orthogonal matrix and

$(w_1(t), \ldots, w_n(t))$ is an n-dimensional Brownian motion, then $(\tilde{w}_1, \ldots, \tilde{w}_n)$ is also an n-dimensional Brownian motion.

8. If $w(t)$ is a Brownian motion then $\alpha w(t/\alpha^2)$ is also a Brownian motion, for any positive constant α.

9. If $w(t)$ is a Brownian motion and $z(t) = \exp[\alpha w(t) - \alpha^2 t/2]$, α positive constant, then

(i) $Ez(t) = 1$.

(ii) $z(t)$ is a martingale; in fact, if $\mathcal{F}_s = \mathcal{F}(w(\lambda), 0 \leq \lambda \leq s)$, then $E(z(t)|\mathcal{F}_s) = z(s)$ if $t > s$.

(iii) $P\{\max_{0\leq s\leq t}[w(s) - \alpha s/2] > \beta\} \leq e^{-\alpha\beta}$ if $\alpha \geq 0$, $\beta > 0$.

10. Let $u_n(t)$ be Brownian motion defined on the same probability space, and suppose that $u_n(t) \xrightarrow{P} u(t)$ for each $t \geq 0$, as $n \to \infty$. Prove that $u(t)$ is a Brownian motion.

11. If $w(t)$ is a Brownian motion and A is a positive constant, then, for any $T > 0$,

$$Ee^{A|w(t)|} \leq C \quad \text{if} \quad 0 \leq t \leq T,$$

where C is a positive constant depending on A, T.

12. Let $\mathcal{F}_t$, $t \geq 0$ be an increasing family of σ-fields and let $x(t)$ be an n-dimensional process such that $x(t)$ is $\mathcal{F}_t$ measurable and $E[x(t+s)|\mathcal{F}_t] = x(t)$ for all $t \geq 0$, $s \geq 0$. Prove that if $\gamma \cdot x(t)$ is a Brownian motion for any $\gamma \in R^n$, $|\gamma| = 1$, then $x(t)$ is an n-dimensional Brownian motion. [*Hint*: Cf. the proof of Theorem 6.2.]

4

The Stochastic Integral

In this chapter we shall define the integral

$$I(T) = \int_0^T f(t)\, dw(t)$$

where $w(t)$ is a Brownian motion and $f(t)$ is a stochastic function, and study its basic properties. One may define

$$I(T) = f(T)w(T) - \int_0^T f'(t)w(t)\, dt$$

if f is absolutely continuous for each ω. However, if f is only continuous, or just integrable, this definition does not make sense.

Since $w(t)$ is nowhere differentiable with probability 1, the integral $\int_0^T f(t)\, dw(t)$ cannot be defined in the usual Lebesgue–Stieltjes sense.

1. Approximation of functions by step functions

We shall call a stochastic process also a stochastic function or, briefly, a function.

Let $w(t)$, $t \geq 0$ be a Brownian motion on a probability space $(\Omega, \mathcal{F}, P)$. Let $\mathcal{F}_t$ $(t \geq 0)$ be an increasing family of σ-fields, i.e., $\mathcal{F}_{t_1} \subset \mathcal{F}_{t_2}$ if $t_1 < t_2$, such that $\mathcal{F}_t \subset \mathcal{F}$, $\mathcal{F}(w(s), 0 \leq s \leq t)$ is in $\mathcal{F}_t$, and

$$\mathcal{F}(w(\lambda + t) - w(t), \lambda \geq 0) \quad \text{is independent of } \mathcal{F}_t$$

for all $t \geq 0$. One can take, for instance, $\mathcal{F}_t = \mathcal{F}(w(s), 0 \leq s \leq t)$.

Let $0 \leq \alpha < \beta < \infty$. A stochastic process $f(t)$ defined for $\alpha \leq t \leq \beta$ is called a *nonanticipative* function with respect to $\mathcal{F}_t$ if:

(i) $f(t)$ is a separable process;

(ii) $f(t)$ is a measurable process, i.e., the function $(t, \omega) \to f(t, \omega)$ from $[\alpha, \beta] \times \Omega$ into R^1 is measurable;

(iii) for each $t \in [\alpha, \beta]$, $f(t)$ is $\mathcal{F}_t$ measurable.

When (iii) holds we say that $f(t)$ is *adapted* to $\mathcal{F}_t$. We denote by $L_w^p[\alpha, \beta]$ $(1 \leqslant p \leqslant \infty)$ the class of all nonanticipative functions $f(t)$ satisfying:

$$P\left\{ \int_\alpha^\beta |f(t)|^p \, dt < \infty \right\} = 1 \qquad \left(P\left\{ \operatorname*{ess\,sup}_{\alpha \leqslant t \leqslant \beta} |f(t)| < \infty \right\} = 1 \text{ if } p = \infty \right).$$

We denote by $M_w^p[\alpha, \beta]$ the subset of $L_w^p[\alpha, \beta]$ consisting of all functions f with

$$E \int_\alpha^\beta |f(t)|^p \, dt < \infty \qquad \left(E\left[\operatorname*{ess\,sup}_{\alpha \leqslant t \leqslant \beta} |f(t)| \right] < \infty \text{ if } p = \infty \right).$$

Definition. A stochastic process $f(t)$ defined on $[\alpha, \beta]$ is called a *step function* if there exists a partition $\alpha = t_0 < t_1 < \cdots < t_r = \beta$ of $[\alpha, \beta]$ such that

$$f(t) = f(t_i) \qquad \text{if} \quad t_i \leqslant t < t_{i+1}, \quad 0 \leqslant i \leqslant r - 1.$$

Lemma 1.1. *Let $f \in L_w^2[\alpha, \beta]$. Then:*

(i) *there exists a sequence of continuous functions g_n in $L_w^2[\alpha, \beta]$ such that*

$$\lim_{n \to \infty} \int_\alpha^\beta |f(t) - g_n(t)|^2 \, dt = 0 \quad a.s.; \tag{1.1}$$

(ii) *there exists a sequence of step functions f_n in $L_w^2[\alpha, \beta]$ such that*

$$\lim_{n \to \infty} \int_\alpha^\beta |f(t) - f_n(t)|^2 \, dt = 0 \quad a.s. \tag{1.2}$$

Proof. Let

$$\rho(t) = \begin{cases} c \exp\left[-1/(1 - t^2) \right] & \text{if} \quad |t| \leqslant 1, \\ 0 & \text{if} \quad |t| > 1 \end{cases}$$

where c is a positive constant determined by the condition $\int_{-\infty}^{\infty} \rho(t) \, dt = 1$.

Define $f(t) = 0$ if $t < \alpha$ and let

$$(J_\epsilon f)(t) = \frac{1}{\epsilon} \int_{\alpha - 1}^{\beta} \rho\left(\frac{t - s - \epsilon}{\epsilon} \right) f(s) \, ds \qquad (2\epsilon < 1). \tag{1.3}$$

Then $J_\epsilon f$ is clearly continuous, and

$$(J_\epsilon f)(t) = \frac{1}{\epsilon}\int_{t-2\epsilon}^{t} \rho\left(\frac{t-s-\epsilon}{\epsilon}\right)f(s)\,ds = \int_{-\epsilon}^{\epsilon} \rho(z)f(t-z-\epsilon)\,dz. \tag{1.4}$$

By Schwarz's inequality,

$$\begin{aligned}\int_\alpha^\beta (J_\epsilon f)^2\,dt &\leq \int_\alpha^\beta \left\{\int_{-\epsilon}^{\epsilon} \rho(z)\,dz \int_{-\epsilon}^{\epsilon} \rho(z)f^2(t-z-\epsilon)\,dz\right\} dt \\ &\leq \int_{-\epsilon}^{\epsilon} \rho(z)\left\{\int_{\alpha-1}^{\beta} f^2(t)\,dt\right\} dz.\end{aligned}$$

Hence

$$\int_\alpha^\beta (J_\epsilon f)^2\,dt \leq \int_\alpha^\beta f^2(t)\,dt. \tag{1.5}$$

For fixed ω for which $\int_\alpha^\beta f^2(t,\omega)\,dt < \infty$, let u_n be nonrandom continuous functions such that $u_n(t) = 0$ if $t < \alpha$ and

$$\int_\alpha^\beta |u_n(t) - f(t,\omega)|^2\,dt \to 0 \qquad \text{if} \quad n\to\infty. \tag{1.6}$$

Since u_n is continuous, it is clear that

$$(J_\epsilon u_n)(t)\to u_n(t) \quad \text{uniformly in} \quad t\in[\alpha,\beta], \qquad \text{as} \quad \epsilon\to 0.$$

Writing

$$\begin{aligned}\int_\alpha^\beta |(J_\epsilon f)(t,\omega) - f(t,\omega)|^2\,dt &\leq \int_\alpha^\beta |J_\epsilon(f(\cdot,\omega) - u_n(\cdot))(t)|^2\,dt \\ &\quad + \int_\alpha^\beta |(J_\epsilon u_n)(t) - u_n(t)|^2\,dt + \int_\alpha^\beta |u_n(t) - f(t,\omega)|^2\,dt\end{aligned}$$

and using (1.5) with f replaced by $f - u_n$, we get, after taking $\epsilon \to 0$,

$$\overline{\lim_{\epsilon\downarrow 0}} \int_\alpha^\beta |(J_\epsilon f)(t,\omega) - f(t,\omega)|^2\,dt \leq 2\int_\alpha^\beta |u_n(t) - f(t,\omega)|^2\,dt.$$

Taking $n\to\infty$ and using (1.6), we obtain

$$\overline{\lim_{\epsilon\downarrow 0}} \int_\alpha^\beta |(J_\epsilon f)(t) - f(t)|^2\,dt = 0 \quad \text{a.s.}$$

Since the integrand on the right-side of (1.4) is a separable process that is $\mathcal{F}_t$ measurable, the integral is also $\mathcal{F}_t$ measurable. Consequently the assertion (i) holds with $g_n = J_{1/n}f$.

To prove (ii), let

$$h_{n,m}(t) = g_n\left(\frac{k}{m}\right) \quad \text{if } \alpha + \frac{k}{m} \leqslant t < \alpha + \frac{k+1}{m} \quad (0 \leqslant k < m(\beta - \alpha)).$$

Then

$$\lim_{m\to\infty} \int_\alpha^\beta |h_{n,m}(t) - g_n(t)|^2\,dt = 0 \quad \text{a.s.} \tag{1.7}$$

Now, for any $\delta > 0$,

$$P\left\{\int_\alpha^\beta |f(t) - g_n(t)|^2\,dt > \frac{\delta}{2}\right\} < \frac{\delta}{2} \qquad \text{for some} \quad n = n_0.$$

From (1.7) with $n = n_0$ we get

$$P\left\{\int_\alpha^\beta |g_{n_0}(t) - h_{n_0,m}(t)|^2\,dt > \frac{\delta}{2}\right\} < \frac{\delta}{2} \qquad \text{for some} \quad m = m_0.$$

Hence

$$P\left\{\int_\alpha^\beta |f(t) - h_{n_0,m_0}(t)|^2\,dt > \delta\right\} < \delta.$$

Taking $\delta = 1/k$ and denoting the corresponding h_{n_0,m_0} by ψ_k, it follows that $\psi_k \in L_w^2[\alpha, \beta]$ and

$$\int_\alpha^\beta |f(t) - \psi_k(t)|^2\,dt \xrightarrow{P} 0.$$

But then there is a subsequence $\{f_n\}$ of $\{\psi_k\}$ satisfying the assertion (ii).

Lemma 1.2. *Let $f \in M_w^2[\alpha, \beta]$. Then:*

(i) *there exists a sequence of continuous functions k_n in $M_w^2[\alpha, \beta]$ such that*

$$E\int_\alpha^\beta |f(t) - k_n(t)|^2\,dt \to 0 \qquad \text{if} \quad n\to\infty; \tag{1.8}$$

(ii) *there exists a sequence of bounded step functions l_n in $M_w^2[\alpha, \beta]$ such that*

$$E\int_\alpha^\beta |f(t) - l_n(t)|^2\,dt \to 0 \qquad \text{if} \quad n\to\infty. \tag{1.9}$$

Proof. Let g_n be as in Lemma 1.1. For any $N > 0$, let

$$\phi_N(t) = \begin{cases} t & \text{if } |t| \leqslant N, \\ Nt/|t| & \text{if } |t| > N. \end{cases}$$

Notice that $|\phi_N(t) - \phi_N(s)| \leq |t - s|$. Therefore

$$\int_\alpha^\beta |\phi_N(f(t)) - \phi_N(g_n(t))|^2\, dt \leq \int_\alpha^\beta |f(t) - g_n(t)|^2\, dt \to 0 \quad \text{a.s.}$$

Since

$$\int_\alpha^\beta |\phi_N(f(t)) - \phi_N(g_n(t))|^2\, dt \leq 4N^2(\beta - \alpha),$$

the Lebesgue bounded convergence theorem gives

$$E\int_\alpha^\beta |\phi_N(f(t)) - \phi_N(g_n(t))|^2\, dt \to 0 \qquad \text{if} \quad n \to \infty. \tag{1.10}$$

Next,

$$E\int_\alpha^\beta |\phi_N(f(t)) - f(t)|^2\, dt \leq 4 \iint_{\{|f|>N\}} |f(t)|^2\, dt\, dP \to 0$$

as $N \to \infty$, since $f \in M_w^2[\alpha, \beta]$. From this and (1.10) it follows that for every positive integer k there are $N = N(k)$ and $n = n(k, N)$ such that

$$E\int_\alpha^\beta |\phi_N(g_n(t)) - f(t)|^2\, dt < \frac{1}{k}.$$

Taking $h_k = \phi_N(g_n)$ where $N = N(k)$, $n = n(N, k)$ and noting that $h_k(t)$ is nonanticipative, the assertion (i) follows.

The proof of (ii) is similar. The l_n are of the form $\phi_N(f_n)$ where the f_n are as in Lemma 1.1. Notice that these are in fact step functions.

2. Definition of the stochastic integral

Definition. Let $f(t)$ be a step function in $L_w^2[\alpha, \beta]$, say $f(t) = f_i$ if $t_i \leq t < t_{i+1}$, $0 \leq i \leq r - 1$ where $\alpha = t_0 < t_1 < \cdots < t_r = \beta$. The random variable

$$\sum_{k=0}^{r-1} f(t_k)[w(t_{k+1}) - w(t_k)]$$

is denoted by

$$\int_\alpha^\beta f(t)\, dw(t)$$

and is called the *stochastic integral* of f with respect to the Brownian motion w; it is also called the *Itô integral.*

Lemma 2.1. *Let f_1, f_2 be two step functions in $L_w^2[\alpha, \beta]$ and let λ_1, λ_2 be real numbers. Then $\lambda_1 f_1 + \lambda_2 f_2$ is in $L_w^2[\alpha, \beta]$ and*

$$\int_\alpha^\beta [\lambda_1 f_1(t) + \lambda_2 f_2(t)]\, dw(t) = \lambda_1 \int_\alpha^\beta f_1(t)\, dw(t) + \lambda_2 \int_\alpha^\beta f_2(t)\, dw(t). \tag{2.1}$$

The proof is left to the reader.

Lemma 2.2. *If f is a step function in $M_w^2[\alpha, \beta]$, then*

$$E \int_\alpha^\beta f(t)\, dw(t) = 0, \tag{2.2}$$

$$E\left| \int_\alpha^\beta f(t)\, dw(t) \right|^2 = E \int_\alpha^\beta f^2(t)\, dt. \tag{2.3}$$

Proof. Since

$$E \int_\alpha^\beta f^2(t)\, dt = \sum_{i=0}^{r-1} Ef^2(t_i)(t_{i+1} - t_i) \tag{2.4}$$

is finite, by assumption, we deduce that $Ef^2(t_i) < \infty$. In particular, $E|f(t_i)| < \infty$. Also, $E|w(t_{i+1}) - w(t_i)| < \infty$. But since $f(t_i)$ is $\mathcal{F}_{t_i}$ measurable whereas $w(t_{i+1}) - w(t_i)$ is independent of $\mathcal{F}_{t_i}$,

$$Ef(t_i)(w(t_{i+1}) - w(t_i)) = Ef(t_i)E(w(t_{i+1}) - w(t_i)) = 0.$$

Summing over i, (2.2) follows.

Next, since $f^2(t_i)$ and $(w(t_{i+1}) - w(t_i))^2$ are independent and have finite expectation, also $f^2(t_i)(w(t_{i+1}) - w(t_i))^2$ has finite expectation. By Schwarz's inequality it follows that

$$E|f(t_k)f(t_i)(w(t_{i+1}) - w(t_i))| < \infty.$$

If $k > i$, then $w(t_{k+1}) - w(t_k)$ is independent of $f(t_k)f(t_i)(w(t_{i+1}) - w(t_i))$. In view of the last inequality and the finiteness of $E|w(t_{k+1}) - w(t_k)|$, we deduce that

$$Ef(t_k)f(t_i)(w(t_{i+1}) - w(t_i))(w(t_{k+1}) - w(t_k)) = 0.$$

Hence

$$E\left|\int_\alpha^\beta f(t)\,dw(t)\right|^2 = \sum_{i=0}^{r-1} Ef^2(t_i)(w(t_{i+1}) - w(t_i))^2$$
$$= \sum_{i=0}^{r-1} Ef^2(t_i)E(w(t_{i+1}) - w(t_i))^2$$
$$= \sum_{i=0}^{r-1} Ef^2(t_i)(t_{i+1} - t_i) = E\int_\alpha^\beta f^2(t)\,dt$$

by (2.4), and (2.3) is proved.

Lemma 2.3. *For any step function f in $L^2_w[\alpha, \beta]$ and for any $\epsilon > 0$, $N > 0$,*

$$P\left\{\left|\int_\alpha^\beta f(t)\,dw(t)\right| > \epsilon\right\} \leqslant P\left\{\int_\alpha^\beta f^2(t)\,dt > N\right\} + \frac{N}{\epsilon^2}. \tag{2.5}$$

Proof. Let

$$\phi_N(t) = \begin{cases} f(t) & \text{if} \quad t_k \leqslant t < t_{k+1} \quad \text{and} \quad \sum_{j=0}^{k} f^2(t_j)(t_{j+1} - t_j) \leqslant N, \\ 0 & \text{if} \quad t_k \leqslant t < t_{k+1} \quad \text{and} \quad \sum_{j=0}^{k} f^2(t_j)(t_{j+1} - t_j) > N, \end{cases}$$

where $f(t) = f(t_j)$ if $t_j \leqslant t < t_{j+1}$; $t_0 = \alpha < t_1 < \cdots < t_r = \beta$. Then $\phi_N \in L^2_w[\alpha, \beta]$ and

$$\int_\alpha^\beta \phi_N^2(t)\,dt = \sum_{j=0}^{\nu} f^2(t_j)(t_{j+1} - t_j)$$

where ν is the largest integer such that

$$\sum_{j=0}^{\nu} f^2(t_j)(t_{j+1} - t_j) \leqslant N, \qquad \nu \leqslant r - 1.$$

Hence

$$E\int_\alpha^\beta \phi_N^2(t)\,dt \leqslant N.$$

Further, $f(t) - \phi_N(t) = 0$ for all $t \in [\alpha, \beta)$ if $\int_\alpha^\beta f^2(t)\,dt \leqslant N$. Therefore

$$P\left\{\left|\int_\alpha^\beta f(t)\,dw(t)\right| > \epsilon\right\} \leqslant P\left\{\left|\int_\alpha^\beta \phi_N(t)\,dw(t)\right| > \epsilon\right\}$$
$$+ P\left\{\int_\alpha^\beta f^2(t)\,dt > N\right\}.$$

Since, by Chebyshev's inequality, the first integral on the right is bounded by

$$\frac{1}{\epsilon^2} E\left|\int_\alpha^\beta \phi_N(t)\, dw(t)\right|^2 < \frac{N}{\epsilon^2},$$

the assertion (2.5) follows.

We shall now proceed to define the stochastic integral for any function f in $L_w^2[\alpha, \beta]$.

By Lemma 1.1 there is a sequence of step functions f_n in $L_w^2[\alpha, \beta]$ such that

$$\int_\alpha^\beta |f_n(t) - f(t)|^2\, dt \xrightarrow{P} 0 \quad \text{if} \quad n \to \infty. \tag{2.6}$$

Hence

$$\lim_{n,m\to\infty} P \int_\alpha^\beta |f_n(t) - f_m(t)|^2\, dt \xrightarrow{P} 0.$$

By Lemma 2.3, for any $\epsilon > 0$, $\rho > 0$,

$$P\left\{\left|\int_\alpha^\beta f_n(t)\, dw(t) - \int_\alpha^\beta f_m(t)\, dw(t)\right| > \epsilon\right\}$$
$$< \rho + P\left\{\int_\alpha^\beta |f_n(t) - f_m(t)|^2\, dt > \epsilon^2\rho\right\}.$$

It follows that the sequence

$$\left\{\int_\alpha^\beta f_n(t)\, dw(t)\right\}$$

is convergent in probability. We denote the limit by

$$\int_\alpha^\beta f(t)\, dw(t)$$

and call it the *stochastic integral* (or the *Itô integral*) of $f(t)$ with respect to the Brownian motion $w(t)$.

The above definition is independent of the particular sequence $\{f_n\}$. For if $\{g_n\}$ is another sequence of step functions in $L_w^2[\alpha, \beta]$ converging to f in the sense that

$$\int_\alpha^\beta |g_n(t) - f(t)|^2\, dt \xrightarrow{P} 0,$$

then the sequence $\{h_n\}$ where $h_{2n} = f_n$, $h_{2n+1} = g_n$ is also convergent to f in the same sense. But then, by what we have proved, the sequence

$$\left\{\int_\alpha^\beta h_n(t)\, dw(t)\right\}$$

is convergent in probability. It follows that the limits (in probability) of $\int_\alpha^\beta f_n\,dw$ and of $\int_\alpha^\beta g_n\,dw$ are equal a.s.

Lemmas 2.1–2.3 extend to any functions from $L_w^2[\alpha, \beta]$:

Theorem 2.4. *Let f_1, f_2 be functions from $L_w^2[\alpha, \beta]$ and let λ_1, λ_2 be real numbers. Then $\lambda_1 f_1 + \lambda_2 f_2$ is in $L_w^2[\alpha, \beta]$ and*

$$\int_\alpha^\beta [\lambda_1 f_1(t) + \lambda_2 f_2(t)]\,dw(t) = \lambda_1 \int_\alpha^\beta f_1(t)\,dw(t) + \lambda_2 \int_\alpha^\beta f_2(t)\,dw(t). \tag{2.7}$$

Theorem 2.5. *If f is a function in $M_w^2[\alpha, \beta]$, then*

$$E \int_\alpha^\beta f(t)\,dw(t) = 0, \tag{2.8}$$

$$E\left|\int_\alpha^\beta f(t)\,dw(t)\right|^2 = E\int_\alpha^\beta f^2(t)\,dt. \tag{2.9}$$

Theorem 2.6. *If f is a function from $L_w^2[\alpha, \beta]$, then, for any $\epsilon > 0$, $N > 0$,*

$$P\left\{\left|\int_\alpha^\beta f(t)\,dw(t)\right| > \epsilon\right\} \leq P\left\{\int_\alpha^\beta f^2(t)\,dt > N\right\} + \frac{N}{\epsilon^2}. \tag{2.10}$$

The proof of Theorem 2.4 is left to the reader.

***Proof of Theorem* 2.5.** By Lemma 1.2 there exists a sequence of step functions f_n in $M_w^2[\alpha, \beta]$ such that

$$E\int_\alpha^\beta |f_n(t) - f(t)|^2\,dt \to 0 \quad \text{if} \quad n \to \infty.$$

This implies that

$$E\int_\alpha^\beta f_n^2(t)\,dt \to E\int_\alpha^\beta f^2(t)\,dt. \tag{2.11}$$

By Lemma 2.2,

$$E\int_\alpha^\beta f_n(t)\,dw(t) = 0, \tag{2.12}$$

$$E\left|\int_\alpha^\beta f_n(t)\,dw(t)\right|^2 = E\int_\alpha^\beta f_n^2(t)\,dt, \tag{2.13}$$

$$E\left|\int_\alpha^\beta f_n(t)\,dw(t) - \int_\alpha^\beta f_m(t)\,dw(t)\right|^2 = E\int_\alpha^\beta |f_n(t) - f_m(t)|^2\,dt \to 0 \quad \text{if} \quad n, m \to \infty. \tag{2.14}$$

From the definition of the stochastic integral,

$$\int_\alpha^\beta f_n(t)\,dw(t) \xrightarrow{P} \int_\alpha^\beta f(t)\,dw(t).$$

Using (2.14) we conclude that actually

$$\int_\alpha^\beta f_n(t)\,dw(t) \to \int_\alpha^\beta f(t)\,dw(t) \qquad \text{in} \quad L^2(\Omega).$$

Hence, in particular,

$$E\int_\alpha^\beta f(t)\,dw(t) = \lim_{n\to\infty} E\int_\alpha^\beta f_n(t)\,dw(t),$$

$$E\left|\int_\alpha^\beta f(t)\,dw(t)\right|^2 = \lim_{n\to\infty} E\left|\int_\alpha^\beta f_n(t)\,dw(t)\right|^2,$$

and using (2.12) and (2.13), (2.11), the assertions (2.8), (2.9) follow.

Proof of Theorem 2.6. By Lemma 1.1 there exists a sequence of step functions f_n in $L_w^2[\alpha, \beta]$ such that

$$\int_\alpha^\beta |f_n(t) - f(t)|^2\,dt \xrightarrow{P} 0. \tag{2.15}$$

By definition of the stochastic integral,

$$\int_\alpha^\beta f_n(t)\,dw(t) \xrightarrow{P} \int_\alpha^\beta f(t)\,dw(t). \tag{2.16}$$

Applying Lemma 2.3 to f_n we have

$$P\left\{\left|\int_\alpha^\beta f_n(t)\,dw(t)\right| > \epsilon'\right\} \leq P\left\{\int_\alpha^\beta f_n^2(t)\,dt > N'\right\} + \frac{N'}{(\epsilon')^2}.$$

Taking $n \to \infty$ and using (2.15), (2.16), we get

$$P\left\{\left|\int_\alpha^\beta f(t)\,dw(t)\right| > \epsilon\right\} \leq P\left\{\int_\alpha^\beta f^2(t)\,dt > N\right\} + \frac{N'}{(\epsilon')^2}$$

for any $\epsilon > \epsilon'$, $N < N'$. Taking $\epsilon' \uparrow \epsilon$, $N' \downarrow N$, (2.10) follows.

Theorem 2.7. *Let f, f_n be in $L_w^2[\alpha, \beta]$ and suppose that*

$$\int_\alpha^\beta |f_n(t) - f(t)|^2\,dt \xrightarrow{P} 0 \qquad \text{as} \quad n \to \infty. \tag{2.17}$$

Then

$$\int_\alpha^\beta f_n(t)\,dw(t) \xrightarrow{P} \int_\alpha^\beta f(t)\,dw(t) \qquad \text{as} \quad n \to \infty. \tag{2.18}$$

Proof. By Theorem 2.6, for any $\epsilon > 0$, $\rho > 0$,

$$P\left\{\left|\int_\alpha^\beta (f_n(t) - f(t))\,dw(t)\right| > \epsilon\right\} \leq P\left\{\int_\alpha^\beta |f_n(t) - f(t)|^2\,dt > \epsilon^2\rho\right\} + \rho.$$

Taking $n\to\infty$ and using (2.17), the assertion (2.18) follows.

The next theorem improves Theorem 2.5.

Theorem 2.8. *Let $f \in M_w^2[\alpha, \beta]$. Then*

$$E\left\{\int_\alpha^\beta f(t)\,dw(t)\middle|\mathcal{F}_\alpha\right\} = 0, \tag{2.19}$$

$$E\left\{\left|\int_\alpha^\beta f(t)\,dw(t)\right|^2 \middle| \mathcal{F}_\alpha\right\} = E\left\{\int_\alpha^\beta f^2(t)\,dt \middle| \mathcal{F}_\alpha\right\} = \int_\alpha^\beta E[f^2(t)|\mathcal{F}_\alpha]\,dt. \tag{2.20}$$

We first need a simple lemma.

Lemma 2.9. *If $f \in L_w^2[\alpha, \beta]$ and ζ is a bounded and $\mathcal{F}_\alpha$ measurable function, then ζf is in $L_w^2[\alpha, \beta]$ and*

$$\int_\alpha^\beta \zeta f(t)\,dw(t) = \zeta \int_\alpha^\beta f(t)\,dw(t). \tag{2.21}$$

Proof. It is clear that ζf is in $L_w^2[\alpha, \beta]$. If f is a step function, then (2.21) follows from the definition of the stochastic integral. For general f in $L_w^2[\alpha, \beta]$, let f_n be step functions in $L_w^2[\alpha, \beta]$ satisfying (2.17). Applying (2.21) to each f_n and taking $n\to\infty$, the assertion (2.21) follows.

Proof of Theorem 2.8. Let ζ be a bounded and $\mathcal{F}_\alpha$ measurable function. Then ζf belongs to $M_w^2[\alpha, \beta]$ and, by Theorem 2.6,

$$E\int_\alpha^\beta \zeta f(t)\,dw(t) = 0.$$

Hence, by (2.21),

$$E\left\{\zeta\int_\alpha^\beta f(t)\,dw(t)\right\} = 0,$$

i.e.,

$$E\left\{\zeta E\left[\int_\alpha^\beta f(t)\,dw(t)|\mathcal{F}_\alpha\right]\right\} = 0.$$

This implies (2.19). The proof of (2.20) is similar.

Theorem 2.10. *If $f \in L_w^2[\alpha, \beta]$ and f is continuous, then, for any sequence Π_n of partitions $\alpha = t_{n,0} < t_{n,1} < \cdots < t_{n,m_n} = \beta$ of $[\alpha, \beta]$ with mesh $|\Pi_n| \to 0$,*

$$\sum_{k=0}^{m_n-1} f(t_{n,k})[w(t_{n,k+1}) - w(t_{n,k})] \xrightarrow{P} \int_\alpha^\beta f(t)\, dw(t) \qquad \text{as} \quad n \to \infty. \tag{2.22}$$

Proof. Introduce the step functions g_n:

$$g_n(t) = f(t_{n,k}) \qquad \text{if} \quad t_{n,k} \leqslant t < t_{n,k+1}, \quad 0 \leqslant k \leqslant m_n - 1.$$

For a.a. ω, $g_n(t) \to f(t)$ uniformly in $t \in [\alpha, \beta)$ as $n \to \infty$. Hence

$$\int_\alpha^\beta |g_n(t) - f(t)|^2\, dt \to 0 \quad \text{a.s.}$$

By Theorem 2.7 we then have

$$\int_\alpha^\beta g_n(t)\, dw(t) \xrightarrow{P} \int_\alpha^\beta f(t)\, dw(t).$$

Since

$$\int_\alpha^\beta g_n(t)\, dw(t) = \sum_{k=0}^{m_n-1} f(t_{n,k})[w(t_{n,k+1}) - w(t_{n,k})],$$

the assertion (2.22) follows.

Lemma 2.11. *Let f, g belong to $L_w^2[\alpha, \beta]$ and assume that $f(t) = g(t)$ for all $\alpha \leqslant t \leqslant \beta$, $\omega \in \Omega_0$. Then*

$$\int_\alpha^\beta f(t)\, dw(t) = \int_\alpha^\beta g(t)\, dw(t) \qquad \textit{for a.a. } \omega \in \Omega_0. \tag{2.23}$$

Proof. Let ψ_k be the step function in $L_w^2[\alpha, \beta]$ constructed in the proof of Lemma 1.1, satisfying

$$\int_\alpha^\beta |f(t) - \psi_k(t)|^2\, dt \xrightarrow{P} 0.$$

Similarly let ϕ_k be step functions in $L_w^2[\alpha, \beta]$ satisfying

$$\int_\alpha^\beta |g(t) - \phi_k(t)|^2\, dt \xrightarrow{P} 0.$$

From the construction in Lemma 1.1 we deduce that we can choose the sequences ϕ_k, ψ_k so that, if $\omega \in \Omega_0$, $\phi_k(t, \omega) = \psi_k(t, \omega)$ for $\alpha \leqslant t \leqslant \beta$. Hence, by the definition ff the integral of a step function,

$$\int_\alpha^\beta \psi_k(t)\, dw(t) = \int_\alpha^\beta \phi_k(t)\, dw(t) \qquad \text{if} \quad \omega \in \Omega_0.$$

Taking $k \to \infty$, the assertion (2.23) follows.

3. The Indefinite Integral

Let $f \in L_w^2[0, T]$ and consider the integral

$$I(t) = \int_0^t f(s)\, dw(s), \qquad 0 \leqslant t \leqslant T \tag{3.1}$$

where, by definition, $\int_0^0 f(s)\, dw(s) = 0$. We refer to $I(t)$ as the indefinite integral of f. Notice that $I(t)$ is $\mathcal{F}_t$ measurable.

If f is a step function, then clearly

$$\int_\alpha^\beta f(s)\, dw(s) + \int_\beta^\gamma f(s)\, dw(s) = \int_\alpha^\gamma f(s)\, dw(s) \quad \text{if} \quad 0 \leqslant \alpha < \beta < \gamma \leqslant T. \tag{3.2}$$

By approximation we find that (3.2) holds for any f in $L_w^2[0, T]$.

Theorem 3.1. *If $f \in M_w^2[0, T]$, then the indefinite integral $I(t)$ $(0 \leqslant t \leqslant T)$ is a martingale.*

Proof. Let $0 \leqslant t' < t \leqslant T$. By (3.2) and Theorem 2.8,

$$E(I(t)|\mathcal{F}_{t'}) = E(I(t')|\mathcal{F}_{t'}) + E\left(\int_{t'}^t f(s)\, ds|\mathcal{F}_{t'}\right) = I(t').$$

Theorem 3.2. *If $f \in L_w^2[0, T]$, then the indefinite integral $I(t)$ $(0 \leqslant t \leqslant T)$ has a continuous version.*

Proof. Suppose first that $f \in M_w^2[0, T]$. Let $\{f_n\}$ be a sequence of step functions in $M_w^2[0, T]$ such that

$$E \int_0^T (f(t) - f_n(t))^2\, dt \to 0 \qquad \text{if} \quad n \to \infty. \tag{3.3}$$

Notice that the indefinite integrals

$$I_n(t) = \int_0^t f_n(s)\, dw(s) \qquad (0 \leqslant t \leqslant T)$$

are continuous functions. By Theorem 2.8, $I_n(t) - I_m(t)$ is a martingale, for each n, m. Hence, by the martingale inequality (Corollary 1.3.3)

$$P\left\{ \sup_{0 \leq t \leq T} |I_n(t) - I_m(t)| > \epsilon \right\}$$
$$\leq \frac{1}{\epsilon^2} E \left| \int_0^T (f_n(s) - f_m(s))\, dw(s) \right|^2$$
$$\leq \frac{1}{\epsilon^2} E \int_0^T |f_n(s) - f_m(s)|^2\, ds \to 0 \qquad \text{if} \quad n, m \to \infty.$$

Taking $\epsilon = 1/2^k$ it follows that for some n_k sufficiently large,

$$P\left\{ \sup_{0 \leq t \leq T} |I_{n_k}(t) - I_m(t)| > \frac{1}{2^k} \right\} < \frac{1}{k^2} \qquad \text{if} \quad m \geq n_k.$$

We can choose the n_k in such a way that $n_k \uparrow$ if $k \uparrow$. Hence,

$$P\left\{ \sup_{0 \leq t \leq T} |I_{n_k}(t) - I_{n_{k+1}}(t)| > \frac{1}{2^k} \right\} < \frac{1}{k^2} \qquad (k = 1, 2, \ldots).$$

Since $\sum k^{-2} < \infty$, the Borel–Cantelli lemma inplies that

$$P\left\{ \sup_{0 \leq t \leq T} |I_{n_k}(t) - I_{n_{k+1}}(t)| > \frac{1}{2^k} \text{ i.o.} \right\} = 0,$$

i.e., for a.a. ω

$$|I_{n_k}(t) - I_{n_{k+1}}(t)| \leq \frac{1}{2^k} \qquad \text{for all} \quad 0 \leq t \leq T, \quad \text{if} \quad k \geq k_0(\omega).$$

But then, with probability one, $\{I_{n_k}(t)\}$ is uniformly convergent in $t \in [0, T]$. The limit $J(t)$ is therefore a continuous function in $t \in [0, T]$ for a.a. ω. Since (3.3) implies that

$$\int_0^t f_n(s)\, dw(s) \to \int_0^t f(s)\, dw(s) \qquad \text{in } L^2(\Omega),$$

it follows that

$$J(t) = \int_0^t f(s)\, dw(s) \quad \text{a.s.}$$

Thus, the indefinite integral has a continuous version.

Consider now the general case where $f \in L_w^2[0, T]$. For any $N > 0$, let

$$\chi_N(z) = \begin{cases} 1 & \text{if} \quad z \leq N, \\ 0 & \text{if} \quad z > N, \end{cases} \tag{3.4}$$

and introduce the function

$$f_N(t) = f(t)\chi_N\left(\int_0^t f^2(s)\, ds \right). \tag{3.5}$$

It is easily checked that f_N belongs to $M_w^2[0, T]$. Hence, by what was already proved, a version of

$$J_N(t) = \int_0^t f_N(s)\, dw(s) \qquad (0 \leqslant t \leqslant T)$$

is a continuous process.

Let

$$\Omega_N = \left\{ \int_0^T f^2(t)\, dt < N \right\}.$$

If $\omega \in \Omega_N$, then $f_N(t) = f_M(t)$ for $0 \leqslant t \leqslant T$, $M > N$. By Lemma 2.11 it follows that for a.a. $\omega \in \Omega_N$

$$J_N(t) = J_M(t) \qquad \text{if} \quad 0 \leqslant t \leqslant T.$$

Therefore

$$\tilde{J}(t) = \lim_{M\to\infty} J_M(t)$$

is continuous in $t \in [0, T]$ for a.a. $\omega \in \Omega_N$.

Since $\Omega_N \uparrow$, $P(\Omega_N) \uparrow 1$ if $N \uparrow \infty$, $\tilde{J}(t)$ $(0 \leqslant t \leqslant T)$ is a continuous process.

But since for each $t \in (0, T]$,

$$P\left\{ \int_0^t |f(s) - f_M(s)|^2\, ds > 0 \right\} = P\left\{ \int_0^t f^2(s)\, ds > M \right\} \to 0$$

as $M \to \infty$, we have, by Theorem 2.7,

$$J_M(t) \xrightarrow{P} \int_0^t f(s)\, dw(s) = I(t).$$

Consequently, $I(t)$ has the continuous version $\tilde{J}(t)$.

Remark. From now on, when we speak of the indefinite integral (3.1) of a function $f \in L^2_w[0, T]$ we always mean a continuous version of it.

Theorem 3.3. *Let $f \in L^2_w[0, T]$. Then, for any $\epsilon > 0$, $N > 0$,*

$$P\left\{ \sup_{0 \leqslant t \leqslant T} \left| \int_0^t f(s)\, dw(s) \right| > \epsilon \right\} \leqslant P\left\{ \int_0^T f^2(t)\, dt > N \right\} + \frac{N}{\epsilon^2}. \qquad (3.6)$$

Proof. With the notation of (3.4), (3.5) we have

$$\begin{aligned} P&\left\{ \sup_{0 \leqslant t \leqslant T} \left| \int_0^t f(s)\, dw(s) \right| > \epsilon \right\} \\ &\leqslant P\left\{ \sup_{0 \leqslant t \leqslant T} \left| \int_0^t f(s)\, dw(s) - \int_0^t f_N(s)\, dw(s) \right| > 0 \right\} \\ &\quad + P\left\{ \sup_{0 \leqslant t \leqslant T} \left| \int_0^t f_N(s)\, dw(s) \right| > \epsilon \right\} \\ &\equiv A + B. \end{aligned}$$

By Theorem 3.1, $\int_0^t f_N(s)\,dw(s)$ is a martingale. Hence, by the martingale inequality,

$$B \leq \frac{1}{\epsilon^2} E\left|\int_0^T f_N(s)\,dw(s)\right|^2 = \frac{1}{\epsilon^2} E\int_0^T (f_N(s))^2\,ds \leq \frac{N}{\epsilon^2}.$$

Next, on the set

$$\Omega_N = \left\{\int_0^T f^2(s)\,ds \leq N\right\}$$

we have $f(t) = f_N(t)$ for $0 \leq t \leq T$. Hence, by Lemma 2.11, for each fixed t in $[0, T]$

$$\int_0^t f(s)\,dw(s) = \int_0^t f_N(s)\,dw(s) \qquad \text{for a.a.} \quad \omega\in\Omega_N.$$

Since both integrals are continuous processes, the last relation holds for all $t\in[0, T]$ and for all $\omega\in\Omega_N'\subset\Omega_N$ where $P(\Omega_N\backslash\Omega_N') = 0$, i.e.,

$$A \leq P(\Omega\backslash\Omega_N) = P\left\{\int_0^T f^2(s)\,ds > N\right\}.$$

Combining the estimates on A, B, (3.6) follows.

Theorem 3.4. *Let f_n, f belong to $L_w^2[0, T]$ and assume that $\int_0^T|f_n - f|^2\,dt \xrightarrow{P} 0$ if $n\to\infty$. Then*

$$\sup_{0\leq t\leq T}\left|\int_0^t f_n(s)\,dw(s) - \int_0^t f(s)\,dw(s)\right| \xrightarrow{P} 0 \qquad \text{if} \quad n\to\infty.$$

This is a consequence of Theorem 3.3.

Theorem 3.5. *Let $f\in M_w^2[0, T]$. Then for any $\lambda > 0$*

$$P\left\{\sup_{0\leq t\leq T}\left|\int_0^t f(s)\,dw(s)\right| > \lambda\right\} \leq \frac{1}{\lambda^2} E\int_0^T f^2(s)\,ds. \tag{3.7}$$

Proof. By Theorems 3.1, 3.2, the indefinite integral of f is a continuous martingale. The inequality (3.7) then follows from the martingale inequality.

Another estimate is given in the following theorem.

Theorem 3.6. *Let $f\in M_w^2[0, T]$. Then*

$$E\left\{\sup_{0\leq t\leq T}\left|\int_0^t f(s)\,dw(s)\right|^2\right\} \leq 4E\left|\int_0^T f(t)\,dw(t)\right|^2 = 4E\int_0^T f^2(t)\,dt. \tag{3.8}$$

Like Theorem 3.5, the present estimate is also a consequence of the fact that $\int_0^t f(s)\,dw(s)$ is a continuous martingale. The general result for martingales is given in the following theorem.

Theorem 3.7. *If $X(t)$ $(0 \leq t \leq T)$ is a separable martingale, then, for any $\alpha > 1$,*

$$E\Big\{\sup_{0 \leq t \leq T} |X(t)|^{\alpha}\Big\} \leq \Big(\frac{\alpha}{\alpha - 1}\Big)^{\alpha} E|X(T)|^{\alpha}. \tag{3.9}$$

The proof follows immediately from the following lemma.

Lemma 3.8. *If $\{X_n\}$ is a submartingale and $X_n \geq 0$ for all n, then, for any $\alpha > 1$,*

$$E\Big[\max_{1 \leq i \leq n} (X_i)^{\alpha}\Big] \leq \Big(\frac{\alpha}{\alpha - 1}\Big)^{\alpha} E(X_n)^{\alpha}. \tag{3.10}$$

Proof. Let $Y = \max_{1 \leq i \leq n} X_i$. For any $\lambda > 0$, set $\chi_k(\lambda) = 1$ if $X_i < \lambda$ for $1 \leq i \leq k-1$, $X_k \geq \lambda$, and $\chi_k(\lambda) = 0$ otherwise. Then $\Sigma_{k=1}^n \chi_k(\lambda) = 1$ if $\lambda \leq Y$ and $\Sigma_{k=1}^n \chi_k(\lambda) = 0$ if $\lambda > Y$. It follows that, for any $\beta > 0$,

$$Y^{\beta} = \beta \int_0^{\infty} \lambda^{\beta - 1}\Big[\sum_{k=1}^{n} \chi_k(\lambda)\Big]\, d\lambda. \tag{3.11}$$

Since $\lambda \chi_k(\lambda) \leq X_k \chi_k(\lambda)$,

$$\lambda^{\alpha - 1} \sum_{k=1}^{n} \chi_k(\lambda) \leq \sum_{k=1}^{n} \lambda^{\alpha - 2} X_k \chi_k(\lambda). \tag{3.12}$$

Observing that $\chi_k(\lambda)$ is measurable with respect to $\mathfrak{F}(X_1, \ldots, X_k)$ and using the submartingale assumption, we get

$$E\{X_n \chi_k(\lambda) | \mathfrak{F}(X_1, \ldots, X_k)\} \geq \chi_k(\lambda) E\{X_n | \mathfrak{F}(X_1, \ldots, X_k)\} \geq \chi_k(\lambda) X_k.$$

Taking the expectation in (3.12) and using the last inequality, we find that

$$E\lambda^{\alpha - 1} \sum_{k=1}^{n} \chi_k(\lambda) \leq E\lambda^{\alpha - 2} \sum_{k=1}^{n} \chi_k(\lambda) X_n.$$

Integrating with respect to λ and using (3.11), we get

$$\frac{1}{\alpha} EY^{\alpha} \leq \frac{1}{\alpha - 1} EY^{\alpha - 1} X_n.$$

By Hölder's inequality,

$$EY^{\alpha} \leq \frac{\alpha}{\alpha - 1} (EY^{\alpha})^{(\alpha - 1)/\alpha} (EX_n^{\alpha})^{1/\alpha},$$

and (3.10) follows.

Remark. It is clear that all the results derived so far in this section regarding the indefinite integral (3.1) extend to indefinite integrals

$$\int_\alpha^t f(s)\,dw(s).$$

Theorem 3.9. *Let $f \in M_w^2[\alpha, \beta]$. Then*

$$E\left\{ \sup_{\alpha \leq t \leq \beta} \left| \int_\alpha^t f(s)\,dw(s) \right|^2 \Big| \mathcal{F}_\alpha \right\} \leq 4E\left\{ \int_\alpha^\beta f^2(t)\,dt \Big| \mathcal{F}_\alpha \right\}. \tag{3.13}$$

The proof is left to the reader.

Definition. Let τ_0 be a random variable, $0 \leq \tau_0 \leq T$. If $f \in L_w^2[0, T]$, we define

$$\int_0^{\tau_0} f(s)\,dw(s)$$

to be the random variable $I(\tau_0)$, where $I(t)$ is given by (3.1).

From Theorems 3.1, 3.2, and 1.3.5 we deduce:

Theorem 3.10. *If $f \in M_w^2[0, \tau]$ and τ is a stopping time with respect to $\mathcal{F}_t$, $0 \leq \tau \leq T$, i.e., $\{\tau \leq t\} \in \mathcal{F}_t$ for all $0 \leq t \leq T$, then the process*

$$\int_0^{\tau \wedge t} f(s)\,dw(s), \qquad 0 \leq t \leq T$$

is a martingale and

$$E \int_0^{\tau \wedge t} f(s)\,dw(s) = 0. \tag{3.14}$$

4. Stochastic integrals with stopping time

If $f \in L_w^2[\alpha, T]$ for all $T > 0$, then we say that f belongs to $L_w^2[\alpha, \infty)$. Similarly we define $f \in M_w^2[\alpha, \infty)$.

Let $f \in L_w^2[0, T]$ and let ζ_1, ζ_2 be random variables, $0 \leq \zeta_1 \leq \zeta_2 \leq T$. We define

$$\int_{\zeta_1}^{\zeta_2} f(t)\,dw(t) = \int_0^{\zeta_2} f(t)\,dw(t) - \int_0^{\zeta_1} f(t)\,dw(t).$$

Lemma 4.1. *Define $\chi_i(t) = 1$ if $t < \zeta_i$, $\chi_i(t) = 0$ if $t \geq \zeta_i$ $(i = 1, 2)$. Then the $\chi_i(t)$ are $\mathcal{F}_t$ measurable and*

$$\int_{\zeta_1}^{\zeta_2} f(t)\,dw(t) = \int_0^T \chi_2(t) f(t)\,dw(t) - \int_0^T \chi_1(t) f(t)\,dw(t). \tag{4.1}$$

Proof. It is enough to prove the lemma in case $\zeta_1 \equiv 0$. It is clear that $\chi_2(t)$ is $\mathfrak{F}_t$ measurable; in fact, it is a nonanticipative function. In case f is a step function and ζ_2 is a simple function, (4.1) follows directly from the definition of the integral. In the general case, let f_n be step functions such that

$$\int_0^T |f_n(t) - f(t)|^2\, dt \xrightarrow{P} 0 \quad \text{if} \quad n \to \infty,$$

and let ζ_{2n} be simple stopping times such that $\zeta_{2n} \downarrow \zeta_2$ everywhere if $n \uparrow \infty$ (cf. Lemma 1.3.4). We have

$$\int_0^{\zeta_{2n}} f_n(t)\, dw(t) = \int_0^T \chi_{2n}(t) f_n(t)\, dw(t) \tag{4.2}$$

where $\chi_{2n}(t) = 1$ if $t < \zeta_{2n}$, $\chi_{2n}(t) = 0$ if $t \geq \zeta_{2n}$. It is clear that $\chi_{2n}(t) \to \chi_2(t)$ for all ω and $t \neq \zeta_2(\omega)$, as $n \to \infty$. Hence

$$\int_0^T |\chi_{2n}(t) - \chi_2(t)|^2 f(t)\, dt \to 0 \quad \text{a.s.} \quad \text{as} \quad n \to \infty,$$

by the Lebesgue bounded convergence theorem. We also have

$$\int_0^T |\chi_{2n}(t)|^2 |f(t) - f_n(t)|^2\, dt \xrightarrow{P} 0 \quad \text{a.s.} \quad \text{as} \quad n \to \infty.$$

Putting these together, we find that

$$\int_0^T |\chi_{2n}(t) f_n(t) - \chi_2(t) f(t)|^2\, dt \xrightarrow{P} 0 \quad \text{as} \quad n \to \infty.$$

Hence, by Theorem 2.7,

$$\int_0^T \chi_{2n}(s) f_n(s)\, dw(s) \xrightarrow{P} \int_0^T \chi_2(s) f(s)\, dw(s). \tag{4.3}$$

By Theorem 3.4,

$$\sup_{0 \leq t \leq T} \left| \int_0^t f_n(s)\, dw(s) - \int_0^t f(s)\, dw(s) \right| \xrightarrow{P} 0 \quad \text{if} \quad n \to \infty.$$

It follows that

$$\int_0^{\zeta_{2n}} f_n(s)\, dw(s) - \int_0^{\zeta_{2n}} f(s)\, dw(s) \xrightarrow{P} 0 \quad \text{if} \quad n \to \infty.$$

Since from the continuity of the indefinite integral we also have

$$\int_0^{\zeta_{2n}} f(s)\, dw(s) \to \int_0^{\zeta_2} f(s)\, dw(s) \quad \text{a.s.} \quad \text{as} \quad n \to \infty,$$

we find that

$$\int_0^{\zeta_{2n}} f_n(s)\, dw(s) \xrightarrow{P} \int_0^{\zeta_2} f(s)\, dw(s) \quad \text{if} \quad n \to \infty.$$

Combining this with (4.2), (4.3), the formula (4.1) (in case $\zeta_1 \equiv 0$) follows.

Theorem 4.2. *Let $f \in M_w^2[0, T]$ and let ζ_1, ζ_2 be stopping times, $0 \leq \zeta_1 \leq \zeta_2 \leq T$. Then*

$$E \int_{\zeta_1}^{\zeta_2} f(t)\, dw(t) = 0, \tag{4.4}$$

$$E\left[\int_{\zeta_1}^{\zeta_2} f(t)\, dw(t)\right]^2 = E \int_{\zeta_1}^{\zeta_2} f^2(t)\, dt. \tag{4.5}$$

Proof. Let $\chi_i(t) = 1$ if $t < \zeta_i$, $\chi_i(t) = 1$ if $t \geq \zeta_i$. By Lemma 4.1,

$$\int_{\zeta_1}^{\zeta_2} f(t)\, dw(t) = \int_0^T (\chi_2(t) - \chi_1(t)) f(t)\, dw(t).$$

Applying Theorem 2.5 to the right-hand side, the assertions (4.4), (4.5) readily follow.

Note that (4.4) follows also from Theorem 3.10.

Theorem 4.2 is a generalization of Theorem 2.5. The next theorem is a generalization of Theorem 2.8.

Theorem 4.3. *Let $f \in M_w^2[0, T]$ and let ζ_1, ζ_2 be stopping times (with respect to $\mathcal{F}_t$), $0 \leq \zeta_1 \leq \zeta_2 \leq T$. Then*

$$E\left\{ \int_{\zeta_1}^{\zeta_2} f(t)\, dw(t) | \mathcal{F}_{\zeta_1} \right\} = 0, \tag{4.6}$$

$$E\left\{ \left| \int_{\zeta_1}^{\zeta_2} f(t)\, dw(t) \right|^2 | \mathcal{F}_{\zeta_1} \right\} = E\left\{ \int_{\zeta_1}^{\zeta_2} f^2(t)\, dt | \mathcal{F}_{\zeta_1} \right\}. \tag{4.7}$$

Here $\mathcal{F}_\zeta$ denotes the σ-field of all events A such that

$$A \cap (\zeta \leq s) \qquad \text{is in } \mathcal{F}_s, \quad \text{for all} \quad s \geq 0.$$

Denote by $\mathcal{F}_\zeta^*$ the σ-field generated by all sets of the form

$$A \cap (\zeta > t), \qquad A \in \mathcal{F}_t, \quad t \geq 0.$$

We shall need the following lemma.

Lemma 4.4. *For any stopping time ζ, $\mathcal{F}_\zeta = \mathcal{F}_\zeta^*$.*

Proof. Let $B = A \cap (\zeta > t)$, $A \in \mathcal{F}_t$. For any $s \geq 0$, let

$$C = B \cap (\zeta \leq s) = A \cap (\zeta > t) \cap (\zeta \leq s).$$

If $s \leq t$, then $C = \varnothing \in \mathcal{F}_s$. If $s > t$, then

$$C = [A^c \cup (\zeta \leq t)]^c \cap (\zeta \leq s)$$

where D^c is the complement of D. Since $A^c \in \mathcal{F}_t$ and $(\zeta \leq t) \in \mathcal{F}_t$, it follows that $[A^c \cup (\zeta \leq t)]^c \in \mathcal{F}_t$. Hence $C \in \mathcal{F}_s$. We have thus proved that $B \cap (\zeta \leq s)$ is in $\mathcal{F}_s$ for all $s \geq 0$, i.e., $B \in \mathcal{F}_\zeta$. It follows that $\mathcal{F}_\zeta^* \subset \mathcal{F}_\zeta$.

Conversely let $B \in \mathcal{F}_\zeta$. Then

$$B_s = B \cap (\zeta \leq s) \qquad \text{is in } \mathcal{F}_s, \quad \text{for all} \quad s \geq 0.$$

Hence $B_s^c = B^c \cup (\zeta > s)$ is in $\mathcal{F}_s$, and since

$$B_s^c = B_s^c \cap (\zeta > s),$$

it follows that B_s^c is one of the sets that generate $\mathcal{F}_\zeta^*$. Therefore $B_s^c \in \mathcal{F}_\zeta^*$. Since $\mathcal{F}_\zeta^*$ is a σ-field, B_s belongs to $\mathcal{F}_\zeta^*$, and also $\lim_{s\uparrow\infty} B_s \in \mathcal{F}_\zeta^*$. But this limit is the set B. Hence $B \in \mathcal{F}_\zeta^*$.

***Proof of Theorem* 4.3.** Let $C = A \cap (\zeta_1 > s)$, $A \in \mathcal{F}_s$. Its indicator function $\chi_C = \chi_A \chi_1(s)$ is $\mathcal{F}_s$ measurable. Consider the function

$$\chi_C(\chi_2(t) - \chi_1(t)) = \chi_A \chi_1(s)(\chi_2(t) - \chi_1(t)).$$

If $s \leq t$, each factor on the left is in $\mathcal{F}_t$, and so is then the product. If $s > t$, then $\chi_1(t) = 1$, so that $\chi_2(t) - \chi_1(t) = 0$. Thus the product is again in $\mathcal{F}_t$. We have thus proved that

$$\chi_C(\chi_2(t) - \chi_1(t)) \qquad \text{is } \mathcal{F}_t \text{ measurable for any set } C. \tag{4.8}$$

Let $B \in \mathcal{F}_{\zeta_1}$. From the proof of Lemma 4.4 we have that B_s^c has the form of the set C for which (4.8) holds. Hence also

$$\chi_{B_s}(\chi_2(t) - \chi_1(t)) \qquad \text{is } \mathcal{F} \text{ measurable.}$$

Taking $s \uparrow \infty$ we conclude that

$$\chi_B(\chi_2(t) - \chi_1(t)) \qquad \text{is } \mathcal{F}_t \text{ measurable.}$$

We can now proceed as in the proof of Theorem 4.2 to prove that

$$E\chi_B \int_{\zeta_1}^{\zeta_2} f(t)\, dw(t) = 0,$$

$$E\left|\chi_B \int_{\zeta_1}^{\zeta_2} f(t)\, dw(t)\right|^2 = E \int_{\zeta_1}^{\zeta_2} \chi_B f^2(t)\, dt,$$

and the assertions (4.6), (4.7) follow.

As an application of Theorem 4.3 we shall prove:

Theorem 4.5. *Let* $f \in L_w^2[0, \infty)$ *and assume that* $\int_0^\infty f^2(t)\, dt = \infty$ *with probability* 1. *Let*

$$\tau(t) = \inf\left\{ s;\ \int_0^s f^2(\lambda)\, d\lambda = t \right\};$$

then the process

$$u(t) = \int_0^{\tau(t)} f(s)\, dw(s)$$

is a Brownian motion.

Let

$$t^*(t) = \int_0^t f^2(s)\, ds.$$

Then τ is the left-continuous inverse of t^*, i.e., $\tau(t) = \min(s, t^*(s) = t)$. t^* is called the *intrinsic time* (or *intrinsic clock*) for $I(t) = \int_0^t f(s)\, dw(s)$. Theorem 4.5 asserts that there is a Brownian motion $u(t)$ such that $u(t^*(t)) = I(t)$.

Proof. It is easily verified that $\tau(t)$ is a stopping time. Notice next that $\mathcal{F}_{\tau(t_1)} \subset \mathcal{F}_{\tau(t_2)}$ if $t_1 < t_2$. Indeed, if $A \in \mathcal{F}_{\tau(t_1)}$, then $A \cap [\tau(t_1) \leqslant s]$ is in $\mathcal{F}_s$ for all $s \geqslant 0$. But since $\tau(t_2) \geqslant \tau(t_1)$,

$$A \cap [\tau(t_2) \leqslant s] = \{A \cap [\tau(t_1) \leqslant s]\} \cap [\tau(t_2) \leqslant s] \qquad \text{is in } \mathcal{F}_s.$$

Thus $A \in \mathcal{F}_{\tau(t_2)}$. We shall now assume that $f \in M_w^2[0, \infty)$ and $f(t) = 1$ if $t > \lambda$, for some $\lambda > 0$. Then $\tau(t_2) < t_2 + \lambda$ and, by Theorem 4.3,

$$E\left\{ \int_{\tau(t_1)}^{\tau(t_2)} f(s)\, dw(s) | \mathcal{F}_{\tau(t_1)} \right\} = 0,$$

$$E\left\{ \left| \int_{\tau(t_1)}^{\tau(t_2)} f(s)\, dw(s) \right|^2 | \mathcal{F}_{\tau(t_1)} \right\} = E\left\{ \int_{\tau(t_1)}^{\tau(t_2)} f^2(s)\, ds | \mathcal{F}_{\tau(t_1)} \right\} = t_2 - t_1.$$

If we prove that $u(t)$ is continuous, then Theorem 3.5.1 implies that $u(t)$ is a Brownian motion. From Theorem 3.6 we deduce that

$$E\left\{ \sup_{\substack{t' \leqslant t \leqslant t' + \epsilon \\ t' \leqslant s \leqslant t' + \epsilon}} |u(t) - u(s)|^2 \right\} \leqslant 4E\left\{ \int_{\tau(t')}^{\tau(t'+\epsilon)} f^2(s)\, ds \right\} = 4\epsilon.$$

Since t' is arbitrary,

$$P\left\{ \sup_{\substack{|t-s| \leqslant \epsilon \\ t \geqslant 0,\, s \geqslant 0}} |u(t) - u(s)| > \epsilon^\alpha \right\} \leqslant \frac{4\epsilon}{\epsilon^{2\alpha}}.$$

Taking $\alpha = \frac{1}{4}$, $\epsilon = 1/k^4$ and using the Borel–Cantelli lemma, we find that

$$P\left\{ \sup_{\substack{|t-s| \leqslant 1/k^4 \\ t \geqslant 0, s \geqslant 0}} |u(t) - u(s)| > 1/k \quad \text{i.o.} \right\} = 0, \qquad \text{i.e.} \quad u \text{ is continuous.}$$

For a general $f \in M_w^2[0, \infty)$, apply the preceding result to f_λ where $f_\lambda(t) = f(t)$ if $t < \lambda$, $f_\lambda(t) = 1$ if $t > \lambda$. Since (by Theorem 4.7 below)

$$\int_0^{\tau_\lambda(t)} f_\lambda(s)\, dw(s) = \int_0^{\tau(t)} f(s)\, ds \quad \text{a.e. on the set } \lambda > \tau(t),$$

where τ_λ is the τ function of f_2, Problem 10, Chapter 3 implies that $u(t)$ is a Brownian motion.

So far we have proved the theorem only in case $f \in M_w^2[0, \infty)$. Now let

$f \in L_w^2[0, \infty)$ and define χ_N, f_N as in (3.4), (3.5). Then $f_N \in M_w^2[0, \infty)$, so that

$$\int_0^{\tau_N(t)} f_N(s)\, dw(s)$$

is a Brownian motion. But $\tau_N(t) = \tau(t)$, $f_N(s) = f(s)$ if $0 < s \leqslant \tau(t)$, $t < N$. Hence (by Theorem 4.7 below)

$$\int_0^{\tau_N(t)} f_N(s)\, dw(s) = \int_0^{\tau(t)} f(s)\, dw(s)$$

if $N > t$. It follows that the function on the right is Brownian.

Corollary 4.6. *Let $f \in L_w^2[0, \infty)$, $|f| \leqslant K$ where K is a constant. Then, for any $\alpha > \frac{1}{2}$,*

$$\frac{1}{t^\alpha} \int_0^t f(s)\, dw(s) \to 0 \quad \text{if} \quad t \to \infty. \tag{4.9}$$

Proof. Let $M > K$ and define

$$t^*(t) = \int_0^t (f(s) + M)^2\, ds,$$

$$a(t) = \int_0^{\tau(t)} (f(s) + M)^2\, dw(s) \qquad (\tau = \text{inverse of } t^*).$$

Since $|f + M| \geqslant M - K > 0$, the function $t^*(t)$ is strictly monotone increasing to ∞ as $t \uparrow \infty$. Hence $a(t)$ is defined for all $t \geqslant 0$ and, by Theorem 4.5, it is a Brownian motion. But then, by the law of the iterated logarithm,

$$\frac{a(t)}{t^\alpha} \to 0 \quad \text{a.s.} \quad \text{if} \quad t \to \infty,$$

or

$$\frac{a(t^*(t))}{(t^*(t))^\alpha} \to 0 \quad \text{a.s.} \qquad \text{if} \quad t \to \infty.$$

Noting that $t^*(t) \leqslant (K + M)^2 t$, we get

$$\frac{1}{t^\alpha} \int_0^t (f(s) + M)\, dw(s) \to 0 \quad \text{a.s.} \qquad \text{if} \quad t \to \infty.$$

Using the fact that $(w(t)/t^\alpha) \to 0$ if $t \to \infty$, the assertion (4.9) follows.

Theorem 4.7. *Let $f \in L_w^2[0, \infty)$, $g \in L_w^2[0, \infty)$, and set*

$$h(t) = \int_0^t f(s)\, dw(s), \qquad k(t) = \int_0^t g(s)\, dw(s).$$

Let τ be a random variable, $\tau \geqslant 0$, and assume that $f(s) = g(s)$ if $s < \tau$. Then $h(t) = k(t)$ a.s. if $t < \tau$.

Proof. Construct, by Lemma 1.1, step functions f_n, g_n which belong to $L_w^2[0, T]$ and which satisfy

$$\int_0^T |f_n(t) - f(t)|^2 \, dt \xrightarrow{P} 0, \qquad \int_0^T |g_n(t) - g(t)|^2 \, dt \xrightarrow{P} 0$$

as $n \to \infty$. That construction is such that

$$f_n(s, \omega) = g_n(s, \omega) \qquad \text{if} \quad s < \tau(\omega) \wedge T. \tag{4.10}$$

Let

$$h_n(t) = \int_0^t f_n(s) \, dw(s), \qquad k_n(t) = \int_0^t g_n(s) \, dw(s).$$

From the definition of the integral of a step function and from (4.10) it easily follows that

$$h_n(t, \omega) = k_n(t, \omega) \qquad \text{if} \quad t < \tau(\omega) \wedge T. \tag{4.11}$$

By Theorem 3.4,

$$\sup_{0 \le t \le T} |h_n(t) - h(t)| \xrightarrow{P} 0, \qquad \sup_{0 \le t \le T} |k_n(t) - k(t)| \xrightarrow{P} 0$$

if $n \to \infty$. Hence, for some subsequence $\{n'\}$,

$$\sup_{0 \le t \le T} |h_{n'}(t) - h(t)| \to 0, \qquad \sup_{0 \le t \le T} |k_{n'}(t) - k(t)| \to 0 \quad \text{a.s.}$$

if $n' \to \infty$. Recalling (4.11), we find that for a.a. ω,

$$h(t, \omega) = k(t, \omega) \qquad \text{if} \quad t < \tau(\omega) \wedge T.$$

The proof is now completed by taking a sequence $T = T_n \uparrow \infty$.

5. Itô's formula

Definition. Let $\xi(t)$ $(0 \le t \le T)$ be a process such that for any $0 \le t_1 < t_2 \le T$

$$\xi(t_2) - \xi(t_1) = \int_{t_1}^{t_2} a(t) \, dt + \int_{t_1}^{t_2} b(t) \, dw(t)$$

where $a \in L_w^1[0, T]$, $b \in L_w^2[0, T]$. Then we say that $\xi(t)$ has *stochastic differential* $d\xi$, on $[0, T]$, given by

$$d\xi(t) = a(t) \, dt + b(t) \, dw(t).$$

Observe that $\xi(t)$ is a nonanticipative function. It is also a continuous process. Hence, in particular, it belongs to $L_w^\infty[0, T]$.

Example 1. If $0 \le t_1 < t_2$ and $\Pi_n = \{t_1 = t_{n,1}, t_{n,2}, \ldots, t_{n,n} = t_2\}$ is a sequence of partitions of $[t_1, t_2]$ with mesh $|\Pi_n| \to 0$ then, by Theorem 2.10,

$$\begin{aligned}\int_{t_1}^{t_2} w(t)\,dw(t) &= \lim_{n\to\infty} \sum_{k=1}^{n-1} w(t_{n,k})(w(t_{n,k+1}) - w(t_{n,k}))\\ &= \tfrac{1}{2} \lim_{n\to\infty} \sum_{k=1}^{n-1} \{[(w(t_{n,k+1}))^2 - (w(t_{n,k}))^2]\\ &\qquad - (w(t_{n,k+1}) - w(t_{n,k}))^2\}\\ &= \tfrac{1}{2}(w(t_2))^2 - \tfrac{1}{2}(w(t_1))^2 - \tfrac{1}{2} \lim_{n\to\infty} \sum_{k=1}^{n-1} (w(t_{n,k+1}) - w(t_{n,k}))^2\end{aligned}$$

where $\lim_{n\to\infty}$ is taken as the limit in probability.

By Theorem 3.3.3, the last limit in probability is equal to $t_2 - t_1$. Hence

$$\int_{t_1}^{t_2} w(t)\,dw(t) = \tfrac{1}{2}(w(t_2))^2 - \tfrac{1}{2}(w(t_1))^2 - \tfrac{1}{2}(t_2 - t_1), \tag{5.1}$$

or

$$d(w(t))^2 = dt + 2w(t)\,dw(t). \tag{5.2}$$

Example 2. By Theorem 2.10,

$$\int_{t_1}^{t_2} t\,dw(t) = \lim_{n\to\infty} \sum_{k=1}^{n-1} t_{n,k}[w(t_{n,k+1}) - w(t_{n,k})] \quad \text{in probability.}$$

Clearly

$$\int_{t_1}^{t_2} w(t)\,dt = \lim_{n\to\infty} \sum_{k=1}^{n-1} w(t_{n,k+1})(t_{n,k+1} - t_{n,k})$$

for all ω for which $w(t, \omega)$ is continuous. The sum of the right-hand sides is equal to

$$\lim_{n\to\infty} \sum_{k=1}^{n-1} [t_{n,k+1} w(t_{n,k+1}) - t_{n,k} w(t_{n,k})] = t_2 w(t_2) - t_1 w(t_1).$$

Hence

$$d(tw(t)) = w(t)\,dt + t\,dw(t). \tag{5.3}$$

Definition. Let $\xi(t)$ be as in the definition above and let $f(t)$ be a function in $L_w^\infty[0, T]$. We define

$$f(t)\,d\xi(t) = f(t)a(t)\,dt + f(t)b(t)\,dw(t).$$

Example 3. $f(t)\,d\xi(t)$ is a stochastic differential $d\eta$, where

$$\eta(t) = \int_0^t f(s)a(s)\,ds + \int_0^t f(s)b(s)\,dw(s).$$

Theorem 5.1. *If* $d\xi_i(t) = a_i(t)\,dt + b_i(t)\,dw(t)$ $(i = 1, 2)$, *then*

$$d(\xi_1(t)\xi_2(t)) = \xi_1(t)\,d\xi_2(t) + \xi_2(t)\,d\xi_1(t) + b_1(t)b_2(t)\,dt. \tag{5.4}$$

The integrated form of (5.4) asserts that, for any $0 \leq t_1 < t_2 \leq T$,

$$\begin{aligned}\xi_1(t_2)\xi_2(t_2) - \xi_1(t_1)\xi_2(t_1) = &\int_{t_1}^{t_2} \xi_1(t)a_2(t)\,dt + \int_{t_1}^{t_2} \xi_1(t)b_2(t)\,dw(t)\\ &+ \int_{t_1}^{t_2} \xi_2(t)a_1(t)\,dt + \int_{t_1}^{t_2} \xi_2(t)b_1(t)\,dw(t)\\ &+ \int_{t_1}^{t_2} b_1(t)b_2(t)\,dt.\end{aligned} \tag{5.5}$$

Proof. Suppose first that a_i, b_i are constants in the interval $[t_1, t_2)$. Then (5.5) follows from (5.2), (5.3). Next, if a_i, b_i are step functions in $[t_1, t_2)$, constants on successive intervals $I_1, I_2, \ldots, I_l$, then (5.5) holds with t_1, t_2 replaced by the end points of each interval I_j. Taking the sum we obtain (5.5).

Consider now the general case. Approximate a_i, b_i by nonanticipative step functions $a_{i,n}$, $b_{i,n}$ in such a way that

$$\int_0^T |a_{i,n}(t) - a_i(t)|\,dt \to 0 \quad \text{a.s.},$$

$$\int_0^T |b_{i,n}(t) - b_i(t)|^2\,dt \to 0 \quad \text{a.s.}$$

Let

$$\xi_{i,n}(t) = \xi_i(0) + \int_0^t a_{i,n}(s)\,ds + \int_0^t b_{i,n}(s)\,dw(s).$$

By Theorem 3.4,

$$\sup_{0 \leq t \leq T} |\xi_{i,n}(t) - \xi_i(t)| \xrightarrow{P} 0 \qquad \text{if} \quad n \to \infty;$$

hence there is a subsequence $\xi_{i,n'}$ such that

$$\xi_{i,n'}(t) \to \xi_i(t) \quad \text{uniformly in} \quad t \in [0, T], \quad \text{a.s.} \tag{5.6}$$

Using (5.6) and Theorem 2.7 it is easily seen that

$$\int_{t_1}^{t_2} \xi_{i,n}(t)b_{j,n}(t)\,dw(t) \xrightarrow{P} \int_{t_1}^{t_2} \xi_i(t)b_j(t)\,dw(t)$$

as $n = n' \to \infty$. Clearly also

$$\int_{t_1}^{t_2} \xi_{i,n}(t)a_{j,n}(t)\,dt \to \int_{t_1}^{t_2} \xi_i(t)a_j(t)\,dt,$$

$$\int_{t_1}^{t_2} b_{1,n}(t)b_{2,n}(t)\,dt \to \int_{t_1}^{t_2} b_1(t)b_2(t)\,dt$$

a.s. Writing (5.5) for $a_{i,n}$, $b_{i,n}$, $\xi_{i,n}$ and taking $n \to \infty$, the assertion (5.5) follows. Since t_1, t_2 are arbitrary, the proof of the theorem is complete.

Theorem 5.2. *Let $d\xi(t) = a(t)\,dt + b(t)\,dw(t)$, and let $f(x, t)$ be a continuous function in $(x, t) \in R^1 \times [0, \infty)$ together with its derivatives f_x, f_{xx}, f_t. Then the process $f(\xi(t), t)$ has a stochastic differential, given by*

$$df(\xi(t), t) = [f_t(\xi(t), t) + f_x(\xi(t), t)a(t) + \tfrac{1}{2}f_{xx}(\xi(t), t)b^2(t)]\,dt + f_x(\xi(t), t)b(t)\,dw(t). \tag{5.7}$$

This is called *Itô's formula.* Notice that if $w(t)$ were continuously differentiable in t, then (by the standard calculus formula for total derivatives) the term $\frac{1}{2}f_{xx}b^2\,dt$ would not appear.

Proof. The proof will be divided into several steps.

Step 1. For any integer $m \geq 2$,

$$d(w(t))^m = m(w(t))^{m-1} + \tfrac{1}{2}m(m-1)(w(t))^{m-2}\,dt. \tag{5.8}$$

Indeed, this follows by induction, using Theorem 5.1.

By linearity of the stochastic differential we then get

$$dQ(w(t)) = Q'(w(t))\,dw(t) + \tfrac{1}{2}Q''(w(t))\,dt \tag{5.9}$$

for any polynomial Q.

Step 2. Let $G(x, t) = Q(x)g(t)$ where $Q(x)$ is a polynomial and $g(t)$ is continuously differentiable for $t \geq 0$. By Theorem 5.1 and (5.9),

$$\begin{aligned} dG(w(t), t) &= f(w(t))dg(t) + g(t)df(w(t)) \\ &= [f(w(t))g'(t) + \tfrac{1}{2}g(t)f''(w(t))]\,dt + g(t)f'(w(t))\,dw(t), \end{aligned}$$

i.e., for any $0 \leq t_1 < t_2 \leq T$,

$$G(w(t_2), t_2) - G(w(t_1), t_1) = \int_{t_1}^{t_2} [G_t(w(t), t) + \tfrac{1}{2}G_{xx}(w(t), t)]\,dt + \int_{t_1}^{t_2} G_x(w(t), t)\,dw(t). \tag{5.10}$$

Step 3. Formula (5.10) remains valid if

$$G(x, t) = \sum_{i=1}^{m} f_i(x)g_i(t)$$

where $f_i(x)$ are polynomials and $g_i(t)$ are continuously differentiable. Now let $G_n(x, t)$ be polynomials in x and t such that

$$G_n(x, t) \to f(x, t),$$

$$\frac{\partial}{\partial x}G_n(x, t) \to f_x(x, t), \qquad \frac{\partial^2}{\partial x^2}G_n(x, t) \to f_{xx}(x, t), \qquad \frac{\partial}{\partial t}G_n(x, t) \to f_t(x, t)$$

uniformly on compact subsets of $x, t) \in R^1 \times [0, \infty)$; see Problem 11 for the proof of the existence of such a sequence. We have

$$G_n(w(t_2), t_2) - G_n(w(t_1), t_1) = \int_{t_1}^{t_2} \left[\frac{\partial}{\partial t} G_n(w(t), t) + \frac{1}{2} \frac{\partial^2}{\partial x^2} G_n(w(t), t) \right] dt + \int_{t_1}^{t_2} \frac{\partial}{\partial x} G_n(w(t), t)\, dw(t). \tag{5.11}$$

It is clear that

$$\int_{t_1}^{t_2} \left[\frac{\partial}{\partial t} G_n(w(t), t) + \frac{1}{2} \frac{\partial^2}{\partial x^2} (w(t), t) \right] dt \to \int_{t_1}^{t_2} [f_t(w(t), t) - \tfrac{1}{2} f_{xx}(w(t), t)]\, dt \quad \text{a.s.},$$

$$\int_{t_1}^{t_2} \left| \frac{\partial}{\partial x} G_n(w(t), t) - f_x(w(t), t) \right|^2 dt \to 0 \quad \text{a.s.}$$

Hence, taking $n \to \infty$ in (5.11), we get the relation

$$f(w(t_2), t_2) - f(w(t_1), t_1) = \int_{t_1}^{t_2} [f_t(w(t), t) + \tfrac{1}{2} f_{xx}(w(t), t)]\, dt + \int_{t_1}^{t_2} f_x(w(t), t)\, dw(t). \tag{5.12}$$

Step 4. Formula (5.12) extends to the process

$$\Phi(w(t), t) = f(\xi_1 + a_1 t + b_1 w(t), t)$$

where ξ_1, a_1, b_1 are random variables measurable with respect to $\mathcal{F}_{t_1}$, i.e.,

$$\Phi(w(t_2), t_2) - \Phi(w(t_1), t_1) = \int_{t_1}^{t_2} [f_t(\tilde{\xi}(t), t) + f_x(\tilde{\xi}(t), t) a_1 + \tfrac{1}{2} f_{xx}(\tilde{\xi}(t), t) b_1^2]\, dt + \int_{t_1}^{t_2} f_x(\tilde{\xi}(t), t) b_1\, dw(t) \tag{5.13}$$

where $\tilde{\xi}(t) = \xi_1 + a_1 t + b_1 w(t)$.

The proof of (5.13) is a repetition of the proof of (5.12) with obvious changes resulting from the formula

$$d(\tilde{\xi}(t))^m = m(\tilde{\xi}(t))^{m-1}[a_1\, dt + b_1\, dw(t)] + \tfrac{1}{2} m(m-1)(\tilde{\xi}(t))^{m-2} b_1^2\, dt, \tag{5.14}$$

which replaces (5.8). The details are left to the reader.

Step 5. If $a(t)$, $b(t)$ are step functions, then

$$f(\xi(t_2), t_2) - f(\xi(t_1), t_1) = \int_{t_1}^{t_2} \left[f_t(\xi(t), t) + f_x(\xi(t), t)a(t) + \tfrac{1}{2} f_{xx}(\xi(t), t)b^2(t) \right] dt + \int_{t_1}^{t_2} f_x(\xi(t), t)b(t)\, dw(t). \tag{5.15}$$

Indeed, denote by $I_1, \ldots, I_k$ the successive intervals in $[t_1, t_2]$ in which a, b are constants. If we apply (5.13) with t_1, t_2 replaced by the end points of I_l, and sum over l, the formula (5.15) follows.

Step 6. Let a_i, b_i be nonanticipative step functions such that

$$\int_0^T |a_i(t) - a(t)|\, dt \to 0 \quad \text{a.s.}, \tag{5.16}$$

$$\int_0^T |b_i(t) - b(t)|^2\, dt \xrightarrow{P} 0, \tag{5.17}$$

and let

$$\xi_i(t) = \xi(0) + \int_0^t a_i(s)\, ds + \int_0^t b_i(s)\, dw(s).$$

Then

$$\sup_{0 < t < T} |\xi_i(t) - \xi(t)| \xrightarrow{P} 0.$$

Hence, for a subsequence $\{i'\}$,

$$\sup_{0 < t < T} |\xi_i(t) - \xi(t)| \to 0 \quad \text{a.s.} \qquad \text{if} \quad i = i' \to \infty. \tag{5.18}$$

This and (5.17) imply that

$$\int_0^T |f_x(\xi_i(t), t)b_i(t) - f_x(\xi(t), t)b(t)|^2\, dt \xrightarrow{P} 0 \qquad \text{if} \quad i = i' \to \infty.$$

It follows that

$$\int_{t_1}^{t_2} f_x(\xi_i(t), t)b_i(t)\, dw(t) \xrightarrow{P} \int_{t_1}^{t_2} f_x(\xi(t), t)b(t)\, dw(t) \qquad \text{if} \quad i = i' \to \infty.$$

It is clear from (5.16)–(5.18) that also

$$\int_{t_1}^{t_2} \left[f_t(\xi_i(t), t) + f_x(\xi_i(t), t)a_i(t) + \tfrac{1}{2} f_{xx}(\xi_i(t), t)(b_i(t))^2 \right] dt$$
$$\xrightarrow{P} \int_{t_1}^{t_2} \left[f_t(\xi(t), t) + f_x(\xi(t), t)a(t) + \tfrac{1}{2} f_{xx}(\xi(t), t)b^2(t) \right] dt$$

if $i = i' \to \infty$.

Writing (5.15) for $a = a_i$, $b = b_i$, $\xi = \xi_i$ and taking $i = i' \to \infty$, the formula (5.15) follows for general a, b. This completes the proof of the theorem.

Theorem 5.3. *Let $d\xi_i(t) = a_i(t)\,dt + b_i(t)\,d\xi$ $(1 \leq i \leq m)$ and let $f(x_1, \ldots, x_m, t)$ be a continuous function in (x, t) where $x = (x_1, \ldots, x_m) \in R^m$, $t \geq 0$, together with its first t-derivative and second x-derivatives. Then $f(\xi_1(t), \ldots, \xi_m(t), t)$ has a stochastic differential, given by*

$$df(X(t), t) = \left[f_t(X(t), t) + \sum_{i=1}^{m} f_{x_i}(X(t), t) a_i(t) + \frac{1}{2} \sum_{i,j=1}^{m} f_{x_i x_j}(X(t), t) b_i(t) b_j(t) \right] dt + \sum_{i=1}^{m} f_{x_i}(X(t), t) b_i(t)\, dw(t), \tag{5.19}$$

where $X(t) = (\xi_1(t), \ldots, \xi_m(t))$.

Formula (5.19) is also called *Itô's formula*. It includes both Theorems 5.1 and 5.2.

The proof is left to the reader (see Problem 15).

Remark. Itô's formula (5.7) asserts that the two processes $f(\xi(t), t) - f(\xi(0), 0)$ and

$$\int_0^t [f_s(\xi(s), s) + f_x(\xi(s), s) a(s) + \tfrac{1}{2} f_{xx}(\xi(s), s) b^2(s)]\, ds + \int_0^t f_x(\xi(s), s) b(s)\, dw(s)$$

are stochastically equivalent. Since they are continuous, their sample paths coincide a.s. Consequently

$$f(\xi(\tau), \tau) - f(\xi(0), 0) = \int_0^\tau [f_t(\xi(t), t) + f_x(\xi(t), t) a(t) + \tfrac{1}{2} f_x(\xi(t), t) b^2(t)]\, dt + \int_0^\tau f_x(\xi(t), t) b(t)\, dw(t) \tag{5.20}$$

for any random variable τ, $0 \leq \tau \leq T$.

If, in particular, τ is a stopping time, when, taking the expectation and using Theorem 3.10, we find that

$$Ef(\xi(\tau), \tau) - Ef(\xi(0), 0) = E \int_0^\tau (Lf)(\xi(t), t)\, dt \tag{5.21}$$

where

$$Lf \equiv f_t + af_x + \tfrac{1}{2}b^2 f_{xx},$$

provided

$$\begin{aligned} b(t)f_x(\xi(t, t)) &\quad \text{belongs to} \quad M_w^2[0, T], \\ (Lf)(\xi(t), t) &\quad \text{belongs to} \quad M_w^1[0, T]. \end{aligned}$$

6. Applications of Itô's formula

Itô's formula will become a standard tool in the sequel. In the present section we give a few straightforward applications. First we need two lemmas.

Lemma 6.1. *If $f \in L_w^p[\alpha, \beta]$ for some $p \geq 1$, then there exists a sequence of step functions f_n in $L_w^p[\alpha, \beta]$ such that*

$$\lim_{n\to\infty} \int_\alpha^\beta |f(t) - f_n(t)|^p\, dt \quad a.s.$$

Proof. The proof is similar to the proof of Lemma 1.1. Instead of (1.5) we now have

$$\int_\alpha^\beta |J_\epsilon f|^p\, dt \leq \int_\alpha^\beta |f|^p\, dt. \tag{6.1}$$

Indeed, from (1.4) and Hölder's inequality,

$$\int_\alpha^\beta |J_\epsilon f|^p\, dt \leq \int_\alpha^\beta \left[\int_{-\epsilon}^{\epsilon} \rho(z)\, dz\right]^{p/q} \left[\int_{-\epsilon}^{\epsilon} \rho(z) |f(t - z - \epsilon)|^p\, dz\right] dt$$

$$\leq \int_{-\epsilon}^{\epsilon} \rho(z) \left[\int_{\alpha-1}^{\beta} |f(t)|^p\, dt\right] dz \qquad \left(\frac{1}{p} + \frac{1}{q} = 1\right)$$

and (6.1) follows. (If $p = 1$, we do not use Hölder's inequality.)

Using (6.1) we can proceed to show that

$$\int_\alpha^\beta |J_\epsilon f - f|^p\, dt \to 0 \qquad \text{if} \quad \epsilon \downarrow 0, \tag{6.2}$$

by the argument given the proof of Lemma 1.1. The rest of the proof is the same as for Lemma 1.1.

Lemma 6.2. *If $f \in M_w^p[\alpha, \beta]$ for some $p > 1$, then there exists a sequence of bounded step functions f_n in $M_w^p[\alpha, \beta]$ such that*

$$E \int_\alpha^\beta |f(t) - f_n(t)|^p \, dt \to 0 \qquad \text{if} \quad n \to \infty.$$

This follows by obvious changes in the proof of Lemma 1.2, exploiting Lemma 6.1.

Theorem 6.3. *Let $f \in M_w^{2m}[0, T]$ where m is a positive integer. Then*

$$E\left[\int_0^T f(t) \, dw(t)\right]^{2m} \leq [m(2m-1)]^m T^{m-1}\left[E \int_0^T f^{2m}(t) \, dt\right]. \tag{6.3}$$

For $m = 1$, there is equality.

Proof. We apply Itô's formula with $f(x, t) = x^{2m}$, $\xi(t) = \int_0^t f(s) \, dw(s)$, and obtain

$$\left[\int_0^t f(s) \, dw(s)\right]^{2m} = 2m \int_0^t \left[\int_0^s f(\lambda) \, dw(\lambda)\right]^{2m-1} f(s) \, dw(s)$$
$$+ m(2m-1)\int_0^t \left[\int_0^s f(\lambda) \, dw(\lambda)\right]^{2m-2} f^2(s) \, ds. \tag{6.4}$$

Suppose $f(t)$ is a bounded step function. From the definition of $\int_0^s f(\lambda) \, dw(\lambda)$ and the fact that

$$E|w(t)|^r < C_{r,t} \qquad (C_{r,t} \text{ constant})$$

for any $t > 0$, $r > 0$, we find that

$$\left[\int_0^s f(\lambda) \, dw(\lambda)\right]^{2m-1} f(s) \qquad \text{belongs to} \qquad M_w^2[0, T].$$

Hence, taking the expectation in (6.4) and using Theorem 2.5, we get

$$E\left[\int_0^T f(s) \, dw(s)\right]^{2m} = m(2m-1)\int_0^T E\left[\int_0^s f(\lambda) \, dw(\lambda)\right]^{2m-2} f^2(s) \, ds. \tag{6.5}$$

By Hölder's inequality, the right-hand side is bounded by

$$m(2m-1)\left\{\int_0^T E\left[\int_0^s f(\lambda) \, dw(\lambda)\right]^{2m} ds\right\}^{(2m-2)/2m} \left\{\int_0^T Ef^{2m}(s) \, ds\right\}^{2/2m}. \tag{6.6}$$

From (6.5) we see that the function

$$t \to E\left[\int_0^t f(s) \, dw(s)\right]^{2m}$$

is monotone increasing in t. Hence

$$\int_0^T E\left[\int_0^s f(\lambda)\, dw(\lambda)\right]^{2m} ds \leq \int_0^T E\left[\int_0^T f(\lambda)\, dw(\lambda)\right]^{2m} ds.$$

Using this to estimate the first expression $\{\cdots\}$ in (6.6), we then obtain from (6.5),

$$E\left[\int_0^T f(s)\, dw(s)\right]^{2m}$$
$$\leq m(2m-1)\left\{TE\left[\int_0^T f(s)\, dw(s)\right]^{2m}\right\}^{(m-1)/m}\left\{\int_0^T Ef^{2m}(s)\, ds\right\}^{1/m},$$

and (6.3) follows.

To prove (6.3) in general, we can take, by Lemma 6.2, a sequence of bounded step functions f_n such that

$$E\int_0^T |f(t) - f_n(t)|^{2m}\, dt \to 0 \quad \text{if} \quad n\to\infty. \tag{6.7}$$

Then

$$\int_0^T f_n(t)\, dw(t) \xrightarrow{P} \int_0^T f(t)\, dw(t).$$

We may assume that the convergence is a.s., for otherwise we can take a subsequence of the f_n. Writing (6.3) for f_n, and using Fatou's lemma and (6.7), the assertion (6.3) follows.

From Theorems 3.7 and 6.3 we obtain:

Corollary 6.4. *If $f \in M_w^{2m}[0, T]$ where m is a positive integer, then*

$$E\left\{\sup_{0\leq t\leq T}\left|\int_0^t f(s)\, dw(s)\right|^{2m}\right\} \leq C_m T^{m-1} E\int_0^T |f(t)|^{2m}\, dt$$

where $C_m = [4m^3/(2m-1)]^m$.

Theorem 6.5. *Let $f \in L_w^2[0, T]$, and let α, β be any positive numbers. Then*

$$P\left\{\max_{0\leq t\leq T}\left[\int_0^t f(\lambda)\, dw(\lambda) - \frac{\alpha}{2}\int_0^t f^2(\lambda)\, d\lambda\right] > \beta\right\} \leq e^{-\alpha\beta}. \tag{6.8}$$

Proof. Consider first the case where f is a bounded step function, and set

$$\xi(t) = \int_0^t f(\lambda)\, dw(\lambda) - \tfrac{1}{2}\int_0^t f^2(\lambda)\, d\lambda,$$
$$\zeta(t) = \exp[\xi(t)].$$

Applying Itô's formula to $f(x) = e^x$ and the process $\xi(t)$, we get

$$\zeta(t) = \zeta(s) + \int_s^t e^{\xi(\lambda)} f(\lambda)\, dw(\lambda).$$

Since f is a bounded step function,

$$e^{\xi(\lambda)} \leq A_0 \prod_{i=1}^{l} e^{A_i |w(t_i)|}$$

for some positive numbers A_i, t_i. It follows that $[\exp(\xi(\lambda))] f(\lambda)$ belongs to $M_w^2[0, T]$. Hence, by Theorem 2.8,

$$E[\zeta(t) | \mathcal{F}_s] = \zeta(s), \tag{6.9}$$

i.e., $\zeta(t)$ is a martingale. Note also that $E\zeta(0) = 1$.

Set

$$\xi_\alpha(t) = \int_0^t f(\lambda)\, dw(\lambda) - \frac{\alpha}{2} \int_0^t f^2(\lambda)\, d\lambda,$$

$$\hat{\xi}_\alpha(t) = \int_0^t \alpha f(\lambda)\, dw(\lambda) - \frac{\alpha^2}{2} \int_0^t f^2(\lambda)\, d\lambda.$$

By what we have already proved, $\exp(\hat{\xi}_\alpha(t))$ is a martingale with expectation 1. Hence, by the martingale inequality,

$$P\Big\{ \max_{0 \leq t \leq T} \xi_\alpha(t) > \beta \Big\} = P\Big\{ \max_{0 \leq t \leq T} [\exp \hat{\xi}_\alpha(t)] > e^{\alpha\beta} \Big\} \leq e^{-\alpha\beta}.$$

This proves (6.8) in case f is a bounded step function.

Now let f be any function in $L_w^2[0, T]$ and take bounded step functions f_n such that

$$\int_0^T |f_n(t) - f(t)|^2\, dt \to 0 \quad \text{a.s.}$$

Theorem 3.4 implies that

$$\max_{0 \leq t \leq T} \left| \int_0^t f_n(s)\, dw(s) - \int_0^t f(s)\, dw(s) \right| \to 0 \quad \text{in probability.}$$

By taking a subsequence, if necessary, we obtain convergence a.s. It follows that a.s.

$$\exp\left[\int_0^t f_n(\lambda)\, dw(\lambda) - \frac{\alpha}{2} \int_0^t f_n^2(\lambda)\, d\lambda \right]$$

$$\to \exp\left[\int_0^t f(\lambda)\, dw(\lambda) - \frac{\alpha}{2} \int_0^t f^2(\lambda)\, d\lambda \right]$$

uniformly with respect to t, $0 \leq t \leq T$. Hence, by writing (6.8) for each f_n and taking $n \to \infty$, the assertion (6.8) follows.

The inequality (6.8) will be referred to as the *exponential martingale inequality*.

Corollary 6.6. *If $f \in L_w^2[0, T]$, then the process*

$$\zeta(t) = \exp\left\{ \int_0^t f(\lambda)\, dw(\lambda) - \tfrac{1}{2} \int_0^t f^2(\lambda)\, d\lambda \right\} \tag{6.10}$$

is a supermartingale, i.e., $-\zeta(t)$ is a submartingale.

Proof. Define

$$\xi_n(t) = \int_0^s f(\lambda)\, dw(\lambda) + \int_s^t f_n(\lambda)\, dw(\lambda) - \tfrac{1}{2} \int_0^s f^2(\lambda)\, d\lambda - \tfrac{1}{2} \int_s^t f_n^2(\lambda)\, d\lambda,$$

$$\zeta_n(t) = \exp(\xi_n(t))$$

for $s \leqslant t \leqslant T$, where the f_n are step functions as in the preceding proof. By the proof of (6.9), $\zeta_n(t)$ is a martingale for $s \leqslant t \leqslant T$. Hence, for any event $A \in \mathcal{F}_s$,

$$\int_A \zeta_n(t)\, dP = \int_A \zeta_n(s)\, dP. \tag{6.11}$$

Notice that $\zeta_n(s) = \zeta(s)$ where $\zeta(t)$ is defined by (6.10). If $n = n' \to \infty$ ($\{n'\}$ a suitable subsequence of $\{n\}$), then $\zeta_n(t) \to \zeta(t)$ a.s. Hence, taking $n \to \infty$ in (6.11), and using Fatou's lemma, we get

$$\int_A \zeta(t)\, dP \leqslant \int_A \zeta(s)\, dP,$$

and the assertion follows.

7. Stochastic integrals and differentials in n dimensions

Let $w(t) = (w_1(t), \ldots, w_n(t))$ be an n-dimensional Brownian motion. Let $\mathcal{F}_t$ ($t \geqslant 0$) be an increasing family of σ-fields such that $w(t)$ is $\mathcal{F}_t$ measurable and $\mathcal{F}(w(\lambda + t) - w(t), \lambda \geqslant 0)$ is independent of $\mathcal{F}_t$, for any $t \geqslant 0$.

We shall say that a matrix of functions belongs to $L_w^p[\alpha, \beta]$ (or to $M_w^p[\alpha, \beta]$) if each of its elements belongs to $L_w^p[\alpha, \beta]$ (or to $M_w^p[\alpha, \beta]$).

Let $b = (b_{ij})$ be an $m \times n$ matrix that belongs to $L_w^2[\alpha, \beta]$. The *stochastic integral* $\int_\alpha^\beta b(t)\, dw(t)$ is an m-vector defined by

$$\int_\alpha^\beta b(t)\, dw(t) = \left\{ \sum_{j=1}^n \int_\alpha^\beta b_{ij}(t)\, dw_j(t) \right\}_{i=1,\ldots,m}$$

where each integral on the right is defined as in Section 2.

If we substitute $\alpha = \int_{t_1}^{t_2} f\, dw_i$, $\beta = \int_{t_1}^{t_2} g\, dw_i$ in the identity $4\alpha\beta = (\alpha + \beta)^2 - (\alpha - \beta)^2$, and use Theorem 2.5, we find that

$$E \int_{t_1}^{t_2} f(t)\, dw_i(t) \int_{t_1}^{t_2} g(t)\, dw_i(t) = E \int_{t_1}^{t_2} f(t) g(t)\, dt \tag{7.1}$$

provided f and g belong to $M_w^2[t_1, t_2]$.

We also have

$$E \int_{t_1}^{t_2} f(t)\, dw_i(t) \int_{t_1}^{t_2} g(t)\, dw_j(t) = 0 \qquad \text{if} \quad i \neq j, \tag{7.2}$$

because the integrals are independent and with zero expectation.

Using (7.1), (7.2) we easily see that if $b = (b_{ij})$ is an $m \times n$ matrix in $M_w^2[t_1, t_2]$, then

$$E\left|\int_{t_1}^{t_2} b(t)\, dw(t)\right|^2 = E\int_{t_1}^{t_2} |b(t)|^2\, dt \tag{7.3}$$

where

$$|b|^2 = \sum_{i=1}^{m} \sum_{j=1}^{n} (b_{ij})^2.$$

Definition. Let $\xi(t)$ be an m-dimensional process for $0 \leq t \leq T$, and suppose that, for any $0 \leq t_1 < t_2 \leq T$,

$$\xi(t_2) - \xi(t_1) = \int_{t_1}^{t_2} a(t)\, dt + \int_{t_1}^{t_2} b(t)\, dw(t)$$

where $a = (a_1, \ldots, a_m)$ and the $m \times n$ matrix $b = (b_{ij})$ belong to $L_w^1[0, T]$ and $L_w^2[0, T]$ respectively. Then we say that $\xi(t)$ has a *stochastic differential* $d\xi(t)$ given by

$$d\xi(t) = a(t)\, dt + b(t)\, dw(t). \tag{7.4}$$

We shall now state *Itô's formula*.

Theorem 7.1. *Let $u(x, t)$ be a continuous function in $(x, t) \in R^m \times [0, \infty)$ together with its derivatives u_t, u_{x_i}, $u_{x_i x_j}$. Let $\xi(t)$ be an m-dimensional process having a stochastic differential*

$$d\xi(t) = a(t)\, dt + b(t)\, dw(t)$$

where $a = (a_1, \ldots, a_m)$ and $b = (b_{ij})$ $(1 \leq i \leq m,\ 1 \leq j \leq n)$ belong to $L_w^1[0, T]$ and $L_w^2[0, T]$ respectively. Then $u(\xi(t), t)$ has a stochastic differential

$$\begin{aligned} du(\xi(t), t) = \Bigg[& u_t(\xi(t), t) + \sum_{i=1}^{m} u_{x_i}(\xi(t), t) a_i(t) \\ & + \frac{1}{2} \sum_{l=1}^{n} \sum_{i,j=1}^{m} u_{x_i x_j}(\xi(t), t) b_{il}(t) b_{jl}(t) \Bigg] dt \\ & + \sum_{l=1}^{n} \sum_{i=1}^{m} u_{x_i}(\xi(t), t) b_{il}(t)\, dw_l(t). \end{aligned} \tag{7.5}$$

We begin with

Lemma 7.2. *If w_1, w_2 are independent Brownian motions, then*

$$d(w_1w_2) = w_1\,dw_2 + w_2dw_1. \tag{7.6}$$

Proof. It is easily seen that $w(t) = (w_1 + w_2)/\sqrt{2}$ is a Brownian motion. Hence, by (5.2)

$$d(w^2) = dt + 2w\,dw.$$

Since $d(w_1^2)$ and $d(w_2^2)$ are also given by the same formula, $d(w_1w_2)$ exists and is equal to

$$d(w^2 - \tfrac{1}{2}w_1^2 - \tfrac{1}{2}w_2^2) = dw^2 - \tfrac{1}{2}\,dw_1^2 - \tfrac{1}{2}\,dw_2^2 = w_1\,dw_2 + w_2\,dw_1.$$

Lemma 7.3. *If $d\xi_i = a_i(t)\,dt + \sum_{j=1}^n b_{ij}(t)\,dw_j$ $(i = 1, 2)$ where $w_1, \ldots, w_n$ are independent Brownian motions, then*

$$d(\xi_1\xi_2) = \xi_1\,d\xi_2 + \xi_2\,d\xi_1 + \sum_{j=1}^{n} b_{1j}b_{2j}.$$

Proof. The proof is similar to the proof of Theorem 5.1. It is based on the special cases (7.6), (5.2), (5.3) and the approximation procedure employed in the proof of Theorem 5.1.

Proof of Theorem 7.1. We begin with the special case of $u(\xi_* + a_*t + b_*w(t), t)$ where $t_1 < t < t_2$ and ξ_*, a_*, b_* are random variables measurable with respect to $\mathcal{F}_{t_1}$. The special case where $u = x_i^k$ follows by induction on m, using Lemma 7.3. The more general case where $u = x_1^{k_1} \cdots x_m^{k_m}$ follows by using the previous special case and Lemma 7.3. The case where $u = g(t)x_1^{k_1} \cdots x_m^{k_m}$ follows by again using Lemma 7.3. Now approximate general u by linear combinations of u's of the last special form.

Once Theorem 7.1 has been proved in the special case where $\xi(t) = \xi_* + a_*t + b_*w(t)$, we can obtain the general case by a proof similar to that of Theorem 5.2.

Let

$$a_{ij} = \sum_{l=1}^{n} b_{il}b_{jl},$$

$$Lu = \tfrac{1}{2}\sum_{i,j=1}^{m} a_{ij}\,\frac{\partial^2 u}{\partial x_i\,\partial x_j} + \sum_{i=1}^{m} a_i\,\frac{\partial u}{\partial x_i} + \frac{\partial u}{\partial t}.$$

Then we can write Itô's formula (7.5) in the form

$$du(\xi(t), t) = Lu(\xi(t), t)\,dt + u_x(\xi(t), t)\cdot b(t)\,dw(t). \tag{7.7}$$

Let us define formally a multiplication table:

$$dw_i\,dt = 0,$$
$$dt\,dt = 0,$$
$$dw_i\,dw_j = 0 \quad \text{if } i \neq j,$$
$$dw_i\,dw_i = dt,$$

so that

$$d\xi_i\,d\xi_j = \sum_{l=1}^{n} b_{il}b_{jl}\,dt. \tag{7.8}$$

Then Itô's formula (7.5) takes the form

$$du(\xi(t), t) = u_t(\xi(t), t)\,dt + \sum_{i=1}^{m} u_{x_i}(\xi(t), t)\,d\xi_i + \frac{1}{2}\sum_{i,j=1}^{m} u_{x_i x_j}(\xi(t), t)\,d\xi_i\,d\xi_j. \tag{7.9}$$

From Itô's formula we obtain (cf. (5.20))

$$u(\xi(\tau), \tau) - u(\xi(0), 0) = \int_0^{\tau} (Lu)(\xi(s), s)\,ds + \int_0^{\tau} u_x(\xi(s), s) \cdot b(s)\,dw(s) \tag{7.10}$$

for any random variable τ, $0 \leq \tau \leq T$. If τ is a stopping time and Lu, $u_x \cdot b$ are in $M_w^1[0, T]$ and $M_w^2[0, T]$ respectively, then (cf. (5.21))

$$Eu(\xi(\tau), \tau) - Eu(\xi(0), 0) = E\int_0^{\tau} Lu(\xi(s), s)\,ds. \tag{7.11}$$

The results of Sections 2–4 extend with obvious changes to stochastic integrals in n dimensions. We shall give here the generalization of Theorem 4.5.

Theorem 7.4. *Let $f = (f_1, \ldots, f_n)$ belong to $L_w^2[0, \infty)$ and suppose that $\int_0^\infty |f|^2\,dt = \infty$ with probability 1, and define*

$$\tau(t) = \inf\left\{ s;\ \int_0^s |f(\lambda)|^2\,d\lambda = t \right\}.$$

Then the process

$$u(t) = \int_0^{\tau(t)} f(s)\,dw(s)$$

is a Brownian motion.

Proof. The proof is similar to the proof of Theorem 4.5 once we have established that if $f \in M_w^2[0, \infty)$ and ζ_1, ζ_2 are bounded stopping times, $\zeta_1 \leq \zeta_2$, then

$$E\left\{ \int_{\zeta_1}^{\zeta_2} f(s)\, dw(s) | \mathcal{F}_{\zeta_1} \right\} = 0,$$

$$E\left\{ \left| \int_{\zeta_1}^{\zeta_2} f(s)\, dw(s) \right|^2 | \mathcal{F}_{\zeta_1} \right\} = E\left\{ \int_{\zeta_1}^{\zeta_2} |f|^2\, ds | \mathcal{F}_{\zeta_1} \right\}$$

a.s. The proof of these formulas is similar to the proof of Theorem 4.3. It employs Theorem 2.8 which remains valid for n-dimensional stochastic integrals.

We conclude this section with an extension of Theorem 6.5 to n dimensions.

Theorem 7.5. *Let $f = (f_1, \ldots, f_n)$ belong to $L_w^2[0, T]$, and let α, β be positive numbers. Then*

$$P\left\{ \max_{0 \leq t \leq T} \left[\int_0^t f(\lambda)\, dw(\lambda) - \frac{\alpha}{2} \int_0^t |f(\lambda)|^2\, d\lambda \right] > \beta \right\} \leq e^{-\alpha\beta}. \quad (7.12)$$

The proof is similar to the proof of Theorem 6.5. First we prove (7.12) in case $f(t)$ is a step function, using the martingale inequality, and then proceed to general f by approximation.

The inequality (7.12) is referred to as the *exponential martingale inequality*.

Corollary 6.6 also extends to the present n-dimensional case, i.e.,

$$\exp\left\{ \int_0^t f\, dw - \tfrac{1}{2} \int_0^t |f|^2\, ds \right\}$$

is a supermartingale.

PROBLEMS

1. Prove (2.20).
2. Prove Theorem 3.9 [*Hint*: Apply Theorem 3.6 to $\xi f(t)$, ξ bounded and $\mathcal{F}_\alpha$ measurable.]
3. Suppose $f \in L_w^2[0, \infty)$ and ζ is a stopping time such that $E \int_0^\zeta f^2(t)\, dt < \infty$. Prove that

$$E \int_0^\zeta f(t)\, dw(t) = 0, \qquad E\left| \int_0^\zeta f(t)\, dw(t) \right|^2 = E \int_0^\zeta f^2(t)\, dt.$$

4. Let

$$\rho(x) = \begin{cases} c \exp[1/(|x|^2 - 1)] & \text{if } |x| < 1 \\ 0 & \text{if } |x| > 1 \end{cases}$$

for $x \in R^n$, where c is a positive constant such that $\int_{R^n} \rho(x)\,dx = 1$. If f is a function locally integrable, then

$$(J_\epsilon f)(x) = \frac{1}{\epsilon^n} \int_{R^n} \rho\left(\frac{x-y}{\epsilon}\right) f(y)\,dy$$

is called a *mollifier* of f. Prove:

(i) $J_\epsilon f$ is in $C^\infty(R^n)$;

(ii) If K is a compact set and Ω a bounded open set containing K, then

$$(J_\epsilon f)(x) = \frac{1}{\epsilon^n} \int_\Omega \rho\left(\frac{x-y}{\epsilon}\right) f(y)\,dy$$

$$= \int_{|z|<1} \rho(z) f(x - \epsilon z)\,dz \qquad (x \in K),$$

provided $\epsilon < \text{dist}(K, R^n \backslash \Omega)$.

(iii) If $f \in L^p(\Omega)$ for some $p \geqslant 1$, then

$$\left\{ \int_K |J_\epsilon f|^p\,dx \right\}^{1/p} \leqslant \left\{ \int_\Omega |f|^p\,dx \right\}^{1/p}.$$

(iv) If $f \in L^p(\Omega)$ for some $p \geqslant 1$, then

$$\int_K |J_\epsilon f - f|^p\,dx \to 0 \qquad \text{if} \quad \epsilon \to 0.$$

5. Let $f(x)$ be a continuous function for $\alpha \leqslant x \leqslant \beta$, and let

$$(P_k f)(x) = \frac{\int_\alpha^\beta \left[1-(x-y)^2\right]^k f(y)\,dy}{\int_{-1}^1 (1-y^2)^k\,dy} \qquad (k = 1, 2, \ldots).$$

Let δ be any positive number. Prove that $(P_k f)(x) \to f(x)$ uniformly in $x \in [\alpha + \delta, \beta - \delta]$ as $k \to \infty$. [*Hint*: $[\int_\epsilon^1 (1-y^2)^k\,dy / \int_0^1 (1-y^2)^k\,dy] \to 0$ if $k \to \infty$, for any $\epsilon > 0$.]

6. Let $f(x)$ be a continuous function in an n-dimensional interval $I \equiv \{x;\ \alpha_i \leqslant x \leqslant \beta_i,\ 1 \leqslant i \leqslant n\}$, and let

$$(P_k f)(x) = \frac{\int_{\alpha_1}^{\beta_1} \cdots \int_{\alpha_n}^{\beta_n} \prod_{i=1}^n \left[1-(x_i-y_i)^2\right]^k f(y)\,dy_n \cdots dy_1}{\left[\int_{-1}^1 (1-y^2)^k\,dy\right]^n}$$

$$(k = 1, 2, \ldots).$$

Let I_0 be any subset lying in the interior of I. Prove that, as $k \to \infty$,

$$(P_k f)(x) \to f(x) \quad \text{uniformly in} \quad x \in I_0.$$

Notice that $P_k f$ is a polynomial. It is called a *polynomial mollifier* of f.

7. If in the preceding problem f belongs to $C^m(I)$ and f vanishes in a neighborhood of the boundary of I, then

$$\frac{\partial^{i_1+\cdots+i_n}}{\partial x_1^{i_1}\cdots\partial x_n^{i_n}}(P_kf)(x)\to\frac{\partial^{i_1+\cdots+i_n}}{\partial x_1^{i_1}\cdots\partial x_n^{i_n}}f(x)\quad\text{if}\quad k\to\infty,$$

uniformly in $x\in I_0$, for any $(i_1,\ldots,i_n)$ such that $0\leqslant i_1+\cdots+i_n\leqslant m$.

8. If $f\in C^m(R^n)$, then there exists a sequence of polynomials Q_k such that, as $k\to\infty$,

$$\frac{\partial^{i_1+\cdots+i_n}}{\partial x_1^{i_1}\cdots\partial x_n^{i_n}}Q_k(x)\to\frac{\partial^{i_1+\cdots+i_n}}{\partial x_1^{i_1}\cdots\partial x_n^{i_n}}f(x)\qquad\text{for}\quad 0\leqslant i_1+\cdots+i_n\leqslant m,$$

uniformly in x in compact subsets of R^n. [*Hint*: Approximate f by functions with compact support, and apply Problem 7 to these functions.]

9. If in the previous problem it is assumed that f, f_{x_i} $(1\leqslant i\leqslant n)$ and $f_{x_ix_j}$ $(2\leqslant i,j\leqslant n)$ are continuous in R^n (instead of $f\in C^m(R^n)$), then

$$Q_k\to f,\qquad \frac{\partial}{\partial x_i}Q_k\to\frac{\partial f}{\partial x_i}\qquad(1\leqslant i\leqslant n),$$

$$\frac{\partial^2}{\partial x_i\,\partial x_j}Q_k\to\frac{\partial^2 f}{\partial x_i\,\partial x_j}\qquad(2\leqslant i,j\leqslant n)$$

uniformly on compact subsets of R^n.

10. Let $f(x,t)$ be a continuous function in $(x,t)\in R^n\times[0,\infty)$ together with its derivatives $f_t, f_{x_i}, f_{x_ix_j}$. Prove that there exists a function F continuous in $(x,t)\in R^n\times R^1$ together with its derivatives $F_t, F_{x_i}, F_{x_ix_j}$, such that $F(x,t)=f(x,t)$ if $x\in R^n$, $t\geqslant 0$.

11. Let $f(x,t)=f(x_1,\ldots,x_n,t)$ be a continuous function in $(x,t)\in R^n\times[0,\infty)$ together with its derivatives $f_t, f_{x_i}, f_{x_ix_j}$. Then there exists a sequence of polynomials $Q_m(x,t)$ such that, as $m\to\infty$,

$$Q_m\to f,\qquad\frac{\partial}{\partial t}Q_m\to f_t,\qquad\frac{\partial}{\partial x_i}Q_m\to f_{x_i},\qquad\frac{\partial^2}{\partial x_i\,\partial x_j}Q_m\to f_{x_ix_j}$$

uniformly in compact subsets. [*Hint*: Combine Problems 9, 10.]

12. Prove (5.8).

13. Prove (5.14) and complete the proof of (5.13).

14. Let $f\in L_w^2[0,\infty)$, $|f|\leqslant K$ (K constant) and let $d\xi(t)=f(t)\,dw(t)$, $\xi(0)=0$ where $w(t)$ is a Brownian motion. Prove:

(i) if $f\leqslant\beta$, then $E|\xi(t)|^2\leqslant\beta^2t$;
(ii) if $f\geqslant\alpha>0$, then $E|\xi(t)|^2\geqslant\alpha^2t$.

15. Prove Theorem 5.3. [*Hint*: Proceed as in the proof of Theorem 5.2, but with

$$\Phi(w(t),t)=f(\xi_{10}+a_1t+b_1w(t),\ldots,\xi_{m0}+a_mt+b_mw(t))$$

where ξ_{i0}, a_i are random variables and the b_i are random n-vectors; cf. Step 4.]

16. Let $\xi(t) = \int_0^t b(t)\,dw(t)$ where b is an $n \times n$ matrix belonging to $L_w^2[0, \infty)$. Suppose that $d\xi_i\,d\xi_j = 0$ if $i \neq j$, $d\xi_i\,d\xi_i = dt$ (see (7.8) for the definition of $d\xi_i\,d\xi_j$), for all $1 \leq i, j \leq n$. Prove that $\xi(t)$ is an n-dimensional Brownian motion. [*Hint*: First proof: Use Theorem 3.6.2. Second proof: Suppose the elements of b are bounded step functions and let $\zeta(t) = \exp[i\gamma \cdot \xi(t) + \gamma^2 t/2]$. By Itô's formula $d\zeta = i\zeta\gamma\,dw$. By Theorem 2.8

$$E[e^{i\gamma \cdot \xi(t)} | \mathcal{F}_s] = e^{i\gamma \cdot \xi(s)} e^{-\gamma^2(t-s)/2}.$$

Use Problem 2, Chapter 3.]

17. Let $\gamma > 0$, $a > 0$, $\tau = \min\{t;\, w(t) = a\}$ where $w(t)$ is one-dimensional Brownian motion. Prove that $P(\tau < \infty) = 1$ and

$$Ee^{-\gamma\tau} = \exp\left(-\sqrt{2\gamma}\, a\right).$$

[*Hint*: For any $c > 0$,

$$P\left[\max_{0 \leq s \leq t} w(s) > c\right] \leq P\left[\max_{0 \leq s \leq t}\left(w(s) - \frac{\alpha}{2} s\right) > \beta\right] < e^{-c^2/2t}$$

where $\alpha = c/t$, $\beta = c/2$. Hence $P(\tau < \infty) = 1$. Since $y(t) = \exp[\gamma w(t) - \gamma^2 t/2]$ is a martingale, so is $y(t \wedge \tau)$. Hence

$$E \exp\left[\gamma w(t \wedge \tau) - \tfrac{1}{2}\gamma^2(t \wedge \tau)\right] = 1.$$

Take $t \uparrow \infty$.]

18. Under the conditions of the previous problem

$$P(\tau \in dt) = \frac{a}{(2\pi t^3)^{1/2}} \exp\left(-\frac{a^2}{2t}\right) dt.$$

[*Hint*: Use the fact (see, for instance, Feller [1]) that if the Laplace transform of two probability distributions concentrated on $[0, \infty)$ coincide, then the probability distributions coincide.]

19. If $w(t)$ is a Brownian motion and $0 \leq y$, $x < y$, then

$$P\left(w(t) \in dx,\ \max_{0 \leq s \leq t} w(s) \in dy\right)$$

$$= \left(\frac{2}{\pi t^3}\right)^{1/2} (2y - x) \exp\left[-\frac{(2y-x)^2}{2t}\right] dx\,dy.$$

[*Hint*: Use Problem 12, Chapter 2 and Theorem 3.6.3 to deduce that

$$P\left[w(t) \in dx,\ \max_{0 \leq s \leq t} w(s) \geq y\right] = \int_0^t P(\tau \in ds) P[w(t-s) + y \in dx]$$

where $\tau = \min\{t;\, w(t) = y\}$, and apply the preceding problem.]

20. Let $f(t)$ be a continuous process in $L^2_w[0, T]$ and let $\Pi_n : t_{n,0} = 0 < t_{n,1} < \cdots < t_{n,n} = T$ be a partition with mesh $|\Pi_n| \to 0$ as $n \to \infty$. Define

$$g_n(t) = \sum_{i=0}^{j-1} f(t_{n,i})(w(t_{n,i+1}) - w(t_{n,i})) + f(t_{n,j})(t - t_{n,j})$$
$$\text{if} \quad t_{n,j} \leq t < t_{n,j+1}.$$

Prove that for some subsequence $\{n'\}$ of $\{n\}$,

$$\sup_{0 \leq t \leq T} \left| \int_0^t f(s)\, dw(s) - g_{n'}(t) \right| \to 0 \quad \text{a.s.} \qquad \text{if} \quad n' \to \infty.$$

21. Let $\sigma(x, t)$ be a measurable function in $(x, t) \in R^n$ such that

$$|\sigma(x, t) - \sigma(\bar{x}, t)| \leq \eta(|x - \bar{x}|), \qquad \eta(\delta) \downarrow 0 \quad \text{if } \delta \downarrow 0,$$

and let $f(t)$ be an n-dimensional continuous process in $L^2_w[0, T]$. Let

$$\sigma_\epsilon(x, t) = \frac{1}{\epsilon} \int_{-1}^{T} \rho\left(\frac{t - s - \epsilon}{\epsilon} \right) \sigma(x, s)\, ds \qquad (2\epsilon < 1)$$

where $\rho(t)$ is defined as in Lemma 1.1 and $\sigma(x, s) = \sigma(x, 0)$ if $-1 < s < 0$. Prove:

(i) $\int_0^T |\sigma(x, t) - \sigma_\epsilon(x, t)|^2\, dt \to 0$ uniformly in x in bounded sets, as $\epsilon \to 0$.

(ii) $\int_0^T |\sigma(f(t), t) - \sigma_\epsilon(f(t), t)|^2\, dt \to 0$ a.s. as $\epsilon \to 0$.

(iii) $\sup_{0 \leq t \leq T} |\int_0^t \sigma(f(s), s)\, dw(s) - \int_0^t \sigma_{\epsilon_n}(f(s), s)\, dw(s)| \to 0$ a.s. for some sequence $\epsilon_n \downarrow 0$.

[*Hint*: for (i), use the uniform continuity in x of $\int \sigma(x, t)\, dt$ and of $\int \sigma_\epsilon(x, t)\, dt$.]

5

Stochastic Differential Equations

1. Existence and uniqueness

If $\sigma = (\sigma_{ij})$ is a matrix, we write $|\sigma|^2 = \sum_{i,j} |\sigma_{ij}|^2$.

Let $b(x, t) = (b_1(x, t), \ldots, b_n(x, t))$, $\sigma(x, t) = (\sigma_{ij}(x, t))_{i,j=1}^n$ and suppose the functions $b_i(x, t)$, $\sigma_{ij}(x, t)$ are measurable in $(x, t) \in R^n \times [0, T]$. If $\xi(t)$ $(0 \leq t \leq T)$ is a stochastic process such that

$$d\xi(t) = b(\xi(t), t)\, dt + \sigma(\xi(t), t)\, dw(t), \tag{1.1}$$

$$\xi(0) = \xi_0 \quad \text{a.s.}, \tag{1.2}$$

then we say that $\xi(t)$ satisfies the *system of stochastic differential equations* (1.1) and the *initial condition* (1.2). Note that it is implicitly assumed that $b(\xi(t), t)$ belongs to $L_w^1[0, T]$ and $\sigma(\xi(t), t)$ belongs to $L_w^2[0, T]$.

Theorem 1.1. *Suppose $b(x, t)$, $\sigma(x, t)$ are measurable in $(x, t) \in R^n \times [0, T]$ and*

$$\begin{aligned} |b(x, t) - b(\bar{x}, t)| \leq K_*|x - \bar{x}|, \quad |\sigma(x, t) - \sigma(\bar{x}, t)| \leq K_*|x - \bar{x}|, \\ |b(x, t)| \leq K(1 + |x|), \quad |\sigma(x, t)| \leq K(1 + |x|) \end{aligned} \tag{1.3}$$

where K_*, K are constants. Let ξ_0 be any n-dimensional random vector independent of $\mathcal{F}(w(t), 0 \leq t \leq T)$, such that $E|\xi_0|^2 < \infty$. Then there exists a unique solution of (1.1), (1.2) *in $M_w^2[0, T]$.*

The assertion of uniqueness means that if $\xi_1(t)$, $\xi_2(t)$ are two solutions of (1.1), (1.2) and if they belong to $M_w^2[0, T]$, then

$$P\{\xi_1(t) = \xi_2(t) \text{ for all } 0 \leq t \leq T\} = 1.$$

Proof. To prove uniqueness, suppose $\xi_1(t)$ and $\xi_2(t)$ are two solutions belonging to $M_w^2[0, T]$. Then

$$\begin{aligned} \xi_1(t) - \xi_2(t) &= \int_0^t [b(\xi_1(s), s) - b_2(\xi_2(s), s)]\, ds \\ &\quad + \int_0^t \sigma(\xi_1(s), s)\, dw(s) - \int_0^t \sigma(\xi_2(s), s)\, dw(s). \end{aligned} \tag{1.4}$$

Set $f_i(s) = \sigma(\xi_i(s), s)$ and note that the stochastic integral $\int_0^t f_i(s)\, dw(s)$ is defined with respect to an increasing family of σ-fields $\mathfrak{F}_t^i$ which may depend on i. If $f_i(s)$ is a step function, for $i = 1, 2$, then using the definition of the stochastic integral we get (cf. the proof of Lemma 4.2.2 and formula (4.7.3))

$$E\left|\int_0^t f_1(s)\, dw(s) - \int_0^t f_2(s)\, dw(s)\right|^2 = E\int_0^t |f_1(s) - f_2(s)|^2\, ds. \quad (1.5)$$

By approximation we find that (1.5) is true for any pair f_1, f_2 of nonanticipative functions with respect to $\mathfrak{F}_t^1$ and $\mathfrak{F}_t^2$ respectively, provided $E\int_0^t |f_i(s)|^2\, ds < \infty$ $(i = 1, 2)$.

Taking the expectation of the squares of the absolute values on both sides of (1.4) and using (1.3) and (1.5) with $f_i(s) = \sigma(\xi_i(s), s)$, we find that

$$E|\xi_1(t) - \xi_2(t)|^2$$
$$\leq 2K_*^2 t \int_0^t E|\xi_1(s) - \xi_2(s)|^2\, ds + 2K_*^2 \int_0^t E|\xi_1(s) - \xi_2(s)|^2\, ds.$$

Thus the function $\phi(t) = E|\xi_1(t) - \xi_2(t)|^2$ satisfies

$$\phi(t) \leq C\int_0^t \phi(s)\, ds, \qquad \phi(0) = 0,$$

where C is a positive constant. Therefore $\phi(t) \equiv 0$, and the assertion of uniqueness is proved.

To prove the existence of a solution we introduce an increasing family of σ-fields $\mathfrak{F}_t$ $(0 \leq t \leq T)$ such that ξ_0 is $\mathfrak{F}_0$ measurable and $w(t)$ is $\mathfrak{F}_t$ measurable, and such that $\mathfrak{F}(w(t+s) - w(t), 0 \leq s \leq T - t)$ is independent of $\mathfrak{F}_t$, for all $t \geq 0$. We can take for instance $\mathfrak{F}_t$ to be the σ-field generated by ξ_0 and $\mathfrak{F}(w(s), s \leq t)$; here we use the assumption that ξ_0 is independent of $\mathfrak{F}(w(s), 0 \leq s \leq T)$.

Define $\xi_0(t) = \xi_0$ and

$$\xi_{m+1}(t) = \xi_0 + \int_0^t b(\xi_m(s), s)\, ds + \int_0^t \sigma(\xi_m(s), s)\, dw(s). \quad (1.6)$$

The inductive assumption is that $\xi_m \in M_w^2[0, T]$ and that

$$E|\xi_{k+1}(t) - \xi_k(t)|^2 \leq \frac{(Mt)^{k+1}}{(k+1)!} \qquad \text{for } 0 \leq k \leq m-1, \quad (1.7)$$

where M is some positive constant (depending only on K, K_*, T).

Since $\xi_0 \in \mathfrak{F}_0$, ξ_{m+1} is well defined if $m = 0$. Further

$$|\xi_1(t) - \xi_0|^2 \leq 2\left|\int_0^t b(\xi_0, s)\, ds\right|^2 + 2\left|\int_0^t \sigma(\xi_0, s)\, dw(s)\right|^2.$$

Taking the expectation and using (1.3), we get

$$E|\xi_1(t) - \xi_0|^2 \leq 2K_*^2 t^2(1 + E|\xi_0|^2) + 2K_*^2 t(1 + E|\xi_0|^2) \leq Mt$$

if $M > 2K^2(T+1)(1+E|\xi_0|^2)$. This implies that $\xi_1 \in M_w^2[0, T]$ and (1.7) holds for $m = 0$.

We now make the inductive assumption for any $m \geq 0$ and prove it for $m+1$.

Since $\xi_m \in M_w^2[0, T]$ it follows, using (1.3), that $b(\xi_m(t), t)$ and $\sigma(\xi_m(t), t)$ belong to $M_w^2[0, T]$. Thus the integrals on the right-hand side of (1.6) are well defined.

Next,

$$|\xi_{m+1}(t) - \xi_m(t)|^2 \leq 2\left|\int_0^t [b(\xi_m(s), s) - b(\xi_{m-1}(s), s)]\, ds\right|^2 + 2\left|\int_0^t [\sigma(\xi_m(s), s) - \sigma(\xi_{m-1}(s), s)]\, dw(s)\right|^2. \quad (1.8)$$

Taking the expectation and using (1.3),

$$E|\xi_{m+1}(s) - \xi_m(s)|^2 \leq 2K_*^2 tE \int_0^t |\xi_m(s) - \xi_{m-1}(s)|^2\, ds + 2K_*^2 E \int_0^t |\xi_m(s) - \xi_{m-1}(s)|^2\, ds.$$

Thus,

$$E|\xi_{m+1}(t) - \xi_m(t)|^2 \leq M \int_0^t E|\xi_m(s) - \xi_{m-1}(s)|^2\, ds$$

if $M > 2K_*^2(T+1)$. Substituting (1.7) with $k = m-1$ into the right-hand side, we get

$$E|\xi_{m+1}(t) - \xi_m(t)|^2 \leq M \int_0^t \frac{(Ms)^m}{m!}\, ds = \frac{(Mt)^{m+1}}{(m+1)!}.$$

Thus (1.7) holds for $k = m$. Since this implies that $\xi_{m+1} \in M_w^2[0, T]$, the proof of the inductive assumption for $m+1$ is complete.

From (1.8) we also have

$$\sup_{0 \leq t \leq T} |\xi_{m+1}(t) - \xi_m(t)|^2 \leq 2TK_*^2 \int_0^T |\xi_m(s) - \xi_{m-1}(s)|^2\, ds + 2 \sup_{0 \leq t \leq T} \left|\int_0^t [\sigma(\xi_m(s), s) - \sigma(\xi_{m-1}(s), s)]\, dw(s)\right|^2.$$

Taking the expectation and using Theorem 4.3.6 and (1.7), we find that

$$E \sup_{0 \leq t \leq T} |\xi_{m+1}(t) - \xi_m(t)|^2 \leq 2K_*^2 T \int_0^T E|\xi_m(s) - \xi_{m-1}(s)|^2\, ds + 8K_*^2 \int_0^T E|\xi_m(s) - \xi_{m-1}(s)|^2\, ds \leq C\frac{(MT)^m}{m!}$$

where $C = 2K_*^2T^2 + 8K_*^2T$. Hence

$$P\left\{\sup_{0 \leq t \leq T} |\xi_{m+1}(t) - \xi_m(t)| > \frac{1}{2^m}\right\} \leq 2^{2m}C\frac{(MT)^m}{m!}.$$

Since $\Sigma[2^m(MT)^m/m!] < \infty$, the Borel–Cantelli lemma implies that

$$P\left\{\sup_{0 \leq t \leq T} |\xi_{m+1}(t) - \xi_m(t)| > \frac{1}{2^m} \text{ i.o.}\right\} = 0.$$

Thus, for almost any ω there is a positive integer $m_0 = m_0(\omega)$ such that

$$\sup_{0 \leq t \leq T} |\xi_{m+1}(t) - \xi_m(t)| \leq \frac{1}{2^m} \quad \text{if} \quad m \geq m_0(\omega).$$

It follows that the partial sums

$$\xi_0 + \sum_{m=0}^{k-1} (\xi_{m+1}(t) - \xi_m(t)) = \xi_k(t)$$

are convergent uniformly in $t \in [0, T]$. Denote the limit by $\xi(t)$. Then $\xi(t)$ is a continuous process. It is clearly also a nonanticipative function and it belongs to $L_w^2[0, T]$. Since for a.a. ω,

$$b(\xi_m(t), t) \to b(\xi(t), t) \quad \text{uniformly in} \quad t \in [0, T],$$
$$\sigma(\xi_m(t), t) \to \sigma(\xi(t), t) \quad \text{uniformly in} \quad t \in [0, T],$$

and hence also

$$\int_0^T |\sigma(\xi_m(t), t) - \sigma(\xi(t), t)|^2 \xrightarrow{P} 0,$$

if we take $m \to \infty$ in (1.6) we obtain the relation

$$\xi(t) = \xi_0 + \int_0^t b(\xi(s), s)\, ds + \int_0^t \sigma(\xi(s), s)\, dw(s). \tag{1.9}$$

Thus $\xi(t)$ is a solution of (1.1), (1.2).

From (1.6) we have

$$E|\xi_{m+1}(t)|^2 \leq 3E|\xi_0|^2 + 3E\left|\int_0^t b(\xi_m(s), s)\, ds\right|^2 + 3E\left|\int_0^t \sigma(\xi_m(s), s)\, dw\right|^2$$
$$\leq C\left(1 + E|\xi_0|^2\right) + C\int_0^t E|\xi_m(s)|^2\, ds$$

where C is some constant depending only on K, T. By induction we then get

$$E|\xi_{m+1}(t)|^2 \leq \left[C + C^2t + C^3\frac{t^2}{2!} + \cdots + C^{m+2}\frac{t^{m+1}}{(m+1)!}\right][1 + E|\xi_0|^2].$$

Therefore

$$E|\xi_{m+1}(t)|^2 \leq C(1 + E|\xi_0|^2)e^{Ct}.$$

Taking $m \uparrow \infty$ and using Fatou's lemma, we conclude that

$$E|\xi(t)|^2 \leq C(1 + E|\xi_0|^2)e^{Ct}. \tag{1.10}$$

This implies that $\xi(t)$ belongs to $M_w^2[0, T]$.

The above method used to prove the existence of a solution $\xi(t)$ is called the *method of successive approximations*; it is modeled after the corresponding proof for ordinary differential equations.

Remark. Very often we shall take the initial value ξ_0 to be a constant function x, i.e., $\xi_0(\omega) = x$ a.s. Notice that this random variable is independent of $\mathcal{F}(w(t), t \geq 0)$.

From (1.9) we obtain

$$\sup_{0 \leq t \leq T} |\xi(t)|^2 \leq 3|\xi_0|^2 + 3\left[\int_0^T |b(\xi(s), s)|\, ds\right]^2 + 3 \sup_{0 \leq t \leq T}\left|\int_0^t \sigma(\xi(s), s)\, dw(s)\right|^2.$$

Taking the expectation and using (1.3) and Theorem 4.3.6, we get

$$E \sup_{0 \leq t \leq T} |\xi(t)|^2 \leq C_0(1 + E|\xi_0|^2) + C_0 \int_0^T E|\xi(s)|^2\, ds$$

where C_0 is a constant depending only on K, T. Making use of (1.10), we obtain

Corollary 1.2. *Under the assumptions of Theorem* 1.1,

$$E\left[\sup_{0 \leq t \leq T} |\xi(t)|^2\right] \leq C^*(1 + E|\xi_0|^2) \tag{1.11}$$

where C^* is a constant depending only on K, T.

2. Stronger uniqueness and existence theorems

Theorem 2.1. *Suppose $b_i(x, t)$, $\sigma_i(x, t)$ are measurable functions in $(x, t) \in R^n \times [0,T]$, for $i = 1, 2$, satisfying*

$$|b_i(x, t) - b_i(\bar{x}, t)| \leq K_*|x - \bar{x}|, \qquad |\sigma_i(x, t) - \sigma_i(\bar{x}, t)| \leq K_*|x - \bar{x}|,$$

$$|b_i(x, t)| \leq K(1 + |x|), \qquad |\sigma_i(x, t)| \leq K(1 + |x|).$$

Let D be a domain in R^n and suppose that

$$b_1(x, t) = b_2(x, t), \quad \sigma_1(x, t) = \sigma_2(x, t) \qquad \text{if } x \in D,\ 0 \leq t \leq T. \tag{2.1}$$

Let $\xi_i(t)$ $(i = 1, 2)$ *be the solution of*

$$d\xi_i(t) = b_i(\xi_i(t), t)\, dt + \sigma_i(\xi_i(t), t), \qquad \xi_i(0) = \xi_{i0}$$

in $M_w^2[0, T]$ *(with the same family of* σ*-fields* $\mathcal{F}_t$*) where* $E|\xi_{i0}|^2 < \infty$. *Assume finally that* $\xi_{10} = \xi_{20}$ *for a.a.* ω *for which either* $\xi_{10}(\omega) \in D$ *or* $\xi_{20}(\omega) \in D$. *Denote by* τ_i *the first time* $\xi_i(t)$ *intersects* $R^n \setminus D$ *if such time* $t \leq T$ *exists, and* $\tau_i = T$ *otherwise. Then*

$$P(\tau_1 = \tau_2) = 1,$$

$$P\Big\{ \sup_{0 \leq s \leq \tau_1} |\xi_1(s) - \xi_2(s)| = 0 \Big\} = 1.$$

Thus, if two stochastic equation have the same coefficients in a cylinder $Q = D \times [0, T]$ and if the initial conditions coincide in D, then the corresponding solutions agree until the first time they both leave D; they first leave D at the same time. This is a local uniqueness theorem. It remains true (with similar proof) for the general domains Q.

Proof. Let $\phi_i(t) = 1$ if $\xi_i(s) \in D$ for all $0 \leq s \leq t$, and $\phi_i(t) = 0$ in all other cases. Then

$$\phi_1(t)(\xi_{10} - \xi_{20}) = 0 \quad \text{a.s.}$$

Hence

$$\begin{aligned}
\phi_1(t)[\xi_1(t) - \xi_2(t)] &= \phi_1(t) \int_0^t [b_1(\xi_1(s), s) - b_2(\xi_1(s), s)]\, ds \\
&\quad + \phi_1(t) \int_0^t [b_2(\xi_1(s), s) - b_2(\xi_2(s), s)]\, ds \\
&\quad + \phi_1(t) \int_0^t [\sigma_1(\xi_1(s), s) - \sigma_2(\xi_1(s), s)]\, dw(s) \\
&\quad + \phi_1(t) \int_0^t [\sigma_2(\xi_1(s), s) - \sigma_2(\xi_2(s), s)]\, dw(s) \\
&= I_1 + I_2 + I_3 + I_4.
\end{aligned}$$

If $\phi_1(t) = 1$, then $b_1(\xi_1(s), s) = b_2(\xi_1(s), s)$. Hence $I_1 = 0$. If $t \leq \tau_1$, then $\sigma_1(\xi_1(s), s) = \sigma_2(\xi_1(s), s)$. Hence, by Theorem 4.4.7, $I_3 = 0$ a.s. if $t \leq \tau_1$. Since also $\phi_1(t) = 0$ when $t > \tau_1$, we have $I_3 = 0$ a.s. Thus

$$\begin{aligned}
\phi_1(t)|\xi_1(t) - \xi_2(t)|^2 &\leq 2\phi_1(t) \left| \int_0^t [b_2(\xi_1(s), s) - b_2(\xi_2(s), s)]\, ds \right|^2 \\
&\quad + 2\phi_1(t) \left| \int_0^t [\sigma_2(\xi_1(s), s) - \sigma_2(\xi_2(s), s)]\, dw(s) \right|^2. \qquad (2.2)
\end{aligned}$$

Since $\phi_1(t) \leq \phi_1(s)$ if $s \leq t$,

$$\phi_1(t)\left|\int_0^t [b_2(\xi_1(s), s) - b_2(\xi_2(s), s)]\, ds\right|^2$$
$$\leq n\left\{\int_0^t \phi_1(s)|\, b_2(\xi_1(s), s) - b_2(\xi_2(s), s)|\, ds\right\}^2. \tag{2.3}$$

Next, if $f \in L_w^2[0, T]$ and $t \in [0, T]$,

$$\phi_1(t)\left|\int_0^t f(s)\, dw(s)\right|^2 \leq \left|\int_0^t \phi_1(s)f(s)\, dw(s)\right|^2. \tag{2.4}$$

Indeed, let $A = \{\phi_1(t) = 1\}$, $B = \{\phi_1(t) = 0\}$. If $\omega \in B$, then the left-hand side of (2.4) vanishes; hence (2.4) is true. If $\omega \in A$, then $\phi_1(s)f(s) = f(s)$ if $s < t$; hence, by Lemma 4.2.11,

$$\int_0^t \phi_1(s)f(s)\, dw(s) = \int_0^t f(s)\, dw(s),$$

and (2.4) again follows.

Using (2.3), (2.4) in (2.2), we find that

$$\phi_1(t)|\xi_1(t) - \xi_2(t)|^2 \leq 2n\left[\int_0^t \phi_1(s)|b_2(\xi_1(s), s) - b_2(\xi_2(s), s)|\, ds\right]^2$$
$$+ 2\left|\int_0^t \phi_1(s)[\sigma_2(\xi_1(s), s) - \sigma_2(\xi_2(s), s)]\, dw(s)\right|^2.$$

Using the Lipschitz continuity of $b_2(x, t)$, $\sigma_2(x, t)$ in x, we get

$$E\phi_1(t)|\xi_1(t) - \xi_2(t)|^2 \leq C\int_0^t E\phi_1(s)|\xi_1(s) - \xi_2(s)|^2\, ds \qquad (C \text{ constant}).$$

This implies that

$$E\phi_1(t)|\xi_1(t) - \xi_2(t)|^2 = 0.$$

Hence, by the continuity of $\xi_1(t)$, $\xi_2(t)$,

$$\sup_{0 \leq t \leq T} \phi_1(t)|\xi_1(t) - \xi_2(t)| = 0 \quad \text{a.s.}$$

It follows that $\xi_1(t, \omega) = \xi_2(t, \omega)$ a.s. if $0 \leq t \leq \tau_1(\omega)$. Consequently, $P(\tau_2 \geq \tau_1) = 1$. Similarly $P(\tau_1 \geq \tau_2) = 1$, and the proof is complete.

We shall now use Theorem 2.1 in order to improve Theorem 1.1.

Theorem 2.2. *Suppose $b(x, t)$, $\sigma(x, t)$ are measurable in $(x, t) \in R^n \times [0, T]$ and*

$$|b(x, t)| \leq K(1 + |x|), \qquad |\sigma(x, t)| \leq K(1 + |x|). \tag{2.5}$$

Suppose further that for any $N > 0$ there is a positive constant K_N such that

$$|b(x, t) - b(\bar{x}, t)| \leq K_N|x - \bar{x}|, \qquad |\sigma(x, t) - \sigma(\bar{x}, t)| \leq K_N|x - \bar{x}| \tag{2.6}$$

if $|x| \leqslant N$, $|\bar{x}| \leqslant N$, $0 \leqslant t \leqslant T$. *Let* ξ_0 *be any n-dimensional random vector independent of* $\mathcal{F}(w(t), 0 \leqslant t \leqslant T)$ *such that* $E|\xi_0|^2 < \infty$. *Then there exists a unique solution* $\xi(t)$ *of* (1.1), (1.2) *in* $M_w^2[0, T]$. *Further*

$$E\Big[\sup_{0 \leqslant t \leqslant T} |\xi(t)|^2\Big] \leqslant C^*(1 + E|\xi_0|^2) \tag{2.7}$$

where C^* *is a constant depending only on* T, K.

Thus, Theorem 1.1 and Corollary 1.2 remain true if the condition (1.3) is replaced by (2.5), (2.6).

Proof. Let

$$\xi_{N0}(\omega) = \begin{cases} \xi_0(\omega) & \text{if} \quad |\xi_0(\omega)| \leqslant N, \\ 0 & \text{if} \quad |\xi_0(\omega)| > N, \end{cases}$$

$$b_N(x, t) = \begin{cases} b(x, t) & \text{if} \quad |x| \leqslant N, \\ b(x, t)(2 - (|x|/N)) & \text{if} \quad N < |x| \leqslant 2N, \\ 0 & \text{if} \quad |x| > 2N, \end{cases}$$

$$\sigma_N(x, t) = \begin{cases} \sigma(x, t) & \text{if} \quad |x| \leqslant N, \\ \sigma(x, t)(2 - (|x|/N)) & \text{if} \quad N < |x| \leqslant 2N, \\ 0 & \text{if} \quad |x| > 2N. \end{cases}$$

By Theorem 1.1 there exists a unique solution $\xi_N(t)$ in $M_w^2[0, T]$ of

$$\begin{aligned} d\xi_N(t) &= b_N(\xi_N(t), t)\, dt + \sigma_N(\xi_N(t), t)\, dw(t), \\ \xi_N(0) &= \xi_{N0}. \end{aligned} \tag{2.8}$$

For definiteness we take $\mathcal{F}_t$ to be the σ-field generated by $w(s)$, $0 \leqslant s \leqslant t$ and ξ_0. By Corollary 1.2,

$$E\Big[\sup_{0 \leqslant t \leqslant T} |\xi_N(t)|^2\Big] \leqslant C^*(1 + E|\xi_0|^2) \tag{2.9}$$

where C^* is a constant independent of N.

Let τ_N be the first time $|\xi_N(t)| \geqslant N$, if such a time $t \leqslant T$ exists, and $\tau_N = T$ otherwise. If $N' > N$, then by Theorem 2.1 $\xi_N(t) = \xi_{N'}(t)$ a.s. if $0 \leqslant t \leqslant \tau_N$. Therefore,

$$\begin{aligned} P\Big\{\sup_{0 \leqslant t \leqslant T} |\xi_N(t) - \xi_{N'}(t)| > 0\Big\} &= P\Big\{\sup_{0 \leqslant t \leqslant T} |\xi_N(t)| > N\Big\} \\ &\leqslant \frac{1}{N^2} E\Big[\sup_{0 \leqslant t \leqslant T} |\xi_N(t)|\Big]^2 \leqslant \frac{C^*}{N^2}(1 + E|\xi_0|^2). \end{aligned}$$

Notice that the set $\Omega_N = \{\sup_{0 \leqslant t \leqslant T} |\xi_N(t)| > N\}$ decreases when N increases, and $P(\Omega_N) \downarrow 0$ if $N \uparrow \infty$. Define

$$\xi(t, \omega) = \lim_{N \to \infty} \xi_N(t, \omega) \qquad \text{if} \quad \omega \not\in \lim \Omega_N.$$

Then, $\xi(t, \omega) = \xi_N(t, \omega)$ if $\omega \not\in \Omega_N$. It is clear that $\xi(t)$ is in $L_w^2[0, T]$.

By Lemma 4.2.11,

$$\int_0^t \sigma(\xi(s), s)\, dw(s) = \int_0^t \sigma_N(\xi_N(s), s)\, dw(s) \qquad \text{a.s.} \qquad \text{if} \quad \omega \not\in \Omega_N.$$

From the stochastic differential equation for $\xi_N(t)$ we then obtain

$$\xi(t) = \xi_0 + \int_0^t b(\xi(s), s)\, ds + \int_0^t \sigma(\xi(s), s)\, dw(s)$$

a.s. if $\omega \not\in \Omega_N$. It follows that $\xi(t)$ is a solution of (1.1), (1.2).

Finally, from (2.9) and Fatou's lemma we obtain (2.7).

It remains to prove uniqueness. Let $\xi_1(t)$, $\xi_2(t)$ be two solutions of (1.1), (1.2) in $M_w^2[0, T]$; ξ_i is nonanticipative with respect to $\mathcal{F}_t^i$. Let $\phi(t) = 1$ if $\sup_{0 \le s \le T} |\xi_i(s)| \le N$ for $i = 1, 2$ and $\phi(t) = 0$ in all other cases. Then (cf. the proof of (2.2))

$$E\big[\phi(t)|\xi_1(t) - \xi_2(t)|^2\big] \le 2E\left\{\phi(t)\left|\int_0^t [b(\xi_1(s), s) - b(\xi_2(s), s)]\, ds\right|^2\right\}$$

$$+\, 2E\left\{\phi(t)\left|\int_0^t \sigma(\xi_1(s), s)\, dw(s) - \int_0^t \sigma(\xi_2(s), s)\, dw(s)\right|^2\right\}$$

$$\equiv I + J. \tag{2.10}$$

It is clear that

$$I \le 2t \int_0^t K_N|\xi_1(s) - \xi_2(s)|^2\phi(s)\, ds. \tag{2.11}$$

Next, if $f_i(s)$ are step functions, nonanticipative with respect to the σ-fields $\mathcal{F}_t^i$ and in $M_w^2[0, T]$, then (cf. the proof of (2.4))

$$E\left\{\phi(t)\left|\int_0^t f_1(s)\, dw(s) - \int_0^t f_2(s)\, dw(s)\right|^2\right\} \le E\int_0^t |f_1(s) - f_2(s)|^2\phi(s)\, ds. \tag{2.12}$$

By approximation, this inequality holds for any f_1, f_2 in $M_w^2[0, T]$; f_i being nonanticipative with respect to $\mathcal{F}_t^i$. Using (2.12) with $f_i(s) = \sigma(\xi_i(s), s))$ and using (2.11), we obtain from (2.10)

$$E\{\phi(t)|\xi_1(t) - \xi_2(t)|^2\} \le (2T + 2)K_N \int_0^t E\phi(s)|\xi_1(s) - \xi_2(s)|^2\, ds.$$

Consequently

$$E\{\phi(t)|\xi_1(t) - \xi_2(t)|^2\} = 0,$$

i.e.,

$$P\{\xi_1(t) \neq \xi_2(t)\} \le P\Big\{\sup_{0 \le s \le T} |\xi_1(s)| > N\Big\} + P\Big\{\sup_{0 \le s \le T} |\xi_2(s)| > N\Big\}.$$

Since $\xi_1(s)$ and $\xi_2(s)$ are continuous process, the right-hand side converges to 0 as $N \to \infty$. Therefore,

$$P\{\xi_1(t) \neq \xi_2(t)\} = 0.$$

Since both processes $\xi_1(t)$ and $\xi_2(t)$ are continuous,

$$P\{\xi_1(t) = \xi_2(t) \text{ for } 0 \leqslant t \leqslant T\} = 1.$$

We conclude this section with an estimate on the higher moments of the solution $\xi(t)$.

Theorem 2.3. *Let $b(x, t)$, $\sigma(x, t)$ be as in Theorem 2.2, and let $E|\xi_0|^{2m} < \infty$ for some positive integer m. Then*

$$E|\xi(t)|^{2m} \leqslant (1 + E|\xi_0|^{2m})e^{Ct}, \tag{2.13}$$

$$E\Big\{\sup_{0 \leqslant s \leqslant t} |\xi(s) - \xi_0|^{2m}\Big\} \leqslant \bar{C}(1 + E|\xi_0|^{2m})t^m, \tag{2.14}$$

where C, $\bar{C}$ are constants depending only on K, T, m.

Proof. By Itô's formula with $f(x, t) = |x|^{2m}$, $\xi_N(t)$,

$$\begin{aligned}|\xi_N(t)|^{2m} = |\xi_N(0)|^{2m} &+ \int_0^t \Big\{ 2m|\xi_N(s)|^{2m-2}\xi_N(s) \cdot b_N(\xi_N(s), s) \\ &+ \sum_{i=1}^n 2m|\xi_N(s)|^{2m-2} a_{ii}^N \\ &+ \sum_{i,j=1}^n 2m(2m-2)|\xi_N(s)|^{2m-4} a_{ij}^N \xi_N^i(s)\xi_N^j(s) \Big\}\, ds \\ &+ \int_0^t 2m|\xi_N(s)|^{2m-2}\xi_N(s) \cdot \sigma_N(\xi(s), s)\, dw(s)\end{aligned} \tag{2.15}$$

where $(a_{ij}^N) = \frac{1}{2}\sigma^N(\xi_N(s), s)(\sigma^N(\xi_N(s), s))^*$ (A^* = transpose of A).

Since

$$\xi_N(s) = \xi_N(0) + \int_0^t b_N(\xi_N(s), s)\, ds + \int_0^t \sigma_N(\xi_N(s), s)\, dw(s)$$

and $E|\xi_N(0)|^2 < \infty$, while $b_N(x, t)$ and $\sigma_N(x, t)$ are bounded, it follows (using Theorem 4.6.3) that $E|\xi_N(s)|^{2m} \leqslant C_N$ for all $0 \leqslant s \leqslant T$, where C_N is a constant. Hence the expectation of the stochastic integral on the right-hand side of (2.15) is equal to 0. We therefore get

$$E|\xi_N(t)|^{2m} \leqslant E|\xi_N(0)|^{2m} + L\int_0^t E\{[1 + |\xi_N(s)|^2]|\xi_N(s)|^{2m-2}\}\, ds,$$

where L is a constant depending only on m, K. Since the integrand on the right is $\leqslant 1 + 2E|\xi_N(s)|^{2m}$, the inequality (2.13) for ξ_N readily follows. Now take $N \uparrow \infty$ to obtain (2.13).

To prove (2.14) notice that for any $\lambda \in [0, T]$

$$\sup_{0 \leq t \leq \lambda} |\xi(t) - \xi_0|^{2m}$$
$$\leq n2^{2m}\left[\int_0^\lambda |b(\xi(s), s)|\, ds\right]^2 + 2^m \sup_{0 \leq t \leq \lambda}\left|\int_0^t \sigma(\xi(s), s)\, dw(s)\right|^{2m}.$$

Taking the expectation and using Corollary 4.6.4 and (2.13) we obtain

$$E \sup_{0 \leq t \leq \lambda} |\xi(t) - \xi_0|^{2m} \leq L\lambda^{m-1} \int_0^\lambda [1 + E|\xi(s)|^{2m}]\, ds \leq L_1\lambda^m(1 + E|\xi_0|^{2m}),$$

where L, L_1 are constants depending only on K, T, m; thus (2.14) holds.

Notice that (2.14) implies that

$$E\left\{\sup_{0 \leq s \leq t} |\xi(s)|^{2m}\right\} \leq C^*(1 + E|\xi_0|^2), \tag{2.16}$$

where C^* is a constant depending only on K, T, m.

Remark. The results proved in this and the previous sections extend immediately to stochastic differential equations (1.1) with initial condition $\xi(s) = \xi_s$ where $s \in [0, T]$. Here ξ_s is assumed to be independent of the σ-field $\mathcal{F}(w(t + s) - w(s), s \leq t \leq T - s)$.

3. The solution of a stochastic differential system as a Markov process

We shall assume:

(A) The n-vector $b(x, t)$ and the $n \times n$ matrix $\sigma(x, t)$ are measurable functions for $(x, t) \in R^n \times [0, \infty)$ and, for any $T > 0$,

$$|b(x, t)| \leq K(1 + |x|), \qquad |\sigma(x, t)| \leq K(1 + |x|), \tag{3.1}$$

$$|b(x, t) - b(\bar{x}, t)| \leq K|x - \bar{x}|, \qquad |\sigma(x, t) - \sigma(\bar{x}, t)| \leq K|x - \bar{x}| \tag{3.2}$$

if $0 \leq t \leq T$, $x \in R^n$, $\bar{x} \in R^n$, where K is a constant depending on T.

By Theorem 1.1 there exists a unique solution in $M_w^2[0, \infty)$ of

$$d\xi(t) = b(\xi(t), t)\, dt + \sigma(\xi(t), t)\, dw(t), \tag{3.3}$$

$$\xi(0) = \xi_0, \tag{3.4}$$

provided ξ_0 is independent of $\mathcal{F}(w(\lambda), \lambda \geq 0)$ and $E|\xi_0|^2 < \infty$. Similarly (see the remark at the end of Section 2), for any $s \geq 0$ there exists a unique solution in $M_w^2[s, \infty)$ of (3.3) and

$$\xi(s) = \xi_s \tag{3.5}$$

provided ξ_s is independent of $\mathcal{F}(w(\lambda + s) - w(s), \lambda \geq 0)$ and $E|\xi_s|^2 < \infty$.

If $\xi_s = x$ a.s., where x is a point in R^n, then we denote the solution of (3.3), (3.5) by $\xi_{x,s}(t)$.

For any Borel set A in R^n and for any $t \geqslant s$, let

$$p(s, x, t, A) = P(\xi_{x,s}(t) \in A). \tag{3.6}$$

Theorem 3.1. *Let* (A) *hold and let* ξ_0 *be independent of* $\mathcal{F}(w(t), t \geqslant 0)$, $E|\xi_0|^2 < \infty$. *Denote by* $\mathcal{F}_t$ *the* σ*-field spanned by* ξ_0 *and* $w(s)$, $0 \leqslant s \leqslant t$. *Then the unique solution* $\xi(t)$ *of* (3.3), (3.4) *satisfies*

$$P(\xi(t) \in A | \mathcal{F}_s) = P(\xi(t) \in A | \xi(s)) = p(s, \xi(s), t, A) \quad \text{a.s.} \tag{3.7}$$

for all $t > s$ *and for any Borel set* A. *Further*, $p(s, x, t, A)$ *is a transition probability function.*

Proof. We can write

$$\xi(t) = \xi(s) + \int_s^t b(\xi(\lambda), \lambda)\, d\lambda + \int_s^t \sigma(\xi(\lambda), \lambda)\, dw(\lambda).$$

By the method of successive approximations, if we set

$$\xi_0(t) = \gamma \qquad (\text{where} \quad \gamma = \xi(s))$$

$$\xi_k(t) = \gamma + \int_s^t b(\xi_{k-1}(\lambda), \lambda)\, d\lambda + \int_s^t \sigma(\xi_{k-1}(\lambda), \lambda)\, dw(\lambda),$$

then $\xi_k(t) \to \xi(t)$ a.s. By induction one can show that each $\xi_k(t)$ is measurable with respect to the σ-field spanned by $\mathcal{F}(w(u + s) - w(s), s \leqslant u \leqslant t)$ and γ. The same is therefore true of $\xi(t)$. More specifically, using Problems 20, 21 of Chapter 4 and Theorem 4.3.4, one can approximate a.s. (and uniformly in t) each $\xi_k(t)$ by a sequence of functions

$$F_m(t, \gamma, w(u_{m,1} + s) - w(s), \ldots, w(u_{m,\mu_m} + s) - w(s))$$

where $0 < u_{m,i} \leqslant t - s$ and $F_m(t, x_0, x_1, \ldots, x_{\mu_m})$ are Borel measurable functions depending only on the functions $b(x, \lambda)$, $\sigma(x, \lambda)$ in the interval $s \leqslant \lambda \leqslant t$. The same therefore holds for $\xi(t)$, i.e., a.s.

$$\xi(t) = \lim_{m\to\infty} F_m(t, \xi(s), w(u_{m,1} + s) - w(s), \ldots, w(u_{m,\mu_m} + s) - w(s)) \tag{3.8}$$

with suitable Borel functions F_m and suitable $u_{m,i}$. In particular,

$$\xi_{x,s}(t) = \lim_{m\to\infty} F_m(t, x, w(u_{m,1} + s) - w(s), \ldots, w(u_{m,\mu_m} + s) - w(s)). \tag{3.9}$$

Now, for any bounded measurable function

$$F(x_0, x_1, \ldots, x_k) = F_0(x_0)F_1(x_1, \ldots, x_k)$$

and for any $u_i \geqslant 0$,

$$E[F(\xi(s), w(u_1 + s) - w(s), \ldots, w(u_k + s) - w(s))|\mathcal{F}_s]$$
$$= F_0(\xi(s))E[F_1(w(u_1 + s) - w(s), \ldots, w(u_k + s) - w(s))|\mathcal{F}_s]$$
$$= F_0(\xi(s))E[F_1(w(u_1 + s) - w(s), \ldots, w(u_k + s) - w(s))]$$

since the random variables $w(u_i + s) - w(s)$ are independent of $\mathcal{F}_s$. Thus,

$$E[F(\xi(s), w(u_1 + s) - w(s), \ldots, w(u_k + s) - w(s))|\mathcal{F}_s]$$
$$= \{EF(x_0, w(u_1 + s) - w(s), \ldots, w(u_k + s) - w(s))\}_{x_0 = \xi(s)}. \quad (3.10)$$

By approximation, (3.10) remains true for any bounded Borel measurable function $F(x_0, x_1, \ldots, x_k)$. Taking, in particular, $F = f(F_m)$ where f is a bounded continuous function and F_m are as in (3.8), (3.9), we conclude that

$$E\{f(\xi(t))|\mathcal{F}_s\} = E\{f(\xi(t))|\xi(s)\} = \phi(x)|_{x=\xi(s)}$$

where $\phi(x) = Ef(\xi_{x,s}(t))$.

Taking a sequence of f's that increase to the indicator function of an open set A in R^n, we obtain the assertions (3.6), (3.7) for any open set A. Since the class of Borel sets for which (3.6), (3.7) hold form a monotone class, and since the open sets belong to this class, this class must contain the σ-field generated by the open sets, i.e., it includes all the Borel sets.

It remains to prove that p is a transition probability function.

From (3.6) it is clear that $p(s, x, t, A)$ is a probability measure in A, for fixed s, x, t. Since $\xi_{x,s}(t)$ is Borel measurable in x (for the same is true of each function in the sequence that converges to $\xi_{x,s}(t)$ by the method of successive approximations), $p(s, x, t, A)$ is also Borel measurable in x, for fixed s, t, A. Thus it remains to verify the Chapman–Kolmogorov equation.

From (3.6),

$$\int p(s, x, t, dy)\psi(y) = \int \psi(\xi_{x,s}(t))\, dP$$

for any bounded Borel measurable function $\psi(y)$. Taking $\psi(y) = p(t, y, \tau, A)$ where $t < \tau$, we get

$$\int p(s, x, t, dy)p(t, y, \tau, A) = \int p(t, \xi_{x,s}(t), \tau, A)\, dP$$
$$= Ep(t, \xi_{x,s}(t), \tau, A) = EP[\xi_{x,s}(\tau) \in A|\xi_{x,s}(t)]$$

where (3.7) has been used in the last equality. Since the expression on the right is equal to

$$EE[\chi_A(\xi_{x,s}(\tau))|\xi_{x,s}(t)] = E\chi_A(\xi_{x,s}(\tau))$$
$$= P(\xi_{x,s}(\tau) \in A) = p(s, x, \tau, A),$$

the Chapman–Kolmogorov equation holds for all $x \in R^n$, $0 \leqslant s < t < \tau$.

Denote by $\mathcal{C}$ the class of all continuous functions $x(\,\cdot\,)$ from $[0, \infty)$ into R^n. Denote by $\mathfrak{M}_t^s$ the smallest σ-field generated by the sets

$$\{x(\,\cdot\,);\, x(u) \in A\}, \qquad s \leq u \leq t$$

where A is any Borel set in R^n. Denote by $\mathfrak{M}_\infty^s$ the smallest σ-field generated by $\{x(\,\cdot\,);\, x(u) \in A\}$, $u \geq s$.

Define a continuous process $X(t) = X(t, x(\,\cdot\,))$ by

$$X(t, x(\,\cdot\,)) = x(t). \tag{3.11}$$

Finally, let

$$P_{x,s}\{x(\,\cdot\,) \in B\} = P\{\omega;\, \xi_{x,s}(\,\cdot\,, \omega) \in B\} \tag{3.12}$$

where B is any set in $\mathfrak{M}_t^s$. Notice that for each $\omega \in \Omega$, $\xi_{x,s}(t) = \xi_{x,s}(t, \omega)$ is a continuous path. This path is the continuous function $\xi_{x,s}(\,\cdot\,, \omega)$ appearing on the right-hand side of (3.12).

It is easily seen that $P_{x,s}$ is a probability measure on $\mathfrak{M}_t^s$. We shall now show that

$$P_{x,s}\{X(t+h) \in A \,|\, \mathfrak{M}_t^s\} = p(t, X(t), t+h, A) \quad \text{a.s.} \tag{3.13}$$

Since

$$P\{\xi_{x,s}(t+h) \in A \,|\, \mathcal{F}(\xi_{x,s}(\lambda), \lambda \leq t)\} = p(t, \xi_{x,s}(t), t+h, A),$$

for any $s \leq t_1 < t_2 < \cdots < t_m \leq t$ and any Borel sets $A_1, \ldots, A_m$, we have

$$\begin{aligned}
P\{\xi_{x,s}(t+h) \in A, \xi_{x,s}(t_1) \in A_1, \ldots, \xi_{x,s}(t_m) \in A_m\} \\
= \int_{\bigcap_{i=1}^m [\xi_{x,s}(t_i) \in A_i]} p(t, \xi_{x,s}(t), t+h, A)\, dP \\
= \int \chi_{A_1}(\xi_{x,s}(t_1)) \cdots \chi_{A_m}(\xi_{x,s}(t_m)) p(t, \xi_{x,s}(t), t+h, A)\, dP.
\end{aligned}$$

Using (3.11), (3.12), we conclude that

$$\begin{aligned}
P_{x,s}[X(t+h) \in A, X(t_1) \in A_1, \ldots, X(t_m) \in A_m] \\
= \int \chi_{A_1}(X(t_1)) \cdots \chi_{A_m}(X(t_m)) p(t, X(t), t+h, A)\, dP_{x,s} \\
= \int_{\bigcap_{i=1}^m [X(t_i) \in A_i]} p(t, X(t), t+h, A)\, dP_{x,s}.
\end{aligned}$$

This implies (3.13).

We have proved:

Theorem 3.2. *Let* (A) *hold. Then* $\{\mathcal{C}, \mathfrak{M}, \mathfrak{M}_t^s, X(t), P_{x,s}\}$ *is a continuous n-dimensional Markov process with the transition probability function* (3.6).

We shall call this Markov process also the *solution of the stochastic differential system* (1.1). The process $X(t)$ will often be denoted also by $\xi(t)$.

Lemma 3.3. *Let* (A) *hold. Then for any* $R > 0$, $T > 0$,

$$E \sup_{\tau \leq t \leq T} |\xi_{x,s}(t) - \xi_{y,\tau}(t)|^2 \leq C(|x - y|^2 + |s - \tau|) \tag{3.14}$$

if $|x| \leq R$, $|y| \leq R$, $0 \leq s \leq \tau \leq T$; *C is a constant depending on* R, T.

Proof. Clearly, if $\tau \leq t$,

$$\xi_{x,s}(t) - \xi_{y,\tau}(t) = \xi_{x,s}(\tau) - y + \int_\tau^t [b(\xi_{x,s}(\lambda), \lambda) - b(\xi_{y,\tau}(\lambda), \lambda)]\, d\lambda$$
$$+ \int_\tau^t [\sigma(\xi_{x,s}(\lambda), \lambda) - \sigma(\xi_{y,\tau}(\lambda), \lambda)]\, dw(\lambda). \tag{3.15}$$

By Theorem 2.3,

$$E|\xi_{x,s}(\tau) - y|^2 \leq 2E|\xi_{x,s}(\tau) - x|^2 + 2|x - y|^2 \leq C_0|\tau - s| + 2|x - y|^2 \tag{3.16}$$

where C_0 is a constant. Taking the expectation of the supremum of the squares of both side of (3.15) and using (3.16), (3.2), we get

$$E \sup_{\tau \leq t \leq t'} |\xi_{x,s}(t) - \xi_{y,\tau}(t)|^2 \leq C_1(|x - y|^2 + |s - \tau|)$$
$$+ C_1 E \int_\tau^{t'} |\xi_{x,s}(\lambda) - \xi_{y,\tau}(\lambda)|^2\, d\lambda$$

where C_1 is a constant. This easily implies (3.14).

Theorem 3.4. *Let* (A) *hold. Then the solution* $\{\mathcal{C}, \mathfrak{M}, \mathfrak{M}_t^s, X(t), P_{x,s}\}$ *of the stochastic differential system* (1.1) *satisfies the Feller property, and therefore also the strong Markov property.*

Proof. If f is a bounded continuous function, then, by Lemma 3.3 and the Lebesgue bounded convergence theorem,

$$Ef(\xi_{y,\tau}(t + \tau)) - Ef(\xi_{x,s}(t + \tau)) \to 0 \quad \text{if} \quad y \to x, \quad \tau \to s.$$

Also,

$$Ef(\xi_{x,s}(t + \tau)) \to Ef(\xi_{x,s}(t + s)) \quad \text{if} \quad \tau \to s.$$

Thus, the function

$$(s, x) \to \int p(s, x, t + s, dy) f(y)$$

is continuous, i.e., the Markov process $\{\mathcal{C}, \mathfrak{M}, \mathfrak{M}_t^s, X(t), P_{x,s}\}$ satisfies the Feller property. Since it is also a continuous process, Corollary 2.2.6 asserts that it possesses the strong Markov property.

Consider a stochastic differential system

$$d\xi(t) = b(\xi(t), t)\, dt + \sigma(\xi(t), t)\, d\tilde{w}(t) \tag{3.17}$$

where $\tilde{w}(t)$ is another n-dimensional Brownian motion. According to Theorem 3.3, there is a Markov process solution $\{\mathcal{C}, \mathfrak{M}, \mathfrak{M}_t^s, X(t), \tilde{P}_{x,s}\}$.

Definition. If for any Brownian motion $\tilde{w}(t)$, $\tilde{P}_{x,s} = P_{x,s}$ for all x, s, then we say that there is *uniqueness in the sense of probability law*. By contrast, the uniqueness asserted in Theorem 1.1 is called *pathwise uniqueness*.

Theorem 3.5. *Let* (A) *hold. Then there is a unique solution of* (1.1) *in the sense of probability law.*

Proof. From the proof of (3.8) one sees that the F_m do not depend on the particular Brownian motion $w(t)$. Hence, for any $s \leq t_1 < t_2 < \cdots < t_k < \infty$ and for any bounded continuous function $\phi(x_1, \ldots, x_k)$, if $\tilde{\xi}(t)$ is a solution of (3.17) with $\tilde{\xi}(s) = \xi(s)$ and if $\xi(s) = x$, then

$$\begin{aligned}
&E\phi(\xi(t_1), \ldots, \xi(t_k)) \\
&\quad = \lim_{m\to\infty} E\phi\big(F_m(t_1, x, w(u_{m,1} + s) - w(s), \ldots, w(u_{m,\mu_m} + s) - w(s)), \\
&\qquad \ldots, F_m(t_k, x, w(u_{m,1} + s) - w(s), \ldots, w(u_{m,\mu_m} + s) - w(s))\big) \\
&\quad = \lim_{m\to\infty} E\phi\big(F_m(t_1, x, \tilde{w}(u_{m,1} + s) - w(s), \ldots, \tilde{w}(u_{m,\mu_m} + s) - \tilde{w}(s)), \\
&\qquad \ldots, F_m(t_k, x, \tilde{w}(u_{m,1} + s) - \tilde{w}(s), \ldots, \tilde{w}(u_{m,\mu_m} + s) - \tilde{w}(s))\big) \\
&\quad = E\phi(\tilde{\xi}(t_1), \ldots, \tilde{\xi}(t_k)).
\end{aligned}$$

Since ϕ is arbitrary, it follows that $P_{x,s} = \tilde{P}_{x,s}$ on cylinder sets in $\mathfrak{M}_t^s$. Therefore also $P_{x,s} = \tilde{P}_{x,s}$ on the whole of $\mathfrak{M}_t^s$.

The results of this section can be extended to the case where the condition (A) is replaced by the weaker condition:

(A′) $b(x, t)$ and $\sigma(x, t)$ are measurable; for every $T > 0$ the inequality (3.1) holds for $0 \leq t \leq T$, $x \in R^n$ with K depending only on T; for every $T > 0$, $R > 0$,

$$|b(x, t) - b(\bar{x}, t)| \leq K_R|x - \bar{x}|, \qquad |\sigma(x, t) - \sigma(\bar{x}, t)| \leq K_R|x - \bar{x}|$$

if $0 \leq t \leq T$, $|x| \leq R$, $|\bar{x}| \leq R$ where K_R is a constant depending on T, R.

Theorem 3.6. *Theorems* 3.1, 3.2, 3.4, 3.5 *remain true if the condition* (A) *is replaced by the weaker condition* (A′).

The proof is left to the reader.

4. Diffusion processes

A continuous n-dimensional Markov process with transition probability function $p(s, x, t, A)$ is called a *diffusion process* if:

(i) for any $\epsilon > 0$, $t \geq 0$, $x \in R^n$,

$$\lim_{h \downarrow 0} \frac{1}{h} \int_{|y-x|>\epsilon} p(t, x, t+h, dy) = 0; \tag{4.1}$$

(ii) there exist an n-vector $b(x, t)$ and an $n \times n$ matrix $a(x, t)$ such that for any $\epsilon > 0$, $t \geq 0$, $x \in R^n$,

$$\lim_{h \downarrow 0} \frac{1}{h} \int_{|y-x|<\epsilon} (y_i - x_i) p(t, x, t+h, dy) = b_i(x, t) \qquad (1 \leq i \leq n), \tag{4.2}$$

$$\lim_{h \downarrow 0} \frac{1}{h} \int_{|y-x|<\epsilon} (y_i - x_i)(y_j - x_j) p(t, x, t+h, dy) = a_{ij}(x, t) \qquad (1 \leq i, j \leq n), \tag{4.3}$$

where $b = (b_1, \ldots, b_n)$, $a = (a_{ij})$.

The vector b is called the *drift coefficient* and the matrix a is called the *diffusion matrix*; when $n = 1$ we call a the *diffusion coefficient.*

Lemma 4.1. *The following conditions imply the conditions* (i), (ii):

(i*) *for some* $\delta > 0$, $t \geq 0$, $x \in R^n$,

$$\lim_{h \downarrow 0} \frac{1}{h} \int_{R^n} |x - y|^{2+\delta} p(t, x, t+h, dy) = 0; \tag{4.4}$$

(ii*) *for any* $t \geq 0$, $x \in R^n$,

$$\lim_{h \downarrow 0} \frac{1}{h} \int_{R^n} (y_i - x_i) p(t, x, t+h, dy) = b_i(x, t) \qquad (1 \leq i \leq n), \tag{4.5}$$

$$\lim_{h \downarrow 0} \frac{1}{h} \int_{R^n} (y_i - x_i)(y_j - x_j) p(t, x, t+h, dy) = a_{ij}(x, t) \qquad (1 \leq i, j \leq n). \tag{4.6}$$

Proof. Using (4.4) we have

$$\int_{|y-x|>\epsilon} p(t, x, t+h, dy) \leq \frac{1}{\epsilon^{2+\delta}} \int_{R^n} |y - x|^{2+\delta} p(t, x, t+h, dy)$$

$$\to 0 \quad \text{if} \quad h \to 0,$$

so that (4.1) holds. By (4.4) we also have, for $j = 1, 2$,

$$\frac{1}{h}\int_{|y-x|>\epsilon}|y-x|^j p(t, x, t+h, dy)$$
$$\leq \frac{1}{\epsilon^{2+\delta-j}}\cdot\frac{1}{h}\int_{R^n}|y-x|^{2+\delta}p(t, x, t+h, dy)\to 0$$

if $h\to 0$. Consequently (4.2), (4.3) hold if and only if (4.5), (4.6) hold.

Theorem 4.2. *Let* (A′) *hold and let* $b(x, t)$, $\sigma(x, t)$ *be continuous in* $(x, t)\in R^n\times[0, \infty)$. *Then the solution of* (1.1) *is a diffusion process with drift* $b(x, t)$ *and diffusion matrix* $a(x, t) = \sigma(x, t)\sigma^*(x, t)$.

Here σ^* is the transpose of σ. Thus, if $a = (a_{ij})$, then $a_{ij} = \sum_{k=1}^n \sigma_{ik}\sigma_{jk}$.

Proof. Since $p(t, x, t+h, A)$ is the probability distribution of the random variable $\xi_{x,t}(t+h)$,

$$Ef(\xi_{x,t}(t+h)-x) = \int_{R^n} f(y-x)p(t, x, t+h, dy) \tag{4.7}$$

for any continuous function $f(z)$ with $|f(z)| \leq K(1+|z|^\alpha)$ for some $K > 0$, $\alpha > 0$; we use here the fact that $E|\xi_{x,t}(t+h)|^\alpha < \infty$ for any $\alpha > 0$ (see Theorem 2.3).

In view of Lemma 4.1 it is therefore sufficient to prove that

$$\frac{1}{h}E|\xi_{x,t}(t+h)-x|^4\to 0 \quad \text{if} \quad h\to 0, \tag{4.8}$$

$$\frac{1}{h}E[\xi_{x,t}(t+h)-x]\to b(x, t) \quad \text{if} \quad h\to 0, \tag{4.9}$$

$$\frac{1}{h}E(\xi^i_{x,t}(t+h)-x)(\xi^j_{x,t}(t+h)-x)\to a_{ij}(x, t) \quad \text{if} \quad h\to 0, \tag{4.10}$$

where $\xi^i_{x,t}$ is the ith component of $\xi_{x,t}$.

From Theorem 2.3,

$$E|\xi_{x,t}(t+h)-x|^4 \leq Kh^2(1+|x|^4) \qquad (K \text{ constant});$$

this gives (4.8).

Next,

$$\frac{1}{h}E[\xi_{x,t}(t+h)-x] = \frac{1}{h}E\int_t^{t+h} b(\xi_{x,t}(\lambda), \lambda)\,d\lambda$$
$$= \int_0^1 Eb(\xi_{x,t}(t+hs), t+hs)\,ds.$$

Consider the random variables $X_h(s, \omega) = b(\xi_{x,t}(t + hs), t + hs)$. Since

$$\int_0^1 E|X_h|^2\, ds \leq C \int_0^1 E[1 + |\xi_{x,t}(t + hs)|^2]\, ds \leq C_1$$

where C, C_1 are constants independent of h, the random variables X_h satisfy the condition of uniform integrability, i.e.,

$$\sup_h \int_{|X_h|>\lambda} |X_h|\, dP \to 0 \qquad \text{if} \quad \lambda \to \infty. \tag{4.11}$$

We also have that $X_h(s, \omega) \xrightarrow{P} b(x, t)$ uniformly with respect to s. Hence, by Lemma 1.3.6,

$$\int_0^1 EX_h(s)\, ds \to b(x, t) \qquad \text{as} \quad h \to 0, \tag{4.12}$$

and the proof of (4.9) is complete.

To prove (4.10) we use Itô's formula (4.7.11) with $u(z, t) = z_i z_j$:

$$\begin{aligned}
&\frac{1}{h}\{E\xi^i_{x,t}(t + h)\xi^j_{x,t}(t + h) - x_i x_j\} \\
&= \frac{1}{h} E \int_t^{t+h} [\xi^i_{x,t}(\lambda) b_j(\xi_{x,t}(\lambda), \lambda) \\
&\quad + \xi^j_{x,t}(\lambda) b_i(\xi_{x,t}(\lambda), \lambda) + a_{ij}(\xi_{x,t}(\lambda), \lambda)]\, d\lambda,
\end{aligned} \tag{4.13}$$

where $x = (x_1, \ldots, x_n)$.

Noting that

$$\int_0^1 E[1 + |\xi_{x,t}(t + hs)|^4]\, ds \leq C_2$$

where C_2 is a constant independent of h, we can apply the argument used to prove (4.12) in order to deduce from (4.13) that, as $h \to 0$,

$$\frac{1}{h} E[\xi^i_{x,t}(t + h)\xi^j_{x,t}(t + h) - x_i x_j] \to x_i b_j(x, t) + x_j b_i(x, t) + a_{ij}(x, t).$$

It follows that

$$\begin{aligned}
&\lim_{h\to 0} \left\{ \frac{1}{h} E(\xi^i_{x,t}(t + h) - x_i)(\xi^j_{x,t}(t + h) - x_j) \right\} \\
&= x_i b_j(x, t) + x_j b_i(x, t) + a_{ij}(x, t) - x_i \lim_{h\to 0} \frac{1}{h} E[\xi^j_{x,t}(t + h) - x_j] \\
&\quad - x_j \lim_{h\to 0} \frac{1}{h} E[\xi^i_{x,t}(t + h) - x_i] \\
&= a_{ij}(x, t),
\end{aligned}$$

where (4.9) has been used in the last equality.

We shall use Theorem 4.2 in order to compute the infinitesimal generators $\mathcal{Q}_s$ of the Markov process solution of (1.1). By definition (see (2.3.7)),

$$(\mathcal{Q}_s f)(x) = \lim_{h \downarrow 0} \frac{(T_{s,s+h}f)(x) - f(x)}{h}$$

where

$$(T_{s,s+h}f)(x) = \int p(s, x, s+h, dy) f(y).$$

Thus,

$$(\mathcal{Q}_s f)(x) = \lim_{h \downarrow 0} \frac{1}{h} \int [f(y) - f(x)] p(s, x, s+h, dy). \tag{4.14}$$

Suppose f is bounded and twice continuously differentiable. Then

$$f(y) - f(x) = \sum_{i=1}^{n} (y_i - x_i) f_{x_i}(x)$$
$$+ \frac{1}{2} \sum_{i,j=1}^{n} (y_i - x_i)(y_j - x_j) f_{x_i x_j}(x) + o(|y - x|^2).$$

Substituting this in the integral in (4.14) for y in a small neighborhood of x, then taking $h \downarrow 0$ and using (4.1)–(4.3), we find that

$$(\mathcal{Q}_s f)(x) = \frac{1}{2} \sum_{i,j=1}^{n} a_{ij}(x, s) \frac{\partial^2 f}{\partial x_i \, \partial x_j} + \sum_{i,j=1}^{n} b_i(x, s) \frac{\partial f}{\partial x_i} \equiv (L_s f)(x). \tag{4.15}$$

The operator $L_s f$ is a partial differential operator. This same operator arises also in Itô's formula:

$$u(\xi(t), t) - u(\xi(s), s) = \int_s^t \left(\frac{\partial u}{\partial \lambda} + L_\lambda u \right)(\xi(\lambda), \lambda) \, d\lambda$$
$$+ \int_s^t u_x(\xi(\lambda), \lambda) \sigma(\xi(\lambda), \lambda) \, dw(\lambda) \tag{4.16}$$

where $\xi(t)$ is any solution of (1.1). We refer to L_s as the differential operator corresponding to the stochastic system (1.1). Using (4.16) with $t = \tau$, one can easily derive Dynkin's formula (2.3.6) when f is twice continuously differen tiable.

5. Equations depending on a parameter

Let $A(x, t)$, $B(x, t)$ be a random n-vector and $n \times n$ matrix respectively, defined for $(x, t) \in R^n \times [s, \infty)$ for some $s > 0$ and assume:

(i) $A(x, t)$, $B(x, t)$ are continuous in (x, t), for each ω;
(ii) $A(x, t)$, $B(x, t)$ are measurable in (x, t, ω);
(iii) $A(x, t)$, $B(x, t)$ are $\mathcal{F}_t$ measurable for each (x, t), where $\mathcal{F}_t$ is an

increasing family of σ-fields such that $w(t)$ is $\mathcal{F}_t$ measurable and $\mathcal{F}(w(t+\lambda)-w(t), \lambda \geq 0)$ is independent of $\mathcal{F}_t$ for all $t \geq 0$;

(iv) there is a constant K such that

$$|A(x, t)| \leq K(1+|x|), \qquad |B(x, t)| \leq K(1+|x|) \quad \text{a.s.,} \tag{5.1}$$

$$|A(x, t) - A(\bar{x}, t)| \leq K|x-\bar{x}|, \qquad |B(x, t) - B(\bar{x}, t)| \leq K|x-\bar{x}| \quad \text{a.s.} \tag{5.2}$$

Let $\phi(t)$ be a function in $M_w^\infty[s, T]$. Consider the equation

$$\xi(t) = \phi(t) + \int_s^t A(\xi(\lambda), \lambda)\, d\lambda + \int_s^t B(\xi(\lambda), \lambda)\, dw(\lambda). \tag{5.3}$$

We refer to it as a *system of stochastic differential equations with random coefficients*.

Theorem 5.1. *If* (i)–(iv) *hold and* $\phi \in M_w^\infty[s, T]$, *then there exists a unique solution* $\xi(t)$ *in* $M_w^2[s, T]$; *further,* $\xi(t)$ *belongs to* $M_w^\infty[s, T]$.

We outline the proof leaving the details to the reader.

One defines by induction

$$\xi_m(t) = \phi(t) + \int_s^t A(\xi_{m-1}(\lambda), \lambda)\, d\lambda + \int_s^t B(\xi_{m-1}(\lambda), \lambda)\, dw(\lambda),$$

$$\xi_0(t) = \phi(t),$$

and shows that the ξ_m are well defined and

$$E|\xi_m(t) - \xi_{m-1}(t)|^2 \leq K_0 \int_s^t E|\xi_{m-1}(s) - \xi_{m-2}(s)|^2\, ds.$$

Since

$$E|\xi_1(t) - \xi_0(t)|^2 \leq K_1 \int_s^t E(1 + |\phi(\lambda)|^2)\, d\lambda,$$

$$E|\xi_m(t) - \xi_{m-1}(t)|^2 \leq \frac{L^m(t-s)^m}{m!} \qquad (L \text{ constant}).$$

Now we proceed as in the proof of Theorem 1.1 to show that $\lim \xi_m(t)$ is a solution. The proof of uniqueness is the same as for Theorem 1.1.

Theorem 5.2. *Let* $A_\alpha(x, t)$, $B_\alpha(x, t)$, $\phi_\alpha(t)$ *satisfy the assumptions of Theorem* 5.1 *for any* $0 \leq \alpha \leq 1$, *with the constant* K (*in* (5.1), (5.2)) *independent of* α, *and with* $\sup_{s \leq t \leq T} E|\phi_\alpha(t)|^2 \leq C$, C *a constant indepen-*

dent of α. Suppose that for any $N > 0$, $t \in [s, T]$, $\epsilon > 0$,

$$\lim_{\alpha \downarrow 0} P\left\{ \sup_{|x| \leq N} |A_\alpha(x, t) - A_0(x, t)| > \epsilon \right\} = 0, \tag{5.4}$$

$$\lim_{\alpha \downarrow 0} P\left\{ \sup_{|x| \leq N} |B_\alpha(x, t) - B_0(x, t)| > \epsilon \right\} = 0. \tag{5.5}$$

Suppose also that

$$\lim_{\alpha \downarrow 0} \sup_{s \leq t \leq T} E|\phi_\alpha(t) - \phi_0(t)|^2 = 0. \tag{5.6}$$

Consider the solutions $\xi_\alpha(t)$ of the equations

$$\xi_\alpha(t) = \phi_\alpha(t) + \int_s^t A_\alpha(\xi_\alpha(\lambda), \lambda)\, d\lambda + \int_s^t B_\alpha(\xi_\alpha(\lambda), \lambda)\, dw(\lambda). \tag{5.7}$$

Then,

$$\sup_{s \leq t \leq T} E|\xi_\alpha(t) - \xi_0(t)|^2 \to 0 \qquad \text{if} \quad \alpha \downarrow 0. \tag{5.8}$$

Proof. Take for simplicity $s = 0$. We can write

$$\xi_\alpha(t) - \xi_0(t) = \eta_\alpha(t) + \int_0^t [A_\alpha(\xi_\alpha(s), s) - A_\alpha(\xi_0(s), s)]\, ds$$
$$+ \int_0^t [B_\alpha(\xi_\alpha(s), s) - B_\alpha(\xi_0(s), s)]\, dw(s)$$

where

$$\eta_\alpha(t) = \phi_\alpha(t) - \phi_0(t) + \int_0^t [A_\alpha(\xi_0(s), s) - A_0(\xi_0(s), s)]\, ds$$
$$+ \int_0^t [B_\alpha(\xi_0(s), s) - B_0(\xi_0(s), s)]\, dw(s).$$

Using (5.1), (5.2) for A_α, B_α, we easily find that

$$E|\xi_\alpha(t) - \xi_0(t)|^2 \leq 3E|\eta_\alpha(t)|^2 + C \int_0^t E|\xi_\alpha(s) - \xi_0(s)|^2\, ds. \tag{5.9}$$

If we prove that

$$\sup_{0 \leq t \leq T} E|\eta_\alpha(t)|^2 \to 0 \qquad \text{as} \quad \alpha \to 0, \tag{5.10}$$

then the assertion (5.8) follows from (5.9).

Now,

$$I_\alpha \equiv E\left| \int_0^t [A_\alpha(\xi_0(s), s) - A_0(\xi_0(s), s)]\, ds \right|^2$$
$$\leq tE \int_0^t |A_\alpha(\xi_0(s), s) - A_0(\xi_0(s), s)|^2\, ds.$$

The last integrand is bounded by $2K(1 + |\xi_0(s)|^2)$, which is an integrable function. In view of (5.4), this integrand also converges to 0, in probability, as $\alpha \to 0$. Hence, by the Lebesgue bounded convergence theorem, $I_\alpha \to 0$ if $\alpha \to 0$.

Next,

$$J_\alpha \equiv E\left| \int_0^t [B_\alpha(\xi_0(s), s) - B_0(\xi_0(s), s)]\, dw \right|^2$$
$$= E \int_0^t |B_\alpha(\xi_0(s), s) - B_0(\xi_0(s), s)|^2\, ds.$$

By the previous argument, $J_\alpha \to 0$ if $\alpha \to 0$.

Finally, making use also of (5.6), the assertion (5.10) follows.

We shall apply Theorems 5.1, 5.2 in studying the behavior of the solution $\xi_{x,s}(t)$ in the parameters x, s. Recall that

$$\xi_{x,s}(t) = x + \int_s^t b(\xi_{x,s}(\lambda), \lambda)\, d\lambda + \int_s^t \sigma(\xi_{x,s}(\lambda), \lambda)\, dw(\lambda). \qquad (5.11)$$

We shall use the notation

$$D_x^\alpha = \frac{\partial^\alpha}{\partial x^\alpha} = \frac{\partial^{\alpha_1 + \cdots + \alpha_n}}{\partial x_1^{\alpha_1} \cdots \partial x_n^{\alpha_n}}, \qquad |\alpha| = \alpha_1 + \cdots + \alpha_n.$$

If $f = f(x, t)$, then the vector $(\partial f/\partial x_1, \ldots, \partial f/\partial x_n)$ is denoted briefly by $D_x f$ or by f_x.

Definition. Let $g(x) = g(x_1, \ldots, x_n)$, $f(x) = f(x_1, \ldots, x_n)$ be random functions for x in some open set. If

$$\int \left| \frac{1}{h} [g(x_1, \ldots, x_{i-1}, x_i + h, x_{i+1}, \ldots, x_n) - g(x_1, \ldots, x_n)] \right.$$
$$\left. - f(x_1, \ldots, x_n) \right|^2 dP \to 0$$

as $h \to 0$, then we say that $g(x)$ has a derivative with respect to x_i in the $L^2(\Omega)$ sense, and the derivative is equal to $f(x)$. We write $(\partial/\partial x_i)g(x) = f(x)$. Similarly one defines the derivative $D^\alpha g(x)$ in the $L^2(\Omega)$ sense, for any $\alpha = (\alpha_1, \ldots, \alpha_n)$.

We shall need the following condition:

$$\partial b/\partial x_i, \quad \partial\sigma/\partial x_i \quad \text{exist and are continuous} \qquad (1 \le i \le n). \qquad (5.12)$$

Theorem 5.3. *If* (A) *and* (5.12) *hold, then the derivatives* $\partial \xi_{x,s}(t)/\partial x_i$ *exist in the* $L^2(\Omega)$ *sense and the functions* $\zeta_i(t) = \partial \xi_{x,s}(t)/\partial x_i$ *satisfy the* *stochastic differential system with random coefficients*

$$\zeta_i(t) = e_i + \int_s^t \zeta_i(\lambda) \cdot b_x(\xi_{x,s}(\lambda), \lambda)\, d\lambda + \int_s \zeta_i(\lambda) \cdot \sigma_x(\xi_{x,s}(\lambda), \lambda)\, dw(\lambda) \tag{5.13}$$

where e_i *is the vector with components* δ_{ij}.

Proof. Take for simplicity $i = 1$, and write $h = (h_1, 0, \ldots, 0)$ where $h_1 \neq 0$. Then

$$\frac{1}{h_1}[\xi_{x+h,s}(t) - \xi_{x,s}(t)] = e_1 + \int_s^t [b(\xi_{x+h,s}(\lambda), \lambda) - b(\xi_{x,s}(\lambda), \lambda)]\, d\lambda$$
$$+ \int_s^t [\sigma(\xi_{x+h,s}(\lambda), \lambda) - \sigma(\xi_{x,s}(\lambda), \lambda)]\, dw(\lambda). \tag{5.14}$$

By Theorem 5.1 there exists a unique solution ζ_1 of (5.13) with $i = 1$. We shall now complete the proof by invoking Theorem 5.2 with $\alpha = h_1$ if $h_1 > 0$ and $\alpha = -h_1$ if $h_1 < 0$. First we note that

$$\frac{1}{h_1}\int_s^t [b(\xi_{x+h,s}(\lambda), \lambda) - b(\xi_{x,s}(\lambda), \lambda)]\, d\lambda$$
$$= \frac{1}{h_1}\int_s^t d\lambda \int_0^1 \frac{d}{d\mu} b(\xi_{x,s}(\lambda) + \mu(\xi_{x+h,s}(\lambda) - \xi_{x,s}(\lambda)), \lambda)\, d\mu$$
$$= \int_s^t \left[\int_0^1 b_x(\xi_{x,s}(\lambda) + \mu(\xi_{x+h,s}(\lambda) - \xi_{x,s}(\lambda)), \lambda)\, d\mu\right]$$
$$\cdot \frac{\xi_{x+h,s}(\lambda) - \xi_{x,s}(\lambda)}{h_1}\, d\lambda.$$

A similar formula holds for the stochastic integral on the right-hand side of (5.14). Therefore, the function

$$\xi_\alpha(t) = \frac{1}{h_1}[\xi_{x+h,s}(t) - \xi_{x,s}(t)]$$

satisfies (5.7) with

$$A_\alpha(z, t) = z \cdot \int_0^1 b_x(\xi_{x,s}(t) + \mu(\xi_{x+h,s}(t) - \xi_{x,s}(t)), t)\, d\mu,$$

$$B_\alpha(z, t) = z \cdot \int_0^1 \sigma_x(\xi_{x,s}(t) + \mu(\xi_{x+h,s}(t) - \xi_{x,s}(t)), t)\, d\mu$$

if $\alpha \neq 0$, and $\xi_0(t) \equiv \zeta_1(t)$ satisfies the same equation with

$$A_0(z, t) = z \cdot b_x(\xi_{x,s}(t), t), \qquad B_0(z, t) = z \cdot \sigma_x(\xi_{x,s}(t), t).$$

Since, by Lemma 3.3,

$$E \sup_t |\xi_{x+h,s}(t) - \xi_{x,s}(t)| \to 0 \qquad \text{if} \quad h \to 0,$$

we conclude, upon using (5.12), that (5.4), (5.5) are satisfied. The assertion of the theorem now follows from Theorem 5.2.

Remark. Notice that $\partial \xi_{x,s}(t)/\partial x_i$ satisfies the stochastic differential system with random coefficients obtained by differentiating formally the stochastic differential system of $\xi_{x,s}(t)$ with respect to x_i.

Theorem 5.4. *Let* (A) *hold and assume that* $D_x^\alpha b(x, t)$, $D_x^\alpha \sigma(x, t)$ *exist and are continuous if* $|\alpha| \leq 2$, *and*

$$|D_x^\alpha b(x, t)| + |D_x^\alpha \sigma(x, t)| \leq K_0(1 + |x|^\beta) \qquad (|\alpha| \leq 2) \tag{5.15}$$

where K_0, β *are positive constants. Then the second derivatives* $\partial^2 \xi_{x,s}(t)/\partial x_i \, \partial x_j$ *exist in the* $L^2(\Omega)$ *sense, and they satisfy the stochastic differential system with random coefficients obtained by applying formally* $\partial^2/\partial x_i \, \partial x_j$ *to* (5.11).

The proof is left to the reader.

Theorem 5.5. *Let* $f(x)$ *be a function with two continuous derivatives, satisfying*

$$|D_x^\alpha f(x)| \leq C(1 + |x|^\beta) \qquad (|\alpha| \leq 2) \tag{5.16}$$

where C, β *are positive constants. Let the conditions of Theorem 5.4 hold, and set*

$$\phi(x) = Ef(\xi_{x,s}(t)). \tag{5.17}$$

Then $\phi(x)$ *has two continuous derivatives; these derivatives can be computed by differentiating the right-hand side of* (5.17) *under the integral sign. Finally,*

$$|D^\alpha \phi(x)| \leq C_0(1 + |x|^\gamma) \qquad \text{if} \quad |\alpha| \leq 2, \tag{5.18}$$

where C_0, γ *are positive constants.*

Proof. We shall prove that

$$\frac{\partial \phi}{\partial x_i} = Ef_x(\xi_{x,s}(t)) \cdot \frac{\partial}{\partial x_i} \xi_{x,s}(t). \tag{5.19}$$

Take for simplicity $i = 1$ and set $h = (h_1, 0, \ldots, 0)$. Then

$$\begin{aligned} \phi(x+h) - \phi(x) &= E[f(\xi_{x+h,s}(t)) - f(\xi_{x,s}(t))] \\ &= E \int_0^1 \frac{d}{d\mu} f(\xi_{x,s}(t) + \mu(\xi_{x+h,s}(t) - \xi_{x,s}(t)))\, d\mu \\ &= E \int_0^1 f_x(\xi_{x,s}(t) + \mu(\xi_{x+h,s}(t) - \xi_{x,s}(t))) \cdot (\xi_{x+h,s}(t) - \xi_{x,s}(t))\, d\mu. \end{aligned}$$

Hence

$$\frac{\phi(x+h)-\phi(x)}{h_1} = E\int_0^1 f_x(\xi_{x,s}(t) + \mu(\xi_{x+h,s}(t)-\xi_{x,s}(t)))\,d\mu \cdot \frac{\xi_{x+h,s}(t)-\xi_{x,s}(t)}{h_1}. \tag{5.20}$$

As $h \to 0$

$$\frac{\xi_{x+h,s}(t)-\xi_{x,s}(t)}{h_1} \to \frac{\partial}{\partial x_1}\xi_{x,s}(t) \quad \text{in } L^2(\Omega). \tag{5.21}$$

We claim that also

$$\psi_h \equiv \int_0^1 f_x(\xi_{x,s}(t) + \mu(\xi_{x+h,s}(t)-\xi_{x,s}(t)))\,d\mu \to f_x(\xi_{x,s}(t)) \quad \text{in } L^2(\Omega) \tag{5.22}$$

as $h \to 0$. Indeed, this follows from Lemma 1.3.6 if we observe that $|\psi_h - f_x(\xi_{x,s}(t)|^2 \to 0$ in probability and

$$E|\psi_h - f_x(\xi_{x,s}(t))|^4 \leq C_1$$

where C_1 is a constant independent of h; the assumption (5.16) is used here.

From (5.20), (5.21), and (5.22) we see that

$$\frac{\phi(x+h)-\phi(x)}{h_1} \to Ef_x(\xi_{x,s}(t)) \cdot \frac{\partial}{\partial x_1}\xi_{x,s}(t) \quad \text{as } h \to 0.$$

Having proved (5.19), one can now check that the right-hand side of (5.19) is a continuous function in x and is bounded by $C_0(1+|x|^\gamma)$ for some positive constants C_0, γ. The same is then true of $\partial\phi/\partial x_i$. The proof that $\phi(x)$ has second derivatives, that these derivatives are obtained by differentiating under the integral sign of (5.17), and that they are continuous and satisfy (5.18), is similar to the proof for the first derivatives; this is left to the reader.

Theorems 5.4, 5.5 extend without difficulty to higher derivatives.

6. The Kolmogorov equation

Consider the function

$$u(x, t) = Ef(\xi_{x,t}(T)).$$

Notice that

$$u(x, t) = E_{x,t}f(\xi(T))$$

where $\{\mathcal{C}, \mathfrak{M}, \mathfrak{M}_t^s, \xi(t), P_{x,s}\}$ is the Markov process that solves the system of stochastic differential equations (1.1).

Theorem 6.1. *If f and b, σ satisfy the conditions of Theorem 5.5, then u_t, u_x, u_{xx} are continuous functions in $(x, t) \in R^n \times [0, T)$ and satisfy*

$$\frac{\partial u}{\partial t} + \sum_{i=1}^{n} b_i \frac{\partial u}{\partial x_i} + \frac{1}{2} \sum_{i,j=1}^{n} a_{ij} \frac{\partial^2 u}{\partial x_i \partial x_j} = 0 \quad \text{in} \quad R^n \times [0, T) \tag{6.1}$$

$$u(x, t) \to f(x) \quad \text{if} \quad t \uparrow T. \tag{6.2}$$

Equation (6.2) is called the *Kolmogorov equation*, or the *backward parabolic equation*.

Proof. Let $g(x)$ be a twice continuously differentiable function satisfying $|D^\alpha g(x)| \leq C(1 + |x|^\beta)$ if $|\alpha| \leq 2$, where C, β are positive constants. By Itô's formula (see (4.16))

$$Eg(\xi_{x,t-h}(t)) - g(x) = E \int_{t-h}^{t} (L_s g)(\xi_{x,t-h}(s), s)\, ds \qquad (h > 0) \tag{6.3}$$

where L_s is the partial differential operator defined in (4.15). The argument used to derive (4.12) can be applied also here to deduce that

$$\frac{1}{h} E \int_{t-h}^{t} (L_s g)(\xi_{x,t-h}(s), s)\, ds \to (L_t g)(x, t) \qquad \text{as} \quad h \to 0.$$

Hence, if we divide both sides of (6.3) by h and let $h \to 0$, we get the formula

$$\lim_{h \downarrow 0} \frac{1}{h} E[g(\xi_{x,t-h}(t)) - g(x)] = (L_t g)(x, t). \tag{6.4}$$

Next, by the Markov property,

$$\begin{aligned} u(x, t-h) &= E_{x,t-h} f(\xi(T)) = E_{x,t-h} E_{x,t-h}(f(\xi(T)) | \mathfrak{M}_t^{t-h}) \\ &= E_{x,t-h} E_{\xi(t),t} f(\xi(T)) = E_{x,t-h} u(\xi(t), t) = Eu(\xi_{x,t-h}(t), t), \end{aligned}$$

so that

$$\frac{u(x, t) - u(x, t-h)}{h} = - \frac{Eu(\xi_{x,t-h}(t), t) - u(x, t)}{h}. \tag{6.5}$$

In view of Theorem 5.5, the function $g(x) = u(x, t)$ has two continuous x-derivatives, bounded by $C_0(1 + |x|^\gamma)$, for some positive constants C_0, γ. Consequently the formula (6.4) can be applied. From (6.5) we then conclude that

$$\lim_{h \downarrow 0} \frac{u(x, t) - u(x, t-h)}{h} = - L_t u(x, t). \tag{6.6}$$

From (6.3) one easily deduces that

$$\frac{1}{h} |Eg(\xi_{x,t-h}(t)) - g(x)| \leq C$$

where C is a constant independent of h. Applying this to the function

$g(x) = u(x, t)$ and then using (6.5), we get

$$|u(x, t) - u(x, t - h)| \leq Ch.$$

It follows that, for fixed x, $u(x, t)$ is absolutely continuous in t. Therefore, $\partial u(x, s)/\partial s$ exists for almost all s and

$$u(x, t) = u(x, 0) + \int_0^t \frac{\partial u(x, s)}{\partial s}\, ds.$$

In view of (6.6),

$$u(x, t) = u(x, 0) - \int_0^t L_s u(x, s)\, ds.$$

But since $L_s u(x, s)$ is a continuous function, it follows that $\partial u(x, t)/\partial t$ exists everywhere and (6.1) holds.

Next,

$$u(x, t) - f(x) = Ef(\xi_{x, t}(T)) - Ef(\xi_{x, T}(T)).$$

Using Lemma 3.3 and Lemma 1.3.6, we find that the right-hand side converges to 0 if $t \uparrow T$. This proves (6.2).

PROBLEMS

1. In Theorem 2.2, replace (2.5) by the inequalities

$$x \cdot b(x, t) \leq K(1 + |x|^2), \qquad |\sigma(x)| \leq K,$$

and prove that the assertions of Theorem 2.2 are still valid. [*Hint*: Apply Itô's formula to $\xi_N^2(t)$. Prove that

$$E \sup_{0 \leq t \leq T} |\xi_N(t)|^2 \leq C,$$

C independent of N.]

2. Prove that if (A′) holds, then for every $x \in R^n$, $s \geq 0$,

$$E|\xi_{y, 0}(t) - \xi_{x, 0}(s)|^2 \leq \eta(|y - x|^2 + |t - s|)$$

where $\eta(r) \to 0$ if $r \to 0$. [*Hint*: Use (2.7).]

3. Prove Theorem 3.6.

4. If $f \in L_w^2[0, T]$ and τ is a stopping time, $0 \leq \tau \leq T$, then

$$\int_\tau^T f(t)\, dw(t) = \int_0^{T-\tau} f(\tau + s)\, d\hat{w}(s)$$

where $\hat{w}(s) = w(s + \tau) - w(\tau)$.

5. If $\{\Omega, \mathcal{F}, \mathcal{F}_t^s, \xi(t), P_{x,s}\}$ is a Markov process with transition probability $p(s, x, t, A)$ and if $g(x, t)$ is a continuous function in $(x, t) \in R^1 \times [0, \infty)$, strictly monotone in x, then $\{\Omega, \mathcal{F}, \mathcal{F}_t^s, g(\xi(t), t), P_{x,s}\}$ is a Markov process with the transition probability $\tilde{p}(s, x, t, A) = p(s, g^{-1}(x, s), t, g^{-1}(A, t))$

where $g^{-1}(x, t)$ is the inverse function to $g(x, t)$.

6. If the Markov process in the preceding problem is a diffusion process with drift $b(x, t)$ and diffusion coefficient $a(x, t)$ and if g_t, g_x, g_{xx} are continuous and $g_x(x, t) \neq 0$ for all (x, t), then the process $\{\Omega, \mathcal{F}, \mathcal{F}_t^s, g(\xi(t), t), P_{x,s}\}$ is also a diffusion process with drift and diffusion coefficients given by:

$$\tilde{b}(x, t) = g_t(y, t) + b(y, t)g_x(y, t) + \tfrac{1}{2}a(y, t)g_{xx}(y, t),$$

$$\tilde{a}(x, t) = a(y, t)(g_x(y, t))^2, \qquad y = g^{-1}(x, t).$$

7. Complete the details of the proof of Theorem 5.1.
8. Prove Theorem 5.4.
9. Complete the proof of Theorem 5.5.
10. Consider a linear stochastic differential equation

$$d\xi(t) = [\alpha(t) + \beta(t)\xi(t)]\,dt + [\gamma(t) + \delta(t)\xi(t)]\,dw(t) \tag{6.7}$$

where the functions α, β, γ, δ are bounded and measurable. Prove:

(i) If $\alpha \equiv 0$, $\gamma \equiv 0$, then the solution $\xi = \xi_0(t)$ is given by

$$\xi_0(t) = \xi_0(0)\exp\left\{\int_0^t [\beta(s) - \tfrac{1}{2}\delta^2(s)]\,ds + \int_0^t \delta(s)\,dw(s)\right\}.$$

(ii) Setting $\xi(t) = \xi_0(t)\zeta(t)$ show that $\xi(t)$ is a solution of (6.7) if and only if

$$\zeta(t) = \zeta(0) + \int_0^t [\xi_0(s)\alpha(s) - \gamma(s)\delta(s)]\,ds + \int_0^t \gamma(s)\xi_0(s)\,ds.$$

Thus the solution of (6.7) is $\xi_0(t)\zeta(t)$ with $\xi(0) = \xi_0(0)\zeta(0)$.

11. If an equation of the form $d\xi = b(x, t)\,dt + \sigma(x, t)\,dw(t)$ can be reduced by a transformation $\eta(t) = f(\xi(t), t)$ to an equation of the form $d\eta(t) = \beta(t)\,dt + \delta(t)\,dw(t)$, then

$$\frac{\partial}{\partial x}\left\{\sigma\left[\frac{\sigma_t}{\sigma^2} - \frac{\partial}{\partial x}\left(\frac{b}{\sigma}\right) + \tfrac{1}{2}\sigma_{xx}\right]\right\} = 0.$$

12. Let $g(x, t)$, $h(x, t)$ be continuous functions in $(x, t) \in R^1 \times [0, T]$ together with their first two x-derivatives, and assume that

$$|D_x^\alpha g(x, t)| + |D_x^\alpha h(x, t)| \leq K_0(1 + |x|^\beta) \qquad \text{if} \quad |\alpha| \leq 2,$$

where K_0, β are positive constants. Let the conditions of Theorem 5.5 also hold, and consider the function

$$u(x, t) = E\exp\left\{i\lambda\int_t^T g(\xi_{x,t}(s), s)\,ds + i\mu\int_t^T h(\xi_{x,t}(s), s)\,dw(s)\right\}f(\xi_{x,t}(T)).$$

Prove that

$$\frac{\partial u}{\partial t} + Lu + i\mu h u_x + \left(i\lambda g - \tfrac{1}{2}\mu^2 h^2\right)u = 0 \qquad \text{if} \quad x \in R^1, \quad 0 \leqslant t < T,$$

$$u(x, t) \to f(x) \qquad \text{if} \quad t \uparrow T.$$

13. Verify that all the results of this chapter remain true for a system of n stochastic differential equations

$$d\xi_i = \sum_{j=1}^{l} \sigma_{ij}(\xi, t)\, dw_j + b_i(\xi, t)\, dt \qquad (1 \leqslant i \leqslant n)$$

where $(w_1, \ldots, w_l)$ is an l-dimensional Brownian motion.

6

Elliptic and Parabolic Partial Differential Equations and Their Relations to Stochastic Differential Equations

1. Square root of a nonnegative definite matrix

As is well known, if a is an $n \times n$ nonnegative definite matrix, then it has a unique nonnegative square root σ, i.e., there is a unique nonnegative definite matrix σ such that $\sigma\sigma = a$. If a depends on a parameter x, say $a = a(x)$, then also $\sigma = \sigma(x)$. We shall be concerned here with the smoothness of $\sigma(x)$ in x, assuming that $a(x)$ varies smoothly with x.

Set $a(x) = (a_{ij}(x))$, $\sigma(x) = (\sigma_{ij}(x))$. The point x varies in an open set G of R^ν.

We shall denote by $C^m(G)$ the class of all functions $f(x)$ having continuous derivatives up to order m in G. If the mth derivatives of $f(x)$ satisfy a Hölder condition with exponent α in compact subsets of G, then we say that f belongs to $C^{m+\alpha}(G)$; here $0 < \alpha \leqslant 1$.

Lemma 1.1. *If $a(x)$ is positive definite for all $x \in G$ and if $a_{ij} \in C^{m+\alpha}(G)$ for all i, j (or if $a_{ij} \in C^m(G)$ for all i, j), then the elements of $\sigma(x)$ belong to $C^{m+\alpha}(G)$ (or to $C^m(G)$).*

Proof. Let G_0 be an open bounded set with closure in G. Let Γ be a smooth contour lying in the half-plane Re $z > 0$ of the complex plane and containing all the eigenvalues of $a(x)$, $x \in G_0$. We claim that if $x \in G_0$, then

$$\sigma(x) = \frac{1}{2\pi} \int_\Gamma \sqrt{z}\,(a(x) - zI)^{-1}\,dz \qquad (I = \text{identity matrix}). \quad (1.1)$$

To prove it, let Γ' be another smooth contour in Re $z > 0$ whose interior contains Γ. Then

$$\begin{aligned}\sigma^2(x) &= \frac{1}{4\pi^2}\int_{\Gamma'}\int_{\Gamma}\sqrt{z}\sqrt{\zeta}\,(a(x)-\zeta I)^{-1}(a(x)-zI)^{-1}\,dz\,d\zeta\\ &= \frac{1}{4\pi^2}\int_{\Gamma'}\left\{\int_{\Gamma}\sqrt{z}\sqrt{\zeta}\,\frac{(a(x)-\zeta I)^{-1}-(a(x)-zI)^{-1}}{\zeta-z}\,dz\right\}d\zeta\\ &= -\frac{1}{4\pi^2}\int_{\Gamma'}\int_{\Gamma}\frac{\sqrt{z}\sqrt{\zeta}\,(a(x)-zI)^{-1}}{\zeta-z}\,dz\,d\zeta\end{aligned}$$

by Cauchy's theorem. Changing the order of integration and using Cauchy's theorem, we get

$$\begin{aligned}\sigma^2(x) &= -\frac{1}{4\pi^2}\int_{\Gamma}(a(x)-zI)^{-1}\sqrt{z}\int_{\Gamma'}\frac{\sqrt{\zeta}}{\zeta-z}\,d\zeta\,dz\\ &= \frac{1}{2\pi i}\int_{\Gamma}(a(x)-zI)^{-1}z\,dz = \frac{1}{2\pi i}\int_{\Gamma}(a(x)-zI)^{-1}a(x)\,dz.\end{aligned}$$

Now modify Γ into the disk $\Gamma_R = \{z;\ |z| = R\}$, R large. Since

$$\|(a(x)-zI)^{-1}-(zI)^{-1}\| < \frac{C}{1+|z|^2} \qquad \text{if} \quad |z| = R,$$

$$\left|\int_{\Gamma_R}\left[(a(x)-zI)^{-1}-(zI)^{-1}\right]dz\right| < \frac{2\pi CR}{1+R^2} \to 0 \qquad \text{if} \quad R\to\infty.$$

It follows that

$$\sigma^2(x) = \lim_{R\to\infty}\left\{\frac{1}{2\pi i}\int_{\Gamma_R}a(x)\,\frac{dz}{z}\right\} = a(x).$$

From (1.1) it is obvious that the elements $\sigma(x)$ are as smooth in G_0 as the elements of $a(x)$. Since G_0 is arbitrary, the proof of the lemma is complete.

Theorem 1.2. *If $a(x)$ is nonnegative definite for all $x\in R^\nu$ and if the $a_{ij}(x)$ belong to $C^2(R^\nu)$, then the $\sigma_{ij}(x)$ are Lipschitz continuous in compact subsets. If, further,*

$$\sup_{x\in R^\nu}\sup_{i,j,k,l}\left|\frac{\partial^2 a_{ij}(x)}{\partial x_k\,\partial x_l}\right| \le M, \tag{1.2}$$

then

$$|\sigma_{ij}(x)-\sigma_{ij}(y)| \le 2\nu\sqrt{\nu}\,\sqrt{M}\,|x-y|. \tag{1.3}$$

Proof. Suppose first that $a(x)$ is positive definite and (1.2) holds. Consider the function $\langle a(x)\xi, \xi\rangle$ where $\langle\ ,\ \rangle$ denotes the scalar product. By Taylor's formula,

$$0 \leq \langle a(x_1, \ldots, x_{k-1}, x_k + h, x_{k+1}, \ldots, x_\nu)\xi, \xi\rangle$$
$$= \langle a(x)\xi, \xi\rangle + h\left\langle \frac{\partial a(x)}{\partial x_k}\xi, \xi\right\rangle + \frac{h^2}{2}\left\langle \frac{\partial^2 a(\tilde{x})}{\partial x_k^2}\xi, \xi\right\rangle$$

for a suitable point $\tilde{x}$. Using (1.2) we get

$$0 \leq \langle a(x)\xi, \xi\rangle + h\left\langle \frac{\partial a(x)}{\partial x_k}\xi, \xi\right\rangle + \frac{h^2}{2}\nu M|\xi|^2.$$

Since the right-hand side is a quadratic polynomial in h taking only nonnegative values,

$$\left\langle \frac{\partial a(x)}{\partial x_k}\xi, \xi\right\rangle^2 \leq 2\nu M|\xi|^2\langle a(x)\xi, \xi\rangle. \tag{1.4}$$

Denoting by $'$ the derivative with respect to x_k and differentiating $\sigma^2(x) = a(x)$ with respect to x_k, we obtain

$$a'(x) = \sigma(x)\sigma'(x) + \sigma'(x)\sigma(x). \tag{1.5}$$

Let $T(x)$ be an orthogonal matrix such that $T(x)\sigma(x)T^{-1}(x)$ is a diagonal matrix $\Lambda(x)$. Multiplying (1.5) by $T(x)$ on the left and by $T^{-1}(x)$ on the right, we get

$$\bar{a}'(x) \equiv T(x)a'(x)T^{-1}(x) = Y(x)\Lambda(x) + \Lambda(x)Y(x) \tag{1.6}$$

where $Y(x) = (y_{ij}(x)) = T(x)\sigma'(x)T^{-1}(x)$. Denoting by $\lambda_i(x)$ the eigenvalues of $\sigma(x)$, we obtain from (1.6),

$$y_{ij} = \frac{\bar{a}'_{ij}}{\lambda_i + \lambda_j} \qquad \text{where} \quad \bar{a}'(x) = (\bar{a}'_{ij}(x)). \tag{1.7}$$

Using the estimate (1.4) we find that

$$\begin{aligned}\langle \bar{a}'(x)\xi, \xi\rangle^2 &= \langle T(x)a'(x)T^{-1}(x)\xi, \xi\rangle^2 = \langle a'(x)T^{-1}(x)\xi, T^{-1}(x)\xi\rangle^2 \\ &\leq 2\nu M|T^{-1}(x)\xi|^2\langle a(x)T^{-1}(x)\xi, T^{-1}(x)\xi\rangle = 2\nu M|\xi|^2\langle\Lambda^2\xi, \xi\rangle.\end{aligned} \tag{1.8}$$

Taking $\xi = (\xi_1, \ldots, \xi_\nu)$ with $\xi_j = 0$ if $j \neq i$, $\xi_i = 1$, we get

$$|\bar{a}'_{ii}(x)|^2 \leq 2\nu M\lambda_i^2(x). \tag{1.9}$$

Next, taking in (1.8) $\xi = (\xi_1, \ldots, \xi_\nu)$ with $\xi_k = 0$ if $k \neq i$, $k \neq j$ (where $i \neq j$) and $\xi_i = \xi_j = 1$, we find that

$$|\bar{a}'_{ii}(x) + 2\bar{a}'_{ij}(x) + \bar{a}'_{jj}(x)|^2 \leq 4\nu M(\lambda_i^2(x) + \lambda_j^2(x)).$$

Hence

$$\left[\frac{\bar{a}'_{ii}(x)}{\lambda_i+\lambda_j}+\frac{2\bar{a}'_{ij}(x)}{\lambda_i+\lambda_j}+\frac{\bar{a}'_{jj}(x)}{\lambda_i+\lambda_j}\right]^2 \leqslant 4\nu M.$$

Recalling (1.9) we get

$$2\left|\frac{\bar{a}'_{ij}(x)}{\lambda_i+\lambda_j}\right| \leqslant \sqrt{2\nu M}+2\sqrt{\nu M}\,. \tag{1.10}$$

Substituting (1.9), (1.10) into (1.7), we obtain

$$|y_{il}(x)| < 2\sqrt{\nu M} \qquad \text{if} \quad 1 \leqslant i, \quad l \leqslant \nu. \tag{1.11}$$

Since $(\sigma'_{ij}(x)) = T^{-1}(x)Y(x)T(x)$ and $T(x) = (T_{ij}(x))$ is orthogonal,

$$|\sigma'_{ij}(x)| \leqslant \left\{\max_{k,l}|y_{kl}(x)|\right\}\sum_{k,l}|T_{ik}(x)T_{lj}(x)| \leqslant 2\nu\sqrt{\nu M}\,. \tag{1.12}$$

Suppose now that $a(x)$ is nonnegative definite and (1.2) holds. Then the matrix $a(x)+\epsilon I$ ($\epsilon > 0$) is positive definite and satisfies (1.2) with M replaced by a constant M_ϵ, $M_\epsilon \downarrow M$ if $\epsilon \downarrow 0$. By what we have already proved, the square root $\sigma^\epsilon(x)$ of $a(x)+\epsilon I$ satisfies

$$|\sigma^\epsilon_{ij}(x) - \sigma^\epsilon_{ij}(y)| \leqslant 2\nu\sqrt{\nu M_\epsilon}\,|x-y|.$$

We can apply the Ascoli–Arzela lemma to conclude that, for some sequence $\epsilon_m \downarrow 0$, $\sigma^{\epsilon_m}(x) \to \sigma(x)$ uniformly on compact sets. This gives (1.3). We have thus completed the proof of the theorem in case (1.2) holds.

If the a_{ij} belong to $C^2(R^\nu)$ but (1.2) is not assumed to hold, then we can slightly modify the previous proof, making use of the fact that for any $R > 0$

$$\sup_{|x|<R}\ \sup_{i,j,k,l}\left|\frac{\partial^2 a_{ij}(x)}{\partial x_k\,\partial x_l}\right| \leqslant M(R) \qquad (M(R)\text{ constant}).$$

Remark 1. If $a(x)$ is nonnegative definite in an open set G and $a_{ij} \in C^2(G)$, then the square root $\sigma(x)$ is Lipschitz continuous in compact subsets of G. Indeed, this follows from the proof of Theorem 1.2.

Remark 2. If $n = 1$ and $a_{11}(x) = |x|^\lambda$, then $\sigma_{11}(x) = |x|^{\lambda/2}$. If $\lambda = 2$, then a_{11} is in C^2 and σ_{11} is Lipschitz continuous, but if $\lambda = 2-\epsilon$ for any small $\epsilon > 0$, then a_{11} is in $C^{2-\epsilon}$ and σ_{11} is not Lipschitz continuous. This example shows that assumption that $a_{ij} \in C^2$ in Theorem 1.2 is sharp.

Remark 3. If $a(x)$ is nonnegative definite in a closed domain G with smooth boundary, and if $a_{ij} \in C^\infty(G)$, there does not exist in general a nonnegative definite matrix σ in G that is Lipschitz continuous and satisfies $\sigma^2 = a$ in G. For example, if $n = 1$ and $G = \{x;\ 0 \leqslant x \leqslant 1\}$, $a_{11}(x) = x$, then $\sigma_{11}(x) = \sqrt{x}$ is not Lipschitz continuous.

2. The maximum principle for elliptic equations

Consider the linear partial differential operator

$$Lu \equiv \frac{1}{2} \sum_{i,j=1}^{n} a_{ij}(x) \frac{\partial^2 u}{\partial x_i\, \partial x_j} + \sum_{i=1}^{n} b_i(x) \frac{\partial u}{\partial x_i} + c(x)u \tag{2.1}$$

with real coefficients defined in an n-dimensional domain D. L is said to be of *elliptic type* (or *elliptic*) at a point x^0 if the matrix $(a_{ij}(x^0))$ is positive definite, i.e., for any real vector $\xi \neq 0$, $\sum a_{ij}(x^0)\xi_i\xi_j > 0$.

Lemma 2.1. *Let $(a_{ij}(x))$ be a nonnegative definite matrix for $x \in D$, and let $c(x) \leqslant 0$. If $u(x)$ is in $C^2(D)$ and if it attains a positive maximum in D at a point x^0, then $Lu(x^0) \leqslant 0$.*

Proof. If A and B are nonnegative definite matrices, then $\operatorname{tr}(AB) \geqslant 0$. Indeed, if λ is an eigenvalue of AB, then $ABx = \lambda x$ for some $x \neq 0$. Taking the scalar product with Bx we see that $\lambda \geqslant 0$. Since $\operatorname{tr}(AB)$ is equal to the sum of the eigenvalues of AB, it follows that $\operatorname{tr}(AB) \geqslant 0$.

Applying the previous result to $A = (a_{ij}(x^0))$, $B = -(u_{x_i x_j}(x^0))$ we conclude that $\sum a_{ij}(x^0) u_{x_i x_j}(x^0) \leqslant 0$. Since also $u_{x_i}(x^0) = 0$, $c(x^0)u(x^0) \leqslant 0$, the assertion $Lu(x^0) \leqslant 0$ follows.

The *weak maximum principle* is the following theorem:

Theorem 2.2. *Let $(a_{ij}(x))$ be nonnegative definite matrix for all x in a bounded domain D, and assume that $c(x) \leqslant 0$ and*

$$\tfrac{1}{2} a_{11}(x)\lambda^2 + b_1(x)\lambda > 0 \qquad \text{for some} \quad \lambda > 0 \quad \text{and for all} \quad x \in D.$$

If $u \in C^2(D) \cap C^0(\bar{D})$ and $Lu \geqslant 0$ in D, and if $\max_{\bar{D}} u(x) > 0$, then

$$\underset{x \in D}{\text{l.u.b.}}\, u(x) \leqslant \max_{x \in \partial D} u(x)$$

where ∂D is the boundary of D.

Thus, if u takes positive values in $\bar{D}$, then the maximum of u in $\bar{D}$ is attained on ∂D; it may also be attained at points of D.

Proof. If the assertion is not true, then there is a point $\bar{x} \in D$ such that

$u(\bar{x}) > \max_{\partial D} u(x)$. Consider the function

$$k(x) = \exp(\lambda x_1^0) - \exp(\lambda x_1)$$

where x_1^0 is a real number such that $x_1 < x_1^0$ for all $x = (x_1, \ldots, x_n)$ in $\bar{D}$. It is clear that $k(x) > 0$ and $Lk(x) < 0$ in D. If ϵ is positive and sufficiently small, then the function $v = u - \epsilon k$ satisfies: $v(\bar{x}) > \max_{\partial D} v(x)$. Hence v assumes a positive maximum in $\bar{D}$ at some point $x^0 \in D$. By Lemma 2.1, $Lv(x^0) \leqslant 0$, i.e., $Lu(x^0) - \epsilon Lk(x^0) \leqslant 0$. Since $Lk(x^0) < 0$, we arrive at the inequality $Lu(x^0) < 0$; this contradicts the assumption that $Lu \geqslant 0$ in D.

Remark. If we apply the weak maximum principle to $-u$, we obtain the following assertion: If the coefficients of L satisfy the conditions imposed in Theorem 2.2, and if $u \in C^2(D) \cap C^0(\bar{D})$, $Lu \leqslant 0$ in D and u assumes a negative minimum in $\bar{D}$, then

$$\operatorname*{g.l.b.}_{x \in D} u(x) \geqslant \min_{x \in \partial D} u(x).$$

Definition. If there is a positive constant μ such that

$$\sum_{i,j=1}^{n} a_{ij}(x)\xi_i\xi_j \geqslant \mu|\xi|^2$$

for all $x \in D$, $\xi \in R^n$, then we say that L is *uniformly elliptic* in D.

The *strong maximum principle* for elliptic operators is the following theorem.

Theorem 2.3. *Let L be a uniformly elliptic operator with bounded coefficients in compact subsets of D, and assume that $c(x) \leqslant 0$ in D. If $u \in C^2(D)$ and $Lu \geqslant 0$ in D, and if $u(x) \not\equiv$ const, then u cannot attain a positive maximum in D.*

Applying this result to $-u$ we conclude that if $Lu \leqslant 0$ in D and $u(x) \not\equiv$ const, then u cannot attain a negative minimum in D.

Proof. We shall assume that u takes a positive maximum M at some point $x^0 \in D$, and derive a contradiction. Since $u \not\equiv$ const, there is a point $x^1 \in D$ such that $u(x^1) < M$. Connect x^1 to x^0 by a continuous curve γ lying in D. Then there is a first point x^2 along γ such that $u(x) < M$ if x lies on γ between x^1 and x^2, and $u(x^2) = M$ (x^2 may coincide with x^0). Take a point x^* on γ between x^1 and x^2 such that the ball $\{x;\ |x - x^*| \leqslant |x^2 - x^*|\}$ lies in D. Let x^{**} be the nearest point of the set $\{x \in D;\ u(x) = M\}$ to x^*. Clearly, the closure of the ball $B^* = \{x;\ |x - x^*| < |x^{**} - x^*|\}$ lies in D, $u(x) < M$ for all $x \in B^*$, and $u(x^{**}) = M$. Take a ball B contained in B^* whose boundary ∂B touches the boundary of B^* at x^{**}. Then $u(x) < M$ for all $x \in B \cup \partial B$, $x \neq x^{**}$. Denote the center of B by $\bar{x}$.

Denote by E a closed ball with center x^{**} and radius $< |x^{**} - \bar{x}|$ lying in D. Its boundary consists of a part E_1 lying in B and a part E_2 lying outside B. On E_1, $u \leq M - \delta$ for some $\delta > 0$.

Set $R = |x^{**} - \bar{x}|$, and consider the function

$$h(x) = \exp[-\alpha|x - \bar{x}|^2] - \exp[-\alpha R^2].$$

It satisfies: $h \geq 0$ in B, $h < 0$ outside B, $h = 0$ on ∂B, and $Lh > 0$ in E if α is sufficiently large. Consider the function $v = u + \epsilon h$ $(\epsilon > 0)$. If ϵ is sufficiently small, then $v < M$ on E_1, $v < M$ on E_2, and $Lv > 0$ in E. By Lemma 2.1, v cannot take a positive maximum in the interior of E. Since however $v(x^{**}) = M > 0$, we get a contradiction.

Consider the problem of finding a solution u of

$$Lu(x) = f(x) \quad \text{in} \quad D, \tag{2.2}$$

$$u(x) = \phi(x) \quad \text{on} \quad \partial D. \tag{2.3}$$

This is called the *Dirichlet problem* or the *first boundary value problem.*

A *barrier* $w_y(x)$ at the point $y \in \partial D$ is a continuous nonnegative function in $\bar{D}$ that vanishes only at the point y and for which $Lw_y(x) \leq -1$.

If there exists a closed ball K such that $K \cap D = \varnothing$, $K \cap \bar{D} = \{y\}$, then

$$w_y(x) = k\{|x^0 - y|^{-p} - |x - y|^{-p}\} \qquad (x^0 = \text{center of } K) \tag{2.4}$$

is a barrier at y provided k and p are sufficiently large.

Theorem 2.4. *Assume that L is uniformly elliptic in D, that* $c(x) \leq 0$, *and that* a_{ij}, b_i, c, f *are uniformly Hölder continuous (exponent* α*) in* $\bar{D}$. *If every point of* ∂D *has a barrier and if* ϕ *is a continuous function on* ∂D, *then there exists a unique solution u in* $C^2(D) \cap C^0(\bar{D})$ *of the Dirichlet problem* (2.2), (2.3).

For the proof of existence the reader is referred to Friedman [1, 2]. The uniqueness follows from the maximum principle.

3. The maximum principle for parabolic equations

Consider the partial differential operator

$$Mu \equiv \frac{1}{2} \sum_{i,j=1}^{n} a_{ij}(x, t) \frac{\partial^2 u}{\partial x_i \, \partial x_j} + \sum_{i=1}^{n} b_i(x, t) \frac{\partial u}{\partial x_i} + c(x, t)u - \frac{\partial u}{\partial t} \tag{3.1}$$

with real coefficients defined in an $(n + 1)$-dimensional domain Q. If $\Sigma a_{ij}(x^0, t^0)\xi_i\xi_j > 0$ for all $\xi \in R^n$, $\xi \neq 0$, then we say that M is of *parabolic type* at (x^0, t^0). M is *uniformly parabolic* in Q if there is a positive constant μ

such that

$$\sum a_{ij}(x, t)\xi_i\xi_j \geq \mu|\xi|^2 \qquad \text{for all} \quad (x, t) \in Q, \quad \xi \in R^n.$$

Suppose Q lies in the strip $0 < t < T$, and define

$$D_T = \{(x, T); (x, T - \delta) \in Q \text{ for all } 0 < \delta < \delta_0, \delta_0 \textit{ depending on } x\},$$

$$Q_0 = Q \cup D_T, \qquad \partial Q = \text{boundary of } Q, \qquad \partial_0 Q = \partial Q \setminus (\text{int } D_T).$$

The *weak maximum principle* is the following theorem:

Theorem 3.1. *Let $(a_{ij}(x, t))$ be nonnegative definite matrix for all (x, t) in a bounded domain Q, and assume that $c(x, t) \leq 0$. If $u \in C^0(\bar{Q})$ and u_{x_i}, $u_{x_i x_j}$, u_t belong to $C^0(Q_0)$, and if $Mu \geq 0$ in Q_0 and $\max_{\bar{Q}} u > 0$, then*

$$\underset{(x,t) \in Q_0}{\text{l.u.b.}}\ u(x, t) \leq \max_{(x,t) \in \partial_0 Q} u(x, t).$$

Proof. If the assertion is false, then, for some $\epsilon > 0$ sufficiently small, the function $v = u - \epsilon t$ takes positive maximum in $\bar{Q}$ at a point (x^0, t^0) in $Q \cup D_T$. If $t^0 = T$, then $v_t(x^0, t^0) \geq 0$; and if $t^0 < T$, then $v_t(x^0, t^0) = 0$. Employing Lemma 2.1 we conclude that $Mv(x^0, t^0) \leq 0$. Since however $Mu \geq 0$ in Q_0 and $M(-\epsilon t) = -c\epsilon t + \epsilon > 0$, we have $Mv(x^0, t^0) > 0$, a contradiction.

For any point $P^0 = (x^0, t^0)$ in Q, denote by $S(P^0)$ the set of all points P in Q that can be connected to P^0 by a simple continuous curve in Q along which the t-coordinate is nondecreasing from P to P^0. We denote by $C(P^0)$ the component, in $t = t^0$, of $Q \cap \{t = t^0\}$ that contains P^0. Notice that $C(P^0) \subset S(P^0)$.

The *strong maximum principle* for parabolic equations states the following.

Theorem 3.2. *Let M be uniformly parabolic in a domain Q, with bounded coefficients, and assume that $c(x, t) \leq 0$. If u, u_{x_i}, $u_{x_i x_j}$, u_t are continuous in Q and $Mu \geq 0$ in Q, and if u takes a positive maximum in Q at a point $P^0 = (x^0, t^0)$, then $u(P) = u(P^0)$ for all $P \in S(P^0)$.*

We need a few lemmas.

Lemma 3.3. *Let $Mu \geq 0$ in Q and suppose that u takes a positive maximum K in Q. Suppose that Q contains a closed solid ellipsoid E:*

$$\sum_{i=1}^{n} \lambda_i (x_i - x_i^*)^2 + \lambda_0 (t - t^*)^2 \leq R^2 \qquad (\lambda_j > 0, \quad R > 0),$$

$u < K$ in the interior of E, and $u(\bar{x}, \bar{t}) = K$ for some point $\bar{P} = (\bar{x}, \bar{t})$ on the boundary ∂E of E. Then $\bar{x} = x^*$ where $x^* = (x_1^*, \ldots, x_n^*)$.

Proof. We may suppose that $\bar{P}$ is the only point on ∂E where $u = K$ since otherwise we confine ourselves to a smaller ellipsoid lying in E and touching ∂E at $\bar{P}$ only. Suppose $\bar{x} \neq x^*$ and let C be an $(n+1)$-dimensional closed ball contained in Q with center $\bar{P}$ and radius $< |\bar{x} - x^*|$. Then

$$|x - x^*| \geq \text{const} > 0 \qquad \text{for all} \quad (x, t) \in C. \tag{3.2}$$

The boundary of C is composed of a part C_1 lying in E and a part C_2 lying outside E. Clearly

$$u \leq K - \delta \qquad \text{on } C_1, \qquad \text{for some constant} \quad \delta > 0.$$

Consider the function

$$h(x, t) = \exp\left\{ -\alpha\left[\sum_{i=1}^{n} \lambda_i (x_i - x_i^*)^2 + \lambda_0 (t - t^*)^2 \right] \right\} - \exp\{-\alpha R^2\}$$

where α is a positive constant. Clearly $h > 0$ in the interior of E, $h = 0$ on ∂E, and $h < 0$ outside E.

Using (3.2) one easily verifies that $Mh > 0$ in C if α is sufficiently large. Consider now the function $v = u + \epsilon h$ in C, where $\epsilon > 0$. If ϵ is sufficiently small, then $v < K$ on both C_1 and C_2. Since $v(\bar{P}) = K$, it follows that v takes its positive maximum in C at an interior point $\tilde{P} = (\tilde{x}, \tilde{t})$. By Lemma 2.1, $Mv \leq 0$ at $\tilde{P}$. Since, however, $Mv = Mu + \epsilon Mh > 0$ in C, we get a contradiction.

Lemma 3.4. *If $Mu \geq 0$ in Q and if u takes a positive maximum in Q at a point $P^0 = (x^0, t^0)$, then $u(P) = u(P^0)$ for all $P \in C(P^0)$.*

Proof. If the assertion is not true, then there exists a point $P^1 = (x^1, t^0)$ in $C(P^0)$ such that $u(P^1) < u(P^0)$. Connect P^1 to P^0 by a simple continuous curve γ lying in $C(P^0)$. Then there exists a point $P^* = (x^*, t^0)$ on γ such that $u(P^*) = u(P^0)$ and $u(\bar{P}) < u(P^0)$ for all $\bar{P}$ on γ between P^1 and P^*. Take one such point $\bar{P} = (\bar{x}, t^0)$ whose distance to the boundary of Q is $\geq 2|\bar{P} - P^*|$.

Since $u(\bar{P}) < u(P^*)$, there exists a sufficiently small interval $\sigma = \{(\bar{x}, t);\ t^0 - \epsilon \leq t \leq t^0 + \epsilon\}$ such that

$$u(P) < u(P^*) \qquad \text{for all} \quad P \in \sigma. \tag{3.3}$$

Consider the family of ellipsoids E_λ:

$$|x - \bar{x}|^2 + \lambda(t - t^0)^2 < R^2 \qquad (\lambda > 0).$$

Take $R^2 = \lambda\epsilon^2$ so that the end points of σ lie on the boundary of E_λ. If $\lambda \to 0$, $E_\lambda \to \sigma$. As λ increases, $E_\lambda \cap \{t = t^0\}$ increases; for sufficiently large λ it will contain the point P^*. Hence, because of (3.3), there is a first value of λ, say $\lambda = \lambda_0$, such that $u < u(P^*)$ in the interior of E_{λ_0} and $u = u(P^*)$ at some boundary point $\hat{P} = (y, \hat{t})$ of E_{λ_0}. Since $u < u(P^*)$ on σ, it is clear that $y \neq \bar{x}$. This contradicts Lemma 3.3.

Lemma 3.5. *Let R be a rectangle*

$$x_i^0 - a_i \leqslant x_i \leqslant x_i^0 + a_i \qquad (i = 1, \ldots, n), \qquad t^0 - a_0 \leqslant t \leqslant t^0$$

contained in Q, and let $Mu \geqslant 0$ *in Q. If u takes a positive maximum in R at the point* $P^0 = (x^0, t^0)$, *where* $x^0 = (x_1^0, \ldots, x_n^0)$, *then* $u(P) = u(P^0)$ *for all* $P \in R$.

Proof. If the assertion is not true, then there is a point $\hat{P}$ in R such that $u(\hat{P}) < u(P^0)$. Since then $u < u(P^0)$ in a neighborhood of $\hat{P}$, we may assume that $\hat{P}$ does not lie on $t = t^0$. On the straight segment γ connecting $\hat{P}$ to P^0 there exists a point P^1 such that $u(P^1) = u(P^0)$ and $u(\bar{P}) < u(P^1)$ for all $\bar{P}$ on γ between $\hat{P}$ and P^1. Without loss of generality we may assume that $P^1 = P^0$ and that $\hat{P}$ lies on $t = t^0 - a_0$, for otherwise we can confine ourselves to a smaller rectangle.

Denote by R_0 the subset of R consisting of all points with $t < t^0$. Since, for every P^1 in R_0, $C(P^1)$ contains a point of γ, and since $u < u(P^0)$ on γ, Lemma 3.4 implies that $u(P^1) < u(P^0)$.

Consider the function

$$h(x, t) = t^0 - t - A|x - x^0|^2 \qquad (A > 0).$$

Clearly $h = 0$ on the paraboloid Π: $t^0 - t = A|x - x^0|^2$, $h < 0$ above Π, and $h > 0$ below Π. Also, as easily seen, $Mh > 0$ in R if $4A\Sigma a_{ii} \leqslant 1$ in R and if the dimensions of R are sufficiently small.

Π divides R into two regions. Denote by R' the lower region. Its upper boundary B' touches $t = t^0$ at the point P^0 only. Hence on the complementary boundary B'' of R' we have $u \leqslant u(P^0) - \delta$ for some $\delta > 0$. It follows that if $\epsilon > 0$ is sufficiently small, then the function $v = u + \epsilon h$ satisfies $v < u(P^0)$ on B''. Next, $v = u < u(P^0)$ at all the points of B' other than P^0. Since $Mv = Mu + \epsilon Mh > 0$ in R', v cannot take a positive maximum in R' at an interior point (by Lemma 2.1). We thus conclude that the maximum of v in the closure of R' is attained at just one point, namely, at P^0. This implies that $\partial v(P^0)/\partial t \geqslant 0$. Since $\partial h(P^0)/\partial t = -1$, we conclude that

$$\partial u(P^0)/\partial t > 0. \tag{3.4}$$

Since however u takes a positive maximum at P^0, Lemma 2.1 implies that $Mu \leqslant -\partial u/\partial t$ at P^0. But since we have assumed that $Mu \geqslant 0$ in Q, $\partial u(P^0)/\partial t \leqslant 0$. This contradicts (3.4).

***Proof of Theorem* 3.2.** Suppose $u \not\equiv u(P^0)$ in $S(P^0)$. Then there exists a point P in $S(P^0)$ such that $u(P) < u(P^0)$. Connect P to P^0 by a simple continuous curve γ lying in $S(P^0)$ such that the t-coordinate is nondecreasing from P to P^0. There exists a point P^1 on γ such that $u(P^1) = u(P^0)$ and $u(\bar{P}) < u(P^1)$ for all $\bar{P}$ on γ lying between P and P^1. Construct a rectangle R:

$$x_i^1 - \alpha \leq x_i \leq x_i^1 + \alpha \qquad (i = 1, \ldots, n), \qquad t^1 - \alpha \leq t \leq t^1$$

where $P^1 = (x_1^1, \ldots, x_n^1, t^1)$ with a sufficiently small α so that R is contained in Q. Applying Lemma 3.5 we conclude that $u \equiv u(P^1)$ in R. This contradicts the definition of P^1.

Let Q be a bounded domain in the $(n + 1)$-dimensional space of variables (x, t). Assume that Q lies in the strip $0 < t < T$ and that $\tilde{B} \equiv \bar{Q} \cap \{t = 0\}$, $\tilde{B}_T \equiv \bar{Q} \cap \{t = T\}$ are nonempty. Let B_T = interior of $\tilde{B}_T$, B = interior of $\tilde{B}$. Denote by S_0 the boundary of Q lying in the strip $0 < t \leq T$, and let $S = S_0 \setminus B_T$. The set $\partial_0 Q = B \cup S$ is called the *normal* (or *parabolic*) *boundary* of Q.

The *first initial-boundary value problem* consists of finding a solution u of

$$Mu(x, t) = f(x, t) \qquad \text{in} \quad Q \cup B_T \tag{3.5}$$

$$u(x, 0) = \phi(x) \qquad \text{on } B, \tag{3.6}$$

$$u(x, t) = g(x, t) \qquad \text{on } S, \tag{3.7}$$

where f, ϕ, g are given functions. We refer to (3.6) as the *initial condition* and to (3.7) as the *boundary condition*. If $g = \phi$ on $\bar{B} \cap \bar{S}$, then the solution u is always understood to be continuous in $\bar{Q}$.

A function $w_R(P)$ $(R \in \bar{B} \cup S)$ is called a *barrier* at the point R if $w_R(P)$ is continuous in $P \in \bar{Q}$, $w_R(P) > 0$ if $P \in \bar{Q}$, $P \neq R$, $w_R(R) = 0$, and $Mw_R \leq -1$ in $Q \cup B_T$.

Suppose Q is a cylinder $B \times (0, T)$, and assume that there exists an n-dimensional closed ball K with center $\bar{x}$ such that $K \cap B = \varnothing$, $K \cap \bar{B} = \{x^0\}$. Then there exists a barrier at each point $R = (x^0, t^0)$ of S $(0 < t^0 \leq T)$, namely,

$$w_R(x, t) = ke^{\gamma t}\left(\frac{1}{R_0^p} - \frac{1}{R^p}\right)$$

where $\gamma \geq c(x, t)$, $R_0 = |x^0 - \bar{x}|$, $R = [|x - \bar{x}|^2 + (t - t^0)^2]^{1/2}$, and k, p are appropriate positive numbers.

Theorem 3.6. *Assume that M is uniformly parabolic in Q, that a_{ij}, b_i, c, f are uniformly Hölder continuous in $\bar{Q}$, and that g, ϕ are continuous functions on $\bar{B}$, $\bar{S}$ respectively and $g = \phi$ on $\bar{B} \cap \bar{S}$. Assume also that there exists a barrier at every point of S. Then there exists a unique solution u of the initial-boundary value problem* (3.5)–(3.7).

The solution u has Hölder continuous derivatives u_{x_i}, $u_{x_i x_j}$, u_t.

For the proof of existence we refer the reader to Friedman [1]. The uniqueness follows from the inequality

$$\max_Q |u| \leqslant e^{\alpha T} \max_{B \cup S} |u| \qquad \text{if } Mu = 0 \text{ and } c(x, t) \leqslant \alpha \qquad \text{in } Q. \quad (3.8)$$

This inequality follows from the weak maximum principle applied to $v = ue^{-\gamma t}$.

4. The Cauchy problem and fundamental solutions for parabolic equations

Let

$$Lu \equiv \frac{1}{2} \sum_{i,j=1}^{n} a_{ij}(x, t) \frac{\partial^2 u}{\partial x_i \, \partial x_j} + \sum_{i=1}^{n} b_i(x, t) \frac{\partial u}{\partial x_i} + c(x, t)u \qquad (4.1)$$

be an elliptic operator in R^n for each $t \in [0, T]$. Consider the parabolic equation

$$Mu \equiv Lu(x, t) - \frac{\partial u(x, t)}{\partial t} = f(x, t) \qquad \text{in} \quad R^n \times (0, T] \qquad (4.2)$$

with the *initial condition*

$$u(x, 0) = \phi(x) \qquad \text{on } R^n. \qquad (4.3)$$

The problem of solving (4.2), (4.3), for given f, ϕ, is called the *Cauchy problem*. The solution is understood to be continuous in $R^n \times [0, T]$ and to have continuous derivatives u_{x_i}, $u_{x_i x_j}$, u_t in $R^n \times (0, T]$.

We first prove uniqueness.

Theorem 4.1. *Let $(a_{ij}(x, t))$ be nonnegative definite and let*

$$|a_{ij}(x, t)| \leqslant C, \qquad |b_i(x, t)| \leqslant C(|x| + 1), \qquad c(x, t) \leqslant C(|x|^2 + 1) \quad (4.4)$$

(C constant). If $Mu \leqslant 0$ in $R^n \times (0, T]$ and

$$u(x, t) \geqslant -B \exp[\beta |x|^2] \qquad in \quad R^n \times [0, T] \qquad (4.5)$$

for some positive constants B, β, and if $u(x, 0) \geqslant 0$ on R^n, then $u(x, t) \geqslant 0$ in $R^n \times [0, T]$.

Proof. Consider the function

$$H(x, t) = \exp\left\{ \frac{k|x|^2}{1 - \mu t} + \nu t \right\} \qquad \left(0 \leqslant t \leqslant \frac{1}{2\mu}\right).$$

One easily checks that, for every $k > 0$, if μ and ν are sufficiently large positive constants, then $MH \leq 0$. Consider the function $v = u/H$ and take $k > \beta$ and μ, ν such that $MH \leq 0$. From (4.5) it follows that

$$\liminf_{|x|\to\infty} v(x, t) \geq 0 \tag{4.6}$$

uniformly in t, $0 \leq t \leq 1/2\mu$. Further, v satisfies

$$\tilde{M}v \equiv \frac{1}{2} \sum a_{ij} \frac{\partial^2 v}{\partial x_i \partial x_j} + \sum \tilde{b}_i \frac{\partial v}{\partial x_i} + \tilde{c}v - \frac{\partial v}{\partial t} = \tilde{f} \qquad \left(0 < t < \frac{1}{2\mu}\right)$$

where $\tilde{f} = (Mu)/H \leq 0$ and

$$\tilde{b}_i = b_i + \sum a_{ij} \frac{\partial H/\partial x_j}{H}, \qquad \tilde{c} = \frac{MH}{H} \leq 0.$$

By (4.6), for any $\epsilon > 0$, $v(x, t) + \epsilon > 0$ if $|x| = R$, $0 \leq t \leq 1/(2\mu)$ provided R is sufficiently large. Also $\tilde{M}(v + \epsilon) \leq \tilde{c}\epsilon \leq 0$ if $0 < t < 1/(2\mu)$ and $v(x, 0) + \epsilon > 0$ if $|x| \leq R$. By the maximum principle, $v(x, t) + \epsilon > 0$ if $|x| \leq R$, $0 < t < 1/2\mu$. Taking $R \to \infty$ and then $\epsilon \downarrow 0$ it follows that $v(x, t) \geq 0$ if $0 \leq t \leq 1/2\mu$. Hence $u(x, t) \geq 0$ if $0 \leq t \leq 1/2\mu$. We can now proceed step-by-step to prove the nonnegativity of u in the strip $0 \leq t \leq T$.

Corollary 4.2. *If $(a_{ij}(x, t))$ is nonnegative definite and* (4.4) *holds, then there exists at most one solution u of the Cauchy problem* (4.2), (4.3) *satisfying*

$$|u(x, t)| \leq B \exp[\beta |x|^2]$$

where B, β are some positive constants.

The next result on uniqueness has different growth conditions on the coefficients of L.

Theorem 4.3. *Let $(a_{ij}(x, t))$ be nonnegative definite and let*

$$|a_{ij}(x, t)| \leq C(|x|^2 + 1), \quad |b_i(x, t)| \leq C(|x| + 1),$$
$$c(x, t) \leq C \quad (C \text{ constant}). \tag{4.7}$$

If $Mu \leq 0$ in $R^n \times (0, T]$ and

$$u(x, t) \geq -N(|x|^q + 1) \qquad \text{in} \quad R^n \times [0, T] \tag{4.8}$$

where N, q are positive constants, and if $u(x, 0) \geq 0$ on R^n, then $u(x, t) \geq 0$ in $R^n \times [0, T]$.

Proof. For any $p > 0$, the function

$$w(x, t) = (|x|^2 + Kt)^p e^{\alpha t}$$

satisfies $Mw < 0$ in $R^n \times [0, T]$ provided K, α are appropriate positive constants. Take $2p > q$ and consider the function $v = u + \epsilon w$ for any $\epsilon > 0$. Then $Mv < 0$. Since $v(x, 0) \geq 0$ and (by (4.8)) $v(x, t) > 0$ if $|x| = R$ (R large), $0 < t < T$, the maximum principle can be applied (to ve^{-Ct}) to yield the inequality $v(x, t) \geq 0$ if $|x| \leq R$, $0 \leq t \leq T$. Taking first $R \to \infty$ and then $\epsilon \to 0$, the assertion follows.

Corollary 4.4. *Let $(a_{ij}(x, t))$ be nonnegative definite and let (4.7) hold. Then there exists at most one solution of the Cauchy problem (4.2), (4.3) satisfying*

$$|u(x, t)| \leq N(1 + |x|^q)$$

where N, q are some positive constants.

Definition. A *fundamental solution* of the parabolic operator $L - \partial/\partial t$ in $R^n \times [0, T]$ is a function $\Gamma(x, t; \xi, \tau)$ defined for all (x, t) and (ξ, τ) in $R^n \times [0, T]$, $t > \tau$, satisfying the following condition:

For any continuous function $f(x)$ with compact support, the function

$$u(x, t) = \int_{R^n} \Gamma(x, t; \xi, \tau) f(\xi)\, d\xi \tag{4.9}$$

satisfies

$$Lu - \partial u/\partial t = 0 \quad \text{if} \quad x \in R^n, \quad \tau < t \leq T, \tag{4.10}$$

$$u(x, t) \to f(x) \quad \text{if} \quad t \downarrow \tau. \tag{4.11}$$

We shall need the conditions:

(A_1) There is a positive constant μ such that

$$\sum a_{ij}(x, t)\xi_i\xi_j \geq \mu|\xi|^2 \qquad \text{for all} \quad (x, t) \in R^n \times [0, T], \quad \xi \in R^n.$$

(A_2) The coefficients of L are bounded continuous functions in $R^n \times [0, T]$, and the coefficients $a_{ij}(x, t)$ are continuous in t, uniformly with respect to (x, t) in $R^n \times [0, T]$.

(A_3) The coefficients of L are Hölder continuous (exponent α) in x, uniformly with respect to (x, t) in compact subsets of $R^n \times [0, T]$; furthermore, the $a_{ij}(x, t)$ are Hölder continuous (exponent α) in x uniformly with respect to (x, t) in $R^n \times [0, T]$.

Theorem 4.5. *Let (A_1)–(A_3) hold. Then there exists a fundamental solution $\Gamma(x, t, \xi, \tau)$ for $L - \partial/\partial t$ satisfying the inequalities*

$$|D_x^m \Gamma(x, t; \xi, \tau)| \leq C(t - \tau)^{-(n+|m|)/2} \exp\left[-c\frac{|x - \xi|^2}{t - \tau}\right] \tag{4.12}$$

for $|m| = 0, 1$, where C, c are positive constants. The functions $D_x^m \Gamma(x, t; \xi, \tau)$ $(0 \leq |m| \leq 2)$ and $D_t \Gamma(x, t; \xi, \tau)$ are continuous in

$(x, t, \xi, \tau) \in R^n \times [0, T] \times R^n \times [0, T]$, $t > \tau$, *and* $L\Gamma - \partial\Gamma/\partial t = 0$ *as a function in* (x, t). *Finally, for any continuous bounded function* $f(x)$,

$$\int_{R^n} \Gamma(x, t; \xi, \tau) f(x)\, dx \to f(\xi) \qquad \text{if} \quad t \downarrow \tau. \tag{4.13}$$

The construction of Γ can be given by the parametrix method; for details regarding the construction of this Γ and the proof of the other assertions of Theorem 4.5, see Friedman [1; Chapter 9]. (The assertion (4.13) follows from Friedman [1, p. 247, formula (2.29); p. 252, formula (4.4), and the estimate on $\Gamma - Z$].)

The following theorem is also proved in Friedman [1]:

Theorem 4.6. *Let* (A_1)–(A_3) *hold. Let* $f(x, t)$ *be a continuous function in* $R^n \times [0, T])$, *Hölder continuous in* x *uniformly with respect to* (x, t) *in compact subsets, and let* $\phi(x)$ *be a continuous function in* R^n. *Assume also that*

$$|f(x, t)| \leqslant A \exp(a|x|^2) \qquad \text{in} \quad R^n \times [0, T], \tag{4.14}$$

$$|\phi(x)| \leqslant A \exp(a|x|^2) \qquad \text{in} \quad R^n \tag{4.15}$$

where A, a *are positive constants. Then there exists a solution of the Cauchy problem* (4.2), (4.3) *in the strip* $0 \leqslant t \leqslant T^*$ *where* $T^* = \min\{T, \bar{c}/a\}$ *and* $\bar{c}$ *is a constant depending only on the coefficients of* L, *and*

$$|u(x, t)| \leqslant A' \exp(a'|x|^2) \qquad \text{in} \quad R^n \times [0, T^*] \tag{4.16}$$

for some positive constants A', a'. *The solution is given by*

$$u(x, t) = \int_{R^n} \Gamma(x, t; \xi, 0)\phi(\xi)\, d\xi - \int_0^t \int_{R^n} \Gamma(x, t; \xi, \tau) f(\xi, \tau)\, d\xi\, d\tau. \tag{4.17}$$

The *formal adjoint operator* M^* of $M = L - \partial/\partial t$, where L is given by (4.1), is given by

$$M^*v = L^*v + \partial v/\partial t,$$

$$L^*v = \frac{1}{2} \sum_{i,j=1}^{n} a_{ij}(x, t) \frac{\partial^2 v}{\partial x_i\, \partial x_j} + \sum_{i,j=1}^{n} b_i^*(x, t) \frac{\partial v}{\partial x_i} + c^*(x, t)v \tag{4.18}$$

where

$$b_i^* = -b_i + \frac{1}{2} \sum_{j=1}^{n} \frac{\partial a_{ij}}{\partial x_j},$$

$$c^* = c - \sum_{i=1}^{n} \frac{\partial b_i}{\partial x_i} + \frac{1}{2} \sum_{i,j=1}^{n} \frac{\partial^2 a_{ij}}{\partial x_i\, \partial x_j}. \tag{4.19}$$

It is assumed here that $\partial a_{ij}/\partial x_j$, $\partial^2 a_{ij}/\partial x_i\, \partial x_j$, $\partial b_i/\partial x_i$ exist and are bounded functions.

Notice that

$$M^*v = \frac{1}{2}\sum_{i,j=1}^{n} \frac{\partial^2}{\partial x_i\,\partial x_j}(a_{ij}v) - \sum_{i=1}^{n}\frac{\partial}{\partial x_i}(b_i v) + cv + \frac{\partial v}{\partial t}. \tag{4.20}$$

Using this form of M^*v, one can easily verify *Green's identity*

$$vMu - uM^*v = \sum_{i=1}^{n}\frac{\partial}{\partial x_i}\left[\frac{1}{2}\sum_{j=1}^{n}\left(va_{ij}\frac{\partial u}{\partial x_j} - ua_{ij}\frac{\partial v}{\partial x_j} - uv\frac{\partial a_{ij}}{\partial x_j}\right) + b_i uv\right] - \frac{\partial}{\partial t}(uv). \tag{4.21}$$

Thus, if u, v have compact support in a domain G, then

$$\int\int_G (vMu - uM^*v)\,dx\,dt = 0. \tag{4.22}$$

Definition. A *fundamental solution* of the operator $L^* + \partial/\partial t$ in $R^n \times [0, T]$ is a function $\Gamma^*(x, t; \xi, \tau)$ defined for all (x, t) and (ξ, τ) in $R^n \times [0, T]$, $t < \tau$, satisfying the following condition:

For any continuous function $g(x)$ with compact support, the function

$$v(x, t) = \int_{R^n} \Gamma^*(x, t; \xi, \tau) g(\xi)\,d\xi$$

satisfies

$$L^*v + \partial v/\partial t = 0 \qquad \text{if} \quad x \in R^n, \quad 0 \leq t < \tau,$$
$$v(x, t) \to g(x) \qquad \text{if} \quad t \uparrow \tau.$$

We shall need the following condition:

(A_4) The functions

$$a_{ij}, \qquad \frac{\partial}{\partial x_i}a_{ij}, \qquad \frac{\partial^2}{\partial x_i\,\partial x_j}a_{ij}, \qquad b_i, \qquad \frac{\partial}{\partial x_i}b_i, \qquad c$$

are bounded functions, and the coefficients of L^* satisfy the conditions (A_2), (A_3).

Theorem 4.7. *If* (A_1)–(A_4) *hold, then there exists a fundamental solution* $\Gamma^*(x, t; \xi, \tau)$ *of* $L^* + \partial/\partial t$, *and*

$$\Gamma(x, t; \xi, \tau) = \Gamma^*(\xi, \tau; x, t) \qquad (t > \tau). \tag{4.23}$$

Proof. The construction of Γ^* can be carried out in the same way as for Γ. Further, Γ^* satisfies inequalities similar to (4.12) and $L^*\Gamma^* + \partial\Gamma^*/\partial t = 0$ as a function of (ξ, τ). Thus it remains to prove (4.23). Consider the functions

$$u(y, \sigma) = \Gamma(y, \sigma; \xi, \tau), \qquad v(y, \sigma) = \Gamma^*(y, \sigma; x, t)$$

for $y \in R^n$, $\tau < \sigma < t$. Integrating Green's identity (4.21) over the domain $|y| < R$, $\tau + \epsilon < \sigma < t - \epsilon$ $(R > 0)$ and using the relations $Mu = 0$, $M^* v = 0$, we obtain

$$\int_{|y|<R} u(y, t-\epsilon)v(y, t-\epsilon)\, dy - \int_{|y|<R} u(y, \tau+\epsilon)v(y, \tau+\epsilon)\, dy = I_{\epsilon, R}$$

where

$$I_{\epsilon, R} = \sum_{i=1}^{n} \int_{\tau+\epsilon}^{t-\epsilon} \int_{|y|=R} \left[\frac{1}{2} \sum_{j=1}^{n} \left(v a_{ij} \frac{\partial u}{\partial y_j} - u a_{ij} \frac{\partial v}{\partial y_j} - uv \frac{\partial a_{ij}}{\partial y_j} \right) + b_i uv \right] \cos(\nu, y_i)\, dS_y\, d\sigma;$$

ν is the outwardly directed normal to $|y| = R$ and dS_y is the surface element on $|y| = R$. Using (4.12) and the corresponding inequalities for Γ^* we find that $I_{\epsilon, R} \to 0$ if $R \to \infty$. Hence

$$\int_{R^n} u(y, t-\epsilon)\Gamma^*(y, t-\epsilon; x, t)\, dy = \int_{R^n} v(y, \tau+\epsilon)\Gamma(y, \tau+\epsilon; \xi, \tau)\, dy. \tag{4.24}$$

Taking $\epsilon \downarrow 0$ the assertion (4.23) follows; cf. Problem 10.

5. Stochastic representation of solutions of partial differential equations

Let L be an elliptic operator in a bounded domain D given by (2.1). Assume that L is uniformly elliptic in D, i.e.,

$$\sum_{i,j=1}^{n} a_{ij}(x)\xi_i\xi_j \geq \mu|\xi|^2 \quad \text{if} \quad x \in D, \quad \xi \in R^n \quad (\mu > 0). \tag{5.1}$$

Assume also that

$$a_{ij},\ b_i \quad \text{are uniformly Lipschitz continuous in } \bar{D}, \tag{5.2}$$

$$c \leq 0,\ c \quad \text{uniformly Hölder continuous in } \bar{D}. \tag{5.3}$$

Assume finally that the boundary ∂D of D is in C^2, so that barriers exist at all the points of ∂D (see Problem 2).

Then, by Theorem 2.4, the Dirichlet problem (2.2), (2.3) has a unique solution u for any given functions f, ϕ satisfying:

$$f \quad \text{is uniformly Hölder continuous in } \bar{D}, \tag{5.4}$$

$$\phi \quad \text{is continuous on } \partial D. \tag{5.5}$$

We shall now represent u in terms of a solution of a stochastic differential system.

By Lemma 1.1 and (1.1) it is clear that the nonnegative definite square root $\sigma(x) = (\sigma_{ij}(x))$ of the matrix $a(x) = (a_{ij}(x))$ is Lipschitz continuous in $\bar{D}$. Extend $\sigma(x)$ and $b(x) = (b_1(x), \ldots, b_n(x))$ into the whole space R^n so that

$$|\sigma(x) - \sigma(y)| \leq C|x - y|, \qquad |b(x) - b(y)| \leq C|x - y|. \tag{5.6}$$

Consider the system of stochastic differential equations

$$d\xi(t) = \sigma(\xi(t))\, dw(t) + b(\xi(t))\, dt. \tag{5.7}$$

Denote by V_ϵ the closed ϵ-neighborhood of ∂D and let $D_\epsilon = D \setminus V_\epsilon$. Let v be a function in $C^2(R^n)$ that coincides with the solution u of (2.2), (2.3) in $D_{\epsilon/2}$, and let τ be any Markov time with respect to the time-homogeneous Markov process solution of (5.7). By Itô's formula,

$$E_x v(\xi(\tau)) \exp\left[\int_0^\tau c(\xi(s))\, ds\right] - v(x)$$
$$= E_x \int_0^\tau [Lv(\xi(t))] \exp\left[\int_0^t c(\xi(s))\, ds\right] dt. \tag{5.8}$$

Take $x \in D_\epsilon$ and $\tau = \tau_\epsilon \wedge T$ where τ_ϵ is the hitting time of V_ϵ. Then $v(\xi(t)) = u(\xi(t))$ for all $0 \leq t \leq \tau_\epsilon \wedge T$. Hence (5.8) holds for $v = u$. Taking $\epsilon \to 0$ and using the Lebesgue bounded convergence theorem, we get

$$u(x) = E_x u(\xi(\tau \wedge T)) \exp\left[\int_0^{\tau\wedge T} c(\xi(s))\, ds\right]$$
$$- E_x \int_0^{\tau \wedge T} f(\xi(t)) \exp\left[\int_0^t c(\xi(s))\, ds\right] dt \tag{5.9}$$

where τ is the exit time from D.

Theorem 5.1. *Let* (5.1)–(5.5) *hold and let ∂D belong to C^2. Then the unique solution u of the Dirichlet problem* (2.2), (2.3) *is given by*

$$u(x) = E_x \phi(\xi(\tau)) \exp\left[\int_0^\tau c(\xi(s))\, ds\right]$$
$$- E_x \int_0^\tau f(\xi(t)) \exp\left[\int_0^t c(\xi(s))\, ds\right] dt \tag{5.10}$$

where τ is the exit time from D.

Proof. If we prove that

$$E_x \tau < \infty \tag{5.11}$$

then, by taking $T \uparrow \infty$ in (5.9) and using the Lebesgue bounded convergence theorem, we get the assertion (5.9).

To prove (5.11) consider the function

$$h(x) = -Ae^{\lambda x_1}$$

for all $x = (x_1, \ldots, x_n)$ in D. If A, λ are sufficiently large (A depending on λ), then

$$\sum a_{ij} h_{x_i x_j} + \sum b_i h_{x_i} \leqslant -1 \qquad \text{in } D.$$

By Itô's formula,

$$E_x h(\xi(\tau \wedge T)) - h(x) \leqslant -E_x(\tau \wedge T).$$

Since $|h(x)| \leqslant K$ in D (K constant), we then have $E_x(\tau \wedge T) \leqslant 2K$. Taking $T \uparrow \infty$ and using the monotone convergence theorem we get $E_x \tau \leqslant 2K$; this proves (5.11).

Remark. The proof of (5.11) requires only that $a_{11}(x) > 0$ in $\bar{D}$. Thus the right-hand side of (5.10) makes sense even if ∂D is not in C^2, L is not elliptic but has bounded coefficients with $c(x) \leqslant 0$, $a_{11}(x) > 0$ in $\bar{D}$, $a = \sigma\sigma^*$, and σ, b are Lipschitz continuous in $\bar{D}$. Recall that, by Theorem 1.2 and Remark 1 following it, if (a_{ij}) is nonnegative definite and in C^2 in some neighborhood of $\bar{D}$ then σ is Lipschitz continuous in $\bar{D}$.

Consider next the initial-boundary value problem (written for t replaced by $T - t$)

$$\begin{aligned} Lu + \partial u/\partial t &= f(x, t) && \text{in} \quad Q = B \times [0, T), \\ u(x, T) &= \phi(x) && \text{on } B, \\ u(x, t) &= g(x, t) && \text{on } S, \end{aligned} \tag{5.12}$$

where B is a bounded domain with C^2 boundary ∂B, $S = \partial B \times [0, T)$, and L is defined by (4.1).

Consider also the system of stochastic differential equations

$$d\xi(t) = \sigma(\xi(t), t)\, dw(t) + b(\xi(t), t)\, dt \tag{5.13}$$

where $(\sigma(x, t))^2 = a(x, t)$ in $\bar{Q}$. The coefficients σ, b here are extensions of $\sigma(x, t)$, $b(x, t)$, originally defined in $\bar{Q}$, such that

$$|\sigma(x, t) - \sigma(y, s)| \leqslant C(|x - y| + |t - s|),$$
$$|b(x, t) - b(y, s)| \leqslant C(|x - y| + |t - s|).$$

We shall assume:

$$\sum a_{ij}(x, t)\xi_i\xi_j \geqslant \mu|\xi|^2 \qquad \text{if} \quad (x, t) \in Q, \quad \xi \in R^n \qquad (\mu > 0),$$

$$\begin{aligned} &a_{ij}, b_i \quad \text{are uniformly Lipschitz continuous in } (x, t) \in \bar{Q}, \\ &c \quad \text{is uniformly Hölder continuous in } (x, t) \in \bar{Q}, \\ &f \quad \text{is uniformly Hölder continuous in } (x, t) \in \bar{Q}, \\ &g \quad \text{is continuous on } \bar{S}, \ \phi \text{ is continuous on } \bar{B} \text{ and} \\ &\qquad g(x, T) = \phi(x) \qquad \text{if} \quad x \in \partial B. \end{aligned} \tag{5.14}$$

By Theorem 3.6 there exists a unique solution u of (5.12).

Theorem 5.2. *Let ∂B belong to C^2 and let* (5.14) *hold. Then the unique solution u of the initial-boundary value problem* (5.12) *is given by*

$$u(x,t) = E_{x,t}g(\xi(\tau),\tau)\exp\left[\int_t^\tau c(\xi(s),s)\,ds\right]\chi_{\tau<T}$$

$$+ E_{x,t}\phi(\xi(T))\exp\left[\int_t^T c(\xi(s),s)\,ds\right]\chi_{\tau=T}$$

$$- E_{x,t}\int_t^\tau f(\xi(s),s)\exp\left[\int_t^s c(\xi(\lambda),\lambda)\,d\lambda\right]ds \tag{5.15}$$

where τ is the first time $\lambda \in [t, T)$ that $\xi(\lambda)$ leaves B if such a time exists and $\tau = T$ otherwise.

The proof of Theorem 5.2 is similar to the proof of Theorem 5.1. Here one applies Itô's formula to

$$u(\xi(\lambda),\lambda)\exp\left[\int_t^\lambda c(\xi(s),s)\,ds\right]. \tag{5.16}$$

Consider next the Cauchy problem

$$\begin{aligned} Lu + \partial u/\partial t &= f(x,t) \quad \text{in } R^n \times [0,T), \\ u(x,T) &= \phi(x) \quad \text{in } R^n \end{aligned} \tag{5.17}$$

where L is given by (4.1).

We shall assume:

(B_1) (i) The functions a_{ij}, b_i are bounded in $R^n \times [0, T]$ and uniformly Lipschitz continuous in (x, t) in compact subsets of $R^n \times [0, T]$.

(ii) The functions a_{ij} are Hölder continuous in x, uniformly with respect to (x, t) in $R^n \times [0, T]$.

(B_2) The function c is bounded in $R^n \times [0, T]$ and uniformly Hölder continuous in (x, t) in compact subsets of $R^n \times [0, T]$.

We shall also assume:

$f(x,t)$ is continuous in $R^n \times [0, T]$, Hölder continuous in x uniformly with respect to $(x, t) \in R^n \times [0, T]$, and

$$|f(x,t)| \leq A(1 + |x|^a) \quad \text{in } R^n \times [0,T], \tag{5.18}$$

$$\phi(x) \text{ is continuous in } R^n \quad \text{and} \quad |\phi(x)| \leq A(1+|x|^a) \tag{5.19}$$

where A, a are positive constants.

Under the conditions (A_1), (B_1), (B_2), (5.18), and (5.19), there exists a unique solution u of the Cauchy problem (5.17) satisfying

$$|u(x,t)| \leq \text{const}(1 + |x|^a). \tag{5.20}$$

Indeed, uniqueness follows from Corollary 4.2. The existence of u follows from Theorem 4.6; the estimate (5.20) on $u(x, t)$ is an easy consequence of the estimate (4.12) (with $m = 0$) on Γ and the assumptions (5.18), (5.19). Using (4.12) with $|m| = 1$ we also get

$$|u_x(x, t)| \leq \text{const}(1 + |x|^a). \tag{5.21}$$

By Lemma 1.1 and (1.1), the nonnegative definite square root $\sigma(x, t)$ of $a(x, t)$ is Lipschitz continuous in (x, t), uniformly in compact subsets of $R^n \times [0, T]$. It is also clear that $\sigma(x, t)$ is uniformly bounded in $R^n \times [0, T]$. Thus (by results in Chapter 5) the stochastic system (5.13) has a unique solution for any initial value $\xi(0) = x$.

Theorem 5.3. *If* (A_1), (B_1), (B_2) *and* (5.18), (5.19) *hold, then the solution of the Cauchy problem* (5.17), (5.20) *is given by*

$$u(x, t) = E_{x,t}\phi(\xi(T)) \exp\left[\int_t^T c(\xi(s), s)\, ds\right]$$
$$- E_{x,t} \int_t^T f(\xi(s), s) \exp\left[\int_t^s c(\xi(\lambda), \lambda)\, d\lambda\right] ds. \tag{5.22}$$

The proof follows by applying Itô's formula to the function in (5.16) and the process (5.13), and then taking the expectation. Since, by (5.21),

$$|u_x(x, t)\sigma(x, t)| \leq \text{const}(1 + |x|^a)$$

and since

$$\sup_{t \leq s \leq T} E_{x,t}|\xi(s)|^m < \infty \qquad \text{for any} \quad m > 0,$$

the stochastic integral occurring in Itô's formula has zero expectation. The assumptions (5.18), (5.19) ensure that the expectations on the right-hand side of (5.22) exist.

Let

$$L_0 = \frac{1}{2} \sum_{i,j=1}^n a_{ij}(x, t) \frac{\partial^2}{\partial x_i\, \partial x_j} + \sum_{i=1}^n b_i(x, t) \frac{\partial}{\partial x_i} \tag{5.23}$$

and denote by $\Gamma_0^*(x, s; y, t)$ the fundamental solution of $L_0 + \partial/\partial s$ $(s < t)$. Taking $f \equiv 0$, $c \equiv 0$ in (5.22), we get

$$u(x, t) = E_{x,t}\phi(\xi(T)). \tag{5.24}$$

By Theorem 4.6 (with t replaced by $T - t$) we can also represent u in the form

$$u(x, t) = \int_{R^n} \Gamma_0^*(x, t; y, T)\phi(y)\, dy.$$

Comparing this with (5.24) we conclude that

$$\int_{R^n} \Gamma_0^*(x, t; y, T)\phi(y)\, dy = E_{x,t}\phi(\xi(T)).$$

Notice that this relation holds for any function ϕ satisfying (5.19).

Similarly, for any $0 \le s < t \le T$,

$$\int_{R^n} \Gamma_0^*(x, s; y, t)\phi(y)\,dy = E_{x,s}\phi(\xi(t)) = \int_{R^n} \phi(y) P_{x,s}(\xi(t) \in dy). \quad (5.25)$$

Since ϕ is arbitrary function satisfying (5.19), we conclude that the transition probability function

$$p(s, x, t, A) = P_{x,s}(\xi(t) \in A) = P(\xi_{x,s}(t) \in A)$$

of the Markov process solution of (5.13) has density, and that the density function is $\Gamma_0^*(s, x, t, y)$. We sum up:

Theorem 5.4. *If* (A_1), (B_1) *hold, then the transition probability function of the solution of the stochastic differential system* (5.13) *has density, i.e.,*

$$P_{x,s}(\xi(t) \in A) = \int_A \Gamma_0^*(x, s; y, t)\,dy \qquad (s < t) \quad (5.26)$$

for any Borel set A, *and* $\Gamma_0^*(x, s; y, t)$ *is the fundamental solution of* $L_0 + \partial/\partial t$ *constructed in Section* 4.

The density function of the transition probability function is called the *transition density function*. From the results of Section 4 we conclude that the transition density function $\Gamma_0^*(x, s; y, t)$ of the solution $\xi(t)$ of (5.13) satisfies in (x, s) the *backward parabolic equation*

$$\frac{\partial}{\partial s}\Gamma_0^*(x, s; y, t) + \frac{1}{2}\sum_{i,j=1}^n a_{ij}(x, s)\frac{\partial^2}{\partial x_i\,\partial x_j}\Gamma_0^*(x, s; y, t) + \sum_{i=1}^n b_i(x, s)\frac{\partial}{\partial x_i}\Gamma_0^*(x, s; y, t) = 0. \quad (5.27)$$

Under the assumptions of Theorem 4.7, $\Gamma_0^*(x, s; y, t)$ also satisfies in (y, t) the *forward parabolic equation*

$$-\frac{\partial}{\partial t}\Gamma_0^*(x, s; y, t) + \frac{1}{2}\sum_{i,j=1}^n \frac{\partial^2}{\partial y_i\,\partial y_j}[a_{ij}(y, t)\Gamma_0^*(x, s; y, t)] - \sum_{i=1}^n \frac{\partial}{\partial y_i}[b_i(y, t)\Gamma_0^*(x, s; y, t)] = 0. \quad (5.28)$$

Notice that under the conditions (A_1), (B_1) the backward equation (5.27) is equivalent to the Kolmogorov equation (5.6.1).

If the coefficients a_{ij}, b_i are independent of t, then $\Gamma_0^*(x, s; y, t) = \Gamma_0^*(x, 0; y, t-s) \equiv \Gamma(x, t-s, y)$; $\Gamma(x, t, y)$ satisfies in (x, t) the parabolic equation

$$-\frac{\partial \Gamma(x, t, y)}{\partial t} + \frac{1}{2}\sum_{i,j=1}^n a_{ij}(x)\frac{\partial^2 \Gamma(x, t, y)}{\partial x_i\,\partial x_j} + \sum_{i=1}^n b_i(x)\frac{\partial \Gamma(x, t, y)}{\partial x_i} = 0. \quad (5.29)$$

Under the conditions of Theorem 4.7, it also satisfies in (y, t) the parabolic equation

$$-\frac{\partial \Gamma(x, t, y)}{\partial t} + \frac{1}{2} \sum_{i,j=1}^{n} \frac{\partial^2}{\partial y_i \, \partial y_j} [a_{ij}(y)\Gamma(x, t, y)]$$

$$-\sum_{i=1}^{n} \left[\frac{\partial}{\partial y_i} b_i(y)\Gamma(x, t, y) \right]. \tag{5.30}$$

PROBLEMS

1. Verify that the function in (2.4) is a barrier for L.

2. If D is a bounded domain with C^2 boundary ∂D, then for any $y \in \partial D$, there exists a closed ball K such that $K \cap D = \varnothing$ and $K \cap \bar{D} = \{y\}$; thus, a barrier exists.

3. Let L be as in Theorem 2.2 and let u be a solution of (2.2), (2.3) where D is a bounded domain. Prove that

$$\max_{\bar{D}} |u| \leq \max_{D} |\phi| + A \max_{\bar{D}} |f|$$

where A is a constant independent of f, ϕ.

4. Let $Lu = y^2 u_{xx} - 2xyu_{xy} + x^2 u_{yy} - xu_x - yu_y$. Verify that any C^2 function $u = f(r)$ (where $r = \sqrt{x^2 + y^2}$) satisfies $Lu = 0$. Hence the weak maximum principle does not hold in any domain $\epsilon^2 < x^2 + y^2 < R^2$. Notice that (a_{ij}) is nonnegative definite, $a_{11} + a_{22} \geq \epsilon^2$, but neither a_{11} nor a_{22} are positive throughout the domain.

5. Let $(a_{ij}(x, t))$ be nonnegative definite, $c(x, t) \leq \alpha$ and $a_{11}(x, t)\lambda^2 + b_1(x)\lambda \geq 1$ in a cylinder $Q = D \times [0, T]$. Denote the diameter of D by d. Show that if u is continuous in $\bar{Q}$ and Mu is bounded in Q (M given by (3.1)), then

$$\max_{\bar{Q}} |u| \leq e^{\alpha T} \Big\{ \max_{\partial_0 Q} |u| + (e^{\lambda d} - 1) \max_{\bar{Q}} |Mu| \Big\}$$

where $\partial_0 Q$ is the normal boundary of Q.

6. Let the coefficients of M satisfy the conditions of either Theorem 4.1 or 4.3. Prove that a fundamental solution Γ is uniquely determined by the requirements:

(i) $\Gamma(x, t; \xi, \tau)$ is continuous in ξ, for fixed x, t, τ;

(ii) for any continuous function f with compact support, the function $u(x, t)$ given by (4.9) satisfies: $|u(x, t)| \leq C(1 + |x|^q)$ in $R^n \times [0, T]$, where C, q are some positive constants (depending on f).

7. Let the coefficients of M satisfy the conditions of either Theorem 4.1 or 4.3 and let Γ be a fundamental solution satisfying (i), (ii) of the preceding problem. Prove that $\Gamma(x, t; \xi, \tau) \geq 0$.

8. The operator $\Delta - \partial/\partial t$ (where Δ is the Laplacian operator $\sum_{i=1}^n \partial^2/\partial x_i^2$) is called the *heat operator*. Prove that

$$\Gamma(x, t; \xi, \tau) = \frac{1}{(2\sqrt{\pi})^n (t-\tau)^{n/2}} \exp\left\{ \frac{|x-\xi|^2}{4(t-\tau)} \right\}$$

is a fundamental solution for the heat operator.

9. Let Γ be the fundamental solution constructed in Theorem 4.5. Prove that, for any $\epsilon > 0$,

$$\int_{|x-\xi|>\epsilon} \Gamma(x, t; \xi, \tau)\, d\xi \to 0 \qquad \text{if} \quad t \downarrow \tau,$$

$$\int_{|x-\xi|>\epsilon} \Gamma(x, t; \xi, \tau)\, dx \to 0 \qquad \text{if} \quad t \downarrow \tau.$$

10. Let $f(x, t)$ be a continuous bounded function in $R^n \times [0, T]$, and let Γ be the fundamental solution constructed in Theorem 4.5. Prove that

$$\int_{R^n} \Gamma(x, t; \xi, \tau) f(\xi, \tau)\, d\xi \to f(x, \tau) \qquad \text{if} \quad t \downarrow \tau,$$

$$\int_{R^n} \Gamma(x, t; \xi, \tau) f(x, \tau)\, dx \to f(\xi, \tau) \qquad \text{if} \quad t \downarrow \tau.$$

[*Hint*: Use the preceding problem.]

11. Let $\Gamma(x, t; \xi, \tau)$ be the fundamental solution constructed in Theorem 4.5. Prove that

$$\Gamma(x, t; \xi, \tau) = \int_{R^n} \Gamma(x, t; y, \sigma)\Gamma(y, \sigma; \xi, \tau)\, dy \qquad (\tau < \sigma < t).$$

12. Give another proof of Theorem 4.1, by applying the maximum principle to $u + \epsilon v$ in $0 \leqslant t \leqslant 1/\alpha$, where

$$v(x, t) = \exp\{ \beta(|x|^2 + 1)e^{\alpha t}\}$$

with suitable constants α, β.

13. Let M^* be an operator of the form $M^*v = \sum \alpha_{ij} v_{x_i x_j} + \sum \beta_i v_{x_i} + \gamma v + \delta v_t$ where α_{ij}, β_i, γ, δ are continuous functions. Prove that if (4.22) holds for any u, v in C^∞ with compact support in G, then M^* is given by (4.20), i.e., M^* is the formal adjoint of M.

14. Prove Theorem 5.2.

15. Consider two stochastic differential systems of n equations

$$d\xi = \sigma(\xi, t)\, dw + b(\xi, t)\, dt, \qquad d\xi' = \sigma'(\xi', t)\, dw' + b'(\xi', t)\, dt$$

where w and w' are n-dimensional Brownian motions. Suppose that the coefficients are uniformly Lipschitz continuous and that $\sigma\sigma^*$ is uniformly positive definite. Prove that if $\sigma\sigma^* \equiv \sigma'(\sigma')^*$, $b \equiv b'$, then the two processes have the same joint distribution functions, given that $\xi(0)$, $\xi'(0)$ have the same distribution functions.

7

The Cameron–Martin–Girsanov Theorem

1. A class of absolutely continuous probabilities

Any nonnegative random variable g defines an absolutely continuous measure P_g whose Radon–Nikodym derivative is g, i.e., $P_g(A) = \int_A g\, dP$ for any measurable set A. In this section we show that the random variable

$$g = \exp\left\{ \int_0^T f(s)\, dw(s) - \tfrac{1}{2}\int_0^T |f(s)|^2\, ds \right\}$$

defines a *probability* P_g, i.e., $Eg = 1$, provided f belongs to $L_w^2[0, T]$ and satisfies a certain growth condition. More precisely:

Theorem 1.1. *Let $f = (f_1, \ldots, f_n)$ belong to $L_w^2[0, T]$ and assume that there exist positive numbers μ, C such that*

$$E \exp[\mu |f(t)|^2] \leqslant C \quad \text{for} \quad 0 \leqslant t \leqslant T. \tag{1.1}$$

Then

$$E \exp\left\{ \int_{t_1}^{t_2} f(s)\, dw(s) - \tfrac{1}{2}\int_{t_1}^{t_2} |f(s)|^2\, ds \right\} = 1 \quad \text{if} \quad 0 \leqslant t_1 < t_2 \leqslant T. \tag{1.2}$$

We first prove two lemmas.

Lemma 1.2. *If in Theorem* 1.1 *the condition* (1.1) *is replaced by the condition*

$$E \exp\left\{ \lambda \int_0^T |f(s)|^2\, ds \right\} < \infty \quad \text{for some} \quad \lambda > 1, \tag{1.3}$$

then the assertion (1.2) *is valid.*

Proof. By Lemma 4.1.1 (see (4.1.5)) there exists a sequence of continuous functions $f_m(t) = (f_{m1}(t), \ldots, f_{mn}(t))$ in $L^2_w[0, T]$ such that

$$\int_0^T |f_m(t) - f(t)|^2 \, dt \xrightarrow{P} 0 \quad \text{as} \quad m \to \infty, \tag{1.4}$$

$$\int_0^T |f_m(t)|^2 \, dt \leq \int_0^T |f(t)|^2 \, dt. \tag{1.5}$$

We may assume that each f_m is bounded, i.e., $|f_m(t, \omega)| \leq C_m$ a.s. in (t, ω), for otherwise we replace f_m by f_{m, N_m} with components $\phi_{N_m}(f_{mi})$, where $\phi_N(r) = r$ if $|r| \leq N$ and $\phi_N(r) = Nr/|r|$ if $|r| > N$, and $N_m \uparrow \infty$ if $m \uparrow \infty$. We claim that

$$E \exp\left\{2\int_{t_1}^{t_2} f_m(s)\, dw(s)\right\} \leq C'_m \qquad \text{for all} \quad 0 \leq t_1 < t_2 \leq T, \tag{1.6}$$

where C'_m is a constant depending on m.

To prove it notice that

$$Ee^{\gamma|w(t)|} = \int_{R^n} e^{\gamma|x|} \frac{e^{-|x|^2/2t}}{\sqrt{2\pi t}} \, dx = e^{\gamma^2 t/2}. \tag{1.7}$$

Let $\Pi_l = \{s^l_1, \ldots, s^l_{k_l}\}$ be a sequence of partitions of $[t_1, t_2]$ with mesh $|\Pi_l| \to 0$. Then

$$\left|\int_{t_1}^{t_2} f_m(s)\, dw(s)\right| = \left|\lim_{l\to\infty} \sum f_m(s^l_i)\left[w(s^l_{i+1}) - w(s^l_i)\right]\right|$$
$$\leq C \varliminf_{l\to\infty} \sum |w(s^l_{i+1}) - w(s^l_i)|$$

where C is a constant (depending on m). By Fatou's lemma,

$$E \exp\left|2\int_{t_1}^{t_2} f_m(s)\, dw(s)\right| \leq \varliminf_{l\to\infty} E \exp\left[2C\sum_i |w(s^l_{i+1}) - w(s^l_i)|\right]$$
$$= \varliminf_{l\to\infty} \prod_i E \exp\left[2C|w(s^l_{i+1}) - w(s^l_i)|\right].$$

By (1.7), the right-hand side is equal to

$$= \varliminf_{l\to\infty} \prod_i \exp\left[2C^2(s^l_{i+1} - s^l_i)\right] = \exp\left[2C^2(t_2 - t_1)\right].$$

Thus (1.6) follows.

Applying Itô's formula to $u = e^x$ and the process

$$\int_{t_1}^t f_m(s)\, dw(s) - \tfrac{1}{2}\int_{t_1}^t |f_m(s)|^2 \, ds,$$

we obtain

$$\exp\left\{\int_{t_1}^{t_2} f_m(s)\,dw(s) - \tfrac{1}{2}\int_{t_1}^{t_2} |f_m(s)|^2\,ds\right\} - 1$$

$$= \int_{t_1}^{t} \exp\left\{\int_{t_1}^{t_2} f_m(s)\,dw(s) - \tfrac{1}{2}\int_{t_1}^{t} |f_m(s)|^2\,ds\right\} f_m(t)\,dw(t). \tag{1.8}$$

Denote the last integral by $\int_{t_1}^{t_2} h(s)\,dw(s)$. By (1.6) and the fact that $|f_m(t)| < C_m$ it follows that

$$E\int_{t_1}^{t_2} |h(t)|^2\,dt < \infty.$$

Hence $E\int_{t_1}^{t_2} h(t)\,dw(t) = 0$. Taking the expectation in (1.8) we then get

$$E\exp\left\{\int_{t_1}^{t_2} f_m(s)\,dw(s) - \tfrac{1}{2}\int_{t_1}^{t_2} |f_m(s)|^2\,ds\right\} = 1. \tag{1.9}$$

We shall prove that

$$I \equiv E\left[\exp\left\{\int_{t_1}^{t_2} f_m(s)\,dw(s) - \tfrac{1}{2}\int_{t_1}^{t_2} |f_m(s)|^2\,ds\right\}\right]^{1+\epsilon} < K \tag{1.10}$$

for some $\epsilon > 0$, where K is a constant independent of m. This would imply that the sequence of random variables

$$X_m \equiv \exp\left\{\int_{t_1}^{t_2} f_m(s)\,dw(s) - \tfrac{1}{2}\int_{t_1}^{t_2} |f_m(s)|^2\,ds\right\}$$

is uniformly integrable. Since, by (1.4),

$$\int_{t_1}^{t_2} f_m(s)\,dw(s) \xrightarrow{P} \int_{t_1}^{t_2} f(s)\,dw(s) \qquad \text{as} \quad m\to\infty,$$

$$X_m \xrightarrow{P} \exp\left\{\int_{t_1}^{t_2} f(s)\,dw(s) - \tfrac{1}{2}\int_{t_1}^{t_2} |f(s)|^2\,ds\right\}.$$

Hence, taking $m\to\infty$ in (1.19), the assertion (1.2) follows.

It remains to prove (1.10). By Hölder's inequality,

$$I = E\left\{\exp\left[(1+\epsilon)\int_{t_1}^{t_2} f_m(s)\,dw(s) - \frac{(1+\epsilon)^3}{2}\int_{t_1}^{t_2} |f_m(s)|^2\,ds\right]\right.$$

$$\left.\cdot\exp\left[\frac{(1+\epsilon)(2\epsilon+\epsilon^2)}{2}\int_{t_1}^{t_2} |f_m(s)|^2\,ds\right]\right\}$$

$$< \left\{E\exp\left[(1+\epsilon)^2\int_{t_1}^{t_2} f_m(s)\,dw(s) - \frac{(1+\epsilon)^4}{2}\int_{t_1}^{t_2} |f_m(s)|^2\,ds\right]\right\}^{1/(1+\epsilon)}$$

$$\cdot\left\{E\exp\left[\frac{(1+\epsilon)^2(2+\epsilon)}{2}\int_{t_1}^{t_2} |f_m(s)|^2\,ds\right]\right\}^{\epsilon/(1+\epsilon)}.$$

By (1.9) the first factor on the right is equal to 1. Using (1.5) we get

$$I \leqslant \left\{ E \exp\left[\frac{(1+\epsilon)^2(2+\epsilon)}{2} \int_{t_1}^{t_2} |f(s)|^2 \, ds \right] \right\}^{\epsilon/(1+\epsilon)} \equiv K.$$

If ϵ is sufficiently small so that $(1+\epsilon)^2(2+\epsilon) < 2\lambda$, then, by (1.3), K is a finite number. This completes the proof of (1.10).

Lemma 1.3. *Under the assumptions of Lemma* 1.2,

$$E\left\{ \exp\left[\int_{t_1}^{t_2} f(s)\, dw(s) - \tfrac{1}{2} \int_{t_1}^{t_2} |f(s)|^2 \, ds \right] \Big| \mathcal{F}_{t_1} \right\} = 1 \quad a.s. \tag{1.11}$$

for any $0 \leqslant t_1 < t_2 \leqslant T$.

Proof. Taking the conditional expectation of both sides of (1.8) with respect to $\mathcal{F}_{t_1}$, we obtain the assertion (1.11) for $f = f_m$. Using (1.10) one can easily justify passage to the limit $m \to \infty$ in the formula (1.11) for $f = f_m$.

Proof of Theorem 1.1. Let $\lambda > 1$. Since $e^{\lambda x}$ is a convex function, Jensen's inequality (see Problem 7, Chapter 1) gives

$$\exp\left[\lambda \int_{t'}^{t''} |f(s)|^2 \, ds \right] = \exp\left[\frac{1}{t''-t'} \int_{t'}^{t''} \lambda(t''-t')|f(s)|^2 \, ds \right]$$
$$\leqslant \frac{1}{t''-t'} \int_{t'}^{t''} \exp\left[\lambda(t''-t')|f(s)|^2 \right] ds \qquad (t' < t'').$$

Hence, if $\lambda(t''-t') < \mu$, then (1.1) implies that

$$E \exp\left[\lambda \int_{t'}^{t''} |f(s)|^2 \, ds \right] < \infty.$$

Consequently, by Lemma 1.3,

$$E\left[\exp\left\{ \int_{t'}^{t''} f(s)\, dw(s) - \tfrac{1}{2} \int_{t'}^{t''} |f(s)|^2 \, ds \right\} \Big| \mathcal{F}_{t'} \right] = 1 \quad \text{a.s.} \tag{1.12}$$

Now let $t_1 = s_1 < s_2 < \cdots < s_m = t_2$ where $\lambda(s_{i+1} - s_i) < \mu$. Then

$$E\left\{ \exp\left[\int_{t_1}^{t_2} f \, dw - \tfrac{1}{2} \int_{t_1}^{t_2} |f|^2 \, ds| \right] \mathcal{F}_{t_1} \right\}$$
$$= E\left\{ \exp\left[\int_{t_1}^{s_{m-1}} f \, dw - \tfrac{1}{2} \int_{t_1}^{s_{m-1}} |f|^2 \, ds \right] \right.$$
$$\left. \cdot\, E \exp\left[\int_{s_{m-1}}^{s_m} f \, dw - \tfrac{1}{2} \int_{s_{m-1}}^{s_m} |f|^2 \, ds \right] \Big| \mathcal{F}_{s_{m-1}} \right\} \Big| \mathcal{F}_{t_1}$$
$$= E \exp\left[\int_{t_1}^{s_{m-1}} f \, dw - \tfrac{1}{2} \int_{t_1}^{s_{m-1}} |f|^2 \, ds \right] \Big| \mathcal{F}_{t_1}$$

by (1.12). Proceeding similarly step-by-step, we arrive at the formula

$$E\left\{\exp\left[\int_{t_1}^{t_2} f(s)\,dw(s) - \tfrac{1}{2}\int_{t_1}^{t_2} |f(s)|^2\,ds\right]\Big|\mathcal{F}_{t_1}\right\} = 1. \quad \text{a.s.} \tag{1.13}$$

Taking the expectation, (1.2) follows.

Corollary 1.4. *Under the assumptions of Theorem* 1.1, *the relation* (1.13) *holds for any* $0 \leqslant t_1 < t_2 \leqslant T$.

Example. If $f(t) = w(t)$, then condition (1.1) is satisfied with $\mu < 1/2T$.

Corollary 1.5. *For any* $f = (f_1, \ldots, f_m)$ *in* $L_w^2[0, T]$,

$$E\exp\left[\int_{t_1}^{t_2} f(s)\,dw(s) - \tfrac{1}{2}\int_{t_1}^{t_2} |f(s)|^2\,ds\right] \leqslant 1, \tag{1.14}$$

$$E\left\{\exp\left[\int_{t_1}^{t_2} f(s)\,dw(s) - \tfrac{1}{2}\int_{t_1}^{t_2} |f(s)|^2\,ds\right]\Big|\mathcal{F}_{t_1}\right\} \leqslant 1 \quad a.s. \tag{1.15}$$

for any $0 \leqslant t_1 < t_2 \leqslant T$.

To prove (1.14) we take $m \to \infty$ in (1.9) and use Fatou's lemma. To prove (1.15) we first note (by taking the conditional expectation in (1.8)) that (1.15) holds (with "=") for $f = f_m$. Now take $m \to \infty$ and use Fatou's lemma.

2. Transformation of Brownian motion

A process $\{w(t), 0 \leqslant t \leqslant T\}$ that satisfies all the conditions imposed on an n-dimensional Brownian motion (including continuity) in the interval $0 \leqslant t \leqslant T$ is called *an n-dimensional Brownian motion in the interval* $[0, T]$.

Let $w(t)$ be an n-dimensional Brownian motion in an interval $[0, T]$. Let $\mathcal{F}_t$ be an increasing family of σ-fields such that $\mathcal{F}(w(\lambda), \lambda \leqslant t)$ is a subset of $\mathcal{F}_t$ and $\mathcal{F}\{w(\lambda + t) - w(t), 0 \leqslant \lambda \leqslant T - t\}$ is independent of $\mathcal{F}_t$, for all $t \in [0, T]$. As in Chapter 3 (where $w(t)$ and $\mathcal{F}_t$ were defined for all $t \geqslant 0$), we can define $L_w^2[0, T]$ and stochastic integrals $\int_0^t f\,dw$ $(0 < t \leqslant T)$ with respect to the present family $\mathcal{F}_t$.

If $\tilde{P}$ is a measure on $(\Omega, \mathcal{F})$ given by

$$\tilde{P}(A) = \int_A f\,dP \qquad (A \in \mathcal{F}),$$

then we write $d\tilde{P}(\omega) = f(\omega)\,dP(\omega)$.

Theorem 2.1. *Let $w(t)$ be an n-dimensional Brownian motion in* $[0, T]$ *and* *let* $\phi = (\phi_1, \ldots, \phi_n)$ *be a function in* $L_w^2[0, T]$. *Define*

$$\zeta_s^t(\phi) = \int_s^t \phi(u)\, dw(u) - \tfrac{1}{2}\int_s^t |\phi(u)|^2\, du, \tag{2.1}$$

$$\tilde{w}(t) = w(t) - \int_0^t \phi(s)\, ds, \tag{2.2}$$

$$d\tilde{P}(\omega) = \exp\left[\zeta_0^T(\phi)\right] dP(\omega). \tag{2.3}$$

If

$$\tilde{P}(\Omega) = 1, \tag{2.4}$$

then $\tilde{w}(t)$, $0 \le t \le T$ *is an n-dimensional Brownian motion in the probability space* $(\Omega, \mathcal{F}, \tilde{P})$.

Recall that in Section 1 we have established sufficient conditions for (2.4) to hold. We have also proved (Corollary 1.5) that $\tilde{P}(\Omega) \le 1$ for any $\phi \in L_w^2[0, T]$.

Theorem 2.1 is due to Girsanov [1]. Cameron and Martin [1] have previously established results of the same nature on nonlinear transformations of Brownian motion. We shall refer to Theorem 2.1 as the *Cameron–Martin–Girsanov theorem.*

We begin with some lemmas.

Lemma 2.2. *If* $\phi \in L_w^2[0, T]$ *and if* $|\phi(t)| \le c$ *a.s., then, for any* $\alpha > 1$,

$$E\exp[\alpha\zeta_s^t(\phi)] \le \exp\left[\frac{\alpha^2 - \alpha}{2}(t - s)c^2\right]. \tag{2.5}$$

Proof. By Corollary 1.5, $E\exp[\zeta_s^t(\alpha\phi)] \le 1$. Hence

$$E\exp[\alpha\zeta_s^t(\phi)] = E\exp[\zeta_s^t(\alpha\phi)]\exp\left[\frac{\alpha^2 - \alpha}{2}\int_s^t |\phi(u)|^2\, du\right]$$

$$\le E\exp[\zeta_s^t(\alpha\phi)]\exp\left[\frac{\alpha^2 - \alpha}{2}(t - s)c^2\right] \le \exp\left[\frac{\alpha^2 - \alpha}{2}(t - s)c^2\right].$$

Denote by $\tilde{E}$ the expectation with respect to $\tilde{P}$.

Lemma 2.3. *Let* $\phi \in L_w^2[0, T]$. *Then for any nonnegative and* $\mathcal{F}_t$ *measurable random variable* η,

$$\tilde{E}\eta \le E\{\eta\exp[\zeta_0^t(\phi)]\}, \qquad 0 \le t \le T. \tag{2.6}$$

If ϕ *satisfies* (2.4), *then*

$$E\exp[\zeta_s^t(\phi)] = 1, \tag{2.7}$$

$$E\{\exp[\zeta_s^t(\phi)] \mid \mathcal{F}_s\} = 1 \quad a.s. \tag{2.8}$$

for all $0 \le s < t \le T$, *and*

$$\tilde{E}\eta = E\{\eta \exp[\zeta_0^t(\phi)]\} \tag{2.9}$$

for any $\mathcal{F}_t$ *measurable random variable* η *for which* $\tilde{E}\eta$ *exists.*

Proof. The assertion (2.6) follows from

$$\begin{aligned}\tilde{E}\eta &= E\eta \exp[\zeta_0^T(\phi)] = EE\{\eta \exp[\zeta_0^T(\phi)] | \mathcal{F}_t\} \\ &= E\eta \exp[\zeta_0^t(\phi)]E\{\exp[\zeta_t^T(\phi)] | \mathcal{F}_t\} \\ &\le E\eta \exp[\zeta_0^t(\phi)],\end{aligned}$$

where (1.15) has been used.

Assume now that (2.4) holds. Using (1.15) we have

$$\begin{aligned}1 = E\exp\left[\zeta_0^T(\phi)\right] &= E\exp[\zeta_0^s(\phi)]E\left\{\exp[\zeta_s^t(\phi)]E\left[\exp\left[\zeta_t^T(\phi)\right] | \mathcal{F}_t\right] | \mathcal{F}_s\right\} \\ &\le E\exp[\zeta_0^s(\phi)]E\left\{\exp[\zeta_s^t(\phi)] | \mathcal{F}_s\right\}.\end{aligned} \tag{2.10}$$

Denote by B_m $(m > 0)$ the set on which

$$E\exp[\zeta_s^t(\phi)] | \mathcal{F}_s \le 1 - 1/m.$$

Since, by (1.15), $E\{\exp[\zeta_s^t(\phi)] | \mathcal{F}_s\} \le 1$ a.s., we conclude from (2.10) that

$$1 \le \int_{\Omega \setminus B_m} \exp[\zeta_0^s(\phi)]\, dP + \left(1 - \frac{1}{m}\right)\int_{B_m} \exp[\zeta_0^s(\phi)]\, dP.$$

But since

$$\int_\Omega \exp[\zeta_0^s(\phi)]\, dP \le 1 \qquad \text{and} \qquad \exp \zeta_0^s(\phi) > 0 \quad \text{a.s.},$$

it follows that $P(B_m) = 0$. Since m is arbitrary, the assertion (2.8) follows. The validity of (2.7) is now obvious, and (2.9) follows from

$$\tilde{E}\eta = E\eta \exp[\zeta_0^t(\phi)]E\{\exp[\zeta_t^T(\phi)] | \mathcal{F}_t\} = E\eta \exp[\zeta_0^t(\phi)].$$

Lemma 2.4. *Let* ϕ *be as in Theorem 2.1 and let* η *be any* $\mathcal{F}_t$ *measurable random variable for which* $\tilde{E}\eta$ *exists. Then*

$$\tilde{E}[\eta | \mathcal{F}_s] = E\{\eta \exp[\zeta_s^t(\phi)] | \mathcal{F}_s\}. \tag{2.11}$$

Proof. Set $\gamma_1 = \exp \zeta_0^s(\phi)$, $\gamma_2 = \exp \zeta_s^t(\phi)$. Let λ be any bounded and $\mathcal{F}_s$ measurable random variable. Then

$$\tilde{E}(\lambda\eta) = \tilde{E}\lambda\tilde{E}(\eta | \mathcal{F}_s). \tag{2.12}$$

We also have

$$\tilde{E}(\lambda\eta) = E(\lambda\eta\gamma_1\gamma_2) = E\lambda\gamma_1 E(\eta\gamma_2 | \mathcal{F}_s). \tag{2.13}$$

If γ is $\mathcal{F}_s$ measurable, then, by (2.8),

$$E\gamma = E\gamma E(\gamma_2 | \mathcal{F}_s) = EE[(\gamma\gamma_2) | \mathcal{F}_s] = E\gamma\gamma_2.$$

Applying this to $\gamma = \lambda\gamma_1 E[(\eta\gamma_2)|\mathcal{F}_s]$, we get from (2.13)

$$\tilde{E}(\lambda\eta) = E\lambda\gamma_1\gamma_2 E[(\eta\gamma_2)|\mathcal{F}_s] = \tilde{E}\lambda E(\eta\gamma_2)|\mathcal{F}_s.$$

Comparing this with (2.12), we get

$$\tilde{E}\lambda\tilde{E}(\eta|\mathcal{F}_s) = \tilde{E}\lambda E(\eta\gamma_2)|\mathcal{F}_s.$$

Since λ is arbitrary bounded random variable that is $\mathcal{F}_s$ measurable, (2.11) follows.

Lemma 2.5. *Let ϕ be as in Theorem 2.1 and let $\{\phi^N\}$ be a sequence of bounded functions in $L_w^2[0, T]$ such that $\zeta_s^t(\phi^N) \to \zeta_s^t(\phi)$ in probability as $N \to \infty$. Then*

$$\int_\Omega |\exp \zeta_s^t(\phi^N) - \exp \zeta_s^t(\phi)|\, dP \to 0 \qquad \textit{if} \quad N \to \infty. \tag{2.14}$$

Proof. By Theorem 1.1,

$$\int_\Omega \exp \zeta_s^t(\phi^N)\, dP = 1. \tag{2.15}$$

Denote by $\Omega_{N,\epsilon}$ the set where

$$|\exp \zeta_s^t(\phi) - \exp \zeta_s^t(\phi^N)| < \epsilon.$$

By assumption, $P(\Omega_{N,\epsilon}) > 1 - \delta_N(\epsilon)$, $\delta_N(\epsilon) \to 0$ if $N \to \infty$. Since $\exp \zeta_s^t(\phi)$ is integrable,

$$\int_{\Omega_{N,\epsilon}} \exp[\zeta_s^t(\phi)]\, dP = 1 - \int_{\Omega\setminus\Omega_{N,\epsilon}} \exp[\zeta_s^t(\phi)]\, dP > 1 - \epsilon \tag{2.16}$$

if N is sufficiently large. Hence

$$\int_{\Omega_{N,\epsilon}} \exp\left[\zeta_s^t(\phi^N)\right] dP > 1 - 2\epsilon.$$

From (2.15) it then follows that

$$\int_{\Omega\setminus\Omega_{N,\epsilon}} \exp\left[\zeta_s^t(\phi^N)\right] dP \leqslant 2\epsilon. \tag{2.17}$$

Since ϕ satisfies (2.7), (2.16) implies that

$$\int_{\Omega\setminus\Omega_{N,\epsilon}} \exp[\zeta_s^t(\phi)]\, dP \leqslant \epsilon. \tag{2.18}$$

Making use of (2.17), (2.18), we find that

$$\begin{aligned}
&\int_\Omega |\exp \zeta_s^t(\phi) - \exp \zeta_s^t(\phi^N)|\, dP \\
&\quad \leqslant \int_{\Omega_{N,\epsilon}} |\exp \zeta_s^t(\phi) - \exp \zeta_s^t(\phi^N)|\, dP \\
&\quad\quad + \int_{\Omega\setminus\Omega_{N,\epsilon}} \left[\exp \zeta_s^t(\phi) + \exp \zeta_s^t(\phi^N)\right] dP > 4\epsilon
\end{aligned}$$

if N is sufficiently large, and the proof of (2.14) is complete.

In proving Theorem 2.1 we shall rely upon Theorem 3.6.2. Since $\tilde{w}(t)$ is a continuous process, it will suffice to prove that, for any $0 \leq s < t \leq T$,

$$\tilde{E}[\tilde{w}_i(t) - \tilde{w}_i(s)]|\mathcal{F}_s = 0 \quad \text{a.s.,} \tag{2.19}$$

$$\tilde{E}[\tilde{w}_i(t) - \tilde{w}_i(s)][\tilde{w}_j(t) - \tilde{w}_j(s)]|\mathcal{F}_s = \delta_{ij}(t-s) \quad \text{a.s.} \tag{2.20}$$

Lemma 2.6. *The assertion of Theorem 2.1 is true if $\phi(t)$ is bounded.*

Proof. We shall verify (2.19), (2.20).

We need the following special cases of the integrated form of Itô's formula $d(\zeta_1\zeta_2) = \zeta_1\, d\zeta_2 + \zeta_2\, d\zeta_1 + d\zeta_1\, d\zeta_2$:

$$\begin{aligned}\int_{t_0}^t f(s)\, dw(s) \int_{t_0}^t g(s)\, dw(s) = \int_{t_0}^t f(s) \cdot g(s)\, ds \\ + \int_{t_0}^t \left(\int_{t_0}^s g(u)\, dw(u)\right) f(s)\, dw(s) \\ + \int_{t_0}^t \left(\int_{t_0}^s f(u)\, dw(u)\right) g(s)\, dw(s),\end{aligned} \tag{2.21}$$

$$\begin{aligned}\int_{t_0}^t f(s)\, dw(s) \int_{t_0}^t h(s)\, ds = \int_{t_0}^t \left(\int_{t_0}^s h(u)\, du\right) f(s)\, dw(s) \\ + \int_{t_0}^t \left(\int_{t_0}^s f(u)\, dw(u)\right) h(s)\, ds.\end{aligned} \tag{2.22}$$

Here f and g are n-dimensional functions in $L_w^2[t_0, t]$ and h is a scalar function in $L_w^1[t_0, t]$.

We shall also need (1.8) with $f_m = \phi$, i.e.,

$$\exp \zeta_{t_1}^{t_2}(\phi) = 1 + \int_{t_1}^{t_2} \exp[\zeta_{t_1}^u(\phi)]\phi(u)\, dw(u).$$

By Lemma 2.4,

$$\begin{aligned}\tilde{E}[\tilde{w}_i(t) - \tilde{w}_i(s)]|\mathcal{F}_s &= E[w_i(t) - w_i(s)] \exp[\zeta_s^t(\phi)]|\mathcal{F}_s \\ &\quad - E\left(\int_s^t \phi_i(u)\, du\right) \exp[\zeta_s^t(\phi)]|\mathcal{F}_s \\ &= E[w_i(t) - w_i(s)]\left[1 + \int_s^t \exp[\zeta_s^u(\phi)]\phi(u)\, dw(u)\right]|\mathcal{F}_s \\ &\quad - E\left(\int_s^t \phi_i(u)\, du\right) \exp[\zeta_s^t(\phi)]|\mathcal{F}_s \\ &= E\left\{\int_s^t \phi_i(u) \exp[\zeta_s^u(\phi)]\, du|\mathcal{F}_s\right\} - E\left(\int_s^t \phi_i(u)\, du\right) \exp[\zeta_s^t(\phi)]|\mathcal{F}_s.\end{aligned} \tag{2.23}$$

In the last equation we have used (2.21) and also the fact that

$$E\left[\int_s^t h(u)\, dw(u)\right]|\mathcal{F}_s = 0$$

if

$$E\int_s^t |h(u)|^2\,du < \infty \qquad (h = (h_1, \ldots, h_n)) \tag{2.24}$$

where, for one stochastic integral,

$$h_i(u) = \int_s^u \exp[\zeta_s^v(\phi)]\phi(v)\,dw(v), \qquad h_j = 0 \quad \text{if} \quad j \neq i$$

and, for another stochastic integral,

$$h(u) = (w_i(u) - w_i(s)) \exp[\zeta_s^u(\phi)]\phi(u).$$

Using the fact that $|\phi| \leq c$ (c constant) and the estimate (2.5), one can easily see that these two functions $h(u)$ indeed satisfy (2.24).

In view of (2.8) the right-hand side of (2.23) is equal to zero. Thus (2.19) holds.

In order to prove (2.20), set

$$\xi(t) = w(t) - w(s), \qquad \eta(t) = \exp \zeta_0^t(\phi).$$

Then

$$\begin{aligned}
&[\tilde{w}_i(t) - \tilde{w}_i(s)][\tilde{w}_j(t) - \tilde{w}_j(s)]\,\eta(t)\\
&= \left[\xi_i(t) - \int_s^t \phi_i(u)\,du\right]\left[\xi_j(t) - \int_s^t \phi_j(u)\,du\right]\eta(t)\\
&= \xi_i(t)\xi_j(t)\left[1 + \int_s^t \eta(u)\phi(u)\,dw(u)\right] - \left[\int_s^t \phi_i(u)\,du\right]\xi_j(t)\eta(t)\\
&\quad - \left[\int_s^t \phi_j(u)\,du\right]\xi_i(t)\eta(t) + \left[\int_s^t \phi_i(u)\,du\right]\left[\int_s^t \phi_j(u)\,du\right]\eta(t)\\
&\equiv \xi_i(t)\xi_j(t) + J_1 + J_2 + J_3 + J_4,
\end{aligned}$$

where J_i ($i = 2, 3, 4$) denotes the $(i + 1)$th term on the right.

Denote by $\cong$ equality moduli stochastic integrals, i.e., $A \cong B$ if $A = B + \sum\int_s^t h_j(u)\,dw_j(u)$. Then, using (2.21), (2.22), we find that

$$\begin{aligned}
J_1 &= \xi_i(t)\int_s^t \left[\int_s^u \eta(v)\phi(v)\,dw(v)\right] dw_j(u) + \xi_i(t)\int_s^t \xi_j(u)\eta(u)\phi(u)\,dw(u)\\
&\quad + \xi_i(t)\int_s^t \eta(u)\phi_j(u)\,du\\
&\cong \delta_{ij}\int_s^t \left[\int_s^u \eta(v)\phi(v)\,dw(v)\right] du\\
&\quad + \int_s^t \xi_j(u)\eta(u)\phi_i(u)\,du + \int_s^t \xi_i(u)\eta(u)\phi_j(u)\,du.
\end{aligned}$$

Next,

$$J_2 = -\left[\int_s^t \phi_i(u)\,du\right]\xi_j(t)\eta(t) \cong -\left[\int_s^t \xi_j(u)\phi_i(u)\,du\right]\eta(t),$$

$$J_3 = -\left[\int_s^t \phi_j(u)\,du\right]\xi_i(t)\eta(t) \cong -\left[\int_s^t \xi_i(u)\phi_j(u)\,du\right]\eta(t).$$

Notice that due to the boundedness of ϕ and Lemma 2.2, all the stochastic integrals $\int_s^t h(u)\,dw(u)$ which were dropped in the final expressions for J_1, J_2, J_3 are such that (2.24) holds. Consequently,

$$\begin{aligned}\tilde{E}[\tilde{w}_i(t) - \tilde{w}_i(s)][\tilde{w}_j(t) - \tilde{w}_j(s)]|\mathcal{F}_s - \delta_{ij}(t-s) \\ = -E\left[\int_s^t [w_i(t) - w_i(u)]\phi_j(u)\,du\right]\eta(t)|\mathcal{F}_s \\ -E\left[\int_s^t [w_j(t) - w_j(u)]\phi_i(u)\,du\right]\eta(t)|\mathcal{F}_s + EJ_4|\mathcal{F}_s. \qquad (2.25)\end{aligned}$$

The first term on the right-hand side is equal to

$$\begin{aligned}- E\left[\int_s^t [\tilde{w}_i(t) - \tilde{w}_i(u)]\phi_j(u)\eta(t)\,du\right]|\mathcal{F}_s \\ -E\left[\int_s^t \left[\int_u^t \phi_i(v)\,dv\right]\phi_j(u)\eta(t)\,du\right]|\mathcal{F}_s. \qquad (2.26)\end{aligned}$$

The first term in (2.26) is equal to

$$\begin{aligned}- E\left\{\int_s^t \{E[\tilde{w}_i(t) - \tilde{w}_i(u)\ \phi_j(u)]\,\eta(t)|\mathcal{F}_u\}\,du\right\}|\mathcal{F}_s \\ = -E\left\{\int_s^t \{\tilde{E}[\tilde{w}_i(t) - \tilde{w}_i(u)]\ \phi_j(u)|\mathcal{F}_u\}\,du\right\}|\mathcal{F}_s \\ = -E\left\{\int_s^t \{\phi_j(u)\tilde{E}[\tilde{w}_i(t) - \tilde{w}_i(u)]\ |\mathcal{F}_u\}\,du\right\}|\mathcal{F}_s = 0\end{aligned}$$

by (2.8) and (2.19). Hence the first term on the right-hand side of (2.25) is equal to the second term in (2.26).

Treating the second term on the right-hand side of (2.25) similarly, we conclude that

$$\begin{aligned}\tilde{E}[\tilde{w}_i(t) - \tilde{w}_i(s)][\tilde{w}_j(t) - \tilde{w}_j(s)]|\mathcal{F}_s - \delta_{ij}(t-s) \\ = E\left\{\int_s^t \frac{d}{du}\left\{\left[\int_u^t \phi_i(v)\,dv\right]\left[\int_u^t \phi_j(v)\,dv\right]\right\}\eta(t)\,du\right\}|\mathcal{F}_s + EJ_4|\mathcal{F}_s \\ = -E\left\{\left[\int_s^t \phi_i(u)\,du\right]\left[\int_s^t \phi_j(u)\,du\right]\eta(t)\right\}|\mathcal{F}_s + EJ_4|\mathcal{F}_s = 0\end{aligned}$$

by the definition of J_4. This completes the proof of (2.20).

Proof of Theorem 2.1. Let $\phi^N = (\phi_1^N, \ldots, \phi_n^N)$ be bounded functions in $L_w^2[0, T]$ such that

$$\int_0^T |\phi^N(t) - \phi(t)|^2\,dt \to 0 \quad \text{a.s.} \qquad \text{as} \quad N \to \infty.$$

Define $\tilde{w}^N(t) = (\tilde{w}_1^N(t), \ldots, \tilde{w}_n^N(t))$ by

$$\tilde{w}^N(t) = w(t) - \int_s^t \phi^N(u)\, du.$$

Notice that

$$\tilde{w}^N(t) \to \tilde{w}(t) \quad \text{a.s.} \tag{2.27}$$

In view of Lemma 2.6, $\tilde{w}^N(t)$ is a Brownian motion in the interval $[0, T]$, with respect to the probability space $(\Omega, \mathcal{F}, \tilde{P}_N)$ where $\tilde{P}_N$ is defined by (2.3) with $\phi = \phi^N$. Consequently, for any $0 \leqslant t_0 < t_1 < \cdots < t_k \leqslant T$ and $\lambda_1, \ldots, \lambda_k$ real,

$$E \exp\left\{ \sqrt{-1} \sum_{j=1}^{k} \lambda_j \left[\tilde{w}_i^N(t_j) - \tilde{w}_i^N(t_{j-1}) \right] \right\} \exp \zeta_0^T(\phi^N)$$

$$= \exp\left\{ - \sum_{j=1}^{k} \tfrac{1}{2} \lambda_j^2 (t_j - t_{j-1}) \right\}, \tag{2.28}$$

and for any $0 \leqslant t, s \leqslant T$ and λ, μ real,

$$E \exp\left\{ \sqrt{-1} \left[\lambda \tilde{w}_i^N(t) + \mu \tilde{w}_j^N(s) \right] \right\} \exp \zeta_0^T(\phi^N)$$

$$= \exp[-(\lambda^2/2)t] \exp[-(\mu^2/2)s] \qquad \text{if} \quad i \neq j. \tag{2.29}$$

We shall need the following fact:

If $\alpha_N \to \alpha$ a.s., $|\alpha_N| \leqslant c$ (c const), and $E|\beta_N - \beta| \to 0$, then $E|\alpha_N \beta_N - \alpha\beta| \to 0$. (2.30)

To prove (2.30) write

$$E|\alpha_N \beta_N - \alpha\beta| \leqslant E|\alpha_N|\,|\beta_N - \beta| + E|\alpha_N - \alpha|\,|\beta|$$
$$\leqslant cE|\beta_N - \beta| + E|\alpha_N - \alpha|\,|\beta|.$$

The first integral on the right converges to zero by assumption, and the second integral converges to zero by the Lebesgue bounded convergence theorem. Thus (2.30) follows.

Taking α_N to be the first exponent on the left-hand side of (2.28) and taking $\beta_N = \exp \zeta_0^T(\phi^N)$, and recalling (2.27) and Lemma 2.5, we see that the assumptions in (2.30) are satisfied. Hence, letting $N \to \infty$ in (2.28) we obtain

$$\tilde{E} \exp\left\{ \sqrt{-1} \sum_{j=1}^{k} \lambda_j [\tilde{w}_i(t_j) - \tilde{w}_i(t_{j-1})] \right\} = \exp\left\{ - \sum_{j=1}^{k} \tfrac{1}{2} \lambda_j^2 (t_j - t_{j-1}) \right\}.$$

This implies (see Problem 2, Chapter 3) that $\tilde{w}_i(t)$, $0 \leqslant t \leqslant T$ is a Brownian motion in the space $(\Omega, \mathcal{F}, \tilde{P})$.

If we let $N \to \infty$ in (2.29) and apply (2.30) (again using (2.27) and Lemma 2.5), we obtain

$$\tilde{E} \exp\left\{ \sqrt{-1} \left[\lambda \tilde{w}_i(t) + \mu \tilde{w}_j(s) \right] \right\} = \exp[-(\lambda^2/2)t] \exp[-(\mu^2/2)s].$$

This proves that $\mathcal{F}\{\tilde{w}_i(t), 0 \leq t \leq T\}$ is independent of $\mathcal{F}\{\tilde{w}_j(t), 0 \leq t \leq T\}$ if $i \neq j$. Thus $\tilde{w}(t)$ is an n-dimensional Brownian motion in the space $(\Omega, \mathcal{F}, \tilde{P})$.

Remark. Since $\tilde{w}^N$ satisfies (2.19), (2.20), the remark at the end of Section 3.5 implies that, if $0 \leq s < t \leq T$,

$$\tilde{E}\exp\Big[i\lambda \cdot \big(\tilde{w}^N(t) - \tilde{w}^N(s)\big)\Big]|\mathcal{F}_s = \exp\Big[-\tfrac{1}{2}|\lambda|^2(t-s)\Big] \quad \text{a.s.} \tag{2.31}$$

If α_N, β_N are as in (2.30) then $E(\alpha_N\beta_N - \alpha\beta)|\mathcal{F}_s \to 0$ in probability, as $N \to \infty$. Using this fact we get, as $N \to \infty$ in (2.31),

$$\tilde{E}\exp[i\lambda \cdot (\tilde{w}(t) - \tilde{w}(s))]|\mathcal{F}_s = \exp\Big[-\tfrac{1}{2}|\lambda|^2(t-s)\Big] \quad \text{a.s.} \tag{2.32}$$

Now, if X has n-dimensional normal distribution, then $E|X|^m \leq c^{m+1}m!$ for some $c > 0$. Hence the series

$$\sum_{m=0}^{\infty} \frac{i^m(\lambda \cdot X)^m}{m!}$$

is absolutely convergent in $L^1(\Omega)$ if $|\lambda| < 1/c$. It follows that

$$E\left[e^{i\lambda \cdot X} - \sum_{m=1}^{l} \frac{i^m(\lambda \cdot X)^m}{m!}\right]|\mathcal{F}_s \xrightarrow{P} 0 \qquad \text{if} \quad l \to \infty. \tag{2.33}$$

(For if $E|\alpha_N - \alpha| \to 0$, then $E(\alpha_N - \alpha)|\mathcal{F}_s \xrightarrow{P} 0$.) If

$$Ee^{i\lambda \cdot X}|\mathcal{F}_s = \exp[-\tfrac{1}{2}|\lambda|^2(t-s)],$$

then (2.33) implies that

$$\sum_{m=0}^{\infty} \frac{i^m}{m!} E(\lambda \cdot X)^m|\mathcal{F}_s = \sum_{m=0}^{\infty} \frac{|\lambda|^{2m}(t-s)^m}{2^m m!} \quad \text{a.s.} \quad \left(|\lambda| < \frac{1}{c}\right). \tag{2.34}$$

In view of (2.32), (2.34) can be applied to $X = \tilde{w}(t) - \tilde{w}(s)$. Comparing coefficients for $m = 1, 2$ we deduce:

$$\tilde{E}(\tilde{w}(t) - \tilde{w}(s))|\mathcal{F}_s = 0 \qquad \text{if} \quad 0 \leq s < t \leq T, \tag{2.35}$$

$$\tilde{E}[\tilde{w}_i(t) - \tilde{w}_i(s)][\tilde{w}_j(t) - \tilde{w}_j(s)]|\mathcal{F}_s = \delta_{ij}(t-s) \qquad \text{if} \quad 0 \leq s < t \leq T. \tag{2.36}$$

3. Girsanov's formula

Let $w(t)$ be an n-dimensional Brownian motion in the interval $0 \leq t \leq T$. In Chapter 4 we have defined the concept of the stochastic integral $\int_0^T f(s)\,dw(s)$ for functions f in $L^2_w[0, T]$, where $L^2_w[0, T]$ is defined with respect to an increasing family of σ-fields $\mathcal{F}_t$ satisfying:

(i) $\mathcal{F}(w(\lambda), \lambda \leq t)$ is in $\mathcal{F}_t$, for any $0 \leq t \leq T$;

(ii) $\mathcal{F}(w(t+\lambda) - w(t), 0 \leq \lambda \leq T - t)$ is independent of $\mathcal{F}_t$, for any $0 \leq t \leq T$.

(More precisely, we have assumed that the $\mathcal{F}_t$ are defined for all $t \geq 0$ and satisfy (i) for all $t \geq 0$, and (ii) with $0 \leq \lambda < \infty$ and for any $t \geq 0$. However we have actually made use only of (i), (ii).)

Suppose we replace (ii) by the weaker condition:

(ii′) For all $0 \leq s < t \leq T$,

$$E[w(t) - w(s)] | \mathcal{F}_s = 0,$$
$$E[w_i(t) - w_i(s)][w_j(t) - w_j(s)] | \mathcal{F}_s = \delta_{ij}(t - s).$$

Then we can still define the stochastic integral and derive all the formulas and estimates as in Chapter 4. In fact we first define the stochastic integral for any step function, next derive Lemmas 4.2.2, 4.2.3 (see Problem 1), and then approximate any f in $L^2_w[0, T]$ by bounded step functions and use the procedure of Section 4.2. All the results of Chapters 4, 5 and of Sections 1, 2 of this chapter remain valid, without any change, for this slightly more general notion of the stochastic integral.

Definition. From now on we shall assume only that the family $\mathcal{F}_t$ satisfies (i) and (ii′). We shall say that $\mathcal{F}_t$ is *adapted* to $w(t)$.

Let $\phi(t)$ be as in Theorem 2.1. In view of (2.35), (2.36), $\mathcal{F}_t$ is adapted to $\tilde{w}(t)$. If $f \in L^2_w[0, T]$, then

$$P\left\{ \int_0^T |f(s, \omega)|^2 \, ds < \infty \right\} = 1.$$

Since $\tilde{P}$ is absolutely continuous with respect to P,

$$\tilde{P}\left\{ \int_0^T |f(s, \omega)|^2 \, ds < \infty \right\} = 1.$$

Hence, if $f \in L^2_w[0, T]$, with any $\mathcal{F}_t$ adapted to $w(t)$, then $f \in L^2_{\tilde{w}}[0, T]$ with the same $\mathcal{F}_t$.

Let f_k be a sequence of step functions in $L^2_w[0, T]$ such that

$$\int_0^T |f_k(s) - f(s)|^2 \, ds \to 0 \quad \text{a.s.} \qquad \text{in } P.$$

Then $f_k \in L^2_{\tilde{w}}[0, T]$ and

$$\int_0^T |f_k(s) - f(s)|^2 \, ds \to 0 \quad \text{a.s.} \qquad \text{in } \tilde{P}.$$

Consequently

$$\int_0^t f_k(s)\,dw(s) \xrightarrow{P} \int_0^t f(s)\,dw(s),$$

$$\int_0^t f_k(s)\,d\tilde{w}(s) \xrightarrow{\tilde{P}} \int_0^t f(s)\,d\tilde{w}(s).$$

Hence, for a subsequence $\{k'\}$,

$$\begin{aligned} &\int_0^t f_{k'}(s)\,dw(s) \to \int_0^t f(s)\,dw(s) \quad \text{a.s.} \quad \text{in } P,\\ &\int_0^t f_{k'}(s)\,d\tilde{w}(s) \to \int_0^t f(s)\,d\tilde{w}(s) \quad \text{a.s.} \quad \text{in } \tilde{P}. \end{aligned} \tag{3.1}$$

From (2.2) we easily see that

$$\int_0^t f_{k'}(s)\,dw(s) = \int_0^t f_{k'}(s)\,d\tilde{w}(s) + \int_0^t f_{k'}(s)\phi(s)\,ds. \tag{3.2}$$

It is also clear that

$$\int_0^t f_{k'}(s)\phi(s)\,ds \to \int_0^t f(s)\phi(s)\,ds \quad \text{a.s.} \quad \text{in } P.$$

Hence, taking $k' \to \infty$ in (3.2) and using (3.1), we get

$$\int_0^t f(s)\,dw(s) = \int_0^t f(s)\,d\tilde{w}(s) + \int_0^t f(s)\phi(s)\,ds \quad \text{if} \quad f \in L_w^2[0, T], \tag{3.3}$$

a.s. in P (or in $\tilde{P}$).

Definition. A stochastic process $x(t)$ $(0 \leq t \leq T)$ is called an *Itô process with respect to* $\{w(t), P, \mathcal{F}_t\}$ (where $\mathcal{F}_t$ is adapted to $w(t)$) *relative to the pair* $\sigma(t)$, $b(t)$ if

$$x(t) = x(0) + \int_0^t b(s)\,ds + \int_0^t \sigma(s)\,dw(s) \qquad \text{for} \quad 0 \leq t \leq T \tag{3.4}$$

where $b = (b_1, \ldots, b_n)$ is in $L_w^1[0, T]$ and $\sigma = (\sigma_{ij})_{i,j=1}^n$ is in $L_w^2[0, T]$.

Theorem 3.1. *Let $x(t)$ $(0 \leq t \leq T)$ be an Itô process with respect to $\{w(t), P, \mathcal{F}_t\}$ relative to the matrix $\sigma(t)$ and the vector $b(t)$. Let $\phi(t) = (\phi_1(t), \ldots, \phi_n(t))$ belong to $L_w^2[0, T]$ and set*

$$\zeta_s^t(\phi) = \int_s^t \phi(u)\,dw(u) - \tfrac{1}{2}\int_s^t |\phi(u)|^2\,du, \tag{3.5}$$

$$\tilde{w}(t) = w(t) - \int_0^t \phi(s)\,ds, \tag{3.6}$$

$$d\tilde{P}(\omega) = \exp\left[\zeta_0^T(\phi)\right] dP(\omega). \tag{3.7}$$

Assume that

$$\tilde{P}(\Omega) = 1. \tag{3.8}$$

Then $\tilde{w}(t)$ is an n-dimensional Brownian motion in the probability space $(\Omega, \mathcal{F}, \tilde{P})$, $\mathcal{F}_t$ is adapted to $\tilde{w}(t)$, and the process $x(t)$ is an Itô process with respect to $\{\tilde{w}(t), \tilde{P}, \mathcal{F}_t\}$ relative to the matrix $\sigma(t)$ and the vector

$$\tilde{b}(t) = b(t) + \sigma(t)\phi(t). \tag{3.9}$$

This follows by combining Theorem 2.1 with (2.35), (2.36), and (3.3).

Denote by $\mathcal{C}_T^n$ the space of continuous functions $x(t)$ from $0 \leq t \leq T$ into R^n. Denote by $\mathfrak{M}_T$ the σ-algebra generated by the sets $\{x(t) \in A\}$ where $0 \leq t \leq T$ and A is any Borel set in R^n.

A continuous n-dimensional process $\xi(t)$ $(0 \leq t \leq T)$ in $(\Omega, \mathcal{F}, P)$ induces a probability μ_p on $(\mathcal{C}_T^n, \mathfrak{M}_T)$ as follows:

Denote by ξ_ω the sample path $t \to \xi(t, \omega)$ $(0 \leq t \leq T)$, and denote by $\mathcal{C}_T^n(\xi)$ the set of all elements ξ_ω, $\omega \in \Omega$. For any set $B \in \mathfrak{M}_T$ define

$$\mu_P(B) = P\{\omega; \xi_\omega \in B\}. \tag{3.10}$$

It is clear that μ_P is a probability, and $\mu_P(B) = \mu_P[B \cap \mathcal{C}_T^n(\xi)]$. In particular

$$\mu_P\{x; x(t_1) \in A_1, \ldots, x(t_k) \in A_k\} = P\{\omega; \xi(t_1, \omega) \in A_1, \ldots, \xi(t_k, \omega) \in A_k\}. \tag{3.11}$$

Suppose now that Q is another probability on $(\Omega, \mathcal{F})$ given by

$$dQ(\omega) = \rho(\omega)\, dP(\omega); \tag{3.12}$$

thus $\int \rho(\omega)\, dP(\omega) = 1$. Introduce the probability μ_Q on $(\mathcal{C}_T^n, \mathfrak{M}_T)$ corresponding to Q:

$$\mu_Q(B) = Q\{\omega; \xi_\omega \in B\}.$$

Lemma 3.2 *Under the foregoing assumptions,*

$$\frac{d\mu_Q}{d\mu_P}(\xi_\omega) = \rho(\omega), \tag{3.13}$$

i.e., for any $B \in \mathfrak{M}_T$, $B \subset \mathcal{C}_T^n(\xi)$,

$$\mu_Q(B) = \int_B \rho(\omega)\, d\mu_P(\xi_\omega). \tag{3.14}$$

Proof. It suffices to verify (3.14) for sets of the form

$$B = \{\xi_\omega; \xi(t_1, \omega) \in A_1, \ldots, \xi(t_k, \omega) \in A_k\}.$$

Let

$$\tilde{B} = \{\omega; \xi(t_1, \omega) \in A_1, \ldots, \xi(t_k, \omega) \in A_k\}.$$

Then, from the definitions of μ_Q and Q,

$$\mu_Q(B) = Q(\tilde{B}) = \int_{\tilde{B}} \rho(\omega)\, dP(\omega). \tag{3.15}$$

Now, (3.10) implies that

$$\int_{\mathcal{C}_T^n(\xi)} f(\omega)\, d\mu_P(\xi_\omega) = \int_\Omega f(\omega)\, dP(\omega)$$

for any bounded measurable function f. Hence, the right-hand side of (3.15) is equal to

$$\int_B \rho(\omega)\, d\mu_P(\xi_\omega),$$

and (3.14) follows.

Corollary 3.3. *Under the assumptions of Lemma 3.2, μ_Q is absolutely continuous with respect to μ_P.*

Proof. Let $\hat{B} \in \mathfrak{M}_T$, $\mu_P(\hat{B}) = 0$. Then $\mu_P(B) = 0$ where $B = \hat{B} \cap \mathcal{C}_T^n(\xi)$. By (3.14), $\mu_Q(B) = 0$. But then $\mu_Q(\hat{B}) = \mu_Q(\hat{B} \cap \mathcal{C}_T^n(\xi)) = \mu_Q(B) = 0$.

We shall now apply the preceding results to solutions of two systems of n stochastic differential equations:

$$d\xi(t) = \sigma(\xi(t), t)\, dw(t) + b(\xi(t), t)\, dt, \tag{3.16}$$

and

$$d\tilde{\xi}(t) = \sigma(\tilde{\xi}(t), t)\, dw(t) + \tilde{b}(\xi(t), t)\, dt. \tag{3.17}$$

We assume:

(G) $\sigma(x, t)$, $b(x, t)$, $\tilde{b}(x, t)$ are measurable in $R^n \times [0, T]$ and $|\sigma(x, t)| \le K(1 + |x|)$, $|b(x, t)| \le K(1 + |x|)$, $|\tilde{b}(x, t)| \le K(1 + |x|)$ where K is a constant. Further, for any $N > 0$,

$$|\sigma(x, t) - \sigma(\bar{x}, t)| \le K_N|x - \bar{x}|, \qquad |b(x, t) - b(\bar{x}, t)| \le K_N|x - \bar{x}|,$$
$$|\tilde{b}(x, t) - \tilde{b}(\bar{x}, t)| \le K_N|x - \bar{x}|$$

if $|x| \le N$, $|\bar{x}| \le N$, $0 \le t \le T$, where K_N is a constant.

We shall also assume that

$$\sigma(x, t) \quad \text{is a nonsingular matrix.} \tag{3.18}$$

Introduce the functions

$$\psi(x, t) = \sigma^{-1}(x, t)[\tilde{b}(x, t) - b(x, t)], \tag{3.19}$$
$$\phi(t) = \psi(\xi(t), t). \tag{3.20}$$

Let $\xi(t)$ be a solution of (3.16). We shall impose the condition that

$$E \exp[\mu|\phi(t)|^2 < C \quad \text{if} \quad 0 < t < T, \qquad \text{for some} \quad \mu > 0, \quad C > 0. \tag{3.21}$$

Then, by Theorem 1.1, ϕ satisfies the conditions imposed in Theorem 3.1. Applying Theorem 3.1 to $x(t) = \xi(t)$ we conclude that

$$\xi(t) = \xi(0) + \int_0^t \sigma(\xi(s), s)\, d\tilde{w}(s) + \int_0^t \tilde{b}(\xi(s), s)\, ds$$

where $\tilde{w}(t)$ is the Brownian motion defined by (3.6). Thus, $\xi(t)$ and $\tilde{\xi}(t)$ are solutions of the same stochastic differential system, except that the Brownian motions are not the same. If $\tilde{\xi}(0) = \xi(0) = x$, then, by Theorem 5.3.5, $\xi(t)$ (in $(\Omega, \mathcal{F}, \tilde{P})$) and $\tilde{\xi}(t)$ (in $(\Omega, \mathcal{F}, P)$) induce the same probability in $(\mathcal{C}_T^n, \mathfrak{M}_T)$.

Denote by μ_ξ the probability induced by $\xi(t)$ and denote by $\mu_{\tilde{\xi}}$ the probability induced by $\tilde{\xi}(t)$, in the space $(\mathcal{C}_T^n, \mathfrak{M}_T)$; both processes are considered here in the same probability space $(\Omega, \mathcal{F}, P)$. Applying Lemma 3.2 and Corollary 3.3 we obtain the following result:

Theorem 3.4. *Let* σ, b, $\tilde{b}$ *satisfy* (G) *and* (3.18) *and let* $\xi(t)$, $\tilde{\xi}(t)$, *be solutions of* (3.16) *and* (3.17) *respectively, with* $\xi(0) = \tilde{\xi}(0) = x$. *If the function* $\phi(t)$ *defined by* (3.20), (3.19) *satisfies* (3.21), *then*

$$\frac{d\mu_{\tilde{\xi}}}{d\mu_\xi}(\xi_\omega) = \exp \zeta_0^T(\phi) \quad a.s. \tag{3.22}$$

where $\zeta_0^T(\phi)$ *is defined by* (3.5).

Example. If $\psi(t, x)$ is bounded function, then (3.21) holds.

We shall refer to either Theorem 3.1 or Theorem 3.4 as the *the Girsanov theorem*, and to (3.22) as *the Girsanov formula*.

PROBLEMS

1. Let $\mathcal{F}_t$ satisfy (i), (ii′) of Section 3. Define the integral $\int_0^T f\, dw$ for any step function f in $L_w^2[0, T]$ ($f = f_i$ on $[t_i, t_{i+1})$) by $\sum f_i[w(t_{i+1}) - w(t_i)]$. Prove:

(a) Lemmas 4.2.2, 4.2.3 hold for any bounded step function in $L_w^2[0, T]$.

(b) If f is any step function ($f = f_i$ on $[t_i, t_{i+1}]$) and g_k is a bounded step function such that $g_k = f_i$ on $[t_i, t_{i+1})$ if $|f_i| < k$ and $g_k = 0$ on $[t_i, t_{i+1})$ if

$|f_i| > k$, then

$$\int_0^T |f - g_k|^2\,dt \xrightarrow{P} 0, \qquad \int_0^T g_k\,dw \xrightarrow{P} \int_0^T f\,dw \qquad \text{if } f \in L_w^2[0, T],$$

$$E\int_0^T |g_k - f|^2\,dt \to 0, \qquad E\left|\int_0^T g_k\,dw - \int_0^T f\,dw\right|^2 \to 0 \qquad \text{if } f \in M_w^2[0, T].$$

(c) Use (a), (b) in order to prove Lemmas 4.2.2, 4.2.3 for any step function f in $L_w^2[0, T]$.

2. Consider a system of n *stochastic functional differential equations*

$$x(t, \omega) = x(0, \omega) + \int_0^t \sigma(x(\,\cdot\,, \omega), s)\,dw(s, \omega) + \int_0^t b(x(\,\cdot\,, \omega), s)\,ds \qquad (3.23)$$

where $b = (b_1, \ldots, b_n)$, $\sigma = (\sigma_{ij})_{i,j=1}^n$, and $b_i(x(\,\cdot\,), t)$, $\sigma_{ij}(x(\,\cdot\,), t)$ are measurable functions on $\mathcal{C}_T^n \times [0, T]$, measurable with respect to $\mathfrak{M}_t$, for each t (where $\mathfrak{M}_t$ is the σ-algebra generated by the sets $\{x(s) \in A\}$, $0 \leqslant s \leqslant t$, A a Borel set in R^n). Assume that

$$|f(x(\,\cdot\,), t)| \leqslant K(1 + \|x(\,\cdot\,)\|),$$
$$|f(x(\,\cdot\,), t) - f(\hat{x}(\,\cdot\,), t)| \leqslant K\|x(\,\cdot\,) - \hat{x}(\,\cdot\,)\|$$

for $f = \sigma_{ij}$ and $f = b_i$, where $\|x(\,\cdot\,)\| = \max_{0 \leqslant t \leqslant T}|x(t)|$.

(i) Prove by the method of successive approximations that for any $x(0, \omega)$ in $L^2(\Omega)$ which is independent of $\mathcal{F}(w(s), 0 \leqslant s \leqslant T)$ there exists a unique solution of (3.23) satisfying $E|x(t, \omega)|^2 \leqslant$ const $(0 \leqslant t \leqslant T)$.

(ii) Prove uniqueness in the sense of probability law.

3. Extend Girsanov's formula to stochastic functional differential equations.

4. Let $b(x)$ be uniformly Lipschitz continuous in $x \in R^1$. Consider the stochastic differential equation $d\xi(t) = dw(t) + b(\xi(t))\,dt$. Prove that

$$P(s, x, t, A) = \int_A \frac{\Phi(s, x, t, y)}{\sqrt{2\pi(t-s)}} \exp\left[-\frac{(y-x)^2}{2(t-s)}\right] dy$$

where

$$\Phi(s, x, t, y) = E \exp\left\{\int_s^t b(\bar{\xi}_{x,s}(u))\,dw(u) - \tfrac{1}{2}\int_s^t b^2(\bar{\xi}_{x,s}(u))\,du\right\} | \bar{\xi}_{x,s}(t)$$

and $\bar{\xi}_{x,s}(t) = x + w(t) - w(s)$.

5. Let $w(t)$ be an n-dimensional Brownian motion. The processes $w(t)$ and $2w(t)$ induce probability measures in $(\mathcal{C}_T^n, \mathfrak{M}_T)$, μ_w and μ_{2w} respectively. Prove that μ_w and μ_{2w} are mutually singular.

6. Let $y = f(x)$ be a local diffeomorphism in R^n. A given stochastic system $dx = \sigma(x)\,dw + b(x)\,dt$ is then transformed, by means of Itô's formula, into $dy = \hat{\sigma}(y)\,dw + \hat{b}(y)\,dt$. Denote by L and $\hat{L}$ the differential operators

corresponding to the first and second stochastic systems, respectively. If $u(x) = \hat{u}(y)$ where $y = f(x)$, prove that $Lu(x) = \hat{L}\hat{u}(y)$; thus the relation between a stochastic differential system and its differential operator is invariant under diffeomorphism.

8

Asymptotic Estimates for Solutions

In this chapter (except for Section 1) we consider the behavior, as $t \to \infty$, of solutions of a stochastic differential system in case the diffusion matrix is nondegenerate for all $(x, t) \in R^n \times [0, \infty)$.

1. Unboundedness of solutions

Consider a system of n stochastic differential equations

$$d\xi(t) = \sigma(\xi(t), t)\, dw(t) + b(\xi(t), t)\, dt \tag{1.1}$$

with initial condition

$$\xi(0) = \xi_0 \tag{1.2}$$

where ξ_0 is independent of $\mathcal{F}\{w(t), t \geq 0\}$ and $E|\xi_0|^2 < \infty$. Set

$$b = (b_1, \ldots, b_n), \qquad \sigma = (\sigma_{ij})_{i,j=1}^n, \qquad a = (a_{ij}), \qquad a_{ij} = \sum_{k=1}^n \sigma_{ik}\sigma_{jk}.$$

If C is the matrix (c_{ij}), we write

$$|C| = \left\{ \sum_{i,j} (c_{ij})^2 \right\}^{1/2}.$$

We shall need the following conditions:

(A_0) $\sigma(x, t)$ and $b(x, t)$ are measurable functions in $R^n \times [0, \infty)$ and, for any $T > 0$, $R > 0$, there are positive constants C_T, $C_{T,R}$ such that

$$|\sigma(x, t)| + |b(x, t)| \leq C_T(1 + |x|)$$
$$\text{if} \quad x \in R^n, \quad 0 \leq t \leq T,$$
$$|\sigma(x, t) - \sigma(\bar{x}, t)| + |b(x, t) - b(\bar{x}, t)| \leq C_{T,R}|x - \bar{x}|$$
$$\text{if} \quad |x| \leq R, \quad |\bar{x}| \leq R, \quad 0 \leq t \leq T.$$

(A_1) For any $R > 0$ there exist positive constants γ_R, Γ_R such that

$$a_{11}(x, t)\Gamma_R + b_1(x, t) \geq \gamma_R \qquad \text{if} \quad |x| \leq R, \quad t \geq 0.$$

This condition is satisfied, for example, if $a_{11}(x, t) \geq \lambda$, $b_1(x, t) \leq C$ where λ, C are positive constants.

Theorem 1.1. *Suppose* (A_0) *and* (A_1) *hold. Let* $\xi(t)$ *be the solution of* (1.1), (1.2). *Then*

$$\limsup_{t\to\infty} |\xi(t)| = \infty \quad a.s. \tag{1.3}$$

Proof. Since $\xi(t)$ is a continuous process, the assertion (1.3) is equivalent to the assertion that

$$\sup_{t>0} |\xi(t)| = \infty \quad \text{a.s.} \tag{1.4}$$

Consider first the case where the range of ξ_0 lies in a bounded set K. Let B be any open ball containing K and denote by $\tau(B)$ the exit time from B, i.e.,

$\tau(B) =$ first $\bar{t}$ such that $\xi(\bar{t}) \notin B$ if such $\bar{t}$ exists,
$\tau(B) = \infty$ if no such $\bar{t}$ exist.

For any $T > 0$ let $\tau_T = \tau(B) \wedge T$. Set $B_T = B \times \{t = T\}$, $S_T = \partial B \times \{0 < t < T\}$ where ∂B is the boundary of B.

Suppose $\phi(x, t)$ is a smooth function satisfying

$$L\phi \equiv \frac{\partial \phi}{\partial t} + \frac{1}{2} \sum_{i,j=1}^{n} a_{ij}(x, t) \frac{\partial^2 \phi}{\partial x_i \, \partial x_j} + \sum_{i=1}^{n} b_i(x, t) \frac{\partial \phi}{\partial t} \leq -1$$
$$\text{in} \quad B \times [0, T], \qquad \phi \geq 0 \quad \text{on} \quad B_T \cup S_T. \tag{1.5}$$

If we apply Itô's formula to $\phi(\xi(t), t)$ and substitute $t = \tau_T$, we get after taking the expectation (cf. Theorem 6.5.2)

$$E\phi(\xi(0), 0) \geq E\tau_T, \tag{1.6}$$

provided the range of $\xi(0)$ is in K.

If $1 + x_1 \leq x_1^0$ for all $(x_1, x_2, \ldots, x_n) \in B$, then we can take

$$\phi(x, t) = A[\exp(\alpha x_1^0) - \exp(\alpha x_1)],$$

where α, A are suitable positive constants. The condition (A_1) is used here in verifying (1.5). Since $\phi \leq C$ where C is a constant independent of T, (1.6) yields $E\tau_T \leq C$.

Now, $\tau_T \uparrow \tau(B)$ if $T \uparrow \infty$. Using the monotone convergence theorem we conclude that $\tau(B)$ satisfies

$$E\tau(B) \leq C. \tag{1.7}$$

In particular,

$$\tau(B) < \infty \quad \text{a.s.} \tag{1.8}$$

Take a strictly increasing sequence of open balls B_m with center 0 and radius $R_m \to \infty$. Let

$$\Omega_m = \left\{ \sup_{t>0} |\xi(t)| \geq R_m \right\}, \qquad \Omega^* = \bigcap_{m=1}^{\infty} \Omega_m.$$

Denote by χ_m the indicator function of B_m and let $\xi_m(t)$ be the solution of (1.1) with the initial condition $\xi_m(0) = \chi_m \xi(0)$. Denote by $\tau_m(B_m)$ the exit time from B_m of the solution $\xi_m(t)$. By Theorem 5.2.1, $\tau_m(B_m) = \tau(B_m)$ a.s. and $\xi_m(t) = \xi(t)$ if $0 \leq t \leq \tau(B_m)$. Hence,

$$P\left\{ \sup_{t>0} |\xi(t)| \geq R_m \right\} = P\left\{ \sup_{t>0} |\xi_m(t)| \geq R_m \right\}$$

and the right-hand side is equal to 1, by (1.8). We conclude that $P(\Omega_m) = 1$. Since the sequence Ω_m is monotone decreasing to $\Omega^* = \bigcap_{m=1}^{\infty} \Omega_m$,

$$P(\Omega^*) = \lim_{m\to\infty} P(\Omega_m) = 1.$$

Now, if $\omega \in \Omega^*$, then $\omega \in \Omega_m$ for all m, so that

$$\sup_{t>0} |\xi(t)| \geq R_m \qquad \text{for all} \quad m, \qquad \text{i.e.,} \qquad \sup_{t>0} |\xi(t)| = \infty.$$

This completes the proof.

From the proof of (1.7) we have:

Corollary 1.2. *Let $\xi(t)$ be a solution of (1.1) with $|\xi(0)| < R$ a.s. Suppose $a_{11}(x, t) \geq \lambda$, $b_1(x, t) \leq C$ if $|x| \leq R$, $t \geq 0$, where λ, C are positive constants. Then*

$$E\tau^R < \infty$$

where τ^R is the exit time from the ball $|x| < R$.

2. Auxiliary estimates

In the sequel we shall assume that $\sigma(x, t)$, $b(x, t)$ are measurable and $\xi(t)$ is any solution of (1.1), (1.2) such that

$$\sup_{0 \leq t \leq T} E|\xi(t)|^2 < \infty \qquad \text{for any} \quad T > 0.$$

If the condition (A_0) holds, then, of course, there exists a unique solution.

Set

$$Lu \equiv \tfrac{1}{2} \sum_{i,j=1}^{n} a_{ij}(x, t) \frac{\partial^2 u}{\partial x_i \, \partial x_j} + \sum_{i=1}^{n} b_i(x, t) \frac{\partial u}{\partial x_i} + \frac{\partial u}{\partial t}.$$

Theorem 2.1. *Assume that*

$$\sum_{i=1}^{n} a_{ii}(x, t) \leq C, \qquad \sum_{i=1}^{n} b_i(x, t) \leq C$$

where C is a positive constant. Then

$$E|\xi(t)|^2 \leq Kt + K' \qquad \textit{for all} \quad t \geq 0, \tag{2.1}$$

where K, K' are positive constants.

Proof. Using Itô's formula with $u(x) = |x|^2$ and $\xi(t)$, and taking the expectation, we get

$$E|\xi(t)|^2 = E|\xi_0|^2 + E\int_0^t g(\xi(s), s)\, ds$$

where

$$g(x, t) = L|x|^2 = \sum a_{ii}(x, t) + 2\sum x_i b_i(x, t).$$

Since, by our assumptions, $g(x, t) \leq C_1$ where C_1 is a positive constant, (2.1) follows.

Notation. We shall denote the eigenvalues of $a(x, t)$ by $\lambda_i(x, t)$ where

$$\lambda_1(x, t) \leq \lambda_2(x, t) \leq \cdots \leq \lambda_n(x, t).$$

Lemma 2.2. *Assume that*

$$\sum_{i,j} |a_{ij}(x, t)| \leq C \qquad \text{for all} \quad x \in R^n, \quad t \geq 0 \qquad (C\ \textit{const}); \tag{2.2}$$

for any $R > 0$ there is a positive constant $\mu(R)$ such that

$$\sum_{i,j} a_{ij}(x, t)\xi_i\xi_j \geq \mu(R)|\xi|^2 \qquad \textit{if} \quad |x| \leq R, \quad t \geq 0, \quad \xi \in R^n, \tag{2.3}$$

$$(1 + |x|)\sum_i |b_i(x, t)| \leq \epsilon(|x|) \qquad \textit{for} \quad x \in R^n, \quad t \geq 0,$$

$$\textit{where} \quad \epsilon(r) \to 0 \quad \textit{if} \ r \to \infty; \tag{2.4}$$

$$\lambda_n(x, t) \geq \gamma > 0 \qquad \textit{for} \quad x \in R^n, \quad t \geq 0 \qquad (\gamma\ \textit{const}). \tag{2.5}$$

Let $\phi(x, t)$ be a bounded measurable function such that $\phi(x, t) = 0$ if $x \notin G$ where G is a compact set. Then, for any $\eta > 0$,

$$E\left|\int_0^t \phi(\xi(s), s)\, ds\right| \leq K_1 + K_2 t^{(1+\eta)/2} \tag{2.6}$$

where K_1, K_2 are positive constants. If further,

$$\lambda_{n-1}(x, t) \geq \gamma' > 0 \qquad \textit{for} \quad x \in R^n, \quad t \geq 0 \qquad (\gamma'\ \textit{const}), \tag{2.7}$$

then (2.6) holds for some $-1 < \eta < 0$.

Proof. Let $r = |x - x^0|$ where x^0 is a point lying outside G. For simplicity we shall assume in what follows that $x^0 = 0$. It is clear that there exists a continuous function $\psi(r)$ such that

$$|\phi(x, t)| \leqslant \psi(r) \tag{2.8}$$

and $\psi(r) = 0$ if $r \leqslant r_0$ or if $r \geqslant r_1$ for some $0 < r_0 < r_1 < \infty$. We shall construct a function $F(r)$ such that the function $f(x) = F(|x|)$ is in $C^2(R^n)$ and

$$(Lf)(x, t) \geqslant \psi(|x|). \tag{2.9}$$

One can easily verify that

$$2Lf(r) = A(x, t)F''(r) + \frac{F'(r)}{r}[B(x, t) - A(x, t) + C(x, t)] \tag{2.10}$$

where

$$A(x, t) = \frac{1}{r^2}\sum_{i,j=1}^{n} a_{ij}(x, t)x_i x_j, \qquad B(x, t) = \sum_{i=1}^{n} a_{ii}(x, t),$$

$$C(x, t) = 2\sum_{i=1}^{n} x_i b_i(x, t).$$

Let $\theta(r)$ be a continuous function, vanishing for $r < r_0/2$ and satisfying, for $r > r_0$,

$$(1 + \theta(r))A(x, t) \leqslant B(x, t) + C(x, t) \tag{2.11}$$

Since, for any $R > 0$, $A(x, t) \geqslant \mu(R) > 0$ if $t \geqslant 0$, $|x| \leqslant R$, and since $B(x, t) \geqslant 0$ and $|C(x, t)|$ is bounded, one can certainly construct such a function $\theta(r)$. Noting that

$$\lambda_1(x, t) \leqslant A(x, t) \leqslant \lambda_n(x, t), \qquad B(x, t) = \lambda_1(x, t) + \cdots + \lambda_n(x, t)$$

and using (2.5) and (2.4), we can construct $\theta(r)$ such that

$$\theta(r) = -\eta \quad \text{if } r \text{ is sufficiently large (for any } \eta > 0). \tag{2.12}$$

If (2.7) holds, then we can take

$$\theta(r) = -\eta \quad \text{if } r \text{ is sufficiently large (for some } -1 < \eta < 0); \tag{2.13}$$

in fact, we can take any η such that

$$-\eta < \gamma'/\delta' \qquad \text{where} \quad \delta' = \limsup_{|x|\to\infty}\Big\{\sup_{t>0}\lambda_n(x, t)\Big\}.$$

Introduce the functions

$$I(s) = \int_{r_0}^{s} \frac{\theta(u)}{u}\, du,$$
$$f(x) = F(r) = \frac{1}{\lambda}\int_0^r e^{-I(s)}\, ds \int_0^s e^{I(u)}\psi(u)\, du \tag{2.14}$$

for some $\lambda > 0$, $r = |x|$. Since F is in C^2 and $F(r) = 0$ if $r \leq r_0$, $f(x)$ is in $C^2(R^n)$. One can easily check that

$$F''(r) + \frac{\theta(r)}{r} F'(r) = \frac{\psi(r)}{\lambda} \tag{2.15}$$

and that $F'(r) \geq 0$. Using (2.10), (2.11), we then get

$$2Lf(x) \geq \left[F''(r) + \frac{\theta(r)}{r} F'(r)\right] A(x, t) \geq A(x, t) \frac{\psi(r)}{\lambda} \geq 2\psi(r)$$

if $2\lambda = \mu(r_1)$ (since $A(x, t) \geq \mu(r_1)$ if $|x| \leq r_1$, $t \geq 0$).

Having proved (2.9), we now use Itô's formula to get

$$E\left|\int_0^t \phi(\xi(s), s)\, ds\right| \leq E\int_0^t \psi(\xi(s))\, ds \leq E\int_0^t (Lf)(\xi(s), s)\, ds$$
$$= EF(|\xi(t)|) - EF(|\xi_0|) \leq EF(|\xi(t)|).$$

Noting that $F(r) \leq Cr^{1+\eta}$, we obtain

$$E\left|\int_0^t \phi(\xi(s), s)\, ds\right| \leq CE|\xi(t)|^{1+\eta} + C \leq C\{E|\xi(t)|^2\}^{(1+\eta)/2} + C$$

for suitable constants C. Using Theorem 2.1, the assertion (2.6) follows.

We shall now extend Lemma 2.2 to the case where $\phi(x, t)$ does not necessarily vanish if $|x|$ is large, but either

$$|\phi(x, t) - \Phi| \leq \frac{C}{(1 + |x|)^\alpha} \qquad (\alpha > 0), \tag{2.16}$$

or

$$|\phi(x, t) - \Phi| \leq \frac{\epsilon(|x|)}{(1 + |x|)^\alpha} \qquad (\alpha \geq 0, \quad \epsilon(r) \to 0 \quad \text{if } r \to 0), \tag{2.17}$$

where Φ is a constant.

Lemma 2.3. *Let the conditions (2.2)–(2.5) hold and let $\phi(x, t)$ be a bounded measurable function satisfying (2.16). Then*

$$E\left|\int_0^t \phi(\xi(s), s)\, ds - \Phi t\right| = o(t^{(1+\eta)/2}) + O(t^{1-\alpha/2}) \tag{2.18}$$

for any $\eta > 0$; if (2.7) also holds, then (2.18) holds for some $-1 < \eta < 0$. If the function $\phi(x, t)$ satisfies (2.17) instead of (2.16), then

$$E\left|\int_0^t \phi(\xi(s), s)\, ds - \Phi t\right| = O(t^{(1+\eta)/2}) + o(t^{1-\alpha/2}) \tag{2.19}$$

with the same η as before.

Proof. Suppose first that (2.17) holds. Then, for any $\epsilon > 0$, there exists an

R_ϵ such that

$$|\phi(x, t) - \Phi| < \frac{\epsilon}{r^\alpha} \qquad \text{if} \quad |x| > R_\epsilon, \quad t \geqslant 0.$$

Let $\chi_\epsilon(x) = 1$ if $|x| < R_\epsilon$, $\chi_\epsilon(x) = 0$ if $|x| \geqslant R_\epsilon$, and write

$$\begin{aligned} \phi(x, t) - \Phi &= [\phi(x, t) - \Phi]\chi_\epsilon(x) + [\phi(x, t) - \Phi][1 - \chi_\epsilon(x)] \\ &= \phi_1(x, t) + \phi_2(x, t). \end{aligned} \tag{2.20}$$

By Lemma 2.2,

$$E\left| \int_0^t \phi_1(\xi(s), s)\, ds \right| = O(t^{(1+\eta)/2}). \tag{2.21}$$

Next, let $\psi(r)$ be a continuous function such that

$$\psi(r) = \begin{cases} 0 & \text{if} \quad 0 < r < r_0/2 \\ \epsilon/r^\alpha & \text{if} \quad r > r_0 \end{cases} \tag{2.22}$$

for some $0 < r_0 < R_\epsilon$. We shall slightly modify the definition of $\theta(r)$ by replacing the condition (2.11) by the stricter condition

$$(1 + \theta(r))A(x, t) \leqslant \rho B(x, t) + C(x, t) \tag{2.23}$$

where ρ is any positive number < 1. We can still satisfy (2.12) for any $\eta = \eta(\rho) > 0$ provided ρ is sufficiently close to 1; if (2.7) holds, then we can even satisfy (2.13) if ρ is sufficiently close to 1.

Let $f(x)$ be defined by (2.14) with $\lambda > 0$ to be determined later. Then, by (2.10), (2.23),

$$\begin{aligned} 2Lf(x) &\geqslant \left[F''(r) + \frac{\theta(r)}{r} F'(r) \right] A(x, t) + \frac{F'(r)}{r}(1 - \rho)B(x, t) \\ &\geqslant \frac{F'(r)}{r}(1 - \rho)B(x, t) \geqslant \frac{c}{\lambda}\, \frac{1}{r}\, e^{-I(r)} \int_{r_0}^r e^{I(s)} \frac{\epsilon}{s^\alpha}\, ds \\ &\geqslant \frac{c}{\lambda}\, \frac{\epsilon}{r^{1+\alpha}}\, e^{-I(r)} \int_{r_0}^r e^{I(s)}\, ds \end{aligned} \tag{2.24}$$

where c is a positive constant $\leqslant (1 - \rho)B(x, t)$. Since $e^{I(r)} \sim r^{-\eta}$ if $r \to \infty$, we find that

$$Lf(x) \geqslant \frac{c_1}{2\lambda}\, \frac{\epsilon}{r^\alpha} \geqslant |\phi_2(x, t)| \qquad \text{if} \quad |x| \geqslant R_\epsilon \tag{2.25}$$

where c_1 is a positive constant independent of ϵ, provided $2\lambda \leqslant c_1$. If $|x| < R_\epsilon$, then clearly $Lf(x) \geqslant 0 = |\phi_2(x, t)|$.

From Itô's formula and (2.25) we obtain

$$\begin{aligned} E\left|\int_0^t \phi_2(\xi(s), s)\, ds\right| &\leq E\int_0^t |\phi_2(\xi(s), s)|\, ds \\ &\leq E\int_0^t (Lf)(\xi(s), s)\, ds \\ &= EF(|\xi(t)|) - EF(|\xi_0|) \leq EF(|\xi(t)|). \end{aligned} \tag{2.26}$$

Since

$$F(r) \leq \begin{cases} C\epsilon r^{2-\alpha} + C & \text{if } 2-\alpha > 1+\eta, \\ C\epsilon r^{1+\eta} + C & \text{if } 2-\alpha < 1+\eta, \end{cases}$$

and since we may assume, without loss of generality, that $\alpha + \eta \neq 1$, we deduce, after using Hölder's inequality and Theorem 2.1,

$$EF(|\xi(t)|) \leq \begin{cases} C\epsilon t^{1-\alpha/2} + C & \text{if } \alpha+\eta < 1, \\ cc\epsilon t^{(1+\eta)/2} + C & \text{if } \alpha+\eta > 1. \end{cases}$$

Substituting this into (2.26) and combining the resulting inequality with (2.21), (2.20), we get

$$E\left|\int_0^t \phi(\xi(s), s)\, ds - \Phi t\right| \leq C_0 t^{(1+\eta)/2} + C_1 \epsilon t^{1-\alpha/2} \tag{2.27}$$

where C_1 is a constant independent of ϵ. Since ϵ is arbitrary, (2.27) implies the assertion (2.19) with η replaced by η', for any $\eta' > \eta$. But this clearly completes the proof of (2.19). The proof of (2.18) (for ϕ satisfying (2.16)) is similar.

We proceed to evaluate (under additional conditions) more precisely the number η occurring in Lemma 2.3. We shall assume that, for $1 \leq i, j \leq n$,

$$a_{ij}(x, t) \to \bar{a}_{ij} \quad \text{as } |x| \to \infty, \text{ uniformly with respect to } t, \tag{2.28}$$

where $\bar{a}_{ij}$ are constants. Let

$$d = \text{number of positive eigenvalues of } \bar{a} = (\bar{a}_{ij}). \tag{2.29}$$

Observe that if (2.3) holds, then (2.5) holds if and only if $d \geq 1$, and (2.7) holds if and only if $d \geq 2$.

Note that the assumptions and assertions of Lemma 2.3 remain unchanged if we perform a nonsingular linear transformation $x \to Ax$. Such a transformation changes $a(x, t)$ into $Aa(x, t)A^*$. We can therefore choose A such that

$$\begin{aligned} \bar{a}_{ij} &= 0 && \text{if } i \neq j, \\ \bar{a}_{ii} &= 1 && \text{if } i = 1, 2, \ldots, d, \\ \bar{a}_{ii} &= 0 && \text{if } i = d+1, \ldots, n; \end{aligned}$$

$d = 0$ means that $\bar{a}_{ii} = 0$ for all i, and $d = n$ means that $\bar{a}_{ii} = 1$ for all i.

If $d \geqslant 2$, then we can take, in the proof of Lemmas 2.2, 2.3, $\theta(r) = \nu$ for any $\nu < 1$ provided r is sufficiently large. This leads to the assertions of Lemmas 2.2, 2.3 with any $-1 < \eta < 0$.

If $d \geqslant 3$, then we can take $\theta(r) = \frac{3}{2}$ provided r is sufficiently large. This leads to the assertion of Lemma 2.2 with $\eta = -1$. If ϕ satisfies (2.16), then, instead of (2.18), we have

$$E\left|\int_0^t \phi(\xi(s), s)\, ds - \Phi t\right| = O(1) + O(t^{1-\alpha/2}) \qquad \text{if} \quad \alpha \neq 2. \tag{2.30}$$

We sum up:

Lemma 2.4. (a) *Let the conditions* (2.2)–(2.4), (2.28) *hold, and let* $d \geqslant 2$. *Let* $\phi(x, t)$ *be a bounded measurable function. If* ϕ *satisfies* (2.16), *then* (2.18) *holds for any* $-1 < \eta < 0$, *and if* ϕ *satisfies* (2.17), *then* (2.19) *holds for any* $-1 < \eta < 0$.

(b) *Let the conditions* (2.2)–(2.4), (2.28) *hold and let* $d \geqslant 3$. *If* $\phi(x, t)$ *is a bounded measurable function satisfying* (2.16), *then* (2.30) *holds.*

3. Asymptotic estimates

Theorem 3.1. *Let the conditions* (2.2)–(2.5) *hold. Then*

$$E|\xi(t)|^2 \geqslant Kt - K' \qquad \textit{for all} \quad t \geqslant 0 \tag{3.1}$$

where K, K' *are positive constants.*

Proof.

$$L|x|^2 = \sum a_{ii}(x, t) + 2\sum x_i b_i(x, t) \geqslant \gamma - \phi(x)$$

where $\phi(x)$ is a bounded measurable and nonnegative function having compact support. By Itô's formula,

$$E|\xi(t)|^2 - E|\xi_0|^2 = E\int_0^t (L|x|^2)(\xi(s), s)\, ds \geqslant \gamma t - E\int_0^t \phi(\xi(s))\, ds.$$

Since, by Lemma 2.2,

$$E\int_0^t \phi(\xi(s))\, ds = o(t)$$

the inequality (3.1) follows.

Remark. Lemma 2.2 and Theorem 3.1 remain true if the condition (2.4) is replaced by the weaker condition

$$(1 + |x|)\sum_i |b_i(x, t)| \leqslant \epsilon_0 \tag{3.2}$$

provided ϵ_0 is a sufficiently small positive constant.

We shall need the following condition:

$$(1+|x|)^{1+\delta}\sum_{i=1}^{n}|b_i(x,t)| \leqslant \epsilon(|x|), \quad \epsilon(r)\to 0 \quad \text{if} \quad r\to 0, \quad \delta \geqslant 0. \tag{3.3}$$

Theorem 3.2. *Let* (2.2), (2.3), (2.28), (3.3) *hold and let* $d \geqslant 1$. *Then*

$$E|\xi(t)|^2 = (\operatorname{tr}\bar{a})t + O(t^{(1+\eta)/2}) + o(t^{1-\delta/2}) \tag{3.4}$$

where $\operatorname{tr}\bar{a} = \Sigma_i \bar{a}_{ii}$ *and* η *is any positive number; if* $d \geqslant 2$, *then* η *is any number* > -1.

Proof. Notice that

$$|L|x|^2 - \operatorname{tr}\bar{a}| \leqslant \epsilon'(|x|)/(1+|x|^\delta)$$

where $\epsilon'(r)\to 0$ if $r\to\infty$. Hence, by Lemmas 2.3 and 2.4(a) with $\phi(x,t) = L|x|^2$, $\Phi = \operatorname{tr}\bar{a}$,

$$\begin{aligned} E|\xi(t)|^2 - E|\xi_0|^2 &= E\int_0^t (L|x|^2)(\xi(s),s)\,ds \\ &= (\operatorname{tr}\bar{a})t + O(t^{(1+\eta)/2}) + o(t^{1-\delta/2}) \end{aligned}$$

where η is any positive number if $d \geqslant 1$ and any number > -1 if $d \geqslant 2$. This yields (3.4).

The special case $\delta = 0$ gives

Corollary 3.3. *Let* (2.2)–(2.4), (2.28) *hold and let* $d \geqslant 1$. *Then*

$$E|\xi(t)|^2 = 2(\operatorname{tr}\bar{a})t + o(t). \tag{3.5}$$

We can now state a key lemma needed for deriving the main results of this section.

Lemma 3.4. *Let the conditions* (2.2), (2.3), *and* (3.3) *hold with* $0 \leqslant \delta < 1$, *and let* $d \geqslant 2$. *Then, for any* $j = 1, 2, \ldots, n$,

$$E\left|\int_0^t b_j(\xi(s),s)\,ds\right|^2 = o(t^{1-\delta}). \tag{3.6}$$

Proof. For any $\epsilon > 0$ we can write

$$|b_j(x,t)| \leqslant g_1(x) + g_2(|x|)$$

where g_i are bounded measurable functions; $g_1(x)$ has compact support,

$$g_2(r) = \epsilon/(1+r)^{1+\delta} \qquad \text{if} \quad r > r_1, \tag{3.7}$$

and $g_2(r) = 0$ if $r \leqslant r_1$ for some $0 < r_1 < \infty$. Clearly

$$E\left|\int_0^t b_j(\xi(s),s)\,ds\right|^2 \leqslant 2E\left|\int_0^t g_1(\xi(s),s)\,ds\right|^2 + 2E\left|\int_0^t g_2(|\xi(s)|)\,ds\right|^2. \tag{3.8}$$

Let x^0 be a point outside the support of g_1 and let $\rho = |x - x^0|$.

Let $F(\rho)$ be the function constructed in the proof of Lemma 2.2 for $\phi = g_1$. Then, for any $-1 < \eta < 0$,

$$F(r) \leq cr^{1+\eta}, \qquad F'(r) \leq cr^{\eta} \qquad (c \text{ positive constant}) \tag{3.9}$$

provided r is sufficiently large. By Itô's formula,

$$\left|\int_0^t g_1(\xi(s))\, ds\right| \leq \int_0^t (LF)(\xi(s), s)\, ds$$
$$\leq F(|\xi(t) - x^0|) - \int_0^t D_x F \cdot \sigma(\xi(s), s)\, dw(s).$$

Consequently

$$E\left|\int_0^t g_1(\xi(s))\, ds\right|^2 \leq Ct^{1+\eta} + C + 2E \int_0^t |D_x F \cdot \sigma|^2\, ds. \tag{3.10}$$

Next, for large r,

$$|D_x F \cdot \sigma|^2 \leq Cr^{2\eta} \qquad \text{for any} \quad -1 < \eta < 0. \tag{3.11}$$

Hence, by Lemma 2.4(a) with $\alpha = -2\eta$,

$$E \int_0^t |D_x F \cdot \sigma|^2\, ds \leq Ct^{1+\eta} + C, \tag{3.12}$$

and (3.10) then yields

$$E\left|\int_0^t g_1(\xi(s))\, ds\right|^2 \leq Ct^{1+\eta} + C. \tag{3.13}$$

To evaluate the integral corresponding to g_2, let $F_0(r)$ be the function $F(r)$ constructed in Lemma 2.3 when $\phi_2 = g_2$. We can take $-1 < \eta < -\delta$. Then

$$F_0(r) \leq C\epsilon r^{1-\delta} + C\epsilon, \qquad F_0'(r) \leq C\epsilon r^{-\delta} + C\epsilon \tag{3.14}$$

where C is a positive constant independent of ϵ. Analogously to (3.10), (3.11) we get

$$E\left|\int_0^t g_2(|\xi(s)|)\, ds\right|^2 \leq C\epsilon t^{1-\delta} + C + 2E \int_0^t |D_x F_0 \cdot \sigma|^2\, ds \tag{3.15}$$

$$|D_x F_0 \cdot \sigma| \leq C\epsilon^2/r^{2\delta}. \tag{3.16}$$

By Lemma 2.4(a),

$$E \int_0^t |D_x F_0 \cdot \sigma|^2\, ds \leq Ct^{(1+\eta)/2} + C\epsilon^2 t^{1-\delta} + C. \tag{3.17}$$

Hence

$$E\left|\int_0^t g_2(|\xi(s)|)\, ds\right|^2 \leq C\epsilon t^{1-\delta} + Ct^{(1+\eta)/2} + C. \tag{3.18}$$

Combining (3.18) with (3.13) and (3.8), and recalling that $\eta < -\delta$, the assertion (3.6) follows.

The following lemma is a variant of Lemma 3.4 in case $d \geq 3$.

Lemma 3.5. *Let the conditions* (2.2), (2.3), (2.28) *hold, let* $d \geq 3$ *and let* (3.3) *hold with some* $\delta > 1$. *Then, for any* $j = 1, \ldots, n$,

$$E\left|\int_0^t b_j(\xi(s), s)\, ds\right|^2 = O(1). \tag{3.19}$$

Proof. The function $F(r)$ occurring in (3.9) is now given by

$$F(r) = \int_{\rho_0}^{r} e^{-I(s)}\, ds \int_{\rho_0}^{s} e^{I(u)}\bar{\gamma}(u)\, du \qquad \text{if} \quad r > \rho_0,$$

where $\bar{\gamma}(r) = 0$ if $r > \rho_1$, for some $0 < \rho_0 < \rho_1 < \infty$. We can take $\eta(r) = \frac{3}{2}$ for large r, so that $I(r) \sim \log r^{3/2}$. Therefore

$$F(r) \leq C, \qquad F'(r) \leq C/r^{3/2}. \tag{3.20}$$

Hence (3.11) is replaced by

$$|D_x F \cdot \sigma|^2 \leq C/r^3. \tag{3.21}$$

Lemma 2.4(b) with $\alpha = 3$ gives

$$E\int_0^t |D_x F \cdot \sigma|^2\, ds = O(1), \tag{3.22}$$

and this estimate is to replace (3.12). We conclude that (instead of (3.13))

$$E\left|\int_0^t g_1(\xi(s))\, ds\right|^2 \leq C. \tag{3.23}$$

Let $0 < r_0 < r_1$ and let $\bar{g}_2(r)$ be a continuous function satisfying

$$\bar{g}_2(r) = \begin{cases} \epsilon/(1+r)^{1+\delta} & \text{if} \quad r > r_0, \\ 0 & \text{if} \quad r < r_0/2. \end{cases}$$

Take

$$F_0(r) = C\int_0^r e^{-I(s)} \int_0^s e^{I(u)}\bar{g}_2(u)\, du \qquad (C > 0).$$

Since $\delta > 1$, F_0 satisfies (instead of (3.14))

$$F_0(r) \leq C, \qquad F_0'(r) \leq C/r^k \qquad k = \min(\delta, \tfrac{3}{2}). \tag{3.24}$$

Using Itô's formula we find that

$$E\left|\int_0^t g_2(|\xi(s)|)\, ds\right|^2 \leq C + E\int_0^t |D_x F_0 \cdot \sigma|^2\, ds. \tag{3.25}$$

Since, by Lemma 2.4(b) with $\alpha = 2k$,

$$E\int_0^t |D_x F_0 \cdot \sigma|^2\, ds \leq C,$$

the right-hand side of (3.25) is bounded by a constant. Combining this with (3.23), (3.8), the assertion (3.19) follows.

Remark. If in Lemma 3.4 we replace the condition (3.3) by

$$\sum_i |b_i(x, t)| \leq C/(1 + |x|)^{1+\delta} \qquad (0 < \delta < 1), \tag{3.26}$$

then

$$E\left|\int_0^t b_j(\xi(s), s)\, ds\right|^2 = O(t^{1-\delta}). \tag{3.27}$$

We shall need the following conditions:

$$\sum_{i,j=1}^{n} |\sigma_{ij}(x, t) - \bar{\sigma}_{ij}| \leq \frac{\epsilon(|x|)}{(1 + |x|)^{\delta}}, \qquad \delta \geq 0, \quad \epsilon(r) \to 0 \quad \text{if} \quad r \to \infty, \tag{3.28}$$

where $\bar{\sigma}_{ij}$ are constants. Note that

$$\bar{a} = \bar{\sigma}\bar{\sigma}^*, \qquad \operatorname{tr} \bar{a} = |\bar{\sigma}|^2.$$

We shall also need the condition

$$\sum_{i,j=1}^{n} |\sigma_{ij}(x, t) - \bar{\sigma}_{ij}| \leq C/(1 + |x|)^{\delta}, \qquad \delta > 0. \tag{3.29}$$

We shall now state the main results of this section.

Theorem 3.6. (a) *Let* (2.2), (2.3), (3.3), (3.28) *hold with* $0 \leq \delta < 1$, *and let* $d \geq 2$. *Then*

$$E|\xi(t) - \bar{\sigma}w(t)|^2 = o(t^{1-\delta}). \tag{3.30}$$

If (3.3), (3.28) *are replaced by* (3.26), (3.29) *and* $0 < \delta < 1$, *then*

$$E|\xi(t) - \bar{\sigma}w(t)|^2 = O(t^{1-\delta}). \tag{3.31}$$

(b) *Let* (2.2), (2.3), (3.26), (3.29) *hold for some* $\delta > 1$, *and let* $d \geq 3$. *Then*

$$E|\xi(t) - \bar{\sigma}w(t)|^2 = O(1). \tag{3.32}$$

Proof. Let (2.2), (2.3), (3.3), (3.28) hold with $0 \leq \delta < 1$, and let $d \geq 2$. Consider the expression

$$E\left|\int_0^t (\sigma - \bar{\sigma})\, dw(s)\right|^2 = E\int_0^t |\sigma - \bar{\sigma}|^2\, ds.$$

Since $|\sigma(x, t) - \bar{\sigma}| \leq \epsilon^2(|x|)/(1 + |x|)^{2\delta}$, Lemma 2.4(a) implies that

$$E\int_0^t |\sigma(\xi(s), s) - \bar{\sigma}|^2\, ds = o(t^{1-\delta}).$$

Writing

$$\xi(t) - \bar{\sigma}w(t) = \xi_0 + \int_0^t [\sigma(\xi(s), s) - \bar{\sigma}]\, dw(s) + \int_0^t b(\xi(s), s)\, ds$$

and using the previous estimate and Lemma 3.4, the assertion (3.30) follows.

The proofs of (3.31), (3.32) are similar. In proving (3.31) we make use of the remark following Lemma 3.5. In proving (3.32) we employ Lemma 3.5.

4. Applications of the asymptotic estimates

We shall need the following conditions:

$$\lim_{|x|\to\infty} \sigma_{ij}(x, t) = \bar{\sigma}_{ij} \quad \text{uniformly with respect to} \quad t, \qquad 1 \leqslant i, j \leqslant n, \tag{4.1}$$

$$\begin{gathered} \sum |b_i(x, t)| \leqslant C \qquad \text{for all} \quad x \in R^n, \quad t \geqslant 0 \quad (C \text{ const}), \\ \lim_{|x|\to\infty} (1 + |x|) \sum_{i=1}^{n} |b_i(x, t)| = 0 \quad \text{uniformly with respect to } t. \end{gathered} \tag{4.2}$$

Theorem 3.6(a) with $\delta = 0$ states:

If (2.2), (2.3), (4.1), (4.2) hold and if $d \geqslant 2$, then

$$E|\xi(t) - \bar{\sigma} w(t)|^2 = o(t). \tag{4.3}$$

From (4.3) we see that, as $t \to \infty$,

$$\frac{\xi(t)}{\sqrt{t}} - \bar{\sigma}\frac{w(t)}{\sqrt{t}} \to 0 \qquad \text{in } L^2;$$

consequently also in probability. This immediately yields the following theorem on convergence in distribution of $\xi(t)$:

Theorem 4.1. *Let the conditions* (2.2), (2.3), (4.1), (4.2) *hold, let* $\bar{\sigma}$ *be nonsingular matrix, and let* $n \geqslant 2$. *Then*

$$\begin{aligned} &\lim_{t\to\infty} P\{\xi(t) < x\sqrt{t}\,\} \\ &= \frac{1}{(2\pi)^{n/2} \det \bar{\sigma}} \int_{-\infty}^{x_1} \cdots \int_{-\infty}^{x_n} \exp\left\{-\tfrac{1}{2}\sum \hat{a}_{ij} y_i y_j\right\} dy_n \cdots dy_1 \end{aligned} \tag{4.4}$$

where $\hat{a}$ *is the inverse matrix to* $\bar{a}$.

Suppose next that (2.2), (2.3) hold and that

$$\sum_{i,j} |\sigma_{ij}(x, t) - \bar{\sigma}_{ij}| \leqslant C/(1 + |x|)^{\delta}, \tag{4.5}$$

$$\sum_i |b_i(x, t)| \leqslant C/(1 + |x|)^{1+\delta} \tag{4.6}$$

for some $0 < \delta < 1$. Suppose $\bar{\sigma}$ is nonsingular and $n \geqslant 2$. Denote by $\hat{\sigma}$ the inverse of $\bar{\sigma}$. Then

$$\begin{aligned} \frac{\hat{\sigma}\xi(t) - w(t)}{\sqrt{2t \log \log t}} &= \frac{\hat{\sigma}}{\sqrt{2t \log \log t}} \int_0^t b(\xi(s), s)\, ds \\ &\quad + \frac{\hat{\sigma}}{\sqrt{2t \log \log t}} \int_0^t [\sigma(\xi(s), s) - \bar{\sigma}]\, dw(s) \\ &\quad + \frac{\hat{\sigma}\xi_0}{\sqrt{2t \log \log t}} \equiv J_1(t) + J_2(t) + J_3(t) \equiv J(t). \end{aligned} \tag{4.7}$$

We shall denote various positive constants by the same symbol C. Let $t_m = m^\lambda$, $\lambda = 4/\delta$, m a positive integer. By the proof of Theorem 3.6(a),

$$P\left\{ \sup_{t_m \leqslant t \leqslant t_{m+1}} |J_1(t)| > \frac{1}{m} \right\} \leqslant P\left\{ \frac{C}{\sqrt{t_m}} \int_0^{t_{m+1}} |b(\xi(s), s)|\, ds > \frac{1}{m} \right\}$$
$$\leqslant \frac{Cm^2}{t_m} E\left[\int_0^{t_{m+1}} |b(\xi(s), s)|\, ds \right]^2$$
$$\leqslant \frac{Cm^2}{t_m} (t_{m+1})^{1-\delta} \leqslant \frac{C}{m^2}.$$

Next, by the martingale inequality and the proof of Theorem 3.6(a),

$$P\left\{ \sup_{t_m \leqslant t \leqslant t_{m+1}} |J_2(t)| > \frac{1}{m} \right\}$$
$$\leqslant P\left\{ \frac{C}{\sqrt{t_m}} \sup_{t_m \leqslant t \leqslant t_{m+1}} \left| \int_0^t [\sigma(\xi(s), s) - \bar{\sigma}]\, dw(s) \right| > \frac{1}{m} \right\}$$
$$\leqslant \frac{Cm^2}{t_m} (t_{m+1})^{1-\delta} \leqslant \frac{C}{m^2}.$$

Since, finally,

$$P\left\{ \sup_{t_m \leqslant t \leqslant t_{m+1}} |J_3(t)| > \frac{1}{m} \right\} \leqslant \frac{Cm^2}{t_m} E|\xi_0|^2 \leqslant \frac{C}{m^2},$$

we conclude that

$$P\left\{ \sup_{t_m \leqslant t \leqslant t_{m+1}} |J(t)| > \frac{3}{m} \right\} \leqslant \frac{C}{m^2}.$$

Applying the Borel–Cantelli lemma we deduce

$$P\left\{ \sup_{t_m \leqslant t \leqslant t_{m+1}} |J(t)| > \frac{3}{m} \text{ i.o.} \right\} = 0;$$

consequently

$$P\{ \lim_{t\to\infty} |J(t)| = 0\} = 1. \tag{4.8}$$

From (4.7) and the law of the iterated logarithm for $w(t)$ (Theorem 3.3.1, Corollary 3.3.2 and Theorem 3.6.1) we obtain:

Theorem 4.2. *Let the conditions* (2.2), (2.3), (4.5), (4.6) *hold for some* $\delta > 0$, *let* $\bar{\sigma}$ *be nonsingular, and let* $n \geqslant 2$. *Then, for any* $i = 1, \ldots, n$,

$$\overline{\lim_{t\to\infty}} \frac{\sum_{j=1}^n \hat{\sigma}_{ij}\xi_j(t)}{\sqrt{2t \log\log t}} = 1 \quad a.s., \tag{4.9}$$

$$\underline{\lim_{t\to\infty}} \frac{\sum_{j=1}^n \hat{\sigma}_{ij}\xi_j(t)}{\sqrt{2t \log\log t}} = 1 \quad a.s. \tag{4.10}$$

Further,

$$\overline{\lim_{t\to\infty}} \frac{|\hat{\sigma}\xi(t)|}{\sqrt{2t\log\log t}} = 1 \quad a.s. \tag{4.11}$$

The next application is to the Cauchy problem:

$$\frac{1}{2}\sum_{i,j=1}^{n} a_{ij}(x)\frac{\partial^2 u}{\partial x_i\,\partial x_j} + \sum_{i=1}^{n} b_i(x)\frac{\partial u}{\partial x_i} - \frac{\partial u}{\partial t} = 0 \qquad \text{if} \quad x\in R^n, \quad t>0, \tag{4.12}$$

$$u(x,0) = f(x) \qquad \text{if} \quad x\in R^n. \tag{4.13}$$

We shall assume:

(i) $\sigma_{ij}(x)$, $b_i(x)$ are bounded functions in R^n, uniformly Lipschitz continuous, and the matrix $(a_{ij}(x))$ is uniformly positive definite.

(ii) For all x, $\bar{x}$ in R^n,

$$|f(x) - f(\bar{x})| \leqslant C|x-\bar{x}|^\gamma \qquad (0<\gamma\leqslant 2, \quad C>0).$$

By Sections 6.4, 6.5, there is a unique solution $u(x,t)$ of (4.12), (4.13) bounded by $O(|x|^\gamma)$ uniformly in t in bounded intervals, and it is given by

$$u(x,t) = Ef(\xi_x(t)) \tag{4.14}$$

where $\xi_x(t)$ is the solution of (1.1) with $\xi(0)=x$.

Let us further assume that

$$\sum_{i,j} |\sigma_{ij}(x) - \delta_{ij}| \leqslant C/(1+|x|)^\delta, \tag{4.15}$$

$$\sum_i |b_i(x)| \leqslant C/(1+|x|)^{1+\delta} \tag{4.16}$$

for some $\delta>0$.

If $a_{ij} = \delta_{ij}$ and $b_i\equiv 0$, then the solution $\tilde{u}$ is given either in the form $Ef(w(t)+x)$, or in terms of the fundamental solution for the heat equation, namely

$$\tilde{u}(x,t) = \frac{1}{(2\pi)^{n/2}t^{n/2}}\int_{R^n}\exp\left\{\frac{|x-y|^2}{2t}\right\}f(y)\,dy. \tag{4.17}$$

From (4.14) and (ii) we get

$$|u(x,t) - \tilde{u}(x,t)| \leqslant C\{E|\xi_x(t) - w(t) - x|^2\}^{\gamma/2}.$$

We can now apply Theorems 3.6(a), 3.6(b) to estimate the right-hand side. A careful review of the proof of (3.31) and (3.32) for $\xi(t)=\xi_x(t)$ shows that

$$|O(t^{1-\delta})| \leqslant C[t^{1-\delta} + |x|^{2(1-\delta)} + 1], \qquad |O(1)| \leqslant C$$

where C is a constant independent of the initial condition $\xi(0)=x$. Hence:

Theorem 4.3. *Let the conditions* (i), (ii), (4.15), (4.16) *hold. If* $n\geqslant 2$,

$0 < \delta < 1$, *then for all* $x \in R^n$, $t \geq 0$,

$$|u(x, t) - \tilde{u}(x, t)| \leq C(1 + t + |x|^2)^{(1-\delta)\gamma/2} \qquad (C \text{ const}). \tag{4.18}$$

If $n \geq 3$ *and* $\delta > 1$, *then for all* $x \in R^n$, $t \geq 0$,

$$|u(x, t) - \tilde{u}(x, t)| \leq C \qquad (C \text{ const}). \tag{4.19}$$

5. The one-dimensional case

In proving Theorem 3.6(a) or even in deriving (4.3), we have assumed that the matrix $\bar{a}$ has at least two positive eigenvalues. To see what may happen when $\bar{a}$ has only one positive eigenvalue, we resort to the case $n = 1$. For simplicity, consider first the equation

$$d\xi(t) = dw(t) + b(\xi(t))\, dt. \tag{5.1}$$

We assume that $b(x)$ is measurable and

$$\int_{-\infty}^{\infty} |b(x)|\, dx < \infty. \tag{5.2}$$

One can construct comparison functions explicitly by solving differential equations of the form

$$Lf = \tfrac{1}{2} f''(x) + b(x) f'(x) = \psi(x). \tag{5.3}$$

Setting $A(x) = 2\int_{-\infty}^{x} b(y)\, dy$, the general solution of (5.3) is given by

$$f(x) = \int_0^x e^{-A(z)}\, dz \left[2 \int_0^t e^{A(y)} \psi(y)\, dy + C_1 \right] + C_2$$

where C_1, C_2 are constants. Using this solution in the proof of Theorem 3.6, one can derive the estimate

$$E|\xi(t) - w(t)|^2 = o(t) \tag{5.4}$$

provided $A(+\infty) = 0$, i.e., provided

$$\int_{-\infty}^{\infty} b(x)\, dx = 0; \tag{5.5}$$

if further

$$|b(x)| \leq C/(1 + |x|)^{1+\delta} \qquad (0 < \delta < 1) \tag{5.6}$$

then

$$E|\xi(t) - w(t)|^2 = O(t^{1-\delta}); \tag{5.7}$$

the details are left to the reader.

Consider next the more general stochastic differential equation

$$d\xi(t) = \sigma(\xi(t))\, dt + b(\xi(t))\, dt. \tag{5.8}$$

We assume that $\sigma(x) > 0$ for all x, and set

$$h(z) = \int_0^z \frac{dy}{\sigma(y)}, \qquad g \text{ inverse to } h.$$

As easily verified, the process $\eta(t) = h(\xi(t))$ satisfies the equation

$$d\eta(t) = dw(t) + \tilde{b}(\eta(t))\, dt \tag{5.9}$$

where

$$\tilde{b}(x) = \frac{b(g(x))}{\sigma(g(x))} - \tfrac{1}{2}\sigma'(g(x)). \tag{5.10}$$

The condition (5.5) for the equation (5.9) becomes

$$\int_{-\infty}^{\infty} \frac{b(x) - \frac{1}{2}\sigma(x)\sigma'(x)}{\sigma^2(x)}\, dx = 0. \tag{5.11}$$

If

$$\left| \frac{b(g(x))}{\sigma(g(x))} - \tfrac{1}{2}\sigma'(g(x)) \right| \leqslant \frac{C}{(1 + |x|)^{1+\delta}} \qquad (\delta > 0) \tag{5.12}$$

and if (5.11) holds, then, by the assertion (5.7) applied to the solution $\eta(t)$ of (5.9),

$$E|\eta(t) - w(t)|^2 \leqslant Ct^{1-\delta} + C.$$

If we further assume that

$$|\bar{\sigma}h(x) - x| \leqslant C(1 + |x|^{1-\mu}) \qquad (0 \leqslant \mu < 1, \bar{\sigma} > 0) \tag{5.13}$$

for some constant $\bar{\sigma}$, then we easily conclude that

$$E|\xi(t) - \bar{\sigma}w(t)|^2 \leqslant Kt^{1-\nu} + K' \qquad \text{for all} \quad t \geqslant 0 \tag{5.14}$$

where $\nu = \min(\delta, \mu)$ and K, K' are positive constants.

Observe that if

$$|\sigma(x) - \bar{\sigma}| \leqslant \frac{C}{(1 + |x|)^{\mu}} \tag{5.15}$$

then (5.13) holds and, further, (5.12) is equivalent to

$$\left| \frac{b(x)}{\sigma(x)} - \tfrac{1}{2}\sigma'(x) \right| \leqslant \frac{C}{(1 + |x|)^{1+\delta}} \tag{5.16}$$

We can therefore state:

Theorem 5.1. *If* (5.15), (5.16), *and* (5.11) *hold, then for any solution* $\xi(t)$ *of* (5.8) *the estimate* (5.14) *holds with* $\nu = \min(\delta, \mu)$.

If (5.15) is replaced by

$$\bar{\sigma} = \lim_{|x|\to\infty} \sigma(x) \quad \text{exists}, \qquad \bar{\sigma} \neq 0,$$

if the left-hand side of (5.12) is in $L^1(-\infty, \infty)$, and if (5.11) holds, then, by applying the assertion (5.4) to the solution $\eta(t)$ of (5.9), we easily obtain

$$E|\xi(t) - \bar{\sigma}w(t)|^2 = o(t). \tag{5.17}$$

The condition (5.11) is essential for the validity of Theorem 5.1. Taking for simplicity the case $\sigma \equiv 1$, where the condition (5.11) reduces to the condition (5.5), we shall prove:

Theorem 5.2. *Let* $b(x)$ *satisfy* (5.2). *If* $\xi(t)$ *is a solution of* (5.1), *then the estimate* (5.4) *holds if and only if* (5.5) *holds.*

Proof. We only have to prove that if

$$\int_{-\infty}^{\infty} b(x)\,dx \neq 0 \tag{5.18}$$

then (5.4) does not hold. We shall assume that (5.18) and (5.4) hold, and derive a contradiction.

Clearly

$$E\xi(t) = E\xi_0 + E\int_0^t b(\xi(s))\,ds. \tag{5.19}$$

A particular solution of (5.3) for $\psi = b$ is

$$f(x) = \int_0^x e^{-A(z)}[e^{A(z)} - 1]\,dz.$$

Since $A(-\infty) = 0$ and (by (5.18)) $A(+\infty) \neq 0$,

$$f(x) = \lambda x^+ + o(|x|) \qquad \text{as} \quad |x|\to\infty \tag{5.20}$$

where λ is a nonvanishing constant.

By Itô's formula

$$E\int_0^t b(\xi(s))\,ds = Ef(\xi(t)) - Ef(\xi_0).$$

Substituting this into (5.19), we get

$$E\xi(t) = E\xi_0 - Ef(\xi_0) + Ef(\xi(t)). \tag{5.21}$$

Since $f(x)$ satisfies a uniform Lipschitz condition,

$$|Ef(\xi(t)) - Ef(w(t))| \leq CE|\xi(t) - w(t)| \qquad (C \text{ const}),$$

and, in view of (5.4),

$$Ef(\xi(t)) = Ef(w(t)) + o(t^{1/2}). \tag{5.22}$$

From (5.20) we have, for any $\epsilon > 0$,

$$|f(x) - \lambda x^+| \leq \epsilon|x| + C(\epsilon) \qquad (C(\epsilon) \text{ const}).$$

Combining this (when $x = w(t)$) with (5.22), we find that

$$|Ef(\xi(t)) - E[\lambda w^+(t)| \leq \epsilon E|w(t)| + C(\epsilon) + o(t^{1/2}) \leq 2\epsilon t^{1/2}$$

if t is sufficiently large. Upon substituting this into (5.21), we obtain

$$E\xi(t) = \lambda E w^+(t) + \theta\epsilon t^{1/2} + C' \qquad (C' \text{ const}, |\theta| \leq 2).$$

But, by (5.4),

$$E\xi(t) = Ew(t) + o(t^{1/2}) = o(t^{1/2}).$$

Consequently,

$$|\lambda| E w^+(t) \leq 3\epsilon t^{1/2} \qquad \text{if} \quad t \text{ is sufficiently large.}$$

Since, however, $Ew^+(t) = (t/2\pi)^{1/2}$, we get a contradiction if $18\pi\epsilon^2 < \lambda^2$.

There is an intuitive reason why for $n = 1$, $\sigma \equiv 1$ the assertion (5.4) cannot hold unless (5.5) is satisfied. In order for the distribution of $\xi(t)/\sqrt{t}$ to approximate the normal distribution as $t \to \infty$ the particles represented by $\xi(t)$, or $\xi(t, \omega)$, must be able to move without significant "resistance" from intervals (α, β) near $+\infty$ to intervals $(-\beta, -\alpha)$. Since in performing this move they must cross the interval $(-\alpha, \alpha)$, they are subject to the influence of the drift term $b(x)$. This drift coefficient will resist the motion if $\int_{-\infty}^{\infty} b(x)\, dx > 0$; thus this inequality cannot take place. Similarly, the reverse inequality cannot take place.

If $n \geq 2$ and $\bar{\sigma}$ is nonsingular, then $\xi(t)$ may move from one n-dimensional interval in a neighborhood G of infinity to another without leaving G. If the drift coefficient $b(x, t)$ is "small" in G (i.e., if $|b(x, t)| \leq C(1 + |x|)^{-1-\mu}$, $\mu > 0$), then there will be negligible resistance by the drift coefficient to the movement of $\xi(t)$ in G. Thus, no condition analogous to (5.5) is required in the case $n \geq 2$.

6. Counterexample

We shall give an example of a system of n equations ($n \geq 1$)

$$d\xi(t) = dw(t) + b(\xi(t))\, dt \tag{6.1}$$

for which

$$b(x) = O\left(\frac{\log|x|}{|x|}\right) \qquad \text{if} \quad |x| \to \infty \tag{6.2}$$

such that the estimate

$$E|\xi(t)|^2 \leq Kt + K' \qquad \text{for all} \quad t \geq 0 \qquad (K, K' \text{ positive constants}) \tag{6.3}$$

is false.

Let $f(x)$ be a function in $C^2(R^n)$ such that

$$f(x) = r^2/\log r \qquad \text{if} \quad r = |x| > 2. \tag{6.4}$$

If $b_i(x) = x_i\beta(r)$ for $|x| > 2$, then

$$\tfrac{1}{2}\,\Delta f(x) + \sum_{i=1}^{n} b_i(x)\,\frac{\partial f(x)}{\partial x_i}$$

$$= \frac{1}{\log r}\left[1 - \frac{m+2}{2}\,\frac{1}{\log r} + \frac{1}{(\log r)^2}\right] + \frac{2r^2}{\log r}\left[1 - \frac{1}{2\log r}\right]\beta(r).$$

Take

$$\beta(r) = \frac{\log r}{2r^2}\,[1 + \gamma(r)]$$

where $\gamma(r)$ is defined so that

$$\tfrac{1}{2}\,\Delta f(x) + \sum_{i=1}^{n} b_i(x)\,\frac{\partial f(x)}{\partial x_i} = 1 \qquad \text{if} \quad |x| > 2. \tag{6.5}$$

It is easily seen that $\gamma(r) = O(1)$. We now define $b_i(x)$ in R^n such that

$$b_i(x) = x_i[\log|x| + \gamma(|x|)]/2|x|^2 \qquad \text{if} \quad |x| > 2; \tag{6.6}$$

we can make the $b_i(x)$ as smooth in R^n as we wish.

By Itô's formula and (6.5),

$$Ef(\xi(t)) = Ef(\xi_0) + E\int_0^t c(\xi(s))\,ds \tag{6.7}$$

where $c(x) - 1 = 0$ if $|x| > 2$.

We wish to apply the proof of Lemma 2.2 to $\phi(x, t) = c(x) - 1$. Here the $b_i(x, t) = b_i(x)$ are not bounded by $\epsilon(|x|)/(1 + |x|)$ where $\epsilon(r) \to 0$ if $r \to \infty$. Nevertheless (2.11) takes the form

$$\tfrac{1}{2}[1 + \theta(r)] \leqslant \tfrac{1}{2}n + \sum_i (x_i - e_i)b_i(x) \tag{6.8}$$

where $r = |x - e|$ and $e = (e_1, \ldots, e_n)$ is a point outside the support of $c(x) - 1$, i.e., $|e| > 2$. In view of (6.6) and the boundedness of $\gamma(r)$ as $r \to \infty$, we can actually construct a continuous function $\theta(r)$ satisfying (6.8) such that, for any $A > 0$, $\lim_{r\to\infty}\theta(r) = A$. Hence, by the proof of Lemma 2.2,

$$E\int_0^t [c(\xi(s)) - 1]\,ds = O(1) \qquad \text{as} \quad t \to \infty.$$

Combining this with (6.7) we get

$$E|f(\xi(t)) - t| \leqslant C_0 \qquad \text{for} \quad t \geqslant 0 \qquad (C_0\ \text{const}). \tag{6.9}$$

For any $\epsilon > 0$ there is a constant B such that

$$|x|^2/\log|x| \leqslant \epsilon|x|^2 + B \qquad \text{if} \quad |x| > 2.$$

Hence

$$f(x) \leqslant \epsilon|x|^2 + C \qquad \text{for all} \quad x \in R^n,$$

where C is a constant depending on ϵ. This implies that

$$Ef(\xi(t)) \leqslant \epsilon E|\xi(t)|^2 + C. \tag{6.10}$$

Now, if (6.3) holds, then from (6.9), (6.10) we obtain

$$t \leqslant \epsilon Kt + \epsilon K' + C + C_0 \qquad \text{for all} \quad t \geqslant 0.$$

But this is impossible if $\epsilon < 1/K$.

If the function $\log|x|$ occurring in (6.2) is replaced by other functions that increase to infinity more mildly, as $|x| \to \infty$, such as $\log\log|x|$, then one can again show by the above method that (6.3) cannot hold. If however $b(x) = O(1/|x|)$, then (6.3) does hold (by Theorem 2.1). Recall that if (3.2) holds, then the estimate (3.1) is also valid.

PROBLEMS

1. All the results of Sections 2, 3 remain valid if the condition (2.4) is replaced by the condition that

$$(1 + |x|)\sum_i |b_i(x, t)| \leqslant \epsilon \qquad \text{for all} \quad x \in R^n, \quad t \geqslant 0$$

provided ϵ is sufficiently small. Check this in the case of Theorem 3.6(a).

2. The methods of Sections 2, 3 can be used to estimate $E|\xi(t) - \hat{\sigma}w(t)|^4$, provided $0 < \delta < \frac{1}{2}$. Show that under the assumptions (2.2), (2.3), (3.26), (3.29) with $0 < \delta < \frac{1}{2}$,

$$E|\xi(t) - \bar{\sigma}w(t)|^4 = O(t^{2(1-\delta)}).$$

3. Prove that if (5.2) and (5.5) hold, then any solution of (5.1) satisfies the estimate (5.4).

4. Prove that if (5.6) and (5.5) hold, then the estimate (5.7) is valid for the solution of (5.1).

5. If

$$\sum a_{ii}(x, t) \leqslant C(1 + |x|^2)^{\mu}, \qquad \left|\sum x_i b_i(x, t)\right| \leqslant C(1 + |x|^2)^{\mu}$$

for some constants $C > 0$, $0 < \mu < 1$, then

$$E|\xi(t)|^{2-2\mu} \leqslant Kt + K' \qquad \text{for all} \quad t \geqslant 0,$$

where K, K' are positive constants.

6. Let $\tilde{b}(x, t) = b(x, t)/(1 + |x|^2)^{\mu}$, $\tilde{\sigma}(x, t) = \sigma(x, t)/(1 + |x|^2)^{\mu/2}$, $\tilde{a} = \tilde{\sigma}\tilde{\sigma}^*$. Let the assumptions of Lemma 2.2 hold for a, b replaced by $\tilde{a}$, $\tilde{b}$. Let $\phi(x)$ be any bounded measurable function with compact support. Prove that if $\xi(t)$ is a solution of (1.1), then

$$E\left|\int_0^t \phi(\xi(s))\,ds\right| \leqslant K_1(1 + t^{(1+\eta)/(2-2\mu)}) \qquad \text{for all} \quad t \geqslant 0,$$

where K_1 is a constant and η is as in Lemma 2.2. [*Hint*: Construct $f(x) = F(r)$ satisfying $Lf(x) \geq (1 + |x|^2)^{\eta}\tilde{\psi}(|x|)$.]

7. Let (2.2)–(2.5) hold with a, b replaced by $\tilde{a}$, $\tilde{b}$ (as in Problem 6), and let $0 \leq \mu < 1$. Prove that

$$E|\xi(t)|^{2-2\mu} \geq Kt - K' \qquad \text{for all} \quad t \geq 0,$$

where K, K' are positive constants.

8. Extend Theorem 5.2 to the case of Eq. (5.8).

9. Let $\xi(t)$ be a solution of (1.1) and assume that $|\sigma(x, t)| \leq C$, $|b(x, t)| \leq C$, $|b(x, t) - \bar{b}| \to 0$ if $t \to \infty$, uniformly with respect to x. Prove that

$$\lim_{t\to\infty} \frac{\xi(t)}{t} = \bar{b} \quad \text{a.s.}$$

10. Consider one-dimensional equations

$$d\xi_i(t) = dw(t) + b_i(\xi_i(t), t) \qquad (i = 1, 2).$$

Assume that $b_1(x, t) < b_2(x, t)$ for all $x \in R^1$, $t \geq 0$. Prove that if $\xi_1(0) = \xi_2(0)$ then $\xi_1(t) < \xi_2(t)$ for all $t > 0$.

11. Let $\xi(t)$ be a solution of a one-dimensional equation

$$d\xi(t) = dw(t) + b(\xi(t), t)\, dt, \qquad \xi(0) = x$$

and assume that

$$b(x, t) \geq \beta(t) \qquad \text{for all} \quad x \in R^1,$$

$$\lim_{t\to\infty} \frac{1}{\sqrt{2t \log\log t}} \int_0^t \beta(s)\, ds > 1.$$

Prove that $\lim_{t\to\infty} \xi(t) = \infty$ a.s.

12. Consider a stochastic differential equation

$$d\xi(t) = \sigma(\xi(t))\, dw(t) + b(\xi(t))\, dt$$

and assume that $\sigma(x) > 0$ in the interval $a \leq x \leq b$. Denote by $\tau_x(a, b)$ the exit time from the interval (a, b), given $\xi(0) = x$. Denote by $p_a(x; b)$ the probability that $\xi(\tau_x(a, b)) = a$. Prove:

(a) If $\frac{1}{2}\sigma(x)u''(x) + b(x)u'(x) = -1$ in $a \leq x \leq b$, and $u(a) = u(b) = 0$ then $E\tau_x(a, b) = u(x)$.

(b) If $\Phi(x) = \exp\{-\int_a^x (2b(z)/\sigma^2(z))\, dz\}$, then

$$u(x) = -\int_a^x 2\Phi(y) \int_a^y \frac{dz}{\sigma^2(z)\Phi(z)}\, dy + \gamma \int_a^b 2\Phi(y) \int_a^y \frac{dz}{\sigma^2(z)\Phi(z)}\, dy,$$

$$\gamma = \int_a^x \Phi(z)\, dz \Big/ \int_a^b \Phi(z)\, dz.$$

(c) If $\frac{1}{2}\sigma^2(x)w''(x) + b(x)w'(x) = 0$ in $a < x < b$, then

$$p_a(x; b) = [w(x) - w(b)]/[w(a) - w(b)].$$

13. If (2.2), (2.3), (3.3) hold with $\delta > 0$, and if (2.5) is replaced by $\lambda_n(x, t) \geq c/(1 + |x|)^\delta$ $(c > 0)$, then the assertion (2.6) of Lemma 2.2 is valid for any $\eta > 0$.

14. If (2.2), (2.3), (2.5), and (3.3) hold with $\delta > 0$, then (2.6) holds with $\eta = 0$. [*Hint*: Take $\theta(r) = -D/r^\beta$ for large r, where β is any positive constant $\leq \delta$ and D is a positive constant.]

15. Suppose, in Lemma 2.3, ϕ satisfies (2.16) and the condition (2.4) is replaced by the stricter condition (3.3) with $\delta > 0$. Prove that

$$E\left|\int_0^t \phi(\xi(s), s)\, ds - \Phi t\right| = O(t^{1/2}) + O(t^{1-(\alpha-\beta)/2})$$

for any $\beta > 0$. [*Hint*: Take $1 - \rho = D/(2r^\beta)$, $\theta(r) = -D/r^\beta$. In (2.14) replace $\psi(u)$ by $\psi(u)u^\beta$; this gives (2.24) with α replaced by $\alpha - \beta$.]

16. Under the same conditions as in Problem 13, if ϕ satisfies (2.16), then

$$E\left|\int_0^t \phi(\xi(s), s)\, ds - \Phi t\right| = O(t^{(1+\eta)/2}) + O(t^{1-(\alpha-\delta)/2})$$

for any $\eta > 0$. [*Hint*: Replace, in (2.14), $\psi(u)$ by $\psi(u)u^\delta$.]

17. Under the assumptions of Theorem 3.2,

$$E[\langle e, \xi(t)\rangle^2] = \langle e, \bar{a}e\rangle + O(t^{(1+\eta)/2}) + o(t^{1-\delta/2})$$

where e is any unit vector and $\langle\ ,\ \rangle$ denotes the scalar product.

9

Recurrent and Transient Solutions

Consider a stochastic differential system of n equations

$$d\xi(t) = \sigma(\xi(t), t)\, dw(t) + b(\xi(t), t)\, dt \tag{0.1}$$

Let G be an open set in R^n. Suppose that for any $x \in G$ and for any open subset V of G,

$$P_x\{\xi(t_m) \in V \quad \text{for a sequence of finite random times } t_m \text{ increasing to } \infty\} = 1. \tag{0.2}$$

Then we say that the solution of (0.1) is a *recurrent process* in G. If $G = R^n$, then we say simply that the solution of (0.1) is a recurrent process.

If, for any $x \in G$,

$$P_x\{\lim_{t\to\infty} |\xi(t)| = \infty\} = 1, \tag{0.3}$$

then we say that the solution of (0.1) is a *transient process* in G. If $G = R^n$, we simply say that the solution of (0.1) is a transient process.

The condition (0.2) says that, given $\xi(0) = x$, the solution $\xi(t)$ "visits" any open subset V of G at a sequence of times increasing to infinity. The condition (0.3) says that, given $\xi(0) = x$, $\xi(t)$ "wanders out to infinity" with probability one.

It is well known (see Itô and McKean [1]) that an n-dimensional Brownian motion is recurrent if $n \leqslant 2$, and transient if $n \geqslant 3$. This fact will follow as a very special case of the results of this chapter.

1. Transient solutions

For simplicity we shall consider only temporally homogeneous processes, i.e., solutions of systems of n stochastic differential equations with time-independent coefficients,

$$d\xi(t) = \sigma(\xi(t))\, dw(t) + b(\xi(t))\, dt. \tag{1.1}$$

Set

$$\sigma = (\sigma_{ij})_{i,j=1}^{n}, \qquad b = (b_1, \ldots, b_n), \qquad a_{ij} = \sum_{k=1}^{n} \sigma_{ik}\sigma_{jk},$$

$$Lu = \frac{1}{2} \sum_{i,j=1}^{n} a_{ij}(x) \frac{\partial^2 u}{\partial x_i \, \partial x_j} + \sum_{i=1}^{n} b_i(x) \frac{\partial u}{\partial x_i}. \tag{1.2}$$

We shall assume:

(A_1) For all $x \in R^n$,

$$\sum_{i=1}^{n} |b_i(x)| + \sum_{i,j=1}^{n} |\sigma_{ij}(x)| \leq C(1 + |x|) \qquad (C \text{ const});$$

for any $R > 0$ there is a positive constant C_R such that

$$\sum_{i=1}^{n} |b_i(x) - b_i(y)| + \sum_{i,j=1}^{n} |\sigma_{ij}(x) - \sigma_{ij}(y)| \leq C_R|x - y|$$

if $|x| < R$, $|y| < R$.

(A_2) The matrix $(a_{ij}(x))$ is positive definite for each $x \in R^n$.

Let

$$A(x, \xi) = \frac{1}{|\xi|^2} \sum_{i,j=1}^{n} a_{ij}(x)\xi_i\xi_j,$$

$$B(x) = \sum_{i=1}^{n} a_{ii}(x),$$

$$C(x, \xi) = 2 \sum_{i=1}^{n} \xi_i b_i(x),$$

and set

$$S(x, \xi) = \frac{B(x) - A(x, \xi) + C(x, \xi)}{A(x, \xi)}, \qquad S(x) = S(x, x). \tag{1.3}$$

We shall need the assumption:

(A_3) There is a positive constant R_0 such that

$$S(x) \geq 1 + \epsilon(|x|) \qquad \text{if} \quad |x| \geq R_0, \tag{1.4}$$

where $\epsilon(r)$ is a continuous function satisfying:

$$\int_{R_0}^{\infty} \frac{1}{t} \exp\left[-\int_{R_0}^{t} \frac{\epsilon(s)}{s} \, ds \right] dt < \infty. \tag{1.5}$$

Notice that (1.5) holds for any of the functions

$$\epsilon(s) = d, \qquad \epsilon(s) = A/s, \qquad \epsilon(s) = B/\log s$$

where d, A, B are positive constants and $B > 1$.

Theorem 1.1. *Let* (A_1)–(A_3) *hold. Then the solution of* (1.1) *is transient, i.e., for any* $x \in R^n$,

$$P_x\{\lim_{t\to\infty} |\xi(t)| = \infty\} = 1. \tag{1.6}$$

Proof. Let $\alpha > 0$. By (A_3), there is a continuous function $\theta(r)$ defined for $r \geq \alpha$ such that

$$S(x) \geq \theta(|x|) \quad \text{if} \quad |x| \geq \alpha, \tag{1.7}$$

$$\theta(r) = 1 + \epsilon(r) \quad \text{if} \quad r \geq R_0. \tag{1.8}$$

We shall construct a function $f(x) = F(r)$, where $r = |x|$, such that $Lf(x) \leq 0$ if $|x| \geq \alpha$. As easily verified,

$$2Lf(x) = A(x, x)F''(r) + \frac{F'(r)}{r}[B(x) - A(x, x) + C(x, x)]. \tag{1.9}$$

Hence, if

$$F'(r) \leq 0 \quad \text{for} \quad r \geq \alpha, \tag{1.10}$$

$$F''(r) + \frac{\theta(r)}{r}F'(r) = 0 \quad \text{for} \quad r \geq \alpha, \tag{1.11}$$

then, by (1.7)–(1.9),

$$Lf(x) \leq 0 \quad \text{if} \quad |x| \geq \alpha. \tag{1.12}$$

Set

$$I(r) = \int_\alpha^r \frac{\theta(s)}{s}\, ds. \tag{1.13}$$

Then a solution of (1.11) is given by

$$F(r) = \int_r^\infty e^{-I(s)}\, ds; \tag{1.14}$$

the integral is convergent, by (1.5), (1.8). Notice that (1.10) is also satisfied. Hence (1.12) holds when F is given by (1.14).

By Itô's formula and (1.12), if $|x| > \alpha$, then

$$E_x F(|\xi(\tau)|) - E_x F(|x|) = E_x \int_0^\tau Lf(\xi(s))\, ds \leq 0 \tag{1.15}$$

where τ is any bounded stopping time such that $|\xi(s)| \geq \alpha$ if $0 \leq s \leq \tau$. Let $\beta > \alpha$, and let $\tau_{\alpha\beta}$ denote the exit time from the shell $\{y;\ \alpha < |y| < \beta\}$. Denote by $P_x(\alpha)$ the probability that $|\xi(\tau_{\alpha\beta})| = \alpha$ (given $\xi(0) = x$) and by

$P_x(\beta)$ the probability that $|\xi(\tau_{\alpha\beta})| = \beta$ (given $\xi(0) = x$). Setting $\tau = T \wedge \tau_{\alpha\beta}$ in (1.15) and taking $T \to \infty$, we get (since $\tau_{\alpha\beta} < \infty$ a.s.; see Theorem 8.1.1),

$$F(\alpha)P_x(\alpha) + F(\beta)P_x(\beta) \leqslant F(|x|).$$

Taking $\beta \to \infty$ and using the fact that $F(\beta) \to 0$ if $\beta \to \infty$, we get

$$\lim_{\beta\to\infty} P_x(\alpha) \leqslant \frac{F(|x|)}{F(\alpha)} . \tag{1.16}$$

Introduce the balls

$$B_\rho = \{ y;\ |y| \leqslant \rho\} \qquad (0 < \rho < \infty)$$

and the event

$$\Omega(\alpha) = \{\xi(t) \text{ hits the ball } B_\alpha \text{ for some } t \geqslant 0\}. \tag{1.17}$$

Then we can write (1.16) in the form:

$$P_x(\Omega(\alpha)) \leqslant \frac{F(|x|)}{F(\alpha)} . \tag{1.18}$$

Denote by t_R the hitting time of the boundary of the ball B_R by $\xi(t)$. By Theorem 8.1.1, if $|x| < R$, then $P_x\{t_R < \infty\} = 1$. Introduce the event

$$\Omega^*(\alpha) = \{\xi(t) \text{ hits the ball } B_\alpha \text{ at a sequence of times increasing to } \infty\}. \tag{1.19}$$

Thus $\Omega^*(\alpha)$ is a subset of $\Omega(\alpha)$.

Let $\alpha < |x| < R$. Since $P_x\{t_R < \infty\} = 1$,

$$P_x(\Omega^*(\alpha)) = P_x\{\xi(t + t_R) \text{ hits } B_\alpha \text{ at a sequence of times increasing to } \infty\}.$$

Using the strong Markov property (see Problem 3) we get

$$P_x(\Omega^*(\alpha)) = E_x P_{\xi(t_R)}(\Omega^*(\alpha)) \leqslant E_x P_{\xi(t_R)}(\Omega(\alpha)) \leqslant E_x \frac{F(R)}{F(\alpha)} = \frac{F(R)}{F(\alpha)} ,$$

where (1.18) has been used. Taking $R \to \infty$, we get $P_x(\Omega^*(\alpha)) = 0$. This means that

$$P_x\Big\{ \varliminf_{t\to\infty} |\xi(t)| \geqslant \alpha \Big\} = 1.$$

Since α is arbitrary, we get

$$P_x\Big\{ \varliminf_{t\to\infty} |\xi(t)| = \infty \Big\} = 1,$$

i.e., (1.6) holds.

We shall now replace the condition (A_3) by:

(A_3') As $|x| \to \infty$,

$$a_{ij}(x) \to a_{ij}^0, \tag{1.20}$$

$$\sum_{i=1}^{n} x_i b_i(x) \to 0, \tag{1.21}$$

where the matrix (a_{ij}^0) has at least three positive eigenvalues.

Theorem 1.2. *Let* (A_1), (A_2), (A_3') *hold. Then the solution of* (1.1) *is transient.*

Proof. We can perform an orthogonal transformation $x \to x'$ in R^n which takes (a_{ij}^0) into (a_{ij}^*), where $a_{ij}^* = 0$ if $i \neq j$, $a_{ii}^* = 1$ if $i = 1, 2, 3$ and $a_{ii}^* = 0$ or 1 if $4 \leqslant i \leqslant n$. In the new coordinates the condition (A_3) holds with $\epsilon(s) = d$ where d is any positive constant < 1. Now apply Theorem 1.1.

2. Recurrent solutions

We shall replace the condition (A_3) by:

(A_4) For any $z \in R^n$ there is a positive constant R_z such that

$$S(x, x - z) \leqslant 1 + \epsilon(|x - z|) \qquad \text{if} \quad |x - z| \geqslant R_z, \tag{2.1}$$

where $\epsilon(r)$ is a continuous function satisfying

$$\int_{R^*}^{\infty} \frac{1}{t} \exp\left[- \int_{R^*}^{t} \frac{\epsilon(s)}{s}\, ds \right] dt = \infty \qquad \text{for some} \quad R^* > 0. \tag{2.2}$$

For simplicity we have taken $\epsilon(r)$ to be independent of z; but the subsequent results are unaffected if $\epsilon(r)$ is allowed to depend on z.

Notice that the function $\epsilon(r) = 1/(\log r)$ satisfies (2.2).

Theorem 2.1. *Let* (A_1), (A_2), (A_4) *hold. Then the solution of* (1.1) *is recurrent, i.e., for any* $x \in R^n$ *and for any ball* $B_\alpha(z) = \{y; |y - z| \leqslant \alpha\}$, $\alpha > 0$

$$P_x\{\xi(t) \text{ hits } B_\alpha(z) \text{ at a sequence of times increasing to } \infty\} = 1. \tag{2.3}$$

Proof. We take, for simplicity, $z = 0$ and write $B_\alpha = B_\alpha(0)$. We shall first construct a function $f(x) = F(r)$ for $r = |x| \geqslant \alpha$ such that

$$Lf(x) \geqslant 0 \qquad \text{if} \quad |x| \geqslant \alpha. \tag{2.4}$$

Let $\theta(r)$ be a continuous function such that

$$S(x) \leqslant \theta(|x|) \qquad \text{if} \quad |x| \geqslant \alpha, \tag{2.5}$$

$$\theta(r) = 1 + \epsilon(r) \qquad \text{if} \quad r \geqslant R_0. \tag{2.6}$$

In view of (1.9), if $F(r)$ satisfies (1.10), (1.11), then (2.4) holds. With the definition (1.13), the function

$$F(r) = -\int_{\alpha}^{r} e^{-I(s)}\, ds \tag{2.7}$$

satisfies both (1.11) and (1.10). In view of (2.6), (2.2),

$$F(r) \rightarrow -\infty \qquad \text{if} \quad r \rightarrow \infty. \tag{2.8}$$

We shall now apply the equality in (1.15) to the present function $F(r)$. Making use of (2.4) and taking $\tau = \tau_{\alpha\beta} \wedge T$, we get, after letting $T \rightarrow \infty$,

$$F(\alpha)P_x(\alpha) + F(\beta)P_x(\beta) - F(|x|) \geqslant 0. \tag{2.9}$$

Taking $\beta \rightarrow \infty$ in (2.9) and using (2.8), we conclude that $P_x(\beta) \rightarrow 0$ if $\beta \rightarrow \infty$. Hence, $P_x(\alpha) = 1 - P_x(\beta) \rightarrow 1$ if $\beta \rightarrow \infty$. This means that

$$P_x(\Omega(\alpha)) = 1 \tag{2.10}$$

where $\Omega(\alpha)$ is defined in (1.17).

For any $\rho > 0$, let $\partial B_\rho = \{y;\ |y| = \rho\}$. Let

$$\alpha < R_1 < R_2 < \cdots < R_m < \cdots, \qquad R_m \rightarrow \infty \qquad \text{if} \quad m \rightarrow \infty.$$

Introduce Markov times:

$$\tau_1 = \text{first time } \xi(t) \text{ hits } B_\alpha;$$

$$\sigma_1 = \text{first time} > \tau_1 \text{ such that } \xi(t) \text{ hits } \partial B_{R_1};$$

in general,

$$\tau_m = \text{first time} > \sigma_{m-1} \text{ such that } \xi(t) \text{ hits } B_\alpha;$$

$$\sigma_m = \text{first time} > \tau_m \text{ such that } \xi(t) \text{ hits } \partial B_{R_m}.$$

By Theorem 8.1.1, on the set where $\tau_m < \infty$ also $\sigma_m < \infty$. By (2.10), $P_x(\tau_1 < \infty) = 1$. Hence $P_x(\sigma_1 < \infty) = 1$, and by the strong Markov property (see Problem 4),

$$P_x(\tau_2 < \infty) = E_x E_{\xi(\sigma_1)} \chi_{\tau_1 < \infty} = 1. \tag{2.11}$$

where (2.10) has been used in the last equality.

We now proceed by induction. Assuming that $\tau_m < \infty$ a.s., we get

$$P_x(\tau_{m+1} < \infty) = E_x E_{\xi(\sigma_m)} \chi_{\tau_1 < \infty} = 1.$$

Now, at each time $t = \tau_m$, $\xi(t)$ hits B_α. Further, since $|\xi(\sigma_m)| = R_m \rightarrow \infty$ as $m \rightarrow \infty$, $\lim_{m\rightarrow\infty} \sigma_m = \infty$; hence also $\lim_{m\rightarrow\infty} \tau_m = \infty$. This completes the proof of (2.3) (in case $z = 0$).

We shall now replace the condition (A_4) by:

(A_4') As $|x|\to\infty$,

$$a_{ij}(x) - a_{ij}^0 = o\left(\frac{1}{\log |x|}\right), \tag{2.12}$$

$$\sum |b_i(x)| = o\left(\frac{1}{|x| \log |x|}\right), \tag{2.13}$$

and the matrix (a_{ij}^0) has precisely two positive eigenvalues.

Theorem 2.2. *Let* (A_1), (A_2), (A_4') *hold. Then, the solution of* (1.1) *is recurrent.*

Proof. We perform an orthogonal transformation $x\to x'$ that takes (a_{ij}^0) into a new matrix (a_{ij}^*) with $a_{ij}^* = 0$ if $i \neq j$ or if $i = j \geqslant 3$, and $a_{ii}^* = 1$ if $i = 1, 2$. In the new coordinates, the condition (A_4) is satisfied with $\epsilon(r) = 1/(\log r)$. Now apply Theorem 2.1.

Remark. Suppose (A_4') is replaced by

$$a_{ij}(x) \to a_{ij}^0 \quad \text{as} \quad |x| \to \infty, \qquad \sum |b_i(x)| = o\left(\frac{1}{|x|}\right) \quad \text{as} \quad |x| \to \infty$$

where the matrix (a_{ij}^0) has precisely one positive eigenvalue. Then the assertion of Theorem 2.1 remains valid, with the same proof; here $\epsilon(r) = -d$ where d is any positive constant < 1.

Example. Consider the case where $n \geqslant 2$, $b_i \equiv 0$, and σ is such that

$$\sigma\sigma^* = (a_{ij}), \qquad a_{ij} = \delta_{ij} + \frac{g(r)}{r^2} x_i x_j \qquad (r = |x|);$$

$g(r)$ is a Lipschitz continuous function vanishing near $r = 0$, and

$$\mu \leqslant g(r) \leqslant M \qquad \text{where} \quad \mu > -1, \quad M < \infty \qquad (\mu, M \text{ const}).$$

The eigenvalues of $(a_{ij}(x))$ are 1 (with multiplicity $n - 1$) and $1 + g$. Hence $(a_{ij}(x))$ is positive definite for all $x \in R^n$. Clearly,

$$S(x) - 1 = \frac{n - 2 - g(r)}{1 + g(r)} \equiv \epsilon(r).$$

Hence, if $g(r)$ is such that $\epsilon(r) = A/(\log r)$ for some $A > 1$ (and all large r), then the assertion of Theorem 1.1 holds. If $g(r)$ is such that $\epsilon(r) = 1/(\log r)$ for all large r, then (A_4) holds with $\epsilon(s) = 1/(\log s) + C/s$ for some positive constant C; consequently the assertion of Theorem 2.1 holds. This example shows that conditions (A_3), (A_4) made in Theorems 1.1, 2.1 are rather sharp.

This example also shows that the behavior asserted in Theorems 1.1 and 2.1 does not depend exclusively on the dimension n. In fact, given any $\epsilon(r)$ which converges to 0 as $r \to \infty$, take

$$g = \frac{n-2-\epsilon}{1+\epsilon} \qquad \text{(for all large } r\text{)}$$

in the above example. Then the behavior of $\xi(t)$ does not depend on n; if $\epsilon(r) \geq A/\log r$ for some $A > 1$, then $\xi(t)$ is transient, whereas if $\epsilon(r) \leq 1/\log r$ then $\xi(t)$ is recurrent.

3. Rate of wandering out to infinity

In this section we return to the situation of Theorem 1.1. We shall assume:

(A_5) $a_{ij}(x)$, $b_i(x)$ are bounded functions in R^n, the $a_{ij}(x)$ are uniformly Hölder continuous in R^n, and, for all $x \in R^n$, $\xi \in R^n$,

$$\sum_{i,j=1}^{n} a_{ij}(x)\xi_i\xi_j \geq \alpha_0|\xi|^2 \qquad (\alpha_0 \text{ positive constant}).$$

We shall also assume that the function $\epsilon(r)$ occurring in the condition (A_3) satisfies, for some r_0 sufficiently large,

$$\epsilon(r) = d \quad \text{if} \quad r \geq r_0 \qquad (d \text{ positive constant}). \tag{3.1}$$

Theorem 3.1. *Let* (A_1), (A_5) *hold and let* (A_3) *hold with* $\epsilon(r)$ *satisfying* (3.1). *Then, for any* $0 < \theta < \frac{1}{2}$, $x \in R^n$,

$$P_x\left\{\frac{|\xi(t)|}{t^\theta} \to \infty \text{ if } t \to \infty\right\} = 1. \tag{3.2}$$

We first prove two lemmas.

Lemma 3.2. *Let* (A_1), (A_5), (A_3), *and* (3.1) *hold. Then there exists a positive constant* C *such that for any* $\alpha > r_0$, $x \in R^n$,

$$P_x\{|\xi(t)| \leq \alpha \textit{ for some } t \geq 0\} \leq C\left(\frac{\alpha}{|x|}\right)^d. \tag{3.3}$$

Proof. Let $R > \alpha$. Consider the Dirichlet problem:

$$Lu_R(x) = 0 \quad \text{if} \quad \alpha < |x| < R,$$
$$u_R = 1 \quad \text{if} \quad |x| = \alpha, \qquad u_R = 0 \quad \text{if} \quad |x| = R.$$

Let $f(x) = F(r)$ $(r = |x|)$ be the function constructed in the proof of Theorem 1.1. Then

$$v(x) = \frac{F(r)}{F(\alpha)} \qquad (r = |x|)$$

satisfies:

$$\begin{aligned} Lv &\leq 0 \quad \text{if} \quad \alpha < |x| < R, \\ v &= 1 \quad \text{if} \quad |x| = \alpha, \qquad v \geq 0 \quad \text{if} \quad |x| = R. \end{aligned}$$

Hence, by the maximum principle,

$$0 \leq u_R(x) \leq v(x). \tag{3.4}$$

By Itô's formula, if $\alpha < |x| < R$,

$$u_R(x) = P_x\{\xi(t) \text{ hits } \partial B_\alpha \text{ before it hits } \partial B_R\}. \tag{3.5}$$

Hence $u_R(x) \uparrow u(x)$ as $R \uparrow \infty$, where

$$u(x) = P_x\{\xi(t) \text{ hits } \partial B_\alpha \text{ for some } t \geq 0\}. \tag{3.6}$$

From (3.4) we get

$$0 \leq u(x) \leq v(x) = \frac{\int_r^\infty \exp[-I(s)]\, ds}{\int_\alpha^\infty \exp[-I(s)]\, ds}.$$

Using (3.1) to estimate the right-hand side, we obtain

$$0 \leq u(x) \leq C\frac{\alpha^d}{r^d} \tag{3.7}$$

where C is a constant independent of α, x.

Now, if $|x| \leq \alpha$, then the assertion (3.3) is trivially true (with $C = 1$). If, on the other hand, $|x| > \alpha$, then the assertion (3.3) follows from (3.6), (3.7).

Lemma 3.3. *Let* (A_1), (A_5), *and* (A_3), (3.1) *hold, with* $d < n$. *Then there is a positive constant* C' *such that, if* $\alpha > r_0$, $|x| < \alpha/4$, $T > 0$,

$$P_x\{|\xi(t)| \leq \alpha \textit{ for some } t \geq T\} \leq C'\left(\frac{\alpha}{T^{1/2}}\right)^d. \tag{3.8}$$

Proof. By the Markov property and Lemma 3.2,

$$\begin{aligned} P_x\{|\xi(t)| \leq \alpha \text{ for some } t \geq T\} &= E_x P_{\xi(T)}\{|\xi(t)| \leq \alpha \text{ for some } t \geq 0\} \\ &= P_x\{|\xi(T)| \leq \alpha\} + E_x\left\{\chi_{|\xi(T)|>\alpha}\left[C\frac{\alpha^d}{|\xi(T)|^d}\right]\right\} \\ &\equiv I + J. \end{aligned} \tag{3.9}$$

Denote by $\Gamma(x, t, y)$ the fundamental solution of the parabolic operator

$L - \partial/\partial t$. By Theorem 6.4.5 and Problem 7, Chapter 6,

$$0 \leqslant \Gamma(x, t, y) \leqslant \frac{M}{t^{n/2}} \exp\left[-\frac{\mu|x-y|^2}{t} \right] \tag{3.10}$$

where M, μ are positive constants.

We can write

$$I = \int_{|y|<\alpha} \Gamma(x, T, y)\, dy,$$

$$J = C\alpha^d \int_{|y|>\alpha} \frac{1}{|y|^d} \Gamma(x, T, y)\, dy.$$

We shall subsequently denote various positive constants by the same symbol C. Substituting $|y - x| = \rho\sqrt{T}$ in the integral I and noting that

$$\rho\sqrt{T} = |y - x| \leqslant 2\alpha \qquad (\text{since } |y| \leqslant \alpha,\ |x| < \alpha),$$

we get

$$I \leqslant C\int_{\rho\sqrt{T} < 2\alpha} \rho^{n-1} e^{-\mu\rho^2}\, d\rho \leqslant C\left(\frac{\alpha}{\sqrt{T}} \right)^n.$$

Substituting $|y - x| = \rho\sqrt{T}$ in the integral J and noting that

$$\rho\sqrt{T} = |y - x| \geqslant \alpha/2 \qquad (\text{since } |y| \geqslant \alpha,\ |x| \leqslant \alpha/2),$$

$$|y| \geqslant |y - x| - |x| \geqslant \rho\sqrt{T}/2 \qquad (\text{since } |x| \leqslant \alpha/4 \leqslant |y - x|/2),$$

we get

$$J \leqslant C\alpha^d \int_{\rho\sqrt{T} > \alpha/2} \frac{\rho^{n-1}}{(\rho\sqrt{T})^d} e^{-\mu\rho^2} d\rho$$

$$\leqslant C\left(\frac{\alpha}{\sqrt{T}} \right)^d \int_{\rho>0} \rho^{n-1-d} e^{-\mu\rho^2}\, d\rho \leqslant C\left(\frac{\alpha}{\sqrt{T}} \right)^d,$$

since $n - 1 - d > -1$. Substituting the estimates for I, J into (3.9), the assertion (3.8) follows.

Proof of Theorem 3.1. Without loss of generality we may assume that $d < n$. We apply Lemma 3.3 with $T = 2^m$, $\alpha = 2^{(m+1)\theta}$ where m is a positive integer such that $|x| < 2^{(m+1)\theta}/4$. We get

$$P_x\{|\xi(t)| \leqslant t^\theta \text{ for some } t,\ 2^m \leqslant t \leqslant 2^{m+1}\}$$

$$\leqslant P_x\{|\xi(t)| \leqslant 2^{(m+1)\theta} \text{ for some } t \geqslant 2^m\} \leqslant C\left[\frac{2^{(m+1)\theta}}{2^{m/2}} \right]^d \leqslant C2^{m(\theta-1/2)d}.$$

Since $\sum 2^{m(\theta-1/2)d} < \infty$, the Borel–Cantelli lemma implies that, with probability 1, the sequence of events

$$\{|\xi(t)| \leqslant t^\theta \text{ for some } t,\ 2^m \leqslant t \leqslant 2^{m+1}\}$$

(where $\xi(0) = x$) occurs only finitely often. Hence

$$P_x\{|\xi(t)| > t^\theta \text{ for all } t \text{ sufficiently large}\} = 1.$$

Since this is true also for θ replaced by any θ', $\theta < \theta' < \frac{1}{2}$, the assertion of the theorem follows.

Corollary 3.4. *Let* (A_1), (A_5), *and* (A_3') *hold. Then, for any* $0 < \theta < \frac{1}{2}$, $x \in R^n$, *the assertion* (3.2) *holds.*

Indeed, perform an orthogonal transformation as in the proof of Theorem 1.2. In the new coordinates the conditions (A_3), (3.1) hold.

Remark 1. By Theorem 3.6.1,

$$P_x\left\{\overline{\lim_{t\to\infty}} \frac{|w(t)|}{\sqrt{t \log\log t}} = 1\right\} = 1.$$

More generally, under the conditions of Theorem 8.4.2,

$$P_x\left\{\overline{\lim_{t\to\infty}} \frac{|\hat{\sigma}\xi(t)|}{\sqrt{t \log\log t}} = 1\right\} = 1 \tag{3.11}$$

where $\hat{\sigma}$ is the inverse of $\lim_{|x|\to\infty}\sigma(x)$. From (3.11) it follows that

$$P_x\left\{\lim_{t\to\infty} \frac{|\xi(t)|}{t^\eta} = 0\right\} = 1 \quad \text{if} \quad \eta > \tfrac{1}{2}.$$

Thus the assertion (3.2) (for any $\theta < \frac{1}{2}$) is rather sharp. However, this assertion can still be strengthened as follows.

Let $g(t)$ be a positive and monotone decreasing function for $t > 0$, satisfying

$$\sum_{m=1}^{\infty} (g(2^m))^d < \infty \tag{3.12}$$

where d is any constant $< n$ such that (3.1) holds. If in the proof of Theorem 3.1 we replace t^θ by $t^{1/2}g(t)$, then we conclude that the events

$$\{|\xi(t)| < t^{1/2}g(t) \text{ for some } t,\ 2^m \leqslant t \leqslant 2^{m+1}\}$$

occur only finitely often. Consequently

$$P_x\left\{\lim_{t\to\infty} \frac{|\xi(t)|}{t^{1/2}g(t)} \geqslant 1\right\} = 1. \tag{3.13}$$

It can be shown (see Dvoretsky and Erdös [1]) that if the function $g(t)$ satisfies

$$\sum_{m=1}^{\infty} \left(g(2^m)\right)^d = \infty \qquad (d = n - 2, n \geqslant 3), \tag{3.14}$$

instead of (3.12), then

$$P_x\left\{ \varlimsup_{t\to\infty} \frac{|w(t)|}{t^{1/2}g(t)} \leqslant 1 \right\} = 1. \tag{3.15}$$

Remark 2. Consider the example at the end of Section 2. If $n = 2$ and $g = -1/(1 + 1/d)$ $(d > 0)$, then the assertion of Theorem 3.1 holds. Thus, even when $n = 2$, a diffusion process $\xi(t)$ may wander out to ∞ at a rate $\geqslant t^\theta$, for any $0 < \theta < \frac{1}{2}$.

4. Obstacles

We shall maintain the condition (A_1) but relax the condition (A_2).

Let G be a closed bounded domain with C^3 connected boundary ∂G, and let $\hat{G} = R^n \backslash G$.

Suppose the diffusion matrix $(a_{ij}(x))$ degenerates on ∂G in such a way that

$$\sum_{i,j=1}^{n} a_{ij}(x)\nu_i\nu_j = 0 \qquad \text{if} \quad x \in \partial G, \tag{4.1}$$

where $\nu = (\nu_1, \ldots, \nu_n)$ is the outward normal at x to ∂G. This condition is equivalent to

$$\sum_{k=1}^{n} \left(\sum_{i=1}^{n} \sigma_{ik}(x)\nu_i \right)^2 = 0,$$

i.e.,

$$\sum_{i=1}^{n} \sigma_{ik}(x)\nu_i = 0 \qquad \text{if} \quad x \in \partial G, \quad 1 \leqslant k \leqslant n. \tag{4.2}$$

The expression $\sum a_{ij}(x)\nu_i\nu_j$ is called the *normal diffusion* to ∂G at x.

Let $\rho(x) = \text{dist}\,(x, \partial G)$ for $x \in \hat{G} \cup \partial G$, and let

$$\hat{G}_\epsilon = \{x \in \hat{G};\ \rho(x) < \epsilon\}.$$

Since ∂G is in C^3, $\rho(x)$ is in $C^2(\hat{G}_{\epsilon_0} \cup \partial G)$ for some $\epsilon_0 > 0$ sufficiently small. Noting that

$$\nu_i(x) = \partial\rho(x)/\partial x_i \qquad x \in \partial G, \quad 1 \leqslant i \leqslant n,$$

we deduce from (4.2) that

$$\sum_{i=1}^{n} \sigma_{ik}(x) \frac{\partial \rho(x)}{\partial x_i} = O(\rho(x)) \qquad \text{as} \quad \rho(x) \to 0, \quad x \in \hat{G}_{\epsilon_0}.$$

Taking the squares, we get

$$\sum_{i,j=1}^{n} a_{ij}(x) \frac{\partial \rho(x)}{\partial x_i} \frac{\partial \rho(x)}{\partial x_j} \leq C(\rho(x))^2 \qquad \text{if} \quad x \in \hat{G}_{\epsilon_0} \tag{4.3}$$

where C is a positive constant.

Let (4.1) hold and suppose $\sigma_{ij} \in C^1(\hat{G}_{\epsilon_0})$. By (4.2), the vector $T_k = (\sigma_{1k}, \ldots, \sigma_{nk})$ is tangent to ∂G. Since the function $\Sigma\, \sigma_{jk}\, \partial\rho/\partial x_j$ vanishes on ∂G, its derivative with respect to the tangent vector of T_k also vanishes on ∂G. Consequently

$$\sum_k \sum_j \sigma_{jk} \frac{\partial}{\partial x_j} \sum_i \sigma_{ik} \frac{\partial \rho}{\partial x_i} = 0.$$

This gives, after using (4.2) once more,

$$\sum_{i,j=1}^{n} a_{ij} \frac{\partial^2 \rho}{\partial x_i\, \partial x_j} = -\sum_{i,j=1}^{n} \frac{\partial a_{ij}}{\partial x_j} \nu_i \qquad \text{on } \partial G. \tag{4.4}$$

The vector with components

$$b_i - \frac{1}{2} \sum_{j=1}^{n} \frac{\partial a_{ij}}{\partial x_j}$$

is called the *Fichera drift*. We shall impose the condition:

$$\sum_{i=1}^{n} b_i \nu_i + \frac{1}{2} \sum_{i,j=1}^{n} a_{ij} \frac{\partial^2 \rho}{\partial x_i\, \partial x_j} > 0 \qquad \text{on } \partial G. \tag{4.5}$$

If $\sigma_{ij} \in C^1(\hat{G}_{\epsilon_0})$ then, by (4.4), this condition is equivalent to

$$\sum_{i=1}^{n} \left(b_i - \frac{1}{2} \sum_{j=1}^{n} \frac{\partial a_{ij}}{\partial x_j} \right) \nu_i > 0 \qquad \text{on } \partial G, \tag{4.6}$$

i.e., the Fichera drift at ∂G points into the exterior of G.

Notice that (4.5) implies that

$$\sum_{i=1}^{n} b_i \frac{\partial \rho}{\partial x_i} + \frac{1}{2} \sum_{i,j=1}^{n} a_{ij} \frac{\partial^2 \rho}{\partial x_i\, \partial x_j} > -c\rho \qquad \text{in } \hat{G}_{\epsilon_0} \tag{4.7}$$

where c is a positive constant.

Theorem 4.1. *Let* (A_1) *and* (4.1), (4.5) *hold. Then*

$$P_x\{\xi(t) \in \hat{G} \text{ for all } t > 0\} = 1 \qquad \text{if} \quad x \in \hat{G}. \tag{4.8}$$

Proof. Let $R(x)$ be a $C^2(\hat{G})$ function such that

$$R(x) = \begin{cases} \rho(x) & \text{if } x \in \hat{G}_{\epsilon_1} \quad (\text{for some } 0 < \epsilon_1 < \epsilon_0), \\ |x| & \text{if } |x| > M, \end{cases}$$

and $\epsilon_1 < R(x) < M$ elsewhere; M is chosen so large that $\hat{G}_{\epsilon_0} \subset \{x;\ |x| < M\}$. (To construct $R(x)$, let

$$\hat{R}(x) = \begin{cases} \rho(x) & \text{if } x \in \hat{G}_{\epsilon_0}, \\ M & \text{if } x \in R^n \setminus \hat{G}_{\epsilon_0} \end{cases}$$

and take $R(x)$ to be a mollifier of $\hat{R}(x)$; see Problem 4, Chapter 4.)

Let $V(x) = 1/(R(x))^\epsilon$ for some $\epsilon > 0$. We have

$$\begin{aligned} LV = & -\epsilon R^{-\epsilon-1} \sum_i b_i \frac{\partial R}{\partial x_i} \\ & + \frac{1}{2} \sum_{i,j} a_{ij} \left\{ \epsilon(\epsilon+1) R^{-\epsilon-2} \frac{\partial R}{\partial x_i} \frac{\partial R}{\partial x_j} - \epsilon R^{-\epsilon-1} \frac{\partial^2 R}{\partial x_i \partial x_j} \right\} \\ = & V \left\{ -\frac{\epsilon}{R} \sum_i b_i \frac{\partial R}{\partial x_i} + \frac{1}{2} \sum_{i,j} \frac{a_{ij}}{R^2} \left[\epsilon(\epsilon+1) \frac{\partial R}{\partial x_i} \frac{\partial R}{\partial x_j} - \epsilon R \frac{\partial^2 R}{\partial x_i \partial x_j} \right] \right\}. \end{aligned} \tag{4.9}$$

In view of (4.3) and (4.7), the right-hand side of (4.9) is bounded by μV if $x \in \hat{G}_{\epsilon_1}$, where μ is a positive constant. In view of the condition (A_1), the same estimate is valid also if $|x| > M$. It follows that

$$LV \leq \mu V \qquad \text{for all } \ x \in \hat{G}. \tag{4.10}$$

Further,

$$V(x) \to \infty \qquad \text{if } \ \rho(x) \to 0, \quad x \in \hat{G}. \tag{4.11}$$

We shall now use (4.10), (4.11) in order to prove (4.8).

Denote by τ_p the hitting time of the set $\hat{G}_{1/p}$ $(p = 1, 2, \dots)$. By Itô's formula,

$$\begin{aligned} e^{-\mu(\tau_p \wedge T)} V(\xi(\tau_p \wedge T)) = & V(x) + \int_0^{\tau_p \wedge T} e^{-\mu s} V_x(\xi(s)) \sigma(\xi(s))\, dw(s) \\ & + \int_0^{\tau_p \wedge T} (LV - \mu V)(\xi(s)) e^{-\mu s}\, ds. \end{aligned} \tag{4.12}$$

Taking the expectation and using (4.10), we get

$$E_x\left[e^{-\mu(\tau_p \wedge T)} \chi_{\rho(\xi(\tau_p \wedge T)) = 1/p} \right] p^\epsilon \leq V(x),$$

and, as $T \uparrow \infty$,

$$E_x e^{-\mu \tau_p} \chi_{\tau_p < \infty} \leq p^{-\epsilon} V(x). \tag{4.13}$$

Now, if (4.8) is not true then the exit time τ from $\hat{G}$ is finite on a set B of positive probability. Since $\tau_p \uparrow \tau$ on B, the monotone convergence theorem gives

$$E_x e^{-\mu\tau}\chi_B \leqslant \lim_{p\to\infty} [p^{-\epsilon}V(x)] = 0.$$

Since $\tau < \infty$ on B, we must then have $P_x(B) = 0$, but this contradicts the definition of B.

Theorem 4.1 motivates the following definition.

Definition. If the conditions (4.1), (4.5) are satisfied, then we say that ∂G is an *obstacle from the outside.*

Suppose (4.5) is replaced by

$$\sum_{i=1}^{n} b_i\nu_i + \frac{1}{2}\sum a_{ij}\frac{\partial^2\rho}{\partial x_i\,\partial x_j} \leqslant 0 \qquad \text{on } \partial G \tag{4.14}$$

with the same function $\rho(x)$ as before. Define $\hat{\rho}(x) = \text{dist}(x, \partial G)$ for $x \in G$, and denote by $\hat{\nu} = (\hat{\nu}_1, \ldots, \hat{\nu}_n)$ the inward normal to ∂G. Then (4.14) holds if and only if

$$\sum_{i=1}^{n} b_i\hat{\nu}_i + \frac{1}{2}\sum_{i,j=1}^{n} a_{ij}\frac{\partial^2\hat{\rho}}{\partial x_i\,\partial x_j} \geqslant 0 \qquad \text{on } \partial G \tag{4.15}$$

where the left-hand side is evaluated by taking the limit as x tends to the boundary from inside G. Using (4.1), (4.15), we can now duplicate the proof of Theorem 4.1, taking $R(x)$ as a $C^2(G)$ function which coincides with $\hat{\rho}(x)$ near ∂G, and taking $V(x) = 1/(R(x))^\epsilon$. We thus conclude:

Corollary 4.2. *If* (A_1) *and* (4.1), (4.14) *hold, then*

$$P_x\{\xi(t) \in \text{int } G \textit{ for all } t > 0\} = 1 \qquad \textit{if} \quad x \in \text{int } G. \tag{4.16}$$

Definitions. If the conditions (4.1), (4.14) are satisfied, then we say that ∂G is an *obstacle from the inside*. If (4.1) holds and if

$$\sum_{i=1}^{n} b_i\nu_i + \frac{1}{2}\sum_{i,j=1}^{n} a_{ij}\frac{\partial^2\rho}{\partial x_i\,\partial x_j} = 0 \qquad \text{on } \partial G, \tag{4.17}$$

then we say that ∂G is a *two-sided obstacle*. In case (4.17) is replaced by either (4.5) or (4.14), we speak of a *one-sided obstacle.*

We shall now consider the case of degeneracy at one point z, and assume:

$$a_{ij}(z) = 0, \quad b_i(z) = 0 \qquad \text{for} \quad 1 \leqslant i, j \leqslant n. \tag{4.18}$$

Notice that $a_{ij}(z) = 0$ for $1 \leqslant i, j \leqslant n$ if and only if $\sigma_{ij}(z) = 0$ for $1 \leqslant i, j \leqslant n$.

Using (4.18) one can show that

$$P_x\{\xi(t) \neq z \quad \text{for all} \quad t > 0\} = 1 \qquad \text{if} \quad x \neq z. \tag{4.19}$$

Indeed, the proof is similar to the proof of Theorem 4.1. Here we take $V(x) = 1/(R(x))^{\epsilon}$ where $R(x) = |x - z|$.

A point z for which (4.18) holds is called a *point obstacle* or a *point of total degeneracy*.

We shall extend the previous considerations to the case where there are several obstacles.

Let $G_1, \ldots, G_k$ be mutually disjoint sets in R^n; if $1 \leqslant j \leqslant k_0$, G_j consists of one point z_j, and, if $k_0 + 1 \leqslant j \leqslant k$, G_j is a closed bounded domain with C^3 connected boundary ∂G_j. Let

$$G = \bigcup_{j=1}^{k} G_j, \qquad \hat{G} = R^n \backslash G$$

$$\rho_j(x) = \text{dist}\,(x, \partial G_j) \qquad \text{if} \quad x \not\in \text{int}\, G_j.$$

We shall assume:

$$a_{ij}(z_h) = 0, \quad b_i(z_h) = 0 \qquad \text{if} \quad 1 \leqslant i, j \leqslant n, \quad 1 \leqslant h \leqslant k_0, \tag{4.20}$$

$$\sum_{i,j=1}^{n} a_{ij}\nu_i\nu_j = 0 \qquad \text{on} \quad \partial G_h, \qquad \text{for} \quad k_0 + 1 \leqslant h \leqslant k, \tag{4.21}$$

$$\sum_{i=1}^{n} b_i\nu_i + \frac{1}{2}\sum_{i,j=1}^{n} a_{ij}\,\frac{\partial^2 \rho_h}{\partial x_i\,\partial x_j} \geqslant 0 \qquad \text{on} \quad \partial G_h, \qquad \text{for} \quad k_0 + 1 \leqslant h \leqslant k. \tag{4.22}$$

Theorem 4.3. *If* (A_1) *and* (4.20)–(4.22) *hold, then*

$$P_x\{\xi(t) \in \hat{G} \text{ for all } t \geqslant 0\} = 1 \qquad \text{if} \quad x \in \hat{G}. \tag{4.23}$$

The proof is similar to the proof of Theorem 4.1. Here one takes $V = 1/R^{\epsilon}$ where $R(x)$ is a C^2 function in $\hat{G}$, satisfying:

$$R(x) = \begin{cases} \rho_h(x) & \text{if} \quad \rho_h(x) < \epsilon_1 \quad (1 \leqslant h \leqslant k), \\ |x| & \text{if} \quad |x| > M \end{cases}$$

and $\epsilon_1 \leqslant R(x) \leqslant M$ elsewhere, where ϵ_1 is sufficiently small and M is sufficiently large.

In the following two sections we shall replace the condition (A_2) by the weaker condition:

(A_2') The matrix $(a_{ij}(x))$ is positive definite for all $x \in \hat{G}$.

We shall also assume that (4.20)–(4.22) hold and

$$\partial G_h \text{ is } C^3 \text{ diffeomorphic to a sphere, for } k_0 + 1 \leqslant h \leqslant k. \tag{4.24}$$

We shall then extend the results of Sections 1, 2. The following lemma will be needed.

Lemma 4.4. *Let* (A_1), (A_2'), *and* (4.24) *hold. Then there exists a continuous function* $R(x)$ *in* R^n *having the following properties*:

(i) $R(x)$ *is in* $C^2(\hat{G})$;

(ii) $R(x) > 0$ *in* $\hat{G}$;

(iii) $R(x) = \rho_h(x)$ *if* $\rho_h(x) \leq \epsilon_0$, $R(x) > \epsilon_0$ *if* $\min_h \rho_h(x) > \epsilon_0$, *for some* ϵ_0 *sufficiently small*;

(iv) $R(x) \to \infty$ *if* $|x| \to \infty$;

(v) $\displaystyle\sum_{i,j=1}^{n} a_{ij}(x) \frac{\partial^2 R(x)}{\partial x_i\, \partial x_j} < 0$ *if* $x \in \hat{G}$, $D_x R(x) = 0$; *there exist precisely* $k - 1$ *points* x *in* $\hat{G}$ *where* $D_x R(x) = 0$.

The number ϵ_0 will be smaller than

$$\min_{i \neq j} \operatorname{dist}(G_i, G_j).$$

Notice that the conditions (4.20)–(4.22) are not assumed in this lemma.

Proof. By the proof of the Schoenflies theorem (see Morse [1]) there is a diffeomorphism $y = f(x)$ of the exterior of $\bigcup_{j=1}^{k} G_j$ onto the exterior of $\bigcup_{j=1}^{k} G_j'$ in R^n, where $G_1', \ldots, G_{k_0}'$ are points situated on the y_1 axis and $G_{k_0+1}', \ldots, G_k'$ are balls with centers on the y_1 axis; the center of G_j' lies to the left of the center of G_{j+1}'. Furthermore, this diffeomorphism preserves the distance functions (to $\bigcup_j G_j$ and to $\bigcup_j G_j'$) as long as the distance is sufficiently small. (The condition (4.24) is needed in order to apply the Schoenflies theorem.) Suppose for simplicity that $k_0 = 0$, $k = 2$. Denote by $(c_1, 0, \ldots, 0)$ the midpoint of the segment connecting the center $(\alpha_1, 0, \ldots, 0)$ of G_1' to the center $(\alpha_2, 0, \ldots, 0)$ of G_2'; it is assumed that $\alpha_1 < \alpha_2$. Construct a positive C^2 function $\phi(y')$ (where $y' = (y_2, \ldots, y_n)$) on the plane $y_1 = c_1$, which increases radially, with $\operatorname{grad} \phi(y') \neq 0$ if $y' \neq 0$, such that $\partial\phi/\partial y_i = 0$, $\partial^2\phi/\partial y_i\, \partial y_j = 0$ $(2 \leq i, j \leq n)$ at $y' = 0$.

Denote by μ_i the radius of G_i'. Construct a C^2 function $\psi(y_1)$, positive for $\alpha_1 + \mu_1 < y_1 < \alpha_2 - \mu_2$, such that

$$\psi(y_1) = \begin{cases} y_1 - \alpha_1 - \mu_1 & \text{for} \quad 0 \leq y_1 - \alpha_1 - \mu_1 < \delta_1, \\ \alpha_2 - \mu_2 - y_1 & \text{for} \quad 0 \leq \alpha_2 - \mu_2 - y_1 < \delta_1 \end{cases}$$

where δ_1 is sufficiently small, and such that $\psi'(y_1) \neq 0$ if $y_1 \neq c_1$ and

$$\psi(c_1) = \phi(0, \ldots, 0), \qquad \psi'(c_1) = 0, \qquad \psi''(c_1) < 0.$$

We now construct a C^2 positive function $\lambda(y)$ for $y \not\in (G_1' \cup G_2')$, which extends the functions ϕ, ψ and the distance function from $G_1' \cup G_2'$ (as long as the distance is sufficiently small). This function is to satisfy:

$$\text{grad } \lambda(y) \neq 0 \quad \text{if} \quad y \neq (c_1, 0, \ldots, 0);$$

$$\text{grad } \lambda(y) = 0, \quad \frac{\partial^2 \lambda(y)}{\partial y_i \, \partial y_j}, \quad = 0 \quad \text{if} \quad (i, j) \neq (1, 1),$$

$$\frac{\partial^2 \lambda(y)}{\partial y_1^2} < 0 \quad \text{at} \quad y = (c_1, 0, \ldots, 0);$$

$$\lambda(y) = |y| \quad \text{if} \quad |y| \text{ is sufficiently large.}$$

The construction of such a function $\lambda(y)$ can be accomplished by introducing a family of curves $\gamma_{y'}$ connecting $(\alpha_1, 0, \ldots, 0)$ to $(\alpha_2, 0, \ldots, 0)$ and intersecting the plane $y_1 = c_1$ orthogonally at (c_1, y'). $\lambda(y)$ is defined along $\gamma_{y'}$ such that its tangential derivative vanishes only at $y_1 = c_1$.

Define $R(x) = \lambda(f(x))$. Clearly $D_x R(x) \neq 0$ if $x \neq x^*$ where $f(x^*)$ is the point $(c_1, 0, \ldots, 0)$. Furthermore, as easily seen,

$$\sum_{i,j=1}^{n} a_{ij}(x) \frac{\partial^2 R}{\partial x_i \, \partial x_j} < 0 \quad \text{at} \quad x = x^*.$$

This completes the proof if $k_0 = 0$, $k = 2$. The proof for any k_0, k is similar.

Lemma 4.5. *Let* (A_1), (A_2') *hold and suppose that*

$$G_h \text{ is } C^3 \text{ diffeomorphic to a closed ball, for } k_0 + 1 \leqslant h \leqslant k. \tag{4.25}$$

Then there exists a function $R(x)$ *satisfying* (i)–(v) *of Lemma* 4.4 *and, in addition,*

(iv′) $R(x) = x$ *if* $|x|$ *is sufficiently large.*

For proof see Problem 8. By Milnor [1], (4.24) implies (4.25) if $n \geqslant 5$; the same is true, by conformal mappings, if $n = 2$.

5. Transient solutions for degenerate diffusion

We shall extend the results of Section 1 to the case the diffusion matrix $(a_{ij}(x))$ degenerates in such a way that (A_2') holds, and G_j, G are as in Lemma 4.4.

For later references we state the following condition:

(B_1) (i) $G_h = \{z_h\}$ if $h = 1, \ldots, k_0$, G_h is a bounded closed domain with C^3 connected boundary if $h = k_0 + 1, \ldots, k$.

(ii) The conditions (4.20)–(4.22) and (4.24) hold.

(iii) The matrix $(a_{ij}(x))$ is nondegenerate for $x \in \hat{G}$, where $\hat{G} = R^n \setminus G$, $G = \bigcup_{h=1}^{k} G_h$.

Let $R(x)$ be the function constructed in Lemma 4.4 or 4.5, and set

$$\mathcal{A} = \frac{1}{2} \sum_{i,j=1}^{n} a_{ij}(x) \frac{\partial R}{\partial x_i} \frac{\partial R}{\partial x_j},$$

$$\mathcal{B} = \sum_{i=1}^{n} b_i(x) \frac{\partial R}{\partial x_i} + \frac{1}{2} \sum_{i,j=1}^{n} a_{ij}(x) \frac{\partial^2 R}{\partial x_i \partial x_j},$$

$$Q = \frac{1}{R}\left(\mathcal{B} - \frac{\mathcal{A}}{R}\right).$$

Then, if $g(x) = \Phi(R(x))$,

$$Lg(x) = \mathcal{A}\left[\Phi''(R) + \frac{1}{R}\Phi'(R)\right] + RQ\Phi'(R). \tag{5.1}$$

If $R(x)$ is as in Lemma 4.5, then the function $S(x)$ defined in (1.3) satisfies

$$S(x) = 1 + \frac{R^2(x)Q(x)}{\mathcal{A}(x)} \qquad \text{if} \quad |x| \text{ is sufficiently large.}$$

As will be proved in Chapter 11, if $\overline{\lim}\ Q(x) < 0$ as $R(x) \to 0$ and as $R(x) \to \infty$, then

$$P_x\left\{\min_{1 \leq h \leq k} [\rho_h(\xi(t))] \to 0 \text{ if } t \to \infty\right\} = 1 \quad \text{if} \quad x \in \hat{G}.$$

Thus, in this case, $\xi(t)$ neither wanders out to ∞ nor visits any open neighborhood in $\hat{G}$ at a sequence of times increasing to ∞. Hence, in order to obtain the type of behavior asserted in Sections 1, 2, we shall have to impose different conditions on $Q(x)$.

As will be shown, in order to generalize the results of Sections 1, 2 to the present case where (B_1) holds, we do not need to change the conditions (A_3), (A_4) near infinity. We only need to impose a condition on $Q(x)$ near $R(x) = 0$. This condition is:

(B_2) For some $0 < \delta_0 \leq \epsilon_0$ there is a continuous function $\epsilon(r)$, defined for $0 < r \leq \delta_0$, such that

$$Q(x) \geq \frac{\mathcal{A}(x)}{R^2(x)} \epsilon(R(x)) \qquad \text{if} \quad 0 < R(x) \leq \delta_0, \tag{5.2}$$

and

$$\int_r^{\delta_0} \frac{1}{s} \exp\left[\int_s^{\delta_0} \frac{\epsilon(t)}{t}\, dt\right] ds \to \infty \qquad \text{if} \quad r \to 0. \tag{5.3}$$

We can take, for example, $\epsilon(s) = -1/[\log(1/s)]$.

Remark. The condition: $\overline{\lim}\ Q(x) < 0$ as $R(x) \to 0$ is a "stability condition," meaning that G attracts $\xi(t)$ near the boundary. The condition (B_2) can be interpreted as a weak "repelling condition."

Theorem 5.1. *Let* (A_1), (B_1), (B_2), *and* (A_3) *hold. Then the solution of* (1.1) *is transient in* $\hat{G}$, *i.e.*,

$$P_x\{\lim_{t\to\infty}|\xi(t)| = \infty\} = 1 \qquad \text{if} \quad x\in\hat{G}. \tag{5.4}$$

Let G be contained in a ball $B_{R_*} = \{y; |y| < R_*\}$.

We shall first prove a lemma.

Lemma 5.2. *Let* (A_1), (B_1), (B_2) *hold and let* $\beta > R_*$. *Then*

$$P_x\{\xi(t) \text{ hits the set } \partial B_\beta \text{ for some } t > 0\} = 1 \qquad \text{if} \quad x\in\hat{G}\cap B_\beta. \tag{5.5}$$

Here B_β is the ball $\{y; |y| < \beta\}$ and ∂B_β is its boundary.

Proof. We first construct a function $g(x) = \Phi(R(x))$ for $x\in\hat{G}\cap B_\beta$, such that

$$Lg(x) < 0 \qquad \text{if} \quad x\in\hat{G}\cap B_\beta. \tag{5.6}$$

Denote by $\zeta_1,\ldots,\zeta_{k-1}$ the points in $\hat{G}$ where $D_xR(x) = 0$. By slightly modifying the proof of Lemma 4.4, we obtain a modified function $R(x)$ for which the points $\zeta_1,\ldots,\zeta_{k-1}$ lie outside the ball B_β. We shall work, in the present proof, with this modified function $R(x)$; it coincides with the original $R(x)$ in the ϵ_0-neighborhood of G.

We claim that there is a continuous function $\theta(r)$ satisfying:

$$1 + \frac{R^2(x)Q(x)}{\mathscr{Q}(x)} \geq \theta(R(x)) \qquad \text{if} \quad x\in\hat{G}\cap B_\beta \tag{5.7}$$

$$\theta(r) = 1 + \epsilon(r) \qquad \text{if} \quad 0 < r \leq \delta_0. \tag{5.8}$$

Indeed, since $\mathscr{Q}(x) \neq 0$ if $x\in\hat{G}$, $D_xR(x) \neq 0$, the left-hand side of (5.7) is a bounded function if $x\in\hat{G}\cap B_\beta$, $\min_h\rho_h(x) \geq \delta_0$. Using the assumption (5.2), the existence of $\theta(r)$ (satisfying (5.7), (5.8)) follows.

Let $\Phi(r)$ be a solution of

$$\Phi''(r) + \frac{\theta(r)}{r}\Phi'(r) = 0 \qquad \text{if} \quad 0 < r < r_0, \tag{5.9}$$

$$\Phi'(r) < 0 \qquad \text{if} \quad 0 < r < r_0 \tag{5.10}$$

where $r_0 = \max_{|x|\leq\beta} R(x)$. Then, upon using (5.1), (5.7) we conclude that $g(x) = \Phi(R(x))$ satisfies (5.6).

A solution of (5.9), (5.10) is given by

$$\Phi(r) = \int_r^{\delta_0} \exp\left[\int_s^{\delta_0} \frac{\theta(t)}{t}\, dt\right] ds. \tag{5.11}$$

In view of (5.3),

$$\Phi(r) \to \infty \quad \text{if} \quad r \to 0. \tag{5.12}$$

By Itô's formula and (5.6),

$$E_x \Phi(|\xi(t)|) - E_x \Phi(|x|) = E_x \int_0^\tau Lg(\xi(s))\, ds \leqslant 0, \tag{5.13}$$

where τ is any bounded stopping time such that $\xi(s) \in \hat{G} \cap B_\beta$ if $0 \leqslant s \leqslant \tau$. Denote by G_ϵ ($\epsilon > 0$) the closed ϵ-neighborhood of G, and denote by $\sigma_{\epsilon\beta}$ the hitting time of the set $G_\epsilon \cup \partial B_\beta$. By Theorem 8.1.1, $P_x(\sigma_{\epsilon\beta} < \infty) = 1$ if $x \not\in G_\epsilon$, $x \in B_\beta$. Denote by $P_x(\epsilon)$ the probability that $\xi(\sigma_\epsilon \beta) \in G_\epsilon$ (given $\xi(0) = x$), and by $P_x(\beta)$ the probability that $\xi(\sigma_{\epsilon\beta}) \in \partial B_\beta$ (given $\xi(0) = x$). Substituting $\tau = \sigma_{\epsilon\beta} \wedge T$ in (5.13), and taking $T \to \infty$, we get

$$\Phi(\epsilon) P_x(\epsilon) + \Phi(\beta) P_x(\beta) \leqslant \Phi(|x|).$$

Taking $\epsilon \to 0$ and using (5.12), we deduce that $P_x(\epsilon) \to 0$ if $\epsilon \to 0$. Hence $P_x(\beta) \to 1$ if $\epsilon \to 0$. But this implies the assertion of the lemma.

Let $R_* < \alpha < R$ and denote by t_R the hitting time of the ball B_R. By Lemma 5.2, $P_x(t_R < \infty) = 1$ if $x \in \hat{G}$. Hence, by the strong Markov property (cf. the proof of Theorem 1.1),

$$P_x(\Omega^*(\alpha)) = E_x P_{\xi(t_R)}(\Omega^*(\alpha)) \leqslant E_x P_{\xi(t_R)}(\Omega(\alpha)),$$

where the notation (1.17), (1.19) is used.

Now, in the domain $\{y;\ |y| > R_*\}$ the matrix $(a_{ij}(y))$ is nondegenerate. Since the condition (A_3) holds, the estimate (1.16) remains valid for $x = \xi(t_R)$. Hence

$$P_x(\Omega^*(\alpha)) \leqslant \frac{F(R)}{F(\alpha)} \qquad (F(r) \to 0 \text{ if } r \to \infty).$$

We can now complete the proof, as in the case of Theorem 1.1, by taking $R \to \infty$ and noting that α can be arbitrarily large.

Theorem 5.3. *Let* (A_1), (B_1), (B_2), *and* (A_3') *hold. Then for any* $x \in \hat{G}$ *the assertion* (5.4) *holds.*

Proof. Proceeding as in the proof of Theorem 5.1, it remains to establish the estimate (1.16). We now perform an orthogonal transformation as in the proof of Theorem 1.2.

6. Recurrent solutions for degenerate diffusion

Theorem 6.1. *Let* (A_1), (B_1), (B_2), (4.25) *and* (A_4) *hold. Then the solution of* (1.1) *is recurrent in* $\hat{G}$, *i.e., for any ball* $B_\alpha(z) = \{y; |y-z| < \alpha\}$, $\alpha > 0$, *lying entirely in* $\hat{G}$, *the assertion* (2.3) *holds.*

Proof. For simplicity we take $z = 0$. Let R be any positive number such that $R > \alpha$ and such that G is contained in the interior of the ball B_R. We shall prove: if $x \in \partial B_R$, then

$$P_x\{\xi(t) \text{ hits the ball } B_\alpha \text{ for some } t \geq 0\} = 1. \tag{6.1}$$

Let $\delta > 0$ be sufficiently small so that $\delta < \epsilon_0$, the closed δ-neighborhood G_δ of G lies in the interior of G_R, and $G_\delta \cap B_\alpha = \emptyset$.

We shall construct a function $f(x) = F(R(x))$ ($R(x)$ as in Lemma 4.5) in $R^n \setminus G_\delta$ such that

$$Lf(x) \geq 0 \qquad \text{if} \quad x \in R^n \setminus G_\delta, \tag{6.2}$$

$$F(r) \to -\infty \qquad \text{if} \quad r \to \infty. \tag{6.3}$$

Notice that at the points ζ_m $(m = 1, \ldots, k-1)$ where $D_x R(x) = 0$, $\mathcal{C} = 0$ and, therefore, by the property (v) of $R(x)$, $Q < 0$. Hence, $Q(x) < 0$ in a neighborhood of each point ζ_m. It follows that there is a continuous function $\theta(r)$, $\delta \leq r < \infty$, such that

$$1 + \frac{R^2 Q}{\mathcal{C}} \leq \theta(R) \qquad \text{if} \quad x \in R^n \setminus G_\delta. \tag{6.4}$$

In view of (A_4), we can choose $\theta(r)$ so that, for all r sufficiently large, $\theta(r) = 1 + \epsilon(r)$ where $\epsilon(r)$ satisfies (2.2). If we now define $F(r)$, for $r \geq \delta$, by

$$F(r) = -\int_\delta^r e^{-I(s)}\,ds, \qquad I(s) = \int_\delta^s \frac{\theta(t)}{t}\,dt$$

then $F'(r) \leq 0$, and (6.2), (6.3) hold. Arguing as in the proof of Theorem 2.1 (following (2.8)) with the present function $f(x) = F(R(x))$ and with the set $\{y; |y| \leq \alpha\}$ replaced by G_δ, we conclude that for any $x \in R^n \setminus G_\delta$,

$$P_x\{\xi(t) \text{ hits the set } G_\delta \text{ for some } t \geq 0\} = 1. \tag{6.5}$$

Let $\beta > R$ and denote by $\hat{G}_{\alpha\beta\delta}$ the domain bounded by ∂G_δ, ∂B_α, ∂B_β. Denote by $\tau_{\alpha\beta}$ the exit time from this domain. By Theorem 8.1.1, $P_x(\tau_{\alpha\beta} < \infty) = 1$ if $x \in \hat{G}_{\alpha\beta\delta}$. Using the strong Markov property we get, for any $x \in \partial B_R$,

$$\begin{aligned} P_x\{\xi(t) \text{ hits } B_\alpha \text{ for some } t \geq 0\} &= P_x\{\xi(t+\tau_{\alpha\beta}) \text{ hits } B_\alpha \text{ for some } t \geq 0\} \\ &= E_x P_{\xi(\tau_{\alpha\beta})}\{\xi(t) \text{ hits } B_\alpha \text{ for some } t \geq 0\} \\ &\geq P_x\{\xi(\tau_{\alpha\beta}) \in \partial B_\alpha\} + P_x\{\xi(\tau_{\alpha\beta}) \in \partial G_\delta\} \\ &\quad \cdot \inf_{y \in \partial G_\delta} P_y\{\xi(t) \text{ hits } B_\alpha \text{ for some } t \geq 0\}. \end{aligned}$$

Denote by $\sigma_{\alpha R}$ the hitting time of $B_\alpha \cup \partial B_R$. By Lemma 5.2, if $y \in \partial G_\delta$, then $P_y(\sigma_{\alpha R} < \infty) = 1$. Hence, by the strong Markov property,

$$\begin{aligned} P_y\{\xi(t) \text{ hits } B_\alpha \text{ for some } t \geqslant 0\} &= P_y\{\xi(t+\sigma_{\alpha R}) \text{ hits } B_\alpha \text{ for some } t \geqslant 0\} \\ &= E_x P_{\xi(\sigma_{\alpha R})}\{\xi(t) \text{ hits } B_\alpha \text{ for some } t \geqslant 0\} \\ &= P_y\{\xi(\sigma_{\alpha R}) \in \partial B_\alpha\} + (1 - P_y\{\xi(\sigma_{\alpha R}) \in \partial B_\alpha\}) \inf_{x \in \partial B_R} P_x\{\xi(t) \text{ hits } B_\alpha \\ &\qquad \text{for some } t \geqslant 0\}. \end{aligned}$$

Combining this with the previous inequality, and setting

$$\begin{aligned} P_x(\alpha) &= P_x\{\xi(t) \text{ hits } B_\alpha \text{ for some } t \geqslant 0\}, \\ \gamma_{\alpha\beta}(x) &= P_x\{\xi(\tau_{\alpha\beta}) \in \partial B_\alpha\}, \\ \gamma_\beta(x) &= P_x\{\xi(\tau_{\alpha\beta}) \in \partial G_\delta\}, \\ \mu(y) &= P_y\{\xi(\sigma_{\alpha R}) \in \partial B_\alpha\}, \end{aligned}$$

we arrive at the inequality

$$P_x(\alpha) \geqslant \gamma_{\alpha\beta}(x) + \gamma_\beta(x) \inf_{y \in \partial G_\delta} \left\{ \mu(y) + [1 - \mu(y)] \inf_{z \in \partial B_R} P_z(\alpha) \right\}. \tag{6.6}$$

Note that $\gamma_{\alpha\beta}(x)$ is the solution $u(x)$ of the Dirichlet problem:

$$\begin{aligned} Lu &= 0 \quad && \text{in } \hat{G}_{\alpha\beta\delta}, \\ u &= 1 && \text{on } \partial B_\alpha \\ u &= 0 && \text{on } \partial G_\delta \cup \partial B_\beta. \end{aligned}$$

Hence, by the strong maximum principle, $\gamma_{\alpha\beta}(x)$ is positive on ∂G_R. Further, $\gamma_{\alpha\beta}(x) \uparrow$ if $\beta \uparrow$. Similar assertions are true for $\gamma_\beta(x)$. Since $\gamma_{\alpha\beta}(x) + \gamma_\beta(x) \leqslant 1$, we conclude that

$$\gamma_\beta(x) \leqslant \theta < 1 \qquad (x \in \partial G_R) \tag{6.7}$$

where θ is a constant *independent* of β.

Notice that (6.5) implies that

$$\gamma_{\alpha\beta}(x) + \gamma_\beta(x) \to 1 \quad \text{if} \quad \beta \to \infty \qquad (x \in \partial G_R). \tag{6.8}$$

Let

$$P_\alpha = \inf_{z \in \partial B_R} P_z(\alpha).$$

Let η be a positive number, and choose x_0 in ∂B_R so that

$$P_\alpha \geqslant P_{x_0}(\alpha) - \eta.$$

Let y_0 be a point in ∂G_δ such that

$$\inf_{y \in \partial G_\delta} \{\mu(y) + [1 - \mu(y)] P_\alpha\} \geqslant \{\mu(y_0) + [1 - \mu(y_0)]\} P_\alpha - \eta.$$

Applying (6.6) with $x = x_0$, we get

$$P_\alpha \geq \gamma_{\alpha\beta}(x_0) + \gamma_\beta(x_0)\{\mu(y_0) + [1 - \mu(y_0)]P_\alpha\} - 2\eta.$$

Hence

$$P_\alpha\{1 - \gamma_\beta(x_0)[1 - \mu(y_0)]\} \geq \gamma_{\alpha\beta}(x_0) + \gamma_\beta(x_0)\mu(y_0) - 2\eta.$$

Taking β sufficiently large and using (6.8) with $x = x_0$, we get

$$P_\alpha\{1 - \gamma_\beta(x_0)[1 - \mu(y_0)]\} \geq \{1 - \gamma_\beta(x_0)[1 - \mu(y_0)]\} - 3\eta.$$

Denote the expression in braces by λ_β. From (6.7) it follows that $\lambda_\beta \geq 1 - \theta > 0$. Hence

$$P_\alpha \geq 1 - \frac{3\eta}{\lambda_\beta} \geq 1 - \frac{3\eta}{1-\theta}.$$

Since η is arbitrary, $P_\alpha = 1$. This implies that $P_\alpha(x) = 1$ for all $x \in \partial G_R$, i.e., (6.1) holds.

Having proved (6.1), we can now easily complete the proof of Theorem 6.1 by the argument given in the proof of Theorem 2.1 (from (2.10) on). Instead of (2.10) we use (6.1), and instead of Theorem 8.1.1 we use Lemma 5.2.

Remark. Theorem 6.1 remains true if the condition (A_4) is replaced by (A_4'), or by the conditions in the remark following Theorem 2.2.

7. The one-dimensional case

Consider the case of one stochastic differential equation

$$d\xi(t) = \sigma(\xi(t))\,dw(t) + b(\xi(t))\,dt. \tag{7.1}$$

We shall assume that the condition (A_1) of Section 1 holds in the present case of $n = 1$, and that $\sigma(x) > 0$ for all x. Let

$$\phi(x) = \int_0^x \exp\left[-\int_0^z \frac{2b(u)}{\sigma^2(u)}\,du\right] dz. \tag{7.2}$$

This is a particular solution of

$$\tfrac{1}{2}\sigma^2 v'' + bv' = 0. \tag{7.3}$$

Notice that $\phi(x)$ is strictly monotone increasing. Set $\phi(\infty) = \lim_{x\to\infty}\phi(x)$, $\phi(-\infty) = \lim_{x\to-\infty}\phi(x)$. These numbers may be finite or infinite.

Theorem 7.1. (a) *If $\phi(-\infty) = -\infty$, then*

$$P_x\left\{\sup_{t>0}\xi(t) = \infty\right\} = 1. \tag{7.4}$$

If $\phi(\infty) = \infty$, *then*

$$P_x\{\inf_{t>0} \xi(t) = -\infty\} = 1. \tag{7.5}$$

(b) *If* $\phi(\infty) = \infty$, $\phi(-\infty) > -\infty$, *then*

$$P_x\{\sup_{t>0} \xi(t) < \infty\} = 1, \tag{7.6}$$

$$P_x\{\lim_{t\to\infty} \xi(t) = -\infty\} = 1. \tag{7.7}$$

(c) *If* $\phi(-\infty) = -\infty$, $\phi(\infty) < \infty$, *then*

$$P_x\{\inf_{t>0} \xi(t) > -\infty\} = 1, \tag{7.8}$$

$$P_x\{\lim_{t\to\infty} \xi(t) = \infty\} = 1. \tag{7.9}$$

(d) *If* $\phi(\infty) < \infty$, $\phi(-\infty) > -\infty$, *then*

$$P_x\{\sup_{t>0} \xi(t) < \infty\} = P_x\{\lim_{t\to\infty} \xi(t) = -\infty\} = \frac{\phi(\infty) - \phi(x)}{\phi(\infty) - \phi(-\infty)}, \tag{7.10}$$

$$P_x\{\inf_{t>0} \xi(t) > -\infty\} = P_x\{\lim_{t\to\infty} \xi(t) = \infty\} = \frac{\phi(x) - \phi(-\infty)}{\phi(\infty) - \phi(-\infty)}, \tag{7.11}$$

$$P_x\{\lim_{t\to\infty} |\xi(t)| = \infty\} = 1. \tag{7.12}$$

It follows that if $\phi(-\infty) = -\infty$ and $\phi(\infty) = \infty$, then the solution of (7.1) is recurrent; in all other cases the solution of (7.1) is transient.

Proof. Let $x_1 < x < x_2$ and denote by $\tau_x[x_1, x_2]$ the first time $\xi(t)$ leaves (x_1, x_2), given $\xi(0) = x$. By Itô's formula (cf. Problem 12(c), Chapter 8),

$$P_x\{\xi(\tau_x[x_1, x_2]) = x_2\} = \frac{\phi(x) - \phi(x_1)}{\phi(x_2) - \phi(x_1)}. \tag{7.13}$$

Noting that

$$P_x\{\sup_{t>0} \xi(t) \geqslant x_2\} \geqslant P_x\{\xi(\tau_x[x_1, x_2]) = x_2\}, \tag{7.14}$$

and taking $x_1 \to -\infty$ in (7.13) we obtain

$$P_x\{\sup_{t>0} \xi(t) \geqslant x_2\} = 1,$$

provided $\phi(-\infty) = -\infty$. Since x_2 can be arbitrarily large,

$$P\{\sup_{t>0} \xi(t) = \infty\} = 1.$$

This proves (7.4). The proof of (7.5), in case $\phi(\infty) = \infty$ is similar.

To prove (b), take $x_1 \to -\infty$ in (7.13) to obtain

$$P_x\Big\{\sup_{t>0} \xi(t) \geq x_2\Big\} = \frac{\phi(x) - \phi(-\infty)}{\phi(x_2) - \phi(-\infty)}.$$

This implies that

$$P\Big\{\sup_{t>0} \xi(t) < x_2\Big\} = \frac{\phi(x_2) - \phi(x)}{\phi(x_2) - \phi(-\infty)}. \tag{7.15}$$

Taking $x_2 \to \infty$ the assertion (7.6) follows.

To prove (7.7), let $y < x$ and denote by τ_y the first time $\xi(t) = y$. By (a), $P_x(\tau_y < \infty) = 1$. Hence, by the strong Markov property, for any $x_2 > y$,

$$P_x\Big\{\sup_{t>0} \xi(t + \tau_y) \geq x_2\Big\} = E_x P_y\Big[\sup_{t>0} \xi(t) \geq x_2\Big] = \frac{\phi(y) - \phi(-\infty)}{\phi(x_2) - \phi(-\infty)}.$$

But

$$P_x\Big\{\sup_{t>0} \xi(t + \tau_y) \geq x_2\Big\} = P_x\Big\{\sup_{t>\tau_y} \xi(t) \geq x_2\Big\} \geq P_x\Big\{\overline{\lim_{t\to\infty}}\, \xi(t) \geq x_2\Big\}.$$

Therefore,

$$P_x\Big\{\overline{\lim_{t\to\infty}}\, \xi(t) \geq x_2\Big\} \leq \frac{\phi(y) - \phi(-\infty)}{\phi(x_2) - \phi(-\infty)}.$$

Taking $y \to -\infty$ we conclude that

$$P_x\Big\{\overline{\lim_{t\to\infty}}\, \xi(t) \geq x_2\Big\} = 0.$$

Here x_2 is any real number. Taking $x_2 \to -\infty$, (7.7) follows.

The proof of (c) is similar to the proof of (b), and will be omitted.

We proceed to prove (d). The relation

$$P_x\Big\{\sup_{t>0} \xi(t) < \infty\Big\} = \frac{\phi(\infty) - \phi(x)}{\phi(\infty) - \phi(-\infty)} \tag{7.16}$$

follows by taking $x_2 \to \infty$ in (7.15). Similarly one proves that

$$P_x\Big\{\inf_{t>0} \xi(t) > -\infty\Big\} = \frac{\phi(x) - \phi(-\infty)}{\phi(\infty) - \phi(-\infty)}. \tag{7.17}$$

If we prove (7.12), then the second equalities in (7.10), (7.11) follow from the equalities in (7.17), (7.16) respectively. Thus it remains to prove (7.12).

For any small $\epsilon > 0$ and large $M > 0$, choose $x_1 < -M$ and $x_2 > M$ so that $x_1 < x < x_2$,

$$P_{x_1}\Big\{\sup_{t\geq 0} \xi(t) < -M\Big\} = \frac{\phi(-M) - \phi(x_1)}{\phi(-M) - \phi(-\infty)} \geq 1 - \epsilon, \tag{7.18}$$

and

$$P_{x_2}\left\{\inf_{t>0} \xi(t) > M\right\} = \frac{\phi(x_2) - \phi(M)}{\phi(\infty) - \phi(M)} > 1 - \epsilon. \tag{7.19}$$

Let $\tau = \tau_x[x_1, x_2]$. Then, by the strong Markov property,

$$P_x\{\xi(t) \not\in [-M, M] \text{ for } t > \tau\} = E_x P_{\xi(\tau)}\{\xi(t) \not\in [-M, M] \text{ for } t > 0\}$$

$$= P_x[\xi(\tau) = x_2] P_{x_2}\left[\inf_{t>0} \xi(t) > M\right] + P_x[\xi(\tau) = x_1] P_{x_1}\left[\sup_{t>0} \xi(t) < -M\right]$$

$$> 1 - \epsilon$$

by (7.18), (7.19). Consequently

$$P_x\left\{\varliminf_{t\to\infty} |\xi(t)| > M\right\} > 1 - \epsilon.$$

Taking first $\epsilon \to 0$ and then $M \to \infty$, the assertion (7.12) follows.

PROBLEMS

1. The solution of (0.1) is recurrent in G if and only if for every $x \in G$ and for any open set V in G, (0.2) holds with "finite random times t_m" replaced by "finite Markov times t_m." [*Hint*: Let W be a closed domain contained in V. Define t_j = first time $> t_{j-1} + 1$ such that $\xi(t)$ hits W.]

2. Let $\xi(t)$ be a continuous process for $t > 0$, with values in R^n. Let $E = \{\xi(t)$ hits a closed set Γ at a sequence of random times t_m increasing to infinity$\}$. Prove that there exists a sequence of bounded, Borel measurable functions $f_k(x_1, \ldots, x_k)$ such that

$$f_k(\xi(t_1), \ldots, \xi(t_k)) \to \chi_E \quad \text{a.s.},$$

and the sequence f_k does not depend on the process $\xi(t)$. [*Hint*: Let $\{y_i\}$ be a dense set in Γ. Show that $E = \{B_m \text{ i.o.}\}$ where

$$B_m = \bigcap_{l=1}^{\infty} \bigcup_{m<r_j<m+1} \bigcup_{i=1}^{\infty} \{|\xi(r_j) - y_i| < 1/l\}$$

where $\{r_j\}$ is the sequence of positive rational numbers; then cf. Problem 9, Chapter 2.]

3. Let Γ be a closed set, τ the hitting time of Γ by the solution $\xi(t)$ of (1.1). Suppose $P_x(\tau < \infty) = 1$. Prove:

$$P_x\{\xi(t+\tau) \text{ hits } \Gamma \text{ at a sequence of times increasing to } \infty\}$$

$$= E_x P_{\xi(\tau)}\{\xi(t) \text{ hits } \Gamma \text{ at a sequence of times increasing to } \infty\}.$$

[*Hint*: Use the preceding problem and the method of solution of Problem 10, Chapter 2.]

4. Prove the first relation in (2.11). [*Hint*: Use the method of proof in Problem 10, Chapter 2.]
5. Let G be a bounded domain with C^3 boundary ∂G. Suppose that (A_1) holds and that the condition (B_2) holds with $R(x)$ replaced by $\rho(x) = \text{dist}(x, \partial G)$, $x \in G$. Suppose further that $(a_{ij}(x))$ is nonsingular for all $x \in \text{int } G$. Prove that $\xi(t)$ is recurrent in int G.
6. Consider (7.1) and set

$$I_1(x) = \int_{-\infty}^{x} \exp\left\{ -\int_0^z \frac{2b(u)}{\sigma^2(u)}\, du \right\} dz,$$

$$I_2(x) = \int_{x}^{\infty} \exp\left\{ -\int_0^z \frac{2b(u)}{\sigma^2(u)}\, du \right\} dz.$$

Prove that if $I_1(x) < \infty$, $I_2(x) < \infty$, then

$$P\left\{ \lim_{t\to\infty} \xi(t) = \infty \right\} = P\left\{ \sup_{t>0} \xi(t) = \infty \right\} = E \frac{I_1(\xi(0))}{I_1(\xi(0)) + I_2(\xi(0))},$$

$$P\left\{ \lim_{t\to\infty} \xi(t) = -\infty \right\} = P\left\{ \inf_{t>0} \xi(t) = -\infty \right\} = E \frac{I_2(\xi(0))}{I_1(\xi(0)) + I_2(\xi(0))}$$

where $\xi(t)$ is any solution of (7.1).
7. Let $x = 0$ be a two-sided obstacle for (7.1), i.e., $\sigma(0) = b(0) = 0$. Assume that $\sigma(x) > 0$ if $x > 0$ and set

$$\bar{\sigma}(x) = \sigma(e^x)e^{-x}, \qquad \bar{b}(x) = b(e^x)e^{-x} + \tfrac{1}{2}\sigma^2(e^x)e^{-2x}.$$

Let $\psi(x)$ be the solution of

$$\tfrac{1}{2}\bar{\sigma}^2\psi'' + \bar{b}\psi' = 0, \qquad \psi(0) = 0, \quad \psi'(0) = 1.$$

Prove that if $\psi(\infty) = \infty$, $\psi(-\infty) > -\infty$, then

$$P_x\left\{ \xi(t) > 0,\ \lim_{t\to\infty} \xi(t) = 0 \right\} = 1 \qquad \text{if} \quad x > 0.$$

8. Prove Lemma 4.5. [*Hint*: Let $k_0 = 0$, $k = 1$. By Palais [1] there is a diffeomorphism $y = f(x)$ of $R^n \setminus (\text{int } G_1)$ onto $|y| \geqslant 1$ such that $f(x) = x$ for x outside a neighborhood of G_1. Take

$$R(x) = \beta\rho_1(x) + (1 - \beta)|f(x)| \tag{*}$$

with suitable β, $\beta = 0$ outside a neighborhood of G_1, $\beta = 1$ in a smaller neighborhood. For $k_0 = 0$, $k = 2$ there exists (by applying Palais [1] twice) a diffeomorphism $y = f(x)$ of $R^n \setminus \text{int}(G_1 \cup G_2)$ onto $R^n \setminus \text{int}(G_1' \cup G_2')$ and G_1', G_2' are as in Lemma 4.4. Replace in (*) $|f(x)|$ by $\lambda(f(x))$ where $\lambda(y)$ is constructed as in the proof of Lemma 4.4.]

Bibliographical Remarks

Chapters 1–3. The material of these chapters is standard. We have used, as sources, Breiman [1], Dynkin [1, 2], Doob [1], Varadhan [1], and McKean [1].

Chapters 4–5. The stochastic integral, Itô's formula, and the basic theory of stochastic differential equations are due to K. Itô [1, 2] (and the references given there). These results are presented in a book form in Doob [1] and, in more detail, in McKean [1] and Gikhman and Skorokhod [1, 2]. The present treatment is based on Gikhman and Skorokhod [2], but it incorporates some results from McKean [1], such as the exponential martingale inequality. Polynomial mollifiers occur in Courant and Hilbert [1]. For more details on mollifiers, see Friedman [2].

Stroock and Varadhan [1] have developed existence and uniqueness theory, for stochastic differential equations, in which the drift coefficient is any bounded measurable function and the diffusion matrix is continuous, bounded, and uniformly positive definite. In [3] they allow the diffusion matrix to degenerate. They have also extended, in [2], their results from [1] to disfusion processes restricted to a given domain Ω, i.e., when the path hits $\partial\Omega$ it is returned into Ω in a prescribed direction. (Some earlier results in this area are described in McKean [1], Gikhman and Skorokhod [1, 2], and in the references given there.)

Stochastic differential equations with $dw(t)$ replaced by the differential of a Poisson process are studied in Gikhman and Skorokhod [2], Komatsu [1], and Stroock [1]. Some applications in stochastic control are given in Wonham [1].

Chapter 6. Theorem 1.2 is due to Freidlin [1] and to Phillips and Sarason [1]. The material of Sections 2, 3, 4 is based on Friedman [1]. The connection between solutions of partial differential equations and stochastic differential equations is discussed in detail in Gikhman and Skorokhod [2], and in an expository article by Freidlin [2].

Chapter 7. The results of this chapter are based on Girsanov [1]. Earlier work in this direction was done by Cameron and Martin [1].

Chapter 8. The results of this chapter are due to Friedman [3]. Theorem 4.1 for $n = 1$ is due to Kulinič [2] (see also Gikhman and Skorokhod [2]) who also proved, for $n = 1$, other related results [1, 3, 4].

Chapter 9. Sections 1–3, 5–6 are due to Friedman [4]. The concept of obstacle and the results of Section 4 are due to Friedman and Pinsky [1]. The results of Section 3, in the special case of Brownian motion, were first proved by Dvoretzky and Erdös [1]. Theorem 7.1 is given in Gikhman and Skorokhod [2].

References

L. Breiman

[1] *Probability*. Addison-Wesley, Reading, Massachusetts, 1968.

R. H. Cameron and W. T. Martin

[1] The transformation of Wiener integrals by nonlinear transformations, *Trans. Amer. Math. Soc.* **66** (1949), 253–283.

R. Courant and D. Hilbert

[1] *Methods of Mathematical Physics*, Vol. 1. Wiley (Interscience), New York, 1953.

J. L. Doob

[1] *Stochastic Processes*. Wiley, New York, 1967.

A. Dvoretsky and P. Erdös

[1] Some problems on random walk in space, *Proc. 2nd Berkeley. Symp. Math. Statist. Probability*, 1950. Univ. of California Press, Berkeley, California, pp. 353–367.

E. B. Dynkin

[1] *Foundation of the Theory of Markov Processes*. Pergamon Press, New York, 1960.
[2] *Markov Processes*, Vols. 1, 2. Springer-Verlag, Berlin, 1965.

W. Feller

[1] *An Introduction to Probability Theory and Its Applications*, Vol. 2. Wiley, New York, 1971.

M. I. Freidlin

[1] On the factorization of nonnegative definite matrices, *Theor. Probability Appl.* **13** (1968), 354–356 [*Teor. Verojatnost. i Primenen* **13** (1968), 375–378].
[2] Markov processes and differential equations, in *Progress in Mathematics*, Vol. 3. Plenum Press, New York, 1969.

A. Friedman

[1] *Partial Differential Equations of Parabolic Type*. Prentice-Hall, Englewood Cliffs, New Jersey, 1964.
[2] *Partial Differential Equations*. Holt, New York, 1969.
[3] Limit behavior of solutions of stochastic differential equations, *Trans. Amer. Math. Soc.* **170** (1972), 359–384.
[4] Wandering out to infinity of diffusion processes, *Trans. Amer. Math. Soc.* **184** (1973), 185–203.

A. Friedman and M. A. Pinsky

[1] Asymptotic stability and spiraling properties of solutions of stochastic equations, *Trans. Amer. Math. Soc.* **186** (1973), 331–358.

I. I. Gikhman and A. V. Skorokhod

[1] *Introduction to the Theory of Random Processes*. Saunders, Philadelphia, 1965.
[2] *Stochastic Differential Equations*. Naukova Dumka, Kiev, 1968. (Springer-Verlag, Vol. 72, New York, 1973).

I. V. Girsanov

[1] On transforming a certain class of stochastic processes by absolutely continuous substitution of measures, *Theor. Probability Appl.* **5** (1960), 285–301 [*Teor. Verojatnost. i Primenen.* **5** (1960), 314–330].

K. Itô

[1] On stochastic differential equations. *Memoirs Amer. Math. Soc.* No. 4 (1961).
[2] *Lectures on Stochastic Processes*. Tata Inst. Fundamental Res., Bombay, 1961.

K. Itô and H. P. McKean

[1] *Diffusion Processes and Their Sample Paths*. Springer-Verlag, Berlin, 1965.

T. Komatsu

[1] Markov processes associated with certain integro-differential operators, *Osaka J. Math.* **10** (1973), 271–303.

G. L. Kulinič

[1] On the limit behavior of the distribution of the solution of a stochastic diffusion equation, *Theor. Probability Appl.* **12** (1967), 497–499 [*Teor. Verojatnost. i Primenen* **12** (1967), 548–551].
[2] Limiting behavior of distributions of a solution of the stochastic diffusion equation, *Ukrain. Mat. Ž.* **19** (1967), 119–125.
[3] Asymptotic normality of the distribution of the solution of a stochastic diffusion equation, *Ukrain. Mat. Ž.* **20** (1968), 396–400.
[4] Limit distributions of a solution of the stochastic diffusion equation, *Theor. Probability Appl.* 13 (1968), 478–482 [*Teor. Verojatnost. i Primenen.* 13 (1968), 502–506].

P. Lévy

[1] *Théorie de l'Addition de Variables Aléatoires*. Gauthier-Villars, Paris, 1937.

H. P. McKean

[1] *Stochastic Integrals*. Academic Press, New York, 1971.

J. Milnor

[1] *Lectures on the h-Cobordism Theorem*. Princeton Math. Notes, Princeton, New Jersey, 1965.

M. Morse

[1] A reduction to the Schoenflies extension problem, *Bull. Amer. Math. Soc.* **66** (1960), 113–115.

P. S. Palais

[1] Extending diffeomorphisms, *Proc. Amer. Math. Soc.* **11** (1960), 274–277.

R. S. Phillips and L. Sarason

[1] Elliptic-parabolic equations of the second order, *J. Math. Mech.* 17 (1967/8), 891–917.

D. W. Stroock

[1] Diffusion processes associated with Lévy generators, *to appear*.

D. W. Stroock and S. R. S. Varadhan

[1] Diffusion processes with continuous coefficients, *Comm. Pure Appl. Math.* Part I, **22** (1969), 345–400; Part II, *ibid.*, 479–530.

[2] Diffusion processes with boundary conditions, *Comm. Pure Appl. Math.* **24** (1971), 147–225.

[3] On degenerate elliptic-parabolic operators of second order and their associated diffusions, *Comm. Pure Appl. Math.* **24** (1972), 651–713.

S. R. S. Varadhan

[1] *Stochastic Processes*. Courant Inst. Math. Sci., New York Univ., New York, 1967–68.

W. M. Wonham

[1] Random differential equations in control theory. *Probabilistic Methods in Applied Mathematics*, Vol. 2, pp. 131–212. Academic Press, New York, 1970.

Stochastic Differential Equations and Applications

Volume 2

Preface–Volume 2

This volume begins with auxiliary results in partial differential equations (Chapter 10) that are needed in the sequel. In Chapters 11 and 12 we study the behavior of the sample paths of solutions of stochastic differential equations in the same spirit as in Chapter 9. Chapter 11 deals with the question whether the paths can hit a given set with positive probability. Chapter 12 is concerned with the stability of paths about a given manifold, and (in case of two dimensions) with spiraling of paths about this manifold.

Chapters 13–15 are concerned with applications to partial differential equations. In Chapter 13 we deal with the Dirichlet problem for degenerate elliptic equations. The results of Chapter 12 play here a fundamental role. In Chapter 14 we consider questions of singular perturbations. Chapter 15 is concerned with the existence of fundamental solutions for degenerate parabolic equations.

Chapters 16 and 17 deal with stopping time problems, stochastic games and stochastic differential games.

This material (except for Chapter 10) appears for the first time in book form. It is based on recent research. We hope that this book will increase and stimulate interest in this emerging area of research which involves stochastic differential equations, partial differential equations, and stochastic control.

I would like to thank Steve Orey for some useful suggestions in connection with the writing of Chapter 14.

General Notation

All functions are real valued, unless otherwise explicity stated.

In Chapter n, Section m the formulas and theorems are indexed by $(m.k)$ and $m.k$ respectively. When in Chapter l, we refer to such a formula $(m.k)$ (or Theorem $m.k$), we designate it by $(n.m.k)$ (or Theorem $n.m.k$) if $l \neq n$, and by $(m.k)$ (or Theorem $m.k$) if $l = n$.

Similarly, when referring to Section m in the same chapter, we designate the section by m; when referring to Section m of another chapter, say n, we designate the section by $n.m$.

Finally, when we refer to conditions (A), (A_1), (B) etc., these conditions are usually stated earlier in the *same* chapter.

10

Auxiliary Results in Partial Differential Equations

1. Schauder's estimates for elliptic and parabolic equations

In this section and in Sections 3 and 4 we state some estimates for solutions of the Dirichlet problem for elliptic equations and for solutions of the initial-boundary value problem for parabolic equations. These estimates do not depend on the fact that the corresponding boundary value problems do in fact have unique solutions; they are therefore called *a priori estimates*. These a priori estimates provide a powerful tool in the theory of partial differential equations. They will be needed in the subsequent chapters.

We begin with the *Schauder estimates* for elliptic operators

$$Lu \equiv \sum_{i,j=1}^{n} a_{ij}(x)u_{x_i x_j} + \sum_{i=1}^{n} b_i(x)u_{x_i} + c(x)u \tag{1.1}$$

in a bounded domain D.

Denote by d_x the distance from a point x of D to the boundary ∂D of D, and set $d_{xy} = \min(d_x, d_y)$. Define

$$H_\alpha(d^k u) = \operatorname*{l.u.b.}_{x,y\in D} d_{xy}^{k+\alpha} \frac{|u(x) - u(y)|}{|x-y|^\alpha},$$

$$|d^k u|_0 = \operatorname*{l.u.b.}_{x\in D} |d_x^k u(x)|,$$

$$|u|_m = \sum_{j=0}^{m} \sum |d^j D^j u|_0$$

where $D^j u$ is the vector whose components are all the jth derivatives of u, and the inner summation on the right is taken over all the components of $D^j u$. Define also

$$|u|_{m+\alpha} = |u|_m + \sum H_\alpha(d^m D^m u) \qquad (0 < \alpha \leqslant 1)$$

where $H_\alpha(d^m D^m u)$ is the vector with components $H_\alpha(d^m w_i)$, w_i varies over the components of $D^m u$, and the summation is taken over the components of $H_\alpha(d^m D^m u)$.

If a function u has m continuous derivatives in D, then we say that u belongs to $C^m(D)$. If the mth derivatives of u are uniformly Hölder continuous (exponent α) in compact subsets of D, then we say that u belongs to $C^{m+\alpha}(D)$.

Theorem 1.1 (Schauder's interior estimates). *Assume that*

$$\sum_{i,j=1}^{n} a_{ij}(x)\xi_i\xi_j \geqslant K_1|\xi|^2 \quad \text{if} \quad x\in D, \quad \xi \in R^n \quad (K_1 > 0), \tag{1.2}$$

$$|a_{ij}|_\alpha \leqslant K_2, \qquad |db_i|_\alpha \leqslant K_2, \qquad |d^2c|_\alpha \leqslant K_2. \tag{1.3}$$

If $Lu = f(x)$ in D and if $|d^2 f|_\alpha < \infty$, $u \in C^{2+\alpha}(D)$ and $|u|_0 < \infty$, then

$$|u|_{2+\alpha} \leqslant K(|u|_0 + |d^2 f|_\alpha) \tag{1.4}$$

where K is a constant depending only on K_1, K_2, n, α.

Next define

$$\overline{H}_\alpha(u) = \operatorname*{l.u.b.}_{x,y\in D} \frac{|u(x) - u(y)|}{|x-y|^\alpha},$$

$$\overline{|u|}_m = \sum_{j=0}^{m} \sum |D^j u|_0 \qquad \text{where} \quad |v|_0 = \operatorname*{l.u.b.}_{x\in D} |v(x)|,$$

$$\overline{|u|}_{m+\alpha} = \overline{|u|}_m + \sum \overline{H}_\alpha(D^m u) \qquad (0 < \alpha \leqslant 1).$$

We shall now assume that ∂D is in $C^{2+\alpha}$, i.e., ∂D can locally be written in the form

$$x_i = h(x_1, \ldots, x_{i-1}, x_{i+1}, \ldots, x_n) \tag{1.5}$$

for some i, where h is in $C^{2+\alpha}$ in some domain. A function ϕ defined on ∂D is said to belong to $C^{2+\alpha}(\partial D)$ if in terms of the local $C^{2+\alpha}$ representations (1.5) of ∂D this function ϕ is in $C^{2+\alpha}$.

It is not difficult to see that, when ∂D is in $C^{2+\alpha}$, the function ϕ is in $C^{2+\alpha}(\partial D)$ if an only if there exists a function Ψ with $\overline{|\Psi|}_{2+\alpha} < \infty$ such that $\Psi = \phi$ on ∂D. We define $\overline{|\phi|}^{*}_{2+\alpha} = \text{l.u.b.}\ \overline{|\Psi|}_{2+\alpha}$ where the "l.u.b." is taken over all such Ψ's.

Theorem 1.2 (Schauder's boundary estimates). *Assume that* (1.2) *holds and that*

$$\overline{|a_{ij}|}_\alpha \leqslant \overline{K}_2, \qquad \overline{|b_i|}_\alpha \leqslant \overline{K}_2, \qquad \overline{|c|}_\alpha \leqslant \overline{K}_2. \tag{1.6}$$

Assume also that ∂D belongs to $C^{2+\alpha}$, $\phi \in C^{2+\alpha}(\partial D)$ and $\overline{|f|}_\alpha < \infty$. If u is *a solution of $Lu = f$ in D, $u = \phi$ on ∂D and if $\overline{|u|}_{2+\alpha} < \infty$, then*

$$\overline{|u|}_{2+\alpha} \leq \bar{K}(\overline{|\phi|}^*_{2+\alpha} + |u|_0 + \overline{|f|}_\alpha) \tag{1.7}$$

where $\bar{K}$ is a constant depending only on K_1, $\bar{K}_2$, α, and D.

For a proof of Theorems 1.1, 1.2 the reader is referred to Agmon *et al.* [1]. Consider next the parabolic operator

$$Lu - \frac{\partial u}{\partial t} \equiv \sum_{i,j=1}^{n} a_{ij}(x, t) \frac{\partial^2 u}{\partial x_i \, \partial x_j} + \sum_{i=1}^{n} b_i(x, t) \frac{\partial u}{\partial x_i} + c(x, t)u - \frac{\partial u}{\partial t} \tag{1.8}$$

with coefficients defined in a bounded domain Q. We assume that Q is bounded by the closure of a domain B on $t = 0$, the closure of a domain B_T on $t = T$ and a manifold S lying in the strip $0 < t < T$.

Set $S_\tau = S \cap \{t \leq \tau\}$. We introduce the distance function

$$d(P, \bar{P}) = \{|x - \bar{x}|^2 + |t - \bar{t}|\}^{1/2} \tag{1.9}$$

where $P = (x, t)$, $\bar{P} = (\bar{x}, \bar{t})$. If $R = (\xi, \tau)$ belongs to Q, we denote by d_R the distance from R to $B \cup S_\tau$, i.e.,

$$d_R = \inf_{P \in B \cup S_\tau} d(R, P).$$

If R, P are any points in Q, we define $d_{RP} = \min(d_R, d_P)$.

Define

$$H_\alpha(d^m u) = \underset{P,R \in Q}{\text{l.u.b.}} \; d_{PR}^{m+\alpha} \frac{|u(P) - u(R)|}{d(P, R)^\alpha},$$

$$|d^m u|_0 = \underset{P \in Q}{\text{l.u.b.}} |d_P^m u(P)|,$$

$$|d^m u|_\alpha = |d^m u|_0 + H_\alpha(d^m u)$$

for any $0 < \alpha \leq 1$, and

$$|u|_{2+\alpha} = |u|_\alpha + \sum |dD_x u|_\alpha + \sum |d^2 D_x^2 u|_\alpha + |d^2 D_t u|_\alpha$$

where $D_x u$ is the vector $(\partial u/\partial x_1, \ldots, \partial u/\partial x_n)$, and the summations are with respect to the components of $D_x u$ and $D_x^2 u$.

We now state the *Schauder interior estimates* for parabolic equations.

Theorem 1.3. *Assume that*

$$\sum_{i,j=1}^{n} a_{ij}(x, t)\xi_i\xi_j \geq K_1|\xi|^2 \quad \text{if} \quad (x, t) \in Q, \quad \xi \in R^n \quad (K_1 > 0), \tag{1.10}$$

$$|a_{ij}|_\alpha \leq K_2, \qquad |db_i|_\alpha \leq K_2, \qquad |d^2 c|_\alpha \leq K_2. \tag{1.11}$$

If $Lu - \partial u/\partial t = f(x, t)$ in Q and if $|d^2 f|_\alpha < \infty$, $|u|_0 < \infty$ and u, $D_x u$, $D_x^2 u$, $D_t u$ are Hölder continuous (exponent α) in compact subsets of Q with respect to the metric (1.9), *then*

$$|u|_{2+\alpha} \leq K(|u|_0 + |d^2 f|_\alpha) \tag{1.12}$$

where K is a constant depending only on K_1, K_2, n, α.

We next define

$$\bar{H}_\alpha(u) = \underset{P,R \in Q}{\text{l.u.b.}} \frac{|u(P) - u(R)|}{d(P, R)^\alpha},$$

$$\overline{|u|}_\alpha = |u|_0 + \bar{H}_\alpha(u),$$

$$\overline{|u|}_{2+\alpha} = \overline{|u|}_\alpha + \sum \overline{|D_x u|}_\alpha + \sum \overline{|D_x^2 u|}_\alpha + \overline{|D_t u|}_\alpha.$$

A function ϕ defined on $B \cup \bar{S}$ is said to belong to $C^{2+\alpha}(B \cup \bar{S})$ if there exists a function Ψ defined on $\bar{Q}$ such that $\overline{|\Psi|}_{2+\alpha} < \infty$ and $\Psi = \phi$ on $B \cup \bar{S}$. We define $\overline{|\phi|}^*_{2+\alpha} = \text{l.u.b.}\ \overline{|\Psi|}_{2+\alpha}$ where the "l.u.b." is taken over all such Ψ's.

The domain Q is said to have the property (E) if for every point P on $\bar{S}$ there is a neighborhood V such that $V \cap \bar{S}$ can be represented in the form

$$x_i = h(x_1, \ldots, x_{i-1}, x_{i+1}, \ldots, x_n, t)$$

for some $1 \leq i \leq n$, and h, $D_x h$, $D_x^2 h$, $D_t h$ are Hölder continuous (exponent α) with respect to the metric (1.9).

We can now state the *Schauder boundary estimates* for parabolic equations.

Theorem 1.4. *Assume that* (1.10) *holds and that*

$$\overline{|a_{ij}|}_\alpha \leq \bar{K}_2, \qquad \overline{|b_i|}_\alpha \leq \bar{K}_2, \qquad \overline{|c|}_\alpha \leq K_2. \tag{1.13}$$

Assume also that Q has the property (E), $\phi \in C^{2+\alpha}(B \cup \bar{S})$ and $\overline{|f|}_\alpha < \infty$. If u is a solution of $Lu - \partial u/\partial t = f(x, t)$ in Q, $u = \phi$ on $B \cup \bar{S}$ and if $\overline{|u|}_{2+\alpha} < \infty$, then

$$\overline{|u|}_{2+\alpha} \leq \bar{K}\left(\overline{|\phi|}^*_{2+\alpha} + \overline{|f|}_\alpha\right) \tag{1.14}$$

where $\bar{K}$ is a constant depending only on K_1, $\bar{K}_2$, α, and Q.

For a proof of Theorems 1.3, 1.4 the reader is referred to Friedman [1].

Remark 1. Theorem 1.1 extends to the case where d_x is the distance from x to a subset Γ of ∂D. Theorem 1.3 extends to the case where d_P is the distance (in the metric (1.9)) from $P = (x, t)$ to the set $\Gamma \cap \{s;\ s \leq t\}$ where Γ is a subset of the normal boundary; see Friedman [1].

Remark 2. Theorem 1.4 can be used to prove existence of solutions u with $\overline{|u|}_{2+\alpha} < \infty$ of the first initial-boundary value problem. Thus, *if* $L, Q, \phi, f,$ *are as in Theorem* 1.4 *and if* $L\phi - \partial\phi/\partial t = f$ *for* $x \in \partial B$, *then there exists a unique solution* u *of*

$$Lu - \frac{\partial u}{\partial t} = f \quad in\ Q, \qquad u = \phi \quad on \quad B \cup \bar{S},$$

and $\overline{|u|}_{2+\alpha} < \infty$; the function $L\phi - \partial\phi/\partial t$ at $(y, 0)$, $y \in \partial B$, is computed by taking an extension Ψ of ϕ into Q with $\overline{|\Psi|}_{2+\alpha} < \infty$, and computing $\lim(L\Psi - \partial\Psi/\partial t)(x, t)$ as $x \to y$, $t \downarrow 0$. For the proof of this result, see Friedman [1].

2. Sobolev's Inequality

We review a few facts used in the standard theory of partial differential equations.

The following notation will be used: $x = (x_1, \ldots, x_n)$ is a variable point in R^n, $D_j = \partial/\partial x_j$, $D^\alpha = D_1^{\alpha_1} \cdots D_n^{\alpha_n}$ where $\alpha = (\alpha_1, \ldots, \alpha_n)$, $\alpha! = \alpha_1! \cdots \alpha_n!$, $|\alpha| = \alpha_1 + \cdots + \alpha_n$, $x^\alpha = x_1^{\alpha_1} \cdots x_n^{\alpha_n}$. If Ω is an open set in R^n, $C^m(\Omega)$ $(C^m(\bar{\Omega}))$ is the set of all real-valued functions continuous (uniformly continuous) in Ω together with their first m derivatives; $C_0^m(\Omega)$ is the subset of $C^m(\Omega)$ consisting of all functions with compact support; $C^\infty(\Omega) = \bigcap_{m=1}^{\infty} C^m(\Omega)$ and $C_0^\infty(\Omega)$ consists of all functions in $C^\infty(\Omega)$ with compact support.

If u, v are locally integrable in Ω and if

$$\int_\Omega u D^\alpha \phi \, dx = (-1)^{|\alpha|} \int v\phi \, dx$$

for all $\phi \in C_0^\infty(\Omega)$, then we say that v is the αth *weak derivative* of u and write: $D^\alpha u = v$ in the weak sense, or $D^\alpha u = v$ (w.d.).

Definition. Let m be a nonnegative integer and let $1 \leqslant p < \infty$. The space $W^{m,p}(\Omega)$ consists of all functions u in the real space $L^p(\Omega)$ whose weak derivatives of all orders $\leqslant m$ exist and belong to $L^p(\Omega)$. The space $W^{m,p}(\Omega)$ is normed by

$$|u|_{m,p}^{\Omega} \equiv \|u\|_{W^{m,p}(\Omega)} = \left\{ \sum_{|\alpha| \leqslant m} \int_\Omega |D^\alpha u(x)|^p \, dx \right\}^{1/p}.$$

It is easy to show that $W^{m,p}(\Omega)$ is a Banach space; if $p = 2$, then it is a Hilbert space.

Theorem 2.1. *Let Ω be a bounded domain with C^2 boundary and let j be a positive integer and p a real number > 1. Then there exists a positive* *constant* ϵ_0 *depending only on* Ω, p, j *such that, for any* $0 < \epsilon < \epsilon_0$,

$$|u|_{j-1,p}^{\Omega} \leq \epsilon |u|_{j,p}^{\Omega} + C|u|_{0,p}^{\Omega} \qquad \textit{for all} \quad u \in C^j(\bar{\Omega}) \tag{2.1}$$

where C *is a constant depending only on* Ω, p, j, ϵ.

For proof see Nirenberg [1] or Friedman [2].

We introduce the notation

$$\underset{\Omega}{\text{l.u.b.}}\,[v]_\alpha = \underset{x,y\in\Omega}{\text{l.u.b.}} \frac{|v(x) - v(y)|}{|x - y|^\alpha},$$

$$|u|_{p,\Omega} = \left\{ \int_\Omega |u(x)|^p \, dx \right\}^{1/p} \qquad \text{if} \quad p > 0,$$

$$|u|_{p,\Omega} = \underset{\Omega}{\text{l.u.b.}}\,|D^h u| \equiv \sum_{|\beta| = h} \underset{\Omega}{\text{l.u.b.}}\,|D^\beta u|$$

if $p < 0$, $h = [-n/p]$, $h + n/p = 0$,

$$|u|_{p,\Omega} = [D^h u]_{\alpha,h} \equiv \sum_{|\beta| = h} \underset{\Omega}{\text{l.u.b.}}\,[D^\beta u]_\alpha$$

if $p<0$, $h=[-n/p]$, $h+n/p<0$ where $-\alpha = h + n/p$.

If $\Omega = R^n$, then we write $|u|_p$ instead of $|u|_{p,\Omega}$.

The *extended Sobolev inequality* in R^n asserts the following.

Theorem 2.2. *Let* q, r *be any numbers satisfying* $1 \leq q, r \leq \infty$ *and let* j, m *be any integers satisfying* $0 \leq j < m$. *If* u *is any function in* $C_0^m(R^n)$, *then*

$$|D^j u|_p \leq C |D^m u|_r^a |u|_q^{1-a} \tag{2.2}$$

where

$$\frac{1}{p} = \frac{j}{n} + a\left(\frac{1}{r} - \frac{m}{n}\right) + (1 - a)\frac{1}{q}$$

for all a *in the interval*

$$\frac{j}{m} \leq a \leq 1,$$

where C *is a constant depending only on* n, m, j, q, r, a, *with the following exception: If* $m - j - n/r$ *is a nonnegative integer, then* (2.2) *is asserted only for* $(j/m) \leq a < 1$.

For proof the reader is referred to Nirenberg [1], Gagliardo [1, 2], or Friedman [2].

From Theorem 2.2 one can derive (see Friedman [2]) the corresponding extended Sobolev inequality in a bounded domain:

Theorem 2.3. *Let Ω be a bounded domain with $\partial\Omega$ in C^m, and let u be any function in $W^{m,r}(\Omega)\cap L^q(\Omega)$, $1 \leqslant r, q \leqslant \infty$. For any integer j, $0 \leqslant j < m$ and for any number a in the interval $j/m \leqslant a \leqslant 1$, set*

$$\frac{1}{p} = \frac{j}{n} + a\left(\frac{1}{r} - \frac{m}{n}\right) + (1-a)\frac{1}{q}.$$

If $m - j - n/r$ is not a nonnegative integer, then

$$|D^j u|_{p,\Omega} \leqslant C\left(|u|^{\Omega}_{m,r}\right)^a \left(|u|^{\Omega}_{0,q}\right)^{1-a}. \tag{2.3}$$

If $m - j - n/r$ is a nonnegative integer, then (2.3) holds for $j/m \leqslant a < 1$. The constant C depends only on Ω, r, q, m, j, a.

We state two special cases which are most useful.

Theorem 2.4. *Let Ω be a bounded domain with boundary $\partial\Omega$ in C^1, and let u be any function $W^{m,r}(\Omega)$, $1 \leqslant r < \infty$. Then, for any integer j, $0 \leqslant j < m$,*

$$|u|^{\Omega}_{j,p} \leqslant C|u|^{\Omega}_{m,r}, \qquad \frac{1}{p} = \frac{j}{n} + \frac{1}{r} - \frac{m}{n} \tag{2.4}$$

provided $p > 0$. The constant C depends only on Ω, m, j, r.

Since we have not assumed that $\partial\Omega$ is in C^m, we cannot deduce Theorem 2.4 as a truly special case of Theorem 2.3. It is a special case of Theorem 2.3 when $j = 0$, $m = 1$. But once (2.4) is known for $j = 0$, $m = 1$, the proof for general j, m follows by induction on m.

Theorem 2.5. *Let Ω be a bounded domain with boundary $\partial\Omega$ in C^1, and let u be a function in $W^{m,p}(\Omega)$ for some $p > 1$. If $m > n/p$, then $u(x)$ has a continuous version (which will be denoted again by $u(x)$) and*

$$\max_{\bar{\Omega}} |u(x)| \leqslant C|u|^{\Omega}_{m,p}, \tag{2.5}$$

$$\underset{x,y\in\Omega}{\text{l.u.b.}} \frac{|u(x) - u(y)|}{|x-y|^{\alpha}} \leqslant C|u|^{\Omega}_{m,p} \tag{2.6}$$

where

$$\frac{1}{k} = \frac{m}{n} - \frac{1}{p}, \quad \alpha = 1 \quad \text{if} \quad \frac{n}{k} > 1, \qquad \alpha = \frac{n}{k} \quad \text{if} \quad \frac{n}{k} < 1;$$

the constant C depends only on Ω, m, p.

If $m = 1$, then Theorem 2.5 is a special case of Theorem 2.3. For $m > 1$, one proceeds by induction on m.

Theorem 2.4 can be used to prove the following *compact imbedding* theorem.

Theorem 2.6. *Let Ω be a bounded domain with boundary $\partial\Omega$ in C^1. Let r be a positive number, $1 \leq r < \infty$, and let j, m be integers, $0 \leq j < m$. If p is any positive number ≥ 1 satisfying*

$$\frac{1}{p} > \frac{j}{n} + \frac{1}{r} - \frac{m}{n},$$

then the imbedding $u \to u$ from $W^{m,r}(\Omega)$ into $W^{j,p}(\Omega)$ is compact.

Thus, from any bounded sequence $\{u_k\}$ in $W^{m,r}(\Omega)$ one can extract a convergent subsequence in $W^{j,p}(\Omega)$.

For proof of Theorem 2.6, see Friedman [2].

3. L^p estimates for elliptic equations

Let L be the elliptic operator (1.1) with coefficients defined in a bounded domain D. We shall assume:

$$\sum_{i,j=1}^{n} a_{ij}(x)\xi_i\xi_j \geq \mu|\xi|^2 \quad \text{if } x \in D, \ \xi \in R^n \quad (\mu > 0); \tag{3.1}$$

$$b_i(x), c(x) \quad \text{are measurable functions,} \tag{3.2}$$

$$\sum_{i=1}^{n} |b_i(x)| + |c(x)| \leq K_1 \quad \text{if } x \in D;$$

$$\text{the } a_{ij}(x) \quad \text{are continuous in } \bar{D}. \tag{3.3}$$

The last condition implies that there is a function $w(\rho)$ ($\rho > 0$) such that

$$\sum_{i,j=1}^{n} |a_{ij}(x) - a_{ij}(y)| \leq w(|x-y|) \quad \text{if } x \in \bar{D}, y \in \bar{D}; \qquad w(\rho) \downarrow 0 \quad \text{if } \rho \downarrow 0. \tag{3.4}$$

Consider the Dirichlet problem

$$Lu(x) = f(x) \quad \text{in } D, \tag{3.5}$$

$$u = 0 \quad \text{on } \partial D. \tag{3.6}$$

If f and the coefficients of L are Hölder continuous and if $c \leq 0$, $\partial D \in C^2$, then by Theorem 2.4 there exists a unique solution of (3.5), (3.6). When the coefficients of L satisfy only (3.1)–(3.3), we have to introduce a weaker concept of solution.

Definition. A function $u(x)$ in $W^{2,p}(D)$ is said to be a *strong solution* of (3.5) if (3.5) holds a.e. in D when the derivatives of u are taken in the weak sense.

Thus the concept of a strong solution is weaker than the concept of a *classical solution* (i.e., a solution u in $C^2(D)$).

There is also a weaker concept of solution: A function in $L^p(D)$ is a *weak solution* of (3.5) if

$$\int_D u(x)L^*\phi(x)\,dx = \int_D f(x)\phi(x)\,dx \qquad \text{for any} \quad \phi\in C_0^\infty(D).$$

This definition requires that the adjoint L^* be defined. This concept will not be used in the future.

Denote by $W_0^{1,p}(D)$ $(p > 1)$ the completion in the space $W^{1,p}(D)$ of the subset $C_0^\infty(D)$. If ∂D is in C^1, then it can be shown that if $u \in C^1(\bar{D})$ and $u = 0$ on ∂D, then $u\in W_0^{1,p}(D)$; conversely, if $u\in W_0^{1,p}(D)$ and $u\in C^0(\bar{D})$, then $u = 0$ on ∂D. For proof in case $p = 2$, see Friedman [2]; the proof for any $p > 1$ is similar. These facts motivate the following

Definition. A function u in $W^{1,p}(\Omega)$ satisfies (3.6) in the *generalized sense* if $u\in W_0^{1,p}(\Omega)$.

Combining the above two definitions, we shall say that u is a *strong solution* of the Dirichlet problem (3.5), (3.6) if $u\in W^{2,p}(D)\cap W_0^{1,p}(D)$ and $Lu = f$ a.s.

We define an operator A with domain $\mathfrak{D}_A = W^{2,p}(D)\cap W_0^{1,p}(D)$ and range in $L^p(D)$ by

$$(Au)(x) = Lu(x).$$

Thus, u is a strong solution of the Dirichlet problem (3.5), (3.6) if and only if $u \in \mathfrak{D}_A$ and $Au = f$ a.e.

Theorem 3.1. *Let D be a bounded domain with boundary ∂D in C^2, and suppose that* (3.1)–(3.3) *hold and that $c \leqslant 0$. Let $1 < p < \infty$. For any $f \in L^p(D)$, there exists a unique strong solution u of the Dirichlet problem* (3.5), (3.6).

Theorem 3.2. *Let D be a bounded domain with boundary ∂D in C^2, and suppose that* (3.1)–(3.3) *hold. Let $1 < p < \infty$. Then there exist positive constants C, Λ such that*

$$|u|_{2,p}^D \leqslant C\big(|Lu|_{0,p}^D + |u|_{0,p}^D\big), \tag{3.7}$$

$$\sum_{j=0}^{2} \lambda^{1-j/2}|u|_{j,p}^D \leqslant C|Lu - \lambda u|_{0,p}^D \qquad \text{if} \quad \lambda \geqslant \Lambda \tag{3.8}$$

for all $u\in W^{2,p}(D)\cap W_0^{1,p}(D)$. If $c \leqslant 0$, then (3.7) *can be replaced by the* *stronger inequality*

$$|u|_{2,p}^D \leq C|Lu|_{0,p}^D, \tag{3.9}$$

and (3.8) *holds with* $\Lambda = 0$. *The constants* C, Λ *depend only on* μ, K_1, *the function* $w(\rho)$ *and the domain* D.

For the proof of Theorems 3.1, 3.2 in case $p = 2$, see Agmon [1] and Friedman [2]. The proof for general p is given in Agmon *et al* [1].

Notice that Theorem 3.1 asserts that the operator A maps $\mathfrak{D}_A$ onto $L^p(D)$. Theorem 3.2 implies the estimate, on the resolvent of A,

$$\|(\lambda I - A)^{-1}\| \leq C_1/\lambda \quad \text{if} \quad \lambda > \Lambda \quad (C_1 \text{ const}).$$

Theorems 3.1, 3.2 can be used to derive regularity theorems for solutions of elliptic equations.

Remark. From the proof of Theorem 3.2 one sees that if ∂D can be covered by a finite number $\leq N$ of local coordinates $x_i = h_i(x_i')$ $(x_i' = (x_1, \ldots, x_{i-1}, x_{i+1}, \ldots, x_n))$ where the h_i have their first two derivatives bounded by a constant K_2, and if every $x \in \partial D$ is contained together with a ν-neighborhood of ∂D in one of the coordinate patches (ν positive constant), then C and Λ depend only on μ, K_1, N, ν, K_2 and the function $w(\rho)$.

4. L^p estimates for parabolic equations

If X is a Banach space with norm $\| \ \|_X$ and $1 < p < \infty$, the space $L^p(\alpha, \beta; X)$ is the space of all functions $u(t)$ from (α, β) into X with finite norm

$$|u|_{L^p(\alpha,\beta;X)} = \left\{ \int_\alpha^\beta (\|u(t)\|_X)^p \, dt \right\}^{1/p}.$$

Similarly one defines the space $L^\infty(\alpha, \beta; X)$.

$C([\alpha, \beta]; X)$ is the space of all continuous functions $u(t)$ from $[\alpha, \beta]$ into X with finite norm

$$|u|_{C([\alpha,\beta];X)} = \max_{\alpha \leq t \leq \beta} \|u(t)\|_X.$$

For functions $u(t)$ from (α, β) into X, the derivative $u'(t) = du(t)/dt$ is defined as the limit in X of the quotient differences $(u(t+h) - u(t))/h$ as $h \to 0$, i.e.,

$$\left\| \frac{u(t+h) - u(t)}{h} - \frac{du(t)}{dt} \right\|_X \to 0 \quad \text{if} \quad h \to 0.$$

We call it the *strong derivative*.

Lemma 4.1. *If $u(t)$ and $u'(t)$ belong to $L^p(\alpha, \beta; X)$ for some $1 < p < \infty$, then $u(t)$ belongs to $C([\alpha, \beta]; X)$.*

More precisely, one can redefine $u(t)$ on a set of Lebesgue measure zero so that the modified function is in $C([\alpha, \beta]; X)$.

The proof is left to the reader (see Problem 12).

Consider the parabolic operator $L - \partial/\partial t$ defined by (1.8), where the coefficients are defined in the closure $\bar{Q}$ of a bounded cylinder $Q = B \times (0, T)$.

Consider the initial-boundary value problem

$$Lu - \partial u/\partial t = f(x, t), \tag{4.1}$$

$$u(x, t) = 0 \quad \text{if} \quad x \in \partial B, \quad 0 < t < T, \tag{4.2}$$

$$u(x, 0) = 0 \quad \text{if} \quad x \in B. \tag{4.3}$$

We shall need the following conditions:

$$\sum_{i,j=1}^{n} a_{ij}(x, t)\xi_i\xi_j \geqslant \mu|\xi|^2 \quad \text{if} \quad (x, t) \in Q \qquad (\mu > 0); \tag{4.4}$$

$b_i(x, t)$, $c(x, t)$ are measurable functions in Q and

$$\sum_{i=1}^{n} |b_i(x, t)| + |c(x, t)| \leqslant K_1; \tag{4.5}$$

$$\sum |a_{ij}(x, t) - a_{ij}(y, s)| \leqslant \eta(|x - y| + |t - s|) \tag{4.6}$$

for all $(x, t) \in \bar{Q}$, $(y, s) \in \bar{Q}$, $\eta(r) \downarrow 0$ if $r \downarrow 0$.

Theorem 4.2. *Let ∂B belong to C^2 and let (4.4)–(4.6) hold. Let $1 < p < \infty$. Then for any $f \in L^p(0, T; L^p(B))$ there exists a unique "strong solution" u of (4.1)–(4.3) in the following sense:*

$$u(t) \in L^p(0, T; W^{2,p}(B)) \cap L^\infty(0, T; W_0^{1,2}(B)), \tag{4.7}$$

$$\frac{du(t)}{dt} \in L^p(0, T; L^p(B)), \tag{4.8}$$

$$\textit{for almost all } t \in (0, T) \textit{ the equation (4.1) holds a.e. in } B \\ (\textit{where the } x\textit{-derivatives are in the weak sense}), \tag{4.9}$$

$$|u(t)|_{L^p(B)} \to 0 \qquad \textit{if} \quad t \downarrow 0. \tag{4.10}$$

Notice, by Lemma 4.1, that (4.7), (4.8) imply that

$$u(t) \in C([0, T]; L^p(B)),$$

i.e., the solution $u(t)$ is a continuous function from $[0, T]$ into $L^p(B)$.

Theorem 4.3. *Let ∂B belong to C^2 and let* (4.4)–(4.6) *hold. Let* $1 < p < \infty$. *Then there exists a constant C depending only on μ, K_1, T, the function $\eta(\rho)$, and the domain B, such that for any $f \in L^p(0, T; L^p(B))$, $p \neq \frac{3}{2}$, the unique strong solution u of* (4.1)–(4.3) *satisfies*:

$$\int_0^T \int_B (|u|^p + |D_x u|^p + |D_x^2 u|^p + |D_t u|^p)\, dx\, dt \leq C \int_0^T \int_B |f|^p\, dx\, dt. \tag{4.11}$$

If $p = \frac{3}{2}$, then an estimate involving a slightly different norm is valid.

For the proof of Theorems 4.2, 4.3 see Solonnikov [1] or Fabes and Riviere [1].

PROBLEMS

1. Let D be a bounded domain in R^n. Let $\{u_m\}$ be a sequence of functions defined in $\bar{D}$ and satisfying $\overline{|u_m|}_{k+\alpha} \leq C$, C constant. Prove that there exists a subsequence $\{u_{m'}\}$ and a function v defined in $\bar{D}$ such that $\overline{|v|}_{k+\alpha} < \infty$ and $\overline{|u_{m'} - v|}_{k+\beta} \to 0$ as $m' \to \infty$, for any $0 < \beta < \alpha$.
2. Prove the same result with the norm $\overline{|\ |}_{k+\alpha}$ replaced by the norm $|\ |_{k+\alpha}$.
3. If in Theorem 1.2, $c(x) \leq 0$, then the Schauder inequality (1.7) reduces to

$$\overline{|u|}_{2+\alpha} \leq \bar{K}(\overline{|\phi|}^*_{2+\alpha} + \overline{|f|}_\alpha) \qquad \text{(with a different constant } \bar{K}).$$

4. Let $L_m = \sum a_{ij}^m(x)\, \partial^2/\partial x_i\, \partial x_j + \sum b_i^m(x)\, \partial/\partial x_i + c^m(x)$ be elliptic operators with $c^m(x) \leq 0$, satisfying the conditions (1.2), (1.6) with constants K_1, $\bar{K}_2$ independent of m, and assume that ∂D belongs to $C^{2+\alpha}$. Let ϕ_m be functions in $C^{2+\alpha}(\partial D)$ satisfying $\overline{|\phi_m|}^*_{2+\alpha} \leq K_3$ where K_3 is independent of m. Let f_m be functions defined on $\bar{D}$ with $\overline{|f_m|}_\alpha \leq K_4$, where K_4 is a constant independent of m. Suppose u_m is a solution of $L_m u_m = f_m$ in D, $u_m = \phi_m$ on ∂D, for each m. Prove: if $a_{ij}^m \to a_{ij}$, $b_i^m \to b_i$, $c^m \to c$, $f_m \to f$ uniformly in D and $\phi_m \to \phi$ uniformly on ∂D, as $m \to \infty$, then $u_m \to u$ uniformly in $\bar{D}$ where u is the solution of $Lu = f(x)$ in D, $u = \phi$ on ∂D and L is given by (1.1).
5. Extend the result of the preceding problem to parabolic equations $L_m u_m - \partial u_m/\partial t = f_m$ in a cylinder Q with $u_m = \phi_m$ on $B \cup \bar{S}$.
6. If $D^\alpha u = v$ in the weak sense and $D^\beta v = w$ in the weak sense, prove that $D^{\alpha+\beta} u = w$ in the weak sense.
7. If u has weak derivative $D_1 u$ and f is continuously differentiable, then $D_1(fu) = fD_1 u + uD_1 f$ in the weak sense.
8. Let A be a compact set in R^n and let B be an open set in R^n, $B \supset A$. Prove that there exists a function $\phi(x)$ in $C^\infty(R^n)$ such that $\phi(x) = 0$ in $R^n \backslash B$, $\phi(x) = 1$ on A, and $0 \leq \phi(x) \leq 1$ elsewhere. [*Hint*: Let $A \subset G \subset \bar{G}$

$\subset B$, G open and bounded, and take ϕ to be the mollifier of χ_G.]

9. Let $1 < p < \infty$. Let Ω be a bounded domain with boundary $\partial\Omega$ in C^m and let Ω_0 be any open set containing $\overline{\Omega}$. Then there exists a constant K depending only on Ω, Ω_0 such that for any $u \in C^m(\overline{\Omega})$ there exists a $\tilde{u}$ in $C_0^m(\Omega_0)$ such that $\tilde{u} = u$ in Ω and

$$|u|_{m,p}^{\Omega_0} \leqslant K|u|_{m,p}^{\Omega}.$$

[*Hint*: If $\Omega = \{x;\ x_n > 0,\ |x| < \rho\}$ and u vanishes in an Ω-neighborhood of the boundary $|x| = \rho$, take

$$\tilde{u}(x_1, \ldots, x_{n-1}, x_n) = \sum_{j=1}^{m+1} c_j u\left(x_1, \ldots, x_{n-1}, -\frac{x_n}{j}\right) \qquad \text{if} \quad x_n < 0.$$

For general Ω, use a partition of unity of $\overline{\Omega}$; each open set in the covering of $\partial\Omega$ is taken so small that its intersection with Ω can be transformed (via the local representation of $\partial\Omega$) into a hemiball with the points of $\partial\Omega$ going into the planar part of the boundary of the hemiball.]

10. Prove that $W^{p,m}(\Omega)$ is a Banach space.

11. Prove that $L^p(\alpha, \beta; X)$ is a Banach space if X is a Banach space.

12. Prove Lemma 4.1. [*Hint*: Let $u_\epsilon(t)$ be a mollifier of $u(t)$. Apply Sobolev's inequality to $f(u_\epsilon(t))$ (f any bounded linear functional on X) to deduce that $|u_\epsilon(t) - u_\epsilon(s)| \leqslant C|t - s|^q$ $(1/p + 1/q = 1)$. Take $\epsilon \downarrow 0$.]

13. Let $u(x)$ be a uniformly Lipschitz continuous function in a bounded domain Ω. Prove that $u \in W^{1,p}(\Omega)$ for any $1 \leqslant p < \infty$.

14. Let $u \in W^{1,2}(\Omega)$. Prove that $D_x u = 0$ a.e. on the set

$$S = \{x \in \Omega;\ u(x) = 0\}.$$

[*Hint*: It suffices to consider $n = 1$. Use the fact that almost every point of S is a Lebesgue point of u_x, and the relation $\int_a^b u_x\, dx = u(b) - u(a)$.]

15. Let $u \in W^{1,2}(\Omega)$. Prove that $|u|$, u^+ belong to $W^{1,2}(\Omega)$.

11

Nonattainability

If $\xi(t)$ is a solution of a stochastic differential system in R^n and if M is a closed set in R^n such that

$$P_x\{\xi(t) \in M \text{ for some } t > 0\} = 0 \qquad \text{whenever} \quad x \notin M,$$

then we say that M is *nonattainable* by the process $\xi(t)$. In Section 9.4 we have shown that (in the present terminology) a two-sided obstacle is non-attainable. The reason for this is that since the normal diffusion and normal component of the Fichera drift both vanish at M, there is "insufficient mobility" for hitting M.

It is well known that an n-dimensional Brownian motion $w(t)$ does not hit a prescribed point $x \neq 0$, with probability 1, if $n > 2$. This is another example of nonattainability of a set M. The reason here is that the set $M = \{x\}$ is "too thin."

In this chapter we shall establish general nonattainability theorems that include, as special cases, the previous two examples.

1. Basic definitions; a lemma

Let M be a k-dimensional C^2 manifold in R^n. At each point $x^0 \in M$, let $N^{k+i}(x^0)$ $(1 \leqslant i \leqslant n-k)$ form a set of linearly independent vectors in R^n which are normal to M and x^0.

Let $a(x)$ be an $n \times n$ matrix, and consider the $(n-k) \times (n-k)$ matrix $\alpha = (\alpha_{ij})$ where

$$\alpha_{ij} = \langle a(x^0)N^{k+i}(x^0), N^{k+j}(x^0)\rangle \qquad (1 \leqslant i, j \leqslant n-k);$$

here $\langle\ ,\ \rangle$ denotes the scalar product in R^n.

Denote the rank of α by $r_{M^\perp}(x^0)$. This number is clearly independent of the choice of the particular set of normals $N^{k+i}(x^0)$.

Definition. The *rank of $a(x)$ orthogonal to M at x^0* is the number $r_{M^\perp}(x^0)$.

If the manifold M has boundary ∂M, then we always take M to be a closed set, i.e., $\overline{M} = M \cup \partial M = M$. If $x^0 \in \partial M$, then by a normal N to M at

x^0 we mean a vector N that is $\lim N(x)$, where $x \in \text{int } M$, $x \to x^0$ and $N(x)$ is normal to M at x. We now define $r_{M^\perp}(x^0)$, for $x^0 \in \partial M$, in the same way as before.

Notice that ∂M is also a manifold, and one can define $r_{(\partial M)^\perp}(x^0)$. Clearly,

$$r_{(\partial M)^\perp}(x^0) \geq r_{M^\perp}(x^0).$$

Notice also that when M consists of just one point x^0, $r_{M^\perp}(x^0)$ is the rank of the matrix $a(x^0)$.

Consider now a diffusion process governed by a system of n stochastic differential equations

$$d\xi(t) = \sigma(\xi(t))\, dw + b(\xi(t))\, dt; \tag{1.1}$$

$\sigma(x)$ is an $n \times n$ matrix $(\sigma_{ij}(x))$, $b(x)$ is a vector $(b_1(x), \ldots, b_n(x))$, and $w(t)$ is an n-dimensional Brownian motion $(w_1(t), \ldots, w_n(t))$.

We assume:

(A_1) $\sigma(x)$ and $b(x)$ satisfy, for all $x \in R^n$,

$$|\sigma(x)| + |b(x)| \leq C(1 + |x|) \qquad (C \text{ const});$$

further, for any $R > 0$ there is a positive constant C_R such that

$$|\sigma(x) - \sigma(y)| + |b(x) - b(y)| \leq C_R|x - y|$$

if $|x| < R$, $|y| < R$.

Introduce the diffusion matrix $a(x) = (a_{ij}(x))$:

$$a(x) = \sigma(x)\sigma^*(x) \qquad [\sigma^*(x) = \text{transpose of } \sigma(x)],$$

and denote the rank of $a(x)$ orthogonal to M at x by $d(x)$, i.e.,

$$d(x) = r_{M^\perp}(x) \qquad \text{for} \quad x \in M. \tag{1.2}$$

Definition. A closed set M in R^n is *nonattainable* by the process $\xi(t)$ if

$$P_x\{\xi(t) \in M \text{ for some } t > 0\} = 0 \qquad \text{for each} \quad x \notin M. \tag{1.3}$$

If (1.3) holds for all x in a set G ($G \cap M = \varnothing$), then we say that M is *nonattainable from* G.

It will be shown later that, roughly speaking, if $d(x) \geq 2$ for all $x \in M$ (M a C^2 manifold), then M is nonattainable. The same assertion is true in some cases when $d(x) \geq 1$ (but not always), provided $n \geq 2$. The interpretation of these results is that M is "too thin" for $\xi(t)$ to hit it.

It will also be shown that when $d(x) \equiv 0$ on M, then the assertion (1.3) is still true provided the normal component of the Fichera drift of $\xi(t)$ vanishes on M. The interpretation of this result is that M is an "obstacle" for the diffusion process $\xi(t)$.

We conclude this section with a lemma that will be useful in reducing the proof of the assertion (1.3) from a global manifold M to a local one.

Let $x^0 \in M$. Then, in a neighborhood of x^0, M can be represented in the form

$$x_{i'} = f_{i'}(x'') \tag{1.4}$$

where i' varies over $n - k$ of the indices $1, 2, \ldots, n$, the coordinates of x'' are $x_{i''}$, and i'' varies over the remaining indices. Suppose for simplicity that i' varies over $k + 1, \ldots, n$, i.e., M is given locally by

$$x_{k+i} = f_{k+i}(x_1, \ldots, x_k) \qquad (i = 1, \ldots, n - k). \tag{1.5}$$

Introduce the mapping

$$\begin{aligned} y_i &= x_i - x_i^0 \qquad (i = 1, \ldots, k), \\ y_{k+i} &= x_{k+i} - f_{k+i}(x_1, \ldots, x_k) \qquad (i = 1, \ldots, n - k) \end{aligned} \tag{1.6}$$

where $x^0 = (x_1^0, \ldots, x_n^0)$. This is a diffeomorphism from a neighborhood $V(x^0)$ of x^0 into a neighborhood V^* of 0 in the y-space. Denote by M^* the image of $M \cap V(x^0)$. Then M^* is given by

$$y_i = 0 \quad (i = 1, \ldots, k), \qquad (y_{k+1}, \ldots, y_n) \in A \tag{1.7}$$

for some set A.

Consider the operator

$$Lu = \frac{1}{2} \sum_{i,j=1}^{n} a_{ij}(x) \frac{\partial^2 u}{\partial x_i \, \partial x_j} + \sum_{i=1}^{n} b_i(x) \frac{\partial u}{\partial x_i}$$

and set $v(y) = u(x)$. Then $Lu(x) = L'v(y)$ where

$$L'v = \frac{1}{2} \sum_{i,j=1}^{n} a_{ij}^*(y) \frac{\partial^2 v}{\partial y_i \, \partial y_j} + \sum_{i=1}^{n} b_i^*(y) \frac{\partial v}{\partial y_i}.$$

It is easily seen that

$$a_{k+i,\,k+j}^*(y) = \langle a(x) N^{k+i}(x), N^{k+j}(x) \rangle$$

where

$$N^{k+i}(x) = \nabla_x g_{k+i}(x), \qquad g_{k+i}(x) = x_{k+i} - f_{k+i}(x_1, \ldots, x_k).$$

Notice that if $x \in M \cap V(x^0)$, then the $N^{k+i}(x)$ $(1 \leq i \leq n - k)$ form a set of linearly independent normal vectors to M at x. Hence

$$d(x) = \operatorname{rank}\left(a_{k+i,\,k+j}^*(x)\right)_{i,j=1}^{n-k} \qquad (x \in M \cap V(x^0)). \tag{1.8}$$

By performing an affine transformation in the space of variables $(y_{k+1}, \ldots, y_n)$ we do not affect the manifold M^* given by (1.7), except for a change in the set A. At the same time, after performing such a transformation we can achieve the conditions

$$\hat{a}_{k+i,\,k+j}(0) = \begin{cases} 1 & \text{if} \quad i = j = k + 1, \ldots, k + d(x^0), \\ 0 & \text{for all other } i, j \quad (1 \leq i, j \leq n - k) \end{cases} \tag{1.9}$$

where $\hat{a}_{k+i,\,k+j}$ are the new $a_{k+i,\,k+j}^*$.

Next, by an affine transformation in the space of variables $(y_1, \ldots, y_k)$ we do not affect the manifold M^*. At the same time we can achieve the additional conditions

$$\tilde{a}_{i,j}(0) = \begin{cases} \eta & \text{if } i = j = 1, \ldots, d^* \quad (\eta > 0), \\ 0 & \text{for all other } i, j \quad (1 \leqslant i, j \leqslant k) \end{cases} \tag{1.10}$$

where η is any given positive number, d^* is the rank of the matrix $(\hat{a}_{ij}(0))_{i,j=1}^k$ and $\tilde{a}_{i,j}$ are the new $\hat{a}_{i,j}$. Notice that d^* can be any number $\geqslant 0$ and $\leqslant k$.

Notation. Let B be any set in R^n and let $x \in R^n$. The distance from x to B will be denoted by $d(x, B)$.

Let $M_V = M \cap V(x^0)$ and let W be a neighborhood of M_V. We shall be interested later in finding a function u satisfying:

$$\begin{aligned} &Lu(x) \leqslant \mu u(x) && \text{if } x \in W \setminus M_V && (\mu \text{ nonnegative constant}), \\ &u(x) \to \infty && \text{if } x \in W \setminus M_V, && d(x, M_V) \to \infty. \end{aligned} \tag{1.11}$$

Suppose after performing the transformation (1.6) and the two affine transformations used above (to get (1.9), (1.10)), we can construct a function $u'(x')$ satisfying (1.11) in the new x'-variable and with the transformed L and M. Then the function $u(x) = u'(x')$ will satisfy (1.11). Consequently, in trying to prove the existence of $u(x)$ satisfying (1.11), we may, without loss of generality, assume that M is given by

$$x_{k+1} = 0, \qquad \ldots, \qquad x_n = 0, \tag{1.12}$$

that $x^0 = 0$, and that

$$a_{k+i,k+j}(0) = \begin{cases} 1 & \text{if } i = j = 1, \ldots, d(0), \\ 0 & \text{for all other } i, j \quad (1 \leqslant i, j \leqslant n - k), \end{cases} \tag{1.13}$$

$$a_{i,j}(0) = \begin{cases} \eta & \text{if } i = j = 1, \ldots, d^* \quad (\eta > 0), \\ 0 & \text{for all other } i, j \quad (1 \leqslant i, j \leqslant k). \end{cases} \tag{1.14}$$

for some $0 \leqslant d^* \leqslant k$.

In the above arguments we have assumed the local representation (1.5). The same arguments also apply, of course, in the general case where M has a local representation of the form (1.4). We sum up:

Lemma 1.1. *In order to find a function u satisfying* (1.11), *we may assume, without loss of generality, that $x^0 = 0$, that M is given by* (1.12), *and that* (1.13), (1.14) *hold.*

From the proof of Lemma 1.1 we obtain:

Lemma 1.2. *Let p be a given positive number. In order to find a function u satisfying*

$$Lu(x) \leqslant -\frac{\mu}{(d(x, M))^p} \quad \text{if} \quad x \in W \setminus M_V \qquad (\mu \text{ positive constant}),$$

$$u(x) \to \infty \quad \text{if} \quad x \in W \setminus M_V, \quad d(x, M_V) \to 0,$$

we may assume, without loss of generality, that $x^0 = 0$, *that M is given by* (1.12) *and that* (1.13), (1.14) *hold.*

2. A fundamental lemma

A function $v(x)$ is said to be piecewise continuous in a region G of R^n if there is in G a finite number of C^1 hypersurfaces $S_1, \ldots, S_l$ and a finite number of C^1 manifolds of dimensions $\leqslant n - 2$, $V_1, \ldots, V_h$, such that:

(i) for any compact subset G_0 of G, $v(x)$ is continuous and bounded on the set $G_0 \setminus (S \cup V)$ where $S = \bigcup_{i=1}^{l} S_i$, $V = \bigcup_{i=1}^{h} V_i$; and

(ii) $v(x)$ $(x \in G \setminus (S \cup V))$ tends to a limit from either side of each S_i.

Let D be an open set in R^n. Denote by ∂D the boundary of D, and by $\bar{D}$ the closure of D. Let

$$\tau = \text{exit time of } \xi(t) \text{ from } D.$$

Let K be a compact subset of $\bar{D}$. For any $\epsilon > 0$, let

$$K_\epsilon = \{x \in D;\, d(x, K) < \epsilon\},$$
$$\hat{K}_\epsilon = K_\epsilon \setminus K.$$

Notice that K need not lie entirely in D, i.e., $K \cap \partial D$ may be nonempty. The following lemma will be fundamental for the subsequent developments.

Lemma 2.1. *Let* (A_1) *hold. Let u be a continuously differentiable function in* $\hat{K}_{\epsilon_0}$, *for some* $\epsilon_0 > 0$, *and let* $D_x^2 u$ *be piecewise continuous in* $\hat{K}_{\epsilon_0}$. *Denote by* $S_1, \ldots, S_l$ *the* $(n-1)$*-dimensional manifolds of discontinuity of* $D_x^2 u$, *and by* $V_1, \ldots, V_h$ *the manifolds of discontinuity of* $D_x^2 u$ *of dimensions* $\leqslant n - 2$. *Let* $S = \bigcup_{i=1}^{l} S_i$, $V = \bigcup_{i=1}^{h} V_i$. *Suppose*

$$Lu(x) \leqslant \mu u(x) \quad \text{if} \quad x \in \hat{K}_{\epsilon_0} \setminus (S \cup V) \quad (\mu \text{ nonnegative constant}), \tag{2.1}$$

$$u(x) \to \infty \quad \text{if} \quad x \in \hat{K}_{\epsilon_0}, \quad d(x, K) \to 0. \tag{2.2}$$

Then, for any $x \in D \setminus K$,

$$P_x\{\xi(t) \in K \text{ for some } 0 \leqslant t < \tau\} = 0. \tag{2.3}$$

This lemma was implicitly proved, by the argument following (9.4.11), in the special case where K is a point or a bounded closed domain, $D = R^n$, $\hat{K}_{\epsilon_0}$ is replaced by R^n, and u is twice continuously differentiable in $R^n \backslash K$.

Proof. Let R, ρ be positive numbers; R will be arbitrarily large and ρ arbitrarily small. Set

$$B_R = \{x; |x| < R\},$$
$$D_\rho = \{x \in D; d(x, \partial D) > \rho\}.$$

R is such that $K \subset B_R$.

Fix a number ϵ_1, $0 < \epsilon_1 < \epsilon_0$ and let

$$0 < \epsilon' < \epsilon < \epsilon_1.$$

Modify and extend u inside $K_{\epsilon'}$ and outside K_{ϵ_1} so as to obtain a function U in D satisfying:

$$\begin{aligned} &U \text{ and } D_x U \quad \text{are continuous in } D; \\ &D_x^2 U \quad \text{is piecewise continuous in } D; \\ &U \quad \text{is positive in } D. \end{aligned} \tag{2.4}$$

Since (by (2.2)) $u(x)$ is positive in some D-neighborhood of K, we can accomplish (2.4) provided ϵ_1 is sufficiently small.

Denote by Σ the set of discontinuities of $D_x^2 U$. Clearly, for any ρ, R,

$$\begin{aligned} |D_x U| + |D_x^2 U| &< C(\rho, R) \qquad \text{if } x \in (D_{\rho/2} \setminus K_{\epsilon_1}) \cap B_{R+1}, \quad x \notin \Sigma, \\ U &> c(\rho, R) \qquad \text{if } x \in (D_{\rho/2} \setminus K_{\epsilon_1}) \cap B_{R+1} \end{aligned} \tag{2.5}$$

where $C(\rho, R)$, $c(\rho, R)$ are positive constants depending on ρ, R, but independent of ϵ'. Since $U = u$ in $K_{\epsilon_1} \backslash K_{\epsilon'}$, we conclude, upon using (2.1) and (2.5), that

$$LU(x) < \mu_{\rho, R} U(x) \qquad \text{if } x \in (D_{\rho/2} \backslash K_{\epsilon'}) \cap B_{R+1}, \quad x \notin \Sigma \tag{2.6}$$

where $\mu_{\rho, R}$ is a positive constant depending on ρ, R, but independent of ϵ'.

Let $p(x)$ be a C^∞ function in R^n, with support in the unit ball $|x| < 1$, such that $p(x) > 0$, $\int_{R^n} p(x)\, dx = 1$. For any $\lambda > 0$, we introduce the mollifier $U_\lambda(x)$ of $U(x)$ defined by (cf. Problem 4, Chapter 4)

$$U_\lambda(x) = \int_{|y-x|<\lambda} U(y) p_\lambda(x-y)\, dy \qquad \left[p_\lambda(x) = \frac{1}{\lambda^n} p\left(\frac{x}{\lambda}\right) \right]. \tag{2.7}$$

We take $\lambda < \rho/2$, $\lambda < \epsilon - \epsilon'$, $x \in D_\rho$. Then $U_\lambda(x)$ is in $C^\infty(D_\rho)$, and

$$D_x U_\lambda(x) = \int_{|y-x|<\lambda} D_y U(y) \cdot p_\lambda(x-y)\, dy. \tag{2.8}$$

Also,

$$D_x^2 U_\lambda(x) = -\int_{|y-x|<\lambda} D_y U(y) \cdot D_y p_\lambda(x-y)\, dy. \tag{2.9}$$

If $d(x, \Sigma) > \lambda$, then clearly

$$D_x^2 U_\lambda(x) = \int_{|y-x|<\lambda} D_y^2 U(y) \cdot p_\lambda(x-y)\, dy. \tag{2.10}$$

Suppose next that $d(x, \Sigma) \leq \lambda$ and $\Sigma \cap \{y;\ |y-x| \leq \lambda\}$ consists of a hypersurface S_1. Then S_1 divides $\{y;\ |y-x| \leq \lambda\}$ into two sets: $S_{1\lambda}$ and $S_{2\lambda}$. Integrating by parts in (2.9) over $S_{1\lambda}$ and $S_{2\lambda}$ separately, and using the continuity of $D_y U$ across S_1, we again get (2.10).

If $\Sigma \cap \{y;\ |y-x| \leq \lambda\}$ consists of manifold V of dimension $\leq n-2$, then we surround V by an η-neighborhood V_η, and split the integral in (2.9) into a part I_1 integrated over $\{y;\ |y-x| < \lambda\} \cap V_\eta$ and a part I_2. In I_2 we integrate by parts so as to obtain

$$I_2 = \int_{W_\eta} D_y^2 U(y) \cdot p_\lambda(x-y)\, dy + O(\eta), \quad W_\eta = \{y;\ |y-x| < \lambda, y \notin V_\eta\}.$$

Taking $\eta \to 0$ in $I_1 + I_2$, (2.10) follows.

Finally, the general case where $d(x, \Sigma) \leq \lambda$ can be handled by combining the above two special cases. Thus (2.10) holds in general.

From (2.7), (2.8), and (2.10) we obtain

$$LU_\lambda(x) - \mu_{\rho,R} U_\lambda(x) = \int_{|y-x|<\lambda} [LU(y) - \mu_{\rho,R} U(y)] p_\lambda(x-y)\, dy.$$

Notice that in $LU(y)$ the coefficients of L are evaluated at the point x. Since these coefficients are Lipschitz continuous, and since U, $D_y U$, $D_y^2 U$ are bounded functions,

$$|LU(y) - (LU)(y)| \leq C|x-y| \leq C\lambda$$

where C is a constant depending on ρ, R, ϵ'. Using (2.6), we get

$$LU_\lambda(x) \leq \mu_{\rho,R} U_\lambda(x) + C\lambda \qquad \text{if} \quad x \in (D_\rho \setminus K_\epsilon) \cap B_R. \tag{2.11}$$

Let

$$\tau^0 = \tau_{\rho,R,\epsilon} = \text{exit time of } \xi(t) \text{ from } (D_\rho \setminus K_\epsilon) \cap B_R,$$

and write, for simplicity, $\mu = \mu_{\rho,R}$. By Itô's formula, if $x \in (D_\rho \setminus K_\epsilon) \cap B_R$, $T > 0$,

$$E_x\{e^{-\mu(\tau^0 \wedge T)} U_\lambda(\xi(\tau^0 \wedge T))\} - U_\lambda(x) = E_x \int_0^{\tau^0 \wedge T} e^{-\mu s}(L-\mu) U_\lambda(\xi(s))\, ds. \tag{2.12}$$

Notice that $\xi(s) \in (D_\rho \setminus K_\epsilon) \cap B_R$ if $0 \leq s < \tau^0 \wedge T$. Hence, by (2.11), the integral on the right-hand side is $\leq CT\lambda$. Taking $\lambda \to 0$ in (2.12) and using the fact that $U_\lambda(y) \to U(y)$ uniformly in $y \in (D_\rho \setminus K_\epsilon) \cap B_R$, we get

$$E_x\{e^{-\mu(\tau^0 \wedge T)} U(\xi(\tau^0 \wedge T))\} - U(x) \leq 0.$$

Since $U > 0$, this yields

$$E_x\left\{e^{-\mu(\tau^0\wedge T)}U(\xi(\tau^0 \wedge T))I_{\{\xi(\tau^0\wedge T)\in\partial K_{\epsilon,\rho}\}}\right\} \leq U(x) \tag{2.13}$$

where $\mu = \mu_{\rho,R}$, $\tau^0 = \tau_{\rho,R,\epsilon}$, $\partial K_{\epsilon,\rho} = \partial K_\epsilon \cap D_\rho$, and ∂K_ϵ is the boundary of K_ϵ.

Noting that

$$U(\xi(\tau^0 \wedge T)) \geq \inf_{\partial K_\epsilon \cap D} u(y) \qquad \text{if} \quad \xi(\tau^0 \wedge T) \in \partial K_{\epsilon,\rho},$$

and taking $T \to \infty$ in (2.13), we get

$$E_x\left\{e^{-\mu\tau^0}I_{\{\tau^0<\infty\}}I_{\{\xi(\tau^0)\in\partial K_{\epsilon,\rho}\}}\right\} \leq U(x)/\left[\inf_{y\in\partial K_\epsilon\cap D} u(y)\right]. \tag{2.14}$$

Suppose now that the assertion (2.3) is false. Then there exists a set G of positive probability such that: if $\omega \in G$, then $\xi(t,\omega) \in K$ for some finite $t = t^*(\omega) < \tau(\omega)$. This implies that for all small ϵ, say $0 < \epsilon < \epsilon^*$, $\xi(s,\omega) \in D_\rho \cap B_R$ if $0 \leq s \leq t_\epsilon$ for some small $\rho > 0$ and large R, $\xi(s,\omega) \notin K_\epsilon$ if $0 \leq s < t_\epsilon$, and $\xi(t_\epsilon,\omega) \in K_\epsilon$; here $t_\epsilon = t_\epsilon(\omega) \leq t^*(\omega)$, ρ and R are independent of ϵ (but they depend on ω) and one can take, for instance, $\epsilon^* = \epsilon_1$ where ϵ_1 is as above.

Setting $\rho_m = 1/m$, $R_m = m$,

$$G_m = G \cap \left\{\tau_{\rho_m,R_m,\epsilon} < \infty;\ \xi(\tau_{\rho_m,R_m,\epsilon}) \in \partial K_{\epsilon,\rho_m} \text{ for all } 0 < \epsilon < \epsilon^*\right\}$$

we then have $G = \bigcup_{m=1}^\infty G_m$. Since $P_x(G) > 0$, it follows that $P_x(G_m) > 0$ for some m. If we take $\rho = \rho_m$, $R = R_m$ in (2.14), and let $\epsilon \to 0$, we obtain, after using (2.2),

$$E_x\{\exp[-\mu_{\rho_m,R_m}\tau_{\rho_m,R_m,\epsilon}] \cdot I_{G_m}\} \to 0 \qquad \text{if} \quad \epsilon \to 0.$$

This implies that for almost all $\omega \in G_m$,

$$\tau_{\rho_m,R_m,\epsilon}(\omega) \to \infty \qquad \text{if} \quad \epsilon \to 0. \tag{2.15}$$

But if $\omega \in G_m$, then

$$\tau_{\rho_m,R_m,\epsilon}(\omega) \leq t^*(\omega) < \infty,$$

which contradicts (2.15), since $P_x(G_m) > 0$.

Remark. The above proof remains valid in case u is continuous in $\hat{K}_{\epsilon_0}$ and has two weak derivatives in $L^\infty(A)$ for any compact subset A of $\hat{K}_{\epsilon_0}$, (2.1) holds almost everywhere, and (2.2) holds. Indeed, the assertions (2.8), (2.10) are then valid by definition of weak derivatives, and the rest of the proof is essentially the same.

3. The case d(x) > 3

When we speak of a manifold M with boundary ∂M, it is always assumed that M is a closed set, i.e., $\partial M \subset M$.

Theorem 3.1. *Let M be a k-dimensional C^2 submanifold of R^n ($0 < k < n - 3$) with C^2 boundary ∂M (∂M may be empty), and let (A_1) hold. Suppose $d(x) > 3$ for each $x \in M$. Then (1.3) holds, i.e., M is nonattainable.*

Proof. If the assertion is not true, then for some $x \not\in M$ there is a point $x^0 \in M$ such that, for any $\delta_0 > 0$,

$$P_x\{\xi(t) \in M \cap B_{\delta_0} \text{ for some } t > 0\} > 0 \tag{3.1}$$

with B_{δ_0} is the closed ball with center x^0 and radius δ_0.

Consider first the case where $x^0 \not\in \partial M$. We want to apply Lemma 2.1 with

$$D = R^n, \qquad K = M \cap B_{\delta_0}.$$

Thus we wish to construct a function u in a δ-neighborhood W_δ of K such that

$$\begin{aligned} Lu(x) &< \mu u(x) \qquad \text{if } \ x \in W_\delta \backslash K \quad (\mu > 0), \\ u(x) &\to \infty \qquad \text{if } \ x \in W_\delta \backslash K, \quad d(x, K) \to 0. \end{aligned} \tag{3.2}$$

In view of Lemma 1.1, we may assume that $x^0 = 0$,

$$K = \{x;\, x_{k+1} = 0, \ldots, x_n = 0, (x_1, \ldots, x_k) \in A\} \tag{3.3}$$

and that the $a_{ij}(x)$ satisfy (1.13), (1.14) with a given arbitrarily small $\eta > 0$. Further, since δ_0 can be taken arbitrarily small, we may assume that A is a k-dimensional cube, say

$$A = A_\epsilon = \{(x_1, \ldots, x_k);\, -\epsilon < x_i < \epsilon \text{ for } i = 1, \ldots, k\} \tag{3.4}$$

and ϵ is sufficiently small. We shall determine later how small ϵ and η are going to be. Also δ can be taken arbitrarily small.

Set $x = (x', x'')$ where $x' = (x_1, \ldots, x_k)$, $x'' = (x_{k+1}, \ldots, x_n)$, and let

$$r = r(x) = |x''|.$$

Thus $r(x)$ is the distance from x to K provided $x' \in A_\epsilon$.

Let

$$u(x) = \phi(r) = \log \frac{1}{r} \qquad \text{if } \ x \in W_\delta \backslash K, \quad x' \in A_\epsilon. \tag{3.5}$$

Then

$$u_{x_i} = -\frac{x_i}{r^2}, \qquad u_{x_i x_j} = -\frac{\delta_{ij}}{r^2} + 2\frac{x_i x_j}{r^2}$$

if $k+1 \leq i, j \leq n$, and $u_{x_i x_j} = 0$ otherwise. Hence, if $d = d(0)$,

$$\sum_{i=1}^{d} a_{k+i,k+i}(0) \frac{\partial^2 u}{\partial x_{k+i}^2} = -\frac{d}{r^2} + 2\frac{x_{k+1}^2 + \cdots + x_{k+d}^2}{r^4} \leq -\frac{1}{r^2}$$

since $d \geq 3$. If $i = j > d$ or if $i \neq j$, $k+1 \leq i, j \leq n$, then

$$\left| a_{k+i,k+j}(x) \frac{\partial^2 u}{\partial x_{k+i}\,\partial x_{k+j}} \right| = |a_{k+i,k+j}(x) - a_{k+i,k+j}(0)| \left| \frac{\partial^2 u}{\partial x_{k+i}\,\partial x_{k+j}} \right|$$

$$\leq C|x| \frac{1}{r^2} \leq \frac{C(\delta+\epsilon)}{r^2}$$

where C is a generic constant. Also

$$\left| [a_{k+i,k+i}(x) - a_{k+i,k+i}(0)] \frac{\partial^2 u}{\partial x_{k+i}^2} \right| \leq \frac{C(\delta+\epsilon)}{r^2} \qquad \text{if} \quad 1 \leq i \leq d.$$

Noting also that $a_{ij} u_{x_i x_j} = 0$ if either $1 \leq i \leq k$ or $1 \leq j \leq k$, and that

$$|b_i u_{x_i}| \leq C|u_{x_i}| \leq C/r,$$

we conclude that

$$Lu \leq -\frac{1}{2r^2} + \frac{C(\delta+\epsilon)}{r^2} < -\frac{1}{4r^2} \qquad \text{if} \quad x \in W_\delta \setminus K \quad x' \in A_\epsilon$$

provided $\delta + \epsilon < 1/4C$.

We next extend the definition of $u(x)$ to the set of points (x', x'') in $W_\delta \setminus K$ where $x' \notin A_\epsilon$. We begin with the subset where

$$x_1 > \epsilon, \qquad -\epsilon \leq x_i \leq \epsilon \qquad \text{if} \quad 2 \leq i \leq k. \tag{3.6}$$

Let $r_1 = r_1(x) = \{(x_1 - \epsilon)^2 + |x''|^2\}^{1/2}$ if $x \in W_\delta \setminus K$, and suppose x' satisfies (3.6). Thus $r_1(x)$ is the distance from x to K. Define $u(x) = \log 1/r_1$ if $x \in W_\delta \setminus K$ and x' satisfies (3.6).

Denote by L' the operator L when $a_{11}(x)$ and the $a_{1,k+i}(x)$, $a_{k+i,1}(x)$ $(1 \leq i \leq d)$ are replaced by 0. Then, by the same calculation as before,

$$L'u(x) < -1/4r_1^2. \tag{3.7}$$

Since $a_{11}(0) = \eta$, $a_{11}(x) < \eta + C(\delta+\epsilon)$ if $x \in W_\delta$. Recalling that $a(x)$ is a nonnegative definite matrix, we also have

$$|a_{1,k+i}(x)| \leq \sqrt{a_{11}(x)}\,\sqrt{a_{k+i,k+i}(x)} \leq C(\eta + \delta + \epsilon)^{1/2} \qquad (x \in W_\delta).$$

Since

$$\left| \frac{\partial^2 u}{\partial x_i\,\partial x_j} \right| \leq \frac{3}{r_1^2},$$

we conclude that

$$|Lu - L'u| \leq \frac{6nC(\eta + \delta + \epsilon)^{1/2}}{r_1^2} \qquad (x \in W_\delta,\ x' \text{ satisfies (3.6)}).$$

Combining this with (3.7) and taking $\eta + \delta + \epsilon$ to be sufficiently small, we get

$$Lu(x) < -1/5r_1^2 \qquad \text{if} \quad x \in W_\delta \setminus K, \quad x' \text{ satisfies (3.6)}.$$

Notice that r and r_1 agree with their first derivatives on the set where $x_1 = \epsilon$. Hence the function $u(x)$ constructed so far is continuously differentiable, and $D_x^2 u$ is piecewise continuous.

Similarly we extend the definition of $u(x)$ to each of the subsets M_i, N_i $(1 \leq i \leq k)$ of $W_\delta \setminus K$ given by

$$M_i = \{x \in W_\delta \setminus K,\ x_i > \epsilon,\ -\epsilon \leq x_j \leq \epsilon \text{ if } 1 \leq j \leq k, j \neq i\},$$

$$N_i = \{x \in W_\delta \setminus K,\ x_i < -\epsilon,\ -\epsilon \leq x_j \leq \epsilon \text{ if } 1 \leq j \leq k, j \neq i\}.$$

Next we extend the definition of $u(x)$ to the subset Γ of $W_\delta \setminus K$ where $x_1 > \epsilon$, $x_2 > \epsilon$. Introducing

$$r_{12}(x) = \{(x_1 - \epsilon)^2 + (x_2 - \epsilon)^2 + |x''|^2\}^{1/2},$$

we define $u(x) = \log(1/r_{12}(x))$. Again we have (if η, δ, ϵ are sufficiently small)

$$Lu < -c/(r_{12})^2 \qquad \text{for some positive constant } c.$$

Notice that the functions r_{12}, r_1, and their first derivatives agree on the set $x_2 = \epsilon$. Similarly the functions r_{12} and $r_2 = [(x_2 - \epsilon)^2 + |x''|^2]^{1/2}$ and their first derivatives agree on the set $x_1 = \epsilon$. Hence, the function $u(x)$ constructed so far is continuously differentiable, and $D_x^2 u$ is piecewise continuous.

We extend the definition of u, in a similar manner, to the subsets of $W_\delta \setminus K$ defined by

$$x_i > \epsilon, \quad x_j > \epsilon, \qquad \text{or} \qquad x_i > \epsilon, \quad x_j < -\epsilon,$$

or

$$x_i < -\epsilon, \quad x_j > \epsilon, \qquad \text{or} \qquad x_i < -\epsilon, \quad x_j < -\epsilon$$

for some $i \neq j$, $1 \leq i, j \leq k$. Then we proceed to define $u(x)$ on sets determined by three inequalities, i.e., $x_i > \epsilon$ or $x_i < -\epsilon$, $x_j > \epsilon$ or $x_j < -\epsilon$, $x_h > \epsilon$ or $x_h < -\epsilon$; etc. The resulting function $u(x)$ is continuously differentiable in the entire set $W_\delta \setminus K$, $D_x^2 u$ is piecewise continuous, and $Lu(x) < 0$ at all the points of $W_\delta \setminus K$ where $D_x^2 u$ exists. Finally, it is clear that $u(x) \to \infty$ if $x \in W_\delta \setminus K$, $d(x, K) \to 0$.

Having constructed u that satisfies (3.2) in the special case where (3.3) and (1.13), (1.14) hold, we appeal to Lemma 1.1 in order to conclude the existence of a continuously differentiable function u, with $D_x^2 u$ piecewise continuous, which satisfies (3.2) in the general case where $K = M \cap B_{\delta_0}$. Applying Lemma 2.1, it follows that

$$P_x\{\xi(t) \in K \text{ for some } t > 0\} = 0 \qquad \text{for any} \quad x \notin K.$$

This, however, contradicts (3.1).

We have assumed so far that $x^0 \notin \partial M$. If $x^0 \in \partial M$, then the proof is similar. The set A_ϵ is simply replaced by its intersection with the half-space $x_1 \geqslant 0$.

4. The case d(x) > 2

We first consider the case where M consists of one point x^0. The number $d(x^0)$ now coincides with the rank of the matrix $a(x^0)$.

Theorem 4.1. *Let* (A_1) *hold and let* $d(x^0) \geqslant 2$. *Then*

$$P_x\{\xi(t) = x^0 \text{ for some } t > 0\} = 0 \qquad \text{for any} \quad x \neq x^0. \tag{4.1}$$

Proof. We may take $x^0 = 0$. We wish to construct a function u such that

$$Lu(x) \leqslant 0 \qquad \text{if} \quad 0 < |x| < \delta, \tag{4.2}$$

$$u(x) \to \infty \qquad \text{if} \quad |x| \to 0 \tag{4.3}$$

where δ is a sufficiently small positive number, and $u(x)$ is in C^2 for $0 < |x| < \delta$. In view of Lemma 2.1, this will complete the proof of (4.1).

Because of Lemma 1.1, we may assume, without loss of generality, that

$$\begin{aligned} a_{ii}(0) &= 1 \qquad \text{if} \quad i = 1, \dots, d \qquad (d \geqslant 2), \\ a_{ij}(0) &= 0 \qquad \text{if} \quad i = j > d \quad \text{or if} \quad i \neq j. \end{aligned} \tag{4.4}$$

We shall take $u(x) = \phi(r)$ where $r = |x|$ and where $\phi(r)$ is defined by

$$\phi'(r) = -r^{-1} e^{r^\theta/\theta}, \qquad \phi(0) = \infty \tag{4.5}$$

for some constant θ, $0 < \theta < 1$. Since (4.3) clearly holds, it remains to verify (4.2). Now

$$u_{x_i} = -\frac{x_i}{r^2} e^{r^\theta/\theta},$$

$$u_{x_i x_j} = \left[-\frac{\delta_{ij}}{r^2} + 2\frac{x_i x_j}{r^4} - \frac{x_i x_j}{r^4} r^\theta \right] e^{r^\theta/\theta}.$$

Using the fact that $d \geqslant 2$, we get

$$\sum_{i=1}^{d} \frac{\partial^2 u}{\partial x_i^2} = \left[-\frac{d}{r^2} + 2\frac{x_1^2 + \cdots + x_d^2}{r^4} - \frac{x_1^2 + \cdots + x_d^2}{r^4} r^\theta \right] e^{r^\theta/\theta}$$

$$\leqslant \left[-2\frac{x_{d+1}^2 + \cdots + x_n^2}{r^4} - \frac{x_1^2 + \cdots + x_d^2}{r^4} r^\theta \right] e^{r^\theta/\theta}$$

$$\leqslant -c\frac{r^\theta}{r^2} \qquad (c = e^{1/\theta})$$

if $r < 1$. On the other hand,

$$|[a_{ij}(x) - a_{ij}(0)]u_{x_i x_j}| \leqslant C|x|\frac{1}{r^2} \leqslant \frac{C}{r},$$

$$|b_i(x)u_{x_i}| \leqslant C|u_{x_i}| \leqslant \frac{C}{r}.$$

Recalling (4.4), we conclude that

$$Lu \leqslant -c\frac{r^\theta}{2r^2} + \frac{C}{r} < 0 \qquad \text{if} \quad 0 < r < \delta$$

and δ is sufficiently small. This completes the proof of (4.2) and thereby also the proof of Theorem 4.1.

We shall now consider the case of a general manifold M (without boundary). By Lemma 1.1, for any $x^0 \in M$ there is a suitable diffeomorphism of a neighborhood of x^0 such that in the new coordinates $W \cap M$ has the form

$$x_{k+1} = 0, \ldots, x_n = 0, \quad x_1^2 + \cdots + x_k^2 \leqslant \delta^2 \qquad (x^0 = 0) \tag{4.6}$$

where W is a neighborhood of x^0, and $a(x)$ satisfies (1.13), (1.14). Set

$$x = (x', x''), \qquad x' = (x_1, \ldots, x_k), \qquad x'' = (x_{k+1}, \ldots, x_n),$$

$$\alpha_{\lambda\mu}(x') = a_{k+\lambda,\, k+\mu}(x', 0) \qquad (1 \leqslant \lambda, \mu \leqslant n-k).$$

Denote by $\alpha(x')$ the $(n-k) \times (n-k)$ matrix $(\alpha_{ij}(x'))$. If $d(x^0) = 2$ and $n - k > 2$, we introduce the $(n-k) \times (n-k)$ symmetric matrix $\alpha_\epsilon^0(x') = (\alpha_{ij}^0(x'))$ $(\epsilon > 0)$, where

$$\alpha_{11}^0(x') = \alpha_{22}^0(x') = (1-\epsilon)\sum_{\lambda=3}^{n-k} \alpha_{\lambda\lambda}(x'), \qquad \alpha_{12}^0(x') = 0,$$

$$\alpha_{1j}^0(x') = -2\alpha_{1j}(x'), \qquad \alpha_{2j}^0(x') = -2\alpha_{2j}(x') \qquad (3 \leqslant j \leqslant n-k),$$

$$\alpha_{ii}(x') = 2 - \epsilon \quad (3 \leqslant i \leqslant n-k), \qquad \alpha_{ij}(x') = 0$$

$$(3 \leqslant i \leqslant j \leqslant n-k, \quad i \neq j).$$

We shall require the condition:

(N_{x^0}) If $d(x^0) = 2$ and $n - k > 2$, then, for some $\epsilon > 0$, the matrix $\alpha_\epsilon^0(x')$ is nonnegative definite for all $|x'|$ sufficiently small.

Definition. Let $n - k > 2$. If at each point $x^0 \in M$ where $d(x^0) = 2$ the condition (N_{x^0}) holds, then we say that the condition (N) is satisfied.

Recall that $\alpha(x')$ is nonnegative definite. Hence $\alpha_{ij}^2 \leq \alpha_{ii}\alpha_{jj}$. It follows that, for any $\epsilon' > 0$,

$$|\alpha_{ij}(x')| \leq (1 + \epsilon')\sqrt{\alpha_{jj}(x')} \qquad \text{if} \quad 1 \leq i \leq 2, \quad 3 \leq j \leq n,$$

provided $|x'|$ is sufficiently small. It is easily seen that if, for some $0 < \theta < \frac{1}{2}$,

$$|\alpha_{ij}(x')| \leq \theta\sqrt{\alpha_{jj}(x')} \qquad \text{if} \quad 1 \leq i \leq 2, \quad 3 \leq j \leq n,$$

for all $|x'|$ sufficiently small, then the matrix $\alpha_\epsilon^0(x')$ is positive definite, for some $\epsilon > 0$, provided $|x'|$ is sufficiently small; hence (N_{x^0}) follows in this case.

Theorem 4.2. *Let M be a k-dimensional C^2 submanifold of R^n $(0 \leq k \leq n - 2)$, and let* (A_1) *hold. Assume also that $a(x)$ is twice continuously differentiable in a neighborhood of M. If $d(x) \geq 2$ and if either $n - k = 2$ or* (N) *holds, then* (1.3) *is satisfied, i.e., M is nonattainable.*

Proof. Consider first the case where M is bounded. Let $x^0 \in M$ and let B_δ be a closed ball with center x^0 and radius δ. We wish to construct a function u in $B_\delta \setminus M$ such that

$$Lu(x) \leq -c(d(x, M))^{\theta - 2} \qquad \text{if} \quad x \in B_\delta \setminus M \quad (c > 0, 0 < \theta < 1), \tag{4.7}$$

$$|D_x u(x)| \leq \frac{C}{d(x, M)} \qquad \text{if} \quad x \in B_\delta \setminus M, \tag{4.8}$$

$$u(x) \to \infty \qquad \text{if} \quad x \in B_\delta \setminus M, \quad d(x, M) \to 0. \tag{4.9}$$

We first consider the case where $x^0 = 0$, $B_\delta \cap M$ is given by (4.6), and (when $n - k \geq 3$) (N_{x^0}) holds. If $d = d(0) \geq 3$, then we can construct u as in the proof of Theorem 3.1 (even with $\theta = 0$). We shall therefore consider only the case $d = 2$.

Let

$$m = n - k, \qquad x'' = (x_{k+1}, \ldots, x_n) = (y_1, \ldots, y_m)$$

and introduce the distance function

$$r(x) = \left\{ \sum_{i,j=1}^{m} b_{ij}(x')y_i y_j \right\}^{1/2}, \qquad b_{ij}(x') = b_{ji}(x'),$$

where the $b_{ij}(x')$ are still to be determined, and $b_{ij}(0) = \delta_{ij}$. Let $\phi(r)$ be the function defined by (4.5). We wish to determine the $b_{ij}(x')$ in such a way that the function

$$u(x) = \phi(r(x))$$

satisfies (4.7)–(4.9), provided δ is sufficiently small.

Clearly,

$$\frac{\partial u}{\partial y_\lambda} = -\frac{1}{r^2}\left(\sum_{i=1}^m b_{i\lambda} y_i\right) e^{r^\theta/\theta},$$

$$\frac{\partial^2 u}{\partial y_\lambda\, \partial y_\mu} = \left[-\frac{1}{r^2} b_{\lambda\mu} + \frac{2}{r^4}\left(\sum_{i=1}^m b_{i\lambda} y_i\right)\left(\sum_{j=1}^m b_{j\mu} y_j\right) - \frac{r^\theta}{r^4}\left(\sum_{i=1}^m b_{i\lambda} y_i\right)\left(\sum_{j=1}^m b_{j\mu} y_j\right)\right] e^{r^\theta/\theta}.$$

Hence

$$\sum_{\lambda,\mu=1}^m \alpha_{\lambda\mu} \frac{\partial^2 u}{\partial x_\lambda\, \partial x_\mu} = \left[-\frac{1}{r^2}\sum_{\lambda,\mu=1}^m \alpha_{\lambda\mu} b_{\lambda\mu} + \frac{2}{r^4}\sum_{\lambda,\mu=1}^m \alpha_{\lambda\mu}\left(\sum_{i=1}^m b_{i\lambda} y_i\right)\left(\sum_{j=1}^m b_{j\mu} y_j\right) - \frac{r^\theta}{r^2}\sum_{\lambda,\mu=1}^m \alpha_{\lambda\mu}\left(\sum_{i=1}^m b_{i\lambda} y_i\right)\left(\sum_{j=1}^m b_{j\mu} y_j\right)\right] e^{r^\theta/\theta}. \tag{4.10}$$

One is tempted to solve the system

$$F_{ij} \equiv b_{ij} \sum_{\lambda,\mu=1}^m \alpha_{\lambda\mu} b_{\lambda\mu} - 2\sum_{\lambda,\mu=1}^m \alpha_{\lambda\mu} b_{i\lambda} b_{j\mu} = -2(\alpha_{ij}(0) - \delta_{ij})$$

in a neighborhood of $x' = 0$, $b_{ij} = \delta_{ij}$, in the form $b_{ij} = b_{ij}(x')$. Unfortunately, the Jacobian vanishes at the point where $x' = 0$, $b_{ij} = \delta_{ij}$. We therefore proceed differently. We define

$$b_{11} = \alpha_{22},\quad b_{22} = \alpha_{11},\quad b_{12} = -\alpha_{12},\qquad b_{jj} = 1 \quad \text{if} \quad 3 \leqslant j \leqslant m,$$
$$b_{ij} = 0 \qquad \text{if} \quad 1 \leqslant i \leqslant j,\quad j \geqslant 3,\quad i \neq j.$$

Set $A = \sum_{\lambda=3}^m \alpha_{\lambda\lambda}$, in case $m \geqslant 3$. One can easily check that $F_{ij} = 0$ if $m = 2$ and $1 \leqslant i \leqslant j \leqslant 2$. If $m > 2$, then

$$F_{11} = \alpha_{22}A,\qquad F_{22} = \alpha_{11}A,\qquad F_{12} = -\alpha_{12}A,$$
$$F_{1j} = -2\alpha_{22}\alpha_{1j} + 2\alpha_{12}\alpha_{2j},\qquad F_{2j} = -2\alpha_{11}\alpha_{2j} + 2\alpha_{12}\alpha_{1j} \quad (3 \leqslant j \leqslant m),$$
$$F_{jj} = \sum_{\lambda=1}^m \alpha_{\lambda\lambda} b_{\lambda\lambda} - 2\alpha_{jj} = 2 + O(|x'|) \qquad \text{if} \quad 3 \leqslant j \leqslant m,$$
$$F_{ij} = -2\alpha_{ij} \qquad \text{if} \quad 3 \leqslant i < j \leqslant m.$$

Suppose $m \geqslant 3$. Using the condition (N_{x^0}) we find that

$$\sum_{i,j=1}^m F_{ij} y_i y_j \geqslant \theta_0(y_3^2 + \cdots + y_m^2) \qquad \text{for some } \theta_0 > 0,$$

provided δ is sufficiently small. Using this in (4.10), and noting that

$$-\sum_{\lambda,\mu=1}^{m} \alpha_{\lambda\mu}\left(\sum_{i=1}^{m} b_{i\lambda}y_i\right)\left(\sum_{j=1}^{m} b_{j\mu}y_j\right) = -y_1^2 - y_2^2 + \sum_{i,j=1}^{m} O(|x'|)y_iy_j,$$

we get

$$\sum_{\lambda,\mu=1}^{m} \alpha_{\lambda\mu}(x')\frac{\partial^2 u}{\partial x_\lambda\,\partial x_\mu}$$
$$\leqslant\left[-\theta_0\frac{y_3^2+\cdots+y_m^2}{r^4} - r^\theta\frac{y_1^2+y_2^2}{r^4} + O(|x'|)\frac{r^\theta}{r^2}\right]e^{r^\theta/\theta}$$
$$\leqslant\left[-\theta_0\frac{r^\theta|y|^2}{r^4} + O(|x'|)\frac{r^\theta}{r^2}\right]e^{r^\theta/\theta} \leqslant -\tfrac{1}{2}\theta_0\frac{r^\theta}{r^2} \tag{4.11}$$

provided δ is sufficiently small. The final inequality is valid (by obvious modifications in the proof) also when $m = 2$.

Next, if $1 \leqslant l, h \leqslant k$, $1 \leqslant i \leqslant m$,

$$\frac{\partial r}{\partial x_l} = O(r),\qquad \frac{\partial^2 r}{\partial x_l\,\partial x_h} = O(r),\qquad \frac{\partial r}{\partial x_{k+i}} = O(1),\qquad \frac{\partial^2 r}{\partial x_l\,\partial x_{k+i}} = O(1).$$

Hence,

$$\frac{\partial u}{\partial x_l} = O(1),\qquad \frac{\partial^2 u}{\partial x_l\,\partial x_h} = O(1),$$
$$\frac{\partial u}{\partial x_{k+i}} = O\left(\frac{1}{r}\right),\qquad \frac{\partial^2 u}{\partial x_l\,\partial x_{k+i}} = O\left(\frac{1}{r}\right).$$

Further,

$$\left|\left[a_{k+\lambda,\,k+\mu}(x',x'') - a_{k+\lambda,\,k+\mu}(x',0)\right]\frac{\partial^2 u}{\partial x_{k+\lambda}\,\partial x_{k+\mu}}\right| \leqslant C|x''|\frac{C}{r^2} = \frac{C}{r}.$$

From (4.11) and the subsequent estimates it follows that

$$Lu \leqslant -\tfrac{1}{2}\theta_0\frac{r^\theta}{r^2} + \frac{C}{r} \leqslant -\frac{cr^\theta}{r^2}\qquad (c>0)$$

provided δ is sufficiently small. Thus (4.7) has been established. The assertions (4.8), (4.9) obviously hold.

Having established (4.7)–(4.9) in the special coordinates where $B_\delta \cap M$ is given by (4.6) and (1.13) holds, we can now return to the original coordinates, and conclude (cf. Lemma 1.2):

For every $y \in M$, there is a ball $B(y, \delta_y)$ with center y and radius δ_y and a

C^2 function $u^y(x)$ defined in $B(y, \delta_y)\backslash M$, such that

$$Lu^y < -c(d(x, M))^{\theta-2} \quad \text{if} \quad x \in B(y, \delta_y)\backslash M \quad (c > 0), \tag{4.12}$$

$$|D_x u^y(x)| \le C/d(x, M) \quad \text{if} \quad x \in B(y, \delta_y)\backslash M, \tag{4.13}$$

$$u^y(x) \to \infty \quad \text{if} \quad x \in B(y, \delta_y)\backslash M, \quad d(x, M) \to 0. \tag{4.14}$$

Cover a small neighborhood W of M by a finite number of balls $B(y, \delta_y)$. Denote these balls by $B_i = B(y_i, \delta_{y_i})$ and the corresponding functions $u^y(x)$ by $u^i(x)$; $1 \le i \le l$.

Let $\{\zeta_i\}$ be a partition of unity subordinate to the covering $\{B_i\}$, and set

$$u_i(x) = \begin{cases} \zeta_i u^i & \text{if} \quad x \in B_i\backslash M, \\ 0 & \text{if} \quad x \notin B_i. \end{cases}$$

Since $\zeta_i = 0$ outside B_i, $u_i(x)$ is in $C^2(W\backslash M)$. Further, by (4.12), (4.13),

$$Lu_i \le \zeta_i L u^i + \frac{C}{d(x, M)} \le -\zeta_i (d(x, M))^{\theta-2} + \frac{C}{d(x, M)}$$

if $x \in B_i\backslash M$. Setting $u = \sum_{i=1}^{l} u_i$, we get

$$Lu \le -\sum_{i=1}^{l} \zeta_i (d(x, M))^{2-\theta} + \frac{C}{d(x, M)} < 0$$

if $x \in W\backslash M$ and $(d(x, M))^{1-\theta} < 1/C$, since $\sum \zeta_i = 1$ on W.

From (4.14) we also have

$$u(x) = \sum_{i=1}^{l} \zeta_i(x) u^i(x) \to \infty \quad \text{if} \quad x \in W\backslash M, \quad d(x, M) \to 0.$$

An application of Lemma 2.1 with $\Omega = R^n$, $K = M$ now yields the assertion of Theorem 4.2, in case M is a bounded set.

Consider next the case where the set M is unbounded. We modify the above construction of u. Thus, instead of a finite covering of M by balls B_i, we now use a countable (but locally finite) covering. Note that the radii of the B_i may decrease to 0 as $i \to \infty$. However, there is still a neighborhood W of M such that

$$Lu(x) < 0 \quad \text{if} \quad x \in W\backslash M,$$

$$u(x) \to 0 \quad \text{if} \quad x \in W\backslash M, \quad d(x, M) \to 0;$$

the last relation holds uniformly in x in bounded subsets. The "thickness" of $W\backslash M$ may go to zero at ∞.

Now, if the assertion (1.3) is false, then there is an event G with $P_x(G) > 0$ such that, if $\omega \in G$, $\xi(t, \omega) \in M$ for some $t = t_\omega < \infty$. Introduce the balls $B_m = \{y; |y| < m\}$, m a positive integer, and the events

$$G_m = \{\omega \in G; \xi(t, \omega) \in B_m \text{ if } 0 \le t \le t_\omega\}.$$

Clearly $G = \bigcup_{m=1}^{\infty} G_m$. Hence there is an m for which $P_x(G_m) > 0$. But this contradicts Lemma 2.1 in the case where $K = M \cap \bar{B}_m$, $\Omega = B_m$.

Corollary 4.3. *Any C^2 $(n-2)$-dimensional manifold in R^n is nonattainable by any diffusion process* (1.1) *with C^2 nondegenerating diffusion matrix, for which* (A_1) *holds.*

Remark. Let M be a manifold with boundary. Suppose that $d(x) \geqslant 3$ if $x \in \partial M$ and $d(x) \geqslant 2$ and (when $n - k \geqslant 3$) (N_x) holds for each $x \in M$. Then M is nonattainable. Indeed, if $x^0 \in \partial M$, then we can construct a function satisfying (4.12)–(4.14) by the proof of Theorem 3.1. If $x^0 \in M \setminus \partial M$, then we can construct u satisfying (4.12)–(4.14) as in the proof of Theorem 4.2. Now use partition of unity (as in the proof of Theorem 4.2) in order to complete the proof.

5. M consists of one point and d = 1

We shall consider the case where M consists of one point $x^0 = 0$ and $d = d(x^0) = 1$. We begin, for simplicity, with the case $n = 2$. Without loss of generality we may take

$$a_{11}(0, 0) > 0, \qquad a_{22}(0, 0) = 0.$$

Since $a_{22}(x, y) \geqslant 0$, we conclude that $\partial a_{22}/\partial x = 0$, $\partial a_{22}/\partial y = 0$ at the origin. Hence, if $a_{22}(x, y)$ is in C^2 in a neighborhood of the origin,

$$a_{22}(x, y) = O(r^2) \qquad \text{where} \quad r^2 = x^2 + y^2.$$

From the inequality $|a_{12}| \leqslant \sqrt{a_{11}}\sqrt{a_{22}}$ we see that $a_{12}(0, 0) = 0$. Hence, if $a_{12}(x, y)$ is continuously differentiable and $a_{22}(x, x)$ is twice continuously differentiable in a neighborhood of the origin, then

$$a(x, y) = \begin{pmatrix} A + o(1) & Mx + Ny + o(r) \\ Mx + Ny + o(r) & Bx^2 + Cxy + Dy^2 + o(r^2) \end{pmatrix}, \qquad A > 0 \tag{5.1}$$

as $r \to 0$. Since the matrix $a(x, y)$ is positive semidefinite,

$$B \geqslant 0, \qquad D \geqslant 0, \qquad M^2 \leqslant AB, \qquad C^2 \leqslant 4BD.$$

We shall assume:

$$B > 0, \tag{5.2}$$

and $|C|$, $|M|$ are "sufficiently small," so that for some $p > 1$, $q > 1$, $p' > 1$,

$q' > 1$, $p_0 > 1$, $q_0 > 1$, where

$$\frac{1}{p} + \frac{1}{q} = 1, \qquad \frac{1}{p'} + \frac{1}{q'} = 1, \qquad \frac{1}{p_0} + \frac{1}{q_0} = 1,$$

and for some $\lambda > 0$, the following inequalities hold:

$$\frac{|C|\lambda}{p} + \frac{2|M|}{p'} < B\lambda, \tag{5.3}$$

$$\frac{|C|}{q} \leqslant D, \tag{5.4}$$

$$\frac{|M|\lambda}{q'} < 2A, \tag{5.5}$$

$$\frac{4|M|\lambda}{q_0} + 2A < B\lambda \tag{5.6}$$

$$\frac{4|M|}{p_0} + B\lambda < 6A. \tag{5.7}$$

Finally, we assume:

$$\text{If} \quad D = 0, \qquad \text{then} \quad a_{22}(x, y) = Bx^2(1 + o(1)). \tag{5.8}$$

Notice that if $|M|$ is sufficiently small so that

$$\frac{4|M|}{q_0} < B, \qquad \frac{2|M|}{p_0} < 3A,$$

and

$$\alpha' \equiv \frac{2A}{B - 4|M|/q_0} < \frac{2(3A - 2|M|/p_0)}{B} \equiv \alpha'',$$

then any λ satisfying $\alpha' < \lambda < \alpha''$ also satisfies (5.6), (5.7).

Regarding the b_i, we require that $b_2(0, 0) = 0$. Hence, if $b_2(x, y)$ is continuously differentiable in a neighborhood of the origin, then

$$b_2(x, y) = c_1 x + c_2 y + o(r). \tag{5.9}$$

Theorem 5.1. *Let* (5.1)–(5.9) *hold. Then, for any* $(x, y) \neq (0, 0)$,

$$P_{(x, y)}\{\xi(t) = 0 \textit{ for some } t > 0\} = 0. \tag{5.10}$$

Proof. Let

$$R(x, y) = x^4 + \mu x^2 y^2 + \lambda y^2$$

where λ is a positive number satisfying (5.3)–(5.7), and μ is a positive constant to be determined later. We shall find a function $u = \Phi(R)$ such that, for some small $\gamma > 0$,

$$L\Phi(R) \leqslant 0 \qquad \text{if} \quad 0 < r < \gamma, \tag{5.11}$$

$$\Phi(R) \to \infty \qquad \text{if} \quad R \to \infty. \tag{5.12}$$

By Lemma 2.1, this will complete the proof of the theorem.

We can write Lu in the form

$$Lu = \alpha\Phi''(R) + \beta\Phi'(R).$$

If we show that

$$\alpha \geqslant 0, \qquad \beta \geqslant \alpha/R \tag{5.13}$$

$$\Phi''(R) + \frac{1}{R}\Phi'(R) = 0, \qquad \Phi'(R) < 0 \tag{5.14}$$

then (5.11) follows. A solution of (5.14) is given by

$$\Phi(R) = \log(1/R).$$

With this $\Phi(R)$, (5.12) is also satisfied. Thus, it remains to verify (5.13).

We shall use the following notation: if E is a constant, then $\hat{E}$ is a function of the form $E(1 + o(1))$.

Now, by direct calculation one finds that

$$\alpha = 16\hat{A}x^6 + 4\hat{B}\lambda^2x^2y^2 + 4\hat{D}\lambda^2y^4 + 4\hat{C}\lambda^2xy^3 + 8M\lambda x^4y$$

$$\beta R = (12\hat{A} + 2\hat{B}\lambda)x^6 + (12\hat{A}\lambda + 2\hat{B}\lambda^2)x^2y^2 + (2D\lambda^2 + 2\hat{A}\lambda\mu + 2c_2\lambda^2)y^4$$
$$+ 2C\lambda^2xy^3 + 2c_1\lambda^2xy^3.$$

Using the inequalities

$$|xy^3| \leqslant \frac{x^2y^2}{p} + \frac{y^4}{q}, \qquad |x^4y| \leqslant \frac{x^2y^2}{p'} + \frac{x^6}{q'}$$

and (5.3)–(5.5), we find that $\alpha \geqslant 0$ (if $D = 0$ we use also (5.8)).

In order to show that $\beta R \geqslant \alpha$, we use the inequalities

$$|xy^3| \leqslant \eta x^2y^2 + \frac{1}{4\eta}y^4, \qquad |x^4y| \leqslant \frac{x^2y^2}{p_0} + \frac{x^6}{q_0},$$

in both α and βR. We then obtain the inequality

$$\beta R - \alpha \geqslant \hat{\gamma}_1x^6 + \hat{\gamma}_2x^2y^2 + \hat{\gamma}_3y^4 \qquad (\hat{\gamma}_i = \gamma_i(1 + o(1))).$$

By (5.6), $\gamma_1 > 0$, and by (5.7), $\gamma_2 > 0$ provided η is sufficiently small. Since μ does not appear in γ_1, γ_2, and since it appears only in the additive term $2\hat{A}\lambda\mu$ of γ_3, we can choose μ so large that $\gamma_3 > 0$. It follows that $\beta R \geqslant \alpha$. We have thus completed the proof of (5.13).

Remark 1. The condition (5.2) is essential for the validity of the assertion of Theorem 5.1. Consider, for example, the system

$$d\xi_1 = dw_1, \qquad d\xi_2 = \sigma(\xi_1, \xi_2)\, dw_2$$

where $\sigma(x_1, 0) = 0$. If $(\xi_1(0), \xi_2(0)) = (\alpha, 0)$, then the solution is $\xi_1(t) = \alpha + w_1(t)$, $\xi_2(t) = 0$. Hence

$$P_{(\alpha, 0)}\{\xi(t) = 0 \text{ for some } t > 0\} = 1.$$

Remark 2. A review of the proof of (5.13) shows that we have actually proved also that $\beta \geq (1 + \delta)\alpha/R$ for some sufficiently small $\delta > 0$. Hence in the above proof we can take

$$\Phi(R) = 1/R^{\delta}.$$

Consider now the case $n \geq 2$. Without loss of generality we may assume that

$$a_{11}(0) > 0, \qquad a_{ii}(0) = 0 \qquad \text{if} \quad 2 \leq i \leq n.$$

If $a_{ii}(x)$ $(2 \leq i \leq n)$ is in C^2 in a neighborhood of 0, then $a_{ii}(x) = O(|x|^2)$. It follows that

$$a_{1i}(x) = O(|x|), \qquad a_{ij}(x) = O(|x|^2) \qquad (2 \leq i, j \leq n).$$

Setting $y_j = x_{j+1}$ $(1 \leq j \leq n-1)$, $m = n - 1$, and assuming that the a_{ij} are in C^2 in a neighborhood of the origin, we then have

$$\begin{aligned}
a_{11} &= A + o(1), \qquad A > 0,\\
a_{1j} &= M_j x_1 + \sum_{l=1}^{m} M_{jl} y_l + o(r) \qquad (2 \leq j \leq n),\\
a_{jj} &= B_j x_1^2 + \sum_{l=1}^{m} C_{jl} x_1 y_l + \sum_{l,k=1}^{m} D_{j,lk} y_l y_k + o(r^2) \qquad (2 \leq j \leq n),\\
a_{ij} &= E_{ij} x_1^2 + \sum_{l=1}^{m} E_{ij,l} x_1 y_l + \sum_{l,k=1}^{m} E_{ij,lk} y_l y_k + o(r^2) \qquad (2 \leq i, j \leq n).
\end{aligned} \tag{5.15}$$

We shall assume:

$$\sum_{j=2}^{n} B_j > 0, \tag{5.16}$$

$$\sum_{l,k,i,j} (D_{i,lk}\delta_{ij} + E_{ij,lk}) y_l y_k y_i y_j \geq c|y|^4 \qquad (c > 0), \tag{5.17}$$

$$|C_{jl}|, \quad |M_j|, \quad |E_{ij}|, \quad |E_{ij,k}| \quad \text{are sufficiently small.} \tag{5.18}$$

Notice that the left-hand side of (5.17) is always ≥ 0. In case (5.17) does not hold, we shall have to impose further restrictions:

if $c = 0$ in (5.17), then $C_{jl} = 0$, $E_{ij,k} = 0$ and the terms $o(r^2)$ occurring in a_{ij}, a_{ij} (in (5.15)) are replaced by $o(x_1^2)$. (5.19)

Theorem 5.2. *Let* (5.15), (5.16) *hold. Assume also that either* (5.17), (5.18) *hold, or* (5.19) *holds and the* $|M_j|$, $|E_{ij}|$ *are sufficiently small. Then,*

$$P_x\{\xi(t) = 0 \text{ for some } t > 0\} = 0 \quad \text{if} \quad x \neq 0. \tag{5.20}$$

The proof is similar to the proof of Theorem 5.1. We now take $u = \Phi(R)$ with Φ as before, but with

$$R(x) = x_1^4 + \mu \sum_{j=1}^{m} x_1^2 y_j^2 + \lambda \sum_{j=1}^{m} y_j^2;$$

λ is a suitable positive number and μ is sufficiently large positive number.

6. The case d(x) = 0

In Section 9.4 we have proved the following theorem:

Theorem 6.1. *Let G be a closed bounded domain in* R^n *with* C^3 *boundary M, and denote by* $\nu = (\nu_1, \ldots, \nu_n)$ *the outward normal to G at M. Let* (A_1) *hold, and assume that*

$$\sum_{i,j=1}^{n} a_{ij}\nu_i\nu_j = 0 \quad \text{on } M, \tag{6.1}$$

$$\langle b, \nu\rangle + \frac{1}{2}\sum_{i,j=1}^{n} a_{ij}\frac{\partial^2\rho}{\partial x_i\,\partial x_j} > 0 \quad \text{on } M \tag{6.2}$$

where $\rho(x) = \operatorname{dist}(x, M)$ *if* $x \not\in \operatorname{int} G$. *Then*

$$P_x\{\xi(t) \in M \text{ for some } t > 0\} = 0 \quad \text{for any} \quad x \not\in G. \tag{6.3}$$

The conditions (6.1), (6.2) are sharp; this is seen from the results in Problems 4–7.

Notice that the condition (6.1) means that $d(x) = 0$ along M. The assertion (6.3) means that M is nonattainable from the exterior of G.

Recall that when the a_{ij} belong to C^1 in a neighborhood of M, the condition (6.2) is equivalent to

$$\sum_{i=1}^{n}\left(b_i - \frac{1}{2}\sum_{j=1}^{n}\frac{\partial a_{ij}}{\partial x_j}\right)\nu_i \geq 0 \qquad \text{on } M. \tag{6.4}$$

The proof of Theorem 6.1 follows by producing a function u satisfying:

$$Lu \leq \mu u \qquad \text{in a } \hat{G}\text{-neighborhood of } M, \quad \hat{G} = R^n \backslash G, \quad \mu > 0,$$
$$u(x) \to \infty \qquad \text{if} \quad x \in \hat{G}, \quad \rho(x) \to 0.$$

Such a function is

$$u(x) = \frac{1}{(\rho(x))^{\epsilon}} \qquad \text{for any} \quad \epsilon > 0. \tag{6.5}$$

Suppose now that G is a bounded, closed, and convex domain, with piecewise C^3 boundary. Thus each point x of the boundary M lies on a finite number of C^3 $(n-1)$-dimensional submanifolds of M, say $M_{i_1}, \ldots, M_{i_s}$. Their intersection is a k-dimensional C^3 manifold through x $(k = n - s)$. Denote by N_x the $(n-k)$-dimensional space of the normals to this submanifold at x.

The function $D_y\rho(y)$ is continuous in a $\hat{G}$-neighborhood W of M. On the other hand, $D_y^2\rho(y)$ is piecewise continuous in W; denote by Σ the set of its discontinuities.

Theorem 6.1 extends to the present case provided (6.1) holds for any $x \in M$, $\nu \in N_x$, and provided (6.2) is replaced by

$$\varliminf_{y\to x} \frac{1}{\rho(y)}\left[\sum_{i=1}^{n} b_i(y)\frac{\partial}{\partial y_i}\rho(y) + \frac{1}{2}\sum_{i,j=1}^{n} a_{ij}(y)\frac{\partial^2}{\partial y_i\,\partial y_j}\rho(y)\right] > -C$$
$$(y \notin G \cup \Sigma,\ C \text{ positive constant}). \tag{6.6}$$

Notice that condition (6.1) for all $\nu \in N_x$ can be interpreted as

$$d_{M^{\perp}}(x) = 0,$$

when the notion of $d_{M^{\perp}}$ is extended in a natural way to the case of a piecewise smooth manifold.

When $\dim N_x = n$, the conditions (6.1) for all $\nu \in N_x$ and (6.6) reduce to

$$a(x) = 0, \qquad b(x) = 0.$$

Suppose next that M is a piecewise C^3 bounded submanifold in R^n, of any dimension k $(1 \leq k \leq n-1)$, with piecewise C^3 boundary ∂M. We can still extend Theorem 6.1 (taking $u(x) = 1/(d(x, M))^{\epsilon}$, $\epsilon > 0$) provided the following conditions hold:

(i) $d(x, M)$ is continuously differentiable and its second derivatives are piecewise continuous in some $\hat{M}$-neighborhood of M; $\hat{M} = R^n \backslash M$; denote by $\hat{\Sigma}$ the set of discontinuities of $D_x^2 d(x, M)$ in $\hat{M}$.

(ii) For any $x \in \operatorname{int} M$, (6.1) holds for all $\nu \in N_x$ (N_x is the space of normals to M at x), and

$$\lim_{y\to x}\frac{1}{d(y,M)}\left[\sum_{i=1}^{n}b_i(y)\frac{\partial}{\partial y_i}d(y,M)\right.$$
$$\left.+\frac{1}{2}\sum_{i,j=1}^{n}a_{ij}(y)\frac{\partial^2}{\partial y_i\,\partial y_j}d(y,M)\right]\geq -C$$
$$(y\not\in M\cup\hat{\Sigma},\ C\text{ positive constant})\tag{6.7}$$

uniformly with respect to x;

(iii) For any $x\in\partial M$, (6.1) holds for all ν normal to ∂M at x, and (6.7) holds.

7. Mixed case

Set

$$d(x)=r_{M^\perp}(x),\qquad d'(x)=r_{(\partial M)^\perp}(x).$$

We shall consider the case where $n=2$, M is an arc, and

$$d(x)=0\quad\text{if } x\in M,\qquad d'(x)=1\quad\text{if } x\in\partial M.\tag{7.1}$$

One can also consider, by the same method, other mixed cases.

The idea for handling the mixed case (7.1) is to form two functions u_1 and u_2 such that:

(i) u_1 is a function constructed for the case $d(x)=0$ (in Section 6);
(ii) u_2 is a function constructed for the case $d'(x)=1$ (in Section 5);
(iii) u_1 and u_2 fit together in a continuously differentiable manner.

For simplicity we take

$$M=\{(x_1,x_2);\ x_1=0,\ 0<x_2<\beta\}.\tag{7.2}$$

The case of a general arc M follows by first performing a local diffeomorphism, mapping the arc onto a linear segment as in (7.2).

Let Ω be a bounded closed domain lying in the half-plane $x_1\geq 0$, with boundary $\partial_1\Omega\cup\partial_2\Omega$, where

$$\partial_1\Omega=\{(x_1,x_2);\ -\alpha<x_1<\alpha,\ x_2=0\}$$

and $\partial_2\Omega$ lies in the half-plane $x_1>0$. We assume that $M\subset\Omega$.

The stochastic differential system is

$$d\xi_i=\sum_{s=1}^{2}\sigma_{is}(\xi)\,dw_s+b_i(\xi)\,dt\qquad(i=1,2).\tag{7.3}$$

Denote by τ the exit time from Ω. In view of the application for the Dirichlet problem (in Section 13.3) we are interested in the process $\xi(t)$ only as long as

$t < \tau$. Thus, we would like to prove that M is nonattainable in time $< \tau$, i.e.,

$$P_x\{\xi(t) \in M \text{ for some } t < \tau\} = 0 \qquad \text{if} \quad x \in \Omega \backslash M. \tag{7.4}$$

First we assume that (6.1), (6.4) hold with respect to both sides of M, i.e., if $a = \sigma\sigma^*$, then

$$a_{11}(0, x_2) = 0 \qquad \text{for} \quad 0 \leq x_2 \leq \beta, \tag{7.5}$$

$$2b_1(0, x_2) - \frac{\partial a_{11}(0, x_2)}{\partial x_1} - \frac{\partial a_{12}(0, x_2)}{\partial x_2} = 0 \qquad \text{if} \quad 0 \leq x_2 \leq \beta. \tag{7.6}$$

If the point $(0, \beta)$ lies on the boundary of Ω, then (7.4) follows from the proof of Theorem 6.1 (when slightly modified). Recall that we apply here Lemma 2.1 with any function

$$u(x) = \frac{c}{(x_1^2)^\epsilon} \qquad (c > 0, \epsilon > 0). \tag{7.7}$$

We shall now consider the case

$$(0, \beta) \in \operatorname{int} \Omega. \tag{7.8}$$

We shall also assume that not all the $a_{ij}(0, \beta)$ $(1 \leq i, j \leq 2)$ vanish. (If they all vanish, then (7.4) again follows from the results of Section 6.)

Assuming the a_{ij} to be in C^2 in a neighborhood of $(0, \beta)$, and recalling (7.5), we then have

$$a(x_1, x_2) = \begin{pmatrix} Bx_1^2 + Cx_1(x_2 - \beta) + D(x_2 - \beta)^2 + o(r^2) & Mx_1 + N(x_2 - \beta) + o(r) \\ Mx_1 + N(x_2 - \beta) + o(r) & A + o(1) \end{pmatrix},$$

$$A > 0, \tag{7.9}$$

where $r^2 = x_1^2 + (x_2 - \beta)^2$. We shall require (cf. (5.9)) that

$$b_1(x_1, x_2) = c_1 x_1 + c_2(x_2 - \beta) + o(r). \tag{7.10}$$

From (7.10), (7.6) it follows that $N = 0$ in (7.9). We finally require that either

$$D > 0, \qquad B > 0, \qquad |C| \text{ is sufficiently small,} \tag{7.11}$$

or

$$D > 0, \qquad B = 0, \qquad C = 0 \qquad \text{and} \qquad a_{11}(x_1, x_2) = Bx_1^2(1 + o(1)). \tag{7.12}$$

Consider the function

$$u(x) = 1/(R(x))^\delta \qquad (\delta > 0) \tag{7.13}$$

where

$$R(x) = (x_2 - \beta)^4 + \mu(x_2 - \beta)^2 x_1^2 + \lambda x_1^2.$$

By Remark 2 at the end of the proof of Theorem 5.1,

$$Lu < 0 \quad \text{if} \quad 0 < x_1^2 + (x_2 - \beta)^2 < \epsilon_0$$

for some $\epsilon_0 > 0$, provided δ is sufficiently small; here μ, λ are suitable positive constants.

Note that the function

$$d(x) = \begin{cases} R(x) & \text{if} \quad x_2 > \beta, \\ \lambda x_1^2 & \text{if} \quad x_2 < \beta \end{cases}$$

is C^1 and piecewise C^2. Recalling (7.7), (7.13), we conclude that the function $u(x) = 1/(d(x))^\delta$ is C^1 and piecewise C^2 in $\Omega \backslash M$, and

$$Lu < 0 \quad \text{for} \quad x \text{ in } (\Omega\backslash M)\text{-neighborhood of } M, \quad x_2 \neq \beta;$$

$$u(x) \to \infty \quad \text{if} \quad x \in \Omega \backslash M, \quad d(x, M) \to 0.$$

Hence, by Lemma 2.1, (7.4) holds. We sum up:

Theorem 7.1. *Let* (7.5), (7.6), (7.9), (7.10) *hold, and let* (7.11) *or* (7.12) *hold. Then* (7.4) *is satisfied.*

PROBLEMS

1. Complete the proof of Theorem 5.2.
2. Let G be a bounded domain with C^1 boundary ∂G. Denote by $\nu = (\nu_1, \ldots, \nu_n)$ the outward normal. Suppose $a_{ij} \in C^1(\overline{G})$, $b_i \in C(\overline{G})$. Let $x^0 \in \partial G$ and let V be a neighborhood of x^0. Consider a transformation $y_i = \psi_i(x_1, \ldots, x_n)$ $(1 \leq i \leq n)$ from V onto V^* which is in C^2 together with its inverse. Denote by W^* the image of $V \cap G$ and by Γ^* the image of $\Gamma = \partial G \cap V$. The outward normal at Γ^* will be denoted by $\tilde{\nu} = (\tilde{\nu}_1, \ldots, \tilde{\nu}_n)$. The operator

$$Lu = \frac{1}{2} \sum a_{ij}(x) \frac{\partial^2 u}{\partial x_i \, \partial x_j} + \sum b_i(x) \frac{\partial u}{\partial x_i}$$

is transformed into

$$\tilde{L}v = \frac{1}{2} \sum \tilde{a}_{ij}(y) \frac{\partial^2 v}{\partial y_i \, \partial y_j} + \sum \tilde{b}_i(y) \frac{\partial v}{\partial y_i} \qquad (v(y) = u(x)).$$

Denote by y^0 the image of x^0, and set $A = \sum a_{ij}\nu_i\nu_j$, $\tilde{A} = \sum \tilde{a}_{ij}\tilde{\nu}_i\tilde{\nu}_j$,

$$I = \sum_i \left(b_i - \frac{1}{2} \sum \frac{\partial a_{ij}}{\partial x_j} \right) \nu_i, \qquad \tilde{I} = \sum \left(\tilde{b}_i - \frac{1}{2} \sum \frac{\partial \tilde{a}_{ij}}{\partial x_j} \right) \tilde{\nu}_i.$$

Prove that sgn $\tilde{A}(y^0) = \text{sgn } A(x^0)$, sgn $\tilde{I}(y^0) = \text{sgn } I(x^0)$.

3. Let (A_1) hold. Let W be a bounded domain and denote by τ_W the exit time from W. Suppose there exists a function v in $C^2(\overline{W})$ such that $Lv \leq -\gamma < 0$ in W. Prove that

$$E_x\tau_W \leq \frac{1}{\gamma} \operatorname*{l.u.b.}_{y\in W} |v(y) - v(x)|.$$

4. Let (A_1) hold. Let G be a bounded domain and let $x^0 \in \partial G$. Let V be a neighborhood of x^0 and denote by τ_W the exit time from $W = V \cap G$. Suppose there exists a function $u(x)$ in $C^2(\overline{W})$ such that $u(x^0) = 0$, $u(x) > 0$ if $x \in \overline{W}\backslash\{x^0\}$, $Lu \leq -\gamma < 0$ in W (i.e., u is a barrier at x^0 with respect to the domain W; cf. Section 6.2). Prove that

$$E_x\tau_W \leq \frac{1}{\gamma} \operatorname*{l.u.b.}_{y\in W} |u(y) - u(x)|, \tag{7.14}$$

$$\lim_{\substack{x\to x^0\\ x\in G}} P_x\{|\xi(\tau_W) - x^0| < \delta\} = 1 \qquad \text{for any} \quad \delta > 0, \tag{7.15}$$

$$\lim_{\substack{x\to x^0\\ x\in G}} P_x\{\tau < \infty, |\xi(\tau) - x^0| < \delta\} = 1 \qquad \text{for any} \quad \delta > 0. \tag{7.16}$$

5. Let (A_1) hold and let G, x^0, y^0, W, W^*, L, $\tilde{L}$ be as in Problem 2. If there exists a function $v(y)$ in $C^2(\overline{W^*})$ such that $\tilde{L}v \leq -\gamma < 0$ in W^*, $v(y^0) = 0$, $v(y) > 0$ if $y \in \overline{W^*}\backslash\{y^0\}$, then the assertions (7.14)–(7.16) hold.

6. Let (A_1) hold and let G be a bounded domain with C^2 boundary. Let $x^0 \in \partial G$, $V_\mu = \{x;\ |x - x^0| < \mu\}$, $W_\mu = V_\mu \cap G$ and denote by τ_μ the exit time from W_μ. If $\sum a_{ij}\nu_i\nu_j > 0$ at x^0, then, for any μ sufficiently small,

$$E_x\tau_\mu \leq C\mu \qquad \text{if} \quad x \in W_\mu \tag{7.17}$$

where C is a constant independent of μ, and (7.15), (7.16) hold with $\tau_W = \tau_\mu$. [*Hint*: If $x_n = \phi(x_1, \ldots, x_{n-1})$ is a representation of $\partial G \cap V_\mu$, perform a transformation $y_i = x_i$ $(1 \leq i \leq n-1)$, $y_n = x_n - \phi(x_1, \ldots, x_n)$ and take $v(y) = y_n(\epsilon - y_n) + \epsilon\sum_{i=1}^{n-1}(y_i - y_i^0)^2$.]

7. Suppose in the preceding problem $\sum a_{ij}\nu_i\nu_j = 0$ on $\partial G \cap V_{\mu_0}$, $a_{ij} \in C^1(\overline{W}_{\mu_0})$ for some $\mu_0 > 0$, and

$$\sum_i \left(b_i - \frac{1}{2}\sum_j \frac{\partial a_{ij}}{\partial x_j}\right)\nu_i > 0 \qquad \text{at} \quad x^0.$$

Prove that (7.15)–(7.17) hold with $\tau_W = \tau_\mu$, μ small. [*Hint*: Show that $\tilde{a}_{nn} = 0$, $\tilde{b}_n < 0$ and take $v(y) = y_n + \epsilon\sum_{i=1}^{n-1}(y_i - y_i^0)^2$.]

8. The assertion of the preceding problem remains true if one assumes that $\sum a_{ij}\nu_i\nu_j$ only vanishes on an open subset S of $\partial G \cap V_{\mu_0}$, and $x^0 \in \bar{S}$.

9. Let (A_1) hold and let G be a bounded domain with C^3 boundary. Let

$x^0 \in \partial G$. Denote by $\rho(x)$ the distance from x to ∂G, if $x \in \bar{G}$. If

$$b \cdot \nu + \frac{1}{2} \sum a_{ij} \frac{\partial^2 \rho}{\partial x_i \, \partial x_j} > 0 \quad \text{at} \quad x^0 \tag{7.18}$$

then the assertions (7.15)–(7.17) hold with $\tau_W = \tau_\mu$, μ small. [*Hint*: Let $x = g(s)$ be a representation of $\partial G \cap V_\mu$ with $g(0) = x^0$. For any $x \in \overline{W}_\mu$ let $g(s(x))$ be the nearest point to x on ∂G. Take $u(x) = \rho(x) + \epsilon |g(s(x)) - x^0|^2$.]

10. Prove the assertion of Problem 6 in case ∂G is in C^3, without resorting to a transformation of coordinates. [*Hint*: Take $u(x) = \rho(x)[\epsilon - \rho(x)] + \epsilon |g(s(x)) - x^0|^2$.]

11. Prove the assertion of Problem 7 in case ∂G is in C^3, without resorting to a change of coordinates.

12. Let K be a compact nonattainable set, and let U be an open set containing K. Denote by τ the exit time from U. Let $u \in C^2(V \backslash K)$ where V is an open set containing $\bar{U}$. Prove Itô's formula

$$u(\xi(\tau \wedge t)) - u(x) = \int_0^{\tau \wedge t} u_x(\xi(s)) \cdot \sigma(\xi(s)) \, dw(s) + \int_0^{\tau \wedge t} (Lu)(\xi(s)) \, ds$$

where $\xi(0) \in U \backslash K$. [*Hint*: Let τ_ϵ be the exit time from $U \backslash K_\epsilon$ where $K_\epsilon = \{x;\ \text{dist}(x, K) < \epsilon\}$. The above formula holds for $\tau_\epsilon \wedge t$. Take $\epsilon \downarrow 0$.]

12

Stability and Spiraling of Solutions

1. Criterion for stability

We denote by $d(x, A)$ the distance from a point x to a set A.

We consider a system of n stochastic differential equations

$$d\xi(t) = \sigma(\xi(t))\, dw(t) + b(\xi(t))\, dt \tag{1.1}$$

and assume, throughout this chapter, that the condition (A_1) of Section 10.1 holds.

Let

$$Lu \equiv \frac{1}{2} \sum_{i,j=1}^{n} a_{ij}(x) \frac{\partial^2 u}{\partial x_i\, \partial x_j} + \sum_{i=1}^{n} b_i(x) \frac{\partial u}{\partial x_i}$$

where $a_{ij} = \sum_{k=1}^{n} \sigma_{ik}\sigma_{jk}$.

Definition. A closed set K is said to be *invariant* with respect to the process defined by (1.1) if

$$P_x\{\xi(t) \in K \text{ for all } t > 0\} = 1 \qquad \text{for all} \quad x \in K;$$

i.e., solutions beginning on K never leave K.

Definition. A nonattainable closed set K is said to be *stable* if for any neighborhood U of K and for any $\epsilon > 0$ there exists a neighborhood U_ϵ of K such that

$$P_x\{\xi(t) \in U \text{ for all } t > 0\} > 1 - \epsilon \qquad \text{for any} \quad x \in U_\epsilon \setminus K.$$

If for any neighborhood U of K and for any $\epsilon > 0$ there exists a neighborhood U_ϵ of K such that

$$P_x\{\xi(t) \in U \ \text{ for all } \ t > 0,\ \lim_{t\to\infty} d(\xi(t), K) = 0\} > 1 - \epsilon \quad \text{for any } x \in U_\epsilon \setminus K,$$

then we say that K is *asymptotically stable*.

Let K be a closed set. Let K' be one of the open connected components of $R^n\backslash K$. Suppose K is nonattainable from the set K', i.e.,

$$P_x\{\xi(t)\in K' \quad \text{for all} \quad t>0\}=1 \qquad \text{if} \quad x\in K'.$$

Then we define the concepts K *stable from* K' and K *asymptotically stable from* K' by replacing in the previous definitions U and U_ϵ by $U\cap K'$, $U_\epsilon\cap K'$. If, in particular, K is the boundary of a bounded domain D with connected boundary, then we can speak of K being *stable* (or asymptotically stable) *from the inside* (i.e., from $K'=D$) or *from the outside* (i.e., from $K'=R^n\backslash\bar{D}$).

It is easily seen (see Problem 2) that if K is stable then K is also an invariant set. If the boundary K of a domain D is stable from the outside, then $\bar{D}$ is an invariant set (see Problem 3).

Definition. Let K be a compact set. Let $v(x)$ be a function in $C^2(U\backslash K)$, where U is some neighborhood of K, satisfying:

$$Lv(x)\leqslant 0 \qquad \text{if} \quad x\in U\backslash K, \tag{1.2}$$

$$\sum_{i,j=1}^{n} a_{ij}(x)\frac{\partial v}{\partial x_i}\frac{\partial v}{\partial x_j}\leqslant C \qquad \text{if} \quad x\in U\setminus K \quad (C \text{ const}), \tag{1.3}$$

$$v(x)>0 \qquad \text{if} \quad x\in U\backslash K, \tag{1.4}$$

$$v(x)\to 0 \qquad \text{if} \quad x\in U\backslash K, \quad d(x,K)\to 0. \tag{1.5}$$

Then we say that $v(x)$ is a *Liapunov function for* K. If the second derivatives of $v(x)$ are only piecewise continuous (and (1.2) is satisfied at the points where the second derivatives exist), then we call $v(x)$ a *piecewise smooth Liapunov function for* K.

Definition. Let K be a compact set. Let $u(x)$ be a function in $C^2(U\backslash K)$, where U is some neighborhood of K, satisfying:

$$Lu(x)\leqslant -1 \qquad \text{if} \quad x\in U\backslash K, \tag{1.6}$$

$$\sum_{i,j=1}^{n} a_{ij}(x)\frac{\partial u}{\partial x_i}\frac{\partial u}{\partial x_j}\leqslant C \qquad \text{if} \quad x\in U\setminus K \quad (C \text{ const}), \tag{1.7}$$

$$u(x)\to -\infty \qquad \text{if} \quad x\in U\backslash K, \quad d(x,K)\to 0. \tag{1.8}$$

Then we call $u(x)$ an S-*function for* K. If the second derivatives of $u(x)$ are only piecewise continuous, then we call $u(x)$ a *piecewise smooth* S-*function for* K.

Let K be a compact set and let K' be one of the open components of $R^n\backslash K$. If (1.2)–(1.5) hold with U replaced by $U\cap K'$, then we speak of *Liapunov function for* K *from* K'. In particular, if K is the boundary ∂G of a bounded domain G with connected boundary, then we speak of a Liapunov function for ∂G from the *outside* if $K'=R^n\backslash\bar{G}$, and from the *inside* if

$K' = G$. Similarly, one defines an S-function for K from K', from the outside, and from the inside.

If u is an S-function, then $v = e^{\lambda u}$ is a Liapunov function provided λ is a sufficiently small positive constant. Indeed, this follows from the identity

$$Le^{\lambda u} = e^{\lambda u}\left\{\lambda Lu + \frac{\lambda^2}{2}\sum a_{ij}\frac{\partial u}{\partial x_i}\frac{\partial u}{\partial x_j}\right\}.$$

This identity also shows that if v is a Liapunov function and if

$$\sum a_{ij}(x)\frac{\partial v}{\partial x_i}\frac{\partial v}{\partial x_j} > \gamma v^2 \qquad (\gamma \text{ positive constant}),$$

then $u = (\log v)/\lambda$ is an S-function provided λ is a sufficiently large positive constant.

Theorem 1.1. *Let K be a compact set and let K' be an open connected component of $R^n \setminus K$. Suppose K is nonattainable from K'. If there exists a piecewise smooth Liapunov function for K from K', then K is stable from K'.*

Proof. For simplicity we take $K' = R^n \setminus K$. Suppose first that there exists a smooth Liapunov function $v(x)$ satisfying (1.2)–(1.5). Let U' be any neighborhood of K contained in U. Denote by τ the exit time for U'. By Itô's formula (see Problem 12, Chapter 11),

$$v(\xi(\tau \wedge T)) = v(x) + \int_0^{\tau\wedge T} v_x \cdot \sigma\, dw + \int_0^{\tau\wedge T} Lv\, ds$$

if $\xi(0) = x \in U' \setminus K$. Here we use the fact that K is nonattainable, so that $\xi(s) \in U' \setminus K$ for all $s < \tau$.

Since

$$|v_x \cdot \sigma|^2 = \sum a_{ij} v_{x_i} v_{x_j} \leq C$$

the expectation of the stochastic integral vanishes. Using also (1.2) we get

$$Ev(\xi(\tau \wedge T)) \leq v(x). \tag{1.9}$$

Taking $T \uparrow \infty$ and using (1.4) we obtain

$$\left[\inf_{y\in\partial U'} v(y)\right] P_x(\tau < \infty) \leq v(x).$$

By (1.5), $v(x) < \epsilon\{\inf_{\partial U'} v\}$ if $d(x, K) < \delta$. Thus

$$P_x(\tau < \infty) \leq \epsilon \qquad \text{if} \quad d(x, K) < \delta,$$

and, consequently, K is stable.

In the above proof we have assumed that v is in $C^2(U \setminus K)$. Suppose now that v is only piecewise smooth. We use mollifiers v_λ as in the proof of

Lemma 11.2.1. By Itô's formula

$$v_\lambda(\xi(\tau_m \wedge T)) = v_\lambda(x) + \int_0^{\tau_m\wedge T} D_x v_\lambda \cdot \sigma\, dw + \int_0^{\tau_m\wedge T} L v_\lambda\, ds$$

where τ_m is the exit time from W_m; here the W_m are open sets satisfying: $W_m \subset W_{m+1}$, $\bigcup_m W_m = U' \setminus K$. Since $Lv_\lambda \leq C\lambda$ in W_m if λ is sufficiently small, where C is a positive constant (cf. the derivation of (11.2.11)),

$$v_\lambda(\xi(\tau_m \wedge T)) \leq v_\lambda(x) + \int_0^{\tau_m\wedge T} D_x v_\lambda \cdot \sigma\, dw + C\lambda.$$

Hence

$$E v_\lambda(\xi(\tau_m \wedge T)) \leq v_\lambda(x) + C\lambda.$$

Taking first $\lambda \downarrow 0$ and then $m \uparrow \infty$, the inequality (1.9) follows. Now proceed as before.

Theorem 1.2. *Let K be a compact set and let K' be an open connected component of $R^n \setminus K$. Suppose K is nonattainable from K'. If there exists an S-function for K from K', then K is asymptotically stable from K'.*

Proof. For simplicity, we take $K' = R^n \setminus K$. Let $u(x)$ be an S-function. Since $v = e^{\lambda u}$ is a Liapunov function (if λ is positive and small), K is stable by Theorem 1.1. To prove asymptotic stability, suppose first that u is in $C^2(U \setminus K)$. Let $\tilde{U}$ be a neighborhood of K whose closure is contained in U. Then we can construct a function $\tilde{u}$ in $C^2(R^n \setminus K)$ that coincides with u on $\tilde{U} \setminus K$, such that $\tilde{u}(x)$ vanishes if $|x|$ is sufficiently large. Using (1.7) we conclude that

$$|D_x\tilde{u} \cdot \sigma|^2 = \sum a_{ij}\tilde{u}_{x_i}\tilde{u}_{x_j} \leq \tilde{C} \qquad (\tilde{C} \text{ const}) \tag{1.10}$$

for all $x \in R^n \setminus K$.

By Itô's formula,

$$\tilde{u}(\xi(t)) = \tilde{u}(x) + \int_0^t \tilde{u}_x \cdot \sigma\, dw + \int_0^t L\tilde{u}\, ds. \tag{1.11}$$

In view of (1.10), Corollary 4.4.6 gives

$$\int_0^t \tilde{u}_x \cdot \sigma\, dw = o(t). \tag{1.12}$$

Let U' be any neighborhood of K contained in $\tilde{U}$. Since K is stable, for any $\epsilon > 0$ there is a neighborhood U_ϵ of K such that

$$P_x\{\xi(t) \in U' \setminus K \quad \text{for all} \quad t > 0\} > 1 - \epsilon \qquad \text{if} \quad x \in U_\epsilon \setminus K. \tag{1.13}$$

Hence, by (1.6),

$$P_x\{Lu(\xi(t)) \leq -1 \qquad \text{for all} \quad t > 0\} > 1 - \epsilon.$$

Using this and (1.12) in (1.11), we get

$$P_x\left\{\xi(t)\in U'\backslash K \quad \text{for all} \quad t > 0,\ \overline{\lim_{t\to\infty}}\ \frac{u(\xi(t))}{t} < -1\right\} > 1-\epsilon. \tag{1.14}$$

In view of (1.8) we conclude that

$$P_x\{\xi(t)\in U'\backslash K \quad \text{for all} \quad t > 0,\ d(\xi(t),K)\to 0 \quad \text{if} \quad t\to\infty\} > 1-\epsilon.$$

Thus K is asymptotically stable.

We has assumed in the above proof that u is in C^2. Suppose now that u is only piecewise smooth. Let K_m be $(1/m)$-neighborhood of K, and $V_m = R^n\backslash K_m$. Introducing mollifiers $\tilde{u}_\lambda$ of u_λ, we have, by Itô's formula

$$\tilde{u}_\lambda(\xi(t\wedge\tau_m)) = \tilde{u}(x) + \int_0^{t\wedge\tau_m} D_x\tilde{u}_\lambda\cdot\sigma\,dw + \int_0^{t\wedge\tau_m} L\tilde{u}_\lambda\,ds,$$

where τ_m is the exit time from K_m and $\lambda < 1/m$. Denote by Ω_x the set occurring in (1.13), i.e.,

$$\Omega_x = \{\xi(t)\in U'\backslash K \quad \text{for all} \quad t > 0\}.$$

If $\omega\in\Omega_x$, $L\tilde{u}_\lambda(\xi(s)) < -1 + C\lambda$ if $0 < s < \tau_m(\omega)$, $x\in U_\epsilon\backslash K$, where C is a positive constant. Hence

$$\tilde{u}_\lambda(\xi(t\wedge\tau_m)) < \tilde{u}(x) + \int_0^{t\wedge\tau_m} D_x\tilde{u}_\lambda\cdot\sigma\,dw - (1-C\lambda)t \quad \text{on} \quad \Omega_x.$$

Taking $\lambda\downarrow 0$ we get

$$\tilde{u}(\xi(t\wedge\tau_m)) < \tilde{u}(x) + \int_0^{t\wedge\tau_m} D_x\tilde{u}\cdot\sigma\,dw - t \qquad \text{on} \quad \Omega_x. \tag{1.15}$$

Let

$$\chi_m(s) = \begin{cases} 1 & \text{if} \quad s < \tau_m, \\ 0 & \text{if} \quad s > \tau_m. \end{cases}$$

Since K is nonattainable, $\lim \tau_m = \infty$ a.s., so that

$$\max_{0<s<t} |\chi_m(s) - 1| \to 0 \qquad \text{if} \quad m\to\infty, \quad \text{a.s.}$$

Hence, for any $f\in L_w^2[0,t]$,

$$\int_0^t |\chi_m f - f|^2\,ds \to 0 \quad \text{a.s.}$$

It follows (using Lemma 4.4.1) that

$$\int_0^{t\wedge\tau_m} f\,dw = \int_0^t \chi_m f\,dw \xrightarrow{P} \int_0^t f\,dw.$$

Applying this to $f = D_x\tilde{u}\cdot\sigma$ we obtain, after taking $m\to\infty$ in (1.15),

$$\tilde{u}(\xi(t)) < \tilde{u}(x) + \int_0^t D_x\tilde{u}\cdot\sigma\,dw - t \quad \text{a.e. on} \quad \Omega_x. \tag{1.16}$$

We now use (1.12) in order to derive from (1.16) the inequality (1.14). This completes the proof of the theorem.

Definition. An asymptotically stable set K is said to be *globally asymptotically stable* if

$$P_x\{\lim_{t\to\infty} d(\xi(t), K) = 0\} = 1 \qquad \text{for any} \quad x \in R^n \setminus K. \tag{1.17}$$

Let K be a closed set and let K' be one of its open connected components. If K is asymptotically stable from K' and if (1.17) holds for any $x \in K'$, then we say that K is *globally asymptotically stable from* K'. If, in particular, K is the boundary of a bounded domain G with connected boundary and $K' = R^n \setminus \overline{G}$ (or $K' = G$), then we say that K is *globally asymptotically stable from the outside* (or *from the inside*).

Definition. Let K be a compact set. Let ϕ be a function in $C^2(R^n \setminus K)$ satisfying:

$$L\phi(x) < 0 \qquad \text{if} \quad x \in R^n \setminus K, \tag{1.18}$$

$$\phi(x) \to \infty \qquad \text{if} \quad x \in R^n \setminus K, \quad |x| \to \infty. \tag{1.19}$$

Then we call $\phi(x)$ a *G-function for K*. If the second derivatives of $\phi(x)$ are piecewise continuous and their set of discontinuities is bounded, then we call $\phi(x)$ a *piecewise smooth G-function for K*.

Let K be a compact set and let K' be one of its open connected components. If (1.18), (1.19) hold for all $x \in K'$, then we call ϕ a (piecewise smooth) *G-function for K from K'*. When K is the boundary of a bounded domain G with connected boundary and K' is $R^n \setminus \overline{G}$ (or G), then we call ϕ a *G-function for K from the outside* (or *from the inside*). When K' is a bounded set, the condition (1.19) is dropped out.

Theorem 1.3. *Let K be a compact set and let K' be an open connected component of $R^n \setminus K$. Suppose K is nonattainable from K'. If there exist piecewise smooth* S-*function and* G-*function for K from K', then K is globally asymptotically stable from K'.*

For simplicity we give the proof in case $K' = R^n \setminus K$. First we establish a lemma.

Lemma 1.4. *Under the conditions of Theorem* 1.3 (*with* $K' = R^n \setminus K$), *for any neighborhood U of K and for any $x \in R^n \setminus K$,*

$$P_x\{\xi(t) \in U \quad \textit{for some} \quad t > 0\} = 1.$$

Proof. Let ϕ be a G-function. For any bounded domain D with $\overline{D} \cap K$

$= \varnothing$, denote by τ^* the exit time from D. If ϕ is in C^2, then, by Itô's formula,

$$E_x\phi(\xi(\tau^* \wedge t)) - \phi(x) = E_x \int_0^{\tau^* \wedge t} L\phi \, ds \leq -\gamma E_x(\tau^* \wedge t)$$

if $x \in D$, where $L\phi(y) \leq -\gamma < 0$ if $y \in D$. Hence

$$\gamma E_x(\tau^* \wedge t) \leq 2\Big[\operatorname*{l.u.b.}_D\Big]|\phi|.$$

Taking $t \uparrow \infty$ we get

$$E_x\tau^* < \frac{2}{\gamma}\Big[\operatorname*{l.u.b.}_D |\phi|\Big] \qquad (x \in D). \tag{1.20}$$

If ϕ is piecewise smooth, we use a mollifier ϕ_λ of ϕ in order to derive (1.20).

Now take $D = B_R \setminus \overline{U}$ where $B_R = \{x;\ |x| < R\}$ and R is sufficiently large. Denote the exit time from D by τ_R. By (1.20)

$$P_x\{\tau_R < \infty\} = 1 \qquad \text{if} \quad x \in B_R \setminus \overline{U}. \tag{1.21}$$

We shall now employ the argument used in the proof of Theorem 9.2.1. If ϕ is in C^2, then, by Itô's formula,

$$E_x\phi(\xi(\tau_R \wedge t)) - \phi(x) = E_x \int_0^{\tau_R \wedge t} L\phi(\xi(s)) \, ds \leq 0.$$

Thus

$$E_x\phi(\xi(\tau_R \wedge t)) \leq \phi(x). \tag{1.22}$$

If ϕ is piecewise smooth, then

$$E_x\phi_\lambda(\xi(\tau_R \wedge t)) \leq \phi_\lambda(x) + C\lambda \qquad (C \text{ const})$$

for a mollifier ϕ_λ of ϕ. Taking $\lambda \downarrow 0$ we obtain (1.22).

Taking $t \uparrow \infty$ in (1.22) and using (1.21) we get

$$E_x\phi(\xi(\tau_R))\chi_{\{\xi(\tau_R) \in \partial U\}} + E_x\phi(\xi(\tau_R))\chi_{\{|\xi(\tau_R)| = R\}} \leq \phi(x).$$

As $R \uparrow \infty$, $\min_{|y| = R} \phi(y) \to \infty$. Hence $P_x\{|\xi(\tau_R)| = R\} \downarrow 0$. Hence, by (1.21), $P_x\{\xi(\tau_R) \in \partial U\} \uparrow 1$. This yields the assertion of the lemma.

Completion of the proof of Theorem 1.3. Since there exists an S-function, K is asymptotically stable. Hence for any neighborhood U of K and for any $\epsilon > 0$, there exists a neighborhood U_ϵ of K such that

$$P_x\{\xi(t) \in U \quad \text{for all} \quad t > 0\} \geq 1 - \epsilon \qquad \text{if} \quad x \in \overline{U}_\epsilon. \tag{1.23}$$

Denote by τ_1 the first time $\xi(t)$ hits $\overline{U}_\epsilon$. Denote by σ_1 the first time $> \tau_1$ that $\xi(t)$ exists from U (if such a time exists). Generally, denote by τ_m the first time $> \sigma_{m-1}$ that $\xi(t)$ hits $\overline{U}_\epsilon$, and denote by σ_m the first time $> \tau_m$ that $\xi(t)$ exits U (if such a time exists).

By Lemma 1.4, $\tau_1 < \infty$ a.s. By the strong Markov property,

$$P_x(\sigma_1 < \infty) = E_x\chi_{\tau_1<\infty}\chi_{\sigma_1<\infty} = E_x\chi_{\tau_1<\infty}P_{\xi(\tau_1)}\{\xi(t) \text{ exits } U\} \leq \epsilon$$

by (1.23). Next, the event $\{\sigma_1 < \infty\}$ coincides with the event $\{\tau_2 < \infty\}$. Indeed, by the strong Markov property,

$$\begin{aligned} P_x\{\tau_2 < \infty, \sigma_1 < \infty\} &= E_x\chi_{\sigma_1<\infty}E_{\xi(\sigma_1)}\{\xi(t) \text{ hits } \overline{U}_\epsilon\} \\ &= E_x\chi_{\sigma_1<\infty} = P_x\{\sigma_1 < \infty\} \end{aligned}$$

where Lemma 1.4 has been used.

Proceeding by induction, we have

$$\begin{aligned} P_x\{\sigma_m < \infty\} &= E_x\chi_{\sigma_m<\infty}\chi_{\tau_m<\infty}\chi_{\sigma_{m-1}} < \infty \\ &= E_x\chi_{\sigma_{m-1}<\infty}\chi_{\tau_m<\infty}P_{\xi(\tau_m)}\{\xi(t) \text{ exits } U\} \\ &\leq \epsilon E_x\chi_{\sigma_{m-1}<\infty}\chi_{\tau_m<\infty} = \epsilon P\{\sigma_{m-1} < \infty\} \leq \epsilon^m, \end{aligned}$$

and

$$\begin{aligned} P_x\{\tau_{m+1} < \infty, \sigma_m < \infty\} &= E_x\chi_{\sigma_m<\infty}P_{\xi(\sigma_m)}\{\xi(t) \text{ hits } \overline{U}_\epsilon\} \\ &= E_x\chi_{\sigma_m<\infty} = P_x\{\sigma_m < \infty\}. \end{aligned}$$

The event $\sigma_{m+1} < \infty$ is well defined when the event $\sigma_m < \infty$ is already defined. We have thus defined, by induction, the events $\sigma_m < \infty$ and established the inequalities

$$P_x\{\sigma_m < \infty\} < 1/\epsilon^m.$$

Taking $\epsilon = \frac{1}{2}$ and using the Borel–Cantelli lemma, we deduce that $P_x\{\sigma_m < \infty \text{ i.o.}\} = 0$. Thus, for a.a. ω there is $m = m(\omega)$ such that $\sigma_m < \infty$, $\sigma_{m+1} = \infty$. By what we have proved above, $\tau_{m+1} < \infty$. Hence $\xi(t) \in U$ if $t > T(\omega)$, $T(\omega) = \tau_{m+1} < \infty$.

Suppose now that there exists a C^2 S-function $u(x)$ for K. Extend it into a C^2 function in $R^n \setminus K$ with bounded support. Denoting this new function again by u, we have, by Itô's formula,

$$u(\xi(t)) = u(x) + \int_0^t u_x \cdot \sigma\, dw + \int_0^t Lu(\xi(s))\, ds.$$

If we take in the above analysis U to be a neighborhood of K for which $Lu(y) \leq -1$ if $y \in U \setminus K$, then we have

$$Lu(\xi(s)) \leq -1 \quad \text{if} \quad s > T(\omega).$$

Using also (1.12) with $\tilde{u} = u$, we conclude that

$$P_x\left\{\xi(t) \in U \quad \text{if} \quad t > T(\omega);\ \overline{\lim_{t\to\infty}}\ \frac{u(\xi(t))}{t} \leq -1\right\} = 1.$$

This gives the assertion

$$P_x\{d(\xi(t), K) \to 0 \quad \text{if} \quad t \to \infty\} = 1.$$

We have assumed so far that the S-function u is in C^2. If u is only piecewise smooth, we use mollifiers u_λ as in the proof of Theorem 1.2 and obtain the inequality (cf. (1.16))

$$u(\xi(t)) < u(x) + \int_0^t u_x \cdot \sigma\, dw + M - [t - T(\omega)] \tag{1.24}$$

where M is a random variable. (For each ω, $M = \text{l.u.b.}\ |Lu(y)|$ where y varies in the set $\xi(s)$, $0 < s < T(\omega)$.) Using (1.24) we can now complete the proof of the theorem as before.

Remark. From the proof of Theorem 1.3 we see that the theorem remains true if instead of assuming that a G-function $\phi(x)$ exists for K from K' we assume that, for any neighborhood V of K, there exists a function ϕ (depending on V) satisfying (1.18), (1.19) for x in $K' \setminus \overline{V}$.

2. Stable obstacles

Let G be a closed bounded domain with C^3 connected boundary ∂G, and let $\hat{G} = R^n \setminus G$. Denote by $\nu = (\nu_1, \ldots, \nu_n)$ the outward normal to ∂G. By Theorem 9.4.1, if

$$\sum_{i,j=1}^n a_{ij}\nu_i\nu_j = 0 \qquad \text{on } \partial G, \tag{2.1}$$

$$\sum_{i=1}^n b_i\nu_i + \frac{1}{2}\sum_{i,j=1}^n a_{ij}\frac{\partial^2\rho}{\partial x_i\, \partial x_j} > 0 \qquad \text{on } \partial G \tag{2.2}$$

then G is nonattainable from the outside. Here $\rho(x) = d(x, G)$ is a function defined in $\hat{G} \cup \partial G$; it belongs to C^2 in a small $(\hat{G} \cup \partial G)$-neighborhood of ∂G.

We now replace (2.2) by

$$\sum_{i=1}^n b_i\nu_i + \frac{1}{2}\sum_{i,j=1}^n a_{ij}\frac{\partial^2\rho}{\partial x_i\, \partial x_j} < 0 \qquad \text{on } \partial G. \tag{2.3}$$

Theorem 2.1. *If* (2.1), (2.3) *hold, then G is an invariant set.*

Proof. Let $R(x)$ be a C^2 function in $R^n \setminus \partial G$ satisfying $R(x) = \rho(x)$ if $x \in \hat{G}$ and $\rho(x)$ is sufficiently small, $R(x) = 0$ if $x \in G$, $R(x) \neq 0$ if $x \notin G$, and $R(x) = \text{const}$ if $|x|$ is sufficiently large. If $R^2(x)$ were in C^2, then, by Itô's formula,

$$E_x R^2(\xi(t)) - R^2(x) = E_x \int_0^t LR^2(\xi(s))\, ds.$$

Using (2.1), (2.3) we find that

$$LR^2(x) = \sum a_{ij} R_{x_i} R_{x_j} + 2R\left\{\tfrac{1}{2} \sum a_{ij} R_{x_i x_j} + \sum b_i R_{x_i}\right\} \leq CR^2$$

if $\rho(x)$ is small, say $\rho(x) < \epsilon_0$, where C is a positive constant; by the definition of $R(x)$ this inequality holds also if $\rho(x) \geq \epsilon_0$. Hence

$$E_x R^2 \xi(t) - R^2(x) \leq C \int_0^t ER^2(\xi(s))\, ds. \tag{2.4}$$

Since $R^2(x)$ is in C^1 and piecewise in C^2, we can establish (2.4), rigorously, using mollifiers.

Now take x in G. Then $R(x) = 0$. Setting $\phi(t) = E_x R^2(\xi(t))$, (2.4) becomes

$$\phi(t) \leq C \int_0^t \phi(s)\, ds, \qquad \phi(0) = 0.$$

Hence $\phi(t) = 0$ for all t, i.e., $R(\xi(t)) = 0$ a.s. for all $t \geq 0$. By the definition of R, then, $\xi(t) \in G$ a.s. for all $t \geq 0$.

We shall now assume that (2.1) holds and that

$$\sum_{i=1}^n b_i \nu_i + \frac{1}{2} \sum_{i,j=1}^n a_{ij} \frac{\partial^2 \rho}{\partial x_i\, \partial x_j} = 0 \qquad \text{on } \partial G. \tag{2.5}$$

Then, by Theorems 9.4.1 and 2.1, G is both nonattainable and invariant. The proofs of these theorems, when slightly modified, establish also the fact that $\hat{G} \cup \partial G$ is nonattainable and invariant. Consequently, if (2.1), (2.5) hold, then ∂G *is nonattainable and invariant.*

We shall next study the asymptotic stability of ∂G from the outside. Introduce the functions

$$\mathcal{A} = \frac{1}{2} \sum_{i,j=1}^n a_{ij}(x) \frac{\partial R}{\partial x_i} \frac{\partial R}{\partial x_j}, \tag{2.6}$$

$$\mathcal{B} = \sum_{i=1}^n b_i(x) \frac{\partial R}{\partial x_i} + \frac{1}{2} \sum_{i,j=1}^n a_{ij}(x) \frac{\partial^2 R}{\partial x_i\, \partial x_j}, \tag{2.7}$$

$$Q = \frac{1}{R}\left(\mathcal{B} - \frac{\mathcal{A}}{R}\right), \tag{2.8}$$

where $R(x) = \rho(x) = d(x, \partial G)$ for $x \in \hat{G} \cup \partial G$, $\rho(x) < \epsilon_0$, where ϵ_0 is sufficiently small. If $u(x) = \Phi(R(x))$, then (cf. (9.5.1))

$$Lu(x) = \mathcal{A}\left[\Phi''(R) + \frac{1}{R}\Phi'(R)\right] + RQ\Phi'(R). \tag{2.9}$$

Consequently,

$$u(x) = \log R(x)$$

in an S-function for ∂G from the outside, provided

$$Q(x) \leq -\theta_0 \quad \text{if} \quad \rho(x) < \epsilon_1 \quad (\theta_0 \text{ positive constant}) \tag{2.10}$$

for some $0 < \epsilon_1 \leq \epsilon_0$.

This condition holds if and only if

$$\overline{\lim_{\rho \downarrow 0}} \left[\frac{\mathcal{B}}{\rho} - \frac{\mathcal{Q}}{\rho^2} \right] < 0 \quad \text{along } \partial G. \tag{2.11}$$

If $b_i \in C^1$, $a_{ij} \in C^2$ in a neighborhood of ∂G, then (2.11) reduces to

$$\left[\frac{\partial}{\partial \rho} \mathcal{B} - \frac{\partial^2 \mathcal{Q}}{\partial \rho^2} \right]_{\rho = 0+} < 0 \quad \text{along } \partial G. \tag{2.12}$$

Consider next the construction of an S-function for ∂G from the inside. We define $\mathcal{Q}$, $\mathcal{B}$, Q as before, but take $R(x) = d(x, \partial G)$ if $x \in G$ and $d(x, \partial G)$ is sufficiently small. We find that $\log R(x)$ is an S-function if $Q(x) \leq -\theta_0 < 0$ when $d(x, \partial G)$ is sufficiently small. If $b_i \in C^1$, $a_{ij} \in C^2$ in a neighborhood of ∂G, then this condition holds if and only if the inequality (2.12) holds. Thus, when (2.10) holds, the ∂G is asymptotically stable both from the outside and from the inside, i.e., it is asymptotically stable.

We sum up:

Theorem 2.2. *Let* (2.1), (2.5) *hold. If* (2.10) *holds, then ∂G is asymptotically stable from the outside. If $b_i \in C^1$, $a_{ij} \in C^2$ in a neighborhood of ∂G, then* (2.12) *holds if and only if* (2.10) *holds and, in that case, ∂G is asymptotically stable.*

Consider now a more general case where

$$G = \bigcup_{j=1}^{k} G_j, \qquad \hat{G} = R^n \setminus G;$$

the G_i are mutually disjoint sets, $G_i = \{z_i\}$ if $1 \leq i \leq k_0$ (where z_i is a point) and G_j is a closed bounded domain with connected C^3 boundary ∂G_j if $k_0 + 1 \leq j \leq k$. We assume that

$$a_{ij}(z_h) = 0, b_i(z_h) = 0 \quad \text{if} \quad 1 \leq i, j \leq n, \quad 1 \leq h \leq k_0, \tag{2.13}$$

$$\sum_{i,j=1}^{n} a_{ij} \nu_i \nu_j = 0 \quad \text{on } \partial G_h, \quad \text{for} \quad k_0 + 1 \leq h \leq k, \tag{2.14}$$

$$\sum_{i=1}^{n} b_i \nu_i - \frac{1}{2} \sum_{i,j=1}^{n} a_{ij} \frac{\partial^2 \rho}{\partial x_i \, \partial x_j} = 0 \quad \text{on } \partial G_h \quad \text{for} \quad k_0 + 1 \leq h \leq k \tag{2.15}$$

where $\rho(x) = d(x, G)$ is defined for $x \in \hat{G} \cup \partial G$. We define the function $Q(x)$ as before, with $R(x) = \rho(x)$.

Set $\partial G_i = \{z_i\}$ if $1 \leq i \leq k_0$, $\partial G = \bigcup_{i=1}^{k} \partial G_i$.

Theorem 2.3. *Let* (2.13)–(2.15) *hold. Then* ∂G *is nonattainable and invariant. If, further,* (2.10) *holds, then* ∂G *is asymptotically stable from* $\hat{G}$. *If* $b_i \in C^1$, $a_{ij} \in C^2$ *in a neighborhood of* $\Gamma \equiv \bigcup_{h=k_0+1}^{k} \partial G_h$, *and if (in addition to* (2.13)–(2.15)) (2.10) *holds, then* ∂G *is asymptotically stable; the condition* (2.10) *for the boundary* ∂G_h, $k_0 + 1 \leq h \leq k$, *is equivalent, in this case, to the condition* (2.12) *along* ∂G_h.

Consider next the global asymptotic stability for ∂G. We shall assume:

$$\left(a_{ij}(x)\right) \quad \text{is positive definite for all} \quad x \in \hat{G}; \tag{2.16}$$

$$R\,\mathcal{B} < \mathcal{A} \qquad \text{if} \quad R(x) \equiv |x|, \quad \text{for all } |x| \text{ sufficiently large.} \tag{2.17}$$

The functions $\mathcal{A}$, $\mathcal{B}$ are defined as in (2.6), (2.7);

$$G_h \quad \text{is } C^2 \text{ diffeomorphic to a closed ball, for} \quad k_0 + 1 \leq h \leq k. \tag{2.18}$$

Theorem 2.4. *If* (2.13)–(2.18) *and* (2.10) *hold, then* ∂G *is globally asymptotically stable from* $\hat{G}$.

Proof. Since we have already constructed an S-function, it sufficies (in view of Theorem 1.3) to construct a piecewise smooth G-function.

We shall employ the function $R(x)$ established in Lemma 9.4.5. In view of property (v) (asserted in that lemma), there is a compact set E containing in its interior the set of points where $D_x R = 0$, such that

$$Q(x) < 0 \qquad \text{if} \quad x \in E. \tag{2.19}$$

For any small $\eta > 0$, let $r_1 = \eta$, $r_2 = 1/\eta$. In view of (2.16)

$$\mathcal{A}(x) \geq \alpha R^2 \qquad \text{if} \quad r_1 \leq R(x) \leq r_2, \quad x \notin E \tag{2.20}$$

where α is a positive constant. We shall construct a continuously differentiable function $\Phi(r)$ for $r > r_1$ whose second derivative $\Phi''(r)$ has a jump discontinuity at r_2 and

$$L\Phi(R(x)) < 0 \qquad \text{if} \quad r_1 < R(x) < \infty, \quad R(x) \neq r_2, \tag{2.21}$$

$$\Phi'(r) > 0, \tag{2.22}$$

$$\Phi(r) \to \infty \qquad \text{if} \quad r \to \infty. \tag{2.23}$$

Let $\theta(r)$ $(r_1 \leq r \leq r_2 + 1)$ be a continuous function satisfying

$$Q(x) \leq \theta(R(x)) \qquad \text{if} \quad r_1 \leq R(x) \leq r_2 \tag{2.24}$$

such that $\theta(R(x)) > 0$ if $x \in E$. We take η so small that E is contained in the set $R(x) \leq r_2$, and (2.17) holds if $R(x) > r_2$.

Define

$$\mu(r) = \exp \int_{r_1}^{r} \frac{1 + \theta(s)/\alpha}{s} \, ds$$

and define $\Phi(r)$ for $r_1 \leqslant r \leqslant r_2$ by

$$\mu(r)\Phi'(r) = \frac{1}{\alpha} \int_{r}^{r_2+1} \frac{\mu(s)}{s^2} \, ds,$$

$$\Phi(r_1) = 0.$$

Then $\Phi'(r) > 0$ and

$$\Phi''(r) + \left(1 + \frac{\theta(r)}{\alpha}\right)\frac{\Phi'(r)}{r} = -\frac{1}{r^2}. \tag{2.25}$$

From (2.9) we have, for $x \notin E$, $r_1 \leqslant R(x) \leqslant r_2$,

$$L\Phi(R(x)) = \mathcal{Q}\left\{\Phi''(R) + \left(1 + \frac{R^2 Q}{\mathcal{Q}}\right)\frac{\Phi'(R)}{R}\right\}$$

$$\leqslant \mathcal{Q}\left\{\Phi''(R) + \left(1 + \frac{\theta(R)}{\alpha}\right)\frac{\Phi'(R)}{R}\right\}$$

where (2.20), (2.24), and (2.22) have been used. In view of (2.25) we then have

$$L\Phi(R(x)) \leqslant -\frac{\mathcal{Q}}{R^2} < 0.$$

If $x \in E$, then by (2.22), (2.25) and the fact that $\theta(R(x)) > 0$,

$$\Phi''(R(x)) + \frac{1}{R(x)}\Phi'(R(x)) < 0.$$

Since also $\Phi'(R) > 0$, $Q(x) < 0$, we obtain from (2.9) the inequality $L\Phi(R(x)) < 0$.

We have thus constructed a C^2 function $\Phi(r)$ satisfying (2.21), (2.22) for $r_1 \leqslant r \leqslant r_2$. Consider next the function

$$\Psi(r) = A \log r + B \qquad (r_2 < r < \infty)$$

where A and B are constants and $A > 0$. Since $\Psi'(r) > 0$, using (2.9) and the assumption (2.17) we see that $L\Psi(R(x)) < 0$ if $R(x) \geqslant r_2$.

If we can choose the constants A, B such that

$$\Psi(r_2) = \Phi(r_2), \qquad \Psi'(r_2) = \Phi'(r_2) \tag{2.26}$$

then by defining $\Phi(r) = \Psi(r)$ for $r > r_2$ we obtain the desired function Φ satisfying (2.21)–(2.23). We solve (2.26) by taking

$$A = r_2\Phi'(r_2), \qquad B = \Phi(r_2) - r_2\Phi'(r_2)\log r_2.$$

The function $\Phi(R(x))$ is a piecewise smooth G-function for the η-neighborhood of G (recall that $r_1 = \eta$). Since η can be arbitrarily small, the remark at the end of Section 1 shows that ∂G is asymptotically stable from $\hat{G}$.

Remark 1. One can easily construct a G-function in the whole domain $\hat{G}$, by extending the above function $\Phi(R(x))$ as $A_0 \log R(x) + B_0$ into the set $0 < R(x) < r_1$, where A_0, B_0 are suitable constants and $A_0 > 0$.

Remark 2. Theorem 2.4 establishes global asymptotic stability from $\hat{G}$. If for a particular G_h $(k_0 + 1 \leq h \leq k)$, there is an i such that $a_{ii}(x) \neq 0$ for all $x \in \operatorname{int} G_h$, then in any compact subset E of G_h there is a function

$$\phi(x) = e^{\alpha x_i^0} - e^{\alpha x_i}$$

satisfying: $L\phi(x) < 0$ if $x \in E$. (Here $x_i < x_i^0$ for all $x = (x_1, \ldots, x_i, \ldots, x_n)$ in E and α is a sufficiently large positive constant.) By the remark at the end of Section 1 it follows that ∂G_h is globally asymptotically stable from int G_h.

3. Stability of point obstacles

We consider the case where $x = 0$ is a point obstacle, and refine the stability theorem derived in the previous section. Consider first the case of a linear stochastic differential system

$$d\xi_i = \sum_{j,s=1}^{n} \sigma_{ij}^s \xi_j \, dw_s + \sum_{j=1}^{n} b_{ij} \xi_j \, dt \qquad (1 \leq i \leq n) \tag{3.1}$$

where σ_{ij}^s, b_{ij} are constants. Set

$$\sigma_{is}(x) = \sum_{j=1}^{n} \sigma_{ij}^s x_j, \qquad a_{ij}(x) = \sum_{s=1}^{n} \sigma_{is}(x)\sigma_{js}(x) = \sum_{s,k,l=1}^{n} \sigma_{ik}^s \sigma_{jl}^s x_k x_l.$$

We shall assume that

$$\sum a_{ij}(x)\xi_i\xi_j \geq \alpha|x|^2|\xi|^2 \quad \text{if } x \in R^n, \ \xi \in R^n, \ x \cdot \xi = 0 \ (\alpha > 0). \tag{3.2}$$

Taking $R(x) = |x|$ in the definition of $Q(x)$ in (2.8), we have

$$Q(x) = \frac{\sum b_{ij} x_i x_j}{|x|^2} + \frac{1}{2} \frac{\sum a_{ii}(x)}{|x|^2} - \frac{\sum a_{ij}(x) x_i x_j}{|x|^4} \equiv Q\left(\frac{x}{|x|}\right). \tag{3.3}$$

If $u(x) = v(\rho, \theta) = v_1(\rho) + v_2(\theta)$ where $\rho = \log|x|$ and $\theta = (\theta_1, \ldots,$

θ_{n-1}) are local coordinates on the sphere S^{n-1}, then (see Problem 5)

$$Lu = \tfrac{1}{2}\sigma^2(\theta)\frac{\partial^2 v}{\partial \rho^2} + \tilde{Q}(\theta)\frac{\partial v}{\partial \rho} + L_0 v \tag{3.4}$$

where

$$\sigma^2(\theta) = \frac{\sum a_{ij}(x)x_i x_j}{|x|^4}, \qquad \tilde{Q}(\theta) = Q\left(\frac{x}{|x|}\right)$$

and L_0 is a nondegenerate elliptic operator on S^{n-1}. The elliptic estimates of Section 10.3 remain valid also on the sphere. Therefore by the elliptic theory (see, for instance, Friedman [2]) the Fredholm alternative holds for the equation $L_0 h = g$. By the maximum principle, the only solutions of the homogeneous equation are constants. Consequently the eigenspace of the adjoint L_0^* is spanned by one function only, say h_0. We normalize it by

$$\int_{S^{n-1}} h_0 \, dA = 1 \tag{3.5}$$

where dA is the surface element. Let

$$Q_0 = \int_{S^{n-1}} Q h_0 \, dA. \tag{3.6}$$

Theorem 3.1. *If* (3.2) *holds, then*

$$\lim_{t\to\infty} \frac{\log|\xi(t)|}{t} = Q_0. \tag{3.7}$$

Thus, if $Q_0 < 0$, then $x = 0$ is globally asymptotically stable.

Proof. Since $-Q + Q_0$ is orthogonal to the homogeneous solutions of $L_0^* w = 0$ (i.e. to h_0), there is a solution g of $L_0 g = -Q + Q_0$ on S^{n-1}. Consider the function

$$u(x) = \log|x| + g. \tag{3.8}$$

If we apply Itô's formula with this function, we obtain, upon using (3.4),

$$\log|\xi(t)| + g\left(\frac{\xi(t)}{|\xi(t)|}\right) = \log|x| + g\left(\frac{x}{|x|}\right) + tQ_0 + k(t) \tag{3.9}$$

where

$$k(t) = \int_0^t \sum_{i,s=1}^n \left(\frac{\xi_i}{|\xi|^2} - \frac{\partial g}{\partial x_i}\right)\sigma_{is}\, dw_s. \tag{3.10}$$

Let

$$f_s(t) = \sum_{i=1}^{n} \left[\frac{\xi_i(t)}{|\xi(t)|^2} - \frac{\partial g}{\partial x_i}\left(\frac{\xi(t)}{|\xi(t)|} \right) \right] \sigma_{is}(\xi(t)).$$

Since $|\partial g/\partial x_i| < \text{const}/|x|$,

$$|f_s(t)| < C \quad \text{a.s. for all} \quad t > 0 \qquad (C \text{ const}). \tag{3.11}$$

By Corollary 4.4.6 it follows that $k(t) = o(t)$. Hence, dividing both sides of (3.9) by t and letting $t \to \infty$, the assertion (3.7) follows.

We now replace (3.2) by a stronger condition of nondegeneracy:

$$\sum a_{ij}(x)\xi_i\xi_j > \alpha|x|^2|\xi|^2 \qquad \text{for all} \quad x \in R^n, \quad \xi \in R^n. \tag{3.12}$$

Theorem 3.2. *If* (3.12) *holds, then for any positive-valued function* $\phi(t) = o(\sqrt{t} \log\log t)$,

$$\overline{\lim_{t\to\infty}} \frac{\log|\xi(t)| - Q_0 t}{\phi(t)} = \infty \quad a.s., \tag{3.13}$$

$$\underline{\lim_{t\to\infty}} \frac{\log|\xi(t)| - Q_0 t}{\phi(t)} = -\infty \quad a.s. \tag{3.14}$$

Proof. Since the function g is homogeneous of degree 0,

$$\sum x_i \frac{\partial g}{\partial x_i} = 0 \qquad \text{by Euler's theorem.}$$

Hence, by (3.12),

$$\sum_{s=1}^{n} |f_s(t)|^2 = \sum a_{ij}\left(\frac{\xi_i}{|\xi|^2} - \frac{\partial g}{\partial x_i} \right)\left(\frac{\xi_j}{|\xi|^2} - \frac{\partial g}{\partial x_j} \right)$$

$$> \alpha|\xi|^2 \sum \left(\frac{\xi_i}{|\xi|^2} - \frac{\partial g}{\partial x_i} \right)^2 > \alpha$$

where $\xi = \xi(t)$ and the argument of g is $\xi(t)/|\xi(t)|$. Recalling (3.11) we conclude that

$$\alpha < \sum |f_s(t)|^2 < \beta \qquad (\beta \text{ const}). \tag{3.15}$$

By Theorem 4.7.4, the process

$$\tilde{w}(t) \equiv \int_0^{\tau(t)} \sum_{s=1}^{n} f_s(\lambda)\, dw_s(\lambda)$$

where

$$\tau(t) = \inf\left\{ \mu;\ \int_0^{\mu} \sum_{s=1}^{n} (f_s(u))^2\,du = t \right\}$$

is a one-dimensional Brownian motion. In view of (3.15)

$$t/\beta \leq \tau(t) \leq t/\alpha. \tag{3.16}$$

Applying the law of the iterated logarithm to $\tilde{w}(t)$ and using (3.16) we deduce that

$$\overline{\lim_{t\to\infty}} \frac{k(t)}{\phi(t)} = \infty \quad \text{a.s.,} \qquad \underline{\lim_{t\to\infty}} \frac{k(t)}{\phi(t)} = -\infty \quad \text{a.s.,}$$

where $k(t)$ is defined in (3.10). Using this in (3.9), the assertions (3.13), (3.14) follow.

Consider now the general case of a nonlinear stochastic differential system. We assume that

$$\sigma_{ij}(0) = 0, \qquad b_i(0) = 0$$

and that σ_{ij}, b_i belong to C^1 in a neighborhood of $x = 0$. Then

$$\sigma_{is}(x) = \sum_{j=1}^{n} \sigma_{ij}^{s} x_j + o(|x|), \qquad b_i(x) = \sum_{j=1}^{n} b_{ij} x_j + o(|x|).$$

Theorem 3.3. *If the σ_{ij}^{s}, b_{ij} are such that* (3.2) *holds, and if $Q_0 < 0$, then $x = 0$ is asymptotically stable.*

Here Q_0 is defined by (3.6) and Q is defined in terms of the σ_{ij}^{s}, b_{ij}. One can easily verify that

$$Q(\theta) = \lim_{r\to 0} \left\{ \frac{\sum x_i b_i(x)}{|x|^2} + \frac{1}{2}\frac{\sum a_{ii}(x)}{|x|^2} - \frac{\sum a_{ij}(x) x_i x_j}{|x|^2} \right\}$$

where (r, θ) are the polar coordinates of x.

The proof of Theorem 3.3 follows from the easily checked fact that $cu(x)$ is an S-function (for the present nonlinear system), where $u(x)$ is defined by (3.8) and c is a suitable positive constant.

4. The method of descent

In the previous sections we have considered the system of n equations (1.1), where $w(t)$ is n-dimensional Brownian motion. However all the results remain valid if $w(t)$ is l-dimensional Brownian motion. In the present section we shall need to work with this slightly more general setting.

Consider a system of $n + 1$ equations

$$d\xi_i = \sum_{s=1}^{l} \sigma_{is}(\xi, \eta)\, dw_s + b_i(\xi, \eta)\, dt \qquad (1 \le i \le n),$$
$$d\eta = \sum_{s=1}^{l} \sigma_{0s}(\xi, \eta)\, dw_s + b_0(\xi, \eta)\, dt \tag{4.1}$$

and suppose that the last equation degenerates on the hyperplane $y = 0$, i.e.,

$$\sigma_{0s}(x, 0) = 0, \qquad b_0(x, 0) = 0 \qquad \text{if} \quad |x| < \delta_0. \tag{4.2}$$

We also assume that the stability condition (2.11) holds with respect to the hyperplane $y = 0$ at $(0, 0)$, i.e.,

$$-\nu \equiv \frac{\partial b_0(0, 0)}{\partial y} - \frac{1}{2} \sum_{s=1}^{l} \left[\frac{\partial \sigma_{0s}(0, 0)}{\partial y} \right]^2 < 0; \tag{4.3}$$

it is assumed that $\partial b_0/\partial y$, $\partial \sigma_{0s}/\partial y$ exist and are continuous in a neighborhood of the origin.

We shall compare the behavior of the solution of (4.1) with the behavior of the solution of the reduced system

$$d\xi_i = \sum_{s=1}^{l} \sigma_{is}(\xi, 0)\, dw_s + b_i(\xi, 0)\, dt \qquad (1 \le i \le n). \tag{4.4}$$

The differential operators corresponding to (4.1) and (4.4) are

$$Lu(x, y) \equiv \frac{1}{2} \sum_{i,j=0}^{n} a_{ij}(x, y) \frac{\partial^2 u}{\partial x_i\, \partial x_j} + \sum_{i=0}^{n} b_i(x, y) \frac{\partial u}{\partial x_i}$$

and

$$L_0 u(x) \equiv \frac{1}{2} \sum_{i,j=1}^{n} a_{ij}(x, 0) \frac{\partial^2 u}{\partial x_i\, \partial x_j} + \sum_{i=1}^{n} b_i(x, 0) \frac{\partial u}{\partial x_i},$$

respectively.

We shall assume that

$$\sigma_{ij}(0, 0) = 0, \qquad b_i(0, 0) = 0 \qquad \text{if} \quad 1 \le i, j \le n. \tag{4.5}$$

Theorem 4.1. *Let* (4.2), (4.3), (4.5) *hold. If* $f(x) = \log|x| + H(x/|x|)$ *is an S-function for the reduced system* (4.4) *for* $x = 0$, *then there exists an S-function for the system* (4.1) *for* $(x, y) = (0, 0)$ *having the form*

$$f(x, y) = \sum_{j=1}^{\infty} \alpha_j \log\{ y^2 + \epsilon_j \exp[2f(x)]\} \tag{4.6}$$

for appropriate nonnegative constants α_j, ϵ_j.

This theorem will be referred to as the *method of descent*.

Proof. Consider a function

$$f_\epsilon(x, y) = \log\left[y^2 + \epsilon \exp(2f(x)) \right] = \log\left[y^2 + \epsilon h(x) \right],$$
$$h(x) = |x|^2 \exp[2H(\theta)].$$

Setting $h_i = \partial h/\partial x_i$, $h_{ij} = \partial^2 h/\partial x_i\, \partial x_j$, $\Psi = y^2 + \epsilon h(x)$, we have

$$\begin{aligned} Q_\epsilon(x, y) &\equiv (Lf_\epsilon)(x, y) \\ &= \tfrac{1}{2} \sum_{i,j=1}^{n} a_{ij}(x, y)\left[\frac{\epsilon h_{ij}}{\Psi} - \epsilon^2 \frac{h_i h_j}{\Psi} \right] + \sum_{i=1}^{n} a_{i0}(x, y)\left[- \frac{2\epsilon y h_i}{\Psi^2} \right] \\ &\quad + \tfrac{1}{2} a_{00}(x, y)\left[\frac{-2y^2 + 2\epsilon h}{\Psi^2} \right] \\ &\quad + \sum_{i=1}^{n} b_i(x, y)\left[\frac{\epsilon h_i(x)}{\Psi} \right] + b_0(x, y)\left[\frac{2y}{\Psi} \right]. \end{aligned} \tag{4.7}$$

If $|x|^2 + y^2 = 1$ and if $\alpha = \min_{|x|<1} h(x)$,

$$\frac{\epsilon}{\Psi} < \frac{\epsilon}{y^2 + \epsilon\alpha|x|^2} = \frac{\epsilon}{\epsilon\alpha + y^2(1 - \epsilon\alpha)} < \frac{1}{\alpha} \qquad \text{if} \quad \epsilon < \frac{1}{\alpha}, \tag{4.8}$$

and

$$\frac{y^2}{\Psi^2} < \frac{y^2}{\left[y^2 + \epsilon\alpha|x|^2 \right]^2} = \frac{y^2}{\left[\epsilon\alpha + y^2(1 - \epsilon\alpha) \right]^2} < 1 \tag{4.9}$$

since the maximum of the third term is attained at $y = 1$.

Consider the first term I on the right-hand side of (4.7). Since, by (4.5), $|a_{ij}(x, y)| < \text{const}(|x|^2 + y^2)$ (if, say, $|x|^2 + y^2 < 1$) and since $h_i = O(|x|)$, $h_{ij} = O(1)$,

$$|I| < C \frac{(|x|^2 + y^2)\epsilon}{\Psi} + C \frac{\epsilon^2 |x|^2}{\Psi} < C' \qquad (C, C' \text{ const})$$

by (4.8), provided $|x|^2 + y^2 < 1$. The other terms on the right-hand side of (4.7) are estimated similarly, using (4.8), (4.9), and the inequalities

$$a_{00}(x, y) < C_1 y^2, \qquad |b_i(x, y)| < C_1\sqrt{|x|^2 + y^2}\,, \qquad |b_0(x, y)| < C_1 |y|$$

which follow from (4.2), (4.5), provided $|x|^2 + y^2 < \mu$, μ sufficiently small. We conclude that

$$Q_\epsilon(x, y) < B^2 \qquad \text{if} \quad |x|^2 + y^2 < \mu, \qquad \epsilon < 1/\alpha \tag{4.10}$$

for some $\mu > 0$, $B > 0$; μ and B are independent of ϵ.

If $y \neq 0$,

$$\lim_{\epsilon \to 0} Q_\epsilon(x, y) = - \frac{y^2 a_{00}(x, y)}{y^4} + \frac{2b_0(x, y)}{y}. \tag{4.11}$$

Using (4.2) we find that the right-hand side is a continuous function (extendable to $y = 0$ by continuity) and its value at $(0, 0)$ is -2ν where ν is defined in (4.3). Since the convergence in (4.11) is uniform in any region $|y| > \delta'$, where $\delta' > 0$, we conclude, after using the fact that $Q_\epsilon(x, y)$ is homogeneous in (x, y), that

$$Q_\epsilon(x, y) < -\nu$$

if

$$|x|^2 + y^2 < \mu^2, \qquad (x, y) \not\in S_\delta \equiv \left\{ |x|^2 + y^2 < \mu^2;\ \tan^{-1} \frac{y}{|x|} < \delta \right\} \tag{4.12}$$

provided $\epsilon = \epsilon(\delta)$ is sufficiently small. Here μ is a sufficiently small positive number, independent of ϵ, δ.

On the other hand, as seen from (4.7),

$$Q_\epsilon(x, 0) = 2L_0 f(x) \leqslant -2 \qquad \text{if} \quad |x| \text{ is small.}$$

Since, for any $\epsilon > 0$, $Q_\epsilon(x, y)$ is continuous in (x, y), we conclude that $Q_\epsilon(x, y) < -1$ if $|x|^2 + y^2 < \mu^2$, $(x, y) \in S_{\delta'}$ provided μ is sufficiently small and provided δ' is sufficiently small. The constant μ can be taken to be independent of ϵ, but $\delta' = \delta'(\epsilon)$.

We conclude that for any small $\delta > 0$ there exist small ϵ and small $\delta' = \delta'(\epsilon)$ such that

$$Q_\epsilon(x, y) \leqslant -A^2 \qquad \text{if} \quad (x, y) \not\in S_\delta \setminus S_{\delta'}, \qquad |x|^2 + y^2 \leqslant \mu^2; \tag{4.13}$$

here μ and A^2 are sufficiently small positive numbers independent of δ, ϵ, δ'.

We now take a sequence of numbers δ_m decreasing to 0 such that the regions $\tilde{S}_m = S_{\delta_m} \setminus S_{\delta'_m}$ are disjoint; for instance, $\delta_{m+1} = \frac{1}{2}\delta'_m$. To each $\delta = \delta_m$ there corresponds an $\epsilon = \epsilon_m$ such that (4.13) holds if $(x, y) \in \tilde{S}_m$.

We now form the function in (4.6) with

$$\sum_{j=1}^{\infty} \alpha_j = 1, \qquad \alpha_j > 0. \tag{4.14}$$

If

$$\sum \alpha_j \log \frac{1}{\epsilon_j} < \infty, \qquad \alpha_j > 0, \tag{4.15}$$

then the series in (4.6) is absolutely and uniformly convergent together with any number of its derivatives in any compact subset lying in the domain $0 < |x|^2 + y^2 < \mu$. Hence,

$$Lf(x, y) = \sum \alpha_j Q_{\epsilon_j}(x, y).$$

If $(x, y) \in \tilde{S}_m$ and $|x|^2 + y^2 < \mu$, then

$$\begin{aligned}(Lf)(x, y) &\leqslant \alpha_m B^2 - A^2 \sum_{j \neq m} \alpha_j = \alpha_m B^2 - A^2(1 - \alpha_m) \\ &= (B^2 + A^2)\alpha_m - A^2 < \frac{A^2}{2} - A^2 = -\frac{A^2}{2}\end{aligned}$$

if

$$\alpha_m < A^2/2(A^2 + B^2). \tag{4.16}$$

On the other hand, if $(x, y) \notin \tilde{S}_m$ for all m, then

$$(Lf)(x, y) < \sum_{j=1}^{\infty} (-A^2\alpha_j) = -A^2.$$

Thus, if the α_j can be chosen to satisfy (4.14)–(4.16), then $Lf(x, y) < -A^2/2$ if $|x|^2 + y^2 < \mu$; thus $cf(x, y)$ is an S-function for the system (4.1), where $c = 2/A^2$.

It remains to construct the α_j satisfying (4.14)–(4.16). Let N be a sufficiently large positive number such that

$$\sum_{j=N+1}^{\infty} \frac{1}{j^2 \log(1/\epsilon_j)} < 1,$$

$$\frac{1}{N^2 \log(1/\epsilon_N)} < \frac{A^2}{2(A^2 + B^2)},$$

$$\frac{1}{N}\left[1 - \sum_{j=N+1}^{\infty} \frac{1}{j^2 \log(1/\epsilon_j)}\right] < \frac{A^2}{2(A^2 + B^2)}.$$

Defining

$$\alpha_j = \begin{cases} \dfrac{1}{N}\left[1 - \displaystyle\sum_{l=N+1}^{\infty} \frac{1}{l^2 \log(1/\epsilon_l)}\right] & \text{if } 1 < j < N, \\ \dfrac{1}{j^2 \log(1/\epsilon_j)} & \text{if } j > N + 1, \end{cases}$$

it is readily seen that (4.14)–(4.16) hold.

5. Spiraling of solutions about a point obstacle

We shall now specialize to the case $n = 2$ and consider the angular behavior of solutions near a stable obstacle. In this section we consider the special case of a point obstacle at $x = 0$. Thus we consider a system

$$d\xi_i = \sum_{s=1}^{l} \sigma_{is}(\xi)\, dw_s + b_i(\xi)\, dt, \qquad i = 1, 2 \tag{5.1}$$

with

$$\begin{aligned} \sigma_{is}(x) &= \sum_{j=1}^{2} \sigma_{isj} x_j + o(|x|) \qquad \text{as } |x| \to 0, \\ b_i(x) &= \sum_{j=1}^{2} b_{ij} x_j + o(|x|) \qquad \text{as } |x| \to 0. \end{aligned} \tag{5.2}$$

We introduce polar coordinates (r, ϕ) by

$$x_1 = r\cos\phi, \qquad x_2 = r\sin\phi.$$

The stochastic differentials dr, $d\phi$ may formally be computed by

$$dr = r_{\xi_1}\,d\xi_1 + r_{\xi_2}\,d\xi_2 + \tfrac{1}{2}a_{11}r_{\xi_1\xi_1}\,dt + a_{12}r_{\xi_1\xi_2}\,dt + \tfrac{1}{2}a_{22}r_{\xi_2\xi_2}\,dt,$$

$$d\phi = \phi_{\xi_1}\,d\xi_1 + \phi_{\xi_2}\,d\xi_2 + \tfrac{1}{2}a_{11}\phi_{\xi_1\xi_1}\,dt + a_{12}\phi_{\xi_1\xi_2}\,dt + \tfrac{1}{2}a_{22}\phi_{\xi_2\xi_2}\,dt.$$

Noting that

$$r_{x_1} = \cos\phi, \qquad r_{x_2} = \sin\phi,$$

$$r_{x_1x_1} = \frac{\sin^2\phi}{r}, \qquad r_{x_1x_2} = -\frac{\sin\phi\cos\phi}{r}, \qquad r_{x_2x_2} = \frac{\cos^2\phi}{r},$$

$$\phi_{x_1} = -\frac{\sin\phi}{r}, \qquad \phi_{x_2} = \frac{\cos\phi}{r},$$

$$\phi_{x_1x_1} = \frac{2\sin\phi\cos\phi}{r^2}, \quad \phi_{x_1x_2} = \frac{\sin^2\phi - \cos^2\phi}{r^2}, \quad \phi_{x_2x_2} = -\frac{2\sin\phi\cos\phi}{r^2},$$

we get

$$\begin{aligned} dr &= \sum_{s=1}^{l} \tilde{\sigma}_s(r,\phi)\,dw_s + \tilde{b}(r,\phi)\,dt, \\ d\phi &= \sum_{s=1}^{l} \tilde{\tilde{\sigma}}_s(r,\phi)\,dw_s + \tilde{\tilde{b}}(r,\phi)\,dt \end{aligned} \tag{5.3}$$

where

$$\begin{aligned} \tilde{\sigma}_s(r,\phi) &= \sigma_{1s}\cos\phi + \sigma_{2s}\sin\phi, \\ \tilde{b}(r,\phi) &= b_1\cos\phi + b_2\sin\phi + \frac{1}{2r}\langle a(x)\lambda^\perp, \lambda^\perp\rangle, \\ \tilde{\tilde{\sigma}}_s(r,\phi) &= -\frac{\sin\phi}{r}\sigma_{1s} + \frac{\cos\phi}{r}\sigma_{2s}, \\ \tilde{\tilde{b}}(r,\phi) &= -\frac{\sin\phi}{r}b_1 + \frac{\cos\phi}{r}b_2 - \frac{1}{r^2}\langle a(x)\lambda, \lambda^\perp\rangle; \end{aligned} \tag{5.4}$$

here

$$\lambda = (\cos\phi, \sin\phi), \qquad \lambda^\perp = (-\sin\phi, \cos\phi)$$

and

$$\langle a(x)\mu, \nu\rangle = \sum a_{ij}(x)\mu_i\nu_j \qquad (\mu = (\mu_1, \mu_2),\ \nu = (\nu_1, \nu_2)).$$

If we substitute (5.4) into (5.3) and make use of (5.2), we find that

$$\begin{aligned} dr &= r\left[\sum_{s=1}^{l} \tilde{\sigma}_s(\phi)\,dw_s + \tilde{b}(\phi)\,dt\right] + \left[\sum_{s=1}^{l} R_s\,dw_s + R_0\,dt\right] \\ d\phi &= \left[\sum_{s=1}^{l} \tilde{\tilde{\sigma}}_s(\phi)\,dw_s + \tilde{\tilde{b}}(\phi)\,dt\right] + \left[\sum_{s=1}^{l} \Theta_s\,dw_s + \Theta_0\,dt\right] \end{aligned} \tag{5.5}$$

where $R_s = o(r)$, $\Theta_s = o(r)$ when $r \to 0$, uniformly for $0 < \phi < 2\pi$.

We proceed to justify (5.5) rigorously. Let $y(t) = (r(t), \phi(t))$ be the diffusion process defined by the solution of the system of stochastic differential equations (5.5) with $r(0) > 0$. By the proof of Theorem 9.4.1 (with $R(x) = r$) we see that the solution $y(t)$ remains in $(0, \infty) \times (-\infty, \infty)$ for all $t > 0$. Define $x(t) = (x_1(t), x_2(t))$ by

$$x_1(t) = r(t) \cos \phi(t), \qquad x_2(t) = r(t) \sin \phi(t). \tag{5.6}$$

Theorem 5.1. *The solution of* (5.1) *can be represented in the form* (5.6) *where* $(r(t), \phi(t))$ *is the solution of* (5.5).

Proof. We have to verify the equations in (5.1) for $x_1(t)$, $x_2(t)$. We resort to an argument of general nature.

Write $x_i(t) = g^i(y)$ where $y = (y_1, y_2)$ and g is the global differentiable transformation $(r, \phi) \to (r \cos \phi, r \sin \phi)$. The stochastic differential of x can be computed by Itô's formula (subscripts following commas denote partial derivatives)

$$dx_i = \sum_k g^i_{,k}\, dy_k + \tfrac{1}{2} \sum_{k,l} g^i_{,kl}\, dy_k\, dy_l. \tag{5.7}$$

On the other hand, by (5.5), (5.6), the stochastic differentials of $(y_1, y_2) = (r, \phi)$ were obtained in terms of the local inverse of g:

$$x \to f(x) = \left(\sqrt{x_1^2 + x_2^2}\, , \tan^{-1} \frac{x_2}{x_1} \right);$$

thus

$$\begin{aligned} dy_k &= \sum_i f^k_{,i}\, dx_i + \tfrac{1}{2} \sum_{i,j} f^k_{,ij}\, dx_i\, dx_j \\ &= \sum_i f^k_{,i}(g(y)) \Big\{ \sum_r \sigma_{ir}(g(y))\, dw_r + b_i(g(y))\, dt \Big\} \\ &\quad + \tfrac{1}{2} \sum_{i,j,r} f^k_{,ij}(g(y)) \sigma_{ir}(g(y)) \sigma_{jr}(g(y))\, dt. \end{aligned} \tag{5.8}$$

If we substitute (5.8) into (5.7), we get (omitting the arguments of g, f, σ, b)

$$\begin{aligned} dx_i &= \sum_k g^i_{,k} \Big\{ \sum_{l,r} f^k_{,l} \sigma_{lr}\, dw_r + \Big[\tfrac{1}{2} \sum_{i,j,r} f^k_{,ij} \sigma_{ir} \sigma_{jr} + \sum f^k_{,i} b_i \Big] dt \Big\} \\ &\quad + \tfrac{1}{2} \Big(\sum_{k,l,p,q,r} g^i_{,kl} f^k_{,p} \sigma_{pr} f^l_{,q} \sigma_{qr} \Big) dt. \end{aligned}$$

Since f and g are inverse functions (locally), it follows (by differentiating once the relation $f(g(y)) = y$) that $\sum_i f^k_{,i} g^i_{,l} = \delta_{kl}$ and (by differentiating once the relation $\sum f^k_{,l} g^i_{,k} = \delta_{il}$) that

$$\sum g^i_{,k} f^k_{,lm} + \sum f^k_{,l} g^i_{,kp} f^p_{,m} = 0.$$

Using these relations in the last expression for dx_i, we arrive at $dx_i = \sum \sigma_{ir}\, dw_r + b_i\, dt$. This completes the proof.

Theorem 5.1 shows that $\phi(t)$, in (5.5), can be identified with the algebraic angle of the solution of (5.1).

Set

$$\sigma(\phi) = \left\{ \sum_{s=1}^{l} (\tilde{\tilde{\sigma}}_s(\phi))^2 \right\}^{1/2}, \qquad b(\phi) = \tilde{\tilde{b}}(\phi) \tag{5.9}$$

and consider the single stochastic equation

$$d\phi = \sigma(\phi)\, dw + b(\phi)\, dt \tag{5.10}$$

where here $w(t)$ is a one-dimensional Brownian motion. This equation has the differential operator

$$L_0 f \equiv \tfrac{1}{2}\sigma^2(\phi) f'' + b(\phi) f'. \tag{5.11}$$

The equation

$$d\phi = \sum_{s=1}^{l} \tilde{\tilde{\sigma}}_s(\phi)\, dw_s + \tilde{\tilde{b}}(\phi)\, dt$$

is an approximation to the second equation in (5.5), and it has the same differential operator (5.11). Thus one expects to study the behavior of the algebraic angle of the solution of (5.1) by analyzing the behavior of the solution of the equation (5.10).

Notice that the $\tilde{\tilde{\sigma}}_s(\phi)$, $\tilde{\tilde{b}}(\phi)$ are trigonometric polynomials, homogeneous of degree 2. In the following analysis, however, we shall not make use of the specific form of $\sigma(\phi)$, $b(\phi)$. We shall only make use of the fact that $\sigma(\phi)$, $b(\phi)$ are periodic functions of period 2π.

We first consider the case where $\sigma(\phi)$ does not vanish.

Theorem 5.2. *Assume that $r(t) \to 0$ a.s. when $t \to \infty$, that $\sigma(\phi) > 0$ for all ϕ, and that*

$$\Lambda = 2 \int_0^{2\pi} \frac{b(\phi)}{\sigma^2(\phi)}\, d\phi \neq 0.$$

Then

$$\lim_{t\to\infty} \frac{\phi(t)}{t} = c \quad a.s.$$

where c is a constant having the same sign as Λ.

Proof. Consider first the case $\Lambda > 0$. It suffices to find a function f such that

$$\tfrac{1}{2}\sigma^2(\phi)f''(\phi) + b(\phi)f'(\phi) = 1 \qquad (-\infty < \phi < \infty), \tag{5.12}$$

$$\lim_{\phi\to\infty} \frac{f(\phi)}{\phi} = \frac{1}{c} \qquad (c \text{ positive constant}), \tag{5.13}$$

$$f(\phi) \quad \text{remains bounded from above as} \quad \phi \to -\infty, \tag{5.14}$$

$$f'(\phi) \quad \text{is bounded.} \tag{5.15}$$

Indeed, if f is such a function, then by Itô's formula,

$$f(\phi(t)) = f(\phi(0)) + \sum_s \int_0^t \tilde{\tilde{\sigma}}_s(r, \phi)f'(\phi)\, dw_s$$
$$+ \int_0^t \left[\frac{1}{2} \sum_s (\tilde{\tilde{\sigma}}_s(r, \phi))^2 f''(\phi) + \tilde{\tilde{b}}(r, \phi)f'(\phi) \right] d\tau. \tag{5.16}$$

Since $|\tilde{\tilde{\sigma}}_s(r(t), \phi(t))f'(\phi(t))| \leqslant$ const, Corollary 4.4.6 gives

$$\lim_{t\to\infty} \frac{1}{t} \int_0^t \sum_s \tilde{\tilde{\sigma}}_s(r, \phi)f'(\phi)\, dw_s = 0 \quad \text{a.s.} \tag{5.17}$$

We now consider the integrand of the second integral on the right-hand side of (5.16). Given $\epsilon > 0$, let $r_0 > 0$ be such that

$$\left| \sum_{s=1}^{l} (\tilde{\tilde{\sigma}}_s(r, \phi))^2 - \sigma^2(\phi) \right| < \epsilon, \qquad |\tilde{\tilde{b}}(r, \theta) - b(\theta)| < \epsilon$$

for $0 < r < r_0$. Let $T_\epsilon = \sup\{t > 0;\ r(t) > r_0\}$. By assumption, $T_\epsilon < \infty$ a.s. From (5.15) and the equation (5.12) we see that f'' is a bounded function. Denote by K a bound on $|f'|$ and $|f''|$. For $t > T_\epsilon$ we then have, by (5.12),

$$\left| \tfrac{1}{2} \sum_s (\tilde{\tilde{\sigma}}_s(r(t), \phi(t)))^2 f''(\phi(t)) + \tilde{\tilde{b}}(r(t), \phi(t))f'(\phi(t)) - 1 \right| < 2\epsilon K.$$

Combining this with (5.17), we conclude from (5.16) that

$$\overline{\lim_{t\to\infty}} \frac{f(\phi(t))}{t} \leqslant 1 + 2\epsilon K,$$

$$\underline{\lim_{t\to\infty}} \frac{f(\phi(t))}{t} \geqslant 1 - 2\epsilon K.$$

This implies that a.s.

$$\lim_{t\to\infty} \frac{f(\phi(t))}{t} = 1;$$

in particular, because of (5.14), $\phi(t) \to \infty$ if $t \to \infty$. Invoking condition (5.13), we then get

$$\lim_{t\to\infty} \frac{\phi(t)}{t} = \lim_{t\to\infty} \frac{\phi(t)}{f(\phi(t))} \frac{f(\phi(t))}{t} = c,$$

which completes the proof of the theorem, subject to the construction of f.

To construct f, let

$$\beta(x) = \exp\left\{2\int_0^x \frac{b(\phi)}{\sigma^2(\phi)}\,d\phi\right\},$$
$$f(x) = \int_0^x \frac{1}{\beta(z)}\int_0^z \frac{2\beta(\phi)}{\sigma^2(\phi)}\,d\phi. \tag{5.18}$$

Clearly f satisfies (5.12). Since $\Lambda > 0$, we may write

$$2\int_0^x \frac{b(z)}{\sigma^2(z)}\,dz = \Lambda\frac{x}{2\pi} + m(x)$$

where

$$m(x) = 2\int_0^{x-[x/2\pi]2\pi} \frac{b(z)}{\sigma^2(z)}\,dz - \Lambda\left(\frac{x}{2\pi} - \left[\frac{x}{2\pi}\right]\right)$$

is a 2π-periodic function. Thus

$$\beta(x) = \exp\{\lambda x + m(x)\} \qquad (\lambda = \Lambda/2\pi).$$

Hence we have

$$f'(x) = \frac{1}{\beta(x)}\int_0^x \frac{\beta(z)}{\sigma^2(z)}\,dz = \int_0^x \frac{\exp\{\lambda z + m(z) - \lambda x - m(x)\}}{\sigma^2(z)}\,dz$$
$$= \int_0^x \frac{\exp\{-\lambda u + m(x-u) - m(x)\}}{\sigma^2(x-u)}\,du \qquad (u = x - z).$$

Since m is a bounded function and σ^2 is bounded below by a positive constant, the last integral $\int_0^x$ differs from the integral $\int_0^\infty$ (with the same integrand) by a quantity which is bounded by

$$\text{const}\int_x^\infty e^{-\lambda u}\,du = \text{const}\; e^{-\lambda x}.$$

Therefore,

$$f'(x) = \int_0^\infty \frac{\exp\{-\lambda u + m(x-u) - m(x)\}}{\sigma^2(x-u)}\,du + O(e^{-\lambda x}). \tag{5.19}$$

Denote the last integral by $G(x)$. Since m and σ^2 are 2π-periodic, the same is true of $G(x)$. Noting that

$$\left| f(x) - \int_0^x G(z)\,dz \right| < \text{const}\int_0^x e^{-\lambda z}\,dz = O(1),$$

we conclude that

$$\lim_{x\to\infty} \frac{f(x)}{x} = \lim_{x\to\infty} \frac{\int_0^x G(z)\,dz}{x} = \frac{1}{2\pi}\int_0^{2\pi} G(z)\,dz.$$

This proves (5.13). The condition (5.15) follows immediately from (5.19). Finally the condition (5.14) follows from the fact that $f'(x) > 0$. We have thus completed the proof of the theorem in case $\Lambda > 0$. The proof in case $\Lambda < 0$ is reduced to the previous case when applied to the process $(r(t), -\phi(t))$.

We now consider the case where $\sigma(\phi)$ is degenerate at a finite number of points.

Theorem 5.3. *Assume that $r(t) \to 0$ a.s. as $t \to \infty$ and that $\sigma(\phi)$ has a finite number of zeros. If $b(\phi) > 0$ at any point ϕ where $\sigma(\phi) = 0$, then*

$$\lim_{t\to\infty} \frac{\phi(t)}{t} = c \quad a.s. \tag{5.20}$$

where c is a positive constant.

Proof. Since $\sigma(\phi)$ vanishes at some points, one cannot find, in general, a regular solution of (5.12). We shall therefore aim at constructing a family of functions f_ϵ ($\epsilon > 0$) in $C^2(-\infty, \infty)$ satisfying:

$$1 - C\epsilon \leqslant L_0 f_\epsilon(x) \leqslant 1 + C\epsilon \qquad \text{if} \quad -\infty < x < \infty \quad (C \text{ const}), \tag{5.21}$$

$$|f_\epsilon'(x)| \leqslant C_1 \qquad \text{if} \quad -\infty < x < \infty \quad (C_1 \text{ const}), \tag{5.22}$$

$$\frac{1}{c} - \epsilon \leqslant \lim_{|x|\to\infty} \frac{f_\epsilon(x)}{x} \leqslant \frac{1}{c} + \epsilon \tag{5.23}$$

where c is a positive constant.

Once the f_ϵ have been constructed, we apply Itô's formula to f_ϵ (cf. (5.16))

$$\begin{aligned} f_\epsilon(\phi(t)) &- f_\epsilon\phi(0) \\ &= \int_0^t (L_0 f_\epsilon)(\phi(s))\, ds + \sum \int_0^t \tilde{\tilde{\sigma}}_j(r(s), \phi(s)) f_\epsilon'(\phi(s))\, dw_j(s) \\ &\quad + \tfrac{1}{2} \int_0^t [\sigma^2(r(s), \phi(s)) - \sigma^2(\phi(s))] f_\epsilon''(\phi(s))\, ds \\ &\quad + \int_0^t [\tilde{\tilde{b}}(r(s), \phi(s)) - b(\phi(s))] f_\epsilon'(\phi(s))\, ds \equiv J_1 + J_2 + J_3 + J_4, \end{aligned}$$

where $\sigma^2(r, \phi) = \sum (\tilde{\tilde{\sigma}}_i(r, \phi))^2$. By Corollary 4.4.6, $J_2 = o(t)$ at $t \to \infty$. Since $r(t) \to 0$ a.s. as $t \to \infty$, we also have from the continuity of $\sigma^2(r, \phi)$, $b(r, \phi)$ at $r = 0$,

$$J_3 = o(t), \qquad J_4 = o(t).$$

Using also (5.21), we conclude that

$$1 - C\epsilon \leqslant \varliminf_{t\to\infty} \frac{f_\epsilon(\phi(t))}{t} \leqslant \varlimsup_{t\to\infty} \frac{f_\epsilon(\phi(t))}{t} \leqslant 1 + C\epsilon.$$

From this and (5.23) it follows that $\phi(t)\to\infty$ if $t\to\infty$, and

$$c_1(\epsilon)(1 - C\epsilon) \le \varliminf_{t\to\infty} \frac{\phi(t)}{t} \le \varlimsup_{t\to\infty} \frac{\phi(t)}{t} \le c_2(\epsilon)(1 + C\epsilon)$$

where

$$\left| \frac{1}{c_i(\epsilon)} - \frac{1}{c} \right| \le \epsilon.$$

Taking $\epsilon\to 0$ we get $\lim_{t\to\infty} [\phi(t)/t] = c$ a.s. Thus it remains to construct the functions f_ϵ.

We shall need a few facts regarding the differential equation

$$\tfrac{1}{2}\sigma^2(x)u'(x) + b(x)u(x) = g \qquad (\alpha < x < \beta). \tag{5.24}$$

We assume that b, g are continuous on $[\alpha, \beta]$ and $\sigma(\alpha) = \sigma(\beta) = 0$. Further, $\sigma(x)$ is Lipschitz continuous in $[\alpha, \beta]$ and does not vanish faster than $\exp[-\epsilon|x - x_0|^{-1}]$ at $x_0 = \alpha, \beta$, for any $\epsilon > 0$.

Lemma 5.4. *Let the foregoing assumptions hold and assume that $\sigma^2(x) > 0$ for $\alpha < x < \beta$, $b(\alpha) > 0$, $b(\beta) > 0$. Then there exists a unique bounded solution $u(x)$ of (5.24) in the interval $\alpha < x < \beta$. This solution satisfies*

$$u(\alpha + 0) = \frac{g(\alpha)}{b(\alpha)}, \qquad u(\beta - 0) = \frac{g(\beta)}{b(\beta)}. \tag{5.25}$$

Furthermore, if g is everywhere positive, then $u(x)$ is everywhere positive.

Proof. Let

$$s(x) = \exp\left\{ -2 \int_{x_0}^{x} \frac{b(y)}{\sigma^2(y)}\, dy \right\} \qquad (\alpha < x_0 < \beta).$$

Then $\frac{1}{2}\sigma^2 s' + bs = 0$, $s(\alpha + 0) = \infty$, $s(\beta - 0) = 0$. Observe that u is a solution of (5.24) if and only if

$$\left(\frac{u}{s}\right)' = \frac{2g}{s\sigma^2}.$$

Consequently the general solution of (5.24) is

$$u(x) = c_1 s(x) + 2s(x) \int_{\alpha}^{x} \frac{2g(y)}{s(y)\sigma^2(y)}\, dy;$$

the integral is convergent since

$$\frac{1}{s(y)} \le \exp\left[-\frac{c}{y - \alpha} \right], \qquad c > 0.$$

If $u(x)$ is a bounded solution, then $u(x)/s(x)\to 0$ if $x \downarrow \alpha$. Consequently

$c_1 = 0$, and

$$u(x) = 2s(x) \int_\alpha^x \frac{g(y)\,dy}{s(y)\sigma^2(y)}. \tag{5.26}$$

It remains to prove (5.25). Since

$$\left(\frac{1}{s}\right)' = -\frac{s'}{s^2} = \frac{2b}{s\sigma^2},$$

an application of l'Hospital's rule gives

$$u(\alpha + 0) = \lim_{x \downarrow \alpha} \frac{g(x)/s(x)\sigma^2(x)}{b(x)/s(x)\sigma^2(x)} = \frac{g(\alpha)}{b(\alpha)}.$$

Similarly, $u(\beta - 0) = g(\beta)/b(\beta)$.

Lemma 5.5. *Let the assumptions of Lemma 5.4 hold and assume that $g \equiv 1$ and that $b(x)$ is constant in a neighborhood of the end points α, β. Then the bounded solution $u(x)$ of (5.24) is in $C^1[\alpha, \beta]$ and $u'(\alpha + 0) = u'(\beta - 0) = 0$.*

Proof. Since

$$\frac{1}{2s(x)} = -\frac{1}{2}\int_\alpha^x \frac{s'}{s^2}\,dy = \frac{1}{2}\int_\alpha^x \frac{2bs}{\sigma^2 s}\,dy = \int_\alpha^x \frac{b}{s\sigma^2}\,dy,$$

we get from (5.26),

$$u(x) = \int_\alpha^x \frac{dy}{s\sigma^2} \Big/ \int_\alpha^x \frac{b\,dy}{s\sigma^2}. \tag{5.27}$$

If $b(x) = b(\alpha)$ for $\alpha < x < \alpha + \delta$, then it follows from (5.27) that $u(x) = 1/b(\alpha)$ for $\alpha < x < \alpha + \delta$. Thus u is continuously differentiable in $[\alpha, \alpha + \delta)$ with $u'(\alpha + 0) = 0$.

Suppose next that $u(x) = b(\beta)$ if $\beta - \delta < x < \beta$. Using (5.27) we see that

$$\begin{aligned}\frac{1}{u(x)} - b(\beta) &= \int_\alpha^{\beta-\delta} \frac{b - b(\beta)}{s\sigma^2}\,dy \Big/ \int_\alpha^x \frac{b}{s\sigma^2}\,dy \\ &= 2s(x)\int_\alpha^{\beta-\delta} \frac{b - b(\beta)}{s\sigma^2}\,dy.\end{aligned}$$

Hence

$$\left|\frac{1}{u(x)} - b(\beta)\right| \leqslant Cs(x) \leqslant C\exp[-c|x - \beta|^{-1}] \tag{5.28}$$

for some $C > 0$, $c > 0$. Since

$$u' = \frac{2u}{\sigma^2}\left(\frac{1}{u} - b\right)$$

we conclude that u' is continuous in $(\beta - \delta, \beta]$ when defined at β by $u'(\beta) = 0$. By the mean value theorem we thus also have $u'(\beta - 0) = 0$. This completes the proof.

We shall now approximate the solution of (5.24) with $g \equiv 1$ by the bounded solutions u_ϵ of

$$\tfrac{1}{2}\sigma^2(x)u_\epsilon'(x) + b_\epsilon(x)u_\epsilon'(x) = 1, \tag{5.29}$$

where

$$b_\epsilon(x) = \begin{cases} b(\alpha) & \text{if } \alpha \leqslant x < \alpha + \delta, \\ b(\alpha) + (x - \alpha - \delta) & \\ \qquad \times \dfrac{b(\alpha + 2\delta) - b(\alpha)}{\delta} & \text{if } \alpha + \delta \leqslant x \leqslant \alpha + 2\delta, \\ b(x) & \text{if } \alpha + 2\delta \leqslant x < \beta - 2\delta, \\ b(\beta) - (\beta - x - \delta) & \\ \qquad \times \dfrac{b(\beta) - b(\beta - 2\delta)}{\delta} & \text{if } \beta - 2\delta \leqslant x < \beta - \delta, \\ b(\beta) & \text{if } \beta - \delta \leqslant x \leqslant \beta \end{cases}$$

where $\delta > 0$ is chosen such that $|b_\epsilon(x) - b(x)| \leqslant \epsilon$. Note that, by Lemma 5.5, u_ϵ is in $C^1[\alpha, \beta]$.

Lemma 5.6. *Let the conditions of Lemma 5.4 hold with $g \equiv 1$. The unique bounded solutions u, u_ϵ of (5.24) (with $g \equiv 1$) and (5.29) satisfy*

$$|u(x) - u_\epsilon(x)| \leqslant C\epsilon \qquad (\alpha < x < \beta) \tag{5.30}$$

where C is a constant independent of ϵ.

Proof. Clearly

$$\tfrac{1}{2}\sigma^2(u - u_\epsilon)' + b(u - u_\epsilon) = (b_\epsilon - b)u_\epsilon.$$

From the proof of Lemma 5.4 we have

$$u - u_\epsilon = 2s(x)\int_\alpha^x \frac{(b_\epsilon - b)u_\epsilon}{s\sigma^2}\,dy. \tag{5.31}$$

Setting

$$K(x, y) = 2s(x)\int_\alpha^x \frac{dy}{s\sigma^2}$$

we then have

$$u_\epsilon(x) \leq u(x) + \epsilon \int_\alpha^x K(x, y)u_\epsilon(y)\, dy.$$

Since $K(x, y) \leq C'$ (C' constant), we obtain, by iteration, $u_\epsilon(x) \leq C'$ l.u.b. $u(x)$, with another positive constant C'. Substituting this into the right-hand side of (5.31), the assertion of the lemma follows.

Completion of the proof of Theorem 5.3. Let $\cdots < x_{-1} < x_0 < x_1 < \cdots$ be the zeros of $\sigma^2(x)$. In each interval (x_k, x_{k+1}) we form a function b_ϵ as in the proof of Lemma 5.6. Denote the corresponding solution of (5.29) by $u_\epsilon^k(x)$, and set $u_\epsilon(x) = u_\epsilon^k(x)$ if $x_k \leq x \leq x_{k+1}$. By Lemma 5.4, $u_\epsilon(x)$ is continuous at the points x_k (with $u_\epsilon(x_k) = 1/b(x_k)$) and by Lemma 5.5 $u_\epsilon'(x)$ is also continuous at the points x_k, with $u_\epsilon'(x_k) = 0$. By uniqueness of the bounded solution and the periodicity of $\sigma^2(x)$, $b_\epsilon(x)$ we deduce that also $u_\epsilon(x)$ is periodic. Set

$$f_\epsilon(x) = \int_0^x u_\epsilon(y)\, dy.$$

Then clearly $f_\epsilon \in C^2(-\infty, \infty)$ and, due to the uniform boundedness of u_ϵ (which follows from (5.30)), the estimate (5.22) is valid. Next, as easily seen,

$$L_0 f_\epsilon = 1 + (b - b_\epsilon)u_\epsilon,$$

and (5.21) thus follows.

To prove (5.23) write

$$f_\epsilon(x) = \int_0^x u(y)\, dy + \int_0^x [u_\epsilon(y) - u(y)]\, dy.$$

Using the periodicity of u and (5.30) we see that

$$\overline{\lim_{|x|\to\infty}} \left| \frac{f_\epsilon(x)}{x} - \frac{1}{2\pi} \int_0^{2\pi} u(y)\, dy \right| \leq C_0\epsilon$$

where C_0 is a constant independent of ϵ. This gives (5.23) with $c > 0$.

Remark. If in Theorem 5.3 we assume that $b(\phi) < 0$ at each point ϕ where $\sigma(\phi) = 0$, then the assertion (5.20) holds with a negative constant c.

6. Spiraling of solutions about any obstacle

We continue to specialize to the case $n = 2$. We shall consider the spiraling of solution about a general obstacle. Consider first the case where G is a closed unit disk and assume that σ_{is}, b_i are continuously differentiable in a neighborhood of $|x| = 1$. We also assume the conditions (2.1), (2.5) so that ∂G is nonattainable and invariant.

Introducing coordinates (r, ϕ) by $x_1 = (r + 1)\cos\phi$, $x_2 = (r + 1)\sin\phi$ we can rewrite the differential system (5.1) in the form

$$\begin{aligned} dr &= r\left[\sum_{s=1}^{l} \tilde{\sigma}_s(\phi)\, dw_s + \tilde{b}(\phi)\, dt\right] + \left[\sum_{s=1}^{l} R_s\, dw_s + R_0\, dt\right], \\ d\phi &= \left[\sum_{s=1}^{l} \tilde{\tilde{\sigma}}_s(\phi)\, dw_s + \tilde{\tilde{b}}(\phi)\, dt\right] + \left[\sum_{s=1}^{l} \theta_s\, dw_s + \theta_0\, dt\right] \end{aligned} \tag{6.1}$$

where $R_s = o(r)$, $\theta_s = o(1)$ if $r \downarrow 0$. The functions $\tilde{\sigma}_s(\phi)$, $\tilde{b}(\phi)$, $\tilde{\tilde{\sigma}}_s(\phi)$, $\tilde{\tilde{b}}(\phi)$ are 2π-periodic continuous functions which are not necessarily trigonometric polynomials. We adhere to the notation

$$\sigma(\phi) = \left\{\sum_{s=1}^{l} (\tilde{\tilde{\sigma}}_s(\phi))^2\right\}^{1/2}, \qquad b(\phi) = \tilde{\tilde{b}}(\phi).$$

Let $y(t) = (r(t), \phi(t))$ be the solution of (6.1) with $r(0) > 0$ and set

$$x_1(t) = (r(t) + 1)\cos\phi(t), \qquad x_2(t) = (r(t) + 1)\sin\phi(t).$$

As in Section 5 one can show that $r(t) > 0$ a.s. for all $t \geq 0$ and $x(t) = (x_1(t), x_2(t))$ is a solution of (5.1). Thus $\phi(t)$ represents the algebraic angle of the solution of (5.1).

Theorem 5.2 now extend word by word to the present case.

If we assume that $\sigma(x)$ has no zeros of infinite order (more precisely, that, for any $\epsilon > 0$, $\sigma(x)$ does not vanish faster than $\exp[-\epsilon|x - x_0|^{-1}]$ at each of its zeros x_0), then Theorem 5.3 also remains valid (with precisely the same proof) in this case.

Consider now the case of a general closed domain G with connected C^3 boundary ∂G.

We introduce new variables (r, ϕ) in a $G^c \equiv R^n \backslash (\text{int } G)$ neighborhood of ∂G, by

$$\phi = 2\pi s/L, \qquad x_1 = f(s) + r\dot{g}(s), \qquad x_2 = g(s) - r\dot{f}(s), \tag{6.2}$$

$0 \leq s \leq L$, $0 \leq r \leq \epsilon_0$ and $\dot{f}^2 + \dot{g}^2 = 1$; L is the length of the boundary ∂G.

Then the stochastic differentials dr, $d\phi$ can be computed in the form

$$\begin{aligned} dr &= \sum_{s=1}^{l} \tilde{\sigma}_s\, dw_s + \tilde{b}\, dt. \\ d\phi &= \sum_{s=1}^{l} \tilde{\tilde{\sigma}}_s\, dw_s + \tilde{\tilde{b}}\, dt. \end{aligned} \tag{6.3}$$

In order to express $\tilde{\sigma}_s$, $\tilde{\tilde{\sigma}}_s$, $\tilde{b}$, $\tilde{\tilde{b}}$ in terms of the σ_{ij}, b_i, we compute dx_i using (6.2) and then compare with the expression for dx_i given by (5.1) (with

$\xi_i = x_i$). After some calculation we arrive at the formulas

$$\tilde{\tilde{\sigma}}_s\,(0,\phi) = \frac{2\pi}{L}\,(\dot{f}\sigma_{1s} + \dot{g}\sigma_{2s})$$
$$\tilde{\tilde{b}}\,(0,\phi) = \frac{2\pi}{L}\left\{(\dot{f}b_1 + gb_2) - (\dot{g}, -\dot{f}\,)\begin{pmatrix} \sum \sigma_{1s}^2 & \sum \sigma_{1s}\sigma_{2s} \\ \sum \sigma_{1s}\sigma_{2s} & \sum \sigma_{2s}^2 \end{pmatrix}\begin{pmatrix} \dot{f} \\ \dot{g} \end{pmatrix}\right\}. \tag{6.4}$$

Set

$$\sigma(\phi) = \left\{\sum_{s=1}^{l} (\tilde{\tilde{\sigma}}_s(0,\phi))^2\right\}^{1/2}, \qquad b(\phi) = \tilde{\tilde{b}}(0,\phi). \tag{6.5}$$

The system (6.3) was defined only locally, i.e., for $0 < r < \epsilon_0$. Suppose we could extend the mapping $(x_1, x_2) \to (r, \phi)$ into a global diffeomorphism Δ from $R^n \setminus G$ into $\{y; |y| > 0\}$, where $y_1 = r\cos\phi$, $y_2 = r\sin\phi$, such that the first derivatives of Δ and of its inverse are bounded near ∞. Then we could apply Theorems 5.2, 5.3 directly to the system (6.3) and conclude:

Theorem 6.1. *Theorems 5.2, 5.3 remain valid for G with $\sigma(\phi)$, $b(\phi)$ given by (6.4), (6.5), provided the conditions (2.1), (2.5) hold and provided the zeros of $\sigma(\phi)$ are of finite order.*

This theorem can be proved without assuming the existence of the above diffeomorphism Δ. Indeed, we first construct a function f satisfying (5.12)–(5.15) (for Theorem 5.2) and functions f_ϵ satisfying (5.21)–(5.23) (for Theorem 5.3). These are functions of ϕ which in turn is a function of s (by (6.2)). Thus, $f = f(s)$, $f_\epsilon = f_\epsilon(s)$. One can check that

$$L_0 f(s) = 1, \qquad 1 - C\epsilon < L_0 f_\epsilon(s) < 1 + C\epsilon, \tag{6.6}$$

where $L_0 v$ is the differential operator Lv restricted to ∂G, i.e., to $r = 0$. Since $r(t) \to 0$ a.s. as $t \to \infty$, we can now proceed to apply Itô's formula to $\tilde{f}(x_1, x_2)$ (for Theorem 5.2) and to $\tilde{f}_\epsilon(x_1, x_2)$ (for Theorem 5.3) and argue in the same way as in the proofs of Theorems 5.2, 5.3; here $\tilde{f}$ and $\tilde{f}_\epsilon$ are C^2 extensions of f and f_ϵ, respectively, into R^n, which are constants along the normal to ∂G in a small neighborhood of ∂G.

Remark 1. The assertion $\phi(t)/t \to c$ can be stated in the following form: Denote by $(r(t), s(t))$ the position of the solution $\xi(t)$ near the boundary ∂G, where $r(t)$ is the distance from ∂G and $s(t)$ is the "algebraic length." (If a point moves along ∂G so that its argument increases (decreases) by 2π, its "algebraic length" increases (decreases) by L.) Then $s(t)/t \to cL/2\pi$ as $t \to \infty$.

Remark 2. If in Theorem 6.1 we assume that $r(t) \to 0$ as $t \to \infty$ for all ω in a set Ω_0, then the assertion $\phi(t)/t \to c$ as $t \to \infty$ is valid a.s. in Ω_0. This is obvious from the proofs of Theorems 5.2, 5.3, 6.1.

7. Spiraling for linear systems

Consider a linear stochastic system

$$d\xi_i = \sum_{j=1}^{2} \sum_{s=1}^{l} \sigma_{ij}^s \xi_j \, dw_s + \sum_{j=1}^{2} b_{ij}\xi_j \, dt \qquad (i = 1, 2) \tag{7.1}$$

in the plane. Introducing polar coordinates $x_1 = r\cos\phi$, $x_2 = r\sin\phi$, we find that

$$d\phi = \sum_{s=1}^{l} \langle \sigma_s\lambda, \lambda^\perp \rangle \, dw_s + \{\langle B\lambda, \lambda^\perp\rangle - \langle a(\lambda)\lambda, \lambda^\perp\rangle\} \, dt \tag{7.2}$$

where

$$\sigma_s = (\sigma_{ij}^s), \qquad B = (b_{ij}), \qquad \lambda = (\cos\phi, \sin\phi), \qquad \lambda^\perp = (-\sin\phi, \cos\phi),$$

$$a(\lambda) = (a_{ij}(\lambda)), \qquad a_{ij}(x) = \sum_{k,l,s} \sigma_{ik}^s \sigma_{jl}^s x_k x_l.$$

Thus the variable r does not enter into the differential equation for ϕ. Consequently $\phi(t)$ *defines a diffusion process*. The differential operator corresponding to it is

$$Lu = \tfrac{1}{2}\sigma^2(x)u'' + b(x)u'$$

where

$$\sigma(\phi) = \left\{ \sum_{s=1}^{l} \langle \sigma_s\lambda, \lambda^\perp\rangle^2 \right\}^{1/2}, \qquad b(\phi) = \langle B\lambda, \lambda^\perp\rangle - \langle a(\lambda)\lambda, \lambda^\perp\rangle. \tag{7.3}$$

We shall now study the behavior of ϕ in cases not covered by Theorems 5.2, 5.3. Suppose first that

$$\sigma(\phi) > 0 \quad \text{if} \quad \alpha < \phi < \beta, \quad \sigma(\alpha) = 0, \quad \sigma(\beta) = 0, \tag{7.4}$$

$$b(\alpha) \geq 0, \qquad b(\beta) \leq 0. \tag{7.5}$$

If $\alpha < \phi(0) < \beta$, then $\alpha < \phi(t) < \beta$ for all $t \geq 0$. To study the limit behavior of $\phi(t)$, let $\alpha < \gamma < \beta$ and introduce the function

$$\pi(x) = \int_\gamma^x \exp\left[-\int_\gamma^z \frac{2b(s)}{\sigma^2(s)} \, ds \right] dz.$$

This is a solution of $L\pi = 0$.

Theorem 7.1. *Let $\phi(0) = x$, $\alpha < x < \beta$ and let* (7.4), (7.5) *hold.*

(i) *If $\pi(\alpha) = -\infty$, $\pi(\beta) = \infty$, then $\alpha < \phi(t) < \beta$ for all $t > 0$ and*

$$\varliminf_{t\to\infty} \phi(t) = \inf_{t>0} \phi(t) = \alpha \quad a.s.,$$

$$\varlimsup_{t\to\infty} \phi(t) = \sup_{t>0} \phi(t) = \beta \quad a.s.$$

(ii) *If* $\pi(\alpha) > -\infty$, $\pi(\beta) = \infty$, *then* $\alpha < \phi(t) < \beta$ *for all* $t > 0$ *and*

$$\lim_{t\to\infty} \phi(t) = \inf_{t>0} \phi(t) = \alpha \quad a.s.,$$
$$\sup_{t>0} \phi(t) < \beta \quad a.s.$$

(iii) *If* $\pi(\alpha) = -\infty$, $\pi(\beta) < \infty$, *then* $\alpha < \phi(t) < \beta$ *for all* $t > 0$ *and*

$$\lim_{t\to\infty} \phi(t) = \sup_{t>0} \phi(t) = \beta \quad a.s.,$$
$$\inf_{t>0} \phi(t) > \alpha \quad a.s.$$

(iv) *If* $\pi(\alpha) > -\infty$, $\pi(\beta) < \infty$, *then* $\alpha < \phi(t) < \beta$ *for all* $t > 0$ *and*

$$P\Big\{\lim_{t\to\infty} \phi(t) = \alpha\Big\} = P\Big\{\inf_{t>0} \phi(t) = \alpha\Big\} = \frac{\pi(\beta) - \pi(x)}{\pi(\beta) - \pi(\alpha)},$$
$$P\Big\{\lim_{t\to\infty} \phi(t) = \beta\Big\} = P\Big\{\sup_{t>0} \phi(t) = \beta\Big\} = \frac{\pi(x) - \pi(\alpha)}{\pi(\beta) - \pi(\alpha)}.$$

The proof is similar to the proof of 9.7.1, with the roles of $-\infty$, $+\infty$ given to the points $\phi = \alpha$ and $\phi = \beta$ respectively. The details are left to the reader.

We next consider the case where (7.4) holds and

$$b(\alpha) = 0, \qquad b(\beta) > 0. \tag{7.6}$$

Notice that if $\phi(0) = x > \alpha$, then $\phi(t) > \alpha$ a.s. for all $t \geq 0$. We denote by $\tau_x[a, c]$ the exit time from (a, c) given $\phi(0) = x \in (a, c)$.

Theorem 7.2. *Let* $\phi(0) = x$, $\alpha < x < \beta$ *and let* (7.4), (7.6) *hold. If* $\pi(\alpha) = -\infty$, *then* $\tau_x[\alpha, \beta] < \infty$ *a.s. and*

$$P\{\phi(\tau_x[\alpha, \beta]) = \beta\} = 1.$$

If $\pi(\alpha) > -\infty$, *then* $\phi(t) > \alpha$ *a.s. and*

$$P\{\phi(\tau_x[\alpha, \beta]) = \beta\} = \frac{\pi(x) - \pi(\alpha)}{\pi(\beta) - \pi(\alpha)},$$
$$P\Big\{\lim_{t\to\infty} \phi(t) = \alpha\Big\} = P\Big\{\inf_{t>0} \phi(t) = \alpha\Big\} = \frac{\pi(\beta) - \pi(x)}{\pi(\beta) - \pi(\alpha)}.$$

Proof. The proof of Theorem 5.3 provides C^2 functions f_ϵ in $[\alpha + \delta, \beta + \delta]$ satisfying $Lf_\epsilon > 1 - \epsilon$; here δ and ϵ are any positive numbers. An application of Itô's formula with f_ϵ (when $\epsilon = \frac{1}{2}$) gives $E\tau_x[\alpha + \delta, \beta + \delta] < \infty$. In particular, $\tau_x[\alpha + \epsilon, \beta] < \infty$ a.s. for any $\epsilon > 0$.

Using this last fact and Itô's formula we get (cf. Problem 12, Chapter 8)

$$P\{\phi(\tau_x[\alpha+\epsilon,\beta])=\beta\}=\frac{\pi(x)-\pi(\alpha+\epsilon)}{\pi(\beta)-\pi(\alpha+\epsilon)},$$

$$P\{\phi(\tau_x[\alpha+\epsilon,\beta])=\alpha+\epsilon\}=\frac{\pi(x)-\pi(\alpha+\epsilon)}{\pi(\beta)-\pi(\alpha+\epsilon)}.$$

Taking $\epsilon\to 0$ the assertions of the theorem readily follow.

The function $(\sigma(\phi))^2$ is a homogeneous polynomial of degree 4 in $(\cos\phi, \sin\phi)$. Consequently, $\sigma(\phi)$ is periodic of period π, and it can have at most two zeros in the interval $[0,\pi)$. The function $b(\phi)$ is also periodic of period π.

The following possibilities may take place:

(i) $\sigma(\phi)$ has two distinct zeros ϕ_1, ϕ_2 in the interval $[0,\pi)$.
(ii) $\sigma(\phi)$ has one zero ϕ_1 in the interval $[0,\pi)$.
(iii) $\sigma(\phi)$ does not vanish in the interval $[0,\pi)$.

If (iii) holds, then Theorem 5.2 can be applied. Suppose (ii) holds. If $b(\phi_1)\neq 0$, then Theorem 5.3 can be applied (see the remark at the end of Section 5). If, on the other hand, $b(\phi_1)=0$, then Theorem 7.1 can be applied.

Suppose finally that (i) holds. If $b(\phi_i)>0$ for $i=1,2$ or $b(\phi_i)<0$ for $i=1,2$ then Theorem 5.3 can be applied. If $b(\phi_1)\gtrdot 0$, $b(\phi_2)\lessdot 0$, then Theorem 7.1 can be applied. If $b(\phi_1)=0$, $b(\phi_2)>0$, then Theorem 7.2 applies. The last possibility is $b(\phi_1)<0$, $b(\phi_2)=0$; in this case an analogue of Theorem 7.2 can be applied.

We shall now compare the behavior of $\phi(t)$ for the stochastic system (7.1) with the behavior of $\phi(t)$ for the deterministic system

$$dx_i=\sum_{j=1}^{2} b_{ij}x_j\,dt \qquad (i=1,2). \tag{7.7}$$

Lemma 7.3. *If for some ϕ, $\sigma(\phi)=0$, then $b(\phi)=\langle B\lambda,\lambda^{\perp}\rangle$.*

Proof. From the definition of $\sigma(\phi)$ we have

$$(\sigma(\phi))^2=\sum_{s=1}^{l}\left(\sum_{i,j=1}^{2}\sigma_{ij}^{s}\lambda_j\lambda_i^{\perp}\right)^2=\sum_{s=1}^{l}(T_{1s}\sin\phi-T_{2s}\cos\phi)^2$$

where $T_{is}=\sigma_{i1}^{s}\cos\phi+\sigma_{i2}^{s}\sin\phi$. Next

$$b(\phi)-\langle B\lambda,\lambda^{\perp}\rangle=-\langle a(\lambda)\lambda,\lambda^{\perp}\rangle=-\sum_{s=1}^{l}\sum_{i,k,j,m=1}^{2}\sigma_{ik}^{s}\lambda_k\sigma_{jm}^{s}\lambda_m\lambda_i\lambda_j^{\perp}$$

$$=-\sum_{s=1}^{l}(T_{1s}\cos\phi+T_{2s}\sin\phi)(-T_{1s}\sin\phi+T_{2s}\cos\phi).$$

Consequently, the right-hand side vanishes whenever $\sigma(\phi) = 0$, and the assertion follows.

Theorem 7.4. *Let* (i) *or* (ii) *hold. If for the deterministic system* (7.7) $\phi(t)\to\infty$ *as* $t\to\infty$, *then the same is true for the stochastic system* (7.1), *i.e., if* $x(t) = (r(t)\cos\phi(t), r(t)\sin\phi(t))$, *then a.s.* $\phi(t)\to\infty$ *as* $t\to\infty$; *in fact* $[\phi(t)/t]\to c$, *c a positive constant. Similarly if* $\phi(t)\to-\infty$ *as* $t\to-\infty$ *for the deterministic system, then, for the stochastic system, a.s.* $[\phi(t)/t]\to -c_0$ *as* $t\to\infty$, *where* c_0 *is a positive constant.*

Proof. In the deterministic case, $|\phi(t)|\to\infty$ as $t\to\infty$ if and only if the origin is a focal point (spiral or vortex). This is the case if and only if the eigenvalues of B are nonreal, i.e., if and only if $\langle B\lambda, \lambda^{\perp}\rangle \neq 0$ for all $\lambda = (\cos, \phi, \sin\phi)$. Now use Lemma 7.3 and Theorem 5.3.

If the eigenvalues of B are real (of the same sign for nodal points, and of different sign for saddle points), then $\langle B\lambda, \lambda^{\perp}\rangle$ does not have a fixed (positive or negative) sign. Nevertheless, the stochastic solutions may still spiral in accordance with Theorems 5.2, 5.3 if either (iii) holds and $\Lambda \neq 0$ or if (ii) holds and $\langle B\lambda_1, \lambda_1^{\perp}\rangle \neq 0$, or if (i) holds and $\langle B\lambda_i, \lambda_i^{\perp}\rangle$ is positive for $i = 1, 2$ or negative for $i = 1, 2$; here $\lambda_i = (\cos\phi_i, \sin\phi_i)$. In all other cases, the stochastic solution does not spiral.

PROBLEMS

1. Let (A_1) hold and let A, B be sets in R^n, A compact, B open, $A\subset B$. Prove that for any $0 \leq t_1 < t_2 < \cdots < t_m$,

$$\varliminf_{x\to y} P_x\{\xi(t_1)\in A, \ldots, \xi(t_m)\in A\} \geq P_y\{\xi(t_1)\in B, \ldots, \xi(t_m)\in B\}.$$

2. A stable set K is invariant. [*Hint*: Let U be any neighborhood of K and ϵ, U_ϵ as in the definition of stability; τ_ϵ is the exit time from U_ϵ, τ the exit time from U. If $x\in K$, $P_x(\tau < \infty) = E_x\chi_{\tau_\epsilon<\infty}P_{\xi(\tau_\epsilon)}(\tau < \infty) < \epsilon$. Another method if K is the boundary of a domain: Use Problem 1.]
3. If the boundary of a domain D is stable from the outside, then $\bar{D}$ is an invariant set. [*Hint*: Cf. Problem 2.]
4. Extend Theorem 2.3 to G convex and with piecewise C^3 boundary, assuming (11.6.6) and (2.13) for any $\nu \in N_x$, $x \in \partial G$ [see Section 11.6 for the definition of N_x.]
5. Verify (3.4). [*Hint*: If $v = v(\theta)$, $L_0 v = \sum \alpha_{\lambda\mu}\, \partial^2 v/\partial\theta_\lambda\, \partial\theta_\mu + \sum \beta_\lambda\, \partial v/\partial\theta_\lambda$ where $\alpha_{\lambda\mu} = \sum a_{ij}(\partial\theta_\lambda/\partial x_i)(\partial\theta_\mu/\partial x_j)$ The inequality $\sum\alpha_{\lambda\mu}\gamma_\lambda\gamma_\mu \geq \alpha|\gamma|^2$ follows from (3.2), noting that $\operatorname{grad}(\sum\gamma_\lambda\theta_\lambda)$ is orthogonal to x.]
6. Extend Theorem 3.1 to the case of an obstacle $\partial G = \{x; |x| = 1\}$.
7. Let G be a convex domain containing the origin, with boundary ∂G

given by $r = g(\phi)$. The function $g(\phi)$ is Lipschitz continuous. Define

$$\sigma(\phi) = \left\{ \sum_{s=1}^{l} [\tilde{\tilde{\sigma}}_s(g(\phi), \phi)]^2 \right\}^{1/2}, \qquad b(\phi) = \tilde{\tilde{b}}(g(\phi), \phi)$$

where the $\tilde{\tilde{\sigma}}_s$, $\tilde{\tilde{b}}$ are as in Section 5. Prove that Theorem 6.1 extends to the present case. (Note that ∂G is not assumed to be in C^3.)

8. Let G be a closed bounded domain with C^3 boundary, and let $\rho(x) = d(x, G)$ if $x \in G$. If (2.1) holds and if

$$\sum b_i \frac{\partial \rho}{\partial x_i} + \frac{1}{2} \sum a_{ij} \frac{\partial^2 \rho}{\partial x_i \, \partial x_j} > 0 \qquad \text{on } \partial G,$$

then $R^n \setminus (\text{int } G)$ is not an invariant set and, consequently, ∂G is not stable from the inside. [*Hint*: If $P_x\{\xi(t) \in G\} = 1$ when $x \in \partial G$, then get a contradiction by using Itô's formula with $\rho^2(x)$.]

9. If for a linear system, with $x = 0$ as obstacle, $Q(x) \geqslant \beta > 0$ for all x, $|x| = 1$, then

$$P_x\left\{ \varliminf_{t\to\infty} \frac{\log |\xi(t)|}{t} \geqslant \beta \right\} = 1, \qquad E_x \log |\xi(t)| > -C + \beta t$$

for any $x \neq 0$, where C is a constant (depending on x).

10. Verify (6.4).

11. Give the details of the proof of Theorem 7.1.

12. For a linear system, the assertions of Theorem 5.2 can be stated as follows:

$$\lim_{t\to\infty} \frac{\phi(t)}{t} = \frac{2\pi}{E(T_1)} \quad \text{a.s.} \qquad \text{if} \quad \Lambda > 0$$

where $T_1 = \inf\{t;\ \phi(t) - \phi_0 = 2\pi\}$, and

$$\lim_{t\to\infty} \frac{\phi(t)}{t} = -\frac{2\pi}{E(T_{-1})} \quad \text{a.s.} \qquad \text{if} \quad \Lambda < 0$$

where $T_{-1} = \inf\{t;\ \phi(t) - \phi_0 = -2\pi\}$. Further, if $\Lambda = 0$, then

$$\varlimsup_{t\to\infty} \phi(t) = \infty, \qquad \varliminf_{t\to\infty} \phi(t) = -\infty \quad \text{a.s.}$$

[*Hint*: If $\Lambda = 0$, cf. Theorem 9.7.1(a). If $\Lambda > 0$, take $\phi_0 = 0$ and define $T_m = \inf\{t;\ \phi(t) = 2m\pi\}$. Show $E(T_1) < \infty$. By the strong law of large numbers $T_m \sim mE(T_1)$.]

13

The Dirichlet Problem for Degenerate Elliptic Equations

1. A general existence theorem

Consider a partial differential operator

$$Lu = \frac{1}{2} \sum_{i,j=1}^{n} a_{ij}(x) \frac{\partial^2 u}{\partial x_i \, \partial x_j} + \sum_{i=1}^{n} b_i(x) \frac{\partial u}{\partial x_i} \tag{1.1}$$

with coefficients defined in the closure $\bar{D}$ of a bounded domain D. It is assumed that the matrix $(a_{ij}(x))$ is nonnegative definite in D. If L is uniformly elliptic, then the Dirichlet problem consists in solving

$$Lu + c(x)u = f(x) \qquad \text{in } D, \tag{1.2}$$

$$u = \phi \qquad \text{on } \partial D. \tag{1.3}$$

This problem has already been studied in Chapter 6. In this chapter we consider the case where L is degenerating on a subset of $\bar{D}$. Our methods will rely upon the theory of stochastic differential equations. We therefore assume:

(A) There exists a uniformly Lipschitz continuous matrix $(\sigma_{ij}(x))$ in R^n such that $a_{ij}(x) = \sum_{k=1}^{n} \sigma_{ik}(x)\sigma_{jk}(x)$ if $x \in \bar{D}$. Further, the vector $b = (b_1, \ldots, b_n)$ is uniformly Lipschitz continuous in R^n.

We can then introduce the stochastic differential equations

$$d\xi = \sigma(\xi)\, dw + b(\xi)\, dt \tag{1.4}$$

whose differential operator is L.

Recall that, by Section 6.1, if the matrix (a_{ij}) belongs to C^2 in a neighborhood of $\bar{D}$, then there exists a matrix $\sigma = (\sigma_{ij})$ as in the condition (A). We can further take $\sigma_{ij}(x) = \delta_{ij}$ if $|x|$ is sufficiently large.

Observe that the elliptic operator, in one dimension,

$$Lu = xu_{xx} \qquad \text{in} \quad 0 < x < 1$$

does not satisfy the condition (A).

The coefficient $c(x)$ in (1.2) will henceforth be subject to the condition:

$$c(x) \leq 0 \quad \text{in } D, \qquad c(x) \quad \text{is Hölder continuous in } \bar{D}. \tag{1.5}$$

In Section 6.5 we have represented the solution u of (1.2), (1.3) (for nondegenerating L) in the form

$$\begin{aligned} u(x) = {} & E_x\phi(\xi(\tau)) \exp\left[\int_0^\tau c(\xi(s))\, ds \right] \\ & - E_x \int_0^\tau f(\xi(t)) \exp\left[\int_0^t c(\xi(s))\, ds \right] dt \end{aligned} \tag{1.6}$$

where τ is the exit time from D. The right-hand side makes sense even if L is degenerate, provided either

$$E_x\tau \leq C \qquad \text{for all} \quad x \in D \qquad (C \text{ const}), \tag{1.7}$$

or

$$c(x) \leq -\gamma < 0 \qquad \text{for all} \quad x \in D \qquad (\gamma \text{ const}). \tag{1.8}$$

When at least one of these conditions is satisfied, it was shown by Stroock and Varadhan [2] that the function $u(x)$, defined by (1.6), is continuous almost everywhere in D; if the condition (B) stated below is also satisfied, then $u(x)$ is continuous everywhere in D.

In this chapter we shall deal with the case where neither (1.7) nor (1.8) is assumed. In order to formulate precisely the Dirichlet problem, we divide the boundary ∂D into four disjoint subsets. Setting $\rho(x) = \operatorname{dist}(x, \partial D)$ for $x \in \bar{D}$ and assuming ∂D to belong to C^3 (so that $\rho \in C^2$ in some $\bar{D}$-neighborhood of ∂D), we define

$$\Sigma_3 = \left\{ x \in \partial D;\ \sum a_{ij}\nu_i\nu_j > 0 \right\},$$

$$\Sigma_2 = \left\{ x \in \partial D;\ \sum a_{ij}\nu_i\nu_j = 0,\ \sum b_i\rho_{x_i} + \tfrac{1}{2}\sum a_{ij}\rho_{x_ix_j} < 0 \right\},$$

$$\Sigma_1 = \left\{ x \in D;\ \sum a_{ij}\nu_i\nu_j = 0,\ \sum b_i\rho_{x_i} + \tfrac{1}{2}\sum a_{ij}\rho_{x_ix_j} > 0 \right\},$$

$$\Sigma_0 = \left\{ x \in D;\ \sum a_{ij}\nu_i\nu_j = 0,\ \sum b_i\rho_{x_i} + \tfrac{1}{2}\sum a_{ij}\rho_{x_ix_j} = 0 \right\}$$

where $\nu = (\nu_1, \ldots, \nu_n)$ is the inward normal. Notice that $\nu = D_x\rho$ on ∂D. Set

$$\Sigma_{23} = \Sigma_2 \cup \Sigma_3.$$

We shall assume:

(B) ∂D is in C^3; Σ_{23} consists of a finite number of connected hypersurfaces; Σ_1 consists of a finite number of connected hypersurfaces, and Σ_0 consists of a finite number of connected hypersurfaces.

Thus, the sets Σ_{23}, Σ_1, Σ_0 are closed sets.

Definition. A point $x^0 \in \partial D$ is called a *regular point* if for any $\delta > 0$

$$\lim_{\substack{x \to x^0 \\ x \in D}} P_x\{\tau < \infty; |\xi(\tau) - x^0| < \delta\} = 1,$$

where τ is the exit time from D. If, in addition, for any μ positive and sufficiently small,

$$E_x \tau_\mu \leq C\mu \qquad \text{if} \quad x \in W_\mu$$

where $W_\mu = \{x \in D; |x - x^0| < \mu\}$ and τ_μ is the exit time from W_μ, then we call x^0 a *strongly regular point.*

From Problems 8 and 9 of Chapter 11 we have that *every point of* Σ_{23} *is a strongly regular point.* On the other hand, the set $\Sigma_0 \cup \Sigma_1$ is nonattainable from D. By Problem 8, Chapter 12, the set Σ_1 is not even stable. Thus, in the event $\tau = \infty$, $\xi(t)$ can only be expected to approach the set Σ_0 as $t \to \infty$, if it approaches the boundary at all.

These considerations indicate that, when L is degenerate, the boundary conditions (1.3) should be replaced by

$$u = \phi \qquad \text{on } \Sigma_{23} \tag{1.9}$$

with perhaps some additional boundary conditions on the set Σ_0. If either (1.7) or (1.8) holds, then as suggested by the previous considerations and formula (1.6), the boundary condition (1.3) should be replaced by just (1.9).

In the next section we shall show, under some conditions, that there exist a finite number of points $\zeta_1, \ldots, \zeta_l$ on Σ_0 such that if $\tau(\omega) = \infty$, then $\xi(t, \omega)$ converges to one of these points. We call these points *distinguished boundary points.*

Setting

$$A_i = \{\tau = \infty, \xi(t) \to \zeta_i \text{ if } t \to \infty\}, \qquad p_i(x) = P_x(A_i), \tag{1.10}$$

we thus have

$$\sum_{i=1}^{l} p_i(x) = P_x\{\tau = \infty\}. \tag{1.11}$$

We shall also show that

$$\zeta_i \text{ is asymptotically stable from } D \qquad (1 \leq i \leq l). \tag{1.12}$$

This implies that $p_i(x) \to 1$ if $x \to \zeta_i$, $x \in D$.

We now add to (1.9) the boundary conditions

$$u(\zeta_i) = g_i \qquad (1 \leq i \leq l) \tag{1.13}$$

where the g_i are given numbers.

Definition. A *classical solution* to the Dirichlet problem

$$\begin{aligned} Lu + cu &= 0 \qquad \text{in } D, \\ u &= \phi \qquad \text{on } \Sigma_{23}, \\ u &= g_i \qquad \text{at } \zeta_i \qquad (1 \le i \le l) \end{aligned} \tag{1.14}$$

is a solution which is in $C^2(D)$ and is continuous on Σ_{23} and at the points ζ_i. Thus

$$u(x) \to \phi(y) \qquad \text{if} \quad x \to y \in \Sigma_{23}, \tag{1.15}$$

$$u(x) \to g_i \qquad \text{if} \quad x \to \zeta_i. \tag{1.16}$$

By applying Itô's formula (cf. the proof of Theorem 6.5.1) one finds that if $u(x)$ is a classical solution of the Dirichlet problem (1.14), then

$$\begin{aligned} u(x) = {} & E_x\left\{\phi(\xi(\tau)) \exp\left[\int_0^\tau c(\xi(t))\,dt\right] I_{\tau<\infty}\right\} \\ & + \sum_{i=1}^{l} g_i E_x\left\{\exp\left[\int_0^\infty c(\xi(t))\,dt\right] I_{A_i}\right\} \end{aligned} \tag{1.17}$$

where I_A is the indicator function of a set A. In order that

$$E_x\left\{\exp\left[\int_0^\infty c(\xi(t))\,dt\right] I_{A_i}\right\} \not\equiv 0$$

we must have $c(\zeta_i) = 0$. We shall actually assume that

$$c(x) = 0 \quad \text{in a neighborhood of } \zeta_i, \qquad 1 \le i \le l. \tag{1.18}$$

We shall now prove that, conversely, the function $u(x)$ given by (1.17) is a classical solution of (1.14) provided the following additional condition holds:

$$(a_{ij}(x)) \quad \text{is positive definite for all} \quad x \in D. \tag{1.19}$$

Theorem 1.1. *Let the conditions* (A), (B), (1.5), (1.11), (1.12), *and* (1.18), (1.19) *hold, and let* ϕ *be a continuous function on* Σ_{23}. *Then there exists a unique classical solution* u *of the Dirichlet problem* (1.14), *and it is given by* (1.17).

Proof. In view of the remark asserting (1.17), a classical solution, if existing, must be given by (1.17). Thus it remains to show that $u(x)$, given by (1.17), is a classical solution. We first verify (1.15).

Let $y \in \Sigma_{23}$, $W_\mu = \{x \in D;\ |x - y| < \mu\}$, τ_μ = exit time from W_μ. Since y

is a strongly regular point,

$$E_x\tau_\mu \leqslant C\mu \qquad \text{if} \quad x \in W_\mu, \qquad 0 < \mu \leqslant \mu_0, \tag{1.20}$$

$$P_x\{\tau < \infty;\ |\xi(\tau) - y| < \lambda\} \to 1 \qquad \text{if} \quad x \to y, \qquad \text{for any} \quad \lambda > 0 \tag{1.21}$$

where C is a constant independent of μ. It follows that

$$P_x(\tau_\mu > \delta) \leqslant \frac{E\tau_\mu}{\delta} \leqslant \frac{C\mu}{\delta} \qquad (x \in W_\mu)$$

for any $\delta > 0$, and

$$\overline{\lim_{x\to y}}\, P_x(\tau > \delta) = \overline{\lim_{x\to y}}\, P_x(\tau_\mu > \delta) \leqslant \frac{C\mu}{\delta}.$$

Since μ is arbitrary,

$$\lim_{x\to y} P_x(\tau > \delta) = 0 \qquad \text{for any} \quad \delta > 0.$$

Since $c(x)$ is a bounded function, we also have

$$\lim_{x\to y} P_x\left\{\left|\exp\left[\int_0^\tau c(\xi(s))\, ds\right] - 1\right| > \delta\right\} = 0 \qquad \text{for any} \quad \delta > 0.$$

An application of the Lebesgue bounded convergence theorem then gives

$$\lim_{x\to y} E_x\left|\exp\left[\int_0^\tau c(\xi(s))\, ds\right] - 1\right| = 0. \tag{1.22}$$

Using (1.21) and the continuity of ϕ, we also have

$$\lim_{x\to y} E_x|\phi(\xi(\tau)) - \phi(y)| = 0.$$

Combining this with (1.22), we find that the first term on the right-hand side of (1.17) converges to $\phi(y)$, as $x\to y$. Each of the other terms on the right-hand side of (1.17) converges to zero, as $x\to y$, since

$$P_x(A_i) \leqslant P_x(\tau = \infty) \to 0$$

by (1.21). Thus the proof of (1.15) is complete.

Let U be a neighborhood of ζ_i such that $c(x) \equiv 0$ in $U \cap D$. Since ζ_i is asymptotically stable from D,

$$P_x\{\xi(t) \in U \text{ for all } t \geqslant 0\} \to 1, \qquad P_x(A_i) \to 1$$

if $x\to\zeta_i$, $x \in D$. Hence, by the Lebesgue bounded convergence theorem,

$$g_i E_x\left\{\exp\left[\int_0^\infty c(\xi(s))\, ds\right] I_{A_i}\right\} \to g_i \qquad \text{if} \quad x \to \zeta_i.$$

All the other terms on the right-hand side of (1.17) converge to zero (as $x\to\zeta_i$), since $P_x(A_i)\to 1$. Thus (1.16) is satisfied.

Before showing that $u(x)$ is a C^2 solution of $Lu + cu = 0$ in D, we prove a lemma.

Lemma 1.2. *The function $p_i(x)$ is in $C^2(D)$ and $Lp_i(x) = 0$ in D.*

Proof. Let N be a ball of radius r_0 with boundary ∂N, such that $\overline{N} = N \cup \partial N$ is contained in D. Since L is nondegenerate in D, the exit time τ_N from N is finite a.s. By a standard argument (see Problem 1), $p_i(y)$ is a Borel measurable function. Hence, by the strong Markov property (see Problem 2), if $x \in N$,

$$p_i(x) = E_x p_i(\xi(\tau_N)) = \int_{\partial N} p_i(y) P_y(\xi(\tau_N) \in dS_y) \tag{1.23}$$

where dS_y is the surface element on ∂N.

Let ψ be a continuous function on ∂N and let v be the solution of the Dirichlet problem

$$Lv = 0 \quad \text{in } N, \qquad v = \psi \quad \text{on } \partial N. \tag{1.24}$$

One can represent v in terms of Green's function (see, for instance, Friedman [1]):

$$v(x) = \int_{\partial N} \psi(y) \frac{\partial G(x, y)}{\partial \nu_y} \, dS_y \tag{1.25}$$

where ν_y is the inward normal. Let $0 < \epsilon < r_0$ and denote by N_ϵ a ball concentric to N with radius $r_0 - \epsilon$. Denote by τ_{N_ϵ} the exit time from N_ϵ. If $x \in N_\epsilon$, then, by Itô's formula,

$$v(x) = E_x v(\xi(\tau_{N_\epsilon})).$$

Taking $\epsilon \to 0$ we arrive at the formula

$$v(x) = E_x v(\xi(\tau_N)) = E_x \psi(\xi(\tau_N)) = \int_{\partial N} \psi(y) P_x(\xi(\tau_N) \in dS_y). \tag{1.26}$$

Comparing this with (1.25) we find that

$$P_x(\xi(\tau_N) \in dS_y) = \frac{\partial G(x, y)}{\partial \nu_y} \, dS_y. \tag{1.27}$$

Using this formula in (1.23), we find that

$$p_i(x) = \int_{\partial N} p_i(y) \frac{\partial G(x, y)}{\partial \nu_y} \, dS_y. \tag{1.28}$$

Since $\partial G(x, y)/\partial \nu_y$ is continuous in $x \in N$ uniformly with respect to $y \in \partial N$, $p_i(x)$ is a continuous function.

Taking $\psi = p_i$ in (1.25) and comparing this with (1.28), we see that $v(x) = p_i(x)$. Thus $p_i(x)$ is a C^2 solution of $Lu = 0$ in D. This completes the proof of the lemma.

Consider now the function

$$q(x) = E_x\{\phi(\xi(\tau))I_{\tau<\infty}\}.$$

By the strong Markov property we have

$$q(x) = E_x q(\xi(\tau_N)).$$

Hence we can prooceed as in the case of $p_i(x)$ to show that $q(x)$ is a C^2 solution of $Lu = 0$ in D. Thus if $c(x) \equiv 0$, then the proof of the theorem is complete. We shall now consider the general case where $c(x) \not\equiv 0$.

Let

$$k(x) = E_x\left\{\exp\left[\int_0^\infty c(\xi(s))\,ds\right]I_{A_1}\right\}.$$

By the strong Markov property

$$k(x) = E_x\left\{\exp\left[\int_0^{\tau_N} c(\xi(t))\,dt\right]k(\xi(\tau_N))\right\}. \tag{1.29}$$

Let $k_m(y)$ be continuous functions on ∂N, uniformly bounded, such that

$$\int_{\partial N} |k_m(y) - k(y)|\,dS_y \to 0 \qquad \text{if} \quad m\to\infty. \tag{1.30}$$

Let v_m be the solution of

$$Lv_m + cv_m = 0 \qquad \text{in } N,$$
$$v_m = k_m \qquad \text{on } \partial N.$$

By the interior Schauder estimates (Section 10.1) we find that there is a subsequence of $\{v_m\}$ (which we again denote by $\{v_m\}$) that is uniformly convergent in compact subsets of N to a solution $v(x)$ of $Lv + cv = 0$.

By Itô's formula we get (cf. the derivation of (1.26))

$$v_m(x) = E_x\left[\exp\left[\int_0^{\tau_N} c(\xi(t))\,dt\right]v_m(\xi(\tau_N))\right]. \tag{1.31}$$

Notice that

$$\begin{aligned} E_x|k(\xi(\tau_N)) - v_m(\xi(\tau_N))| &= \int_{\partial N} |k(y) - k_m(y)|P(\xi(\tau_N)\in dS_y) \\ &= \int_{\partial N} |k(y) - k_m(y)|\,\frac{\partial G(x,y)}{\partial \nu_y}\,dS_y \to 0 \\ &\qquad \text{if} \quad m\to\infty \end{aligned}$$

by (1.30). Therefore, by (1.29), (1.31),

$$|v_m(x) - k(x)| < E_x|k(\xi(\tau_N)) - v_m(\xi(\tau_N))| \to 0 \qquad \text{if} \quad m\to\infty.$$

It follows that $k(x) = v(x)$. Since N is arbitrary, $k(x)$ is in $C^2(D)$ and $Lk + ck = 0$ in D.

Similarly one can prove that the first term on the right-hand side of (1.17) is in $C^2(D)$ and satisfies the equation $Lu + cu = 0$. Thus the proof of Theorem 1.1 is complete.

2. Convergence of paths to boundary points

Suppose (A), (B) hold. We wish to find sufficient conditions for (1.11), (1.12).

For any C^2 function $\rho(x)$, define

$$\mathcal{A}_\rho(x) = \tfrac{1}{2}\sum a_{ij}(x)\rho_{x_i}(x)\rho_{x_j}(x), \qquad \mathcal{B}_\rho(x) = \sum b_i(x)\rho_{x_i}(x) + \tfrac{1}{2}\sum a_{ij}(x)\rho_{x_ix_j}(x),$$

$$Q_\rho(x) = \frac{1}{\rho}\left[\mathcal{B}_\rho(x) - \frac{1}{\rho}\mathcal{A}_\rho(x)\right].$$

Denote by $d(x, A)$ the distance from a point x to a set A. For any set $A \subset \Sigma_0$, we denote by A^γ or $(A)^\gamma$ the set $\{x \in D, d(x, A) < \gamma\}$; in particular, Σ_0^ϵ is A^ϵ when $A = \Sigma_0$.

Denote by $\Sigma_{23,\delta}$ the manifold consisting of all points $x \not\in D$ with $d(x, \Sigma_{23}) = \delta$. Denote by D_δ the domain bounded by $\Sigma_0, \Sigma_1, \Sigma_{23,\delta}$. Notice that on Σ_{23} we have either $\mathcal{A}_\rho > 0$ or $\mathcal{B}_\rho < 0$, where $\rho(x) = d(x, \Sigma_{23})$. This implies that, for any $c > 0$,

$$Q_{\rho_\delta}(x) \leq -c \quad \text{if} \quad x \in D_\delta, \quad \rho_\delta(x) < \epsilon' \qquad (\rho_\delta(x) = d(x, \Sigma_{23,\delta})) \tag{2.1}$$

for any sufficiently small δ, ϵ'.

Next, for any $c > 0$,

$$Q_\rho(x) \geq c \quad \text{if} \quad x \in D, \quad \rho(x) < \epsilon'' \qquad (\rho(x) = d(x, \Sigma_1)) \tag{2.2}$$

for some $\epsilon'' > 0$.

We now make the assumption that

$$Q_\rho(x) \leq -\theta_0 < 0 \quad \text{if} \quad x \in D, \quad \rho(x) < \epsilon^* \qquad (\rho(x) = d(x, \Sigma_0)) \tag{2.3}$$

for some $\theta_0 > 0$, $\epsilon^* > 0$.

Next we assume:

(G) There exists a function $R(x)$ in $C^2(\bar{D})$ such that $R(x) = d(x, \Sigma_{23})$ if $d(x, \Sigma_{23}) \leq \epsilon_0$, $R(x) = 1 - d(x, \Sigma_1)$ if $\Sigma_1 \neq \varnothing$ and $d(x, \Sigma_1) \leq \epsilon_0$, and $\epsilon_0 < R(x) < 1 - \epsilon_0$ elsewhere in D. Further, $D_xR(x)$ vanishes only at a finite number of points z_i in $\bar{D}$, and $\sum a_{ij}R_{x_ix_j} > 0$ at these points. Finally, $\mathcal{A}_R(x) \neq 0$ if $x \neq z_i$, $x \in D$.

If (1.19) holds and if each connected component of $R^n \setminus D$ is C^2 diffeomorphic to a closed ball, then the method of proof of Lemma 9.4.5 shows that the condition (G) holds if either (i) Σ_1 is the outer boundary of D, or (ii) $\Sigma_1 = \varnothing$ and $\Sigma_0 \neq \varnothing$.

Theorem 2.1. *Let the conditions* (A), (B), (G), *and* (2.3) *hold. Then* $\tau = \infty$ *implies a.s. that* $d(\xi(t), \Sigma_0) \to 0$ *if* $t \to \infty$.

Proof. By Theorem 12.2.2, for any neighborhood U of Σ_0 there is a neighborhood U_ϵ of Σ_0 such that

$$P_x\{\xi(t) \in U \text{ for all } t \geq 0,\ d(\xi(t), \Sigma_0) \to 0 \text{ if } t \to \infty\} > 1 - \epsilon \qquad \text{if} \quad x \in U_\epsilon \cap D. \tag{2.4}$$

Next we construct a G-function ψ. From the condition (G) one can deduce that the same condition holds also with respect to D_δ, if δ is sufficiently small. Denote the corresponding R function by R_δ. For any $\epsilon > 0$ we wish to construct a function

$$u_\delta(x) = \Phi(R_\delta(x)) \qquad \text{in} \quad C^2(D_\delta \backslash \Sigma_0^\epsilon)$$

such that $Lu_\delta(x) \leq -\nu < 0$ in $D_\delta \backslash \Sigma_0^\epsilon$. The construction of $\Phi(r)$ is similar to the construction given in the proof of Theorem 12.2.4. Here one makes use of the inequalities (2.1), (2.2) when one takes

$$\Phi(r) = A_1 \log r + B_1 \qquad \text{near} \quad r = 0, \quad A_1 > 0,$$

$$\Phi(r) = A_2 \log(1 - r) + B_2 \qquad \text{near} \quad r = 1, \quad A_2 > 0.$$

Using this function one can show that for any neighborhood V of Σ_0,

$$\tau = \infty \quad \text{implies a.s. that} \quad \xi(t) \text{ hits } V. \tag{2.5}$$

Indeed, take ϵ in the construction of $u_\delta(x)$ so that V contains an ϵ-neighborhood of Σ_0. Denoting by τ^* the hitting time of V, we then have by Itô's formula,

$$u_\delta(\xi(\tau^* \wedge \tau \wedge t)) - u_\delta(x) = \int_0^{t^* \wedge \tau \wedge t} D_x u_\delta \cdot \sigma\, dw + \int_0^{\tau^* \wedge \tau \wedge t} Lu_\delta\, ds. \tag{2.6}$$

If $\tau(\omega) = \infty$ and $\tau^*(\omega) = \infty$ on a set B of positive probability, then the right-hand side of (2.6) is $\leq o(t) - \nu t$ (a.s. for $\omega \in B$). But this is impossible since the left-hand side of (2.6) is bounded.

We can now use (2.4), (2.5) in order to complete the proof of Theorem 2.1 by the argument used in the proof of Theorem 12.1.3.

Remark. Theorem 2.1 remains valid if the conditions (2.3) and (G) are replaced by weaker conditions, namely:

Suppose $\Sigma_0 = \Sigma_0^- \cup \Sigma_0^+$ where Σ_0^-, Σ_0^+ are disjoint sets, each consisting of a finite number of connected manifolds. The inequality (2.3) holds with $\rho(x) = d(x, \Sigma_0^-)$, whereas, for some $\lambda > 0$,

$$\mathfrak{B}_\rho + \frac{1}{\lambda}\mathfrak{A}_\rho > 0 \qquad \text{if} \quad x \in D, \quad \rho(x) < \epsilon^*, \qquad \text{where} \quad \rho(x) = d(x, \Sigma_0^+).$$

The condition (G) holds with Σ_1 replaced by $\Sigma_1 \cup \Sigma_0^+$ and with $R(x) = \lambda - d(x, \Sigma_1 \cup \Sigma_0^+)$ if $d(x, \Sigma_1 \cup \Sigma_0^+) \leq \epsilon_0$.

In fact, under these conditions we can again construct a G-function $\Phi(R_\delta(x))$ with the same $\Phi(r)$ as before, and thus the rest of the proof of Theorem 2.1 remains the same.

Let u be a C^2 function on Σ_0. Extend it into a small $\bar{D}$-neighborhood of Σ_0 by defining it to be constant along normals. Denote by $\tilde{u}$ the extended function.

Definition. The operator

$$(L_0 u)(y) = L\tilde{u}(x)|_{x=y} \qquad (y \in \Sigma_0)$$

is called the *restriction* of L to Σ_0.

Notice that for any $\epsilon > 0$ there is a $\delta > 0$ such that

$$|L_0 u(y) - L\tilde{u}(x)| < \epsilon \qquad \text{if} \quad x \in \bar{D}, \quad |x - y| < \delta.$$

We shall assume:

$$\begin{array}{l}\text{There exist points } \zeta_1, \ldots, \zeta_l \text{ on } \Sigma_0 \text{ such that} \\ \text{(i) } L \text{ is totally degenerate at each } \zeta_i, \\ \text{and (ii) there exists an S-function for each } \zeta_i \text{ from } D.\end{array} \tag{2.7}$$

Set

$$K = \{\zeta_1, \ldots, \zeta_l\}.$$

For any $\rho > 0$, let $\Sigma_{0,\rho} = \{y \in \Sigma_0, d(y, K) > \rho\}$. We shall need another assumption:

$$\begin{array}{l}\text{For any } \rho > 0 \text{ sufficiently small, there exists a} \\ \text{function } \phi_0 \text{ in } C^2(\Sigma_{0,\rho}) \text{ such that } L_0\phi_0 \leq -1.\end{array} \tag{2.8}$$

Theorem 2.2. *Let the conditions* (A), (B), (G), (2.3), (2.7), (2.8) *hold. Then, for any* $x \in D$,

$$\tau = \infty \qquad \textit{implies a.s. that} \qquad d(\xi(t), K) \to 0 \qquad \textit{if} \quad t \to \infty. \tag{2.9}$$

Proof. By Theorem 2.1, for any $\lambda' > 0$,

$$\tau = \infty \qquad \text{implies a.s. that} \qquad \xi(t) \text{ hits } \Sigma_0^{\lambda'} \text{ in finite time } \tau_{\lambda'}. \tag{2.10}$$

Since Σ_0 is stable from D, we also have, for any $\epsilon > 0$, $\lambda > 0$,

$$P_y\{\xi(t) \text{ exits } \Sigma_0^{\lambda} \text{ in finite time}\} < \epsilon \qquad \text{if} \quad y \in \Sigma_0^{2\lambda'} \tag{2.11}$$

provided λ' is sufficiently small.

Let ϕ_0 be as in (2.8). Extend ϕ_0 into D by defining it as constant along normals. Then, for any $\rho' > \rho$, the extended function ϕ_0 satisfies, in some region $(\Sigma_{0,\rho'})^\lambda$, the inequality $L\phi_0 \le -\frac{1}{2}$, provided λ is sufficiently small. But then (cf. the proof of Lemma 12.1.4), if $\xi(0) = x \in (\Sigma_{0,\rho'})^\lambda$, $\xi(t)$ exits $(\Sigma_{0,\rho'})^\lambda$ with probability 1. Combining this fact with (2.10), (2.11), and using the strong Markov property, we get, for any $x \in D$,

$$\begin{aligned} P_x\{\tau = \infty, \xi(t) \text{ does not hit } \mu\text{-neighborhood of } K\} \\ = E_x \chi_{\tau=\infty} P_{\xi(\tau_{\lambda'})}\{\xi(t) \text{ does not hit } \mu\text{-neighborhood of } K\} \\ = E_x \chi_{\tau=\infty} P_{\xi(\tau_{\lambda'})}\{\xi(t) \text{ exits } \Sigma_0^\lambda\} < \epsilon \end{aligned}$$

where $\mu = \rho' + \lambda$, and λ' depends on ϵ, λ. Since ϵ is arbitrary, we get:

$$P_x\{\tau = \infty, \xi(t) \text{ does not hit } \mu\text{-neighborhood of } K\} = 0. \tag{2.12}$$

Note that μ can be any positive number. Using (2.12) and the condition (2.7), we can now employ the proof of Theorem 12.1.3 in order to complete the proof of Theorem 2.2.

3. Application to the Dirichlet problem

Theorem 3.1. *Let the conditions* (A), (B), (G), (1.5), (1.18), (1.19), *and* (2.3), (2.7), (2.8) *hold, and let ϕ be a continuous function on Σ_{23}. Then there exists a unique classical solution of the Dirichlet problem* (1.14), *and it is given by* (1.17).

Indeed, in view of Theorem 1.1, we only have to verify the conditions (1.11), (1.12). But (1.11) follows from Theorem 2.2 and (1.12) follows from the condition (2.7).

In order to apply Theorem 3.1 one has to verify the conditions (G), (2.7), (2.8). As for (G), see the remark following the definition of this condition.

The condition (2.8) is satisfied if L_0 is nondegenerate on $\Sigma_0 \setminus K$, with

$$\phi_0(x) = -A \exp[\alpha|x - \zeta_1|^2]$$

where A, α are sufficiently large positive numbers (depending on ρ). We shall see later that, when $n = 2$, the condition (2.8) is satisfied also in some cases where L_0 degenerates at some points of $\Sigma_0 \setminus K$.

As for (2.7), the construction of an S-function for ζ_i from D was already studied in Chapter 12. Thus, if

$$Q_{\rho_i}(x) \le -\theta_0 \qquad \text{if} \quad \rho_i(x) = |x - \zeta_i| < \epsilon_0 \tag{3.1}$$

for some $\theta_0 > 0$, $\epsilon_0 > 0$, then there exists an S-function, namely, $-\log \rho_i(x)$.

A more delicate sufficient condition for the existence of an S-function for

ζ_i can be obtained by the method of descent (Theorem 12.4.1), assuming:

σ_{jk}, b_j are continuously differentiable and a_{jk} are twice continuous differentiable in a neighborhood of ζ_i. (3.2)

We perform a diffeomorphism $x \to y$ from a neighborhood V of $x = \zeta_i$ onto a neighborhood W of $y = 0$ (ζ_i is mapped into 0) such that $V \cap D$ is mapped into $y_n > 0$ and $V \cap \partial D$ is mapped into $y_n = 0$. The stochastic differential equations take the form

$$dy_j = \sum_{k=1}^{n} \tilde{\sigma}_{jk}(y)\, dw_k(t) + \tilde{b}_j(y)\, dt \qquad (1 \leq j \leq n). \tag{3.3}$$

Set $y' = (y_1, \ldots, y_{n-1})$. The condition (2.3) implies (see Problem 9)

$$\frac{\partial \tilde{b}_n}{\partial y_n} - \frac{1}{2} \sum_{k=1}^{n} \left(\frac{\partial \tilde{\sigma}_{nk}}{\partial y_n} \right)^2 < 0 \qquad \text{at} \quad y = 0.$$

This is precisely the condition (12.4.3) (in the notation of Theorem 12.4.1). In view of Theorem 12.4.1, if there exists an S-function for

$$dy_j = \sum_{k=1}^{n} \tilde{\sigma}_{jk}(y', 0)\, dw_k(t) + \tilde{b}_j(y', 0)\, dt \qquad (1 \leq j \leq n-1) \tag{3.4}$$

about $y' = 0$ having the form $f(y') = \log|y'| + H(y'/|y'|)$, then there exists an S-function for ζ_i.

If $n = 2$, then (3.4) reduces to

$$dy_1 = \sum_{k=1}^{2} \tilde{\sigma}_k(y_1)\, dw_k(t) + \tilde{b}(y_1)\, dt.$$

In this case, $f(y_1) = \log|y_1|$ is an S-function if and only if

$$\frac{d}{dy_1} \tilde{b}(0) < \frac{1}{2} \frac{d^2}{dy_1^2} \sigma^2(0) \tag{3.5}$$

where $\sigma^2 = \tilde{\sigma}_1^2 + \tilde{\sigma}_2^2$.

For the remainder of this section we specialize to the case $n = 2$. For simplicity we assume that Σ_0 consists of just one simple closed curve. Let

$$x_1 = f(s), \quad x_2 = g(s) \qquad (0 \leq s \leq L)$$

be a C^3 representation of Σ_0, with $(f'(s))^2 + (g'(s))^2 \equiv 1$.

Denote by $\rho(x)$ the distance from x (in $\bar{D}$) to Σ_0, and introduce coordinates (y_1, y_2) in a $\bar{D}$-neighborhood of Σ_0 by $y_1 = s$, $y_2 = \rho(x)$. As easily verified,

$$x_1 = f(y_1) + y_2 g'(y_1), \qquad x_2 = g(y_1) - y_2 f'(y_1) \tag{3.6}$$

where $y_2 = \rho(x)$ and $(f(y_1), g(y_1))$ is the nearest point on Σ_0 to $x = (x_1, x_2)$.

The curve Σ_0 is mapped into $y_2 = 0$, and the original stochastic system can formally be written in the form (3.3) with $n = 2$. Set

$$\tilde{\sigma}(s) = \left\{ \sum_{k=1}^{2} (\tilde{\sigma}_{1k}(s, 0))^2 \right\}^{1/2}, \qquad \tilde{b}(s) = \tilde{b}_1(s, 0).$$

Notice that the transformation (3.6) "flattens" the boundary Σ_0 entirely.

Comparing the transformation (3.6) with the transformation (12.6.2), it is clear that

$$\tilde{\sigma}(s) = \sigma\left(\frac{2\pi s}{L} \right), \qquad \tilde{b}(s) = b\left(\frac{2\pi s}{L} \right)$$

where $\sigma(\phi)$ are defined in (12.6.4), (11.6.5).

Denote by $\tilde{L}$ the differential operator corresponding to (3.3). Then one easily sees that the restriction $\tilde{L}_0$ of $\tilde{L}$ to $y_2 = 0$ is given by

$$\tilde{L}_0 v(s) \equiv \tfrac{1}{2} (\tilde{\sigma}(s))^2 v''(s) + \tilde{b}(s) v'(s).$$

A point $x = (f(s), g(s))$ of Σ_0 is called *nondegenerate* if $\tilde{\sigma}(s) > 0$, and *degenerate* if $\tilde{\sigma}(s) = 0$. A degenerate point $x^0 = (f(s), g(s))$ is called a *shunt* if $\tilde{b}(s) \neq 0$. If a degenerate point is not a shunt, i.e., if $\tilde{\sigma}(s) = 0$, $\tilde{b}(s) = 0$, then the point is called a *stable trap* if

$$Q_0(s) \equiv \lim_{t \to s} \left\{ \frac{\tilde{b}(t)}{t - s} - \frac{\tilde{\sigma}^2(t)}{(t - s)^2} \right\}$$

is negative, and an *unstable trap* if $Q_0(s) > 0$. Notice that $x^0 = (f(s), g(s))$ is a stable trap if and only if $\tilde{L}_0 \log |t - s| \leq -\mu < 0$ for all t with $|t - s|$ small, or if and only if

$$L_0 \log\left[(x_1 - f(s))^2 + (x^2 - g(s))^2 \right]^{1/2} \leq -\nu < 0$$

for all $x = (x_1, x_2)$ on Σ_0 with $|x - x^0|$ sufficiently small. Similarly, $x^0 = (f(x), g(s))$ is unstable trap if and only if $\tilde{L}_0 \log|t - s| \geq \mu > 0$ for all t with $|t - s|$ small.

We shall make the following assumption:

(S) There are a finite number of stable traps $\zeta_i = (f(s_i), g(s_i))$ $(1 \leq i \leq l)$. Between two consecutive points ζ_i, ζ_{i+1} (where $\zeta_{l+1} = \zeta_1$) there is at most a finite number of degenerate points $\eta_{i,j} = (f(s_{i,j}), g(s_{i,j}))$, $1 \leq j \leq M_i$, each being a shunt, and, for each i, the numbers

$$\tilde{b}(s_{i,j}), \qquad 1 \leq j \leq M_i,$$

have the same sign.

From previous considerations it then follows (cf. (3.5)) that the condition (2.7) holds. Also the condition (2.8) holds. That is (since this condition is invariant under the diffeomorphism (3.6)), for each $i = 1, \ldots, l$ and for each small $\rho > 0$ there is a function $\phi_0(s)$ in $s_i + \rho \leq s \leq s_{i+1} - \rho$, satisfying $\tilde{L}_0\phi_0 < -1$. (Here $s_{l+1} = L + s_1$.) Indeed, if f_ϵ is the function constructed in the proof of Theorem 12.5.3 (cf. (12.5.21)–(12.5.23)) with ϵ sufficiently small, then take $\phi_0 = -2f_\epsilon$.

Appealing to Theorem 3.1, we can now state:

Theorem 3.2. *Let* $n = 2$ *and assume that* (A), (B), (G), (1.5), (1.18), (1.19), (2.3), (3.2) (*for* $1 \le i \le l$), *and* (S) *hold. Then, for any continuous function* ϕ *on* Σ_{23} *and numbers* $g_1, \ldots, g_l$ *there exists a unique classical solution of the Dirichlet problem* (1.14), *and it is given by* (1.17).

Let $n = 2$ and consider now the situation where the matrix $(a_{ij}(x))$ degenerates along arcs γ_i $(1 \le i \le l)$; each arc initiates at ζ_i and terminates at a point $\eta_i \in D$, and it lies entirely in D, with the exception of its initial point ζ_i. We shall call γ_i a *boundary spoke*.

Let us assume that

$$\gamma_i \quad \text{is nonattainable from } D. \tag{3.7}$$

Recall that in Section 11.7 we have stated sufficient conditions for (3.7) to hold.

Let N be a small neighborhood of ζ_i. Then $(D \cap N)\setminus\gamma_i$ consists of two regions: N_i^+ and N_i^-. If

$$x \to \zeta_i, \qquad x \in N_i^+$$

then we write $x \to \zeta_i^+$. Similarly we define the concept of $x \to \zeta_i^-$. The boundary conditions

$$u(\zeta_i^+) = g_i^+, \qquad u(\zeta_i^-) = g_i^-$$

are understood in the sense that

$$u(x) \to g_i^+ \quad \text{if } x \to \zeta_i^+, \qquad u(x) \to g_i^- \quad \text{if } x \to \zeta_i^-. \tag{3.8}$$

Consider now the Dirichlet problem

$$\begin{aligned} Lu + cu &= 0 \qquad \text{in } D,\\ u &= \phi \qquad \text{on } \Sigma_{23},\\ u(\zeta_i^+) = g_i^+, \qquad u(\zeta_i^-) &= g_i^- \qquad (1 \le i \le l). \end{aligned} \tag{3.9}$$

Set $\gamma = \bigcup_{i=1}^{l} (\gamma_i \cup \{\eta_i\})$. By a classical solution of (3.9) we shall mean a function u in $C^2(D\setminus\gamma)$ which satisfies $Lu + cu = 0$ in $D\setminus\gamma$ and which satisfies the boundary conditions: (i) (3.8), and (ii) $u(x) \to \phi(y)$ if $x \to y \in \Sigma_{23}$.

The consideration leading to the proof of Theorem 3.2 gives:

Theorem 3.3. *Let all the conditions of Theorem* 3.2 *hold, with the exception of* (1.19). *Assume also that* $(a_{ij}(x))$ *is nondegenerate in* $D\setminus\gamma$, *and that* (3.7) *holds. Then there exists a unique classical solution of the Dirichlet* *problem* (3.9), *and it is given by*

$$u(x) = E_x\left\{\phi(\xi(\tau)) \exp\left[\int_0^\tau c(\xi(t))\,dt\right] I_{\tau<\infty}\right\}$$
$$+ \sum_{i=1}^{l} g_i^+ E_x\left\{\exp\left[\int_0^\infty c(\xi(t))\,dt\right] I_{A_i^+}\right\}$$
$$+ \sum_{i=1}^{l} g_i^- E_x\left\{\exp\left[\int_0^\infty c(\xi(t))\,dt\right] I_{A_i^-}\right\}$$

where

$$A_i^+ = \{\tau = \infty, \xi(t)\to\zeta_i^+\}, \qquad A_i^- = \{\tau = \infty, \xi(t)\to\zeta_i^-\}.$$

Theorems 3.1, 3.2 can be generalized to include unstable traps, provided there are boundary spokes initiating at these points that are nonattainable from D; see Friedman and Pinsky [3].

PROBLEMS

1. Prove that $p_i(x)$ is a Borel measurable function. [*Hint*: Let $\{t_j\}$ be dense in $[0, \infty)$, $\{q_{il}\}$ dense in D_i, $D_i = \{x \in D, d(x, \partial D) > 1/i\}$. Let

$$B = \bigcup_i \bigcap_j \bigcup_{m>2i} \bigcup_l \{|\xi(t_j) - q_{il}| < 1/m\},$$

$$C = \bigcup_l \bigcap_m \bigcup_{t_k>m} \{|\xi(t_k) - \zeta_i| < 1/l\}.$$

Prove that $A_i = B \cap C$.]

2. Prove the first relation in (1.23). [*Hint*: Cf. the proof of Problem 10, Chapter 2.]

3. Suppose $\partial D = \Sigma_{23}$, ∂D is in C^3, and (A), (1.5), (1.7), (1.19) hold. Let f be a Hölder continuous function in $\bar{D}$. Prove that there is a unique C^2 solution of

$$Lu + cu = f \quad \text{in } D, \qquad u = \phi \quad \text{on } \partial D,$$

and it is given by (6.5.10).

4. Suppose that (A), (1.5), (1.19) hold and $\Sigma_{23} = \partial D$, ∂D in C^3. Prove that:
   (i) $P_x\{\tau < \infty\} = 1$ for any $x \in D$;
   (ii) there is a unique classical solution of $Lu + cu = 0$ in D, $u = \phi$ on ∂D and it is given by

$$u(x) = E_x\left\{\phi(\xi(\tau)) \exp\left[\int_0^\tau c(\xi(s))\,ds\right]\right\}.$$

[*Hint*: For (i), let $\Gamma_\epsilon = \{x \in D, d(x, \partial D) = \epsilon\}$, τ_ϵ = hitting time of Γ_ϵ. $P_x(\tau < \infty) = E_x P_{\xi(\tau_\epsilon)}(\tau < \infty)$. Estimate $P_y(\tau < \infty)$ for y near ∂D by Problems 8, 9 of Chapter 11.]

5. Let (A) hold and assume that $\Sigma_{23} = \partial D$, ∂D in C^3. Assume that the exit time τ from D is finite for every $\xi(0) = x \in D$, and $P_x(\tau > t_0) \leq \beta < 1$ for all $x \in D$, where t_0 is some positive number. Let $\phi(x)$ be a Lipschitz continuous function on ∂D. Prove that the function $u(x) = E_x\phi(\xi(\tau))$ is uniformly Hölder continuous in D. [*Hint*: Let $\tau_z = \tau$ given $\xi(0) = z$, $\bar{\tau} = \tau_x \wedge \tau_y$, $A_x = \{\tau_x \leq \tau_y\}$, $A_y = \{\tau_y < \tau_x\}$. On the set A_x, $u(\xi_y(\tau_x)) = E_{\xi_y(\tau_x)}\phi(\xi(\tau))$ where $\xi_y(t)$ is $\xi(t)$ when $\xi(0) = y$. Show that

$$\begin{aligned}|u(x) - u(y)| &\leq E|u(\xi_x(\bar{\tau})) - u(\xi_y(\bar{\tau}))|\\ &\leq E|\phi(\xi_x(\tau_x)) - E_{\xi_y(\tau_x)}\phi(\xi(\tau))|\chi_{A_x}\\ &\quad + E|E_{\xi_x(\tau_y)}\phi(\xi(\tau)) - \phi(\xi_y(\tau_y))|\chi_{A_y}\end{aligned}$$

Use the barrier $v_b(x)$ of Problems 8, 9 of Chapter 11 to show

$$E_x|\xi(\tau) - b| \leq C_1 v_b(x) \leq C_2|x - b|.$$

Hence deduce

$$|u_x - u(y)| \leq C_3 E|\xi_x(\bar{\tau}) - \xi_y(\bar{\tau})|.$$

By the strong Markov property, $P_x(\bar{\tau} > t) \leq P_x(\tau > t) \leq ce^{-\alpha t}$ for some $c > 0$, $\alpha > 0$. Finally, if χ_m is the indicator function of $(m \leq \bar{\tau} < m + 1)$,

$$\begin{aligned}E|\xi_x(\bar{\tau}) - \xi_y(\bar{\tau})|^\lambda &\leq \sum E\Big\{\sup_{m \leq s \leq m+1}|\xi_x(s) - \xi_y(s)|^\lambda\chi_m\Big\}\\ &\leq \sum\left[E\Big\{\sup_{0 \leq s \leq m+1}|\xi_x(s) - \xi_y(s)|^{2\lambda}\Big\}^{1/2} P(\bar{\tau} > m + 1)\right]^{1/2}\\ &\leq C_4|x - y|^\lambda\end{aligned}$$

if λ is sufficiently small, since

$$E\Big\{\sup_{0 \leq s \leq t}|\xi_x(s) - \xi_y(s)|^{2\lambda}\Big\} \leq E\Big\{\sup_{0 \leq s \leq t}|\xi_x(s) - \xi_y(s)|^2\Big\}^\lambda \leq e^{\lambda At}|x - y|^{\lambda}]$$

6. Let D be a domain with C^3 boundary ∂D, and let $R(x) = \text{dist}(x, \partial D)$. Denote by $(\nu_1, \ldots, \nu_n)$ the normal to ∂D, and by $(t_1, \ldots, t_n)$ any tangent to ∂D. Prove that $\sum \partial^2 R/(\partial x_i\, \partial x_j)t_i\nu_j = 0$. [*Hint*: Differentiate $\sum(\partial R/\partial x_i)^2 = 1$.]

7. Let D be a domain with C^3 boundary ∂D, and let V be a neighborhood of a point $x^0 \in \partial D$. Let $R(x)$ be a C^2 function in $\bar{D} \cap V$ satisfying:

(i) $R > 0$ in $D \cap V$,

(ii) $R = 0$ on $\partial D \cap V$,

(iii) $D_x R \neq 0$ on $\partial D \cap V$,

(iv) $\displaystyle\sum \frac{\partial^2 R}{\partial x_i\, \partial x_j} t_i\nu_j = 0$ at x^0, where the ν_i, t_i are defined as in Problem 6. Suppose also that x^0 is an interior point of Σ_0, that σ_{ij}, b_i are in $C^1(\bar{D} \cap V)$, and the a_{ij} are in $C^2(\bar{D} \cap V)$. Set

$$Q_R(x) = L(\log R) = -\frac{\mathcal{A}_R(x)}{R^2} + \frac{\mathcal{B}_R(x)}{R},$$

$$Q_R(x^0) = \lim_{x \to x^0} Q_R(x).$$

Prove

$$\lim_{x \to x^0} \frac{\mathcal{A}_R(x)}{R^2} \quad \text{is independent of the function } R.$$

[*Hint*: Suppose $n = 2$ and, without loss of generality, $x^0 = (0, 0)$, ∂D is given by $x_2 = f(x_1)$ with $f'(0) = 0$. Then $R(x_1, x_2) = \lambda x_2 + Cx_1^2 + Dx_2^2 + o(|x|^2)$, $\lambda \neq 0$, and $\sigma_{is}(x) = \sigma_{is}^0 + \sigma_{is}^1 x_1 + \sigma_{is}^2 x_2 + o(|x|^2)$. The condition $\sum \sigma_{is}\, \partial R/\partial x_i \to 0$ as x approaches Σ_0 gives $\sigma_{2s}^0 = 0$ and (taking $x_2 = O(x_1^2)$) $\lambda\sigma_{2s}^1 + 2C\sigma_{1s}^0 = 0$; hence

$$\sum \sigma_{is} \frac{\partial R}{\partial x_i} = \lambda\sigma_{2s}^2 x_2 + o(|x|^2), \qquad (\mathcal{A}_R/R^2) \to \tfrac{1}{2}\left(\sum \sigma_{2s}^2\right)^2.]$$

8. Under the same assumptions as in the preceding problem show that $\lim_{x \to x^0} \mathcal{B}_R/R$ is independent of the function R; consequently $Q_R(x^0)$ is also independent of R. [*Hint:*

$$\mathcal{B}_R = \sum \left(b_i - \tfrac{1}{2} \sum \frac{\partial a_{ij}}{\partial x_j}\right) \frac{\partial R}{\partial x_i} = \sum \beta_i \frac{\partial R}{\partial x_i},$$

$$\beta_i = \beta_i^0 + \sum \beta_{ij} x_j + o(|x|^2).$$

$\mathcal{B}_R \to 0$ as x approaches Σ_0 gives $\beta_2^0 = 0$ and (with $x_2 = O(x_1^2)$) $\lambda\beta_{21} + 2C\beta_1^0 = 0$. Hence $\mathcal{B}_R = \lambda\beta_{22}x_2 + o(|x|)$, $(\mathcal{B}_R/R) \to \beta_{22}$.]

9. Let $x_n = f(x_1, \ldots, x_{n-1})$ be a representation of ∂D in a neighborhood V of $x^0 \in D$, $x_n > f(x_1, \ldots, x_{n-1})$ in $D \cap V$. Perform a transformation

$$y_i = x_i \quad (1 \le i \le n-1), \qquad y_n = x_n - f(x_1, \ldots, x_{n-1}).$$

Then $D \cap V$ is mapped into $y_n > 0$, and $\partial D \cap V$ is mapped into $y_n = 0$. Let y^0 be the image of x^0. Assume that x^0 is an interior point of Σ_0, σ_{ij}, b_i belong to $C^1(\bar{D} \cap V)$, $a_{ij} \in C^2(\bar{D} \cap V)$, and let $R(x) = \operatorname{dist}(x, \partial D)$, $x \in \bar{D}$. Define $Q_R(x^0)$ as in Problem 7, and define, analogously,

$$\tilde{Q}_\rho(y^0) = \lim_{y \to y^0} \tilde{L}(\log \rho(y))$$

where $\tilde{L}\tilde{u}(y) = Lu(x)$ if $\tilde{u}(y) = u(x)$. Let $R^*(y) = y_n$ (i.e., the distance to the image of ∂D, for small $|y|$), and let $\tilde{R}(y) = R(x)$. Prove

$$Q_R(x^0) = \tilde{Q}_{R^*}(y^0);$$

thus, $Q_R(x^0)$ (where $R(x)$ is the distance to the boundary) does not change by a transformation that "flattens" the boundary. [*Hint*: Take $n = 2$ and, without loss of generality, $x^0 = 0$, $x_2 = f(x_1)$, $f(0) = f'(0) = 0$, and check that

$$\frac{\partial^2 \tilde{R}(y^0)}{\partial y_1 \, \partial y_2} = \frac{\partial^2 R(x^0)}{\partial x_1 \, \partial x_2}.$$

The latter vanishes by Problem 6. Apply Problem 8.]

14

Small Random Perturbations of Dynamical Systems

1. The functional $I_T(\phi)$

Consider a system of n stochastic differential equations

$$d\xi^\epsilon(t) = b(\xi^\epsilon(t))\,dt + \epsilon\sigma(\xi^\epsilon(t))\,dw(t) \qquad (\epsilon > 0) \tag{1.1}$$

where $\sigma = (\sigma_{ij})$ is $n \times n$ matrix and $b = (b_1, \ldots, b_n)$. Set

$$a_{ij} = \sum_{k=1}^{n} \sigma_{ik}\sigma_{jk},$$

and let a be the matrix (a_{ij}), i.e., $a = \sigma\sigma^*$ where σ^* = transpose of σ. In this chapter we shall study the behavior of $\xi^\epsilon(t)$ as $\epsilon \to 0$. This behavior will depend on the behavior of solutions of the dynamical system

$$d\xi^0(t) = b(\xi^0(t)). \tag{1.2}$$

The system (1.1) can be considered as a small random perturbation of (1.2), with randomness expressed by a diffusion term $\epsilon\sigma\,dw$.

We shall make the following assumption throughout this chapter:

(A) $a(x)$, $b(x)$ are uniformly Lipschitz continuous in compact subsets of R^n, and

$$|a(x)| \leqslant M, \qquad |a(x) - a(y)| \leqslant M|x - y|^\alpha,$$
$$|b(x)| \leqslant M, \qquad |b(x) - b(y)| \leqslant M|x - y|^\alpha,$$

for all x, y in R^n, where M, α are positive constants and $0 < \alpha \leqslant 1$, and

$$\sum_{i,j=1}^{n} a_{ij}(x)\xi_i\xi_j \geqslant \mu|\xi|^2 \qquad (\mu \text{ positive constant}) \tag{1.3}$$

for all $x \in R^n$, $\xi \in R^n$.

Notice that, since (1.3) holds, $a(x)$ is Hölder (or Lipschitz) continuous if and only if $\sigma(x)$ is Hölder (or Lipschitz) continuous.

The process (1.1) determines a time-homogeneous Markov process with probabilities P_x^ϵ and expectations E_x^ϵ which depend on the parameter ϵ. However, for brevity, we shall often write these probabilities and expectations simply as P_x and E_x respectively.

For x, y in R^n, set

$$\rho(x, y) = |x - y|.$$

Denote by C_{T_1, T_2} the space of all continuous functions $\phi(t)$, $T_1 \leq t \leq T_2$, with range in R^n, and set

$$\rho_{T_1, T_2}(\phi, \psi) = \max_{T_1 \leq t \leq T_2} \rho_t(\phi(t), \psi(t))$$

if ϕ, ψ belong to C_{T_1, T_2}. If Φ is a subset of C_{T_1, T_2}, let

$$d_{T_1, T_2}(\psi, \Phi) = \inf_{\phi \in \Phi} \rho_{T_1, T_2}(\psi, \phi).$$

If $\phi \in C_{T_1, T_2}$, we set $I_{T_1, T_2}(\phi) = \infty$ in case ϕ is not absolutely continuous, and

$$I_{T_1, T_2}(\phi) = \int_{T_1}^{T_2} \left\| \frac{d\phi}{dt} - b(\phi(t)) \right\|^2 dt$$

in case ϕ is absolutely continuous; here

$$\begin{aligned} \left\| \frac{d\phi}{dt} - b(\phi(t)) \right\|^2 &= \left| \sigma^{-1}(\phi(t)) \left[\frac{d\phi}{dt} - b(\phi(t)) \right] \right|^2 \\ &= \left[\frac{d\phi}{dt} - b(\phi(t)) \right]^* a^{-1}(\phi(t)) \left[\frac{d\phi}{dt} - b(\phi(t)) \right]. \end{aligned}$$

We write

$$\begin{aligned} C_T &= C_{0, T}, \\ \rho_T(\phi, \psi) &= \rho_{0, T}, \\ d_T(\psi, \Phi) &= d_{0, T}(\psi, \Phi), \\ I_T(\phi) &= I_{0, T}(\phi). \end{aligned}$$

Lemma 1.1. *A function $F(t)$ in a bounded interval $a \leq t \leq b$ can be written in the form*

$$F(t) = \int_a^t f(s)\, ds \qquad \textit{with} \quad f \in L^p(a, b) \qquad (1 < p < \infty) \tag{1.4}$$

if and only if

$$M_F \equiv \sup_{a < t_0 < \cdots < t_m < b} \sum_{k=1}^{m} \frac{|F(t_k) - F(t_{k-1})|^p}{(t_k - t_{k-1})^{p-1}}$$

is finite, and in this case

$$M_F = \int_a^b |f(s)|^p \, ds. \tag{1.5}$$

Proof. Suppose (1.4) holds. By Hölder's inequality,

$$\begin{aligned} |F(t_k) - F(t_{k-1})|^p &\leq \left| \int_{t_{k-1}}^{t_k} f(s)\, ds \right|^p \\ &\leq (t_k - t_{k-1})^{p-1} \int_{t_{k-1}}^{t_k} |f(s)|^p \, ds. \end{aligned}$$

Hence

$$M_F \leq \int_a^b |f(s)|^p \, ds. \tag{1.6}$$

Suppose conversely that $M_F < \infty$. Then, for any sequence of disjoint intervals (α_k, β_k) in (a, b),

$$\begin{aligned} \sum |F(\beta_k) - F(\alpha_k)| &\leq \left[\sum \frac{|F(\beta_k) - F(\alpha_k)|^p}{(\beta_k - \alpha_k)^{p-1}} \right]^{1/p} \left[\sum (\beta_k - \alpha_k) \right]^{(p-1)/p} \\ &\leq M_F \left[\sum (\beta_k - \alpha_k) \right]^{(p-1)/p} \end{aligned}$$

by Hölder's inequality and the definition of M_F. It follows that F is absolutely continuous, i.e., $F(t) = \int_a^t f(s)\, ds$ for some $f \in L^1(a, b)$.

Take a sequence of partitions Π_n of (a, b) into intervals $(\alpha_{n, l-1}, \alpha_{n, l})$ of equal length $(b - a)/2^n$ $(l = 1, \ldots, 2^n)$. For any function g in $L^p(a, b)$, define step functions g_n by

$$g_n = \frac{1}{\alpha_{n,l} - \alpha_{n,l-1}} \int_{\alpha_{n,l-1}}^{\alpha_{n,l}} g(s)\, ds \quad \text{in} \quad (\alpha_{n,l-1}, \alpha_{n,l}).$$

Write $g_n = J_n g$. Then the operator J_n acts on functions in $L^p(a, b)$ somewhat like a mollifier (see Chapter 4, Problem 4), namely,

(i) $\int_a^b |J_n g|^p \, ds \leq \int_a^b |g|^p \, ds$;
(ii) $J_n g \to g$ uniformly in (a, b) if g is continuous in $[a, b]$.

From (ii) it follows that

(ii′) $\int_a^b |J_n g - g|^p \, ds \to 0$ if g is continuous in $[a, b]$.

Using (i), (ii′) one can then establish the fact that

$$\int_a^b |J_n g - g|^p \, ds \to 0 \quad \text{as} \quad n\to\infty, \qquad \text{for any} \quad g \in L^p(a, b). \tag{1.7}$$

We return to the function $f = F'$. Since $f \in L^1(a, b)$, we can apply (1.7) with $p = 1$. Thus,

$$\int_a^b |f_n - f| \, ds \to 0 \quad \text{if} \quad n\to\infty \qquad (f_n = J_n f).$$

Hence

$$|f_{n'}| \to |f| \quad \text{a.s.} \qquad \text{for a subsequence} \quad n' \to \infty. \tag{1.8}$$

Notice next that

$$f_n = \frac{F(\alpha_{n,l}) - F(\alpha_{n,l-1})}{\alpha_{n,l} - \alpha_{n,l-1}} \qquad \text{in} \quad (\alpha_{n,l-1}, \alpha_{n,l}).$$

Hence

$$\int_a^b |f_n|^p \, ds \le M_F.$$

Taking $n = n' \to \infty$ and using (1.8) and Fatou's lemma, we find that $f \in L^p(a, b)$ and

$$\int_a^b |f|^p \, ds \le \varliminf_{n'\to\infty} \int_a^b |f_{n'}|^p \, ds \le M_F.$$

Recalling (1.6), the assertion (1.5) follows, and the proof of the lemma is complete.

Remark. If $F = (F_1, \ldots, F_n)$ and M_F is defined as before, with

$$|F(\beta) - F(\alpha)| = \left\{ \sum_{i=1}^n (F_i(\beta) - F_i(\alpha))^2 \right\}^{1/2},$$

then, by Lemma 1.1, (1.4) holds with $f = (f_1, \ldots, f_n)$, $f_i \in L^p(a, b)$ if and only if $M_F < \infty$. A look at the proof of (1.5) (for $n = 1$) shows that the equality (1.5) remains valid for any $n \ge 1$.

Lemma 1.2. *The function $I_T(\phi)$ is lower semicontinuous, i.e., if $\phi_m \to \phi$ in C_T, then $I_T(\phi) \le \varliminf_{m\to\infty} I_T(\phi_m)$.*

Proof. We may assume that $\varliminf I_T(\phi_m) < \infty$. We may further assume that $\lim I_T(\phi_m)$ exists (otherwise we proceed with a subsequence $\phi_{m'}$ for which

$\lim I_T(\phi_{m'}) = \underline{\lim}\, I_T(\phi_m))$. From the boundedness of the sequence $I_T(\phi_m)$ we immediately deduce that

$$\int_0^T |\dot{\phi}_m(t)|^2\, dt \leq K < \infty \qquad \text{for all } m. \tag{1.9}$$

Let $\{m'\}$ be a subsequence such that

$$\underline{\lim}_{m\to\infty} \int_0^T |\dot{\phi}_m(t)|^2\, dt = \lim_{m'\to\infty} \int_0^T |\dot{\phi}_{m'}(t)|^2\, dt.$$

If $0 \leq t_1 < \cdots < t_l \leq T$, then, by Lemma 1.1 (with $p = 2$) and the remark following it,

$$\sum_{k=1}^{l} \frac{|\phi(t_k) - \phi(t_{k-1})|^2}{t_k - t_{k-1}} = \lim_{m'\to\infty} \sum_{k=1}^{l} \frac{|\phi_{m'}(t_k) - \phi_m(t_{k-1})|^2}{t_k - t_{k-1}}$$

$$\leq \lim_{m'\to\infty} \int_0^T |\dot{\phi}_{m'}(t)|^2\, dt = \underline{\lim}_{m\to\infty} \int_0^T |\dot{\phi}_m(t)|^2\, dt.$$

Taking the supremum with respect to $t_1, \ldots, t_l$ and l, and using Lemma 1.1 and the remark following it, and (1.9), we conclude that ϕ is absolutely continuous, $\dot{\phi} \in L^2(0, T)$, and

$$\int_0^T |\dot{\phi}(t)|^2\, dt \leq \underline{\lim}_{m\to\infty} \int_0^T |\dot{\phi}_m(t)|^2\, dt. \tag{1.10}$$

Take any partition $0 = \alpha_0 < \alpha_1 < \cdots < \alpha_l = T$ of $(0, T)$ with mesh η. Set $\sigma^{-1}(t) = \sigma^{-1}(\phi(t))$. By (1.10) applied to $\sigma^{-1}(\alpha_i)\phi(t)$,

$$\int_{\alpha_i}^{\alpha_{i+1}} |\sigma^{-1}(\alpha_i)\dot{\phi}(t)|^2\, dt \leq \underline{\lim}_{m\to\infty} \int_{\alpha_i}^{\alpha_{i+1}} |\sigma^{-1}(\alpha_i)\dot{\phi}_m(t)|^2\, dt.$$

Summing over i,

$$\sum \int_{\alpha_i}^{\alpha_{i+1}} |\sigma^{-1}(\alpha_i)\dot{\phi}(t)|^2\, dt \leq \underline{\lim}_{m\to\infty} \sum \int_{\alpha_i}^{\alpha_{i+1}} |\sigma^{-1}(\alpha_i)\dot{\phi}_m(t)|^2\, dt. \tag{1.11}$$

Since $\sigma^{-1}(t)$ is continuous,

$$|\sigma^{-1}(\alpha_i) - \sigma^{-1}(t)| \leq \gamma(\eta) \qquad (\alpha_i < t < \alpha_{i+1})$$

where $\gamma(\eta) \to 0$ if $\eta \to 0$. Using (1.9) and the fact that $\dot{\phi} \in L^2(0, T)$, we obtain from (1.11),

$$\int_0^T |\sigma^{-1}(\phi(t))\dot{\phi}(t)|^2\, dt \leq \underline{\lim}_{m\to\infty} \int_0^T |\sigma^{-1}(\phi(t))\dot{\phi}_m(t)|^2\, dt + A\gamma(\eta)$$

where A is a constant independent of η. Since η is arbitrary,

$$\int_0^T |\sigma^{-1}(\phi(t))\dot{\phi}(t)|^2\, dt \leq \underline{\lim}_{m\to\infty} \int_0^T |\sigma^{-1}(\phi(t))\dot{\phi}_m(t)|^2\, dt. \tag{1.12}$$

Since $\phi_m \to \phi$ uniformly in $[0, T]$,

$$|\sigma^{-1}(\phi(t)) - \sigma^{-1}(\phi_m(t))| \leq \epsilon_m, \qquad \epsilon_m \to 0 \quad \text{if} \quad m \to \infty.$$

Using this and (1.9) in (1.12), we find that

$$\int_0^T |\sigma^{-1}(\phi(t))\dot{\phi}(t)|^2\,dt \le \varliminf_{m\to\infty} \int_0^T |\sigma^{-1}(\phi_m(t))\dot{\phi}_m(t)|^2\,dt. \tag{1.13}$$

Next, using (1.9) it is clear that

$$\int_0^T [a^{-1}(\phi_m(t))b(\phi_m(t)) - a^{-1}(\phi(t))b(\phi(t))] \cdot \dot{\phi}_m(t)\,dt \to 0$$

if $m \to \infty$. By integration by parts we find that

$$\int_0^T \gamma(t) \cdot \dot{\phi}_m(t)\,dt \to \int_0^T \gamma(t) \cdot \dot{\phi}(t)\,dt \qquad \text{as} \quad m \to \infty,$$

for any absolute continuous function $\gamma(t)$ with $\dot{\gamma} \in L^2(0, T)$. Using these observations and (1.13) we deduce that

$$\begin{aligned} I_T(\phi) &= \int_0^T \{|\sigma^{-1}(\phi(t))\dot{\phi}(t)|^2 - 2a^{-1}(\phi(t))b(\phi(t)) \cdot \dot{\phi}(t) \\ &\qquad + a^{-1}(\phi(t))b(\phi(t)) \cdot b(\phi(t))\}\,dt \\ &\le \varliminf_{m\to\infty} \int_0^T \{|\sigma^{-1}(\phi_m(t))\dot{\phi}_m(t)|^2 - 2a^{-1}(\phi_m(t))b(\phi_m(t)) \cdot \dot{\phi}_m(t) \\ &\qquad + a^{-1}(\phi_m(t))b(\phi_m(t)) \cdot b(\phi_m(t))\}\,dt \\ &= \varliminf_{m\to\infty} I_T(\phi_m). \end{aligned}$$

Lemma 1.3. *Let K be a compact set in R^n, and let A be a positive number. Denote by $\Phi_{K,A}$ the class of all functions ϕ in C_T with $\phi(0) \in K$ and $I_T(\phi) \le A$. Then $\Phi_{K,A}$ is a compact subset of C_T.*

Proof. If $0 \le t < t + h \le T$,

$$|\phi(t+h) - \phi(t)| = \left|\int_t^{t+h} \dot{\phi}(s)\,ds\right| \le \sqrt{h}\left\{\int_t^{t+h} |\dot{\phi}(s)|^2\,ds\right\}^{1/2}.$$

Since $I_T(\phi) \le A$ implies

$$\int_0^T |\dot{\phi}(t)|^2\,dt \le B$$

where B is a constant independent of ϕ, we conclude that

$$|\phi(t+h) - \phi(t)| \le \sqrt{B}\,\sqrt{h} \qquad \text{if} \quad \phi \in \Phi_{K,A}.$$

Thus, $\Phi_{K,A}$ is a class of equicontinuous functions. These functions are also uniformly bounded, since $\phi(0) \in K$ and K is bounded. By the lemma of Ascoli–Arzela, every sequence in $\Phi_{K,A}$ has a convergent subsequence in C_T. Thus, if $\Phi_{K,A}$ is also closed then it is compact, and the proof of the lemma is complete. To prove that $\Phi_{K,A}$ is closed let $\phi_m \in \Phi_{K,A}$, $\phi_m \to \phi$ in C_T. Then clearly $\phi(0) = \lim \phi_m(0) \in K$, and, by Lemma 1.2, $I_T(\phi) \leqslant A$. It follows that $\phi \in \Phi_{K,A}$.

Corollary 1.4. *Consider the function $I_T(\phi)$ over the set Ψ consisting of all $\phi \in C_T$ with $\phi(0) \in K$, K a compact set. Then $I_T(\phi)$ attains a minimum on Ψ.*

Proof. Let $J = \inf_{\phi\in\Psi} I_T(\phi)$. Then there is a sequence ϕ_m in Ψ with $I_T(\phi_m) \to J$. Since $I_T(\phi_m) \leqslant J + 1$ for all m sufficiently large, and $\phi_m(0) \in K$, we can apply Lemma 1.3 to deduce that $\phi_{m'} \to \hat{\phi}$ in C_T, $\hat{\phi} \in \Psi$ where $\{\phi_{m'}\}$ is a subsequence of $\{\phi_m\}$. By Lemma 1.2, $I_T(\hat{\phi}) \leqslant \underline{\lim}\, I_T(\phi_{m'}) \leqslant J$. Hence $\hat{\phi}$ is a minimum point for $I_T(\phi)$ on Ψ.

2. The first Ventcel–Freidlin estimate

Theorem 2.1. *Let* (A) *hold. For any $\delta > 0$, $x \in R^n$, and for any $\phi \in C_T$, with $\phi(0) = x$,*

$$\underline{\lim}_{\epsilon\to 0} \{2\epsilon^2 \log[P_x(\rho_T(\xi^\epsilon, \phi) < \delta)]\} \geqslant -I_T(\phi); \tag{2.1}$$

more precisely, if $\int_0^T (1 + |d\phi/ds|^2)\, ds \leqslant K < \infty$, then

$$P_x(\rho_T(\xi^\epsilon, \phi) < \delta) > \frac{1}{2} \exp\left\{ -\frac{I_T(\phi)}{2\epsilon^2} - \frac{C\delta^\alpha K}{2\epsilon^2} - \frac{(4CK)^{1/2}}{\epsilon} \right\} \tag{2.2}$$

provided $C_0\epsilon^2 T \leqslant \frac{1}{4}\delta^2$, where C, C_0 are positive constants depending only on M, μ.

Proof. The function $\eta^\epsilon(t) = \xi^\epsilon(t) - \phi(t)$ (with $\xi^\epsilon(0) = x$) satisfies

$$\eta^\epsilon(t) = \epsilon \int_0^t \sigma(\eta^\epsilon(s) + \phi(s))\, dw(s) + \int_0^t [b(\eta^\epsilon(s) + \phi(s)) - \dot{\phi}(s)]\, ds, \tag{2.3}$$

where $\dot{\phi} = d\phi/ds$. We shall compare this process with the process ζ^ϵ given by

$$\zeta^\epsilon(t) = \epsilon \int_0^t \sigma(\zeta^\epsilon(s) + \phi(s))\, dw(s). \tag{2.4}$$

By Girsanov's formula (Theorem 7.3.4)

$$\frac{d\mu_{\eta^\epsilon}}{d\mu_{\zeta^\epsilon}}(\zeta^\epsilon) = \exp\left\{ \frac{1}{\epsilon} \int_0^T h(s)\, dw(s) - \frac{1}{2\epsilon^2} \int_0^T |h(s)|^2\, ds \right\} \equiv \rho \tag{2.5}$$

where

$$h(s) = \sigma^{-1}(\zeta^\epsilon(s) + \phi(s))(b(\zeta^\epsilon(s) + \phi(s)) - \dot{\phi}(s)).$$

Here we think of $\zeta^\epsilon(t)$ as the process $X(t)$ of continuous functions on

$(C^n_T, \mathfrak{M}_T)$ with probability P_x and expectation E_x, and we think of $\eta^\epsilon(t)$ as the same process $X(t)$ defined on the same measure space, but with probability P^*_x and expectation E^*_x. Then (2.5) gives

$$E^*_x\Phi(\eta^\epsilon(t_1), \ldots, \eta^\epsilon(t_m)) = E_x[\Phi(\zeta^\epsilon(t_1), \ldots, \zeta^\epsilon(t_m))\rho]$$

for any bounded measurable function $\Phi(x_1, \ldots, x_m)$ where each x_i varies in R^n and $0 \leqslant t_1 < \cdots < t_m \leqslant T$. Taking a sequence of Φ's such that $\Phi(X(t_1), \ldots, X(t_m))$ decreases to

$$\operatorname{sgn}\Big[\delta - \sup_{0 \leqslant t \leqslant T} |X(t)|\Big]^+,$$

we get

$$P_x\Big\{\sup_{0 \leqslant s \leqslant T} |\eta^\epsilon(s)| < \delta\Big\} = E^*_x\Big\{\chi_{\{\sup_{0 \leqslant s \leqslant T} |\zeta^\epsilon(s)| < \delta\}}\rho\Big\}. \tag{2.6}$$

Notice that E^*_x is actually the expectation E_x with respect to the Markov process $\zeta^\epsilon(t)$. Denoting this expectation by E_x (this should cause no confusion), we can rewrite (2.6) in the form

$$P_x\{\rho_T(\xi^\epsilon, \phi) < \delta\} = E_x\left\{\chi_{\{\sup_{0 \leqslant s \leqslant T} |\zeta^\epsilon(s)| < \delta\}} \frac{d\mu_{\eta^\epsilon}}{d\mu_{\zeta^\epsilon}}(\zeta^\epsilon)\right\}. \tag{2.7}$$

Now, if $|\zeta^\epsilon(s)| < \delta$ for $0 \leqslant s \leqslant T$, then

$$\left|\int_0^T |h|^2\, ds - \int_0^T \|b(\phi(s)) - \dot\phi(s)\|^2\, ds\right| \leqslant C\delta^\alpha K \tag{2.8}$$

where C is a constant depending only on M, μ.

By Chebyshev's inequality,

$$P_x\left\{\exp\left|\frac{1}{\epsilon}\int_0^T h(s)\, dw(s)\right| < e^{-a/\epsilon}\right\} = P_x\left\{-\int_0^T h(s)\, dw(s) > a\right\}$$

$$\leqslant \frac{1}{a^2}\int_0^T E_x|h(s)|^2\, ds \leqslant \frac{C}{a^2}K \tag{2.9}$$

with another constant C depending on M, μ. We can take C to be the same constant as in (2.8).

Applying the martingale inequality to the solution $\zeta^\epsilon(t)$ of (2.4) we get

$$P_x\Big\{\sup_{0 \leqslant s \leqslant T} |\zeta^\epsilon(s)| \geqslant \delta\Big\} \leqslant \frac{1}{\delta^2} E|\zeta^\epsilon(T)|^2 \leqslant \frac{C_0\epsilon^2 T}{\delta^2} \tag{2.10}$$

where C_0 is a constant depending only on M.

Taking $a^2 = 4CK$ in (2.9) and ϵ such that $C_0\epsilon^2 T < \delta^2/4$, we see that the set where

$$\exp\left\{\frac{1}{\epsilon}\int_0^T h(s)\, dw(s)\right\} > e^{-a/\epsilon} \qquad \text{and} \qquad \sup_{0 \leqslant s \leqslant T} |\zeta^\epsilon(s)| < \delta$$

has probability $> \frac{1}{2}$. From (2.5), (2.8) we then conclude that the right-hand side of (2.7) is larger than

$$\frac{1}{2}\exp\left\{-\frac{1}{2\epsilon^2}\int_0^T \|b(\phi(s)) - \dot{\phi}(s)\|^2\,ds\right\}\exp\left\{-\frac{C\delta^\alpha K}{2\epsilon^2}\right\}\exp\left\{-\frac{\sqrt{4CK}}{\epsilon}\right\},$$

and (2.2) is thereby proved.

Let $\{\mathcal{C}, \mathfrak{M}, \mathfrak{M}_t, \omega(t), P_x^\epsilon\}$ be the Markov process corresponding to $\xi^\epsilon(t)$ and let $\Omega = \mathcal{C}_T$ where $\mathcal{C}_T$ is the space of all continuous functions ω defined on $0 \leqslant t \leqslant T$ with values $\omega(t)$ in R^n. Let G be any open set in Ω, and let $G_x = \{\omega \in G;\ \omega(0) = x\}$.

If $\varphi \in G_x$, then

$$\{\xi^\epsilon(0) = x, \rho_T(\xi^\epsilon, \varphi) < \delta\} \subset \{\xi^\epsilon(0) = x, \xi^\epsilon \in G\}$$

provided δ is sufficiently small. Applying Theorem 2.1, we get

$$\varliminf_{\epsilon\to 0}\left[2\epsilon^2 \log P_x^\epsilon(G)\right] \geqslant -I_T(\varphi)$$

for any $\varphi \in G_x$. Hence:

Theorem 2.2. *Let* (A) *hold. Then for any* $x \in R^n$ *and for any open set* $G \subset \mathcal{C}_T$,

$$\varliminf_{\epsilon\to 0}\left[2\epsilon^2 \log P_x^\epsilon(G)\right] \geqslant -\inf_{\omega\in G_x} I_T(\omega) \tag{2.11}$$

where $G_x = \{\omega \in G;\ \omega(0) = x\}$.

Notice that Theorem 2.2 contains the assertion (2.1).

3. The second Ventcel–Freidlin estimate

Let Γ and Δ be disjoint sets in R^n; Γ is compact and Δ is a finite disjoint union of closed domains with C^2 boundary $\partial\Delta$.

For any set A, write $\rho(x, A) = \inf_{y\in A} \rho(x, y)$.

Denote by $\Phi_{x,T}(\Gamma, \Delta)$ the set of all $\phi \in C_T$ satisfying:

(1) $\phi(0) = x$;
(2) ϕ intersects Γ;
(3) if $\tilde{t} = \min\{t;\ \phi(t) \in \Gamma\}$, then $\phi(s) \notin \Delta$ for all $0 \leqslant s \leqslant \tilde{t}$, i.e., ϕ intersects Γ before it can possibly intersect Δ.

Let I_0 be a positive number and denote by Φ_0 the subset of $\Phi_{x,T}(\Gamma, \Delta)$ consisting of all ϕ with $I_T(\phi) \leqslant I_0$.

Theorem 3.1. *Let* (A) *hold. For any* $\delta > 0$, $x \in R^n\backslash\Delta$,

$$\varlimsup_{\epsilon\to 0}\{2\epsilon^2 \log[P_x(d_T(\xi^\epsilon, \Phi_0) > \delta, \xi^\epsilon \in \Phi_{x,T}(\Gamma, \Delta))]\} \leqslant -I_0. \tag{3.1}$$

If the set Φ_0 is empty, then the condition $d_T(\xi^\epsilon, \Phi_0) > \delta$ is trivially true, i.e., (3.1) becomes

$$\overline{\lim_{\epsilon\to 0}} \{2\epsilon^2 \log[P_x(\xi^\epsilon \in \Phi_{x,T}(\Gamma, \Delta))]\} \leq -I_0.$$

Proof. We shall give the proof only in case the set Φ_0 is nonempty; the proof in case Φ_0 is empty can be obtained by minor modifications.

Let $0 < \mu < \frac{1}{2}\operatorname{dist}(\Delta, \Gamma)$, $J > 0$, and define

$$\Gamma_\mu = \{x \in R^n;\ \rho(x, \Gamma) < \mu\},$$

$$\Phi_{\mu,J} = \{\phi \in \Phi_{x,T}(\Gamma_\mu, \Delta);\ I_T(\phi) \leq J\}.$$

We claim that for every $\phi \in \Phi_{\mu,J}$ there exists a function ψ in $\Phi_{x,T}(\Gamma, \Delta)$ such that

$$|I_T(\phi) - I_T(\psi)| \leq \tilde{C}(J+1)\mu, \tag{3.2}$$

$$\rho_T(\phi, \psi) \leq \tilde{C}(J+1)\mu, \tag{3.3}$$

where $\tilde{C}$ is a constant independent of J, μ.

Indeed, let s_0 denote the first time ϕ intersects Γ_μ. If

$$\tilde{\psi}(s) = \begin{cases} \phi(s) & \text{if } 0 \leq s < s_0, \\ \text{straight segment of length } \mu \text{ connecting } \phi(s_0) \text{ to } \Gamma \text{ and} \\ \text{then traced back to } \phi(s_0), \text{ for } s_0 < s < s_0 + 2\mu, \\ \phi(s - 2\mu) & \text{if } s_0 + 2\mu < s < T + 2\mu, \end{cases}$$

$$\psi(s) = \tilde{\psi}\left(\frac{s(T+2\mu)}{T}\right),$$

then (3.2), (3.3) hold.

Let r_0, h_0 be small positive numbers to be determined later. Define Markov times: $\tau_0 = 0$ and

$$\tau_i = (\tau_{i-1} + h_0) \wedge \inf\{t > \tau_{i-1};\ |\sigma^{-1}(\xi^\epsilon(\tau_{i-1}))[\xi^\epsilon(t) - \xi^\epsilon(\tau_{i-1}) - b(\xi^\epsilon(\tau_{i-1}))(t - \tau_{i-1})]| = r_0\}. \tag{3.4}$$

By Theorem 8.1.1, the τ_i are all finite. Further, $\tau_i \uparrow \infty$ if $i \uparrow \infty$. Indeed, if $\tau_i \uparrow \tau$ and $\tau(\omega) < \infty$ on a set of positive probability, then, by taking $t = \tau_i$, $i \to \infty$ in the equality under the "inf" in (3.4) we get a contradiction.

Let ν be the positive-integer random variable such that $\tau_{\nu-1} < T \leq \tau_\nu$. We construct a polygonal curve $l^\epsilon(t)$ by

$$l^\epsilon(t) = \xi^\epsilon(\tau_{i-1}) + \frac{t - \tau_{i-1}}{\tau_i - \tau_{i-1}}(\xi^\epsilon(\tau_i) - \xi^\epsilon(\tau_{i-1})) \qquad (\tau_{i-1} \leq t \leq \tau_i)$$

for $1 \leq i \leq \nu$.

It is clear that for any $\delta_0 > 0$,

$$|\xi^\epsilon(t) - \xi^\epsilon(\tau_{i-1})| < \delta_0 \qquad \text{if} \quad \tau_{i-1} \leq t \leq \tau_i \tag{3.5}$$

provided h_0, r_0 are sufficiently small (depending on M, μ, δ_0). Hence, for any $\delta_1 > 0$,

$$|l^\epsilon(t) - \xi^\epsilon(t)| < \delta_1 \qquad (0 \leqslant t \leqslant T) \tag{3.6}$$

if h_0, r_0 are sufficiently small (depending on M, μ, δ_1).

Suppose

$$d_T(\xi^\epsilon, \Phi_0) > \delta, \qquad \xi^\epsilon \in \Phi_{x,T}(\Gamma, \Delta). \tag{3.7}$$

Using (3.6) we conclude that for any small $\mu > 0$, $\nu > 0$,

$$l^\epsilon \in \Phi_{x,T}(\Gamma_\mu, \Delta^\nu), \qquad \rho(l^\epsilon, \xi^\epsilon) < \delta/2$$

provided h_0, r_0 are sufficiently small; here Δ^ν is the subset of Δ consisting of all points x with $\rho(x, R^n \backslash \Delta) > \nu$.

Suppose $I_T(l^\epsilon) \leqslant J$. Then, by (3.2), (3.3) (with Δ replaced by Δ^ν), there exists a curve ψ in $\Phi_{x,T}(\Gamma, \Delta^\nu)$ such that

$$I_T(\psi) \leqslant J + \tilde{C}(J+1)\mu, \qquad \rho_T(\psi, l^\epsilon) \leqslant \tilde{C}(J+1)\mu.$$

Further, since $\partial\Delta$ is in C^2, we can modify ψ into a curve ψ^* in C_T such that

$$\psi^* \in \Phi_{x,T}(\Gamma, \Delta), \qquad I_T(\psi^*) \leqslant I_T(\psi) + C\nu, \qquad \rho_T(\psi^*; \psi) \leqslant C\nu,$$

where $C = C_0(J+1)$ and C_0 is a constant independent of J, ν. Indeed, let Ω_0 be a δ^*-neighborhood of $\partial\Delta$, δ^* small. Let Λ be a diffeomorphism of R^n onto R^n such that Λ maps $R^n \backslash \Delta^\nu$ onto $R \backslash \Delta$ and its Jacobian is equal to the identity $+ O(\nu)$. Λ can be constructed by "pushing" the points of Ω_0 along the normals to $\partial\Delta$ in an appropriate smooth manner, while leaving the points of $R^n \backslash \Omega_0$ unchanged. Now take $\psi^*(t) = \Lambda\psi(t)$.

If $J + \tilde{C}(J+1)\mu + C\nu \leqslant I_0$ and $\tilde{C}(J+1)\mu + C\nu < \delta/2$, then $I_T(\psi^*) \leqslant I_0$ (so that $\psi^* \in \Phi_0$) and $\rho_T(\xi^\epsilon, \psi^*) < \delta$; i.e., $d_T(\xi^\epsilon, \Phi_0) < \delta$. Since this contradicts (3.7), we must have

$$I_T(l^\epsilon) > J, \qquad J = I_0 - h, \tag{3.8}$$

where h can be taken arbitrarily small (if μ, ν are arbitrarily small).

We have proved that (3.7) implies (3.8). Hence

$$P_x[\rho_T(\xi^\epsilon, \Phi_0) > \delta, \xi^\epsilon \in \Phi_{x,T}(\Gamma, \Delta)] \leqslant P_x[I_T(l^\epsilon) > J] \qquad (J = I_0 - h). \tag{3.9}$$

We proceed to evaluate the right-hand side.

Clearly,

$$I_T(l^\epsilon) \leqslant \sum_{i=1}^{\nu-1} \int_{\tau_{i-1}}^{\tau_i} \|\dot{l}^\epsilon(t) - b(l^\epsilon(t))\|^2\, dt + \int_{\tau_{\nu-1}}^{T} \|\dot{l}^\epsilon(t) - b(l^\epsilon(t))\|^2\, dt. \tag{3.10}$$

Since

$$\dot{l}^\epsilon(t) = \frac{\xi^\epsilon(\tau_i) - \xi^\epsilon(\tau_{i-1})}{\tau_i - \tau_{i-1}} \qquad (\tau_{i-1} < t < \tau_i),$$

we can write

$$\begin{aligned}\|\dot{l}^\epsilon(t) - b(l^\epsilon(t))\|^2 &= |\sigma^{-1}(l^\epsilon(t))[\dot{l}^\epsilon(t) - b(l^\epsilon(t))]|^2 \\ &= \Bigg| \frac{\sigma^{-1}(l^\epsilon(t))[\xi^\epsilon(\tau_i) - \xi^\epsilon(\tau_{i-1}) - b(\xi^\epsilon(\tau_{i-1}))(\tau_i - \tau_{i-1})]}{\tau_i - \tau_{i-1}} \\ &\qquad + \sigma^{-1}(l^\epsilon(t))[b(\xi^\epsilon(\tau_{i-1})) - b(l^\epsilon(t))]\Bigg|^2.\end{aligned}$$

Since, by (3.6), $\sigma^{-1}(l^\epsilon(t)) = \sigma^{-1}(\xi^\epsilon(\tau_{i-1}))(1 + \theta)$ where $|\theta| < C\delta_1^\alpha$ (and $C = C(M, \mu)$ is a constant), the numerator is bounded by $(1 + \kappa)r_0$, for any given $\kappa > 0$, provided $\delta_1 = \delta_1(\kappa)$ is sufficiently small. The second term under the absolute value sign on the right is bounded (by (3.5)) by any given positive number β_0, provided δ_0 is sufficiently small. We thus have, for $\tau_{i-1} < t < \tau_i$,

$$\begin{aligned}\|\dot{l}^\epsilon(t) - b(l^\epsilon(t))\|^2 &< \left[\frac{r_0(1+\kappa)}{\tau_i - \tau_{i-1}} + \beta_0\right]^2 \\ &< \frac{r_0^2(1+\kappa)^2}{(\tau_i - \tau_{i-1})^2} + \frac{2r_0(1+\kappa)}{\tau_i - \tau_{i-1}}\sqrt{\kappa}\,\frac{\beta_0}{\sqrt{\kappa}} + \beta_0^2 \\ &= \frac{r_0^2(1+\kappa)^3}{(\tau_i - \tau_{i-1})^2} + \beta_0^2\left(1 + \frac{1}{\kappa}\right).\end{aligned}$$

Consequently, by (3.10),

$$I_T(l^\epsilon) < (1+\kappa)^3 \sum_{i=1}^{\nu} \frac{r_0^2}{\tau_i - \tau_{i-1}} + \left(1 + \frac{1}{\kappa}\right)\beta_0^2 T.$$

Hence, for any β $(0 < \beta < 1)$,

$$\begin{aligned}P_x\{I_T(l^\epsilon) > J\} &< P_x\left\{(1+\kappa)^3 \sum_{i=1}^{\nu} \frac{r_0^2}{\tau_i - \tau_{i-1}} > J - \left(1 + \frac{1}{\kappa}\right)\beta_0^2 T\right\} \\ &< \frac{E_x\left\{\exp \dfrac{(1-\beta)(1+\kappa)^3}{2\epsilon^2} \displaystyle\sum_{i=1}^{\nu} \frac{r_0^2}{\tau_i - \tau_{i-1}}\right\}}{\exp \dfrac{(1-\beta)J - (1-\beta)(1 + 1/\kappa)(\beta_0^2 T)}{2\epsilon^2}}. \qquad (3.11)\end{aligned}$$

The constants κ, β will be chosen so that $(1 - \beta)(1 + \kappa)^4 < 1$. Hence, if

$$1 - \beta' = (1 - \beta)(1 + \kappa)^3, \qquad 1 - \beta'' = (1 - \beta)(1 + \kappa)^4,$$

then $\beta' > 0$, $\beta'' > 0$. By Hölder's inequality,

$$E_x\left\{\exp \frac{(1-\beta)(1+\kappa)^3}{2\epsilon^2} \sum_{i=1}^{\nu} \frac{r_0^2}{\tau_i - \tau_{i-1}}\right\}$$

$$= \sum_{m=1}^{\infty} E_x\left\{\chi_{\{\nu=m\}} \exp\left[\frac{1-\beta'}{2\epsilon^2} \sum_{i=1}^{m} \frac{r_0^2}{\tau_i - \tau_{i-1}}\right]\right\}$$

$$\leqslant \sum_{m=1}^{\infty} [P_x(\nu = m)]^{\kappa/(1+\kappa)}$$

$$\times\left[E_x \exp\left[\frac{(1-\beta')(1+\kappa)}{2\epsilon^2} \sum_{i=1}^{m} \frac{r_0^2}{\tau_i - \tau_{i-1}}\right]\right]^{1/(1+\kappa)}. \tag{3.12}$$

It is elementary to verify that if

$$\sum_{i=1}^{m} \alpha_i < T, \qquad \alpha_i > 0,$$

then

$$\sum_{i=1}^{m} \frac{1}{\alpha_i} > \frac{m^2}{T}.$$

Using this inequality with $\alpha_i = \tau_i - \tau_{i-1}$, we find that

$$P_x(\nu = m) = P_x(\tau_{m-1} < T \leqslant \tau_m) \leqslant P_x(\tau_{m-1} < T)$$

$$= P_x\{(\tau_1 - \tau_0) + (\tau_2 - \tau_1) + \cdots + (\tau_{m-1} - \tau_{m-2}) < T\}$$

$$\leqslant P_x\left\{\frac{1}{\tau_1 - \tau_0} + \frac{1}{\tau_2 - \tau_1} + \cdots + \frac{1}{\tau_{m-1} - \tau_{m-2}} > \frac{m^2}{T}\right\}$$

$$\leqslant \frac{E_x \exp\left[\dfrac{1-\beta''}{2\epsilon^2} \displaystyle\sum_{i=1}^{m} \frac{r_0^2}{\tau_i - \tau_{i-1}}\right]}{\exp\left[\dfrac{(1-\beta'')r_0^2 m^2}{2\epsilon^2 T^2}\right]}. \tag{3.13}$$

Setting

$$F_m(x) = E_x \exp\left[\frac{1-\beta''}{2\epsilon^2} \sum_{i=1}^{m} \frac{r_0^2}{\tau_i - \tau_{i-1}}\right], \tag{3.14}$$

we then have, by (3.11)–(3.13),

$$P_x\{I_T(l^\epsilon) > J\} < \sum_{m=1}^{\infty} \frac{F_m(x)}{\exp(A + B_m)},$$

$$A = \frac{(1-\beta)J - (1-\beta)(1+1/\kappa)\beta_0^2 T}{2\epsilon^2}, \qquad B_m = \frac{(1-\beta'')r_0^2 m^2 \kappa}{2\epsilon^2 T^2(\kappa+1)}. \tag{3.15}$$

We shall estimate

$$F_1(x) = E_x \exp \frac{1-\beta''}{2\epsilon^2} \frac{r_0^2}{\tau_1}.$$

Consider the process

$$\eta^\epsilon(t) = \frac{1}{\epsilon} \sigma^{-1}(\xi^\epsilon(0))[\xi^\epsilon(t) - \xi^\epsilon(0) - b(\xi^\epsilon(0))t].$$

It satisfies

$$d\eta^\epsilon(t) = b^\epsilon(\eta^\epsilon(t), t)\, dt + \sigma^\epsilon(\eta^\epsilon(t), t)\, dw(t)$$

where

$$b^\epsilon(x,t) = \epsilon^{-1}\sigma^{-1}(\xi^\epsilon(0))[b(\xi^\epsilon(0) + b(\xi^\epsilon(0))t + \epsilon\sigma(\xi^\epsilon(0))x) - b(\xi^\epsilon(0))],$$

$$\sigma^\epsilon(x,t) = \sigma^{-1}(\xi^\epsilon(0))[\sigma(\xi^\epsilon(0) + b(\xi^\epsilon(0))t + \epsilon\sigma(\xi^\epsilon(0)x)].$$

Denote by τ_η the exit time of $\eta^\epsilon(t)$ from the (r_0/ϵ)-neighborhood of 0. Then

$$\tau_1 = h_0 \wedge \tau_\eta.$$

We shall compare the process η^ϵ with a process ζ^ϵ defined as follows: Let $\hat{\zeta}^\epsilon(t)$ be the solution of

$$\hat{\zeta}^\epsilon(t) = \int_0^t \sigma^\epsilon(\hat{\zeta}^\epsilon(s), s)\, dw(s)$$

and denote by τ_ζ the exit time of $\hat{\zeta}^\epsilon(t)$ from the (r_0/ϵ)-neighborhood of 0. Let $\zeta^\epsilon(t)$ be the solution of

$$\zeta^\epsilon(t) = \int_0^t \sigma^\epsilon(\zeta^\epsilon(s), s)\, dw(s) + \int_0^t \chi_{(s>\tau_\zeta)} b^\epsilon(\zeta^\epsilon(s), s)\, ds.$$

By Theorem 5.5.1, this equation has a unique solution. By the proof of local uniqueness (Theorem 5.2.1), a.s. $\zeta^\epsilon(t) = \hat{\zeta}^\epsilon(t)$ for all $t < \tau_\zeta$. Notice that $\zeta^\epsilon(t)$ has zero drift until its exit time (which is τ_ζ) from the (r_0/ϵ)-neighborhood of 0; thereafter it has the same drift as the process $\eta^\epsilon(t)$.

We now apply Girsanov's formula (Theorem 7.3.1 and Lemma 7.3.2) and obtain (cf. the argument in the paragraph following (2.5))

$$P_x\{\tau_\eta < h\} = E_x\{\chi_{(\tau_\zeta < h)}\rho\} \tag{3.16}$$

where

$$\rho = \frac{d\mu_{\eta^\epsilon}}{d\mu_{\zeta^\epsilon}}(\zeta^\epsilon)$$

is given by

$$\rho = \exp\left\{\int_0^{h\wedge\tau_\zeta} B^\epsilon(\zeta^\epsilon(t), t)\, dw(t) - \frac{1}{2}\int_0^{h\wedge\tau_\zeta} |B^\epsilon(\zeta^\epsilon(t), t)|^2\, dt\right\},$$

where

$$\begin{aligned} B^\epsilon(x, t) &= [\sigma^\epsilon(x, t)]^{-1} b^\epsilon(x, t) \\ &= \epsilon^{-1}\sigma^{-1}(\xi^\epsilon(0) + b(\xi^\epsilon(0)t + \epsilon\sigma(\xi^\epsilon(0))x)[b(\xi^\epsilon(0) + b(\xi^\epsilon(0))t \\ &\quad + \epsilon\sigma(\xi^\epsilon(0))x) - b(\xi^\epsilon(0))]. \end{aligned}$$

Set

$$\rho_\kappa = \exp\left\{\frac{1}{\kappa}\int_0^{h\wedge\tau_\zeta} B^\epsilon(\zeta^\epsilon(t), t)\, dw(t) - \frac{1}{2\kappa^2}\int_0^{h\wedge\tau_\zeta} |B^\epsilon(\zeta^\epsilon(t), t)|^2\, dt\right\}.$$

By Theorem 7.1.1, $E_x\rho_\kappa = 1$. Using this fact and Hölder's inequality, we get from (3.16)

$$\begin{aligned} P_x\{\tau_\eta < h\} &\le [P_x\{\tau_\zeta < h\}]^{1-\kappa}[E_x\rho^{1/\kappa}]^\kappa \\ &= [P_x\{\tau_\zeta < h\}]^{1-\kappa} \exp\left[\left(\frac{1}{\kappa^2} - \frac{1}{\kappa}\right)\frac{h\hat{\gamma}^2}{2}\right] \end{aligned}$$

where $\hat{\gamma}$ is a bound on $B^\epsilon(\zeta^\epsilon(t), t)$, for $0 \le t \le \tau_\zeta \wedge h$. Since $\hat{\gamma} \le C(h + r_0)^\alpha/\epsilon$, we get

$$P_x\{\tau_\eta < h\} \le [P_x\{\tau_\zeta < h\}]^{1-\kappa} \exp\left\{\left(\frac{1}{\kappa^2} - \frac{1}{\kappa}\right)C\,\frac{h(h_0^2 + r_0^2)^\alpha}{2\epsilon^2}\right\} \tag{3.17}$$

where C is a constant depending only on M, μ.

To evaluate the right-hand side we shall need the following lemma:

Lemma 3.2. *Let $z(t)$ be a diffusion process in R^n with drift 0 and diffusion matrix $a(x, t)$. Let $h > 0$, $R > 0$ and suppose that*

$$\sum_{i,j=1}^n |a_{ij}(x, t) - \delta_{ij}| \le \kappa, \qquad \kappa < 1.$$

Denote by τ_R the exit time from the ball $\{x;\ |x| < R\}$. Then

$$P_0\{\tau_R \le h\} \le C(\kappa)\exp\left[-\frac{(1-\kappa)^2 R^2}{2h}\right]$$

where $C(\kappa)$ is a constant depending only on κ, n.

It is tacitly assumed that $a = \sigma\sigma^*$ where $\sigma(x, t)$ is uniformly Lipschitz continuous in bounded sets, and $|\sigma(x, t)| \leq \text{const}(1 + |x|)$.

Proof. Let $0 < \theta < 1$, and choose points $e_1, \ldots, e_l$ on the unit sphere such that

$$|x| \geq 1 \qquad \text{implies} \qquad x \cdot e_j \geq \theta \qquad \text{for at least one } j.$$

For any $\Lambda > 0$, let $\lambda_j = \Lambda e_j$. If

$$\sup_{0 \leq t \leq h} |z(t, \omega)| \geq R$$

then $|z(t_0, \omega)| \geq R$ for at least one $t_0 = t_0(\omega)$ in $[0, h]$. Hence

$$z(t_0, \omega) \cdot \lambda_{j_0} \geq \theta R \Lambda \qquad \text{for at least one } j_0 = j_0(t_0, \omega).$$

Consequently,

$$\sup_{0 \leq t \leq h} z(t, \omega) \cdot \lambda_j \geq \theta R \Lambda \qquad \text{for at least one } j.$$

We therefore have

$$P_0\Big\{\sup_{0 \leq t \leq h} |z(t, \omega)| \geq R\Big\} \leq \sum_{j=1}^{l} P_0\Big\{\sup_{0 \leq t \leq h} z(t, \omega) \cdot \lambda_j \geq \theta R \Lambda\Big\}. \tag{3.18}$$

Set

$$g_j(t) = \lambda_j \cdot z(t) - \tfrac{1}{2}\left\langle \int_0^t [a(z(t), t)\, dt] \cdot \lambda_j, \lambda_j \right\rangle.$$

Then

$$P_0\Big\{\sup_{0 \leq t \leq h} z(t, \omega) \cdot \lambda_j \geq \theta R \Lambda\Big\} \leq P_0\Big\{\sup_{0 \leq t \leq h} g_j(t) \geq \theta R \Lambda - \frac{\Lambda^2}{2}(1 + \kappa)h\Big\}$$
$$\leq \exp\left[-\theta R \Lambda + \frac{\Lambda^2}{2}(1 + \kappa)h\right]$$

by the exponential martingale inequality (Theorem 4.7.5). Taking $\theta = [1 + \kappa + (1 - \kappa)^2]/2$, $\Lambda = R/h$ and using (3.18), the assertion of the lemma follows.

We shall now apply Lemma 3.2 to $\zeta^\epsilon(t)$, with $h \leq h_0$, $R = r_0/\epsilon$. If h_0, r_0 are sufficiently small, then the diffusion matrix $a^\epsilon(x, t) = (a_{ij}^\epsilon(x, t))$ satisfies

$$\sum_{i,j=1}^{n} |a_{ij}^\epsilon(x, t) - \delta_{ij}| \leq \kappa, \qquad \kappa \text{ as in (3.17)},$$

if $0 \leq t \leq h$, $|x| \leq R$. Hence, from (3.17) and Lemma 3.2,

$$P_x\{\tau_\eta \leq h\} \leq (C(\kappa))^{1-\kappa}$$
$$\times \exp\left[\left(\frac{1}{\kappa^2} - \frac{1}{\kappa}\right) C\, \frac{h(h_0^2 + r_0^2)^\alpha}{2\epsilon^2}\right] \exp\left[-\frac{(1 - \kappa)^3 r_0^2}{2\epsilon^2 h}\right]. \tag{3.19}$$

We can now estimate $F_1(x)$ by writing

$$F_1(x) = E_x\left[\exp \frac{1-\beta''}{2\epsilon^2} \frac{r_0^2}{h_0 \wedge \tau_\eta}\right]$$
$$= \int_0^{h_0+0} \exp \frac{(1-\beta'')r_0^2}{2\epsilon^2 h}\, d_h P\{h_0 \wedge \tau_\eta < h\}$$

and integrating by parts. We get

$$F_1(x) \leq \exp\left[\frac{(1-\beta'')r_0^2}{2\epsilon^2 h_0}\right] + (C(\kappa))^{1-\kappa} \exp\left[\left(\frac{1}{\kappa^2} - \frac{1}{\kappa}\right)\frac{Ch_0(h_0^2+r_0^2)^\alpha}{2\epsilon^2}\right]$$
$$\times \int_0^{h_0+0} \exp\left\{-\frac{(1-\kappa)^3 r_0^2}{2\epsilon^2 h} + \frac{(1-\beta'')r_0^2}{2\epsilon^2 h}\right\} \frac{(1-\beta'')r_0^2}{2\epsilon^2 h^2}\, dh.$$

If we choose κ so small (with respect to β) that

$$(1-\kappa)^3 > 1 - \beta'' = (1-\beta)(1+\kappa)^4,$$

then we find that

$$F_1(x) < C' \exp\left[\frac{\gamma}{2\epsilon^2}\right] \equiv \Gamma, \quad \text{where} \quad \gamma = \frac{(1-\beta'')r_0^2}{h_0} + \frac{Ch_0}{\kappa^2}(h_0^2+r_0^2)^\alpha \tag{3.20}$$

and C' is a constant depending on κ, β''.

By the strong Markov property, if $c = (1-\beta'')r_0^2/(2\epsilon^2)$,

$$F_m(x) = E_x \exp\left[\sum_{i=1}^{m} \frac{c}{\tau_i - \tau_{i-1}}\right]$$
$$= E_x\left\{\exp\left[\sum_{i=1}^{m-1} \frac{c}{\tau_i - \tau_{i-1}}\right] E_{\xi^\epsilon(\tau_{m-1})} \exp\left[\frac{c}{\tau_1}\right]\right\}$$
$$< \Gamma E_x\left\{\exp\left[\sum_{i=1}^{m-1} \frac{c}{\tau_i - \tau_{i-1}}\right]\right\} = \Gamma F_{m-1}(x),$$

where (3.20) has been used. Hence, by induction,

$$F_m(x) < \Gamma^m. \tag{3.21}$$

Substituting this in (3.15), we get

$$P_x\{I_T(l^\epsilon) \geq J\} < e^{-A} \sum_{m=0}^{\infty} \exp\left[\frac{m\gamma}{2\epsilon^2} - \frac{(1-\beta'')r_0^2\kappa m^2}{2\epsilon^2 T^2(\kappa+1)}\right].$$

We break the series into two sums

$$\sum_{m=0}^{m'} + \sum_{m=m'+1}^{\infty} \tag{3.22}$$

where

$$m' = \left[\frac{\gamma T^2(\kappa + 1)}{(1 - \beta'')r_0^2\kappa} + 1 \right].$$

The first sum can be estimated by

$$\sum_{m=0}^{m'} \exp\left[\frac{m\gamma}{2\epsilon^2} \right] \leq m' \exp\left[\frac{m'\gamma}{2\epsilon^2} \right].$$

If ϵ is sufficiently small, depending on h_0, r_0, we get the bound

$$c_0 \exp\left[\frac{c\gamma^2}{r_0^2\epsilon^2} \right]$$

where c, c_0 are positive constants (depending on β'', κ). The second sum in (3.22) is bounded by 1 if ϵ is sufficiently small. Recalling the definition of γ in (3.20), we find that

$$P_x\{I_T(l^\epsilon) > J\} \leq e^{-A}\left\{1 + c_0 \exp\left[\frac{cr_0^2}{h_0^2\epsilon^2} + \frac{ch_0^2(h_0^2 + r_0^2)^\alpha}{r_0^2\epsilon^2} \right]\right\} \quad (3.23)$$

with a different constant c, depending on β'', κ.

Recalling the definition of A in (3.15) we see that if we first choose β sufficiently small (depending on h, where h is as in (3.8), (3.9)) and then β_0 sufficiently small (depending on β, h) and finally h_0, r_0 sufficiently small (depending on β, β_0, κ, h) such that also

$$\frac{r_0}{h_0}, \quad \frac{h_0}{r_0}(h_0^2 + r_0^2)^{\alpha/2}$$

are sufficiently small (depending on c in (3.23)), then we get from (3.23)

$$P_x\{I_T(l^\epsilon) > J\} \leq e^{-(J-h)/2\epsilon^2}$$

provided ϵ is sufficiently small. Substituting this into (3.9), we find that

$$\overline{\lim_{\epsilon\to 0}}\, [2\epsilon^2 \log P_x\{d_T(\xi^\epsilon, \Phi_0) > \delta, \xi^\epsilon \in \Phi_{x,T}(\Gamma, \Delta)\}] \leq -I_0 + 2h.$$

Since h is arbitrary, the assertion (3.1) follows.

Denote by $\Phi^*_{x,T}(\Gamma, \Delta)$ the subset of $\Phi_{x,T}(\Gamma, \Delta)$ consisting of all the curves ϕ that actually do intersect the set Δ (of course, only after intersecting Γ at some preceding time). Let Φ_0^* be the subset of $\Phi^*_{x,T}(\Gamma, \Delta)$ consisting of all ϕ with $I_T(\phi) \leq I_0$.

Later we shall need the following variant of Theorem 3.1.

Theorem 3.3. *Let* (A) *hold. For any* $\delta > 0$, $x \in R^n \backslash \Delta$,

$$\overline{\lim_{\epsilon\to 0}}\, \{2\epsilon^2 \log[P_x\{d_T(\xi^\epsilon, \Phi_0^*) > \delta, \xi^\epsilon \in \Phi^*_{x,T}(\Gamma, \Delta)\}]\} \leq -I_0. \quad (3.24)$$

Proof. Arguing as before we conclude that if

$$\begin{aligned} &d_T(\xi^\epsilon, \Phi_0^*) < \delta, \\ &\xi^\epsilon \in \Phi_{x,T}^*(\Gamma, \Delta) \end{aligned} \tag{3.25}$$

and if $I_T(l^\epsilon) \leqslant J$, then there is a curve ψ^* in $\Phi_{x,T}(\Gamma, \Delta)$ such that

$$\rho_T(\xi^\epsilon, \psi^*) < k,$$
$$I_T(\psi^*) \leqslant J + h$$

where k, h are any given positive and small numbers, provided h_0, r_0 are sufficiently small.

Modify ψ^* into $\psi^{**} \in \Phi_{x,T}^*(\Gamma, \Delta)$ as follows: If $\psi^* \in \Phi_{x,T}^*(\Gamma, \Delta)$, take $\psi^{**} = \psi^*$. If $\psi^* \notin \Phi_{x,T}^*(\Gamma, \Delta)$, then notice that, since $\xi^\epsilon \in \Phi_{x,T}^*(\Gamma, \Delta)$, $\xi^\epsilon(t_1) \in \Gamma$ and $\xi^\epsilon(t_2) \in \Delta$ for some $0 \leqslant t_1 < t_2 \leqslant T$. Therefore $\psi^*(t_1)$ and $\psi^*(t_2)$ belong to k-neighborhoods of Γ and Δ respectively. Modify ψ^* into ψ^{**} so that ψ^{**} intersects Γ at $t_1 + O(k)$ and intersects Δ for the first time at $t_2 + O(k)$, and

$$|I_T(\psi^*) - I_T(\psi^{**})| \leqslant C_1 k,$$
$$\rho_T(\psi^*, \psi^{**}) \leqslant C_1 k$$

(cf. the modification from ϕ into ψ in the proof of Theorem 3.1).

We conclude that, if k is sufficiently small,

$$\rho_T(\xi^\epsilon, \psi^{**}) < \delta,$$
$$I_T(\psi^{**}) < J + 2h.$$

Since $\psi^{**} \in \Phi_{x,T}^*(\Gamma, \Delta)$, we get a contradiction to the first part of (3.25) if $J + 2h \leqslant I_0$. Therefore, if $I_T(l^\epsilon) \leqslant J$, then $J \geqslant I_0 - 2h$, i.e., (3.25) implies that $I_T(l^\epsilon) \geqslant I_0 - 2h$.

We can now proceed precisely as in the proof of Theorem 3.1.

Let D be a bounded domain with C^2 boundary, and let Γ be a closed ball in D. Let $\Delta = R^n \setminus D$. Denote by $\tilde{\Phi}_{x,T}(\Gamma, \Delta)$ the set of all curves ϕ in C_T such that $\phi(0) = x$, $\phi(t) \in \Delta$ for some $0 \leqslant t < T$, $\phi(T) \in \Gamma$. Let $\tilde{\Phi}_0$ be the subset of $\tilde{\Phi}_{x,T}(\Gamma, \Delta)$ consisting of all curves ϕ with $I_T(\phi) \leqslant I_0$; I_0 a given positive number.

Theorem 3.4. *Let* (A) *hold. For any* $\delta > 0$, $x \in D \setminus \Gamma$,

$$\overline{\lim_{\epsilon \to 0}} \{2\epsilon^2 \log[P_x\{d_T(\xi^\epsilon, \tilde{\Phi}_0) > \delta, \xi^\epsilon \in \tilde{\Phi}_{x,T}(\Gamma, \Delta)\}]\} - \leqslant I_0.$$

The proof proceeds similarly to the proof of Theorem 3.3. One shows that

if

$$d_T(\xi^\epsilon, \tilde{\Phi}_0) < \delta, \qquad \xi^\epsilon \in \tilde{\Phi}_{x,T}(\Gamma, \Delta), \qquad I_T(l^\epsilon) \leq J,$$

then there exists a curve $\psi \in \tilde{\Phi}_{x,T}(\Gamma, \Delta)$ with $\rho_T(\xi^\epsilon, \psi) < k$, $I_T(\psi) \leq J + h$ (k, h are small); in this proof one employs a smooth deformation of neighborhoods of ∂D and $\partial \Gamma$; cf. the proof of Theorem 3.1. Details are left to the reader.

Theorem 3.5. *Let* (A) *hold. Then for any* $x \in R^n$ *and for any closed set* C *in* $\mathcal{C}_T$,

$$\overline{\lim_{\epsilon \to 0}} \left[2\epsilon^2 \log P_x^\epsilon(C)\right] \leq - \inf_{\omega \in C_x} I_T(\omega) \tag{3.26}$$

where $C_x = \{\omega \in C;\ \omega(0) = x\}$.

This theorem is a complement to Theorem 2.2. Strictly speaking, it does not contain Theorems 3.1, 3.3, 3.4. These theorems, however, can be deduced from Theorem 3.5 applied to appropriate sequences of sets C. As we shall see, the proof of Theorem 3.5 is not much different from the proof of Theorem 3.1.

Proof. Let δ be any small positive number. Denote by C_δ the δ-neighborhood of C, i.e., $\omega \in C_\delta$ if and only if there exists an $\omega' \in C$ such that $\rho_T(\omega, \omega') < \delta$. Let $C_{\delta, x} = \{\omega \in C_\delta;\ \omega(0) = x\}$,

$$j_0 = \inf_{\omega \in C_x} I_T(\omega), \qquad j_\delta = \inf_{\omega \in C_{\delta, x}} I_T(\omega). \tag{3.27}$$

Let $\varphi_\delta \in C_{\delta, x}$ be such that $I_T(\varphi_\delta) < j_\delta + \delta$, and choose $\tilde{\varphi}_\delta \in C_x$ such that $\rho_T(\tilde{\varphi}_\delta, \varphi_\delta) \to 0$ if $\delta \to 0$. Since $I_T(\varphi_\delta) \leq c$, c independent of δ, from any sequence $\{\tilde{\delta}_m\}$ ($\tilde{\delta}_m \to 0$) we can extract a subsequence $\{\delta_m\}$ such that $\varphi_{\delta_m} \to \varphi$ in C_T. But then, by Lemma 1.2,

$$I_T(\varphi) \leq \underline{\lim}\, I_T(\varphi_{\delta_m}) \leq \underline{\lim}\, j_{\delta_m}.$$

We also have $\rho_T(\varphi, \tilde{\varphi}_{\delta_m}) \to 0$, $\tilde{\varphi}_{\delta_m} \in C_x$. Since C is closed, we deduce that $\varphi \in C_x$. Therefore,

$$j_0 = \inf_{\omega \in C_x} I_T(\omega) \leq I_T(\varphi) \leq \underline{\lim}\, j_{\delta_m}.$$

Choosing $\tilde{\delta}_m$ such that $j_{\tilde{\delta}_m} \to \underline{\lim}_{\delta \to 0}\, j_\delta$, we get

$$j_0 \leq \underline{\lim}_{\delta \to 0}\, j_\delta.$$

Since, obviously, $j_\delta \leq j_0$ for any δ, we conclude that

$$\lim_{\delta \to 0} j_\delta = j_0 = \inf_{\omega \in C_x} I_T(\omega). \tag{3.28}$$

Using the same notation l^ϵ as in the proof of Theorem 3.1, we have, for any $\delta > 0$,

$$\{\xi^\epsilon \in C_x\} \subset \{l^\epsilon \in C_{\delta, x}\} \subset \{I_T(l^\epsilon) > j_\delta\}$$

provided h_0, r_0 are sufficiently small (depending on μ, M, δ). Hence,

$$P_x^\epsilon(C_x) < P_x[I_T(l^\epsilon) > J], \qquad J = j_0 - h \tag{3.29}$$

where $h = j_0 - j_\delta \to 0$ if $\delta \to 0$, by (3.28). Thus (3.29) holds for any small $h > 0$ provided h_0, r_0 are sufficiently small. We can now proceed precisely as in the proof of Theorem 3.1.

4. Application to the first initial-boundary value problem

Let D be a bounded domain in R^n with C^2 boundary ∂D. Let

$$L_\epsilon u = \frac{\epsilon^2}{2} \sum_{i,j=1}^{n} a_{ij}(x) \frac{\partial^2 u}{\partial x_i \partial x_j} + \sum_{i=1}^{n} b_i(x) \frac{\partial u}{\partial x_i}. \tag{4.1}$$

Definition. A function $q_\epsilon(t, x, y)$ defined and continuous for $x \in D$, $y \in D$, $t > 0$ is called *Green's function* for $L_\epsilon - \partial/\partial t$ in the cylinder $Q = D \times (0, \infty)$ if for any continuous function $f(x)$ in $\bar{D}$ vanishing on ∂D,

$$u(x, t) = \int_D q_\epsilon(t, x, y) f(y)\, dy$$

is the classical solution of

$$\begin{aligned} L_\epsilon u - \frac{\partial u}{\partial t} &= 0 \quad \text{in } Q, \\ u(x, 0) &= f(x) \quad \text{if } x \in D, \\ u(x, t) &= 0 \quad \text{if } x \in \partial D, \ t > 0; \end{aligned} \tag{4.2}$$

note that u is continuous in $\bar{Q}$ and u_t, $D_x u$, $D_x^2 u$ are continuous in Q.

By the maximum principle it follows that Green's function, if existing, is unique. Indeed, if q_ϵ and $\bar{q}_\epsilon$ are two Green's functions, then from the uniqueness of the solution of (4.2) it follows that

$$\int_D [q_\epsilon(t, x, y) - \bar{q}_\epsilon(t, x, y)] f(y)\, dy = 0.$$

Since this is true for any continuous f with support in D, $q_\epsilon(t, x, y) = \bar{q}_\epsilon(t, x, y)$.

One can construct the Green function $q_\epsilon(t, x, y)$ in the form

$$q_\epsilon(t, x, y) = p_\epsilon(t, x, y) + V_\epsilon(t, x, y)$$

where $p_\epsilon(t, x, y)$ is the fundamental solution $\Gamma(x, t; y, 0)$ of $L_\epsilon - \partial/\partial t$ occurring in Section 6.4. The function V_ϵ is defined, for each y, as the solution of

$$\left(L_\epsilon - \frac{\partial}{\partial t}\right) V_\epsilon = 0 \qquad \text{in } Q$$

$$V_\epsilon(0, x, y) = 0 \qquad \text{if } x \in D,$$

$$V_\epsilon(t, x, y) = -p_\epsilon(t, x, y) \qquad \text{if } x \in \partial D, \quad t > 0.$$

One can prove (see Friedman [1]) that q_ϵ, $D_x q_\epsilon$, $D_x^2 q_\epsilon$ and $D_t q_\epsilon$ are continuous in (t, x, y) for $t > 0$, $x \in \bar{D}$, $y \in \bar{D}$. By the maximum principle

$$q_\epsilon(t, x, y) \leqslant p_\epsilon(t, x, y). \tag{4.3}$$

In Section 5 we shall study the behavior of the fundamental solution p_ϵ, as $\epsilon \to 0$. In Section 6 we shall study the behavior of q_ϵ as $\epsilon \to 0$. In the present section we study the behavior, as $\epsilon \to 0$, of the solution u_ϵ of the initial-boundary value problem

$$\partial u_\epsilon / \partial t = L_\epsilon u_\epsilon \qquad \text{if } x \in D, \quad t > 0$$

$$u_\epsilon(x, 0) = 0 \qquad \text{if } x \in D, \tag{4.4}$$

$$u_\epsilon(x, t) = 1 \qquad \text{if } x \in \partial D, \quad t > 0.$$

By a solution u of (4.4) we understand a function u that has continuous derivatives $D_x u$, $D_x^2 u$, $D_t u$ in the cylinder $Q = D \times (0, \infty)$, that is continuous at all the points of $\bar{Q}$ with the exception of $\{(x, 0); x \in \partial D\}$, that is bounded in Q, and that satisfies (4.4).

One can show (see Problems 1, 2) that there exists a unique solution of (4.4), and (see Problem 3)

$$u_\epsilon(x, t) = P_x\{\tau^\epsilon < t\} \tag{4.5}$$

where τ^ϵ is the exit time of $\xi^\epsilon(t)$ from D.

Let

$$I(t, x, \partial D) = \inf_{\phi \in \Psi_t} I_t(\phi)$$

where Ψ_t consists of all functions ϕ in C_t satisfying: $\phi(0) = x$, $\min_{0 < s < t} \rho(\phi(s), \partial D) = 0$.

Theorem 4.1. *Let* (A) *hold. Then, for any* $x \in D$, $t > 0$,

$$\lim_{\epsilon \to 0} [2\epsilon^2 \log u_\epsilon(x, t)] = -I(t, x, \partial D). \tag{4.6}$$

Proof. Denote by D_δ ($\delta > 0$) a δ-neighborhood of D. For any $h_1 > 0$ there exists a $\delta > 0$ sufficiently small such that the following is true: There is a curve ϕ in C_t with $\phi(0) = x$ such that

$$I_t(\phi) < I(t, x, \partial D) + h_1$$

and $\phi(s)$ intersects the boundary of D_δ at some time $s < t$. By Theorem 2.1,

$$P_x\{\tau^\epsilon < t\} > P_x\{\rho_t(\xi^\epsilon, \phi) < \delta\} \geqslant \exp\left\{-\frac{I(t, x, \partial D) + h_1 + h}{2\epsilon^2}\right\}$$

for any $h > 0$, provided ϵ is sufficiently small. From this and from (4.5) we get

$$\varliminf_{\epsilon\to 0} \left[2\epsilon^2 \log u_\epsilon(x, t)\right] \geqslant -I(t, x, \partial D).$$

It remains to prove that

$$\varlimsup_{\epsilon\to 0} \left[2\epsilon^2 \log u_\epsilon(x, t)\right] \leqslant -I(t, x, \partial D). \tag{4.7}$$

Notice that the class Ψ_t introduced in the definition of $I(t, x, \partial D)$ coincides with $\Phi_{x,t}(\partial D, \varnothing)$, in the notation of Theorem 3.1. Denote by Φ_0 the subset of Ψ_t consisting of all ϕ with $I_T(\phi) \leqslant I_0$. If $I_0 = I(t, x, \partial D) - h_1$ $(h_1 > 0)$ then Φ_0 is empty. Hence, by Theorem 3.1,

$$\varlimsup_{\epsilon\to 0} \{2\epsilon^2 \log[P_x\{\xi^\epsilon \in \Phi_{x,t}(\partial D, \varnothing)\}]\} \leqslant -I(t, x, \partial D) + h_1.$$

Since

$$\{\tau^\epsilon \leqslant t\} \subset \{\xi^\epsilon \in \Phi_{x,t}(\partial D, \varnothing)\},$$

the assertion (4.7) follows.

5. Behavior of the fundamental solution as $\epsilon \to 0$

In this section and in the following one we assume:

(A′) The condition (A) holds and, in addition, the b_i are continuously differentiable and

$$\left|\frac{\partial b_i(x)}{\partial x_j} - \frac{\partial b_i(\bar{x})}{\partial x_j}\right| \leqslant M|x - \bar{x}|^\alpha.$$

Denote by $p_\epsilon(t, x, y)$ the fundamental solution $\Gamma(x, t; y, 0)$ (constructed in Section 6.4) of the Cauchy problem for the parabolic equation

$$\frac{\partial u}{\partial t} = L_\epsilon u \qquad (x \in R^n, t > 0), \tag{5.1}$$

where L_ϵ is given by (4.1). It was proved by Aronson [1] that (when (A′) holds)

$$p_\epsilon(t, x, y) \leq \frac{A_0}{\epsilon^n t^{n/2}} \exp\left\{ -\frac{c|\phi(t, x) - y|^2}{\epsilon^2 t} \right\} \quad \text{if} \quad t < T^*, \tag{5.2}$$

$$p_\epsilon(t, x, y) \geq \frac{A_1}{\epsilon^n t^{n/2}} \exp\left\{ -\frac{\gamma|\phi(t, x) - y|^2}{\epsilon^2 t} \right\}$$
$$- \frac{A_2}{\epsilon^{n-2\alpha} t^{n/2-\alpha}} \exp\left\{ -\frac{c|\phi(t, x) - y|^2}{\epsilon^2 t} \right\} \quad \text{if} \quad t < t^*; \tag{5.3}$$

T^* is any positive number, t^* is a sufficiently small positive number, A_0, A_1, A_2, c, γ are positive constants, and $\phi(t, x)$ is the solution of

$$d\phi/dt = b(\phi), \qquad \phi(0) = x.$$

The proof of (5.2), (5.3) is obtained by a careful analysis of a variant of the parametrix method. If $b_i \equiv 0$, then (5.2), (5.3) are obtained immediately from the explicit formula for the fundamental solution for the heat equation.

Let

$$I_t(x, y) = \inf_\phi I_t(\phi)$$

where ϕ varies in C_t, $\phi(0) = x$, $\phi(t) = y$.

We shall prove:

Theorem 5.1. *Let* (A′) *hold. Then, for any x, y in R^n and $t > 0$,*

$$\lim_{\epsilon \to 0} [2\epsilon^2 \log p_\epsilon(t, x, y)] = -I_t(x, y). \tag{5.4}$$

Proof. We first prove that

$$\overline{\lim_{\epsilon \to 0}} [2\epsilon^2 \log p_\epsilon(t, x, y)] \leq -I_t(x, y), \tag{5.5}$$

i.e., for any $h > 0$,

$$p_\epsilon(t, x, y) \leq \exp\left\{ \frac{-I_t(x, y) + h}{2\epsilon^2} \right\} \tag{5.6}$$

if ϵ is sufficiently small.

Let

$$\tilde{C}_\delta = \{(z, s); s \leq t, |\phi(t - s, z) - y| \leq \delta\}.$$

Note that if $(z, s) \in \partial\tilde{C}_\delta$, then the trajectory of $dx/dt = b(x)$ which is at z at time 0 will be at a distance δ from y at time $t - s$.

If $\phi(t, x) = y$, then (5.5) is a consequence of (5.2). We shall therefore assume that $\phi(t, x) \neq y$. Then $(x, 0)$ does not belong to $\tilde{C}_\delta$ for any δ sufficiently small. Let

$$\tau_\delta^\epsilon = \inf\{\bar{t} > 0; (\xi^\epsilon(\bar{t}), \bar{t}) \in \tilde{C}_\delta\}, \qquad u_\delta^\epsilon(x, s) = P_x\{\tau_\delta^\epsilon < s\}.$$

By the proof of Theorem 4.1 we have,

$$\varlimsup_{\epsilon\to 0}\left[2\epsilon^2 \log u_\delta^\epsilon(x, t)\right] \leq -i_\delta \tag{5.7}$$

where

$$i_\delta = \inf_{\phi\in\Phi_\delta} I_t(\phi), \quad \Phi_\delta = \left\{\phi \in C_t;\ \phi(0) = x,\ \min_{0\leq s\leq t} \operatorname{dist}((\phi(s), s), \partial\tilde{C}_\delta) = 0\right\}.$$

By the strong Markov property,

$$\begin{aligned} p_\epsilon(t, x, B) &= E\{\chi_{\{\tau_\delta^\epsilon = s \text{ for some } s<t\}}[P_x[\xi^\epsilon(t-s) \in B | \xi^\epsilon(0) = x]]_{x=\xi^\epsilon(s)}\} \\ &\leq E\left\{\chi_{\{\tau_\delta^\epsilon = s \text{ for some } s<t\}} \sup_{(z,s)\in\partial\tilde{C}_\delta} [P_z[\xi^\epsilon(t-s)B | \xi^\epsilon(0) = z]]\right\} \\ &= \int_0^t P(\tau_\delta^\epsilon \in ds) \sup_{(z,s)\in\partial\tilde{C}_\delta} p_\epsilon(t-s, z, B) \end{aligned} \tag{5.8}$$

for any Borel set B such that $B \times \{t\} \subset \tilde{C}_\delta$; here

$$p_\epsilon(t, x, B) = \int_B p_\epsilon(t, x, y)\, dy.$$

Hence,

$$p_\epsilon(t, x, y) \leq \int_0^t u_\delta^\epsilon(x, ds) \sup_{z,s} p_\epsilon(t-s, z, y). \tag{5.9}$$

Since $|\phi(t-s, z) - y| = \delta$ in the last integral, (5.2) gives

$$p_\epsilon(t-s, z, y) \leq \frac{A_0}{\epsilon^n (t-s)^{n/2}} \exp\left\{-\frac{c\delta^2}{\epsilon^2(t-s)}\right\}.$$

Substituting this into (5.9), we get

$$p_\epsilon(t, x, y) \leq \frac{A(\delta)}{\epsilon^n} \int_0^t u_\delta^\epsilon(x, ds) = \frac{A(\delta)}{\epsilon^n} u_\delta^\epsilon(x, t)$$

where $A(\delta)$ is a constant depending only on δ. Recalling (5.7), we conclude that

$$\varlimsup_{\epsilon\to 0} [2\epsilon^2 \log p_\epsilon(t, x, y)] \leq -i_\delta \qquad \text{for any} \quad \delta > 0. \tag{5.10}$$

As $\delta \downarrow 0$, $i_\delta \uparrow i_0$. Hence, in order to prove (5.5), it suffices to show that $i_0 \geq I_t(x, y)$. Since $I_t(\phi)$ is lower semicontinuous, we find that there exists a curve $\phi \in \Phi_0$ with $I_t(\phi) = i_0$. Let

$$s = \inf\{\bar{t};\ (\phi(\bar{t}), \bar{t}) \in \tilde{C}_0\}.$$

Since ϕ is a minimum for I_t over Φ_0, we must have $d\phi/d\lambda = b(\phi)$ if

$s < \lambda < t$. But since $(\phi(\bar{t}), \bar{t}) \in \tilde{C}_0$, it then follows that $\phi(t) = y$. Consequently, $I_t(x, y) \leqslant I_t(\phi) = i_0$.

We next have to prove that

$$\varliminf_{\epsilon \to 0} [2\epsilon^2 \log p_\epsilon(t, x, y)] \geqslant -I_t(x, y), \tag{5.11}$$

i.e., for any $h_0 > 0$,

$$p_\epsilon(t, x, y) > \exp\left[\frac{-I_t(x, y) - h_0}{2\epsilon^2}\right] \tag{5.12}$$

provided ϵ is sufficiently small.

Let $\zeta \in R^n$, $\theta > 0$, $|\zeta - y| \leqslant h$ where $0 < h < 1$, and consider the curve ψ:

$$\psi(s) = \phi(-s, y) \qquad \text{for} \quad 0 \leqslant s \leqslant \theta.$$

Let M be a positive number (to be determined later). Suppose first that

$$|\zeta - \psi(\theta)| > M\epsilon. \tag{5.13}$$

Let $0 < \lambda < \theta$ and consider the function

$$w(z, s) = p_\epsilon(s + \lambda, z, y) \qquad \text{for} \quad z \in B_{\lambda+s}, \quad 0 < s < \theta - \lambda$$

where $B_s = \{z; |z - \psi(s)| > M\epsilon\}$. If $z \in \partial B_{\lambda+s}$, $|z - \psi(s + \lambda)| = M\epsilon$; hence

$$|\phi(s + \lambda, z) - \phi(s + \lambda, \psi(s + \lambda))| \leqslant CM\epsilon \qquad (C \text{ const}).$$

But since

$$\phi(s + \lambda, \psi(s + \lambda)) = \phi(s + \lambda, \phi(-s - \lambda, y)) = \phi(0, y) = y,$$

we get

$$|\phi(s + \lambda, z) - y| \leqslant CM\epsilon.$$

Using (5.3) we conclude that

$$w(z, s) = p_\epsilon(s + \lambda, z, y) \geqslant \frac{C_\lambda}{\epsilon^n}$$
$$\text{if} \quad z \in \partial B_{\lambda+s}, \quad 0 < s < \theta - \lambda, \quad \theta < t^*, \tag{5.14}$$

where C_λ is a positive constant depending on λ, provided ϵ is sufficiently small (depending on λ).

We also have

$$w(z, 0) > 0 \qquad \text{if} \quad z \in B_\lambda. \tag{5.15}$$

Let

$$G = G(M) = \bigcup_{0 < s < \theta - \lambda} (B_{s+\lambda} \times \{s\}),$$
$$\partial_0 G = \partial_0 G(M) = \bigcup_{0 < s < \theta - \lambda} (\partial B_{s+\lambda} \times \{s\}).$$

Consider the solution $u_\epsilon(z, s)$ of

$$\begin{aligned} \partial u_\epsilon/\partial t &= L_\epsilon u_\epsilon \quad \text{if} \quad z \in B_{s+\lambda}, \quad 0 < s < \theta - \lambda, \\ u_\epsilon(z, 0) &= 0 \quad \text{if} \quad z \in B_\lambda, \\ u_\epsilon(z, s) &= 1 \quad \text{if} \quad z \in \partial B_{\lambda+s}, \quad 0 < s < \theta - \lambda. \end{aligned} \tag{5.16}$$

Assume first that $\partial_0 G$ is in C^2. Then there exists a unique solution u_ϵ of (5.16). By Itô's formula,

$$u_\epsilon(z, \theta - \lambda) = P_z(\tau < \theta - \lambda) \quad \text{if} \quad z \in B_{(\theta-\lambda)+\lambda} = B_\theta, \tag{5.17}$$

where τ is the first time the path $(\theta - \lambda - s, \xi^\epsilon(s))$ hits the set $\partial_0 G(M)$. Notice that, by (5.13), $\zeta \in B_\theta$; hence (5.17) holds, in particular, for $z = \zeta$.

Let ψ_0 be a straight line in $C_{\theta-\lambda}$ connecting ζ to $\psi(\lambda)$, i.e., $\psi_0(0) = \zeta$, $\psi_0(\theta - \lambda) = \psi(\lambda)$. Since $|\zeta - \psi(\lambda)| \leq |\zeta - y| + |\psi(\lambda) - y| \leq h + (\text{const})\lambda$,

$$I_{\theta-\lambda}(\psi_0) \leq \frac{\tilde{c}\hat{h}^2}{\theta - \lambda} + \tilde{c}(\theta - \lambda) \qquad (\tilde{c} \text{ positive constant}), \quad \hat{h}^2 = h^2 + \lambda^2. \tag{5.18}$$

Clearly

$$\{\tau < \theta - \lambda\} \supset \{\rho_{\theta-\lambda}(\xi^\epsilon, \psi_0) < M\epsilon\}.$$

Hence, by Theorem 2.1 and (5.17) (with $z = \zeta$)

$$\begin{aligned} u_\epsilon(\zeta, \theta - \lambda) &\geq P_\zeta\{\rho_{\theta-\lambda}(\xi^\epsilon, \psi_0) < M\epsilon\} \\ &\geq \tfrac{1}{2} \exp\left\{ -\frac{I_{\theta-\lambda}(\psi_0)}{2\epsilon^2} - \frac{C\delta^\alpha}{2\epsilon^2} - \frac{(4CK)^{1/2}}{\epsilon} \right\} \end{aligned}$$

where $K = \int_0^{\theta-\lambda}(1 + |d\psi_0/ds|^2)\, ds$, $\delta = M\epsilon$, provided $C_0\epsilon^2\theta \leq \delta^2/4$. Notice that the last inequality means that $C_0\theta \leq M^2/4$. We now choose M so that this last inequality holds. Making use also of (5.18), we get

$$u_\epsilon(\zeta, \theta - \lambda) \geq \exp\left\{ -\frac{(\tilde{c} + 1)(\hat{h}^2 + \theta^2)}{(\theta - \lambda)\epsilon^2} \right\} \tag{5.19}$$

provided ϵ is sufficiently small.

We now apply the maximum principle in order to compare $u_\epsilon(z, s)$ with $p_\epsilon(s + \lambda, z, y)$ for $z \in B_{s+\lambda}$, $0 < s \leq \theta - \lambda$. Using (5.14), (5.15), we get

$$p_\epsilon(s + \lambda, z, y) \geq \frac{C_\lambda}{\epsilon^n} u(z, s). \tag{5.20}$$

Taking, in particular, $s = \theta - \lambda$, $z = \zeta$, $\lambda = \theta/2$ and using (5.19), we obtain

$$p_\epsilon(\theta, \zeta, y) \geq \frac{C(\theta)}{\epsilon^n} \exp\left\{ -\frac{c(h^2 + \theta^2)}{\theta\epsilon^2} \right\} \qquad (|\zeta - y| \leq h, \theta < t^*), \tag{5.21}$$

for all ϵ sufficiently small, where $C(\theta)$ is a positive constant depending on θ ($C(\theta) \to 0$ if $\theta \to 0$), and $c = 3(\tilde{c} + 1)$.

We have proved (5.21) assuming that $\partial_0 G(M)$ is in C^2. If this is not the case, we replace G by a region G' with $G(M') \subset G' \subset G(M)$ such that its lateral boundary $\partial_0 G'$ is in C^2 and $(\theta - \lambda, \zeta) \in G'$. We then proceed as before, replacing G and $\partial_0 G$ everywhere by G' and $\partial_0 G'$ respectively.

So far we have proved (5.21) under the restriction (5.13). If $|\zeta - \psi(\theta)| \leqslant M\epsilon$, then $|\phi(\theta, \zeta) - y| \leqslant CM\epsilon$ for some constant C. But then (5.21) follows immediately from (5.3), provided $0 < \theta < t^*$ and ϵ is sufficiently small (depending on θ).

In order to prove (5.11) we shall need, in addition to (5.21), the following result:

Let G be a ball of radius $|G|$ and center z. Then, for any $s > 0$ and for any $h' > 0$,

$$P_x\{\xi^\epsilon(s - \nu) \in G\} \geqslant \exp\left\{ -\frac{I_s(x, z)}{2\epsilon^2} - \frac{h'}{2\epsilon^2} \right\} \tag{5.22}$$

for any $\nu > 0$ sufficiently small, provided ϵ is sufficiently small.

To prove (5.22) let ψ be a curve in C_s such that

$$\psi(0) = x, \qquad \psi(s - \nu) = z, \qquad I_s(\psi) < I_s(x, z) + \frac{h'}{2}$$

for some $\nu > 0$ sufficiently small. Clearly

$$\{\xi^\epsilon(s - \nu) \in G\} \supset \{\rho_{s-\nu}(\xi^\epsilon, \psi) < |G|\},$$

and (5.22) follows upon taking P_x of both sides and applying Theorem 2.1.

We proceed to prove (5.11), using the semigroup property

$$p_\epsilon(t, x, y) = \int_{R^n} p_\epsilon(t - \mu, x, z) p_\epsilon(\mu, z, y)\, dz \tag{5.23}$$

This implies that

$$p_\epsilon(t, x, y) \geqslant \int_G p_\epsilon(t - \mu, x, z) p_\epsilon(\mu, z, y)\, dz$$

where $G = \{z; |z - y| < h\}$, and μ, h are any given small positive numbers.

By (5.21),

$$p_\epsilon(\mu, z, y) \geqslant \beta(\mu) \exp\left\{ -\frac{\beta(h^2 + \mu^2)}{\epsilon^2 \mu} \right\}$$

provided ϵ is sufficiently small, where $\beta(\mu), \beta$ are positive constants independent of h, ϵ (and β is also independent of μ). Hence

$$p_\epsilon(t, x, y) \geqslant \gamma_0 \beta(\mu) \exp\left\{ -\frac{\beta(h^2 + \mu^2)}{\epsilon^2 \mu} \right\} P_x\{\xi^\epsilon(t - \mu) \in G\}, \tag{5.24}$$

where γ_0 is a positive constant.

By (5.22) with $s = t$ and $\nu = \mu$,

$$P_x\{\xi^\epsilon(t-\mu) \in G\} > \exp\left\{-\frac{I_t(x, y) + h'}{2\epsilon^2}\right\}$$

if ϵ is sufficiently small. Recalling (5.24), we get

$$\varliminf_{\epsilon\to 0}\,[2\epsilon^2 \log p_\epsilon(t, x, y)] > -\frac{\beta(h^2 + \mu^2)}{\mu} - I_t(x, y) - h'.$$

Notice that μ is independent of the parameter h. Taking first $h \to 0$, then $\mu \to 0$, and finally $h' \to 0$, the assertion (5.11) follows.

6. Behavior of Green's function as $\epsilon \to 0$

Let D be a bounded domain in R^n with C^2 boundary ∂D. Denote by $q_\epsilon(t, x, y)$ the Green function of $L_\epsilon - \partial/\partial t$ in the cylinder $Q = D \times (0, \infty)$, and set

$$q_\epsilon(t, x, A) = \int_A q(t, x, y)\, dy \tag{6.1}$$

for any Borel set A in D. Let

$$u(x, t) = \int_D q_\epsilon(t, x, y) f(y)\, dy$$

where f is continuous in $\bar{D}$ and vanishes on ∂D. Applying Itô's formula to $u(x, t-s)$ and $\xi^\epsilon(s)$, we find that

$$u(x, t) = E_x\{f(\xi^\epsilon(t))\chi_{\tau > t}\},$$

where τ is the exit time from D. Since f is arbitrary, we conclude that

$$q_\epsilon(t, x, A) = P_x\{\xi^\epsilon(s) \in D \text{ for } 0 < s < t, \xi^\epsilon(t) \in A\}.$$

For any x, y in D, let

$$I_t^D(x, y) = \inf_\phi I_t(\phi) \tag{6.2}$$

where ϕ varies over the function in C_t satisfying: $\phi(0) = x$, $\phi(t) = y$, and $\phi(s) \in D$ for $0 < s < t$.

Theorem 6.1. *Let* (A′) *hold. Then*

$$\lim_{\epsilon\to 0}\,[2\epsilon^2 \log q_\epsilon(t, x, y)] = -I_t^D(x, y). \tag{6.3}$$

Proof. We first prove that

$$\varlimsup_{\epsilon\to 0}\,[2\epsilon^2 \log q_\epsilon(t, x, y)] < -I_t^D(x, y). \tag{6.4}$$

Let

$\tilde{C}_\delta = \{(z, s); 0 \le s \le t, |\phi(t-s, z) - y| < \delta, \phi(t-u) \in D \text{ for } u \in [0, s]\}$.

Introduce

$$\tilde{\tau}_\delta^\epsilon = \inf\{\bar{t}; \xi^\epsilon(s) \in D \text{ for all } 0 \le s \le \bar{t}, (\xi^\epsilon(\bar{t}), \bar{t}) \in \partial \tilde{C}_\delta\}.$$

Notice that if $(z, s) = (\xi^\epsilon(\bar{t}), \bar{t})$, then $|\phi(t-s, z) - y| = \delta$, so that

$$\begin{aligned} q_\epsilon(t-s, z, y) &\le p_\epsilon(t-s, z, y) \\ &\le \frac{A_0}{\epsilon^n (t-s)^{n/2}} \exp\left[-\frac{c\delta^2}{\epsilon^2 (t-s)}\right] \\ &\le \frac{A(\delta)}{\epsilon^n} \qquad (A(\delta) \text{ const}). \end{aligned} \tag{6.5}$$

By the strong Markov property we get (cf. (5.8), (5.9))

$$q_\epsilon(t, x, y) \le \int_0^t P(\tilde{\tau}_\delta^\epsilon \in ds) \sup_{(z, s)} q_\epsilon(t-s, z, y)(|\phi(t-s, t) - y| = \delta).$$

Using (6.5) we obtain

$$q_\epsilon(t, x, y) \le \frac{A(\delta)}{\epsilon^n} P(\tilde{\tau}_\delta^\epsilon \le t). \tag{6.6}$$

Let $\tilde{\Phi}_\delta = \{\phi \in C_t; \phi(0) = x, (\phi(s), s) \in \partial \tilde{C}_\delta$ for some $0 < s \le t$; if $\bar{s} = \inf s$ such that $(\phi(s), s) \in \partial \tilde{C}_\delta$, then $\phi(\lambda) \in D$ if $0 \le \lambda < \bar{s}\}$,

$$\tilde{i}_\delta = \inf_{\phi \in \tilde{\Phi}_\delta} I_t(\phi).$$

By slightly modifying the first part of the proof of Theorem 4.1, we get

$$\overline{\lim_{\epsilon \to 0}} \, [2\epsilon^2 \log P(\tilde{\tau}_\delta^\epsilon \le t)] \le -\tilde{i}_\delta.$$

Using this in (6.6), we conclude that

$$\overline{\lim_{\epsilon \to 0}} \, [2\epsilon^2 \log q_\epsilon(t, x, y)] \le -\tilde{i}_\delta. \tag{6.7}$$

As $\delta \downarrow 0$, $\tilde{i}_\delta \uparrow i$ for some i. Since D is not closed, a compactness argument is not available here. However we can still prove that $i \ge I_t^D(x, y)$ as follows:

For any $\epsilon' > 0$, there is a $\delta > 0$ and $\phi \in \tilde{\Phi}_\delta$ such that

$$I_t(\phi) < \tilde{i}_\delta + \epsilon', \qquad |\phi(t) - y| < \delta;$$

δ can be taken arbitrarily small. But then ϕ can be modified into $\tilde{\phi}$ such that $\tilde{\phi} \in \Phi_\delta$, $I_t(\tilde{\phi}) < \tilde{i}_\delta + 2\epsilon'$, $\tilde{\phi}(t) = y$. Hence

$$I_t^D(x, y) < \tilde{i}_\delta + 2\epsilon' \le i + 2\epsilon'.$$

Since ϵ' is arbitrary, $I_t^D(x, y) \le i$. Taking $\delta \to 0$ in (6.7) and using the last inequality, (6.4) follows.

We shall next prove that

$$\lim_{\epsilon\to 0} [2\epsilon^2 \log q_\epsilon(t, x, y)] \geq -I_t^D(x, y). \tag{6.8}$$

We shall need the following lemma:

Lemma 6.2. *Let $\rho(y, \partial D) \geq \delta_0 > 0$. If $z \in D$, $|z - y| < \delta$ and δ, t are positive and sufficiently small (depending on δ_0), then*

$$q_\epsilon(t, z, y) \geq \beta_0(t) \exp\left[-\frac{\beta(\delta^2 + t^2)}{\epsilon^2 t}\right] \tag{6.9}$$

for all ϵ sufficiently small (depending on δ, t), where $\beta_0(t)$ is a positive constant depending on t but not on δ, ϵ, and β is a positive constant independent of t, δ, ϵ.

Proof. Denote by $\hat{\tau}$ the exit time from D. By the strong Markov property, if $z \in D$ and B is a Borel subset of D,

$$\begin{aligned} p_\epsilon(t, z, B) &= q_\epsilon(t, z, B) + P_z\{\hat{\tau} < t, \xi^\epsilon(t) \in B\} \\ &= q_\epsilon(t, z, B) + E_z \chi_{(\hat{\tau} < t)} p_\epsilon(t - \hat{\tau}, \xi^\epsilon(\hat{\tau}), B). \end{aligned}$$

Hence

$$q_\epsilon(t, z, y) \geq p_\epsilon(t, z, y) - \sup_{\hat{\tau}(\omega) < t} p_\epsilon(t - \hat{\tau}(\omega), \xi^\epsilon(\hat{\tau}, \omega), y). \tag{6.10}$$

If $\hat{\tau}(\omega) < t$, then $|\xi^\epsilon(\hat{\tau}(\omega)) - y| \geq \delta_0$. Hence

$$|\phi(t - \hat{\tau}(\omega), \xi^\epsilon(\hat{\tau}(\omega))) - y| > \tfrac{1}{2}\delta_0$$

provided t is sufficiently small, say $t \in (0, t_0)$. But then, by (5.2),

$$p_\epsilon(t - \hat{\tau}(\omega), \xi^\epsilon(\hat{\tau}, \omega), y) \leq C \exp[-k/\epsilon^2 t]$$

where C, k are positive constants depending on δ_0. Using also (5.21), we deduce from (6.10) that, if $t_0 < t^*$,

$$q_\epsilon(t, z, y) \geq \frac{C(t)}{\epsilon^n} \exp\left[-\frac{c(\delta^2 + t^2)}{\epsilon^2 t}\right] - C_1 \exp\left[-\frac{k}{\epsilon^2 t}\right]$$

where C_1 is a positive constant. Thus, if $c(\delta^2 + t^2) < k/2$ and ϵ is sufficiently small then (6.10) follows.

The next fact we need is the following: Let A be a ball with center ζ and radius $|A|$, contained in D. Then, for any $h' > 0$ and $s \in (0, t]$

$$q_\epsilon(s - \nu, x, A) \geq \exp\left\{-\frac{I_s^D(x, \zeta) + h'}{2\epsilon^2}\right\} \tag{6.11}$$

if $\nu > 0$ is sufficiently small, provided ϵ is sufficiently small.

To prove it, let ψ be a curve in C_s such that $\psi(0) = x$, $\psi(s - \nu) = \zeta$, $\psi(\lambda) \in D$ for $0 \leq \lambda \leq s - \nu$, and

$$I_s(\psi) \leq I_s^D(x, \zeta) + \tfrac{1}{2}h'.$$

Then, $\rho(\psi(\lambda), \partial D) \geq c_0 > 0$ if $0 \leq \lambda \leq s - \nu$, where c_0 is some positive constant. Hence,

$$\{\xi^\epsilon(\lambda) \in D \text{ for } 0 \leq \lambda \leq s - \nu, \xi^\epsilon(s - \nu) \in A\} \supset \{\rho_{s-\nu}(\xi^\epsilon, \psi) < \delta_1\}$$

provided $\delta_1 < \min(c_0, |A|)$. Using Theorem 2.1, (6.11) follows.

We shall now prove (6.8). By the semigroup property of q_ϵ (see Problem 5),

$$\begin{aligned} q_\epsilon(t, x, y) &= \int_D q_\epsilon(t - \mu, x, z) q_\epsilon(\mu, z, y)\, dz \\ &\geq \int_G q_\epsilon(t - \mu, x, z) q_\epsilon(\mu, z, y)\, dz \end{aligned} \tag{6.12}$$

where $G = \{z;\ |z - y| < \delta\}$. By (6.9),

$$q_\epsilon(\mu, z, y) \geq \beta_0(\mu) \exp\left[-\frac{\beta(\delta^2 + \mu^2)}{\epsilon^2 \mu} \right] \tag{6.13}$$

provided μ and δ are sufficiently small, say $\mu \leq \mu^*$, $\delta < \delta^*$, and ϵ is sufficiently small, and $\tilde{\beta}_0(\mu)$, $\tilde{\beta}$ are positive constants.

By (6.11) with $s = t$

$$\int_G q_\epsilon(t - \nu, x, z)\, dz = q_\epsilon(t - \nu, x, G) \geq \exp\left\{ -\frac{I_t^D(x, y) + h'}{2\epsilon^2} \right\}; \tag{6.14}$$

here we can take $\nu = \mu$. Using (6.13), (6.14) in (6.12), we get

$$q_\epsilon(t, x, y) \geq C(t, \delta, h') \exp\left\{ -\frac{I_t^D(x, y) + h' + c\delta^2/\mu + c\mu}{2\epsilon^2} \right\}$$

where c is a positive constant (independent of δ, μ). Hence

$$\varliminf_{\epsilon \to 0} [2\epsilon^2 \log q_\epsilon(t, x, y)] \geq -I_t^D(x, y) - h' - \frac{c\delta^2}{\mu} - c\mu.$$

Taking $\delta \to 0$, then $\mu \to 0$, and finally $h' \to 0$, the assertion (6.8) follows.

Let

$$I_t^{\bar{D}}(x, y) = \inf_\phi I_t(\phi) \tag{6.15}$$

where ϕ varies over the subset $\tilde{\Phi}$ of C_t of functions ϕ satisfying: $\phi(0) = x$, $\phi(t) = y$, and $\phi(s) \in \bar{D}$ for $0 \leq s \leq t$. Clearly

$$I_t^{\bar{D}}(x, y) \leq I_t^D(x, y).$$

Since, however, ∂D is in C^2, it is easily seen that

$$I_t^{\bar{D}}(x, y) = I_t^D(x, y)$$

(cf. the construction of ψ^* in the proof of Theorem 3.1). Notice that there actually exists a function $\bar{\phi}$ in $\tilde{\Phi}$ such that

$$I_t^{\bar{D}}(x, y) = I_t(\bar{\phi}).$$

Theorem 6.3. (i) *Let* (A′) *hold. If* $I_t(x, y) < I_t^{\bar{D}}(x, y)$, *then*

$$\frac{q_\epsilon(t, x, y)}{p_\epsilon(t, x, y)} \to 0 \qquad \text{if} \quad \epsilon \to 0. \tag{6.16}$$

(ii) *Let* (A′) *hold. If* $I_t(x, y) = I_t^{\bar{D}}(x, y)$ *and for every* $\bar{\phi}$ *in* $\tilde{\Phi}$ *for which* $I_t(x, y) = I_t(\bar{\phi})$ *we have:* $\bar{\phi}(s) \in D$ *for all* $0 \leqslant s \leqslant t$, *then*

$$\frac{q_\epsilon(t, x, y)}{p_\epsilon(t, x, y)} \to 1 \qquad \text{if} \quad \epsilon \to 0. \tag{6.17}$$

Proof. The assertion (i) is a consequence of Theorems 5.1, 6.1. To prove (ii), denote by $\tilde{\tau}$ the first time $\xi^\epsilon(s)$ reaches the sphere S_δ, with center y and radius δ, after hitting ∂D at some previous time. If such time does not exist, set $\tilde{\tau} = \infty$. By the strong Markov property we get

$$p_\epsilon(t, x, y) - q_\epsilon(t, x, y) = E_x \chi_{(\tilde{\tau}<t)} p_\epsilon(t - \tilde{\tau}, \xi(\tilde{\tau}), y). \tag{6.18}$$

Since $|\xi(\tilde{\tau}) - y| = \delta > 0$, we have, by (5.2),

$$p_\epsilon(t - \tilde{\tau}, \xi(\tilde{\tau}), y) \leqslant \frac{C}{\epsilon^n}. \tag{6.19}$$

Denote by Φ the set of all ϕ in C_t with $\phi(0) = x$, such that $\phi(t) \in S_\delta$ and $\phi(s') \in \partial D$ for some $0 < s' < t$. If $\phi \in \Phi$, then, by the assumptions of (ii) and Lemma 1.2,

$$I_t(\phi) > I_t(x, y) + 2c$$

where c is a positive constant (independent of ϕ), provided δ is sufficiently small (independently of ϕ). Hence, by Theorem 3.3,

$$P_x\{\tilde{\tau} \leqslant t\} \leqslant \exp\left\{-\frac{I_t(x, y) + c}{2\epsilon^2}\right\}.$$

Using this and (6.19) in (6.18), we get

$$\frac{p_\epsilon(t, x, y)}{q_\epsilon(t, x, y)} - 1 \leqslant \frac{C}{\epsilon^n} \exp\left\{-\frac{I_t(x, y) + c}{2\epsilon^2}\right\} \frac{1}{q_\epsilon(t, x, y)}.$$

Since, by Theorem 6.1, the right-hand side converges to zero as $\epsilon \to 0$, the proof of (6.17) is complete.

7. The problem of exit

Let D be a bounded domain with C^2 boundary ∂D. We shall assume, in addition to (A), that

(B_1) $b \cdot \nu < 0$ along ∂D, where ν is the outward normal.

It follows that every solution of the dynamical system

$$\frac{dx}{dt} = b(x) \tag{7.1}$$

with $x(0) \in D$ remains in D for all $t > 0$.

Notice that the solution $\xi^\epsilon(t)$ of (1.1) leaves D in finite time τ^ϵ. The *problem of exit* is concerned with the behavior of the set $\xi^\epsilon(\tau^\epsilon)$, as $\epsilon \to 0$. This problem will be studied in this section and in the following one.

Lemma 7.1. *Let* (A) *hold and let* K *be a compact set in* R^n. *Suppose that every solution of* (7.1) *with* $x(0) \in K$ *leaves* K *in finite time. Then there exist positive constants* α, T_0 *such that*: (i) *If* $T > T_0$, $\phi \in C_T$ *and* $\phi(t) \in K$ *for all* $0 \leq t \leq T$, *then* $I_T(\phi) > \alpha(T - T_0)$; (ii) *If* τ_K^ϵ *is the exit time of* $\xi^\epsilon(t)$ *from* K, *then, for all* $\epsilon > 0$ *sufficiently small,*

$$P_x\{\tau_K^\epsilon > T\} \leq \exp\left\{-\frac{\alpha(T - T_0)}{2\epsilon^2}\right\} \qquad \text{if} \quad T > T_0.$$

Proof. For any $x \in K$, let $\tau(x)$ be the exit time from K of the solution of (7.1) with $x(0) = x$. By assumption, $\tau(x) < \infty$. Since $\tau(x)$ is an upper semicontinuous function, it achieves a maximum T_1 on K. Let $T_0 = T_1 + 1$, and consider the class Φ of all functions ϕ in C_{T_0} with values in K. This is a closed set in C_{T_0}. By Lemma 1.2, $I_{T_0}(\phi)$ attains a minimum A on the set Φ. Since no solutions of (7.1) are among the elements of Φ, we must have $A > 0$. Thus, $I_{T_0}(\phi) \geq A > 0$ for any $\phi \in \Phi$. But then also $I_{s,s+T_0}(\phi) \geq A$ for any $\phi \in C_{s,s+T_0}$ for which $\phi(t) \in K$ for all $s \leq t \leq s + T_0$. It follows that, if $\phi(t) \in K$ for $0 \leq t \leq T$,

$$I_T(\phi) \geq I_{T_0}(\phi) + I_{T_0, 2T_0}(\phi) + \cdots + I_{(\nu-1)T_0, \nu T_0}(\phi) \geq \nu A$$

where $\nu = [T/T_0]$. Hence

$$I_T(\phi) \geq A\left[\frac{T}{T_0}\right] \geq A\left(\frac{T}{T_0} - 1\right) = \alpha(T - T_0)$$

if $\alpha = A/T_0$.

To prove (ii) notice first that if δ is a sufficiently small positive number, then every solution of (7.1) with $x(0)$ in K leaves a closed δ-neighborhood K_δ of K in finite time. In fact, this follows from the continuous dependence of the solution of (7.1) upon the initial condition. We fix such a small δ, and denote by α, T_0 the positive constants asserted in (i), when K is replaced by K_δ.

Let $x \in K$ and denote by Φ_0 the class of all $\phi \in C_T$ such that $\phi(0) = x$ and $I_T(\phi) < \alpha(T - T_0)$. Then each $\phi \in \Phi_0$ exits K_δ at some time $< T$. Hence,

$$P_x\{\tau_K^\epsilon > T\} \leqslant P_x\{\rho_T(\xi^\epsilon, \Phi_0) > \delta\}.$$

Applying Theorem 3.1 to the right-hand side, we get

$$P_x\{\tau_K^\epsilon > T\} \leqslant \exp\left\{-\frac{\alpha(T - T_0) - h}{2\epsilon^2}\right\}$$

for any $h > 0$, provided ϵ is sufficiently small. This completes the proof of (ii).

For any x, y in $\bar{D}$, let

$$I(x, y) = \inf_{t>0} \inf_{\phi \in \Phi_t} I_t(\phi)$$

where Φ_t is the subset of C_t consisting of all functions ϕ satisfying $\phi(0) = x$, $\phi(t) = y$, $\phi(s) \in D$ for all $0 < s < t$. It is easily seen that $I(x, y)$ is Lipschitz continuous in $(x, y) \in \bar{D} \times \bar{D}$.

If $I(x, y) = 0$ and $I(y, x) = 0$, then we say that x is *equivalent* to y, and write $x \sim y$.

A point ζ is said to be in the *ω-limit set* of a solution $x(t)$ of (7.1) if there exists a sequence $t_m \uparrow \infty$ such that $x(t_m) \to \zeta$. The ω-limit set of any solution is clearly a closed set. Notice that if ζ, η belong to the ω-limit set of a solution of (7.1) then $\zeta \sim \eta$.

Set

$$I(x, A) = \inf_{y \in A} I(x, y).$$

We shall assume:

(B_2) (i) There exists a finite number of disjoint compact sets $K_1, \ldots, K_l$ in D such that the ω-limit set of each solution of (4.1) with $x(0)$ in $D \setminus (\bigcup_{i=1}^{l} K_i)$ is contained in one of the sets K_i.

(ii) If $x \in K_i$, $z \in K_j$, then $x \sim z$ if $i = j$ and $x \not\sim z$ if $i \neq j$.

(iii) For every μ-neighborhood $K_i(\mu)$ of K_i (μ sufficiently small) and for every pair of points x, y on $\partial K_i(\mu)$, the boundary of $K_i(\mu)$, there exists a curve $\phi(s)$ $(0 \leqslant s \leqslant \beta)$ such that $\phi(0) = x$, $\phi(\beta) = y$, $\phi(s) \in D \setminus K_i(\frac{1}{2}\mu)$ if $0 \leqslant s \leqslant \beta$, and $I_\beta(\phi) \leqslant \eta(\mu)$, where $\eta(\mu) \downarrow 0$ if $\mu \downarrow 0$.

The last condition is clearly satisfied for a set K_i consisting of one point only. It is also generally satisfied for all the K_i in case $n = 2$ and $b(x)$ has only a finite number of zeros; this follows from the Poincaré-Bendixon theory (see Coddington and Levinson [1]). Finally, all the subsequent developments remain valid (with few obvious changes) if in the condition (B_2) (iii) we replace the restriction that $\phi(s) \in D \setminus K_i(\frac{1}{2}\mu)$ by the restriction that $\phi(s) \in K \setminus K_i(\mu')$ for some $\mu' = \mu'(\mu) > 0$.

For any $x \in K_i$, let $V_i = I(x, \partial D)$. This number is clearly independent of the choice of x in K_i. In view of (B_1), $V_i > 0$; see Problem 11.

Let $x \in K_i$. Consider all sequences $\{\psi_k\}$, $\psi_k \in C_{T_k}$, $\psi_k(t) \in D$ for $0 \leq t < T_k$, $\psi_k(0) = x$, $\psi_k(T_k) \in \partial D$, $I_{T_k}(\psi_k) \to V_i$. Denote by Σ_x the set of all limit points of the sequences $\{\psi_k(T_k)\}$. This is a subset of ∂D. Since, as easily seen, this set is independent of x in K_i, we shall denote it by Σ_i. Notice that $V_i = \min_{y \in \partial D} I(x, y)$, $\Sigma_i = \{y;\ y \in \partial D \text{ and } I(x, y) = V_i\}$ for any $x \in K_i$.

Set $\Sigma = \Sigma_1 \cup \cdots \cup \Sigma_l$.

Denote by τ^ϵ the first time $\xi^\epsilon(t)$ leaves D. Since the solution of (7.1) does not leave D in finite time, it is a priori not clear at all how the set $\{\xi^\epsilon(\tau^\epsilon)\}$ behaves as $\epsilon \to 0$.

Theorem 7.2. *Let* (A), (B_1), (B_2) *hold. Then* $\rho(\xi^\epsilon(\tau^\epsilon), \Sigma) \to 0$ *in probability, as* $\epsilon \to 0$.

The proof given below is based on Theorems 2.1, 3.1. Another (somewhat different) proof which does not require the condition (B_2)(iii) is outlined in Problem 18; it is based on extensions of Theorems 2.2, 3.5 (see Problems 16, 17).

In the next section we shall generalize Theorem 7.2 to the case where instead of the condition (B_1) it is assumed that no solution of (7.1) with $x(0) \in D$ leaves D in finite time.

Proof of Theorem 7.2. For clarity we consider first the case $l = 1$. Set $K = K_1$, $V = V_1$.

For $\nu > 0$, let $\mathcal{E}_\nu$ be a ν-neighborhood of K, and denote its boundary by $\partial \mathcal{E}_\nu$. Let $\gamma^+ = \partial \mathcal{E}_\mu$. Let $\mathcal{E}'_\mu$ be a finite union of domains with C^2 boundary γ^- such that $\mathcal{E}_{\mu/8} \subset \mathcal{E}'_\mu \subset \mathcal{E}_{\mu/4}$. For any $d > 0$, if μ is sufficiently small (depending on d), then the following is true: For any $x \in \gamma^+$ there exists a curve $\phi(s)$, $0 \leq s \leq \hat{T}$ for some $\hat{T} > 0$, such that

$$\begin{gathered} \phi(0) = x, \qquad \phi(s) \in D \setminus \mathcal{E}'_\mu \quad \text{if} \quad 0 < s < \hat{T}, \quad \phi(\hat{T}) \in \partial D, \\ I_{\hat{T}}(\phi) \leq V + d/4. \end{gathered} \tag{7.2}$$

In fact, let $\psi(s)$ $(\alpha \leq s \leq \beta)$ be a curve such that $\psi(\alpha) \in K$, $\psi(\beta) \in \partial D$, $\psi(s) \in D$ for $\alpha \leq s \leq \beta$, and $I_{\alpha,\beta}(\psi) < V + d/8$. Let $s = \gamma_0 \in [\alpha, \beta)$ be the last time when $\psi(s)$ intersects γ^+. By (B_2) (iii) there exists a curve $\chi(s)$ $(0 \leq s \leq \alpha_1)$ connecting x to $\psi(\gamma_0)$ and lying outside $\mathcal{E}_{\mu/2}$ with $I_{\alpha_1}(\psi) < \eta(\mu)$. The curve

$$\phi(s) = \begin{cases} \chi(s) & \text{if} \quad 0 \leq s \leq \alpha_1, \\ \psi(s - \alpha_1 + \gamma_0) & \text{if} \quad \alpha_1 < s < \beta - \gamma_0 + \alpha_1 \end{cases}$$

satisfies (7.2) provided μ is sufficiently small, i.e., provided $\eta(\mu) < d/8$. From now on we take μ such that $\eta(\mu) < d/8$.

Consider all the curves ϕ satisfying (7.2). Each ϕ is defined in an interval $[0, \hat{T}]$, where $\hat{T}$ may vary from one ϕ to another. However, by Lemma 7.1, $\hat{T} < T_*$ where T_* is independent of ϕ. Let $T = T_* + 1$. Extend each $\phi(s)$ (originally defined for $0 \leq s \leq \hat{T}$) into $\hat{T} < s \leq T$ so that $I_T(\phi) \leq V + d/4$. One can choose, in particular, an extension of $\phi(t)$ which is the solution of $dx/dt = b(x)$. Denote by H_x the set of all possible extensions of all the curves satisfying (7.2). Then clearly

$$H_x = \{\phi \in \Phi_{x,T}(\partial D, \gamma^-); I_T(\phi) \leq V + d/4\}.$$

Define the Markov time

$$\tau^- = \inf\{t; \xi^\epsilon(t) \in \gamma^-\}.$$

By Lemma 7.1, if $T = T(d, \mu)$ is sufficiently large (which we may assume), then

$$P_x\{\tau^- > T, \tau^\epsilon > T\} \leq \exp\left\{-\frac{V+d}{2\epsilon^2}\right\} \tag{7.3}$$

provided ϵ is sufficiently small.

Define the events

$$\begin{aligned} A_1 &= A_1(x) = \{\tau^\epsilon < \tau^-, \tau^\epsilon > T\}, \\ A_2 &= A_2(x) = \{\tau^\epsilon < T \wedge \tau^-, d_T(\xi^\epsilon, H_x) < \lambda\}, \\ A_3 &= A_3(x) = \{\tau^\epsilon < T \wedge \tau^-, d_T(\xi^\epsilon, H_x) > \lambda\}, \end{aligned}$$

where λ is a positive number, to be determined later. Then,

$$P_x\{\tau^\epsilon < \tau^-\} = P_x(A_1) + P_x(A_2) + P_x(A_3). \tag{7.4}$$

By (7.3),

$$P_x(A_1) \leq \exp\left\{-\frac{V+d}{2\epsilon^2}\right\}. \tag{7.5}$$

Lemma 7.3. *For any $\lambda > 0$ and for any $\phi \in H_x$, denote by $s_{\lambda d\mu\phi}$ the first time $\rho(\phi(t), \partial D) = \lambda$; denote by $t_{\lambda d\mu\phi}$ the last time $\rho(\phi(t), \partial D) = \lambda$, and denote by $\sigma_{\lambda d\mu\phi}$ the first time when $\phi(t) \in \partial D$. Then*

$$\sup_{x\in\gamma^+} \sup_{\phi\in H_x} [t_{\lambda d\mu\phi} - s_{\lambda d\mu\phi}] \to 0 \quad \textit{if} \quad d\to 0, \quad \mu\to 0, \quad \lambda\to 0, \tag{7.6}$$

$$\sup_{x\in\gamma^+} \sup_{\phi\in H_x} \sup_{s_{\lambda d\mu\phi} \leq t \leq t_{\lambda d\mu\phi}} \rho(\phi(t), \Sigma) \to 0 \quad \textit{if} \quad d\to 0, \quad \mu\to 0, \quad \lambda\to 0. \tag{7.7}$$

Note that we may always take T in the definition of H_x so large that $t_{\lambda d\mu\phi}$ exists, if λ is sufficiently small.

Proof. If (7.6) is false, there exist sequences $d_m \to 0$, $\mu_m \to 0$, $\lambda_m \to 0$, $\phi_m \in H_{x_m}$

(with $T = T_m$) for which

$$t_{\lambda_m d_m \mu_m \phi_m} - s_{\lambda_m d_m \mu_m \phi_m} \geq 2\epsilon_0 > 0 \qquad (\epsilon_0 \text{ const}). \tag{7.8}$$

Set $\alpha_m = s_{\lambda_m d_m \mu_m \phi_m}$, $\beta_m = \sigma_{\lambda_m d_m \mu_m \phi_m}$, $\gamma_m = t_{\lambda_m d_m \mu_m \phi_m}$. Consider the curves $\psi_m(t) = \phi_m(t + \beta_m)$ for $0 \leq t \leq T_m - \beta_m$. We may assume that $T_m - \beta_m > \epsilon_0$. Observe that $I_{\beta_m, T_m}(\phi_m) \to 0$ if $m \to \infty$. Consequently, $I_{\epsilon_0}(\psi_m) \to 0$. Hence there exists a subsequence $\psi_{m'}(t)$ (for simplicity we take $m' = m$) which is uniformly convergent to $\tilde{\psi}(t)$ $(0 \leq t \leq \epsilon_0)$ and $I_{\epsilon_0}(\tilde{\psi}) = 0$, i.e., $d\tilde{\psi}/dt = b(\tilde{\psi})$. Since $\tilde{\psi}(0) \in \partial D$, the condition (B_1) implies that $\tilde{\psi}(\epsilon_0) \in D$. Recalling that $\psi_m(\epsilon_0) \to \tilde{\psi}(\epsilon_0)$ and $i_{\alpha_m, T_m}(\phi_m) \to 0$, we conclude (since, by Problem 11, $I(\tilde{\psi}(\epsilon_0), \partial D) > 0$) that $\psi_m(t)$, for $t \geq \epsilon_0$, does not intersect any given sufficiently small neighborhood of ∂D, provided m is sufficiently large. Thus, $\gamma_m - \beta_m \leq \epsilon_0$ if m is sufficiently large. From (7.8) it then follows that $\beta_m - \alpha_m \geq \epsilon_0$.

Next,

$$I_{\alpha_m, \beta_m}(\phi_m) \to 0 \qquad \text{as} \quad m \to \infty, \tag{7.9}$$

for, otherwise, we can easily exhibit a function $\phi(s)$, in some space C_{T_0}, with $\phi(0) \in K$, $\phi(T_0) \in \partial D$, $\phi(s) \in D$ for $0 \leq s \leq T_0$ such that $I_{T_0}(\phi) < V$ (and this contradicts the definition of V).

Consider the functions

$$\psi_m(t) = \phi_m(t + \beta_m - \epsilon_0) \qquad \text{for} \quad 0 \leq t \leq \epsilon_0.$$

Since $I_{\epsilon_0}(\psi_m) \leq C$, there is a subsequence $\{\psi_{m'}\}$ which is convergent to some $\tilde{\psi}$, uniformly in $t \in [0, \epsilon_0]$. In view of (7.9), we also have $I_{\epsilon_0}(\tilde{\psi}) = 0$, i.e.,

$$\frac{d\tilde{\psi}}{dt} = b(\tilde{\psi}(t)). \tag{7.10}$$

Since $\tilde{\psi}(t) \in \bar{D}$ for $0 \leq t < \epsilon_0$, $\tilde{\psi}(\epsilon_0) \in \partial D$,

$$-\nu \cdot \frac{d\tilde{\psi}(\epsilon_0)}{dt} = \frac{d}{dt} \rho(\tilde{\psi}(t), \partial D)|_{t=\epsilon_0} \leq 0$$

where ν is the outward normal to ∂D at $\tilde{\psi}(\epsilon_0)$. On the other hand, from (7.10) and (B_1) we get

$$-\nu \cdot \frac{d\tilde{\psi}(\epsilon_0)}{dt} = -\nu \cdot b(\tilde{\psi}(\epsilon_0)) > 0,$$

a contradiction. This proves the assertion (7.6).

To prove (7.7) we first show that

$$\sup_{x \in \gamma^+} \sup_{\phi \in H_x} \rho(\phi(\sigma_{\lambda d \mu \phi}), \Sigma) \to 0 \qquad \text{if} \quad d \to 0, \quad \mu \to 0, \quad \lambda \to 0. \tag{7.11}$$

If (7.11) is false, then there exist sequences $\lambda_m \to 0$, $d_m \to 0$, $\mu_m \to 0$, $\phi_m \in H_{x_m}$

(ϕ_m is a curve in C_{T_m}) such that $I_{T_m}(\phi_m) \to V$ as $m \to \infty$, but

$$\rho\big(\phi_m(\sigma_{\lambda_m d_m \mu_m \phi_m}), \Sigma\big) \geq \epsilon_1 > 0 \qquad \text{for all } m.$$

But then one can easily construct curves $\tilde{\phi}_m$ with $\tilde{\phi}_m(0) \in K$, $\tilde{\phi}_m(s) \in D$ for $0 \leq s < T_m^*$, $\tilde{\phi}_m(T_m^*) \in \partial D$, $\rho(\tilde{\phi}_m(T_m^*), \Sigma) \geq \epsilon_1/2$ and $I_{T_m^*}(\tilde{\phi}_m) \to V$. This however contradicts the definition of Σ.

Since $I_T(\phi) \leq C$ for all $\phi \in H_x$, $x \in \gamma^+$, the functions $\phi(t)$ of H_x satisfy a Hölder condition with coefficient and exponent which are independent of d, μ, λ, x. Hence the assertion (7.7) is a consequence of (7.6), (7.11).

We return to the proof of Theorem 7.2. Clearly,

$$P_x(A_3) = P_x\{d_T(\xi^\epsilon, H_x) > \lambda, \xi^\epsilon \in \Phi_{x,T}(\partial D, \gamma^-)\}.$$

Since γ^- is in C^2, we can apply Theorem 3.1, and get

$$P_x(A_3) \leq \exp\left\{-\frac{V + d/4 - k}{2\epsilon^2}\right\} \qquad \text{for any} \quad k > 0 \tag{7.12}$$

provided ϵ is sufficiently small.

To estimate $P_x(A_2)$, let ϕ^* be a curve in C_{t_0} such that $\phi^*(0) \in K$, $\phi^*(s) \in D$ for $0 \leq s < t_0$, $\phi^*(t_0) \in \partial D$ and $I_{t_0}(\phi^*) < V + d/20$. Denote by t^* the last time $\phi(t)$ intersects γ^+. Construct a continuous curve $\phi(t)$ as follows: For $0 \leq t \leq t_1$, $\phi(t)$ connects x ($x \in \gamma^+$ is given) to $x^* = \phi^*(t^*)$ by a curve lying outside $\mathcal{E}_{\mu/2}$, and $I_{t_1}(\phi) < \eta(\mu)$ (its existence is assured by (B_2) (iii)). For $t_1 < t < t_2$, $\phi(t) = \phi^*(t + t^* - t_1)$ where $t_2 + t^* - t_1 = t_0$. Notice, by Lemma 7.1, that $t_0 - t^*$ is bounded by a constant depending on d, μ. Without loss of generality we may take the number $T = T(d, \mu)$ to be such that $T \geq t_2$. We now define $\phi(t)$ for $t_2 \leq t \leq T$ as a solution of $d\phi/dt = b(\phi)$.

Notice that $\phi \in H_x$, $\rho(\phi(s), \mathcal{E}_\mu') \geq \mu/4$ for $0 \leq s \leq T$, and

$$I_T(\phi) < V + \frac{d}{10} \qquad \text{provided} \qquad \eta(\mu) \leq \frac{d}{20}. \tag{7.13}$$

Denote by $\hat{t}$ the (unique) time that $\phi(t) \in \partial D$. We now modify ϕ into $\hat{\phi}$ as follows: $\hat{\phi}(t) = \phi(t)$ for $0 \leq t \leq \hat{t}$. During the interval $\hat{t} < t < \hat{t} + h$, $\hat{\phi}(t)$ traces a line segment from $\phi(\hat{t})$ to a point ζ in $R^n \setminus D$ satisfying $\rho(\zeta, \partial D) = \rho(\zeta, \phi(\hat{t})) = h$ (if h is sufficiently small, the point ζ is on the normal to ∂D at $\phi(\hat{t})$). During the interval $\hat{t} + h < t < \hat{t} + 2h$, $\hat{\phi}(t)$ traces back the line segment from ζ to $\phi(\hat{t})$. Finally, for $\hat{t} + 2h < t < T + 2h$, $\hat{\phi}(t)$ proceeds along the previous path ϕ, i.e., $\hat{\phi}(t + 2h) = \phi(t)$ if $\hat{t} < t < T$. Let

$$\tilde{\phi}(s) = \hat{\phi}\left(\frac{(T + 2h)s}{T}\right).$$

If h is sufficiently small, then (cf. (3.2), (3.3) in the proof of Theorem 3.1)

$$I_T(\tilde{\phi}) \leq V + \frac{d}{10} + C_1 h, \qquad \rho_T(\phi, \tilde{\phi}) < C_1 h, \tag{7.14}$$

where C_1 is a positive constant. We fix h so that

$$C_1 h \le \frac{d}{10}, \quad C_1 h < \frac{\lambda}{2}. \tag{7.15}$$

If $\rho_T(\xi^\epsilon, \tilde{\phi}) < \lambda/2$, the $\rho(\xi^\epsilon, \phi) < \lambda$. Hence $d_T(\xi^\epsilon, H_x) < \lambda$.

If $\rho_T(\xi^\epsilon, \tilde{\phi}) < \min(h, \mu/4)$, then also $\tau^\epsilon < T \wedge \tau^-$. Hence

$$P_x(A_2) \ge P_x\{\rho_T(\xi^\epsilon, \tilde{\phi}) < \epsilon^*\}$$

where $\epsilon^* = \frac{1}{2}\min(h, \mu/4, \lambda/2)$. Using Theorem 2.1, we get

$$P_x(A_2) \ge \exp\left\{-\frac{V + d/5 + k}{2\epsilon^2}\right\} \qquad \text{for any} \quad k > 0, \tag{7.16}$$

provided ϵ is sufficiently small.

Combining (7.16), (7.12), (7.5) with (7.4), we get, after taking $k < d/40$,

$$\begin{gathered} P_x\{\tau^\epsilon < \tau^-\} = P_x(A_2)[1 + \gamma(x, \epsilon)], \\ 0 \le \gamma(x, \epsilon) \le \gamma(\epsilon), \qquad \gamma(\epsilon) \to 0 \qquad \text{if} \quad \epsilon \to 0. \end{gathered} \tag{7.17}$$

Now, if $\omega \in A_2$ then $\rho_T(\xi^\epsilon(\cdot, \omega), \phi) < \lambda$ for some $\phi \in H_x$. If $\phi(\sigma_\lambda) \in \partial D$ then, by part (7.6) of Lemma 7.3, the first time $\tilde{s}_\lambda$ when $\rho(\phi(s), \partial D) = \lambda$ and the last time $\tilde{t}_\lambda$ when $\rho(\phi(s), \partial D) = \lambda$ satisfy

$$\tilde{s}_\lambda < \sigma_\lambda < \tilde{t}_\lambda, \qquad \tilde{t}_\lambda - \tilde{s}_\lambda \to 0 \qquad \text{if} \quad d \to 0, \quad \mu \to 0, \quad \lambda \to 0$$

(and $\eta(\mu) \le d/20$) uniformly with respect to $\phi \in H_x$ and $x \in \gamma^+$. Since $\xi^\epsilon(\tau^\epsilon(\omega)) \in \partial D$ and $\rho(\xi^\epsilon(\tau^\epsilon(\omega)), \phi(\tau^\epsilon(\omega))) < \lambda$, it follows that $\tilde{s}_\lambda < \tau^\epsilon(\omega) < \tilde{t}_\lambda$. Applying part (7.7) of Lemma 7.3, we conclude that $\rho(\phi(\tau^\epsilon(\omega)), \Sigma) \to 0$ if $d \to 0$, $\mu \to 0$, $\lambda \to 0$, uniformly with respect to ω in A_2. Hence, if we fix $\mu = \mu(d)$, λ such that

$$\eta(\mu) = \frac{d}{20}, \qquad \lambda = d,.$$

we get:

$$\rho(\xi^\epsilon(\tau^\epsilon(\omega)), \Sigma) \le \zeta \qquad (\zeta = \zeta(d))$$

if ϵ is sufficiently small, where $\zeta = \zeta(d) \to 0$ if $d \to 0$. We have thus proved that, if $x \in \gamma^+$,

$$\begin{gathered} P_x\{\tau^\epsilon < \tau^-\} = P_x\{\tau^\epsilon < \tau^-, \rho(\xi^\epsilon(\tau^\epsilon), \Sigma) \le \zeta\}\{1 + \Lambda_{\epsilon, x}(\zeta)\}, \\ 0 \le \Lambda_{\epsilon, x}(\zeta) \le \Lambda(\zeta) \end{gathered} \tag{7.18}$$

if ϵ is sufficiently small (depending on ζ), and $\Lambda(\zeta) \to 0$ if $\zeta \to 0$.

Notice that the set $\gamma^+ = \partial \mathcal{E}_\mu$ and $\zeta = \zeta(d)$ both depend on d.

Let $\tau^+ = \inf\{t;\ t > \tau^-, \xi^\epsilon(t) \in \gamma^+\}$.

Now, if $\tau^- < \tau^\epsilon$, then $\tau^+ < \tau^\epsilon$. Therefore, by the strong Markov property and (7.18), if $x \in \gamma^+$, then

$$\begin{aligned} P_x\{\rho(\xi^\epsilon(\tau^\epsilon), \Sigma) < \zeta\} &= P_x\{\tau^\epsilon < \tau^-, \rho(\xi^\epsilon(\tau^\epsilon), \Sigma) < \zeta\} \\ &\quad + P_x\{\tau^\epsilon > \tau^+, \rho(\xi^\epsilon(\tau^\epsilon), \Sigma) < \zeta\} \\ &= P_x(\tau^\epsilon < \tau^-)[1 - \tilde{\Lambda}_{\epsilon, x}(\zeta)] \\ &\quad + E_x\chi_{\tau^+ < \tau^\epsilon}[E_x\chi_{\rho(\xi^\epsilon(\tau^\epsilon), \Sigma) < \tau}]_{x = \xi^\epsilon(\tau^+)} \end{aligned}$$

where $0 < \tilde{\Lambda}_{\epsilon, x}(\zeta) < \tilde{\Lambda}(\zeta)$, $\tilde{\Lambda}(\zeta) \to 0$ if $\zeta = \zeta(d)$, $d \to 0$.

Let

$$\gamma_\zeta = \inf_{x \in \gamma^+} P_x\{\rho(\xi^\epsilon(\tau^\epsilon), \Sigma) < \zeta\},$$

and take, for any $\eta > 0$, a particular point $x \in \gamma^+$ such that

$$\gamma_\zeta > P_x\{\rho(\xi^\epsilon(\tau^\epsilon), \Sigma) < \zeta\} - \eta.$$

Then

$$\gamma_\zeta + \eta > P_x(\tau^\epsilon < \tau^-)[1 - \tilde{\Lambda}(\zeta)] + P_x(\tau^- < \tau^\epsilon) \cdot \gamma_\zeta,$$

i.e.,

$$\gamma_\zeta + \frac{\eta}{P_x(\tau^\epsilon < \tau^-)} > 1 - \tilde{\Lambda}(\zeta).$$

Since $P_y(\tau^\epsilon < \tau^-) > \rho > 0$ (ρ depends on ϵ, d) for all $y \in \gamma^+$, and since η is arbitrary, we get $\gamma_\zeta > 1 - \tilde{\Lambda}(\zeta)$, provided ϵ is sufficiently small. Thus, for every $x \in \gamma^+$,

$$P_x\{\rho(\xi^\epsilon(\tau^\epsilon), \Sigma) > \zeta\} < \tilde{\Lambda}(\zeta) \qquad \text{if } \epsilon \text{ is sufficiently small.} \tag{7.19}$$

Now let x be any point in D. Denote by $\tilde{\tau}$ the first time $\xi^\epsilon(t)$ hits δ-neighborhood Γ_δ of ∂D. We take δ such that $K \cap \overline{\Gamma}_\delta = \varnothing$. By the strong Markov property,

$$P_x\{(\xi^\epsilon(\tau^\epsilon), \Sigma) < \zeta\} = E_x[E_x\chi_{\rho(\xi^\epsilon(\tau^\epsilon), \Sigma) < \zeta}]_{x = \xi^\epsilon(\tilde{\tau})}. \tag{7.20}$$

Let $\hat{\tau}$ be the first time $\xi^\epsilon(t)$ hits γ^+, given $\xi^\epsilon(0) \in \Gamma_\delta \cap D$. Because of ($B_2$) (i) and the fact that for any $T > 0$, $\delta_0 > 0$

$$P\Big\{\sup_{0 < t < T}\{|\xi^\epsilon(t) - \xi^0(t)| > \delta_0\}\Big\} \to 0 \qquad \text{as} \quad \epsilon \to 0,$$

we have

$$P_x(\hat{\tau} < \tau^\epsilon) = 1 - M_x(\epsilon), \qquad 0 < M_x(\epsilon) < M(\epsilon) \to 0 \qquad \text{if} \quad \epsilon \to 0.$$

Using this inequality and (7.19), we get from (7.20), upon employing the strong Markov property, that

$$P_x^\epsilon\{\rho(\xi^\epsilon(\tau^\epsilon), \Sigma) > \zeta\} < M(\epsilon) + \tilde{\Lambda}(\zeta) \qquad \text{if} \quad \epsilon \text{ is sufficiently small}.$$

This completes the proof of Theorem 7.2 in case $l = 1$.

Consider the case of $K_1, \ldots, K_l$. Denote by $K_i(\mu)$ the μ-neighborhood of K_i, and denote its boundary by $\partial \mathcal{E}_{i,\mu}$. Let $\mathcal{E}'_{i,\mu}$ be a finite union of domains with C^2 boundary γ_i^- such that $\mathcal{E}_{i,\mu/8} \subset \mathcal{E}'_{i,\mu} \subset \mathcal{E}_{i,\mu/4}$.

Let

$$\gamma^+ = \bigcup_{i=1}^{l} \partial \mathcal{E}_{i,\mu}, \quad \gamma^- = \bigcup_{i=1}^{l} \gamma_i^-.$$

Define τ^-, τ^+ as before, and define H_x for $x \in \gamma^+$ by defining it for $x \in \partial \mathcal{E}_{i,\mu}$ using V_i, K_i instead of V, K. The curves $\phi(t)$ of H_x do not intersect $\bigcup_{i=1}^{l} \mathcal{E}'_{i,\mu}$ for all $t \leqslant \bar{t}$ where $\bar{t}$ is the first time for which $\phi(\bar{t}) \in \partial D$. The previous estimates of $P_x(A_l)$ remain valid, and the proof for the present general case then follows as in the case $l = 1$.

8. The problem of exit (continued)

In this section we replace the condition (B_1) by the weaker condition:

(B_1') Any solution of (7.1) with $x(0) \in D$ remains in D for all $t \geqslant 0$.

It is then natural to replace the condition (B_2) by a weaker condition in which one of the sets K_i is allowed to intersect ∂D (and its V_i is then equal to zero). For simplicity we consider first the case $l = 1$ and take K_1 to consist of one point, say ζ, lying on ∂D. Thus, we assume:

(B_2') (i) The ω-limit set of each solution of (7.1) with $x(0) \in D$ coincides with ζ;

(ii) $\zeta \in \partial D$;

(iii) ζ is an asymptotically stable equilibrium point of the system $dx/dt = b(x)$ in the sense that $b(\zeta) = 0$ and all the eigenvalues of the matrix $[\partial b/\partial x]_{x=\zeta}$ have negative real parts.

Theorem 8.1. *Let* (A), (B_1'), (B_2') *hold. Then, for any* $\delta > 0$,

$$P_x^\epsilon\{\rho(\xi^\epsilon(\tau^\epsilon), \zeta) > \delta\} \to 0 \quad \text{if} \quad \epsilon \to 0. \tag{8.1}$$

Proof. For any $\rho > 0$, denote by D_ρ the ρ D-neighborhood of ζ. Let $0 < \nu < \mu$; μ and ν are small numbers to be determined later. Given $x \in D$ consider the solution $\xi^0(t)$ of (7.1) with $\xi^0(0) = x$. By (B_1'), (B_2'), $\xi^0(t)$ lies in D for all $t > 0$ and it intersects $D_{\nu/2}$ at some first time t_ν.

Now, for any $T > 0$, $\delta_0 > 0$

$$P_x\Big\{\sup_{0 \leqslant t \leqslant T} |\xi^\epsilon(t) - \xi^0(t)| > \delta_0\Big\} \to 0 \quad \text{if} \quad \epsilon \to 0.$$

Hence, for any $\eta_1 > 0$,

$$P_x\{\xi^\epsilon(t) \text{ remains in } D \text{ for } 0 \leq t \leq t_\nu;\ \xi^\epsilon(t) \text{ hits } \partial D_\nu \text{ at some first time } \tau_\nu \leq t_\nu\} > 1 - \eta_1, \tag{8.2}$$

provided ϵ is sufficiently small.

Denoting the exit time from D by τ^ϵ, and using (8.2) and the strong Markov property, we then have

$$\begin{aligned} P_x\{\rho(\xi^\epsilon(\tau^\epsilon), \zeta) > \mu\} &= P_x\{\tau^\epsilon < \tau_\nu, \rho(\xi^\epsilon(\tau^\epsilon), \zeta) > \mu\} \\ &\quad + P_x\{\tau^\epsilon > \tau_\nu, \rho(\xi^\epsilon(\tau^\epsilon), \zeta) > \mu\} \\ &< \eta_1 + E_x\{\chi_{\tau^\epsilon > \tau_\nu}[E_x\chi_{\rho(\xi^\epsilon(\tau^\epsilon), \zeta) > \mu}]_{x = \xi(\tau_\nu)}\}. \end{aligned}$$

If we prove that the function

$$w(x) = E_x\{\chi_{\rho(\xi^\epsilon(\tau^\epsilon), \zeta) > \mu}\} \tag{8.3}$$

satisfies

$$w(x) \leq \eta(\nu, \mu) \quad (x \in \partial D_\nu \cap D) \quad \text{where} \quad \eta(\nu, \mu) \to 0 \quad \text{if} \quad \nu \to 0, \tag{8.4}$$

then we conclude that

$$P_x\{\rho(\xi^\epsilon(\tau^\epsilon), \zeta) > \mu\} < \eta_1 + \eta(\nu, \mu) < 2\eta_1$$

if ν is sufficiently small (so that $\eta(\nu, \mu) < \eta_1$), and the proof of Theorem 8.1 is complete.

To prove (8.4) notice that $w(x)$ satisfies:

$$\begin{aligned} &L_\epsilon w(x) = 0 \quad \text{in } D, \\ &w(x) = 0 \quad \text{if} \quad x \in \partial D, \quad \rho(x, \zeta) < \mu, \\ &w(x) = 1 \quad \text{if} \quad x \in \partial D, \quad \rho(x, \zeta) > \mu. \end{aligned} \tag{8.5}$$

For any $\rho > 0$, let $\Gamma_\rho = \partial D \cap \partial D_\rho$, $\Gamma'_\rho = D \cap \partial D_\rho$. Suppose $u(x)$ satisfies

$$\begin{aligned} L_\epsilon u &\leq 0 \quad \text{in} \quad D_\mu, \\ u &\geq 0 \quad \text{on} \quad \Gamma_\mu, \\ u &\geq 1 \quad \text{on} \quad \Gamma'_\mu, \end{aligned} \tag{8.6}$$

and

$$u(x) \leq \eta(\nu, \mu) \quad \text{on } \Gamma'_\nu. \tag{8.7}$$

By the maximum principle, $w \leq 1$ in D and, in particular, $w \leq 1$ on Γ'_μ. Comparing w with u, by means of the maximum principle, we conclude that

$$w(x) \leq u(x) \qquad \text{if} \quad x \in \partial D_\nu \cap D.$$

Thus (8.4) would follow from (8.7).

Set $Mu = \Sigma a_{ij} u_{x_i x_j}$. If a function u satisfies

$$Mu < 0 \quad \text{in} \quad D_\mu, \tag{8.8}$$

$$b \cdot u_x < 0 \quad \text{in} \quad D_\mu, \tag{8.9}$$

$$u > 0 \quad \text{on } \Gamma_\mu, \qquad u > 1 \quad \text{on } \Gamma'_\mu, \qquad u(\zeta) = 0, \tag{8.10}$$

then u clearly satisfies (8.6), (8.7). Thus it remains to find a solution of (8.8)–(8.10).

For simplicity we shall now assume that $\zeta = 0$.

Suppose we perform a nonsingular linear map $y = Px$. The stochastic system (1.1) becomes

$$\begin{aligned} d\eta^\epsilon(t) &= P\,d\xi^\epsilon(t) = \epsilon P\sigma(\xi^\epsilon(t))\,dw + Pb(\xi^\epsilon(t))\,dt \\ &= \epsilon P\sigma(P^{-1}(\eta^\epsilon(t)))\,dw + Pb(P^{-1}(\eta^\epsilon(t)))\,dt \\ &\equiv \epsilon\tilde{\sigma}(\eta^\epsilon(t))\,dw + \tilde{b}(\eta^\epsilon(t))\,dt \end{aligned}$$

where

$$\tilde{b}(y) = PBP^{-1}y + o(|y|), \qquad B = [\partial b/\partial x]_{x=\zeta=0}.$$

By assumption,

$$\text{Re}\,\lambda_i < 0, \qquad \text{where} \quad \lambda_i \text{ are the eigenvalues of } B.$$

Notice that the transformation $y = Px$ does not change the assumptions and assertions, except for changing Γ_μ, Γ'_μ, D_μ into $\tilde{\Gamma}_\mu$, $\tilde{\Gamma}'_\mu$, $\tilde{D}_\mu$, respectively. Thus, it suffices to prove the assertions (8.8)–(8.10) in the y-space with Γ_μ, Γ'_μ, D_μ replaced by $\tilde{\Gamma}_\mu$, $\tilde{\Gamma}'_\mu$, $\tilde{D}_\mu$. For simplicity we take $\Gamma_\mu = \tilde{\Gamma}_\mu$, $\Gamma'_\mu = \tilde{\Gamma}'_\mu$, $D_\mu = \tilde{D}_\mu$; since we shall take $u > 0$ away from 0, the general case follows by minor changes.

We choose P so that $P^{-1}BP$ has the Jordan canonical form. Consider the function

$$\Phi(y) = \left\{ \sum_{j=1}^{n} \alpha_j y_j^2 \right\}^k \qquad (\alpha_j \text{ are positive constants}, \quad k > 0).$$

Set $\tilde{B} = PBP^{-1}$. We shall take later $u(y) = Cf(\Phi(y))$ with $f(z)$ such that $f'(z) > 0$ and C a positive constant. Then, (8.9) holds (in the y-space) if

$$\tilde{B}y \cdot \Phi_y < 0 \qquad \text{when} \quad y \neq 0, \tag{8.11}$$

i.e., if

$$\sum \alpha_i \tilde{b}_{ij} y_i y_j < 0 \qquad \text{when} \quad y \neq 0 \qquad (\tilde{B} = (\tilde{b}_{ij})).$$

Writing explicitly the $\tilde{b}_{ij}$, one can quickly determine how to choose the α_i so that (8.11) is satisfied. Thus, if the first $l \times l$ block is given by $\tilde{b}_{ii} = \lambda$, $\tilde{b}_{ij} = 1$ if $j = i + 1$, then we take, step by step, α_2/α_1 sufficiently large,

α_3/α_2 sufficiently large, ..., α_l/α_{l-1} sufficiently large. If, on the other hand, the first $l \times l$ block is given by

$$\begin{bmatrix} \lambda & -\mu & 1 & 0 & & & & 0 & \\ \mu & \lambda & 0 & 1 & & & & & \\ & & \lambda & -\mu & 1 & 0 & & & \\ & & \mu & \lambda & 0 & 1 & & & \\ & & & & \vdots & & & & \\ & & & & & & & \lambda & -\mu \\ & 0 & & & & & & \mu & \lambda \end{bmatrix}$$

then we take $\alpha_1 = \alpha_2$, $\alpha_3 = \alpha_4, \ldots, \alpha_{l-1} = \alpha_l$ and choose, step by step, α_3/α_1 sufficiently large, α_5/α_3 sufficiently large, etc.

As mentioned before, we shall take $u(y) = Cf(\Phi(y))$ with $f'(z) > 0$. If $f(0) = 0$, then (8.10) is satisfied for a suitable $C > 0$. Thus, it remains then to verify (8.8), i.e.,

$$Mu \equiv f'(\Phi)\sum a_{ij}\Phi_{y_iy_j} + f''(\Phi)\sum a_{ij}\Phi_{y_i}\Phi_{y_j} < 0. \tag{8.12}$$

It is easily seen that (8.12) is a consequence of

$$\gamma f'(\Phi) + kf''(\Phi)\Phi = 0, \qquad f'(\Phi) > 0, \tag{8.13}$$

where γ is a positive constant depending only on the a_{ij}, α_i. A solution of (8.13) with $f(0) = 0$ is given by

$$f(z) = z^{1-\gamma/k}, \qquad k > \gamma.$$

Having thus completed the construction of u, the proof of Theorem 8.1 is complete.

We shall now state a result which includes both Theorem 7.1 and Theorem 8.1. We shall assume that (B_1') holds, but replace (B_2) and $B_2')$ by:

(B_2'') There are disjoint compact subsets $K_1, \ldots, K_l$ of D and a point ζ on ∂D such that the ω-limit set of each solution of (7.1) with $x(0) \in D\setminus(\bigcup_{i=1}^{l} K_i)$ either coincides with ζ or it is contained in one of the sets K_i. Further, the conditions (B_2) (ii), (iii) hold for the K_i and the condition (B_2') (iii) holds for ζ.

Set $\Sigma' = \Sigma \cup \{\zeta\}$, where Σ is defined as in Theorem 7.1.

Theorem 8.2. *Let* (A), (B_1'), (B_2'') *hold. Then, for any* $\delta > 0$,

$$P_x\{\rho(\xi^\epsilon(\tau^\epsilon), \Sigma') > \delta\} \to 0 \quad \textit{if} \quad \epsilon \to 0 \quad (x \in D). \tag{8.14}$$

Proof. The proof follows by combining the proofs of Theorems 7.1 and 8.1. If $x \in D \cap \Gamma_\delta$ (Γ_δ is a δ-neighborhood of ∂D, as in the paragraph following

(7.19)), τ_ν is defined as in (8.2), and $\tilde{\tau}$ = first time $\xi^\epsilon(t)$ hits γ^+, then, for any $\eta > 0$,

$$P_x\{\rho(\xi^\epsilon(\tau^\epsilon), \Sigma') > \eta\} \leqslant M(\epsilon) + E_x\chi_{\tau_\nu < \tau^\epsilon \wedge \tilde{\tau}}[E_x\chi_{\rho(\xi^\epsilon(\tau^\epsilon), \zeta) > \eta}]_{x = \xi^\epsilon(\tau_\nu)}$$

$$+ E_x\chi_{\tilde{\tau} < \tau^\epsilon \wedge \tau_\nu}[E_x\chi_{\rho(\xi^\epsilon(\tau^\epsilon), \Sigma) > \eta}]_{x = \xi^\epsilon(\tilde{\tau})} \qquad (8.15)$$

where $M(\epsilon)\to0$ if $\epsilon\to0$. An estimate of the from (7.19) holds. (The curves ϕ defined analogously to (7.2) do not intersect $[\bigcup_{i=1}^{l} \mathcal{S}_{i,\mu}']\cup D_\nu$; ν is small with respect to d, say $\nu \leqslant \nu_0(d)$.) Using (7.19) and (8.4), we find that the right-hand side of (8.15) is smaller than $M(\epsilon) + \lambda$, for any given $\lambda > 0$, provided ν is chosen sufficiently small (depending on d, λ) and ϵ is sufficiently small (depending on ν, d, λ). Thus, if $x\in\Gamma_\delta\cap D$,

$$P_x\{\rho(\xi^\epsilon(\tau^\epsilon), \Sigma') > \eta\} \leqslant \Lambda_0(\eta), \qquad \Lambda_0(\eta)\to0 \quad \text{if} \quad \eta\to0. \qquad (8.16)$$

The proof of (8.14) (for any $x\in D$) now follows from (8.16) as in the case of Theorem 7.1.

Remark. Suppose in Theorem 8.1 the point ζ is replaced by a k-dimensional manifold S contained in ∂D and S is asymptotically stable with respect to the system $dx/dt = b(x)$ in a sense similar to that of (B$_2'$) (iii) (i.e., all the eigenvalues of the matrix induced by $\partial b/\partial x$ on the orthogonal complement of S, at each point of S, have negative real parts). Then we can extend Theorem 8.1 and its proof. The function $\Phi(y)$ is now taken to be linear combination with suitable positive coefficients of the squares of the $n - k$ variables normal to S. Theorem 8.2 can also be extended to this case.

9. Application to the Dirichlet problem

Let (A) hold and let D be a bounded domain in R^n with C^2 boundary ∂D. Consider the Dirichlet problem

$$\begin{aligned} L_\epsilon u_\epsilon &= 0 \qquad \text{in} \quad D, \\ u_\epsilon &= g \qquad \text{on} \quad \partial D, \end{aligned} \qquad (9.1)$$

where g is a continuous function on ∂D.

The solution u_ϵ of (9.1) has the form

$$u_\epsilon(x) = E_x g(\xi^\epsilon(\tau^\epsilon))$$

where τ^ϵ is the exit time from D. We shall need the following condition:

(M)   Every solution of (7.1) with $x(0) = x\in D$ exits D in finite time $\tau^0(x)$, and $b \cdot \nu > 0$ at the point of exit, where ν is the outward normal to ∂D.

Theorem 9.1. *Let* (A) *and* (M) *hold. Then, for any* $x \in D$,

$$u_\epsilon(x) \to g(\xi^0(\tau^0(x))) \quad \textit{if} \quad \epsilon \to 0, \tag{9.2}$$

where $\xi^0(t)$ *is the solution of* (7.1) *with* $\xi^0(0) = x$.

Proof. For any small $\gamma > 0$, $\xi^0(t)$ remains in a compact subset of D for all $0 < t < \tau^0(x) - \gamma$, and it exits a neighborhood of $\bar{D}$ at some time t in the interval $\tau^0(x) < t < \tau^0(x) + \gamma$. Since, for any $\mu > 0$,

$$P_x\left\{ \sup_{0 < t < \tau^0(x)+\gamma} |\xi^\epsilon(t) - \xi^0(t)| > \mu \right\} \to 0 \quad \text{if} \quad \epsilon \to 0, \tag{9.3}$$

it follows that

$$P_x\{|\tau^\epsilon - \tau^0(x)| < \gamma\} \to 1 \quad \text{if} \quad \epsilon \to 0.$$

Consequently, for any sequence $\epsilon = \epsilon_m \downarrow 0$, there is a subsequence $\epsilon = \epsilon'_m \downarrow 0$ such that

$$\tau^\epsilon \to \tau^0(x) \quad \text{a.s.} \quad \text{if} \quad \epsilon = \epsilon'_m \downarrow 0. \tag{9.4}$$

From (9.3) with $\epsilon = \epsilon'_m$ it follows that there is a subsequence $\epsilon = \epsilon''_m$ for which

$$\sup_{0 < t < \tau^0(x)+\gamma} |\xi^\epsilon(t) - \xi^0(t)| \to 0 \quad \text{a.s.} \quad \text{as} \quad \epsilon = \epsilon''_m \downarrow 0.$$

Together with (9.4) we find that

$$\xi^\epsilon(\tau^\epsilon) \to \xi^0(\tau^0(x)) \quad \text{a.s.} \quad \text{if} \quad \epsilon = \epsilon''_m \downarrow 0.$$

Hence, by the Lebesgue bounded convergence theorem,

$$u_\epsilon(x) = E_x g(\xi^\epsilon(\tau^\epsilon)) \to g(\xi^0(\tau^0(x))) \quad \text{if} \quad \epsilon = \epsilon''_m \downarrow 0.$$

Since the limit is independent of the original sequence ϵ_m, the assertion (9.2) follows.

Consider now the case where (M) does not hold, and, in fact, $\tau^0(x) = \infty$ for all $x \in D$. From Theorems 7.2, 8.1, 8.2, we immediately obtain:

Theorem 9.2. (i) *If* (A), (B_1), (B_2) *hold, and if* $g(y) = \gamma$ *for all* $y \in \Sigma$, *then for all* $x \in D$,

$$u_\epsilon(x) \to \gamma \quad \textit{if} \quad \epsilon \to 0. \tag{9.5}$$

(ii) *If* (A), (B'_1), (B'_2) *hold, then, for all* $x \in D$,

$$u_\epsilon(x) \to g(\zeta) \quad \textit{if} \quad \epsilon \to 0.$$

(iii) *If* (A), (B'_1), (B''_2) *hold and if* $g(y) = \gamma$ *for all* $y \in \Sigma'$, *then* (9.5) *holds for all* $x \in D$.

10. The principal eigenvalue

Let (A) hold and set

$$Lu = \frac{1}{2} \sum_{i,j=1}^{n} a_{ij}(x) \frac{\partial^2 u}{\partial x_i \partial x_j} + \sum_{i=1}^{n} b_i(x) \frac{\partial u}{\partial x_i}. \tag{10.1}$$

Let D be a bounded domain in R^n with C^2 boundary ∂D. Consider the eigenvalue problem

$$\begin{aligned} -Lu &= \lambda u \quad \text{in } D, \\ u &= 0 \quad \text{on } \partial D. \end{aligned} \tag{10.2}$$

If $a_{ij} = \delta_{ij}$ and $b = (b_1, \ldots, b_n)$ is a gradient of a function, then L will be self-adjoint in a suitable space. However, in general, L is not self-adjoint. From the general theory of elliptic operators (see Agmon [1] or Friedman [2]) it is known that L has a sequence of eigenvalues, nonreal in general. However, by the general theory of positive operators (see Krasnoselskii [1]), there does exist at least one positive eigenvalue. Further, if λ_0 is the smallest positive eigenvalue, then the corrseponding space of eigenfunctions consists of multiples of a function $\phi_0(x)$ which is positive throughout D. The eigenvalue λ_0 is called the *principal eigenvalue*. It is known (see Protter and Weinberger [1]) that Re $\lambda \geq \lambda_0$ for any eigenvalue λ of (10.2).

We shall give in this section a probabilistic characterization of the principal eigenvalue λ_0, in terms of the exit time τ from D of the solution of

$$d\xi(t) = \sigma(\xi(t))\, dw(t) + b(\xi(t))\, dt. \tag{10.3}$$

Theorem 10.1. *Let* (A) *hold and define*

$$\Lambda = \sup\left\{\lambda \geq 0;\ \sup_{x \in D} E_x e^{\lambda\tau} < \infty\right\}.$$

Then $\lambda_0 = \Lambda$.

We shall need the following lemma.

Lemma 10.2. *Let u be a solution of the elliptic equation*

$$Lu + cu = f \qquad \text{in a bounded domain } N \tag{10.4}$$

with the boundary condition

$$u = \phi \qquad \text{on } \partial N. \tag{10.5}$$

Suppose $a_{11}\alpha^2 + b_1\alpha \geq 1$, $0 \leq x_1 \leq d$ *for all* $x = (x_1, \ldots, x_n)$ *in* N, *and suppose that* $\max_{\bar N} c(x) > 0$ *and*

$$\rho \equiv \Big(\max_{\bar N} c\Big)(e^{\alpha d} - 1) < 1. \tag{10.6}$$

Then

$$\max_{\bar N}|u| \le \Big\{\max_{\partial N}|\phi| + \Big(\max_{\bar N}|f|\Big)(e^{\alpha d} - 1)\Big\}/(1-\rho). \tag{10.7}$$

Proof. Let $c^- = \min(c, 0)$ and write

$$Lu + c^- u = (c^- - c)u + f \equiv \hat f.$$

The function

$$v = \Big(\max_{\bar N}|\hat f|\Big)(e^{\alpha d} - e^{\alpha x_1}) + \max_{\partial D}|\phi|$$

satisfies $Lv + c^- v \le -|\hat f|$ in N, $v \ge |\phi|$ on ∂N. Hence, by the maximum principle

$$\max_{\bar N}|u| \le \max_{\bar N}|v|$$
$$\le \Big(\max_{\bar N}|f|\Big)(e^{\alpha d} - 1) + \max_{\partial N}|\phi| + \Big(\max_{\bar N} c\Big)(e^{\alpha d} - 1)\Big(\max_{\bar N}|u|\Big),$$

and (10.7) follows.

By the general theory of elliptic operators (see, for instance, Friedman [2]), the Fredholm alternative holds. Thus, if L satisfies (A) and c is Hölder continuous, and if there is at most one solution for the Dirichlet problem (10.4), (10.5), then there does in fact exist a solution (for any continuous ϕ and Hölder continuous f).

If c is any given function in D, then the condition (10.6) is satisfied (after translation of the origin) in any ball N of sufficiently small parameter. Consequently, the Dirichlet problem (10.4), (10.5) has, in this case, at most one solution. Appealing to the remarks of the preceding paragraph, we can assert:

Corollary 10.3. *Let* (A) *hold. For any* $\mu > 0$ *there exists a unique solution of the Dirichlet problem*

$$Lu + \mu u = f \quad \text{in } N \quad (f \text{ Hölder continuous in } \bar N),$$
$$u = \phi \quad \text{on } \partial N \quad (\phi \text{ continuous on } \partial N),$$

for any ball N *lying in* D, *provided the radius of* N *is sufficiently small, depending on* μ.

Proof of Theorem 10.1. Suppose $\mu < \Lambda$, and consider the Dirichlet problem

$$Lu + \mu u = f \quad \text{in} \quad D \qquad \left(f \text{ Hölder continuous in } \bar{D}\right), \tag{10.8}$$

$$u = \phi \quad \text{on} \quad \partial D \qquad (\phi \text{ continuous on } \partial D). \tag{10.9}$$

A natural candidate for a solution is

$$u(x) = E_x\{e^{\mu\tau}\phi(\xi(\tau))\} + E_x\left\{\int_0^\tau e^{\mu s}f(\xi(s))\,ds\right\}.$$

Since $\mu < \Lambda$, this function is well defined. To prove that $u(x)$ is a solution of (10.8), we resort to the argument used in the proof of Theorem 13.1.1 (in showing that $u(x)$, given by (13.1.17) is a solution of $Lu + cu = 0$). Here we use balls N with radius sufficiently small, as in Corollary 10.3.

To prove that u satisfies (10.9), notice that

$$E_x|e^{\mu\tau}\phi(\xi(\tau))|^{1+\epsilon} \leq C_1 E_x e^{\mu(1+\epsilon)\tau} \leq C$$

if $\mu(1+\epsilon) < \Lambda$, where C_1, C are constants independent of x. Since every point $y \in \partial D$ is a regular point, we can apply Lemma 1.3.6 to conclude that

$$E_x e^{\mu\tau}\phi(\xi(\tau)) \to \phi(y) \qquad \text{if} \quad x \to y.$$

Similarly,

$$E_x \int_0^\tau e^{\mu s}f(\xi(s))\,ds \to 0 \qquad \text{if} \quad x \to y,$$

and (10.9) is proved.

We have thus proved that if $\mu < \Lambda$, then the Dirichlet problem (10.8), (10.9) has a solution for any f, ϕ. Consequently, μ is not an eigenvalue. But if every $\mu \in (0, \Lambda)$ is not an eigenvalue, we must have $\lambda_0 \geq \Lambda$.

To prove the converse, let $\mu < \lambda_0$. Then the Dirichlet problem

$$Lu + \mu u = 0 \qquad \text{in } D,$$
$$u = 1 \qquad \text{on } \partial D \tag{10.10}$$

has a solution u. We claim that

$$u > 0 \qquad \text{in } D. \tag{10.11}$$

Indeed, denoting by ϕ_0 a positive eigenfunction corresponding to λ_0, we have

$$L\phi_0 + \mu\phi_0 < 0 \quad \text{in } D, \qquad \phi_0 > 0 \quad \text{in } D. \tag{10.12}$$

Let $D_\delta = \{x \in D;\ \rho(x, \partial D) > \delta\}$. Since $u = 1$ on ∂D, there is a sufficiently small δ such that

$$u(x) > 0 \qquad \text{in} \quad D \setminus D_\delta. \tag{10.13}$$

Next

$$Lu + \mu u = 0 \quad \text{in } D_\delta, \qquad u > 0 \quad \text{on } \partial D_\delta. \tag{10.14}$$

Writing $u = v\phi_0$ we find that

$$\frac{1}{\phi_0}(L + \mu)u = \sum a_{ij} \frac{\partial^2 v}{\partial x_i \, \partial x_j} + \sum \tilde{b}_i \frac{\partial v}{\partial x_i} + \tilde{c}v \equiv \tilde{L}v$$

where

$$\tilde{b}_i = \frac{2}{\phi_0} \sum_j a_{ij} \frac{\partial \phi_0}{\partial x_j} + b_i, \qquad \tilde{c} = \frac{1}{\phi_0}(L + \mu)\phi_0.$$

Since by (10.14), $\tilde{L}v = 0$ in D_δ and $v > 0$ on ∂D_δ, and since, by (10.12), $\tilde{c} < 0$, we can apply the maximum principle to deduce that $v > 0$ in D_δ. Therefore $u > 0$ in $\bar{D}_\delta$. Combining this with (10.13), the assertion (10.11) follows.

Let γ be a positive constant such that

$$u(x) \geq \gamma \qquad \text{if} \quad x \in \bar{D}. \tag{10.15}$$

By Itô's formula,

$$E_x\left\{u(\xi(\tau \wedge t))e^{\mu(\tau \wedge t)}\right\} = u(x) \leq C \qquad (C \text{ const})$$

for all $x \in D$. Exploiting (10.15), we get

$$E_x e^{\mu(\tau \wedge t)} \leq C/\gamma.$$

Taking $t \uparrow \infty$ we find that $E_x e^{\mu\tau} \leq C/\gamma$ for all $x \in D$. Consequently, $\mu \leq \Lambda$. We have thus proved that if $\mu < \lambda_0$, then $\mu \leq \Lambda$. This implies that $\lambda_0 \leq \Lambda$.

11. Asymptotic behavior of the principal eigenvalue

Consider the eigenvalue problem

$$-L_\epsilon u = \lambda u \qquad \text{in } D,$$
$$u = 0 \qquad \text{on } \partial D,$$

and denote by λ_ϵ the principal eigenvalue. We shall study the behavior of λ_ϵ as $\epsilon \to 0$, under the assumption (A) and the assumptions (B_1), (B_2) (i), (ii) (defined in Section 7). Recall that

$$V_i = I(x, \partial D) \qquad \text{when} \quad x \in K_i.$$

Set

$$V^* = \max\{V_1, \ldots, V_l\}, \qquad V_* = \min\{V_1, \ldots, V_l\}.$$

Theorem 11.1. *Let* (A), (B_1) *and* (B_2) (i), (ii) *hold. Then*

$$\overline{\lim_{\epsilon\to 0}}\{-2\epsilon^2 \log \lambda_\epsilon\} \leq V^*, \tag{11.1}$$

$$\underline{\lim}_{\epsilon\to 0}\{-2\epsilon^2 \log \lambda_\epsilon\} \geq V_* \tag{11.2}$$

Corollary 11.2. *If* $V^* = V_*$ *(which is the case when* $l = 1$*), then*

$$\lim_{\epsilon\to 0}\{-2\epsilon^2 \log \lambda_\epsilon\} = V^*.$$

Proof of (11.1). By Theorem 4.1, for any $h > 0$

$$\exp\left\{\frac{-I(t, x, \partial D) - h}{2\epsilon^2}\right\} < P_x\{\tau^\epsilon < t\} < \exp\left\{\frac{-I(t, x, \partial D) + h}{2\epsilon^2}\right\} \tag{11.3}$$

provided ϵ is sufficiently small, say $0 < \epsilon \leq \epsilon_0$. Here τ^ϵ is the exit time from D. A careful review of the proof of (11.3) shows that ϵ_0 can be taken to be independent of $x \in D$ and $t \in (0, T]$, for any fixed $T > 0$; it may however depend on T.

Set $V(x) = I(x, \partial D)$. From the definition of $I(t, x, \partial D)$ we conclude that, when (B_1) holds,

$$I(t, x, \partial D) \to V(x) \qquad \text{if} \quad t \to \infty \tag{11.4}$$

for any $x \in D$. Using the facts that $V(x)$ is uniformly continuous in D and $I(t, x, \partial D)$ is continuous in $x \in D$, uniformly with respect to $(t, x) \in [1, \infty) \times D$, we conclude that the convergence in (11.4) is uniform with respect to $x \in D$.

For any $x \in D$, the ω-limit set of the solution of (7.1) with $x(0) = x$ lies in some set K_i. It follows that $V(x, z) = 0$ for any $z \in K_i$. Hence

$$V(x) \leq V(x, z) + V(z) = V_i.$$

Recalling (11.4) and the first inequality in (11.3), we get

$$P_x\{\tau^\epsilon < t\} > \exp\left\{\frac{-V_i - 2h}{2\epsilon^2}\right\}$$

if t is sufficiently large, and ϵ is sufficiently small (depending on h, t, but independently of x). Consequently, for some $T > 0$ sufficiently large and for all $x \in D$,

$$P_x\{\tau^\epsilon \geq T\} < 1 - \exp\left[-\frac{\alpha}{2\epsilon^2}\right], \qquad \alpha = V^* + 2h \tag{11.5}$$

for all ϵ is sufficiently small (depending on T, h).

By the strong Markov property (see Problem 11, Chapter 2),

$$P_x\{\tau^\epsilon > mT\} < \left\{1 - \exp\left[-\frac{\alpha}{2\epsilon^2}\right]\right\}^m$$

for any positive integer m. Hence

$$\begin{aligned} E_x e^{\lambda\tau^\epsilon} &= \int_0^\infty e^{\lambda t} P(\tau^\epsilon \in dt) \\ &< e^{\lambda T} + \sum_{m=1}^\infty (e^{\lambda(m+1)T} - e^{\lambda mT}) P_x(\tau^\epsilon > mT) \\ &< e^{\lambda T} + c \sum_{m=1}^\infty e^{\lambda mT}\left[1 - \exp\left(-\frac{\alpha}{2\epsilon^2}\right)\right]^m \qquad (c \text{ const}). \end{aligned}$$

The right-hand side is finite and bounded by a constant independent of x if

$$\lambda T + \log\left[1 - \exp\left(-\frac{\alpha}{2\epsilon^2}\right)\right] < 0,$$

i.e., if

$$\lambda T < \left[\exp\left(-\frac{\alpha}{2\epsilon^2}\right)\right](1 + o(1)) \qquad (\epsilon \to 0).$$

Hence, if

$$-2\epsilon^2 \log \lambda > \alpha + h = V^* + 3h$$

for all ϵ sufficiently small (depending on h), then

$$E_x e^{\lambda\tau^\epsilon} < C < \infty \qquad \text{for all} \quad x \in D \qquad (C \text{ const}).$$

By Theorem 10.1 it then follows that $\lambda_\epsilon > \lambda$. Taking λ such that

$$-2\epsilon^2 \log \lambda = V^* + 4h$$

we conclude that

$$-2\epsilon^2 \log \lambda_\epsilon < V^* + 4h$$

if ϵ is sufficiently small. Since h is arbitrary, the assertion (11.1) follows.

Proof of (11.2). Let $h > 0$ be any positive number. Let N_j be a small neighborhood of K_j such that $N_j \subset D$ and

$$|V(x) - V_j| < \tfrac{1}{2}h \qquad \text{if} \quad x \in N_j.$$

From (11.4) it follows that there exists a $T^* > 0$ sufficiently large such that

$$|I(t, x, \partial D) - V_j| < h \qquad \text{if} \quad t > T^*, \quad x \in \bar{N}_j.$$

Let $N = \bigcup_{j=1}^{l} N_j$. Then, from the second inequality in (11.3) we deduce that for any $T^{**} > T^*$

$$P_x\{\tau^\epsilon < T^{**}\} < \exp\left[-\frac{\beta}{2\epsilon^2}\right] \qquad \text{for all} \quad x \in \bar{N} \qquad (\beta = V_* - 2h) \tag{11.6}$$

provided ϵ is sufficiently small (depending on T^{**}, but independently of x).

For $x \in R^n$, denote by $\rho(x)$ the distance from x to $\bar{D}$. Denote by D_δ ($\delta > 0$) the set of all points x with $\rho(x) < \delta$. If δ_0 is sufficiently small, then, for any point $x' \in D_{\delta_0} \backslash \bar{D}$, there is a unique point x on ∂D such that $|x' - x| = \rho(x')$. Denote by ∂D_δ ($\delta > 0$) the set of all points x^δ in D_{δ_0} with $\rho(x^\delta) = \delta$. Each point x^δ is the end point of a unique normal segment of length δ initiating at some point $x \in \partial D$, and $|x^\delta - x| = \rho(x^\delta) = \delta$. The normal to ∂D_δ at x^δ is on the same ray as the normal to ∂D at x.

Denote by $\tau(x)$ the time it takes the solution of (7.1) with $x(0) = x$ to enter the set N. By (B_1), (B_2) (i), $\tau(x) < \infty$ for all $x \in \bar{D}$. Since Theorem 11.1 depends on properties of $b(x)$ in $\bar{D}$ only, we may modify the definition of $b(x)$ outside $\bar{D}$ so that $\tau(x)$ remains finite for all $x \in \bar{D}_{\delta_0}$ and, moreover,

$$b(x) \cdot \nu(x) > 0 \qquad \text{if} \quad x \in \bar{D}_{\delta_0} \backslash D, \tag{11.7}$$

$$b(x) \cdot \nu(x) > 1/\eta \qquad \text{if} \quad x \in \bar{D}_{\delta_0} \backslash D_{\delta_0/2} \tag{11.8}$$

where $\nu(x)$, for $x \in \partial D_\delta$, is the inward to ∂D_δ, and η is an arbitrarily given positive number.

Lemma 11.3. *For any $A > 0$, $B > 0$ there exist positive numbers $\gamma = \gamma(A)$, $\eta = \eta(A)$ (depending on A but independent of B) such that if we take $\eta = \eta(A)$ in (11.8), then the following holds for any absolutely continuous curve $\phi(t)$ with $\phi(0) = x \in D$:*

(i) *if $I_{0,B+\gamma}(\phi) < A$, then $\phi(t) \in D_{\delta_0}$ for all $0 < t < B + \gamma$; and*
(ii) *if $\phi(t) \not\in N$ for $B < t < B + \gamma$, then*

$$I_{0,B+\gamma}(\phi) > A. \tag{11.9}$$

Proof. Consider the differential system

$$\frac{d\phi}{dt} - b(\phi) = f(t) \qquad \text{for} \quad t > 0, \qquad \phi(0) = x \tag{11.10}$$

where

$$\int_0^\infty |f(t)|^2\,dt < \bar{A}, \qquad x \in D \tag{11.11}$$

and $\bar{A} = A_0 A$, $A_0 = \Sigma_{i,j} \sup |a_{ij}|$. We claim that if (11.8) holds with $\eta = \delta_0/\bar{A}$, then

$$\phi(t) \in D_{\delta_0} \quad \text{for all} \quad t > 0. \tag{11.12}$$

Indeed, otherwise there is an interval $(s, \bar{t})$ such that

$$\delta_0/2 < \rho(\phi(t)) < \delta_0 \quad \text{if} \quad s < t < \bar{t},$$
$$\rho(\phi(s)) = \delta_0/2, \qquad \rho(\phi(\bar{t})) = \delta_0.$$

In this interval

$$\frac{d}{dt}\rho(\phi(t)) = \frac{d\phi}{dt} \cdot \nabla\rho(\phi(t)) = -b(\phi(t)) \cdot \nu(\phi(t)) - f(t) \cdot \nu(\phi(t))$$

by (11.10), since $\nabla\rho(x) = -\nu(x)$ if $x \in \partial D_\delta$. Hence, by (11.8), (11.11),

$$\tfrac{1}{2}\delta_0 = \rho(\phi(\bar{t})) - \rho(\phi(s)) \leq -\frac{\bar{t}-s}{\eta} + \int_s^{\bar{t}} |f(t)|\, dt$$

$$\leq -\frac{\bar{t}-s}{\eta} + \int_s^{\bar{t}} \left(\frac{1}{\eta} + \frac{\eta}{4}|f(t)|^2\right) dt \leq \frac{\eta}{4}\bar{A} = \tfrac{1}{4}\delta_0,$$

a contradiction. This completes the proof of (11.12) and, clearly, also the proof of (i).

For any $x \in \bar{D}_{\delta_0}$, $\tau(x)$ is finite. Since $\tau(x)$ is upper semicontinuous,

$$\zeta_0 = \max_{x \in \bar{D}_{\delta_0}} \tau(x) < \infty.$$

Let $\zeta = \zeta_0 + 1$. We claim that for any absolutely continuous $\phi(t)$, $\mu \leq t \leq \mu + \zeta$ with $\phi(\mu) \in \bar{D}_{\delta_0}$ and $\mu \geq 0$, the following is true:

$$\text{if} \quad \phi(t) \not\in N \quad \text{for} \quad \mu \leq t \leq \mu + \zeta, \quad \text{then} \quad I_{\mu,\mu+\zeta}(\phi) \geq \nu > 0 \tag{11.13}$$

where ν is a constant independent of μ.

Suppose (11.13) is false. By the lower semicontinuity of $I_{\mu,\mu+\zeta}(\phi)$ (see Lemma 3.2) we deduce the existence of a particular ϕ for which $I_{\mu,\mu+\zeta}(\phi) = 0$. But then ϕ is a solution of (7.1), whereas $\phi(\mu) \in \bar{D}_{\delta_0}$, $\phi(t) \not\in N$ for $\mu \leq t \leq \mu + \zeta$. This contradicts the definition of ζ.

Notice that since a_{ij}, b_i are functions independent of t, the constant ν occurring in (11.13) is independent of t.

We shall now prove the assertion (ii) of Lemma 11.3 with $\gamma = j_0\zeta$ where j_0 is any positive integer such that $j_0\nu > A$. Suppose $\phi(t) \not\in N$ for $B \leq t \leq B + \gamma$, $\phi(0) = x \in D$, and suppose that (3.3) is not satisfied, i.e.,

$$I_{0,B+\gamma}(\phi) \leq A.$$

Then, by (i), $\phi(\mu) \in D_{\delta_0}$ if $0 \le \mu \le B + \gamma$. Hence, by (11.13),

$$I_{B+(j-1)\zeta, B+j\zeta}(\phi) \geq \nu \qquad \text{for} \quad j = 1, 2, \ldots, j_0.$$

It follows that

$$I_{B, B+\gamma}(\phi) = I_{B, B+j_0\zeta}(\phi) \geq j_0\nu > A,$$

a contradiction.

Completion of the proof of (11.2). Fix

$$A = V_* + 1, \qquad \eta = \eta(A) = \delta_0/\bar{A}.$$

Fix also T so that, by (11.6),

$$P_x\{\tau^\epsilon < \tfrac{5}{4}T\} < \exp\left[-\frac{\beta}{2\epsilon^2}\right] \qquad \text{if} \quad x \in N \qquad (\beta = V_* - 2h), \quad (11.14)$$

and so that the assertions of Lemma 11.3 hold with $\gamma = \frac{1}{4}T$ and with $\bar{N}$ replaced by some open set N' containing $\bigcup_{j=1}^{l} K_j$ and whose closure lies in N. Note that (11.14) holds for all ϵ sufficiently small (independently of x).

Let

$$\delta = \inf\{|x - y|;\, x \in N',\, y \notin N\}.$$

Denote by τ_N^ϵ the first time in the interval $T \le t \le \frac{5}{4}T$ when $\xi^\epsilon(t)$ hits $\bar{N}$ if such a time exists, and set $\tau_N^\epsilon = \frac{5}{4}T$ if $\xi^\epsilon(t) \notin \bar{N}$ for all $T \le t \le \frac{5}{4}T$.

For any $x \in D$, let $\Phi(x)$ be the set of all continuous curves $\phi(t)$, $0 \le t \le \frac{5}{4}T$ with $\phi(0) = x$, such that

$$I_{0, 5T/4}(\phi) < A.$$

Then, by Lemma 11.3, each $\phi(t)$ in $\Phi(x)$ intersects N' at some time $t \in [T, \frac{5}{4}T]$. Hence

$$P_x\{\xi^\epsilon(\tau_N^\epsilon) \notin \bar{N}\} \le P_x\{\rho_{0, 5T/4}(\xi^\epsilon, \Phi(x)) > \delta/2\}.$$

By Theorem 3.1, the right-hand side is $\le \exp[-B/2\epsilon^2]$ where $B = A - \frac{1}{2}$, provided ϵ is sufficiently small. Hence

$$P_x\{\xi^\epsilon(\tau_N^\epsilon) \in \bar{N}\} > 1 - e^{-B/2\epsilon^2} \qquad (x \in D). \qquad (11.15)$$

Set $\tau = \tau^\epsilon$, $\bar{\tau} = \tau_N^\epsilon$. Then, by the strong Markov property,

$$\begin{aligned} P_x\{\tau^\epsilon \geq 2T\} &= E_x I_{(\tau \geq 2T)} \\ &= E_x I_{\tau \geq \bar{\tau}} P_{\xi^\epsilon(\bar{\tau})}\{\tau \geq 2T - \bar{\tau}\} \geq E_x I_{\tau \geq \bar{\tau}} P_{\xi^\epsilon(\bar{\tau})}\{\tau \geq T\}. \end{aligned}$$

Using (11.14), (11.15) we find that

$$P_x\{\tau^\epsilon > 2T\} > (1 - e^{-B/2\epsilon^2})(1 - e^{-\beta/2\epsilon^2})E_x I_{\tau > 5T/4} \qquad \text{if} \quad x \in D, \tag{11.16}$$

$$P_x\{\tau^\epsilon > 2T\} > (1 - e^{-B/2\epsilon^2})(1 - e^{-\beta/2\epsilon^2})^2 \qquad \text{if} \quad x \in \bar{N}. \tag{11.17}$$

Similarly, by induction,

$$\begin{aligned} P_x\{\tau^\epsilon > (m+1)T\} &= E_x I_{\tau > \bar{\tau}} P_{\xi^\epsilon(\bar{\tau})}\{\tau > (m+1)T - \bar{\tau}\} \\ &> E_x I_{\tau > \bar{\tau}} P_{\xi^\epsilon(\bar{\tau})}\{\tau > mT\} \\ &> (1 - e^{-B/2\epsilon^2})^m (1 - e^{-\beta/2\epsilon^2})^m E_x I_{\tau > 5T/4} \end{aligned}$$

if $x \in D$. In particular, if $x \in \bar{N}$,

$$P_x\{\tau^\epsilon > (m+1)T\} > (1 - e^{-B/2\epsilon^2})^m (1 - e^{-\beta/2\epsilon^2})^{m+1} \equiv \Delta_m.$$

It follows that, for $x \in N$,

$$\begin{aligned} E_x e^{\lambda\tau^\epsilon} &> \sum_{m=1}^\infty [e^{\lambda(m+1)T} - e^{\lambda mT}] P_x\{\tau^\epsilon > (m+1)T\} \\ &> c \sum_{m=1}^\infty e^{\lambda mT}\Delta_m = \infty \qquad (c > 0) \end{aligned}$$

provided

$$\lambda T + \overline{\lim_{m\to\infty}} \frac{1}{m} \log \Delta_m > 0.$$

Since

$$\overline{\lim_{m\to\infty}} \left(\frac{1}{m} \log \Delta_m \right) > -2(e^{-B/2\epsilon^2} + e^{-\beta/2\epsilon^2})$$

if ϵ is sufficiently small, we conclude that

$$E_x e^{\lambda\tau^\epsilon} = \infty \qquad \text{if} \quad x \in N \tag{11.18}$$

provided

$$\tfrac{1}{2}\lambda T = e^{-B/2\epsilon^2} + e^{-\beta/2\epsilon^2}.$$

By Theorem 10.1, $\lambda_\epsilon < \lambda$. Hence

$$-2\epsilon^2 \log \lambda_\epsilon > -2\epsilon^2 \log \lambda.$$

Recalling that $B = V_* + \frac{1}{2}$, $\beta = V_* - 2h$, we deduce from (11.18) that $-2\epsilon^2 \log \lambda \to V_* - 2h$ if $h < \frac{1}{4}$, $\epsilon \to 0$. Consequently,

$$\underline{\lim}_{\epsilon\to 0} [-2\epsilon^2 \log \lambda_\epsilon] > V_* - 2h.$$

Since h is arbitrary, the proof of (11.2) is complete.

PROBLEMS

1. Prove that there exists a solution v of (4.4). [*Hint*: Let $\zeta(t)$ be continuous, $\zeta(t) = 0$ if $t < \frac{1}{2}$, $\zeta(t) = 1$ if $t > 1$, $0 < \zeta(t) < 1$ if $\frac{1}{2} < t < 1$, and consider

$$\frac{\partial v_m}{\partial t} = L_\epsilon v_m,$$

$$v_m(x, 0) = 0 \quad \text{if } x \in D, \qquad v_m(x, t) = \zeta(mt) \quad \text{if } x \in \partial D, \quad t > 0.$$

By Section 6.3 there exists a unique solution v_m. By Schauder's estimates $v_m \to u_\epsilon$ for a subsequence $m = m' \to \infty$, and $\partial u_\epsilon / \partial t = L_\epsilon u_\epsilon$. If q belongs to the normal boundary and w_q is a barrier, apply the maximum principle to $(Aw_q + v_m(q) + \delta) \pm v_m$ to deduce that u_ϵ satisfies the desired boundary conditions.]

2. The solution of (4.4) is unique. [*Hint*: If v_1, v_2 are two solutions and $v = v_1 - v_2$, then $|v| \leq$ const. For any $\delta > 0$,

$$v(x, t) = \int q_\epsilon(t - \delta, x, y) v(\delta, y)\, dy.$$

Take $\delta \downarrow 0$.]

3. Prove (4.5). [*Hint*: Apply Itô's formula to v_m and use the monotone convergence theorem.]

4. Prove Theorem 3.4.

5. Prove (6.12). [*Hint*: Use the uniqueness for the first initial-boundary value problem.]

6. Let (A), (B_1) hold and suppose the ω-limit set of any solution of (7.1) with $x(0) \in D$ coincides with the origin 0. Let $V(y) = I(0, y)$ where $I(x, y)$ is defined as in Section 7. Assume that $b(x) = -\nabla U(x) + \gamma(x)$ where $U(0) = 0$, $U \in C^2(\bar{D})$, $\gamma(x) \cdot \nabla U(x) = 0$, and that $a_{ij}(x) = \delta_{ij}$. Prove:

(i) $V(y) \geq 4U(y)$ for any $y \in D$;

(ii) If $y = \psi(T)$ for some $T > 0$ where $\dot{\psi} = \nabla U(\psi) + \gamma(\psi)$ if $0 < t < T$, $\psi(0) = 0$, then $V(y) = 4U(y)$;

(iii) If the solution of $\dot{\psi} = \nabla U(\psi) + \gamma(\psi)$, $\psi(0) = 0$ exits D at a point $y_0 \in \partial D$ and $U(y_0) < U(y)$ for any $y \in \partial D$, $y \neq y_0$, then the set Σ occurring in Theorem 7.2 coincides with $\{y_0\}$.

7. Let (A) hold. Consider the Dirichlet problem

$$\begin{aligned} \partial v_\epsilon / \partial t &= L_\epsilon v_\epsilon + c v_\epsilon \quad && \text{if } x \in D, \quad t > 0, \\ v_\epsilon(x, 0) &= 0 && \text{if } x \in D, \\ v_\epsilon(x, t) &= \psi(x) && \text{if } x \in \partial D, \quad t > 0 \end{aligned} \tag{11.19}$$

where ψ is continuous on ∂D, ∂D is in C^2, c is Hölder continuous, and $c \leq 0$. Prove that it has a unique solution, where the concept of solution is the same as for (4.4).

8. Suppose for some $x \in D$, $t > 0$, $I(t, x, \partial D) \equiv \inf\{I_t(\phi);\ \phi \in \Psi_t\}$ is actually a minimum, and the minimum is attained for a unique curve $\hat{\phi}$ in C_t. Denote by τ^ϵ the exit time of $\xi^\epsilon(t)$ from D. Let

$$A_\delta = \{\tau^\epsilon \leq t,\ \rho_t(\xi^\epsilon, \hat{\phi}) < \delta\}, \qquad \delta > 0.$$

Prove that $P_x(A_\delta) = P_x(\tau^\epsilon \leq t)(1 + o(1))$, where $o(1) \to 0$ if $\delta \to 0$. [*Hint*: If $\rho_t(\chi, \hat{\phi}) \geq \delta/2$, $\chi \in \Psi_t$, then $I_t(\chi) \geq I_t(\hat{\phi}) + \lambda$ where $\lambda > 0$ and is independent of χ; otherwise $\hat{\phi}$ is not unique. Let $\Phi_0 = \{\chi \in C_t, \chi(0) = x, \min_{0 \leq s \leq t} \rho(\chi(s), \partial D) = 0, I_t(\chi) \leq I_t(\hat{\phi}) + \lambda/2$. Then $\chi \in \Phi_0$ implies $\rho_t(x, \hat{\phi}) \leq \delta/2$. Deduce that $A_\delta \subset \{\xi^\epsilon(s)$ intersects ∂D for some $0 \leq s \leq t$, and $\rho_t(\xi^\epsilon, \Phi_0) \geq \delta/2\}$. By Theorem 3.1,

$$P(A_\delta) \leq \exp\left[\left(-I_t(\hat{\phi}) + \frac{\lambda}{2} - k\right)/2\epsilon^2\right] \qquad \text{for any } \ k > 0,$$

if ϵ is small. Next modify $\hat{\phi}$ into $\tilde{\phi}$ which penetrates a distance h into $R^n \setminus D$ and is such that $I_t(\tilde{\phi}) \leq I_t(\hat{\phi}) + \lambda/4$, $\rho_t(\hat{\phi}, \tilde{\phi}) \leq ch$. If $\rho(\xi^\epsilon, \tilde{\phi}) < \delta^*$ and δ^*, h are sufficiently small, then $\rho_t(\xi^\epsilon, \phi) < \delta$ and (if $\delta^* < h$) ξ^ϵ exits D at some time $s < t$. Therefore, $\{\rho_t(\xi^\epsilon, \tilde{\phi}) < \delta^*\} \subset A_\delta$. Apply Theorem 2.1.]

9. Let the assumptions of Problems 7, 8 hold. Denote by $\hat{\tau}$ the time ϕ exits D. Assume that $b \cdot \nu \neq 0$ on ∂D, where ν is the normal to ∂D. Prove that, if $\omega \in A_\delta$,

$$\left|\psi(\hat{\phi}(\hat{\tau})) \exp\left[\int_0^{\hat{\tau}} c(\hat{\phi}(s))\, ds\right] - \psi(\xi^\epsilon(\tau^\epsilon)) \exp\left[\int_0^{\tau^\epsilon} c(\xi^\epsilon(s))\, ds\right]\right| \leq \lambda(\delta),$$

where $\lambda(\delta) \to 0$ if $\delta \to 0$; $\lambda(\delta)$ is independent of ω. [*Hint*: If $\hat{\tau} = t$, then $\hat{\tau} - \gamma < \tau^\epsilon \leq t = \hat{\tau}$ where $\gamma \to 0$ if $\delta \to 0$. If $\hat{\tau} < t$, then $d\hat{\phi}(s)/ds = b(\hat{\phi}(s))$ if $s > \hat{\tau}$. Since $b \cdot \nu \neq 0$ on ∂D, $\hat{\tau} - \gamma < \tau^\epsilon \leq \hat{\tau} + C\delta$.]

10. Let the assumptions of Problems 7, 8 hold and assume that $b \cdot \nu \neq 0$ along ∂D. Denote by $\hat{\tau}$ the time $\hat{\phi}$ exits D. Prove that

$$\lim_{\epsilon \to} \frac{v_\epsilon(x, t)}{u_\epsilon(x, t)} = \psi(\hat{\phi}(\hat{\tau})) \exp\left[\int_0^{\hat{\tau}} c(\hat{\phi}(s))\, ds\right],$$

where u_ϵ is the solution of (4.4). [*Hint*: Setting $\psi(\xi^\epsilon(\tau^\epsilon \wedge t)) = 0$ if $\tau^\epsilon > t$, we have

$$v_\epsilon(x, t) = E_x\left\{\psi(\xi^\epsilon(\tau^\epsilon \wedge t)) \exp\left[\int_0^{\tau^\epsilon \wedge t} c(\xi^\epsilon(s))\, ds\right]\right\}.$$

Use Problems 7–9.]

11. Let (A), (B_1) hold and let $x \in D$. Prove that $I(x, \partial D) > 0$. [*Hint*: If $I_{T_m}(\phi_m) \to 0$, $\phi_m(0) = x$, $T_m \to \infty$, let σ_m be the last time ϕ_m enters a δ-neighborhood of ∂D before intersecting ∂D, δ small. Apply Lemma 1.2 to $\psi_m(t) = \phi_m(t + \sigma_m)$, $0 \leq t \leq 1$.]
12. Extend Theorem 9.1 to $L_\epsilon u + cu = 0$, $c \leq 0$.
13. Let D, E be bounded domains with C^2 boundary. Assume also that (A) holds. Denote by λ_D, λ_E the principal eigenvalues corresponding to the domains D and E respectively. Prove that if $D \subset E$, then $\lambda_D \geq \lambda_E$.
14. Let D be a bounded domain with C^2 boundary, and let (A) hold. Assume that there is a point ζ in D which is a stable equilibrium point for $\dot{x} = b(x)$. Denote by λ_ϵ the principal eigenvalue of $-L_\epsilon$ with respect to D. Prove that there is a positive constant β such that $\lambda_\epsilon \geq \exp[-\beta/\epsilon^2]$ for all ϵ sufficiently small.
15. Let (A) hold and let D be a bounded domain with C^2 boundary. Denote by λ_ϵ the principal eigenvalue of $-L_\epsilon$ with respect to D. Suppose that every solution of $\dot{x} = b(x)$ with $x(0) \in \bar{D}$ exits $\bar{D}$ at some finite time. Prove that $\lim_{\epsilon \to 0} \lambda_\epsilon = \infty$. [*Hint*: Verify that $P_x(\tau^\epsilon > T) \to 0$ if $\epsilon \to 0$, for some large T, and deduce that $E_x e^{\lambda \tau^\epsilon} \leq C < \infty$ for any $\lambda > 0$, provided ϵ is sufficiently small.]
16. Let (A) hold. Let $\{\mathcal{C}, \mathfrak{M}, \mathfrak{M}_t, \omega(t), P_x^\epsilon\}$ be the Markov process corresponding to $\xi^\epsilon(t)$, and let $\Omega = \mathcal{C}_T$ where $\mathcal{C}_T$ is the space of all continuous functions ω defined on $0 \leq t \leq T$ with values $\omega(t)$ in R^n. Let G be any open set in Ω. Prove

$$\liminf_{\epsilon \to 0,\, y \to x} \left[2\epsilon^2 \log P_y^\epsilon(G)\right] \geq -\inf_{\omega \in G_x} I_T(\omega)$$

where $G_x = \{\omega \in G;\ \omega(0) = x\}$.
17. Under the assumptions and notation of the preceding problem, for any closed set C of Ω,

$$\limsup_{\epsilon \to 0,\, y \to x} \left[2\epsilon^2 \log P_y^\epsilon(C)\right] \leq -\inf_{\omega \in C_x} I_T(\omega)$$

where $C_x = \{\omega \in C;\ \omega(0) = x\}$.
18. We outline a proof of Theorem 7.2 which does not require the condition (B_2)(iii). Some details are left to the reader. Take for simplicity $l = 1$. Let N be a neighborhood of Σ and let S_1 be a neighborhood of K with smooth boundary Γ_1 such that

$$V(z, y) < V + \frac{\eta}{6} \quad \text{if} \quad z \in \Gamma_1, \quad y \in \Sigma,$$

$$V(z, y) > V + \frac{7\eta}{8} \quad \text{if} \quad z \in \Gamma_1, \quad y \in N_0 \qquad (N_0 = \partial D \setminus N). \tag{11.20}$$

Choose T^* and a neighborhood S_2 of K with smooth boundary Γ_2 and with

$\bar{S}_2 \subset S_1$ such that for any $z \in \Gamma_1$ there is a curve $\varphi_z(t)$, $0 \leq t \leq T_z$ connecting z to Σ in time $T_z \leq T^*$ and

$$I_{T_z}(\varphi_z) \leq V + \frac{\eta}{5},$$

$$\varphi_z \quad \text{intersects} \quad \Gamma_1 \cup \Gamma_2 \quad \text{at most } m_0 \text{ times,}$$

where m_0 is a positive integer independent of z. If T is sufficiently large,

$$I_T(\varphi) \geq V + \eta \tag{11.21}$$

for any $\varphi \in C_T$, $\varphi(t) \in D \setminus S_2$ if $0 < t < T$. Take $T > T^*$ and extend φ_z to $0 \leq t \leq T$ so that

$$I_T(\varphi_z) < V + \frac{\eta}{4}, \tag{11.22}$$

and φ_z intersects $\Gamma_1 \cup \Gamma_2$ at most m times; m independent of z. Let

$$\tau_0 = \inf[s;\ \xi^\epsilon(s) \in \Gamma_2],$$
$$\sigma_k = \inf[s;\ s \geq \tau_{k-1},\ \xi^\epsilon(s) \in \Gamma_1],$$
$$\tau_k = \inf[s;\ s \geq \sigma_k,\ \xi^\epsilon(s) \in \Gamma_2],$$

$$E_0 = \{\tau^\epsilon < \tau_0\}, \qquad \text{and, for} \quad j \geq 1, \qquad E_j = \{\sigma_{(j-1)m+1} < \tau^\epsilon < \tau_{jm}\},$$

$$A_j = \{\xi^\epsilon(t) \text{ visits } \partial D \cap N \text{ between } \sigma_{(j-1)m+1} \text{ and } \tau_{jm}\},$$

$$B_j = \{\xi^\epsilon(t) \text{ visits } N_0 \text{ between } \sigma_{(j-1)m+1} \text{ and } \tau_{jm}\}.$$

Let $\alpha(\omega) = \inf\{j;\ \omega \in E_j\}$. Suffices to show $P_x^\epsilon[\omega \in B_{\alpha(\omega)}] \to 0$ if $\epsilon \to 0$. Let $\mathcal{F}_j = \mathcal{F}_{\sigma_{jm+1}}$. If $x \in \Gamma_1$

$$\begin{aligned} P_x^\epsilon(B_j|\mathcal{F}_{j-1}) &= \text{conditional probability that in at most} \\ &\quad\ m \text{ trips from } \Gamma_1 \text{ to } \Gamma_2 \text{ the process visits } N_0 \\ &\leq m \sup_{y \in \Gamma_1} P_y^\epsilon[\text{path visits } N_0 \text{ before visiting } \Gamma_2], \end{aligned}$$

$$P_y^\epsilon[\cdots] = P_y^\epsilon[\ldots, \tau^\epsilon < T] + P_y^\epsilon[\ldots, \tau^\epsilon \geq T].$$

Use Problem 17 and (11.20), (11.21) to conclude

$$P_x^\epsilon(B_j|\mathcal{F}_{j-1}) \leq \exp\left(-\frac{V + \frac{7}{8}\eta - h}{2\epsilon^2}\right) \qquad \text{for any} \quad h > 0,$$

uniformly in $x \in \Gamma_2$ and in small ϵ. Use Problem 16 and (11.22) to deduce

$$\begin{aligned} P_x^\epsilon(A_j|\mathcal{F}_{j-1}) &\geq \inf_{y \in \Gamma_1} P_y^\epsilon[\text{path visits } \partial D \cap N \text{ before it} \\ &\quad\ \text{makes } m \text{ visits from } \Gamma_1 \text{ to } \Gamma_2] \\ &\geq \exp\left(-\frac{V + \frac{1}{4}\eta + h}{2\epsilon^2}\right) \qquad \text{for any} \quad h > 0. \end{aligned}$$

Hence

$$\frac{P_x^\epsilon(B_j|\mathcal{F}_{j-1})}{P_x^\epsilon(A_j|\mathcal{F}_{j-1})} < \delta(\epsilon)\to 0 \qquad \text{if} \quad \epsilon\to 0.$$

If $x\in\Gamma_1$, use the strong Markov property to deduce

$$\begin{aligned}
P_x^\epsilon[\omega\in B_{\alpha(\omega)}] &= \sum_0^\infty P_x^\epsilon[\alpha(\omega) = j, B_j] \\
&< P_x^\epsilon(E_0) + \sum_1^\infty P_x^\epsilon[\alpha(\omega) > j, B_j] \\
&< P_x^\epsilon(E_0) + \delta(\epsilon)\sum_1^\infty P_x^\epsilon[\alpha(\omega) > j, A_j] \\
&= P_x^\epsilon(E_0) + \delta(\epsilon)\sum_1^\infty P^\epsilon[\alpha(\omega) = j, A_j] \\
&< P_x^\epsilon(E_0) + \delta(\epsilon)\sum_1^\infty P_x^\epsilon[\alpha(\omega) = j] \\
&< P_x^\epsilon(E_0) + \delta(\epsilon)\to 0 \qquad \text{if} \quad \epsilon\to 0.]
\end{aligned}$$

19. If (A) holds, then for any $T > 0$, $\delta > 0$ there exist positive constants ϵ_0, β such that

$$P_x\left\{\rho_T(\xi^\epsilon, \xi^0) > \delta\right\} < \exp\left[-\frac{\beta}{2\epsilon^2}\right].$$

[*Hint:* Apply Problem 17 with $C = \{\phi \in C_T;\ \phi(0) = x,\ \rho_T(\phi, \xi^0) > \delta\}$.]

15

Fundamental Solutions for Degenerate Parabolic Equations

1. Construction of a candidate for a fundamental solution

Consider a partial differential operator

$$Lu \equiv \frac{1}{2} \sum_{i,j=1}^{n} a_{ij}(x) \frac{\partial^2 u}{\partial x_i \, \partial x_j} + \sum_{i=1}^{n} b_i(x) \frac{\partial u}{\partial x_i} \qquad (a_{ij} = a_{ji}) \tag{1.1}$$

and suppose that

$$a_{ij} = \sum_{k=1}^{n} \sigma_{ik}\sigma_{jk}$$

for some $n \times n$ matrix $\sigma = (\sigma_{ij})$. Set $b = (b_1, \ldots, b_n)$ and introduce the system of n stochastic differential equations

$$d\xi(t) = \sigma(\xi(t))\, dw(t) + b(\xi(t))\, dt. \tag{1.2}$$

In Section 6.4 we have introduced the concept of fundamental solution and asserted the existence, smoothenss, and certain bounds for a fundamental solution Γ. The underlying assumptions were that $(a_{ij}(x))$ is uniformly positive definite and a_{ij}, b_i are bounded and uniformly Hölder continuous. We have also proved that if σ_{ij}, b_i are Lipschitz continuous, uniformly in compact sets, then, for any Borel set A,

$$P_x(\xi(t) \in A) = \int_A \Gamma(x, t, y)\, dy \tag{1.3}$$

where $\Gamma(x, t, y) = \Gamma(x, t; y, 0)$. Let L^* be the adjoint of L. If the coefficients of L^* are uniformly Hölder continuous, then it was proved that

$$\Gamma, \quad D_x\Gamma, \quad D_x^2\Gamma, \quad D_t\Gamma, \quad D_y\Gamma, \quad D_y^2\Gamma$$

are continuous and

$$L\Gamma - \frac{\partial \Gamma}{\partial t} = 0 \qquad \text{as a function in} \quad (x, t),$$

$$L^*\Gamma - \frac{\partial \Gamma}{\partial t} = 0 \qquad \text{as a function in} \quad (y, t).$$

In this chapter we consider the case where L is a degenerate elliptic operator. Thus the standard construction of a fundamental solution Γ breaks down. The concept of a fundamental solution Γ, if taken in the sense of (1.3), still makes sense, and we shall, in fact, prove that such a fundamental solution exists for a class of operators which degenerate on obstacles. The following condition will be needed:

(A) The functions

$$a_{ij}(x), \quad \frac{\partial}{\partial x_\lambda} a_{ij}(x), \quad \frac{\partial^2}{\partial x_\lambda\, \partial x_\mu} a_{ij}(x), \quad b_i(x), \quad \frac{\partial}{\partial x_\lambda} b_i(x)$$

are uniformly Hölder continuous in compact subsets of R^n.

Let S be a closed subset of R^n, and assume:

(B_S) The matrix $a_{ij}(x)$) is positive definite for any $x \notin S$, and positive semidefinite for any $x \in S$.

In the present section we shall construct a function $K(x, t, \xi)$ as a limit of fundamental solutions $K_\epsilon(x, t, \xi)$ for the parabolic equations

$$L_\epsilon u - \frac{\partial u}{\partial t} = 0, \qquad \text{where} \quad L_\epsilon = Lu + \epsilon \sum_{i=1}^{n} \frac{\partial^2 u}{\partial x_i^2} \qquad (\epsilon > 0). \quad (1.4)$$

In the following sections we shall show, under some conditions on S and on the coefficients of L, that a fundamental solution, if defined by (1.3), coincides with $K(x, t, \xi)$, at least away from S.

Let

$$B_m = \{x;\ |x| < m\}, \qquad m = 1, 2, \ldots .$$

Denote by $G_{m,\epsilon}(x, t, \xi)$ the Green function for (1.4) in the cylinder $Q_m = B_m \times (0, \infty)$ (see Section 14.4). Thus $G_{m,\epsilon}(x, t, \xi)$, its first t-derivative and its second x-derivatives are continuous in (x, t, ξ) for $x \in \bar{B}_m$, $t > 0$, $\xi \in \bar{B}_m$, and as a function of (x, t),

$$L_\epsilon G_{m,\epsilon}(x, t, \xi) - \frac{\partial}{\partial t} G_{m,\epsilon}(x, t, \xi) = 0 \quad \text{if} \quad (x, t) \in Q_m \qquad (\xi \text{ fixed in } B_m),$$

$$G_{m,\epsilon}(x, t, \xi) \to 0 \qquad \text{if} \quad t \to 0, \quad x \neq \xi, \quad x \in B_m,$$

$$G_{m,\epsilon}(x, t, \xi) = 0 \qquad \text{if} \quad t > 0, \quad x \in \partial B_m.$$

Finally, for any continuous function $f(\xi)$ with support in B_m, the function

$$u(x, t) = \int_{B_m} G_{m,\epsilon}(x, t, \xi) f(\xi)\, d\xi$$

satisfies:

$$L_\epsilon u(x, t) = 0 \quad \text{in } Q_m,$$
$$u(x, t) \to f(x) \quad \text{if} \quad t \to 0, \quad x \in B_m,$$
$$u(x, t) = 0 \quad \text{if} \quad t > 0, \quad x \in \partial B_m.$$

Denote by L^*, L_ϵ^* the adjoint operators of L, L_ϵ respectively. Denote by $G_{m,\epsilon}^*(x, t, \xi)$ the Green function for the equation

$$L_\epsilon^* u - \partial u/\partial t = 0$$

in Q_m. It can be shown (see Problem 1) that

$$G_{m,\epsilon}(x, t, \xi) = G_{m,\epsilon}^*(\xi, t, x). \tag{1.5}$$

It follows that as a function of (ξ, t),

$$L_\epsilon^* G_{m,\epsilon}(x, t, \xi) - \frac{\partial}{\partial t} G_{m,\epsilon}(x, t, \xi) = 0 \quad \text{if} \quad (\xi, t) \in Q_m \quad (x \text{ fixed in } B_m).$$

Lemma 1.1. *Let* (A) *hold. Then*: (i)

$$0 < G_{m,\epsilon}(x, t, \xi) < G_{m+1,\epsilon}(x, t, \xi) \text{ if } (x, t) \in Q_m, \quad \xi \in B_m, \tag{1.6}$$

$$\lim_{m\to\infty} G_{m,\epsilon}(x, t, \xi) \equiv K_\epsilon(x, t, \xi) \text{ is finite for all } x \in R^n, \quad t > 0, \quad \xi \in R^n. \tag{1.7}$$

(ii) *The functions*

$$K_\epsilon(x, t, \xi), \quad \frac{\partial}{\partial x_\lambda} K_\epsilon(x, t, \xi), \quad \frac{\partial^2}{\partial x_\lambda\, \partial x_\mu} K_\epsilon(x, t, \xi), \quad \frac{\partial}{\partial t} K_\epsilon(x, t, \xi)$$

are continuous in (x, t, ξ) *for* $x \in R^n$, $t > 0$, $\xi \in R^n$; *for any continuous function* $f(\xi)$ *with compact support, the function*

$$u(x, t) = \int_{R^n} K_\epsilon(x, t, \xi) f(\xi)\, d\xi \tag{1.8}$$

satisfies

$$\begin{aligned} L_\epsilon u - \frac{\partial u}{\partial t} &= 0 \quad \text{if} \quad x \in R^n, \quad t > 0, \\ u(x, t) &\to f(x) \quad \text{if} \quad t \to 0. \end{aligned} \tag{1.9}$$

(iii) *The functions*

$$\frac{\partial}{\partial \xi_\lambda} K_\epsilon(x, t, \xi), \quad \frac{\partial^2}{\partial \xi_\lambda\, \partial \xi_\mu} K_\epsilon(x, t, \xi)$$

are continuous in (x, t, ξ) *for* $x \in R^n$, $t > 0$, $\xi \in R^n$; *for any continuous function* $g(x)$ *with compact support, the function*

$$v(\xi, t) = \int_{R^n} K_\epsilon(x, t, \xi) g(x)\, dx \tag{1.10}$$

satisfies

$$\begin{aligned} L_\epsilon^* v - \frac{\partial v}{\partial t} &= 0 \qquad \textit{if} \quad \xi \in R^n, \quad t > 0, \\ v(\xi, t) &\to g(\xi) \qquad \textit{if} \quad t \to 0. \end{aligned} \tag{1.11}$$

In view of (1.9), $K_\epsilon(x, t, \xi)$ is a fundamental solution for (1.4).

Proof. The inequalities in (1.6) are an easy consequence of the maximum principle. In fact, for any continuous and nonnegative function $f_k(\xi)$ with support in B_m,

$$0 \leq \int_{B_m} G_{m,\epsilon}(x, t, \xi) f_k(\xi)\, d\xi \leq \int_{B_{m+1}} G_{m+1,\epsilon}(x, t, \xi) f_k(\xi)\, d\xi$$

by the maximum principle. Taking a sequence $\{f_k\}$ converging to the Dirac measure at ξ^0, the inequalities in (1.6), at $\xi = \xi^0$, follow.

Again, by the maximum principle,

$$\int_{B_m} G_{m,\epsilon}(x, t, \xi)\, d\xi \leq 1. \tag{1.12}$$

Similarly

$$\int_{B_m} G_{m,\epsilon}(x, t, \xi)\, dx \leq 1. \tag{1.13}$$

Now fix a positive integer m. Denote by $\partial/\partial T_\zeta$ the inward conormal derivative to ∂B_m at ζ, i.e., the derivative in the direction of the vector $\sum a_{ij}\nu_j$ $(1 \leq i \leq n)$ where ν is the inward normal. By Green's formula (see Problem 2), for any positive integer $k, k > m$,

$$\begin{aligned} G_{k,\epsilon}(x, t, \xi) = &\int_{B_m} G_{m,\epsilon}(x, s, \zeta) G_{k,\epsilon}(\zeta, t - s, \xi)\, d\zeta \\ &+ \int_0^s \int_{\partial B_m} \frac{\partial}{\partial T_\zeta} G_{m,\epsilon}(x, \sigma, \zeta) \cdot G_{k,\epsilon}(\zeta, t - s + \sigma, \xi)\, dS_\zeta\, d\sigma \end{aligned} \tag{1.14}$$

for any $0 < s < t$, $x \in B_m$, $\xi \in B_m$. Taking $s = t/2$ and using the estimates (see Problem 3)

$$G_{m,\epsilon}(x, t/2, \zeta) \leq C_m \qquad (x \in B_m, \xi \in B_m), \tag{1.15}$$

$$\left| \frac{\partial}{\partial T_\zeta} G_{m,\epsilon}(x, \sigma, \zeta) \right| \leq C_m \qquad (\zeta \in \partial B_m, x \in K, 0 < \sigma < s) \tag{1.16}$$

where K is a compact subset of B_m (C_m depends on m, ϵ, t, K), we get

$$G_{k,\epsilon}(x, t, \xi) \leq C_m \int_{B_m} G_{k,\epsilon}\left(\zeta, \frac{t}{2}, \xi\right) d\zeta + C_m \int_{t/2}^{t} \int_{\partial B_m} G_{k,\epsilon}(\zeta, \sigma, \xi)\, dS_\zeta\, d\sigma$$
$$\leq C_m + C_m \int_{t/2}^{t} \int_{\partial B_m} G_{k,\epsilon}(\zeta, \sigma, \xi)\, dS_\zeta\, d\sigma, \tag{1.17}$$

where (1.13) has been used. If we replace the ball B_m by a ball $B_{m+\lambda}$ ($0 < \lambda < 1$) with center 0 and radius $m + \lambda$, and Green's function $G_{m,\epsilon}$ by the corresponding Green function $G_{m+\lambda,\epsilon}$, then the constants $C_{m+\lambda}$ will remain bounded, independently of λ. In fact, this can be verified as follows: If $|x - \zeta| \geq c > 0$, $0 < s \leq T$, or if $0 < c_0 \leq s \leq T$, the inequality

$$G_{m+\lambda,\epsilon}(x, s, \zeta) \leq C \qquad (C \text{ depends on } c, c_0, \epsilon, T \text{ but not on } \lambda) \tag{1.18}$$

follows from the construction of $G_{m+\lambda,\epsilon}$. For fixed x, the function

$$v(\zeta, s) = G_{m+\lambda,\epsilon}(x, s, \zeta)$$

satisfies

$$\begin{aligned} L_\epsilon^* v - \frac{\partial v}{\partial s} &= 0 \qquad \text{in } B_{m+\lambda} \times (0, \infty), \\ v(\zeta, s) &= 0 \qquad \text{if } \zeta \in \partial B_{m+\lambda}, \quad s > 0. \end{aligned} \tag{1.19}$$

By (1.18), if x varies in a compact set K, $K \subset B_m$, if $0 < s < T$ and if ζ varies in a $B_{m+\lambda}$-neighborhood V of $\partial B_{m+\lambda}$ such that $K \cap \bar{V} = \emptyset$, then $v \leq C$. Using this fact and (1.19), we deduce (see Problem 4)

$$\left| \frac{\partial}{\partial \zeta_i} G_{m+\lambda,\epsilon}(x, s, \zeta) \right| \leq C \tag{1.20}$$

if $x \in K$, $0 < s < T$, $\zeta \in V$. From this inequality and (1.18) we see that, analogously to (1.17), we have

$$G_{k,\epsilon}(x, t, \xi) \leq C_{m+\lambda} + C_{m+\lambda} \int_{t/2}^{t} \int_{\partial B_{m+\lambda}} G_{k,\epsilon}(\zeta, \sigma, \xi)\, dS_\zeta\, d\sigma, \quad C_{m+\lambda} \leq C_m^* \tag{1.21}$$

where the constant C_m^* is independent of λ, provided $x \in K$, $\xi \in B_m$, $t > 0$. The constant C_m^* may depend on t. However, as the proof of (1.21) shows, if $t_0 \leq t \leq T_0$ where $t_0 > 0$, $T_0 > 0$, then C_m^* can be taken to depend on t_0, T_0, but not on t.

Integrating both sides of (1.21) with respect to λ, $0 < \lambda < 1$, we get

$$G_{k,\epsilon}(x, t, \xi) \leq C_m^* + C_m^* \int_{t/2}^{t} \int_{D_m} G_{k,\epsilon}(\zeta, \sigma, \xi)\, d\zeta\, d\sigma,$$

where D_m is the shell $\{x;\ m < |x| < m + 1\}$. Using (1.13), we conclude that

$$G_{k,\epsilon}(x, t, \xi) \leq C_m^{**} \qquad \text{if } x \in K, \quad \xi \in B_m, \quad t_0 \leq t \leq T_0 \tag{1.22}$$

where C_m^{**} is a constant independent of k. Combining this with (1.6), the assertion (1.7) follows.

The inequality (1.22) for m replaced by $m+1$ and $K = \bar{B}_m$ shows that the family $\{G_{k,\epsilon}(x, t, \xi)\}$ (for $k > m$) is uniformly bounded for $x \in B_m$, $\xi \in B_m$, $t_0 < t < T_0$. We can employ the Schauder-type interior estimates, considering the $G_{k,\epsilon}$ first as functions of (x, t) and then as functions of (ξ, t). We conclude that there is a subsequence that is uniformly convergent to a function $G_\epsilon(x, t, \xi)$ with the corresponding derivatives

$$\frac{\partial}{\partial x_\lambda}, \quad \frac{\partial^2}{\partial x_\lambda \, \partial x_\mu}, \quad \frac{\partial}{\partial t}, \quad \frac{\partial}{\partial \xi_\lambda}, \quad \frac{\partial^2}{\partial \xi_\lambda \, \partial \xi_\mu}, \tag{1.23}$$

in compact subsets of $\{(x, t, \xi); x \in B_m, t_0 < t < T_0, \xi \in B_m\}$. Since however the entire sequence $\{G_{k,\epsilon}(x, t, \xi)\}$ is convergent to $K_\epsilon(x, t, \xi)$, the same is true of the entire sequence of each of the partial derivatives of (1.23). It follows that the function $K_\epsilon(x, t, \xi)$ and its derivatives

$$\frac{\partial}{\partial x_\lambda} K_\epsilon, \quad \frac{\partial^2}{\partial x_\lambda \, \partial x_\mu} K_\epsilon, \quad \frac{\partial}{\partial t} K_\epsilon, \quad \frac{\partial}{\partial \xi_\lambda} K_\epsilon, \quad \frac{\partial^2}{\partial \xi_\lambda \, \partial \xi_\mu} K_\epsilon$$

are continuous in (x, t, ξ) for x, ξ in R^n and $t > 0$. Further, as a function of (x, t),

$$L_\epsilon K_\epsilon - \frac{\partial}{\partial t} K_\epsilon = 0 \qquad (\xi \text{ fixed}),$$

and as a function of (ξ, t)

$$L_\epsilon^* K_\epsilon - \frac{\partial}{\partial t} K_\epsilon = 0 \qquad (x \text{ fixed}).$$

Consequently, the functions u, v defined in (1.8), (1.10) satisfy the parabolic equations of (1.9), (1.11), respectively. It remains to show that

$$u(x, t) \to f(x) \quad \text{if} \quad t \to 0, \tag{1.24}$$

$$v(x, t) \to g(x) \quad \text{if} \quad t \to 0. \tag{1.25}$$

Note that (1.6), (1.7), (1.12), (1.13) imply that

$$\int_{R^n} K_\epsilon(x, t, \xi)\, d\xi \leq 1, \qquad \int_{R^n} K_\epsilon(x, t, \xi)\, dx \leq 1. \tag{1.26}$$

We proceed to prove (1.24). Let the support of f be contained in some ball B_m. Suppose first that $f \in C^3$. For $k > m$, consider the functions

$$u_k(x, t) = \int_{R^n} G_{k,\epsilon}(x, t, \xi) f(\xi)\, d\xi.$$

The uniform convergence of $\{G_{k,\epsilon}(x, t, \xi)\}$ to $K_\epsilon(x, t, \xi)$ implies that $u_k(x, t)$

$\to u(x, t)$ for any $x \in R^n$, $t > 0$. Notice next that

$$|u_k(x, t)| \le (\sup|f|) \int_{R^n} G_{k,\epsilon}(x, t, \xi)\, d\xi \le \sup|f|,$$

$$u_k(x, 0) = f(x) \qquad \text{is a } C^3 \text{ function.}$$

Hence the Schauder-type boundary estimates for the parabolic operator $L_\epsilon - \partial/\partial t$ (see Remark 1 at the end of Section 10.1) imply that the sequence $\{u_k(x, t)\}$ is uniformly convergent (with its second x-derivatives) for $x \in B_m$, $t \ge 0$. It follows that $u(x, t)$ $(t > 0)$ has a continuous extension $u(x, 0)$ to $t = 0$ and

$$u(x, 0) = \lim_{k\to\infty} u_k(x, 0) = f(x).$$

If f is only assumed to be continuous, let f_i be C^3 functions such that

$$\gamma_i \equiv \sup_{x \in R^n} |f_i(x) - f(x)| \to 0 \qquad \text{if} \quad i \to \infty,$$

and such that the support of each f_i is in B_m. Then

$$\int_{B_m} |K_\epsilon(x, t, \xi)[f_i(\xi) - f(\xi)]|\, d\xi \le \gamma_i \int_{B_m} K_\epsilon(x, t, \xi)\, d\xi \le \gamma_i$$

by (1.26). Also, by what we have already proved,

$$\delta_i(t) \equiv \left| \int_{B_m} K_\epsilon(x, t, \xi) f_i(\xi)\, d\xi - f_i(x) \right| \to 0 \qquad \text{if} \quad t \to 0 \qquad (i \text{ fixed}).$$

It follows that

$$\overline{\lim_{t\to 0}}\, |u(x, t) - f(x)| \le 2\gamma_i + \lim_{t\to 0} \delta_i(t) = 2\gamma_i.$$

Since $\gamma_i \to 0$ if $i \to \infty$, the assertion (1.24) follows. The proof of (1.25) is similar. This completes the proof of Lemma 1.1.

Theorem 1.2. *Let* (A), (B_S) *hold. Then there exists a sequence* $\epsilon_m \downarrow 0$ *such that, as* $m \to \infty$,

$$K_{\epsilon_m}(x, t, \xi) \to K(x, t, \xi) \tag{1.27}$$

together with the first two x*-derivatives, the first two* ξ*-derivatives, and the first* t*-derivative uniformly for all* x, ξ *in* E, $\delta < t < 1/\delta$, *where* E *is any compact set in* R^n *such that* $\text{E} \cap \text{S} = \varnothing$, *and* δ *is any positive number,* $0 < \delta < 1$.

Proof. Let E_0 be a compact set that does not intersect S.

Let B_λ $(0 \le \lambda \le 1)$ be a family of bounded open sets such that $\bar{B}_\lambda \subset B_{\lambda'}$ if $\lambda < \lambda'$, $E_0 \subset B_0$, $\bar{B}_1 \cap \text{S} = \varnothing$, and such that as λ varies from 0 to 1 the boundary ∂B_λ covers simply a finite disjoint union D of domains, and $dx = \rho\, dS^\lambda\, d\lambda$, where dS^λ is the surface element of ∂B_λ and ρ is a positive continuous function. It is assumed that each ∂B_λ consists of a finite number of C^3 hypersurfaces.

Taking $k \to \infty$ in (1.14) and using the monotone convergence theorem, we obtain the relation (1.14) with $G_{k,\epsilon}$ replaced by K_ϵ. This relation holds also with B_m replaced by B_λ and $G_{m,\epsilon}$ replaced by Green's function $G_{\epsilon,\lambda}$ of $L_\epsilon - \partial/\partial t$ in the cylinder $B_\lambda \times (0, \infty)$. The estimates (cf. (1.18), (1.20))

$$G_{\epsilon,\lambda}(x, t, \zeta) < C_\epsilon \qquad (x \in E_0,\ \zeta \in B_\lambda,\ t_0 < t < T_0), \quad (1.28)$$

$$\left| \frac{\partial}{\partial T_\zeta} G_{\epsilon,\lambda}(x, t, \zeta) \right| < C_\epsilon \qquad (x \in E_0,\ \zeta \in \partial B_\lambda,\ 0 < t < T) \quad (1.29)$$

hold, where $t_0 > 0$, $T_0 < \infty$. Since $(a_{ij}(x))$ is positive definite for $x \in \bar{B}_1$, the constants C_ϵ can be taken to be independent of both ϵ and λ; the proof is similar to the proof of (1.18), (1.20). It follows that if $x \in E_0$, $\xi \in E_0$, $t_0 < t < T_0$,

$$\begin{aligned} K_\epsilon(x, t, \xi) &< C^* \int_{B_\lambda} K_\epsilon\left(\zeta, \frac{t}{2}, \zeta\right) d\xi \\ &\quad + C^* \int_0^{t/2} \int_{\partial B_\lambda} K_\epsilon\left(\zeta, \frac{t}{2} + \sigma, \xi\right) dS_\zeta^\lambda \, d\sigma \\ &< C^* + C^* \int_0^{t/2} \int_{\partial B_\lambda} K_\epsilon\left(\zeta, \frac{t}{2} + \sigma, \xi\right) dS_\zeta^\lambda \, d\sigma \end{aligned} \quad (1.30)$$

where C^* is a constant independent of ϵ, λ; (1.26) has been used here. Integrating with respect to λ and using (1.26), we find that

$$K_\epsilon(x, t, \xi) < C \qquad (C \text{ independent of } \epsilon). \quad (1.31)$$

This bound is valid x, ξ in E_0 and $t \in [t_0, T_0]$; the constant C depends on E_0, t_0, T_0, but not on ϵ.

From the Schauder-type interior estimates applied to $K_\epsilon(x, t, \xi)$ first as a function of (x, t) and then as a function of (ξ, t) we conclude, upon using (1.31), that

$$K_\epsilon(x, t, \xi), \quad \frac{\partial}{\partial x_\lambda} K_\epsilon(x, t, \xi), \quad \frac{\partial^2}{\partial x_\lambda \, \partial x_\mu} K_\epsilon(x, t, \xi),$$

$$\frac{\partial}{\partial t} K_\epsilon(x, t, \xi), \quad \frac{\partial}{\partial \xi_\lambda} K_\epsilon(x, t, \xi), \quad \frac{\partial^2}{\partial \xi_\lambda \, \partial \xi_\mu} K_\epsilon(x, t, \xi)$$

satisfy a uniform Hölder condition in (x, t, ξ) when $x \in E'$, $\xi \in E'$, $t_0 + \delta < t < T_0 - \delta$ for any $\delta > 0$, where E' is any set in the interior of E_0; the Hölder constants are independent of ϵ (since $(a_{ij}(x))$ is positive definite for $x \in E_0$). Since E_0, t_0, T_0 are arbitrary, we conclude, by diagonalization, that there is a sequence $\{\epsilon_m\}$, $\epsilon_m \to 0$ if $m \to \infty$, such that

$$K(x, t, \xi) \equiv \lim_{m \to \infty} K_{\epsilon_m}(x, t, \xi)$$

exists, and the convergence is uniform together with the convergence of the respective first two x-derivatives, first two ξ-derivatives, and first t-derivative, for all x, ξ in any compact set E, $E \cap S = \varnothing$, and for all t, $\delta < t < 1/\delta$, where δ is any positive number.

Corollary 1.3. *The function $K(x, t, \xi)$ satisfies*: (i) *as a function of* (x, t), $LK(x, t, \xi) - \partial K(x, t, \xi)/\partial t = 0$, *and* (ii) *as a function of* (ξ, t), $L^*K(x, t, \xi) - \partial K(x, t, \xi)/\partial t = 0$, *for all* $x \notin S$, $\xi \notin S$, $t > 0$.

2. Interior estimates

We denote by D_x the vector $(\partial/\partial x_1, \ldots, \partial/\partial x_n)$.

Lemma 2.1. *Let* (A), (B_S) *hold. Let B be a bounded domain with C^2 boundary ∂B, and let $\bar{B} \cap S = \varnothing$. Denote by $G_{B,\epsilon}(x, t, \xi)$ the Green function of $L_\epsilon - \partial/\partial t$ in the cylinder $B \times (0, \infty)$. Then, for any compact subset B_0 of B and for any $\epsilon_0 > 0$, $T > 0$,*

$$G_{B,\epsilon}(x, t, \xi) \leq \frac{C}{t^{n/2}} \quad \text{if} \quad (x, \xi) \in (B \times B_0) \cup (B_0 \times B), \quad 0 < t \leq T, \tag{2.1}$$

$$G_{B,\epsilon}(x, t, \xi) \leq Ce^{-c/t} \quad \text{if} \quad (x, \xi) \in (B \times B_0) \cup (B_0 \times B), \quad |x - \xi| \geq \epsilon_0, \quad 0 < t \leq T, \tag{2.2}$$

$$|D_x G_{B,\epsilon}(x, t, \xi)| \leq Ce^{-c/t} \quad \text{if} \quad (x, \xi) \in B \times B_0, \quad |x - \xi| \geq \epsilon_0, \ 0 < t < T, \tag{2.3}$$

$$|D_\xi G_{B,\epsilon}(x, t, \xi)| \leq Ce^{-c/t} \quad \text{if} \quad (x, \xi) \in B_0 \times B, \quad |x - \xi| \geq \epsilon_0, \ 0 < t < T, \tag{2.4}$$

where C, c are positive constants depending on B, B_0, ϵ_0, T but independent of ϵ.

Proof. We write (cf. Section 14.4)

$$G_{B,\epsilon}(x, t, \xi) = \Gamma_\epsilon(x, t, \xi) + V_\epsilon(x, t, \xi) \tag{2.5}$$

where $\Gamma_\epsilon(x, t, \xi)$ is a fundamental solution for $L_\epsilon - \partial/\partial t$ in a cylinder $Q = B' \times (0, \infty)$ and B' is an open neighborhood of $\bar{B}$ such that its closure does not intersect S. Since L is nondegenerate outside S, the construction of Γ_ϵ can be carried out as in Friemdan [1] and (cf. (6.4.12))

$$|\Gamma_\epsilon(x, t, \xi)| + |D_x\Gamma_\epsilon(x, t, \xi)| \leq Ce^{-c/t} \qquad \text{if} \quad |x - \xi| \geq \epsilon_0 > 0, \quad 0 < t < T; \tag{2.6}$$

the positive constants C, c can be taken to be independent of ϵ. Notice also that

$$|\Gamma_\epsilon(x, t, \xi)| \leq \frac{C}{t^{n/2}} \qquad \text{if} \quad 0 < t < T. \tag{2.7}$$

By the methods of Friedman [1] one can actually also prove that

$$|D_x^2\Gamma_\epsilon(x, t, \xi)| + |D_t\Gamma_\epsilon(x, t, \xi)| \leq Ce^{-c/t} \qquad \text{if} \quad |x - \xi| \geq \epsilon_0 > 0, \quad 0 < t < T. \tag{2.8}$$

The points (x, ξ) in (2.6)–(2.8) vary in B'.

The function $V_\epsilon(x, t, \xi)$, for fixed ξ in B, satisfies

$$L_\epsilon V_\epsilon - \frac{\partial}{\partial t} V_\epsilon = 0 \qquad \text{if} \quad x \in B, \quad 0 < t < T,$$

$$V_\epsilon(x, t, \xi) = -\Gamma_\epsilon(x, t, \xi) \qquad \text{if} \quad x \in \partial B, \quad 0 < t < T,$$

$$V_\epsilon(x, 0, \xi) = 0 \qquad \text{if} \quad x \in B.$$

If ξ remains in a compact subset E of B, then, by (2.6) and the maximum principle,

$$|V_\epsilon(x, t, \xi)| \leq Ce^{-c/t} \qquad (x \in B, \quad \xi \in E, \quad 0 < t < T). \tag{2.9}$$

This inequality together with (2.5)–(2.7) imply (2.1), (2.2) for $(x, \xi) \in B \times B_0$. Since similar inequalities hold for Green's function $G^*_{B,\epsilon}(x, t, \xi)$ of $L^*_\epsilon - \partial/\partial t$, and since $G_{B,\epsilon}(x, t, \xi) = G^*_{B,\epsilon}(\xi, t, x)$, the inequalities (2.1), (2.2) follow also when $(x, \xi) \in B_0 \times B$.

From (2.6), (2.8) we see that for any ξ in a compact subset E of B there is a function $f(x, t)$ that coincides with $-\Gamma_\epsilon(x, t, \xi)$ for $x \in \partial B$, $0 < t < T$, and which satisfies

$$|f(x, t)| + |D_t f(x, t)| + |D_x f(x, t)| + |D_x^2 f(x, t)| \leq C^* e^{-c/t} \qquad (x \in B, \quad 0 < t < T)$$

where C^* is a constant independent of ξ, ϵ. We use here the fact that ∂B is in C^2. Notice that

$$\begin{aligned} L_\epsilon(V_\epsilon - f) - \frac{\partial}{\partial t}(V_\epsilon - f) &= -L_\epsilon f + \frac{\partial f}{\partial t} \equiv \tilde{f}, \\ |\tilde{f}(x, t)| \leq C^{**}e^{-c/t} &\qquad (x \in B, \quad 0 < t < T), \end{aligned} \tag{2.10}$$

$$V_\epsilon - f = 0 \qquad \text{if} \quad x \in \partial B, \quad 0 < t < T \qquad \text{or if} \quad x \in B, \quad t = 0;$$

the constant C^{**} is independent of ϵ. Using this one can show (see Problem 5) that

$$|D_x V_\epsilon(x, t, \xi)| \leq C_1 C^{**} e^{-c/t} \qquad \text{if} \quad x \in B, \quad 0 < t < T, \tag{2.11}$$

where C_1 is a constant independent of ϵ. Recalling (2.5), (2.6), the assertion

(2.3) follows. A similar inequality holds for Green's function $G^*_{B,\epsilon}$; since $G_{B,\epsilon}(x, t, \xi) = G^*_{B,\epsilon}(\xi, t, x)$, this inequality gives (2.4).

Theorem 2.2. *Let* (A), (B_S) *hold. Let E be any compact subset in R^n such that $E \cap S = \varnothing$ and let ϵ_0, T be any positive numbers. Then*

$$K(x, t, \xi) < \frac{C}{t^{n/2}} \quad \text{if} \quad x \in E, \quad \xi \in E, \quad 0 < t < T, \tag{2.12}$$

$$K(x, t, \xi) < Ce^{-c/t} \quad \textit{if} \quad x \in E, \quad \xi \in E, \quad |x - \xi| > \epsilon_0, \quad 0 < t < T, \tag{2.13}$$

where C, c are positive constants.

Proof. Let B_λ $(0 < \lambda < 1)$ be an increasing family of bounded open sets with C^2 boundary, as in the proof of Theorem 1.2. Let F be a compact subset of B_0. Recall that $\bar{B}_1 \cap S = \varnothing$. We proceed as in the proof of Theorem 1.2 to employ the relation (1.14) with B_m replaced by B_λ and with $G_{m,\epsilon}$ replaced by $G_{B_\lambda,\epsilon}$:

$$G_{k,\epsilon}(x, t, \xi) = \int_{B_\lambda} G_{B_\lambda,\epsilon}(x, s, \zeta) G_{k,\epsilon}(\zeta, t - s, \xi)\, d\zeta$$
$$+ \int_0^s \int_{\partial B_\lambda} \frac{\partial}{\partial T_\zeta} G_{B_\lambda,\epsilon}(x, \sigma, \zeta) \cdot G_{k,\epsilon}(\zeta, t - s + \sigma, \xi)\, dS_\zeta^\lambda\, d\sigma. \tag{2.14}$$

From the proof of Lemma 2.1 we see that the estimates (2.1)–(2.4) hold for $G_{B_\lambda,\epsilon}$ with constants C, c independent of λ. Using (2.1), (2.4) for $B = B_\lambda$ in (2.14), we obtain, after applying the inequality (1.13) for $m = k$, integrating with respect to λ $(0 < \lambda < 1)$ and applying once more (1.13) with $m = k$,

$$G_{k,\epsilon}(x, t, \xi) < \frac{C}{t^{n/2}} \quad \text{provided} \quad x \in F, \quad \xi \in F, \quad 0 < t < T.$$

Taking $k \to \infty$, we get

$$K_\epsilon(x, t, \xi) < \frac{C}{t^{n/2}} \quad \text{if} \quad x \in F, \quad \xi \in F, \quad 0 < t < T. \tag{2.15}$$

Taking $\epsilon = \epsilon_m \to \infty$, the inequality (2.12) follows.

To prove (2.13), let A, F be disjoint compact domains, $(A \cup F) \cap S = \varnothing$, and let ∂F be in C^2. Consider the function

$$v_\epsilon(x, t) = K_\epsilon(x, t, \xi) \quad \text{for} \quad x \in F, \quad 0 < t < T \qquad (\xi \text{ fixed in } A).$$

Denote by $G_{F,\epsilon}(x, t, \xi)$ the Green function of $L_\epsilon - \partial/\partial t$ in $F \times (0, \infty)$. By Lemma 2.1,

$$|D_\zeta G_{F,\epsilon}(x, t, \zeta)| < Ce^{-c/t} \quad \text{if} \quad \zeta \in \partial F, \quad x \in F_0, \quad 0 < t < T, \tag{2.16}$$

where F_0 is any compact subset in the interior of F.

We have the following representation for $v_\epsilon(x, t)$:

$$v_\epsilon(x, t) = \int_0^t \int_{\partial F} \frac{\partial}{\partial T_\zeta} G_{F,\epsilon}(x, s, \zeta) \cdot v_\epsilon(\zeta, s)\, dS_\zeta\, ds$$

$$(x \in \operatorname{int} F, \quad 0 < t < T). \tag{2.17}$$

Indeed, this formula is valid for $v_{k,\epsilon}(x, t) \equiv G_{k,\epsilon}(x, t, \xi)$ since $v_{k,\epsilon}(x, 0) = 0$. Taking $k \to \infty$ and using the monotone convergence theorem, (2.17) follows.

Substituting the estimates (2.15), (2.16) into the right-hand side of (2.17), we obtain

$$v_\epsilon(x, t) \leq \frac{C'}{t^{n/2}} e^{-c'/t} \leq Ce^{-c/t}$$

where C', c', C, c are positive constants independent of ϵ. Taking $\epsilon = \epsilon_m \to 0$, the assertion (2.13) follows.

3. Boundary estimates

We shall need the condition:

(C) There is a finite number of disjoint sets $G_1, \ldots, G_{k_0}, G_{k_0+1}, \ldots, G_k$ such that each G_i $(1 \leq i \leq k_0)$ consists of one point z_i and each G_j $(k_0 + 1 \leq j \leq k)$ is a bounded closed domain with C^3 connected boundary ∂G_j. Further,

$$a_{ij}(z_l) = 0, \qquad b_i(z_l) = 0 \qquad \text{if} \quad 1 \leq l \leq k_0; \quad 1 \leq i, j \leq n, \tag{3.1}$$

$$\sum_{i,j=1}^n a_{ij}(x)\nu_i\nu_j = 0 \qquad \text{for} \quad x \in \partial G_j \qquad (k_0 + 1 \leq j \leq k), \tag{3.2}$$

$$\sum_{i=1}^n \left(b_i(x) - \frac{1}{2} \sum_{j=1}^n \frac{\partial a_{ij}(x)}{\partial x_j} \right) \nu_i \leq 0 \qquad \text{for} \quad x \in \partial G_j \qquad (k_0 + 1 \leq j \leq k) \tag{3.3}$$

where $\nu = (\nu_1, \ldots, \nu_n)$ is the outward normal to ∂G_j at x.

Let $\Omega = \bigcup_{j=1}^k G_j$, $\hat{\Omega} = R^n \setminus \Omega$, $\partial G_j = G_j = \{z_j\}$ if $1 \leq j \leq k_0$, $\partial\Omega = \bigcup_{j=1}^k \partial G_j$. In this section, and in Sections 6–10, we shall assume that

$$S = \partial\Omega. \tag{3.4}$$

Let $\{N_m\}$ be a sequence of domains with C^3 boundary ∂N_m, such that $\bar{N}_m \subset N_{m+1} \subset \hat{\Omega}$, $\bigcup_m N_m = \hat{\Omega}$. We take N_m such that ∂N_m consists of two disjoint parts: $\partial_1 N_m$ which lies in $(1/m)$-neighborhood of $\partial\Omega$ and $\partial_2 N_m$ which is the sphere $|x| = m$.

Denote by $G_m(x, t, \xi)$ the Green function for $L - \partial/\partial t$ in $N_m \times (0, \infty)$. By arguments similar to those used in the proofs of Lemma 1.1 and Theorem

2.2, we have:

$$0 \leq G_m(x, t, \xi) \leq G_{m+1}(x, t, \xi), \tag{3.5}$$

$$G(x, t, \xi) = \lim_{m\to\infty} G_m(x, t, \xi) \quad \text{is finite} \tag{3.6}$$

for all x, ξ in $\hat{\Omega}$, $t > 0$. Further

$$G_m(x, t, \xi) \leq \frac{C}{t^{n/2}} \qquad \text{if} \quad x \in E, \quad \xi \in E, \quad 0 < t < T, \tag{3.7}$$

$$G_m(x, t, \xi) \leq Ce^{-c/t} \qquad \text{if} \quad x \in E, \quad \xi \in E, \quad |x - \xi| \geq \epsilon_0, \quad 0 < t < T, \tag{3.8}$$

$$G(x, t, \xi) \leq \frac{C}{t^{n/2}} \qquad \text{if} \quad x \in E, \quad \xi \in E, \quad 0 < t < T, \tag{3.9}$$

$$G(x, t, \xi) \leq Ce^{-c/t} \qquad \text{if} \quad x \in E, \quad \xi \in E, \quad |x - \xi| \geq \epsilon_0, \quad 0 < t < T, \tag{3.10}$$

where E is any compact set such that $E \subset \hat{\Omega}$, T, and ϵ_0 are any positive numbers, and C, c are positive constants depending on E, ϵ_0, T but independent of m. We also have, by the strong maximum principle, that $G(x, t, \xi) > 0$ if $x \in \hat{\Omega}$, $t > 0$, $\xi \in \hat{\Omega}$. Finally,

$$LG(x, t, \xi) - \frac{\partial}{\partial t} G(x, t, \xi) = 0 \qquad \text{if} \quad x \in \hat{\Omega}, \quad t > 0 \qquad (\xi \text{ fixed in } \hat{\Omega}), \tag{3.11}$$

$$L^*G(x, t, \xi) - \frac{\partial}{\partial t} G(x, t, \xi) = 0 \qquad \text{if} \quad \xi \in \hat{\Omega}, \quad t > 0 \qquad (x \text{ fixed in } \hat{\Omega}). \tag{3.12}$$

Notice that in proving (3.5)–(3.12) we do not use the conditions (3.1)–(3.3).

Denote by $R(x)$ the distance from $x \in \hat{\Omega}$ to the set Ω. This function is in C^2 in some $\hat{\Omega}$-neighborhood of $\partial\Omega$ and also up to the boundary $\bigcup_{j=k_0+1}^{k} \partial G_j$.

Theorem 3.1. *Let* (A), (B_S), (C), *and* (3.4) *hold. Let E be any compact subset of $\hat{\Omega}$. Then, for any $T > 0$ and for any $\rho > 0$ sufficiently small, there are positive constants C, γ such that*

$$G(x, t, \xi) \leq C \exp\left\{-\frac{\gamma}{t}(\log R(x))^2\right\} \tag{3.13}$$

if $\xi \in E$, $x \in \hat{\Omega}$, $R(x) < \rho$, $0 < t < T$.

Corollary 3.2. *If in Theorem* 3.1 *the condition* (3.3) *is replaced by the* *condition*

$$\sum_{i=1}^{n}\left[b_i(x) - \tfrac{1}{2}\sum_{j=1}^{n}\frac{\partial a_{ij}(x)}{\partial x_j}\right]\nu_i \geq 0 \quad \textit{for} \quad x \in \partial G_j \qquad (k_0 + 1 \leq j \leq k), \tag{3.14}$$

then

$$G(x, t, \xi) \leq C \exp\left\{-\frac{\gamma}{t}(\log R(\xi))^2\right\} \tag{3.15}$$

if $x \in E$, $\xi \in \hat{\Omega}$, $R(\xi) < \rho$, $0 < t < T$.

The point of these results will become obvious when, in Section 6, we shall prove that

$$K(x, t, \xi) = G(x, t, \xi) \qquad \text{if} \quad x \in \hat{\Omega}, \quad \xi \in \hat{\Omega}, \quad t > 0.$$

Proof of Theorem 3.1. For any $\epsilon > 0$, denote by M_ϵ the set of all points $x \in \hat{\Omega}$ for which $R(x) < \epsilon$, and by Γ_ϵ the set of all points $x \in \hat{\Omega}$ with $R(x) = \epsilon$. The number ϵ is such that $E \cap \overline{M}_\epsilon = \emptyset$ and $R(x)$ is in $C^2(M_\epsilon)$; later we shall impose another restriction on the size of ϵ (depending only on the coefficients of L).

Let $M_{\epsilon,m} = M_\epsilon \cap N_m$. Its boundary $\partial M_{\epsilon,m}$ consists of Γ_ϵ and of $\partial_1 N_m$ (the "inner" boundary of N_m), provided m is sufficiently large, say $m \geq m_0(\epsilon)$.

For $m \geq m_0(\epsilon)$, consider the function

$$v(x, t) = G_m(x, t, \xi) \qquad \text{for} \quad x \in M_{\epsilon,m}, \qquad 0 < t < T \qquad (\xi \text{ fixed in } E).$$

If $x \in \partial_1 N_m$, $v(x, t) = 0$. If $x \in \Gamma_\epsilon$, $0 < t < T$, then, by (3.8),

$$0 \leq v(x, t) \leq Ce^{-c/t}.$$

Finally, $v(x, 0) = 0$ if $x \in M_{\epsilon,m}$. We shall compare $v(x, t)$ with a function of the form

$$w(x, t) = C \exp\left\{-\frac{\gamma}{t}(\log R(x))^2\right\} \qquad (\gamma(\log \epsilon)^2 \leq c) \tag{3.16}$$

where γ is a sufficiently small positive constant independent of m. Notice that $w(x, 0) = 0$ if $x \in M_{\epsilon,m}$, $w(x, t) \geq 0$ if $x \in \partial_1 N_m$, and $w(x, t) \geq Ce^{-c/t}$ if $x \in \Gamma_\epsilon$, $0 \leq t \leq T$. Hence, if we can show that

$$Lw - w_t < 0 \qquad \text{for} \quad x \in M_{\epsilon,m}, \qquad 0 < t < T, \tag{3.17}$$

then, by the maximum principle,

$$G_m(x, t, \xi) \equiv v(x, t) \leq w(x, t).$$

Taking $m \to \infty$, the assertion (3.13) follows.

To prove (3.17), set $\Phi = 1/w$. Then

$$w_{x_i} = -\frac{1}{\Phi}\frac{2\gamma}{t}\frac{\log R}{R}R_{x_i},$$

$$w_{x_ix_j} = \frac{1}{\Phi}\left\{\frac{4\gamma^2}{t^2}\frac{(\log R)^2}{R^2}R_{x_i}R_{x_j} - \frac{2\gamma}{t}\frac{1}{R^2}R_{x_ix_j} + \frac{2\gamma}{t}\frac{\log R}{R^2}R_{x_i}R_{x_j}\right.$$
$$\left. - \frac{2\gamma}{t}\frac{\log R}{R}R_{x_ix_j}\right\},$$

$$-w_t = -\frac{1}{\Phi}\frac{\gamma}{t^2}(\log R)^2.$$

Hence

$$[Lw - w_t]\Phi$$
$$= \frac{2\gamma^2}{t^2}\frac{(\log R)^2}{R^2}\sum a_{ij}R_{x_i}R_{x_j} - \frac{\gamma}{t}\frac{1}{R^2}\left(1 + \log\frac{1}{R}\right)\sum a_{ij}R_{x_i}R_{x_j}$$
$$+ \frac{\gamma}{t}\frac{1}{R}\left(\log\frac{1}{R}\right)\sum a_{ij}R_{x_ix_j} + \frac{2\gamma}{t}\frac{1}{R}\left(\log\frac{1}{R}\right)\sum b_iR_{x_i} - \frac{\gamma}{t^2}(\log R)^2. \tag{3.18}$$

Setting

$$\mathscr{A} = \tfrac{1}{2}\sum a_{ij}R_{x_i}R_{x_j}, \qquad \mathscr{B} = \sum b_iR_{x_i} + \tfrac{1}{2}\sum a_{ij}R_{x_ix_j},$$

we find that

$$(Lw - w_t)\Phi = \frac{4\gamma^2}{t^2}\frac{(\log R)^2}{R^2}\mathscr{A} - \frac{2\gamma}{t}\frac{1 + \log(1/R)}{R^2}\mathscr{A}$$
$$+ \frac{2\gamma}{t}\frac{\log(1/R)}{R}\mathscr{B} - \frac{\gamma}{t^2}(\log R)^2. \tag{3.19}$$

By (3.1), (3.2), $\mathscr{A} = 0$ on $\partial\Omega$. Since $\mathscr{A} > 0$ elsewhere, we conclude that

$$\mathscr{A} < C_0R^2 \quad \text{if} \quad 0 < R(x) < 1 \qquad (C_0 \text{ positive constant}). \tag{3.20}$$

When $\mathscr{A} = 0$ we have (cf. 9.4.4))

$$\sum a_{ij}R_{x_ix_j} = -\sum \frac{\partial a_{ij}}{\partial x_j}\nu_i \qquad \text{on } \partial\Omega.$$

Recalling (3.1)–(3.3) we deduce that $\mathscr{B} < 0$ on $\partial\Omega$, so that

$$\mathscr{B} < C_0R \quad \text{if} \quad 0 < R(x) < 1 \qquad (C_0 \text{ positive constant}). \tag{3.21}$$

Now, if γ is sufficiently small, then, by (3.20),

$$\frac{4\gamma^2}{t^2}\,\frac{(\log R)^2}{R^2}\,\mathcal{Q} < \frac{1}{2}\,\frac{\gamma}{t^2}\,(\log R)^2.$$

Since also

$$-\frac{2\gamma}{t}\,\frac{1+\log(1/R)}{R^2}\,\mathcal{Q} < 0 \qquad \text{if} \quad R(x) < \epsilon, \quad \epsilon < 1,$$

we conclude from (3.19) that

$$(Lw - w_t)\Phi < \frac{2\gamma}{t}\,\frac{\log(1/R)}{R}\,\mathcal{B} - \frac{1}{2}\,\frac{\gamma}{t^2}\,(\log R)^2.$$

Using (3.21) we see that if ϵ is sufficiently small, then (3.17) holds.

Proof of Corollary 3.2. The formal adjoint of Lu is

$$L^*u = \frac{1}{2}\sum a_{ij}\,\frac{\partial^2 u}{\partial x_i\,\partial x_j} + \sum \tilde{b}_i\,\frac{\partial u}{\partial x_i} + \tilde{c}u$$

where

$$\tilde{b}_i = -b_i + \frac{1}{2}\sum \frac{\partial a_{ij}}{\partial x_j}, \qquad \tilde{c} = \frac{1}{2}\sum \frac{\partial^2 a_{ij}}{\partial x_i\,\partial x_j} - \sum \frac{\partial b_i}{\partial x_i}. \tag{3.22}$$

Since

$$\tilde{b}_i - \frac{1}{2}\sum \frac{\partial a_{ij}}{\partial x_j} = -\left(b_i - \frac{1}{2}\sum \frac{\partial a_{ij}}{\partial x_j}\right),$$

the condition (3.14) implies the condition (3.3) for L^*. The proof of (3.17) remains valid for L^* (with a minor change due to the term $\tilde{c}w$). We conclude that Green's function $G_m^*(x, t, \xi)$ corresponding to $L^* - \partial/\partial t$ in $N_m \times (0, \infty)$ satisfies:

$$G_m^*(x, t, \xi) \leq w(x, t) \qquad (x \in M_{\epsilon, m}, \quad 0 < t < T, \quad \xi \in E).$$

Recalling that $G_m(x, t, \xi) = G_m^*(\xi, t, x)$ and taking $m \to \infty$, the assertion (3.15) follows.

We shall now assume that

$$\mathcal{Q}(x) = O(R^{p+1}) \qquad \text{as} \quad R = R(x) \to 0, \tag{3.23}$$

where p is a positive number, $p > 1$.

Theorem 3.3. *Let* (A), (B_S), (C), (3.4), *and* (3.23) *hold. Let E be any compact subset of $\hat{\Omega}$. Then, for any $T > 0$ and for any $\rho > 0$ sufficiently* *small, there are positive constants* C, γ *such that*

$$G(x, t, \xi) < C \exp\left\{ -\frac{\gamma}{t} (R(x))^{1-p} \right\} \tag{3.24}$$

if $\xi \in E$, $x \in \hat{\Omega}$, $R(x) < \rho$, $0 < t < T$.

Corollary 3.4. *If in Theorem* 3.3 *the condition* (3.3) *is replaced by the condition* (3.14), *then*

$$G(x, t, \xi) < C \exp\left\{ -\frac{\gamma}{t} (R(\xi))^{1-p} \right\} \tag{3.25}$$

if $x \in E$, $\xi \in \hat{\Omega}$, $R(\xi) < \rho$, $0 < t < T$.

Proof of Theorem 3.3. We proceed as in the proof of Theorem 3.1, but with a different function $w(x, t)$. First we consider the interval $0 < t < \delta$ (δ is small and will be determined later), and take

$$w(x, t) = C \exp\left\{ -\frac{\gamma}{t} (R(x))^{1-p} \right\}. \tag{3.26}$$

If we prove that, for any $\gamma > 0$ sufficiently small and independent of m, (3.17) holds for $x \in M_{\epsilon, m}$, $0 < t < \delta$, then the inequality (3.24), for $0 < t < \delta$, follows as in the proof of Theorem 3.1. To prove (3.17), set $\Phi = 1/w$. Then

$$w_{x_i} = \frac{1}{\Phi} \frac{\gamma}{t} \frac{p-1}{R^p} R_{x_i},$$

$$w_{x_i x_j} = \frac{1}{\Phi} \left\{ \frac{\gamma^2 (p-1)^2}{t^2 R^{2p}} R_{x_i} R_{x_j} - \frac{\gamma p (p-1)}{t R^{p+1}} R_{x_i} R_{x_j} + \frac{\gamma (p-1)}{t R^p} R_{x_i x_j} \right\},$$

$$-w_t = \frac{1}{\Phi} \left\{ -\frac{\gamma}{t^2} \frac{1}{R^{p-1}} \right\}.$$

Hence

$$(Lw - w_t)\Phi = \frac{\gamma^2 (p-1)^2}{t^2} \frac{\mathcal{A}}{R^{2p}} - \frac{\gamma p (p-1)}{t} \frac{\mathcal{A}}{R^{p+1}} + \frac{\gamma (p-1)}{t} \frac{\mathcal{B}}{R^p} - \frac{\gamma}{t^2 R^{p-1}}. \tag{3.27}$$

If γ is sufficiently small, then, by (3.23),

$$\frac{\gamma^2 (p-1)^2}{t^2} \frac{\mathcal{A}}{R^{2p}} < \frac{1}{3} \frac{\gamma}{t^2} \frac{1}{R^{p-1}}.$$

By (3.21),

$$\frac{\gamma(p-1)}{t}\frac{\mathcal{B}}{R^p} < \frac{1}{3}\frac{\gamma}{t^2}\frac{1}{R^{p-1}}$$

if $0 < t < \delta$ and σ is sufficiently small. From (3.27) we then conclude that (3.17) holds if $0 < t < \delta$. As mentioned above, this implies (3.24) for $0 < t < \delta$. In order to prove (3.24) for $\delta < t < T$ we introduce another comparison function, namely,

$$w^0(x, t) = \hat{C}\exp\left\{-\frac{\hat{\gamma}}{(t+1)^\lambda}(R(x))^{1-p}\right\}$$

where $\hat{C}$, $\hat{\gamma}$, λ are positive numbers. With $\Phi = 1/w^0$, we have

$$\left(Lw^0 - w_t^0\right)\Phi = \frac{\hat{\gamma}^2(p-1)^2}{(t+1)^{2\lambda}}\frac{\mathcal{A}}{R^{2p}} - \frac{\hat{\gamma}p(p-1)}{(t+1)^\lambda}\frac{\mathcal{A}}{R^{p+1}} + \frac{\hat{\gamma}(p-1)}{(t+1)^\lambda}\frac{\mathcal{B}}{R^p} - \frac{\hat{\gamma}\lambda}{(t+1)^{\lambda+1}}\frac{1}{R^{p-1}}. \tag{3.28}$$

We choose λ (independently of $\hat{\gamma}$) sufficiently large so that

$$\lambda > 1, \qquad (p-1)\frac{\mathcal{B}}{R} < \frac{1}{3}\frac{\lambda}{T+1};$$

this is possible by (3.21). With λ fixed we next choose $\hat{\gamma}$ sufficiently small so that

$$\frac{\hat{\gamma}(p-1)^2}{(\delta+1)^{\lambda-1}}\frac{\mathcal{A}}{R^{p+1}} < \frac{1}{3}\lambda. \tag{3.29}$$

It then follows from (3.28) that $Lw^0 - w_t^0 \leq 0$ if $x \in M_{\epsilon, m}$, $\delta < t < T$. Notice that if $\hat{\gamma}$ is sufficiently small and $\hat{C}$ is sufficiently large (both independent of m), then, by (3.8),

$$G_m(x, t, \xi) \leq w^0(x, t) \qquad (\xi \text{ fixed in } E) \tag{3.30}$$

if $x \in \Gamma_\epsilon$, $0 < t < T$. The same inequality clearly holds also if $x \in \partial_1 N_m$, $t > 0$ and, by what we have already proved above, for $x \in M_{\epsilon, m}$, $t = \delta$. Hence, we can apply the maximum principle and conclude that (3.30) holds for $x \in M_{\epsilon, m}$, $\delta < t < T$. Taking $m \to \infty$, the proof of (3.24), for $\delta < t < T$, follows.

The proof of Corollary 3.4 is obtained by applying the proof of Theorem 3.3 to the equation $L^*u - \partial u/\partial t = 0$; cf. the proof of Corollary 3.2. The details are left to the reader.

Remark. Set $\Omega_0 = \text{int}\,\Omega$. In Theorems 3.1, 3.3 and their corollaries we were concerned with Green's function $G(x, t, \xi)$ for x, ξ in $\hat{\Omega}$. Similarly one can construct a Green function $G_0(x, t, \xi)$ for x, ξ in Ω_0. If (A), (B_S), (C), and (3.4) hold with ν (in (C)) being the inward normal to ∂G_j at x ($k_0 + 1 \leq j \leq k$), then (3.13) holds with $G(x, t, \xi)$ replaced by $G_0(x, t, \xi)$; $\xi \in E$, $x \in \Omega_0$, $0 < t < T$, $\text{dist}(x, \partial\Omega) < \rho$, where E is any compact subset of Ω_0. Similarly, if (3.3) is replaced by (3.14) (ν the inward normal), then (3.15) holds with $G(x, t, \xi)$ replaced by $G_0(x, t, \xi)$; $x \in E$, $\xi \in \Omega_0$, $0 < t < T$, $\text{dist}(\xi, \partial\Omega) < \rho$. The assertions of Theorem 3.3 and Corollary 3.4 also extend to $G_0(x, t, \xi)$. Note that $G_0(x, t, \xi) = 0$ if $x \in G_j$, $\xi \in G_h$, and $j \neq h$.

4. Estimates near Infinity

In this section we replace the conditions (C), (3.4) by the much weaker condition:

$$S \quad \text{is a compact set.} \tag{4.1}$$

Let $\hat{S} = R^n \backslash S$.

Theorem 4.1. *Let* (A), (B_S), *and* (4.1) *hold. Assume also that*

$$\sum_{i,j=1}^{n} a_{ij}(x)x_i x_j \leq C_0(1 + |x|^4), \tag{4.2}$$

$$-\left[\sum_{i=1}^{n} x_i b_i(x) + \frac{1}{2}\sum_{i=1}^{n} a_{ii}(x)\right] \leq C_0(1 + |x|^2) \tag{4.3}$$

where C_0 is a positive constant. Let E be any bounded subset of $\hat{S}$. Then, for any $T > 0$ and for any ρ sufficiently large, there are positive constants C, γ such that

$$K(x, t, \xi) \leq C \exp\left\{-\frac{\gamma}{t}(\log|x|)^2\right\} \tag{4.4}$$

if $\xi \in E$, $|x| > \rho$, $0 < t < T$.

Notice that the closure of E may intersect S.

Corollary 4.2. *If in Theorem* 4.1 *the condition* (4.3) *is replaced by the conditions*

$$\sum_{i=1}^{n} x_i b_i(x) + \tfrac{1}{2}\sum_{i=1}^{n} a_{ii}(x) \leq C_0(1 + |x|^2), \tag{4.5}$$

$$\frac{1}{2}\sum_{i,j=1}^{n} \frac{\partial^2 a_{ij}(x)}{\partial x_i\,\partial x_j} - \sum_{i=1}^{n} \frac{\partial b_i(x)}{\partial x_i} \leq [\log(2 + |x|)]^2\eta(|x|)$$

$$(\eta(r)\to 0 \qquad \text{if} \quad r\to\infty), \tag{4.6}$$

then

$$K(x, t, \xi) \leq C \exp\left\{ -\frac{\gamma}{t} (\log |\xi|)^2 \right\} \tag{4.7}$$

if $x \in E$, $|\xi| > \rho$, $0 < t < T$.

Proof of Theorem 4.1. Consider first the case where $\bar{E} \cap S = \varnothing$. For any $\rho > 0$, m a positive integer, let

$$N_{m,\rho} = \{x; \rho < |x| < m\}, \qquad \Delta_\rho = \{x; |x| = \rho\}, \qquad \Delta_m = \{x; |x| = m\}.$$

The number ρ is sufficiently large (to be determined later), whereas $m > \rho$. The boundary of $N_{m,\rho}$ then consists of the spheres Δ_ρ, Δ_m. Proceeding similarly to the proof of Theorem 3.1, we shall compare the function $v(x, t) = G_{m,\epsilon}(x, t, \xi)$ (ξ fixed in E) with a function $w(x, t)$ in the cylinder $N_{m,\rho} \times (0, T)$. We take

$$w(x, t) = C \exp\left\{ -\frac{\gamma}{t} (\log |x|)^2 \right\} \tag{4.8}$$

where C, γ are positive constants. It is clear that (3.19) holds with $R(x) = |x|$, L replaced by L_ϵ, a_{ij} replaced by $a_{ij}^\epsilon = a_{ij} + \epsilon\delta_{ij}$, where

$$\mathcal{A} = \frac{1}{2|x|^2} \sum a_{ij}^\epsilon(x) x_i x_j,$$

$$\mathcal{B} = \frac{1}{|x|} \left[\sum x_i b_i(x) + \tfrac{1}{2} \sum a_{ii}^\epsilon(x) \right] - \frac{1}{2|x|^3} \sum a_{ij}^\epsilon(x) x_i x_j.$$

By (4.2), (4.3) we have, for all $R(x) = |x|$ sufficiently large,

$$\mathcal{A} \leq C_0 R^2, \qquad -\mathcal{B} \leq C_0 R \qquad (C_0 \text{ positive constant}).$$

Now choose γ so small that

$$\frac{4\gamma^2}{t^2} \frac{(\log R)^2}{R^2} \mathcal{A} \leq \frac{1}{3} \frac{\gamma}{t^2} (\log R)^2. \tag{4.9}$$

Next choose ρ such that if $R(x) = |x| > \rho$,

$$-\frac{2\gamma}{t} \frac{1 + \log(1/R)}{R^2} \mathcal{A} = \frac{2\gamma}{t} \frac{\log R - 1}{R^2} \mathcal{A} < \frac{1}{3} \frac{\gamma}{t^2} (\log R)^2, \tag{4.10}$$

$$\frac{2\gamma}{t} \frac{\log(1/R)}{R} \mathcal{B} = -\frac{2\gamma}{t} \frac{\log R}{R} \mathcal{B} < \frac{1}{3} \frac{\gamma}{t^2} (\log R)^2 \tag{4.11}$$

for all $0 < t < T$. It follows that $L_\epsilon w - w_t < 0$ if $x \in N_{m,\rho}$, $0 < t < T$.

Notice that ρ was chosen independently of γ. Hence with ρ fixed, we can further decrease γ (if necessary) so that

$$v(x, t) \leq w(x, t) \qquad \text{if} \quad x \in \Delta_\rho, \quad 0 < t < T$$

for some positive constant C (in 4.8)). The last inequality evidently holds also if $x \in \Delta_m$, $0 < t < T$ or if $x \in N_{m,\rho}$, $t = 0$. Applying the maximum principle, we get

$$G_{m,\epsilon}(x, t, \xi) = v(x, t) \leq w(x, t) \qquad \text{if} \quad x \in N_{m,\rho}, \quad 0 < t < T.$$

From this the assertion (4.4) follows by taking first $m \to \infty$ and then $\epsilon = \epsilon_m \to 0$.

So far we have proved (4.4) only in case $\bar{E} \cap S = \emptyset$. Now let E be any bounded set disjoint to S. Let Σ be a sphere whose interior Δ contains both E and S. From what we have proved so far we know that if $x \in N_{m,\rho}$, then

$$G_{m,\epsilon}(x, t, \xi) \leq w(x, t) \tag{4.12}$$

if $\xi \in \Sigma$, $0 < t < T$. Now, as a function of (ξ, t) the function $w(x, t)$ satisfies

$$\left(L_\epsilon^* - \frac{\partial}{\partial t}\right)w = \left[\tilde{c}(\xi) - \frac{\gamma}{t^2}(\log|x|)^2\right]w(x, t) < 0$$

if ρ is sufficiently large and $\xi \in \Delta$, $0 < t < T$. Hence, by the maximum principle, (4.12) holds also for $\xi \in \Delta$, $0 < t < T$. Taking $m \to \infty$ and then $\epsilon = \epsilon_m \to 0$, the inequality (4.4) follows.

Proof of Corollary 4.2. We apply the proof of Theorem 4.1 to the adjoint L^* of L (cf. the proof of Corollary 3.2). Since (4.9)–(4.11) remain valid (with $\mathfrak{B}$ replaced by $-\mathfrak{B}$) with the factor $\frac{1}{3}$ on the right-hand sides replaced by $\frac{1}{4}$, it remains to show that

$$\tilde{c}(x) < \frac{1}{4}\frac{\gamma}{t^2}(\log R)^2,$$

where $\tilde{c}$ is defined in (3.22). In view of (4.6), this inequality holds if $0 < t < T$, provided ρ is sufficiently large and $R(x) = |x| > \rho$.

Suppose next that (4.2) is replaced by

$$\sum_{i,j=1}^{n} a_{ij}(x)x_i x_j \leq C_0(1 + |x|^{4-p}) \qquad (0 < p \leq 2). \tag{4.13}$$

Then we can use, for $0 < t < \delta$, the comparison function

$$w(x, t) = C \exp\left\{-\frac{\gamma}{t}|x|^p\right\}. \tag{4.14}$$

In fact one easily verifies that $L_\epsilon w - w_t < 0$ for $x \in N_{m,\rho}$, $0 < t < \delta$, provided γ and δ are sufficiently small. For $\delta < t < T$ we use the comparison function

$$w^0(x, t) = C \exp\left\{-\frac{\hat{\gamma}}{(t+1)^\lambda}|x|^p\right\}. \tag{4.15}$$

Choosing first λ sufficiently large, and then $\hat{\gamma}$ sufficiently small, we find that $L_\epsilon w^0 - \partial w^0/\partial t < 0$ if $x \in N_{m,\rho}$, $\delta < t < T$.

With the aid of these comparison functions we obtain:

Theorem 4.3. *Let* (A), $(\hat{B}_S)$, (4.1), (4.13), *and* (4.3) *hold. Let* E *be any bounded subset of* $\hat{S}$. *Then, for any* $T > 0$ *and for any* ρ *sufficiently large, there are positive constants* C, γ *such that*

$$K(x, t, \xi) \leq C \exp\left\{-\frac{\gamma}{t}|x|^p\right\} \tag{4.16}$$

if $\xi \in E$, $|x| > \rho$, $0 < t < T$.

Corollary 4.4. *If in Theorem* 4.3 *the condition* (4.3) *is replaced by the conditions* (4.5) *and*

$$\frac{1}{2}\sum_{i,j=1}^{n} \frac{\partial^2 a_{ij}(x)}{\partial x_i\,\partial x_j} - \sum_{i=1}^{n} \frac{\partial b_i(x)}{\partial x_i} \leq (1 + |x|^p)\eta(|x|)$$

$$(\eta(r) \to 0 \quad \text{if} \quad r \to \infty), \tag{4.17}$$

then

$$K(x, t, \xi) \leq C \exp\left\{-\frac{\gamma}{t}|\xi|^p\right\} \tag{4.18}$$

if $x \in E$, $|\xi| > \rho$, $0 < t < T$.

The proof of the corollary is obtained by applying the proof of Theorem 4.3 (with the same comparison functions w, w^0 as in (4.14), (4.15)) to L^*.

Remark. Denote by $\tilde{S}$ the unbounded component of $R^n \setminus S$. One can construct the function $G(x, t, \xi)$, for x, ξ in $\tilde{S}$ and $t > 0$, in the same way that we have constructed $G(x, t, \xi)$ for x, ξ in $\hat{\Omega}$, $t > 0$, as a limit of Green's functions $G_m(x, t, \xi)$ (cf. the remark at the end of Section 3). Using the same comparison functions as in Theorems 4.1, 4.3 and Corollaries 4.2, 4.4, we can estimate the functions $G_m(x, t, \xi)$ and, consequently, also $G(x, t, \xi)$. The estimates on G are the same as for K, except that now $\bar{E} \cap S$ is required to be empty.

5. Relation between K and a diffusion process

If the symmetric matrix $(a_{ij}(x))$ is nonnegative definite and the a_{ij} belong to $C^2(R^n)$, then (see Section 6.1) there exists an $n \times n$ matrix $\sigma(x) = (\sigma_{ij}(x))$

which is Lipschitz continuous, uniformly in compact subsets of R^n, such that

$$\sigma(x)\sigma^*(x) = (a_{ij}(x)) \qquad [\sigma^* = \text{transpose of } \sigma]$$

i.e., $\Sigma\sigma_{ik}(x)\sigma_{jk}(x) = a_{ij}(x)$. If

$$\sum_{i=1}^{n} a_{ii}(x) \leq C(1 + |x|^2), \tag{5.1}$$

then, clearly,

$$|\sigma(x)| \leq C(1 + |x|) \tag{5.2}$$

with a different constant C. Conversely, (5.2) implies (5.1) and, in fact, implies

$$\sum_{i,j=1}^{n} |a_{ij}(x)| \leq C(1 + |x|^2).$$

We shall now assume that (5.1) holds and, in addition,

$$\sum_{i=1}^{n} |b_i(x)| \leq C(1 + |x|). \tag{5.3}$$

Set $b = (b_1, \ldots, b_n)$. Since we always assume that (A) holds, the functions $\sigma(x)$, $b(x)$ are uniformly Lipschitz continuous in compact subsets of R^n.

Consider the system of n stochastic differential equations (1.2), and set

$$P(t, x, A) = E_x(\xi(t) \in A) \tag{5.4}$$

for any Borel set A in R^n.

Definition. If there is a function $\Gamma(x, t, \xi)$ defined for all x, ξ in R^n and $t > 0$ and Borel measurable in ξ for fixed (x, t), such that

$$P(t, x, A) = \int_A \Gamma(x, t, \xi)\, d\xi \tag{5.5}$$

for any Borel set A in R^n and for any $x \in R^n$, $t > 0$, then we call $\Gamma(x, t, \xi)$ *the fundamental solution* of the parabolic equation

$$Lu - \frac{\partial u}{\partial t} = 0. \tag{5.6}$$

Note that $\Gamma(x, t, \xi)$, if existing, is uniquely determined, for each (x, t), almost everywhere in ξ. Note also that for any bounded Borel measurable function $f(\xi)$ with compact support,

$$E_x[f(\xi(t))] = \int_{R^n} \Gamma(x, t, \xi) f(\xi)\, d\xi. \tag{5.7}$$

As mentioned in Section 1, if $(a_{ij}(x))$ is uniformly positive definite and if the a_{ij}, b_i are locally Lipschitz continuous and uniformly Hölder continuous, then the fundamental solution as defined here is a fundamental solution as defined in Section 6.4.

Theorem 5.1. *Let* (A), (B_S), *and* (5.1), (5.3) *hold. Then*

$$\lim_{\epsilon\to 0} K_\epsilon(x, t, \xi) \quad \textit{exists for all} \quad x \notin S, \quad \xi \notin S, \quad t > 0, \tag{5.8}$$

and the function $K(x, t, \xi) = \lim_{\epsilon\to 0} K_\epsilon(x, t, \xi)$ *satisfies*

$$P_x(\xi(t) \in A) = \int_A K(x, t, \xi)\, d\xi \tag{5.9}$$

for any Borel set A *with* $A \cap S = \varnothing$.

Proof. In Section 1 we have proved that there is a sequence $\{\epsilon_m\}$ converging to zero such that

$$K_{\epsilon_m}(x, t, \xi) \to K(x, t, \xi) \quad \text{as} \quad m \to \infty \tag{5.10}$$

for all $x \notin S$, $\xi \notin S$, $t > 0$; the convergence is uniform when x, ξ vary in any compact set E, $E \cap S = \varnothing$, and t varies in any interval $(\delta, 1/\delta)$, $\delta > 0$. The same proof shows that any sequence $\{\epsilon'_m\}$ converging to zero has a subsequence $\{\epsilon''_m\}$ such that

$$K_{\epsilon''_m}(x, t, \xi) \to M(x, t, \xi) \quad \text{as} \quad m \to \infty$$

for some function M, and the convergence is uniform in the same sense as before. If we can show that $M(x, t, \xi) = K(x, t, \xi)$, then the assertion (5.8) follows.

If we show that

$$P_x(\xi(t) \in A) = \int_A M(x, t, \xi)\, d\xi \tag{5.11}$$

for any bounded Borel set A, $\bar{A} \cap S = \varnothing$, then, by applying this to the particular sequence $\{\epsilon_m\}$ we derive (5.11) with M replaced by K. Consequently, $M = K$ (so that (5.8) is true) and (5.9) holds. Thus, in order to complete the proof of the theorem it remains to verify (5.11).

For any $\epsilon > 0$, consider the stochastic differential system

$$d\xi^\epsilon(t) = \sigma^\epsilon(\xi^\epsilon(t))\, dw(t) + b(\xi^\epsilon(t))\, dt$$

where σ^ϵ is such that $\sigma^\epsilon(\sigma^\epsilon)^* = (a_{ij} + \epsilon^2\delta_{ij})$; here $(\sigma^\epsilon)^* =$ transpose of σ^ϵ. For any continuous function f with compact support, the function

$$\int K_\epsilon(x, t, \xi) f(\xi)\, d\xi \tag{5.12}$$

is a bounded solution of

$$L_\epsilon u - \frac{\partial u}{\partial t} = 0 \quad \text{if} \quad x \in R^n, \quad 0 < t < T, \qquad u(x, 0) = f(x) \quad \text{if} \quad x \in R^n. \tag{5.13}$$

Indeed, this is true of $\int G_{m,\epsilon}(x, t, \xi) f(\xi)\, d\xi$ and, by the boundedness of the $G_{m,\epsilon}$ (cf. the proof of Theorem 4.1) and the Schauder estimates, also for the function in (5.12). By Theorem 6.5.4, also

$$E_x f(\xi^\epsilon(t)) \qquad (0 < t < T)$$

is a bounded solution of (5.13). Hence, by the uniqueness of bounded solutions for the Cauchy problem (Corollary 6.4.4)

$$E_x f(\xi^\epsilon(t)) = \int K_\epsilon(x, t, \xi) f(\xi)\, d\xi. \tag{5.14}$$

Taking a sequence of f's which converges to the indicator function of a Borel set A, we obtain

$$P_x(\xi^\epsilon(t) \in A) = \int_A K_\epsilon(x, t, \xi)\, d\xi. \tag{5.15}$$

Since (by Section 6.1) $\sigma^\epsilon(x) \to \sigma(x)$ uniformly on compact sets, as $\epsilon \to 0$, it follows (for instance, by Theorem 5.5.2) that

$$E_x|\xi^\epsilon(t) - \xi(t)|^2 \to 0 \qquad \text{if} \quad \epsilon \to 0. \tag{5.16}$$

Suppose now that A is a ball of radius R and denote by B_ρ $(\rho > 0)$ the ball of radius ρ concentric with A. From (5.16) it follows that if $\rho < R < \rho'$, then

$$\overline{\lim_{\epsilon \to 0}}\, P_x(\xi^\epsilon(t) \in B_\rho) \leqslant P_x(\xi(t) \in A),$$

$$\underline{\lim_{\epsilon \to 0}}\, P_x(\xi^\epsilon(t) \in B_{\rho'}) \geqslant P_x(\xi(t) \in A).$$

By (5.15) and Theorem 1.2 we also have

$$P_x(\xi^\epsilon(t) \in B_{\rho'} \setminus B_\rho) = \int_{B_{\rho'} \setminus B_\rho} K_\epsilon(x, t, \xi)\, d\xi \leqslant C(\rho' - \rho)$$

provided ρ' is sufficiently close to R (so that $\overline{B_{\rho'}} \cap S = \varnothing$), where C is a constant independent of ϵ. From the last three relations we deduce that

$$P_x(\xi^\epsilon(t) \in A) \to P_x(\xi(t) \in A) \qquad \text{if} \quad \epsilon \to 0. \tag{5.17}$$

Taking $\epsilon = \epsilon_m'' \to 0$, the right-hand side of (5.15) converges to the right-hand side of (5.11). If A is a ball, then, by (5.17), the left-hand side of (5.15) converges to the left-hand side of (5.11). We have thus established (5.11) in case A is a ball with $\bar{A} \cap S = \varnothing$. But then (5.11) follows also for any Borel set A with $A \cap S = \varnothing$.

Theorem 5.2. *Let* (A), (B_S), (4.1), *and* (5.1), (5.3) *hold. Then, for any* $x \in S$,

$$K(x, t, \xi) \equiv \lim_{\epsilon \to 0} K_\epsilon(x, t, \xi) \tag{5.18}$$

exists for all $\xi \notin S$, $t > 0$; *the convergence is uniform with respect to* (ξ, t) *in compact subsets of* $(R^n \setminus S) \times [0, \infty)$, *and* (5.9) *holds for any Borel set* A *with* $A \cap S = \varnothing$. *Finally, for any disjoint compact sets* M, E *in* R^n *with* $S \subset M$, *and for any* $T > 0$,

$$K(x, t, \xi) \leqslant Ce^{-c/t} \qquad \textit{for all} \quad x \in M, \quad \xi \in E, \quad 0 < t < T \tag{5.19}$$

where C, c are positive constants depending on M, E, T.

Proof. Let E be a compact set, $E \cap S = \varnothing$, and let M be a bounded neighborhood of S such that $\bar{M} \cap E = \varnothing$. For fixed ξ in E, consider the function

$$v_\epsilon(x, t) = K_\epsilon(x, t, \xi) \qquad \text{for} \quad x \in M, \quad 0 < t \leqslant T.$$

If $x \in \partial M$, $0 < t < T$, then, by the results of Sections 1, 2,

$$0 \leqslant v_\epsilon(x, t) \leqslant Ce^{-c/t}$$

where C, c are positive constants independent of ξ, ϵ. Further

$$v_\epsilon(x, 0) = 0 \qquad \text{if} \quad x \in M,$$

$$L_\epsilon v_\epsilon - \frac{\partial v_\epsilon}{\partial t} = 0 \qquad \text{if} \quad x \in M, \quad t > 0.$$

Hence, by the maximum principle,

$$0 \leqslant v_\epsilon(x, t) \leqslant Ce^{-c/t} \qquad \text{if} \quad x \in M, \quad 0 \leqslant t \leqslant T,$$

i.e.,

$$0 \leqslant K_\epsilon(x, t, \xi) \leqslant Ce^{-c/t} \qquad \text{if} \quad x \in M, \quad 0 \leqslant t \leqslant T, \quad \xi \in E. \tag{5.20}$$

Fix x in S and consider the function

$$\phi_\epsilon(\xi, t) = K_\epsilon(x, t, \xi) \qquad \text{for} \quad \xi \in E, \quad 0 \leqslant t \leqslant T.$$

By (5.20) this function is bounded. Since $\phi_\epsilon(\xi, 0) = 0$ if $\xi \in E$, and

$$L_\epsilon^* \phi_\epsilon - \frac{\partial}{\partial t} \phi_\epsilon = 0 \qquad \text{if} \quad \xi \in E, \quad 0 < t \leqslant T,$$

and since L^* is nondegenerate for $\xi \in E$, we can apply the Schauder-type estimates in order to conclude the following:

For any sequence $\{\epsilon_m'\}$ converging to 0, there is a subsequence $\{\epsilon_m^*\}$ such that $\{\phi_{\epsilon_m^*}\}$ is convergent to some function $\phi(\xi, t) = \hat{K}(x, t, \xi)$, together with the first t-derivative and the first two ξ-derivatives, uniformly for ξ in any set interior to E in t in $[0, T]$. By diagonalization, there is a subsequence $\{\epsilon_m''\}$ of $\{\epsilon_m^*\}$ for which

$$K_{\epsilon_m''}(x, t, \xi) \to \hat{K}(x, t, \xi) \qquad \text{for all} \quad \xi \in R^n \setminus S, \quad t > 0;$$

the first t-derivatives and the first two ξ-derivatives also converge, and the convergence is uniform for (ξ, t) in compact subsets of $(R^n \setminus S) \times [0, \infty)$.

Notice that the sequence $\{\epsilon_m''\}$ may depend on the parameter x. Now let A be a Borel set such that $\bar{A} \cap S = \varnothing$. Taking, in (5.15), $x \in S$ and $\epsilon = \epsilon_m'' \to 0$, and noting (upon using (5.20)) that the proof of (5.17) remains valid for $x \in S$, we conclude that

$$P_x(\xi(t) \in A) = \int_A \hat{K}(x, t, \xi)\, d\xi.$$

Thus, $\hat{K}(x, t, \xi)$ is independent of the particular sequence $\{\epsilon'_m\}$ with which we started. It follows that (5.18) holds. The other assertions of the theorem now follow immediately; in particular, (5.19) follows from (5.20).

From the above proof we see that, for fixed x in S,

$$L^* K(x, t, \xi) - \frac{\partial}{\partial t} K(x, t, \xi) = 0 \quad \text{if} \quad \xi \not\in S, \quad t > 0.$$

Theorem 5.3. *Let* (A), (B_S), (4.1), *and* (5.1), (5.3) *hold. Then for any disjoint compact sets* M, E *in* R^n *with* $S \subset M$, *and for any* $T > 0$

$$K_\epsilon(x, t, \xi) \leq Ce^{-c/t} \quad \text{for all} \quad x \in E, \quad \xi \in M, \quad 0 < t < T, \tag{5.21}$$

$$K(x, t, \xi) \leq Ce^{-c/t} \quad \text{for all} \quad x \in E, \quad \xi \in M \setminus S, \quad 0 < t < T, \tag{5.22}$$

where C, c *are positive constants depending on* M, E, T.

Indeed, we apply the argument which led to (5.20) to L^*, $K^*_\epsilon(x, t, \xi)$ instead of L, $K_\epsilon(x, t, \xi)$. We then get

$$K^*_\epsilon(x, t, \xi) \leq Ce^{-c/t}$$

if $x \in M$, $\xi \in E$, $0 < t < T$. Since $K^*_\epsilon(x, t, \xi) = K_\epsilon(\xi, t, x)$, (5.21) follows. Recalling that $K_\epsilon(\xi, t, x) \to K(\xi, t, x)$ as $\epsilon \to 0$, provided $\xi \not\in S$, $x \not\in S$, (5.22) also follows.

6. The behavior of $\xi(t)$ near S

In Section 3 we have introduced the condition (C). In this section we shall need also other similar conditions:

(C_0) The condition (C) holds with one exception, namely, the condition (3.3) is omitted.

(C′) The condition (C) holds with one exception, namely, the inequality (3.3) is replaced by the inequality (3.14).

(C*) The condition (C) holds with one exception, namely, the inequality (3.3) is replaced by equality, i.e.,

$$\sum_{i=1}^{n} \left(b_i(x) - \frac{1}{2} \sum_{j=1}^{n} \frac{\partial a_{ij}(x)}{\partial x_j} \right) \nu_i = 0 \quad \text{for} \quad x \in \partial G_j \quad (k_0 + 1 \leq j \leq k). \tag{6.1}$$

(C**) There is a finite number of disjoint closed bounded domains G_j

$(1 \leq j \leq k)$ with C^3 connected boundary ∂G_j, such that

$$\sum_{i,j=1}^{n} a_{ij}(x)\nu_i\nu_j = 0 \qquad \text{for} \quad x \in \partial G_j \qquad (1 \leq j \leq k), \tag{6.2}$$

$$\sum_{i=1}^{n} \left(b_i(x) - \tfrac{1}{2} \sum_{j=1}^{n} \frac{\partial a_{ij}(x)}{\partial x_j} \right) \nu_i > 0 \qquad \text{for} \quad x \in \partial G_j \qquad (1 \leq j \leq k) \tag{6.3}$$

where $\nu = (\nu_1, \ldots, \nu_n)$ is the outward normal to ∂G_j at x.

We shall also need the following condition:

(A_0) The inequalities (5.2), (5.3) hold and $\sigma(x)$, $b(x)$ are uniformly Lipschitz continuous in compact subsets of R^n. Finally, the matrix $a = \sigma\sigma^*$ is continuously differentiable in R^n.

Notice that if (A), (B_S), and (5.1), (5.3) hold, then the condition (A_0) is satisfied.

Theorem 6.1. *Let* (A_0), (C^*) *hold. Then, for any* $1 \leq j \leq k$,

$$P_x\{\xi(t) \in \partial G_j \quad \text{for all } t > 0\} = 1 \qquad \text{if} \quad x \in \partial G_j. \tag{6.4}$$

Thus, each set ∂G_j is an invariant set. This result was already established in Section 12.2. We shall give here a somewhat more direct proof.

Proof. Since (6.4) is obvious if $x = z_j$, $1 \leq j \leq k_0$, it remains to consider the case where $k_0 + 1 \leq j \leq k$.

Let $R(x)$ be a function such that $R(x) = \text{dist}(x, \partial G_j)$ if x is in a small $\hat{\Omega}$-neighborhood of ∂G_j; $R(x) = -\text{dist}(x, \partial G_j)$ if x is in a small Ω-neighborhood of ∂G_j; $R(x) \neq 0$ if $x \notin \partial G_j$; $R(x) = \text{const}$ if $|x|$ is sufficiently large, and $R(x)$ is in $C^2(R^n)$. Then

$$\begin{aligned} LR^2(x) &= \sum a_{ij} R_{x_i} R_{x_j} + 2R \left\{ \tfrac{1}{2} \sum a_{ij} R_{x_i x_j} + \sum b_i R_{x_i} \right\} \\ &\equiv 2\mathcal{A} + 2R\,\mathcal{B} \leq CR^2, \end{aligned}$$

since $\mathcal{A} = O(R^2)$, $|\mathcal{B}| = O(R)$ if R is small, and $\mathcal{A} = \mathcal{B} = 0$ if $|x|$ is large. By Itô's formula,

$$E_x R^2(\xi(t)) - R^2(x) = E_x \int_0^t LR^2(\xi(s))\, ds \leq CE_x \int_0^t R^2(\xi(s))\, ds.$$

If $x \in \partial G_j$, then $R(x) = 0$. Setting $\phi(t) = E_x R^2(\xi(t))$, we then have

$$\phi(t) \leq C \int_0^t \phi(s)\, ds, \qquad \phi(0) = 0.$$

Hence $\phi(t) = 0$ for all t, i.e., $R^2(\xi(t)) = 0$ a.s. This proves (6.4).

Theorem 6.2. *Let* (A_0), (C^{**}) *hold. Then, for any* $t > 0$,

$$P_x(\xi(t) \in G_j) = 0 \quad \text{if} \quad x \in \partial G_j \quad (1 \leq j \leq k). \tag{6.5}$$

This motivates us to call $\partial\Omega$ a *strictly one-sided obstacle, from the side* $\hat{\Omega}$, when the condition (C^{**}) holds.

Set

$$\rho(x) = \operatorname{dist}(x, \partial\Omega).$$

We shall first establish the following lemma.

Lemma 6.3. *Let* (A_0), (C_0) *hold. Then*

$$E_x\rho^2(\xi(t)) \leq Ct^2 \quad \text{if} \quad x \in \partial\Omega, \quad 0 < t < 1 \quad (C \text{ const}). \tag{6.6}$$

Proof. Since $\rho(\xi(t)) \equiv 0$ if $x = z_j$ $(1 \leq j \leq k_0)$, it remains to prove (6.6) in case $x \in \partial_0\Omega$, where

$$\partial_0\Omega = \bigcup_{j=k_0+1}^{k} \partial G_j.$$

Set

$$\rho_0(x) = \operatorname{dist}(x, \partial_0\Omega).$$

Let $M(x)$ be a C^2 function in R^n such that

$$M(x) = \begin{cases} \rho_0(x) & \text{if } x \text{ is in a small } \hat{\Omega}\text{-neighborhood of } \partial_0\Omega, \\ -\rho_0(x) & \text{if } x \text{ is in a small } \Omega\text{-neighborhood of } \partial_0\Omega, \\ |x| & \text{if } |x| \text{ is sufficiently large,} \end{cases}$$

and $M(x) \neq 0$ if $x \notin \partial_0\Omega$. If $x \in \partial_0\Omega$, then, by Itô's formula,

$$M(\xi(t)) = \int_0^t M_x \cdot \sigma \, dw + \int_0^t LM \, ds.$$

Squaring both sides and taking the expectation, we obtain

$$E_x M^2(\xi(t)) \leq CE_x \int_0^t |M_x \cdot \sigma|^2 \, ds + CE_x\left(\int_0^t |LM| \, ds\right)^2. \tag{6.7}$$

Near $\partial_0\Omega$,

$$|M_x \cdot \sigma|^2 = \sum_i \left(\sum_j \sigma_{ij} \frac{\partial \rho_0}{\partial x_j}\right)^2 = O(\rho^2) = O(M^2),$$

by (3.2), and near ∞,

$$|M_x \cdot \sigma|^2 = O(|x|^2) = O(M^2)$$

by (5.2). Next, for $|x|$ large

$$|LM| \leqslant C|x| = CM$$

by (5.2), (5.3), and for $|x|$ in a bounded set,

$$|LM| \leqslant C.$$

Using all these estimates in (6.7), and using Schwarz's inequality, we get

$$E_x M^2(\xi(t)) \leqslant C \int_0^t E_x M^2(\xi(s))\, ds + Ct \int_0^t E_x M^2(\xi(s))\, ds + Ct^2.$$

By iteration we then obtain

$$E_x M^2(\xi(t)) \leqslant Ct^2,$$

and this implies (6.6).

Proof of Theorem 6.2. For any $\epsilon > 0$, let $G_{i,\epsilon}$ be the set of points $x \in G_i$ with $\rho(x) < \epsilon$. The boundary $\partial G_{i,\epsilon}$ of $G_{i,\epsilon}$ consists of ∂G_i and $\partial' G_{i,\epsilon}$; the latter is the set of all points x in G_i with $\rho(x) = \epsilon$. Denote by τ_ϵ the hitting time of $\partial' G_{i,\epsilon}$.

Let ϵ_0 be a small positive number, so that $\rho \in C^2$ in G_{i,ϵ_0}. Let

$$\Psi(x) = \begin{cases} -\rho^2(x) & \text{if } x \in G_{i,\epsilon_0}, \\ 0 & \text{if } x \notin G_i. \end{cases} \tag{6.8}$$

Then $D_x\Psi$ is continuous, and $D_x^2\Psi$ is piecewise continuous, with discontinuity of the first kind across ∂G_i.

Define

$$\mathcal{A} = \tfrac{1}{2} \sum a_{ij} \rho_{x_i} \rho_{x_j}, \qquad \mathcal{B} = \tfrac{1}{2} \sum a_{ij} \rho_{x_i x_j} + \sum b_i \rho_{x_i}$$

for $x \in G_{i,\epsilon_0}$. Then

$$L\Psi(x) = -(2\mathcal{A} + 2\rho\mathcal{B}) \qquad \text{if } x \in G_{i,\epsilon_0}.$$

Hence, by (6.2), (6.3), if ϵ_0 is sufficiently small, then

$$L\Psi(x) \geqslant \begin{cases} \beta\rho(x) & \text{if } x \in G_{i,\epsilon_0} \quad (\beta \text{ positive constant}), \\ 0 & \text{if } x \notin G_i. \end{cases} \tag{6.9}$$

One can justify the use of Itô's formula for $\Psi(\xi(t))$ (cf. the proof of Lemma 11.2.1). Recalling (6.8), (6.9) and taking $0 < \epsilon < \epsilon_0$, we then get

$$0 \geqslant E_x \Psi(\xi(t \wedge \tau_\epsilon)) = E_x \int_0^{t \wedge \tau_\epsilon} L\Psi(\xi(s))\, ds \geqslant 0 \qquad (x \in \partial G_i).$$

Hence $E_x \Psi(\xi(t \wedge \tau_\epsilon)) = 0$; by (6.8) this implies

$$P_x(\xi(t \wedge \tau_\epsilon) \in \partial' G_{i,\epsilon}) = 0,$$

i.e.,

$$P_x(\tau_\epsilon > t) = 1.$$

Since this is true for any $t > 0$, $P_x(\tau_\epsilon = \infty) = 1$, i.e.,

$$P_x(\xi(t) \in G_i \setminus G_{i,\epsilon}) = 0.$$

Since this is true for any $0 < \epsilon < \epsilon_0$,

$$P_x(\xi(t) \in \text{int } G_i) = 0 \qquad (x \in \partial G_i). \tag{6.10}$$

Thus, in order to complete the proof of Theorem 6.2 it remains to show that

$$P_x(\xi(t) \in \partial\Omega) = 0 \qquad \text{if} \quad x \in \partial\Omega, \quad t > 0. \tag{6.11}$$

Let $\Psi(x)$ be a C^2 function in $\hat{\Omega} \in \partial\Omega$ such that

$$\Psi(x) = \begin{cases} \rho(x) & \text{if} \quad 0 \leqslant \rho(x) < r_1, \\ 1 & \text{if} \quad \rho(x) > 1, \end{cases}$$

where $0 < r_1 < 1$, and $\Psi(x) > 0$ if $\rho(x) > 0$. If r_1 is sufficiently small, then, by (6.2), (6.3), $L\Psi(x) \geqslant \alpha_0 > 0$ if $\rho(x) < r_1$. Hence, for all $x \in \hat{\Omega} \cup \partial\Omega$,

$$L\Psi(x) \geqslant \alpha_0 - C_1\Psi(x) \qquad (C_1 \text{ positive constant}). \tag{6.12}$$

Notice also that for all $x \in \hat{\Omega} \cup \partial\Omega$,

$$L\Psi(x) \leqslant \alpha_1 \qquad (\alpha_1 \text{ positive constant}). \tag{6.13}$$

Observe that $\Psi(x)$ has a C^2 extension into an Ω-neighborhood of $\partial\Omega$. By (6.10) and the nonattainability of Ω,

$$P_x\{\exists t > 0 \text{ such that } \xi(t) \not\in \hat{\Omega} \cup \partial\Omega\} = 0 \qquad \text{if} \quad x \in \partial\Omega.$$

Hence, if $x \in \partial\Omega$, we can apply Itô's formula to get

$$E_x\Psi(\xi(t)) = \int_0^t E_x[L\Psi(\xi(s))]\, ds. \tag{6.14}$$

Using (6.12)–(6.14), we find that

$$E_x\Psi(\xi(t)) \leqslant \alpha_1 t,$$

$$E_x\Psi(\xi(t)) \geqslant \alpha_0 t - C_1 E_x \int_0^t \Psi(\xi(s))\, ds.$$

Hence,

$$\alpha_0 t \leqslant E_x\Psi(\xi(t)) + \tfrac{1}{2}\alpha_1 C_1 t^2.$$

Consequently

$$\tfrac{1}{2}\alpha t \leqslant E_x\rho(\xi(t)), \qquad \text{if} \quad 0 < t < t^* \qquad (x \in \partial\Omega) \tag{6.15}$$

provided t^* is sufficiently small and α is any positive constant such that $\alpha\Psi(x) \leqslant \alpha_0\rho(x)$ for all $x \in \hat{\Omega}$.

Set

$$\delta_x(t) = P_x(\xi(t) \in \partial\Omega).$$

Then, by (6.15) and Lemma 6.3,

$$\begin{aligned}\tfrac{1}{2}\alpha t &< E_x\{\chi_{\xi(t)\in\hat{\Omega}}\rho(\xi(t))\}\\ &< \{E_x\chi_{\xi(t)\in\hat{\Omega}}\}^{1/2}\{E_x\rho^2(\xi(t))\}^{1/2}\\ &< C\{1-\delta_x(t)\}^{1/2}t.\end{aligned}$$

It follows that

$$\alpha/2C < (1-\delta_x(t))^{1/2},$$

i.e.,

$$\delta_x(t) < \delta = 1 - \frac{\alpha^2}{4C^2} < 1 \qquad \text{if} \quad 0 < t < t^*. \tag{6.16}$$

By the Markov property, if $t = s + r$ where s, r are positive numbers smaller than t^*,

$$\begin{aligned}P_x(\xi(t)\in\partial\Omega) = E_x\{\chi_{\xi(s)\in\partial\Omega}P_{\xi(s)}(\xi(r)\in\partial\Omega)\}\\ + E_x\{\chi_{\xi(s)\in\hat{\Omega}}P_{\xi(s)}(\xi(r)\in\partial\Omega)\}.\end{aligned}$$

The second term vanishes by the nonattainability of Ω. Applying (6.16) to the first term, we get

$$P_x(\xi(t)\in\partial\Omega) < \delta E_x\{\chi_{\xi(s)\in\partial\Omega}\} = \delta P_x(\xi(s)\in\partial\Omega) < \delta^2.$$

Similarly,

$$P_x(\xi(t)\in\partial\Omega) < \delta^m$$

for any m, if $t < t^*m$. Taking $m\to\infty$, the assertion (6.11) follows.

We shall now establish a relation between the functions $K(x, t, \xi)$ and $G(x, t, \xi)$, $G_0(x, t, \xi)$.

Theorem 6.4. *If* (A), (B_S), (C′), (3.4), *and* (5.1), (5.3) *hold, then*

$$K(x, t, \xi) = G(x, t, \xi) \qquad \text{if} \quad x\in\hat{\Omega}, \quad \xi\in\hat{\Omega}, \quad t > 0. \tag{6.17}$$

If (A), (B_S), (C), (3.4), *and* (5.1), (5.3) *hold, then*

$$K(x, t, \xi) = G_0(x, t, \xi) \qquad \text{if} \quad x\in\Omega_0, \quad \xi\in\Omega_0, \quad t > 0. \tag{6.18}$$

The function G was constructed in Section 2, and the function G_0 was defined at the end of Section 3.

Proof. Let $f(x)$ be a continuous nonnegative function with support in a compact Borel set A, $A\subset\hat{\Omega}$. Choose m so large that $A\subset N_m$, and consider the function

$$u_m(x, t) = \int_A G_m(x, t, y)f(y)\, dy. \tag{6.19}$$

It satisfies

$$Lu_m - \frac{\partial u_m}{\partial t} = 0 \quad \text{if} \quad x \in N_m, \quad t > 0,$$
$$u_m(x, 0) = f(x) \quad \text{if} \quad x \in N_m,$$
$$u_m(x, t) = 0 \quad \text{if} \quad x \in \partial N_m, \quad t > 0.$$

Using Itô's formula, we get

$$u_m(x, t) = E_x\{u(\xi(\tau_m), t - \tau_m)\} = E_x\{f(\xi(\tau_m))\chi_{\tau_m = t}\}$$

where τ_m is the first time the process $(s, \xi(s))$ hits the set $\{\partial N_m\{(0, t]\} \cup \{N_m \times \{t\}\}$. If (C′) holds, then Ω is nonattainable, so that $\tau_m \to t$ a.s. as $m \to \infty$. Hence

$$\lim_{m \to \infty} u_m(x, t) = E_x f(\xi(t)) = \int_A K(x, t, y) f(y)\, dy,$$

by Theorem 5.1. Since, on the other hand, by (6.19)

$$\lim_{m \to \infty} u_m(x, t) = \int_A G(x, t, y) f(y)\, dy,$$

the assertion (6.17) holds. The proof of (6.18) is similar.

Theorem 6.5. *If* (A), (B_S), (C′), (3.4), *and* (5.1), (5.3) *hold, then*

$$K(x, t, \xi) = 0 \quad \text{if} \quad x \in \hat{\Omega}, \quad \xi \in \Omega_0. \tag{6.20}$$

If (A), (B_S), (C), (3.4), *and* (5.1), (5.3) *hold, then*

$$K(x, t, \xi) = 0 \quad \text{if} \quad x \in \Omega_0, \quad \xi \in \hat{\Omega}. \tag{6.21}$$

Indeed, this follows from Theorem 5.1 and the nonattainability of Ω (when (C′) holds) or the nonattainability of $\hat{\Omega}$ (when (C) holds).

7. Existence of a generalized solution in the case of a two-sided obstacle

We consider in this section the case where $\partial\Omega$ is a two-sided obstacle, i.e., (C*) holds. We shall also assume:

(D) Denote by L_i the restriction (as defined in Section 13.2) of the elliptic operator L to the manifold ∂G_i, $k_0 + 1 \leq i \leq k$. Then, each L_i is elliptic on ∂G_i.

Thus, in local coordinates $\theta_1, \ldots, \theta_{n-1}$ of ∂G_i,

$$L_i = \sum_{\lambda,\mu=1}^{n-1} \alpha_{\lambda\mu}^i(\theta) \frac{\partial^2}{\partial\theta_\lambda\, \partial\theta_\mu} + \sum_{\lambda=1}^{n-1} \beta_\lambda^i \frac{\partial}{\partial\theta_\lambda}$$

and the $(n-1) \times (n-1)$ matrix $(\alpha_{\lambda\mu}^i(\theta))$ is positive definite for each θ.

A fundamental solution for $L_j - \partial/\partial t$ is a function $\hat{K}_j(x, t, \xi)$ defined for x, ξ in ∂G_j and $t > 0$ and having the following property: For any continuous function f on ∂G_j, the function

$$\hat{u}(x, t) = \int_{\partial G_j} \hat{K}_j(x, t, y) f(y)\, dS_y^j \tag{7.1}$$

satisfies

$$L_j \hat{u}(x, t) - \frac{\partial \hat{u}(x, t)}{\partial t} = 0 \quad \text{if } x \in \partial G_j, \quad t > 0,$$

$$\hat{u}(x, 0) = f(x) \quad \text{if } x \in \partial G_j.$$

Here dS_y^j is the surface element of ∂G_j.

The existence of $\hat{K}_j$ can be established by the same parametrix method by which one proves the existence of the fundamental solution of Section 6.4; for details, see, for instance, S. Itô [1].

For $x \in \partial G_i$, denote by $K_i(x, t, d\xi)$ $(k_0 + 1 \leqslant i \leqslant k)$ the measure supported on ∂G_i with density $\hat{K}_i(x, t, \xi)\, dS_\xi^i$. For $1 \leqslant i \leqslant k_0$, let

$$K_i(z_i, t, d\xi) = \text{the Dirac measure concentrated at } \xi = z_i.$$

Now define $K(x, t, \xi) = 0$ if $x \notin \partial\Omega$, $\xi \in \partial\Omega$, $t > 0$, and set

$$\Gamma(x, t, d\xi) = \begin{cases} K(x, t, \xi)\, d\xi & \text{if } x \notin \partial\Omega, \quad t > 0, \\ K_i(x, t, d\xi) & \text{if } x \in \partial G_i, \quad t > 0 \quad (k_0 + 1 \leqslant i \leqslant k), \\ K_i(z_i, t, d\xi) & \text{if } t > 0 \quad (1 \leqslant i \leqslant k_0). \end{cases} \tag{7.2}$$

In view of Theorems 6.4 and 6.5,

$$\Gamma(x, t, d\xi) = \begin{cases} G(x, t, \xi)\, d\xi & \text{if } x \in \hat{\Omega}, \quad \xi \in \hat{\Omega}, \quad t > 0, \\ G_0(x, t, \xi)\, d\xi & \text{if } x \in \Omega_0, \quad \xi \in \Omega_0, \quad t > 0, \\ 0 & \text{if } x \in \hat{\Omega}, \quad \xi \in \Omega_0, \quad t > 0 \\ & \text{or } x \in \Omega_0, \quad \xi \in \hat{\Omega}, \quad t > 0. \end{cases}$$

Theorem 7.1. *Let* (A), (B_S), (C^*), (3.4), (D), *and* (5.1), (5.3) *hold. Then, for any Borel set A in* R^n,

$$P_x(\xi(t) \in A) = \int_A \Gamma(x, t, dy). \tag{7.3}$$

Definition. $\Gamma(x, t, d\xi)$ is called *the generalized fundamental solution* for the parabolic equation (5.6).

For $x \notin \partial\Omega$, it is given by $K(x, t, \xi)\, d\xi$, and for $x \in \partial\Omega$ it is a certain measure supported on $\partial\Omega$.

Proof of Theorem 7.1. Consider first the case where $x \notin \partial\Omega$. If $A \cap (\partial\Omega) = \emptyset$, then (7.3) is a consequence of Theorem 5.1. If $A \subset \partial\Omega$, then both sides of (7.3) vanish. The truth of (7.3) for any Borel set A follows from the preceding special cases, upon writing $A = (A \cap \partial\Omega) \cup (A \setminus \partial\Omega)$.

Consider next the case where $x \in \partial\Omega$. If $x \in \partial G_j$ and $1 \leq j \leq k_0$, then $x = z_j$ and, by the definition of Γ,

$$\int_A \Gamma(z_j, t, d\xi) = \begin{cases} 1 & \text{if } z_j \in A, \\ 0 & \text{if } z_j \notin A. \end{cases}$$

On the other hand, by Lemma 6.1,

$$P_{z_j}(\xi(t) \in A) = \begin{cases} 1 & \text{if } z_j \in A, \\ 0 & \text{if } z_j \notin A. \end{cases}$$

Thus (7.3) follows. If $x \in \partial G_j$ and $k_0 + 1 \leq j \leq k$, then, by Theorem 6.1, $\xi(t)$ remains on ∂G_j for all $t > 0$. Extend the function $\hat{u}(x, t)$ defined in (7.1) for $x \in \partial G_j$ so that it remains constant along the outward normals to ∂G. Call the extended function u. Then, on ∂G_j, $Lu = L_j \hat{u}$ (by the definition of L_j). By Itô's formula applied to $u(\xi(s), t - s)$ and the fact that $\xi(s) \in \partial G_j$ if $x \in \partial G_j$, we then have

$$E_x \hat{u}(\xi(t), 0) - \hat{u}(x, t) = E_x \int_0^t (L_j - \partial/\partial s)\hat{u}(\xi(s), t - s)\, ds = 0,$$

i.e., $\hat{u}(x, t) = E_x(f(\xi(t)))$. Comparing this with (7.1), we conclude that

$$E_x f(\xi(t)) = \int_{\partial G_j} \hat{K}_j(x, t, y) f(y)\, dS_y^j.$$

This implies that

$$P_x(\xi(t) \in B) = \int_B \hat{K}_j(x, t, y)\, dS_y^j \tag{7.4}$$

for any Borel set B in ∂G_j.

Again, by Theorem 6.1,

$$P_x(\xi(t) \in A) = P_x[\xi(t) \in (A \cap \partial G_j)]$$

for any Borel set A in R^n. Using (7.4) with $B = A \cap \partial G_j$, we get

$$P_x(\xi(t) \in A) = \int_{A \cap \partial G_j} \hat{K}_j(x, t, \xi)\, dS_\xi^j = \int_A \Gamma(x, t, d\xi)$$

where the definition of Γ has been used in the last step. We have thus completed the proof of the theorem.

Remark 1. The estimates derived in Section 2 for the functions G, G_0 are, by Theorem 6.4, estimates on Γ.

Remark 2. We have assumed in Theorem 7.1 that the L_i $(k_0 + 1 \leq i \leq k)$ are nondegenerate elliptic operators on ∂G_i. Suppose now that a particular L_i

degenerates along a C^3 $(n-2)$-dimensional manifold Δ, $\Delta \subset G_i$, and that Δ is a two-sided obstacle. Then we can analyze the generalized fundamental solution $\hat{K}_i$ on ∂G_i by the same procedure as in Theorem 7.1. Thus, if the restriction of L_i to Δ is nondegenerate, then $\hat{K}_i(x, t, d\xi)$ will be (on ∂G_i) of the form $K_i(x, t, \xi)\, dS_\xi^i$ if $x \not\in \Delta$; for $x \in \Delta$ it is given by some measure supported on Δ. (If Δ consists of one point z, then this measure is the Dirac measure concentrated at z.) If the restriction of L_i to Δ degenerates on an $(n-2)$-dimensional manifold, then we can further explore the situation by the method of Theorem 7.1. Thus, in general, the measure $\hat{K}_i$ may consist of densities distributed on submanifolds of ∂G_i of any dimension l, $0 < l < n-2$.

8. Existence of a fundamental solution in the case of a strictly one-sided obstacle

We shall now replace the condition (C*) by the condition (C**). We define

$$\Gamma(x, t, \xi) = K(x, t, \xi) \qquad \text{if} \quad x \in R^n, \quad t > 0, \quad \xi \not\in \partial\Omega. \tag{8.1}$$

For definiteness we also set $\Gamma(x, t, \xi) = 0$ if $x \in R^n$, $t > 0$, $\xi \in \partial\Omega$. Notice, by Theorem 6.5, that

$$\Gamma(x, t, \xi) = 0 \qquad \text{if} \quad x \in \hat{\Omega}, \quad t > 0, \quad \xi \in \Omega_0.$$

By Theorem 6.4,

$$\Gamma(x, t, \xi) = G(x, t, \xi) \qquad \text{if} \quad x \in \hat{\Omega}, \quad t > 0, \quad \xi \in \hat{\Omega}.$$

Thus, the boundary estimates derived in Section 3 apply to Γ.

Theorem 8.1. *Let* (A), (B_S), (C**), (3.4), *and* (5.1), (5.3) *hold. Then* $\Gamma(x, t, \xi)$ *is the fundamental solution of the parabolic equation* (5.6).

Proof. We have to verify the relation

$$P_x(\xi(t) \in A) = \int_A K(x, t, y)\, dy \tag{8.2}$$

for any Borel set A. Consider first the case where $x \not\in \partial\Omega$. For any $\delta > 0$, let V_δ be the δ-neighborhood of $\partial\Omega$.

If δ is sufficiently small, then $x \not\in V_\delta$. Using Theorem 5.3, we get

$$\int_{A \cap V_\delta} K_\epsilon(x, t, \xi)\, d\xi < C \int_{A \cap V_\delta} d\xi < C\delta, \qquad \int_{A \cap V_\delta} K(x, t, \xi)\, d\xi < C\delta.$$

Recalling that for each δ fixed,

$$\int_{A \setminus V_\delta} K_\epsilon(x, t, \xi)\, d\xi \to \int_{A \setminus V_\delta} K(x, t, \xi)\, d\xi \qquad \text{if} \quad \epsilon \to 0,$$

we conclude that

$$\int_A K_\epsilon(x, t, \xi)\, d\xi \to \int_A K(x, t, \xi)\, d\xi \qquad \text{if} \quad \epsilon \to 0. \tag{8.3}$$

Using the estimate (5.21) of Theorem 5,3 and the estimate (2.15), we can argue as in the proof of (5.17) to deduce the relation

$$P_x(\xi^\epsilon(t) \in A) \to P_x(\xi(t) \in A) \qquad \text{if} \quad \epsilon \to 0 \tag{8.4}$$

provided A is a ball. Taking $\epsilon \to 0$ in (5.15) and using (8.3), (8.4), the relation (8.2) follows in case A is a ball. This relation is therefore valid also for any Borel set A.

Consider next the case where $x \in \partial\Omega$. By Theorem 5.2,

$$\int_{A \setminus V_\delta} K_\epsilon(x, t, \xi)\, d\xi \to \int_{A \setminus V_\delta} K(x, t, \xi)\, d\xi \qquad \text{if} \quad \epsilon \to 0. \tag{8.5}$$

Suppose A is a ball. By the proof of Theorem 5.2 (cf. (5.20)), $K_\epsilon(x, t, \xi) \leqslant C$ if ξ belongs to a small neighborhood of $A \setminus V_\delta$. Hence, the argument used to prove (5.17) can be applied also here to deduce that

$$P_x(\xi^\epsilon(t) \in A \setminus V_\delta) \to P_x(\xi(t) \in A \setminus V_\delta) \quad \text{if} \quad \epsilon \to 0. \tag{8.6}$$

Taking $\epsilon \to 0$ in (5.15) (with A replaced by $A \setminus V_\delta$) and using (8.5), (8.6), we get

$$P_x(\xi(t) \in A / V_\delta) = \int_{A \setminus V_\delta} K(x, t, \xi)\, d\xi \tag{8.7}$$

for any $\delta > 0$. Since $K(x, t, \xi) \geqslant 0$ for all ξ, the monotone convergence theorem yields

$$\lim_{\delta \to 0} \int_{A \setminus V_\delta} K(x, t, \xi)\, d\xi = \int_A K(x, t, \xi)\, d\xi. \tag{8.8}$$

Using Theorem 6.2 we also have

$$\lim_{\delta \to 0} P_x(\xi(t) \in A \setminus V_\delta) = P_x(\xi(t) \in A \setminus \partial\Omega) = P_x(\xi(t) \in A). \tag{8.9}$$

Taking $\delta \to 0$ in (8.7) and using (8.8), (8.9), the assertion (8.2) follows in case A is a ball. But then (8.2) clearly holds also for any Borel set A.

Remark 1. From Theorem 6.2 and (8.2) it follows that

$$K(x, t, \xi) = 0 \qquad \text{if} \quad x \in \partial\Omega, \quad t > 0, \quad \xi \in \Omega. \tag{8.10}$$

From Theorem 6.2, $P_x(\xi(t) \in \Omega) = 1$ if $x \in \partial\Omega$. Hence, by the strong maximum principle, $K(x, t, \xi) > 0$ if $x \in \partial\Omega$, $t > 0$, $\xi \in \Omega$. If A is a closed ball in $\hat{\Omega}$, and A' is a closed ball in the interior of A, then (cf. the proof of Lemma 10.2 below)

$$\lim_{x \in \Omega, x \to y} P_x(\xi(t) \in A) \geqslant P_y(\xi(t) \in A') = \int_{A'} K(y, t, \xi)\, d\xi > 0$$

if $y \in \partial\Omega$. It follows that

$$P_x(\xi(t) \in A) > 0 \qquad \text{if} \quad x \in \Omega, \quad \text{dist}(x, \partial\Omega) < \epsilon_0$$

for some ϵ_0 small. Applying the strong maximum principle to $\int_A K(x, t, \xi)\, d\xi$, as a function of (x, t), we conclude that

$$\int_A K(x, t, \xi)\, d\xi > 0 \qquad \text{if} \quad x \in \Omega, \quad t > 0.$$

Applying once more the strong maximum principle, to $K(x, t, \xi)$ as a function of (ξ, t) we conclude that

$$K(x, t, \xi) > 0 \qquad \text{if} \quad x \in \Omega, \quad t > 0, \quad \xi \in \hat{\Omega} \tag{8.11}$$

***Remark* 2.** Theorem 8.1 extends without difficulty to the case where the condition (C**) is replaced by the more general condition where the inequality (6.3) holds for $j = 1, \ldots, l$ and the reverse inequality holds for $j = l + 1, \ldots, k$. In case $n = 1$ we can just assume that each G_i consists of one point z_i and either $a(z_i) = 0$, $b(z_i) > 0$ or $a(z_i) = 0$, $b(z_i) < 0$.

***Remark* 3.** One can easily combine cases of strictly one-sided obstacles with two-sided obstacles.

***Remark* 4.** Theorem 8.1 extends to the case where S is any compact subset of R^n such that

$$P_x\{\xi(t) \in S\} = 0 \quad \text{for all} \quad x \in R^n, \quad t > 0. \tag{8.12}$$

Let S be a C^1 manifold dimension k $(0 \leqslant k \leqslant n-1)$, and denote by $d(x)$ $(x \in S)$ the rank of the linear operator $(a_{ij}(x))$ restricted to the linear space normal to S at x. By Theorem 11.3.1, if

$$d(x) \geqslant 3 \quad \text{for all} \quad x \in S, \tag{8.13}$$

then (8.12) holds for all $x \notin S$. We claim that (8.12) holds also for $x \in S$. To prove it note, by Theorem 12.3.1, that

$$P_x\{\xi(t) \in S \setminus V_\delta\} = 0 \qquad \text{if} \quad t > 0,$$

for any δ-neighborhood V_δ of x. Hence $P_x(\xi(t) \in S \setminus \{x\}) = 0$. Thus, it remains to prove that

$$P_x\{\xi(t) = x\} = 0 \qquad \textit{if} \quad t > 0 \quad (x \in S). \tag{8.14}$$

Suppose for simplicity that $x = 0$. Let $\rho(x)$ be a function in $C^2(R^n)$ such that

$$\rho(x) = \begin{cases} |x|^2 & \text{if} \quad |x| \text{ is small,} \\ 1 & \text{if} \quad |x| \text{ is large,} \end{cases}$$

and $\rho(x) > 0$, if $x \neq 0$. Since $\Sigma a_{ii}(0) > 0$,

$$\gamma_0 - C_0\rho(x) \leqslant L\rho(x) \leqslant \gamma_1 \qquad (x \in R^n) \tag{8.15}$$

where γ_0, C_0, γ_1 are positive constants. By Itô's formula,

$$E_0\rho(\xi(t)) = E_0\int_0^t L\rho(\xi(x))\,ds \leq \gamma_1 t,$$

$$E_0\rho(\xi(t)) = E_0\int_0^t L\rho(\xi(s))\,ds \geq \gamma_0 t - C_0E_0\int_0^t \rho(\xi(s))\,ds.$$

Hence

$$\gamma_0 t \leq E_0\rho(\xi(t)) + C_0\int_0^t \gamma_1 s\,ds = E_0\rho(\xi(t)) + \tfrac{1}{2}C_0\gamma_1 t^2.$$

It follows that

$$\gamma' t \leq E_0\rho(\xi(t)) \qquad (\gamma' \text{ positive constant})$$

if t is sufficiently small, say $t < t^*$. Hence

$$\gamma t \leq E_0|\xi(t)|^2 \quad \text{if} \quad t < t^* \qquad (\gamma \text{ positive constant}). \tag{8.16}$$

Setting $\delta_x(t) = P_x(\xi(t) = 0)$, we then have

$$\begin{aligned}\gamma t &\leq E_0\{\chi_{\xi(t)\neq 0}|\xi(t)|^2\} \leq \{E_0\chi_{\xi(t)\neq 0}\}^{1/2}\{E_0|\xi(t)|^4\}^{1/2}\\ &\leq C\{1-\delta_0(t)\}^{1/2}t,\end{aligned}$$

since $E_0|\xi(t)|^4 < Ct^2$. Hence

$$\delta_0(t) \leq \delta < 1 \quad \text{if} \quad 0 < t < t^* \qquad (\delta \text{ const}).$$

We can now proceed to establish (8.14) by the argument following (6.16).

The assertion (8.12) can be proved also in cases where $d(y) \geq 2$ for all $y \in S$. For $x \notin S$, one applies Theorems 12.4.1, 12.4.2. If $x \in S$, we cannot reduce the proof of (8.12) to that of proving (8.14) as before; instead, we proceed directly to prove (8.12) by the argument used to prove (8.14), employing a positive function

$$\tilde{\rho}(\xi) = (\text{dist}(\xi, S))^2 \quad \text{if} \quad \text{dist}(\xi, S) \text{ is small}, \qquad \tilde{\rho}(\xi) = 1 \quad \text{if} \quad |\xi| \text{ is large},$$

instead of $\rho(\xi)$. Note that also $\tilde{\rho}$ satisfies the differential inequalities of (8.15).

9. Lower bounds on the fundamental solution

In Theorem 3.1 we have derived the bound

$$G(x, t, \xi) \leq C\exp\left\{-\frac{c}{t}(\log R(x))^2\right\} \qquad (C > 0, \quad c > 0) \tag{9.1}$$

if ξ varies in a compact set E of $\hat{\Omega}$, $0 < t < T$, $x \in \hat{\Omega}$, and $R(x)$ is sufficiently small. Recall that the condition (C) was assumed in that theorem.

We shall now assume that the condition (C′) holds and that

$$\sum_{i,j=1}^{n} a_{ij}(x)R_{x_i}R_{x_j} \geq \alpha R^2 \qquad (\alpha \text{ positive constant}) \tag{9.2}$$

for all x in some $\hat{\Omega}$-neighborhood of $\partial\Omega$, where $R(x) = \text{dist}(x, \partial\Omega)$. We shall then derive the estimate

$$G(x, t, \xi) \geq N \exp\left\{-\frac{\nu}{t}(\log R(x))^2\right\} \qquad (N > 0, \nu > 0) \tag{9.3}$$

for some positive constants N, ν, for all $\xi \in E$, $0 < t < T$, $x \in \hat{\Omega}$, provided $R(x)$ is sufficiently small.

To do this, we compare (for fixed $\xi \in E$) the function

$$v(x, t) = G(x, t, \xi) \qquad (x \in \hat{\Omega}, 0 < R(x) < \epsilon, 0 < t < T)$$

with a function $w(x, t)$ of the form

$$w(x, t) = N \exp\left\{-\frac{\nu}{t}(\log R(x))^2\right\},$$

where ϵ is sufficiently small, N is sufficiently small, and ν is sufficiently large. We fix ϵ such that $\epsilon < 1$, $\text{dist}(x, \xi) \geq c_0 > 0$ if $\xi \in E$, $x \in \hat{\Omega}$ and $R(x) < \epsilon$, and such that $R(x)$ is in C^2 if $x \in \hat{\Omega}$, $R(x) < \epsilon$. Fix m so large that N_m (defined in Section 3) contains the set where $x \in \hat{\Omega}$, $R(x) = \epsilon$. By a result of Aronson [2],

$$G_m(x, t, \xi) > w(x, t) \qquad \text{if} \quad x \in \hat{\Omega}, \quad R(x) = \epsilon, \quad 0 < t < T$$

provided N is sufficiently small and ν is sufficiently large.

Since $G(x, t, \xi) \geq G_m(x, t, \xi)$, we have

$$v(x, t) > w(x, t) \qquad \text{if} \quad x \in \hat{\Omega}, \quad R(x) = \epsilon, \quad 0 < t < T.$$

Notice also that

$$v(x, 0) = w(x, 0) = 0 \qquad \text{if} \quad x \in \hat{\Omega}, \quad 0 < R(x) < \epsilon,$$

$$\varliminf_{R(x)\to 0} [v(x, t) - w(x, t)] = \varliminf_{R(x)\to 0} v(x, t) \geq 0 \qquad \text{if} \quad 0 < t < T.$$

Hence, if

$$Lw - w_t > 0 \qquad \text{for} \quad x \in \hat{\Omega}, \quad 0 < R(x) < \epsilon, \quad 0 < t < T, \tag{9.4}$$

then the maximum principle can be applied; it yields the assertion (9.3). Now, the left-hand side of (9.4) can be expressed by (3.19) with $\gamma = \nu$. Since, by (C′), $\mathcal{B}/R > -C$, it is clear that if ν is sufficiently large, then the first term on the right-hand side (with $\gamma = \nu$) dominates the negative contribution of each of the remaining terms. Thus (9.4) holds.

Similarly one can prove that, when (9.2) and the condition (C) hold,

$$G(x, t, \xi) \geq N \exp\left\{-\frac{\nu}{t}(\log R(\xi))^2\right\} \quad (N > 0, \quad \nu > 0) \tag{9.5}$$

provided $x \in E$, $0 < t < T$, $\xi \in \hat{\Omega}$, $R(\xi) < \epsilon$. We can thus state:

Theorem 9.1. *Let* (A), (B_S), (C′), (3.4) *and* (9.2) *hold. Let E be any compact subset of* $\hat{\Omega}$. *Then, for any* $T > 0$ *and for any* $\rho > 0$ *sufficiently small, there are positive constants* N, ν *such that* (9.3) *holds if* $\xi \in E$, $x \in \hat{\Omega}$, $R(x) < \rho$, $0 < t < T$. *If the condition* (C′) *is replaced by the condition* (C), *then* (9.5) *holds for* $x \in E$, $\xi \in \hat{\Omega}$, $R(\xi) < \rho$, $0 < t < T$.

If the condition (9.2) is replaced by the weaker condition

$$\sum a_{ij}(x) R_{x_i} R_{x_j} \geq \alpha R^{p+1} \qquad (\alpha > 0, \quad p > 1) \tag{9.6}$$

for all x in some $\hat{\Omega}$-neighborhood of $\partial\Omega$, then we can establish, instead of (9.3), (9.5), the inequalities

$$G(x, t, \xi) \geq N \exp\left\{-\frac{\nu}{t}(R(x))^{1-p}\right\},$$

$$G(x, t, \xi) \geq N \exp\left\{-\frac{\nu}{t}(R(\xi))^{1-p}\right\}. \tag{9.7}$$

respectively (for x, t, ξ in the same sets as before).

Finally, lower bounds at ∞, supplementary to the upper bounds derived in Section 4, can also be obtained using the above comparison function $w(x)$ with $R(x) = |x|$.

10. The Cauchy problem

Consider the Cauchy problem

$$\begin{aligned} Lu - u_t &= 0 \quad \text{if} \quad x \in R^n, \quad t > 0, \\ u(x, 0) &= f(x) \quad \text{if} \quad x \in R^n, \end{aligned} \tag{10.1}$$

where $f(x)$ is a bounded Borel measurable function and L is a degenerate elliptic operator (1.1). We define the solution of this problem to be the function

$$u(x, t) = E_x f(\xi(t)). \tag{10.2}$$

When the matrix $(a_{ij}(x))$ is uniformly positive definite, a_{ij}, b_i are bounded and uniformly Hölder continuous, and $f(x)$ is continuous, the function $u(x, t)$ is a classical solution of the Cauchy problem (see Section 6.4).

The purpose of this section is to investigate the continuity of $u(x, t)$ when $(a_{ij}(x))$ is degenerate and f is continuous or just measurable.

Theorem 10.1. *Let σ_{ij}, b_i be uniformly Lipschitz continuous in compact subsets of R^n and let* (5.2), (5.3) *hold. If $f(x)$ is a bounded continuous function, then $u(x, t)$ is continuous in $(x, t) \in R^n \times [0, \infty)$ and $u(x, 0) = f(x)$.*

Proof. We shall use the inequality (see Problem 2, Chapter 5)

$$E|\xi_y(t) - \xi_x(s)|^2 \leq \eta(|x - y|^2 + |t - s|) \qquad (\eta(r) \to 0 \quad \text{if} \quad r \to 0) \quad (10.3)$$

where $\xi_z(t)$ is the solution $\xi(t)$ of (1.2) with $\xi_z(0) = z$. By the Lebesgue bounded convergence theorem we then get

$$Ef(\xi_y(t)) \to Ef(\xi_x(s)) \qquad \text{if} \quad y \to x, \quad t \to s.$$

This proves the continuity of $u(y, t)$ at (x, s); $x \in R^n$, $s \geq 0$. Notice that $u(x, 0) = E_x f(\xi(0)) = f(x)$.

We now consider the more general case where $f(x)$ is Borel measurable. When (a_{ij}) is uniformly positive definite and a fundamental solution $\Gamma(x, t, \xi)$ can be constructed as in Section 6.4, the function $u(x, t)$ of (10.2) coincides with

$$\int \Gamma(x, t, \xi) f(\xi)\, d\xi;$$

since one can then show (using continuity properties of Γ) that this latter function is continuous in (x, t) in $R^n \times (0, \infty)$, the same is then true of $u(x, t)$. We shall prove there a similar result in case (a_{ij}) is degenerate.

Lemma 10.2. *Let σ_{ij}, b_i be uniformly Lipschitz continuous in compact subsets of R^n and let* (5.2), (5.3) *hold. Let A be a bounded domain with C^1 boundary and suppose that $P_x(\xi(s) \in \partial A) = 0$ for some $x \in R^n$, $s > 0$. Then the function*

$$(y, t) \to P_y(\xi(t) \in A)$$

is continuous at the point $(y, t) = (x, s)$.

Proof. From (10.3) it follows that

$$\overline{\lim_{y \to x, t \to s}} P_y(\xi(t) \in A) \leq P_x(\xi(s) \in A_\delta) \qquad \text{for any } \delta > 0,$$

where A_δ is a δ-neighborhood of A. Taking $\delta \to 0$, we get

$$\overline{\lim_{y \to x, t \to s}} P_y(\xi(t) \in A) \leq P_x(\xi(s) \in A \cup \partial A) = P_x(\xi(s) \in A).$$

Similarly,

$$\underline{\lim_{y \to x, t \to s}} P_y(\xi(t) \in A) \geq P_x(\xi(s) \in A),$$

and the proof is complete.

Theorem 10.3. *Let $f(x)$ be a bounded Borel measurable function in R^n, and let* (4.6) *and the assumptions of Theorem* 8.1 *hold. Then the solution $u(x, t)$ is continuous in $(x, t) \in R^n \times (0, \infty)$.*

Proof. If A is a bounded domain with C^1 boundary, then, by Theorem 8.1,

$$P_x(\xi(t) \in \partial A) = \int_{\partial A} K(x, t, \xi)\, d\xi = 0 \qquad (t > 0).$$

Thus, by Lemma 10.2, the function

$$(x, t) \to P_x(\xi(t) \in A) \tag{10.4}$$

is continuous in $R^n \times (0, \infty)$.Consider now the special case where f has compact support. For any $\epsilon > 0$, let $g(x)$ be a simple function such that

$$\sup|g| \leq 1 + \sup|f|,$$

$g(x) = \alpha_i$ (α_i constant) if $x \in A_i$ $(1 \leq i \leq l)$, $A_i \cap A_j = \emptyset$ if $i \neq j$, $\bigcup_{i=1}^{l} A_i$ contains the support of f, each A_i is bounded, $g(x) = 0$ if $x \notin \bigcup_{i=1}^{l} A_i$ and $|f(x) - g(x)| < \epsilon$ almost everywhere. Let B_i be bounded domains with C^1 boundary such that $B_i \supset A_i$ and the Lebesgue measure of $\bigcup_{i=1}^{l}(B_i \setminus A_i)$ is less than ϵ.

Then, for all (x, t), (x', t'),

$$\left| \int_{R^n} K(x, t, \xi) g(\xi)\, d\xi - \int_{R^n} K(x, t, \xi) f(\xi)\, d\xi \right| < \epsilon,$$

$$\left| \int_{R^n} K(x', t', \xi) g(\xi)\, d\xi - \int_{R^n} K(x', t', \xi) f(\xi)\, d\xi \right| < \epsilon.$$

Further, if $(x', t') \to (x, t)$, $t > 0$,

$$\overline{\lim} \left| \int_{R^n} K(x', t', \xi) g(\xi)\, d\xi - \int_{R^n} K(x, t, \xi) g(\xi)\, d\xi \right|$$

$$\leq (1 + \sup|f|) \left\{ \overline{\lim} \int_E K(x', t', \xi)\, d\xi + \int_E K(x, t, \xi)\, d\xi \right\}$$

by (10.4), where $E = \bigcup_{i=1}^{l}(B_i \setminus A_i)$. From the proof of Lemma 10.2,

$$\overline{\lim} \int_E K(x', t', \xi)\, d\xi \leq \int_{E_\delta} K(x, t, \xi)\, d\xi$$

where E_δ is any δ-neighborhood of E.

Putting these estimates together, we conclude that if $(x, t) \to (x, t)$, $t > 0$, then

$$\overline{\lim}\, |u(x' t') - u(x, t)| \leq 2\epsilon + 2(1 + \sup|f|) \int_{E_\delta} K(x, t, \xi) d\xi.$$

Since ϵ and δ are arbitrary, the left-hand side can be made arbitrarily small. consequently u is continuous at (x, t).

Consider now the general case where f is a bounded measurable function. Let

$$f_m(x) = \begin{cases} f(x) & \text{if } |x| \leqslant m, \\ 0 & \text{if } |x| > m. \end{cases}$$

Denote the solution of the Cauchy problem corresponding to f_m by u_m. By what we have already proved, each u_m is continuous. By Corollary 4.2, $u_m \to u$ uniformly on compact subsets. Consequently, u is continuous.

Consider next the case of two-sided obstacle, where only a generalized fundamental solution exists. We first take

$$f(x) = \chi_A(x), \tag{10.5}$$

the characteristic function of a set A. We assume:

(E) A is a bounded domain with C^1 boundary, and it intersects precisely one of the sets ∂G_i; further, $k_0 + 1 \leqslant i \leqslant k$ and the intersection $\partial A \cap \partial G_i$ is a C^1 $(n-2)$-dimensional hypersurface.

Theorem 10.4. *Let the assumptions of Theorem* 7.1 *and* (10.5), (E) *hold. Then the solution* $u(x, t)$ *is continuous in* $(x, t) \in R^n \times (0, \infty)$.

Proof. It is enough to prove the continuity of $u(y, t)$ at $y \in \partial\Omega$. In view of Lemma 10.2, it suffices to prove that

$$P_y(\xi(t) \in \partial A) = 0 \qquad \text{if} \quad y \in \partial G_j, \quad t > 0. \tag{10.6}$$

In view of Theorem 6.1 the left-hand side of (10.6) vanishes if $j \neq i$. If $j = i$, then, by Theorems 6.1, 7.1,

$$P_y(\xi(t) \in \partial A) = P_y\{\xi(t) \in (\partial A \cap G_i)\} = \int_{\partial A \cap G_i} \hat{K}_i(x, t, \xi)\, dS_\xi^i = 0.$$

Thus the proof is complete.

Corollary 10.5. *Let the assumption of Theorem* 7.1 *hold and let* $f(x)$ *be any bounded Borel measurable function, continuous at all the points of* $\partial\Omega$. *Then* $u(x, t)$ *is continuous in* $(x, t) \in R^n \times (0, \infty)$.

The proof is left to the reader (see Problem 8).

Remark. If f is a bounded continuous function in R^n, then $u(x, t)$ is continuous (by Theorem 10.1). Let

$$\tilde{f}(x) = \begin{cases} f(x) & \text{if} \quad x \neq z_i, \\ f_i & \text{if} \quad x = z_i \qquad (f_i \neq f(z_i)) \end{cases}$$

for some $i, 1 \leqslant i \leqslant k_0$. Denote by $\tilde{u}$ the solution corresponding to $\tilde{f}$. Then

$u(x, t) = u(x, t)$ if $x \neq z_i$, but

$$\tilde{u}(z_i, t) = f_i \neq f(z_i) = u(z_i, t).$$

Consequently, $\tilde{u}(x, t)$ is discontinuous at the points (z_i, t), $t > 0$. On the other hand, if S is as in Remark 4 at the end of Section 8, so that (8.12) holds, then Theorem 10.1 remains valid even if one changes the definition of $f(x)$, in an arbitrary manner, on the set S. Further, the solution $u(x, t)$ $(t > 0)$ does not change when one changes the definition of f on S.

PROBLEMS

1. Prove the relation (1.5). [*Hint:* Use Green's identity with $u(y, \sigma) = G_{m,\epsilon}(y, \sigma, \xi)$, $v(y, \sigma) = G^*_{m,\epsilon}(y, \sigma, x)$ in $B_m \times (\epsilon < \sigma < t - \epsilon)$ and take $\epsilon \to 0$; cf. the proof of Theorem 6.4.7.]
2. Prove (1.14). [*Hint:* Use (6.4.21) with $u = G_{k,\epsilon}$, $v = G_{m,\epsilon}$ in $B_m \times (\epsilon < s < t - \epsilon)$ and take $\epsilon \to 0$.]
3. Prove (1.15), (1.16). [*Hint:* Recall that $G_{m,\epsilon} = \Gamma_\epsilon + V_\epsilon$ where Γ_ϵ is a fundamental solution, and use the maximum principle to estimate V_ϵ, $D_\zeta V_\epsilon$.]
4. Prove (1.20). [*Hint:* Let $v_y(x)$ be a barrier ($L_\epsilon v_y(x) \leq -1$). Show that $C_0 v_y(\zeta) - v(\zeta, s) \geq 0$ if $\zeta \in V$, and $= 0$ at $\zeta = y$.]
5. Prove that if $V_\epsilon - f$ satisfies (2.10), then (2.11) holds. [*Hint:* If $v_y(x)$ is a barrier, show that
$$|V_\epsilon(x, t, \xi) - f(x, t)| \leq Ce^{-c/t} v_y(x).$$
Hence $|D_y V_\epsilon(y, t, \xi)| \leq C_1 e^{-c/t}$ for $y \in \partial B$. Now use Green's formula
$$V_\epsilon(x, t, \xi) = \int \left(\Gamma_\epsilon \frac{\partial}{\partial T} V_\epsilon - V_\epsilon \frac{\partial}{\partial T} \Gamma_\epsilon \right) \quad \text{in} \quad B \times (0, t).]$$
6. Prove that if in the assumptions of Theorem 9.1 we replace (9.2) by (9.6), then (9.7) holds.
7. If A contains in its interior the point z_i but does not intersect the other sets G_j, $j \neq i$, then the assertion of Theorem 10.4 is again valid.
8. Prove Corollary 10.5. [*Hint*: Approximate $f(x)$ uniformly by simple functions $\Sigma c_j \chi_{A_j}$ where A_j are bounded closed domains and either $A_j \cap \partial\Omega = \emptyset$, or A_j satisfies the condition (E), or A_j contains in its interior one point z_i and does not intersect any G_l, $l \neq i$.]
9. The assertion of Theorem 10.4 is false if $\partial A \cap \partial G_j$ contains a set of positive surface area, or if $A = \{z_i\}$ for some $1 \leq i \leq k_0$.
10. Consider the equation $u_t = x^2 u_{xx} + b(x) u_x$ with either $b(x) = x$ or $b(x) = 0$. Use the transformation $x' = \log x$ in order to compute the fundamental solution Γ_0. Verify directly the general properties of Γ_0, proved in this chapter, from the explicit formula for Γ_0.

16

Stopping Time Problems and Stochastic Games

Part I. The Stationary Case

1. Statement of the problem

Consider a system of n stochastic differential equations

$$dx(t) = b(x(t))\,dt + \sigma(x(t))\,dw(t) \tag{1.1}$$

where $b = (b_1, \ldots, b_n)$ and σ is an $n \times n$ matrix (σ_{ij}). We assume:

(A_1) For all $x \in R^n$,

$$|b(x)| + |\sigma(x)| \leq C(1 + |x|) \qquad (C \text{ const}),$$

and for any $R > 0$ there is a constant C_R such that

$$|b(x) - b(y)| + |\sigma(x) - \sigma(y)| \leq C_R|x - y|$$

if $x \in R^n$, $y \in R^n$, $|x| \leq R$, $|y| \leq R$.

For any nonempty closed set $A \subset R^n$, denote by t_A the hitting time of the set A by $x(t)$, i.e.,

$$t_A = \inf\{t;\, t \geq 0,\, x(t) \in A\}.$$

For any set $B \subset R^n$, denote by B^c the complement of B in R^n.

Let Ω be a nonempty domain in R^n. (In particular, one can take $\Omega = R^n$.) Denote by $\partial\Omega$ the boundary of Ω, and let $\overline{\Omega} = \Omega \cup \partial\Omega$.

Let E, F be given closed subsets of $\overline{\Omega}$, such that

$$\partial\Omega \subset E, \qquad \partial\Omega \subset F.$$

We denote by $\mathcal{C}_x$ the set of all a.s. finite-valued stopping times of the process $x(t)$, given $x(0) = x$.

We denote by $\mathcal{Q}_x$ the subset of $\mathcal{C}_x$ consisting of all stopping times σ for which

$$\sigma < t_{\Omega^c}, \qquad x(\sigma) \in E \quad \text{a.s.}$$

Similarly we shall denote by $\mathcal{B}_x$ the subset of $\mathcal{C}_x$ consisting of all stopping times τ for which

$$\tau < t_{\Omega^c}, \qquad x(\tau) \in F \quad \text{a.s.}$$

In the sequel it is always assumed that the initial point $x(0) = x$ is in $\bar{\Omega}$.

Notice that $\sigma \equiv 0$ is in $\mathcal{A}_x$ if and only if $x \in E$. If $\Omega = R^n$, then $\sigma \in \mathcal{A}_x$ if and only if $P_x(\sigma < \infty, x(\sigma) \in E) = 1$ and $\tau \in \mathcal{B}_s$ if and only if $P_x(\tau < \infty, x(\tau) \in F) = 1$. In case $E = F = \bar{\Omega}$, then $\mathcal{A}_x = \mathcal{B}_x$; also, $\sigma \in \mathcal{A}_x$ if and only if $P_x(\sigma < \infty, \sigma < t_{\Omega^c}) = 1$.

Let $f(x)$, $\psi_1(x)$, $\psi_2(x)$ be given functions defined on $\bar{\Omega}$, and let α be a nonnegative number. We introduce two functionals:

$$J_x(\sigma) = E_x\left\{ \int_0^\sigma e^{-\alpha t} f(x(t))\, dt + e^{-\alpha\sigma}\psi_1(x(\sigma)) \right\}, \tag{1.2}$$

$$J_x(\sigma, \tau) = E_x\left\{ \int_0^{\sigma\wedge\tau} e^{-\alpha t} f(x(t))\, dt + e^{-\alpha\sigma}\psi_1(x(\sigma))\chi_{\sigma<\tau} + e^{-\alpha\tau}\psi_2(x(\tau))\chi_{\tau\leq\sigma} \right\}. \tag{1.3}$$

Here χ_A denotes the indicator function of a set A, $\sigma \wedge \tau = \min(\sigma, \tau)$, σ varies in $\mathcal{A}_x$ and τ varies in $\mathcal{B}_x$. The nonnegative number α will be called the *discount coefficient*.

Definition. The functional (1.2) will be called the *cost functional* and the functional (1.3) will be called the *payoff functional*.

First we introduce the problem corresponding to the cost functional. We are interested in the quantity

$$V(x) = \inf_{\sigma \in \mathcal{A}_x} J_x(\sigma), \tag{1.4}$$

which we call the *optimal cost*.

Definition. If there exists a stopping time $\hat{\sigma} \in \mathcal{A}_x$ such that

$$V(x) = J_x(\hat{\sigma}), \tag{1.5}$$

then we say that $\hat{\sigma}$ is an *optimal stopping time, given* $x(0) = x$ (or, *for* x). If there exists a closed subset $\hat{E}$ of E such that $t_{\hat{E}}$ is an optimal stopping time for all $x \in \bar{\Omega}$, then we say that $\hat{E}$ is an *optimal stopping set* and that $\bar{\Omega} \setminus \hat{E}$ is an *optimal domain of continuation*.

The recipe for finding $V(x)$ is then to continue with the stochastic process $x(t)$ as long as $x(t)$ is in $\bar{\Omega} \setminus \hat{E}$ and to stop immediately upon hitting the set $\hat{E}$.

The problem of studying the optimal stopping sets will be called the *stopping time problem.*

If instead of (1.4) we consider

$$V(x) = \sup_{\tau \in \mathfrak{B}_x} J_x(\tau),$$

then we call $J_x(\tau)$ the *reward functional* and $V(x)$ the *optimal reward.* If we replace J_x by $-J_x$, then the study of this problem reduces to the study of the preceding problem; we shall therefore not pursue it further.

Next we introduce a problem corresponding to the payoff functional (1.3). We consider a scheme whereby, for a given $x \in \bar{\Omega}$, a player P_1 chooses any stopping time $\sigma \in \mathfrak{A}_x$ and a player P_2 chooses any stopping time $\tau \in \mathfrak{B}_x$. The resulting payoff, that P_1 pays to P_2, is $J_x(\sigma, \tau)$. Thus, the aim of P_1 is to minimize $J_x(\sigma, \tau)$, and the aim of P_2 is to maximize $J_x(\sigma, \tau)$. We shall call this scheme the *stochastic game* associated with (1.1), (1.3) and denote it by G_x. We shall denote the collection $\{G_x;\, x \in \bar{\Omega}\}$ by G, and call it the stochastic game associated with (1.1), (1.3) in $\bar{\Omega}$.

If

$$\inf_{\sigma \in \mathfrak{A}_x} \sup_{\tau \in \mathfrak{B}_x} J_x(\sigma, \tau) = \sup_{\tau \in \mathfrak{B}_x} \inf_{\sigma \in \mathfrak{A}_x} J_x(\sigma, \tau), \tag{1.6}$$

then we say that the stochastic game G_x has *value*, and the common number in (1.6) is called the *value of the game* G_x. We shall denote it by $V(x)$.

Suppose there exist stopping times σ_x^*, τ_x^* in $\mathfrak{A}_x$ and $\mathfrak{B}_x$, respectively, such that

$$J_x(\sigma_x^*, \tau) \leq J_x(\sigma_x^*, \tau_x^*) \leq J_x(\sigma, \tau_x^*) \tag{1.7}$$

for all $\sigma \in \mathfrak{A}_x$, $\tau \in \mathfrak{B}_x$. Then we call (σ_x^*, τ_x^*) a *saddle point* of G_x. It is then clear that the game G_x has value, and

$$V(x) = J_x(\sigma_x^*, \tau_x^*). \tag{1.8}$$

Suppose there exist closed sets $\hat{E} \subset E$ and $\hat{F} \subset F$ such that the pair

$$\sigma_x^* = t_{\hat{E}}, \qquad \tau_x^* = t_{\hat{F}}$$

forms a saddle point of G_x, for any $x \in \bar{\Omega}$. Then we say that the pair $(t_{\hat{E}}, t_{\hat{F}})$ is a saddle point for G, and we call the pair $(\hat{E}, \hat{F})$ a *saddle point of sets* for G.

The study of the stopping time problem is similar to (but simpler than) the study of stochastic games. We shall adopt the following course: we first study in detail stochastic games, and then state briefly the corresponding results for the stopping time problem, leaving the proofs for the reader.

In Part I of this chapter (Sections 1–8), the coefficients of the stochastic system and the coefficients f, ψ_1, ψ_2 occurring in the payoff are independent of t. We refer to this case as the *stationary case*. The problem of finding a saddle point (or optimal stopping set) will be reduced to a problem of solving an elliptic variational inequality (see Sections 4, 7). In Part II (Sections 9–13) we shall deal with time-dependent coefficients (i.e., with the *nonstationary* case). In that case the stochastic problem reduces to a problem of solving a parabolic variational inequality.

2. Characterization of saddle points

We shall need the following conditions:

(A_2) The sets $\mathcal{A}_x$, $\mathcal{B}_x$ are nonempty, for any $x \in \Omega$.

(A_3) $f(x)$, $\psi_1(x)$, and $\psi_2(x)$ are continuous and bounded functions in $\bar{\Omega}$, and

$$E_x \int_0^{\sigma \wedge \tau} e^{-\alpha t} |f(x(t))| \, dt < \infty \tag{2.1}$$

for all $x \in \Omega$, $\sigma \in \mathcal{A}_x$, $\tau \in \mathcal{B}_x$.

Denote by L the elliptic operator corresponding to the diffusion process (1.1), that is,

$$Lu \equiv \frac{1}{2} \sum_{i,j=1}^{n} a_{ij}(x) \frac{\partial^2 u}{\partial x_i \, \partial x_j} + \sum_{i=1}^{n} b_i(x) \frac{\partial u}{\partial x_i}$$

where $a = (a_{ij}) = \sigma\sigma^*$ (σ^* = transpose of σ).

Lemma 2.1. *Let* (A_1) *hold. Suppose there exists a function* u *in* $C(\bar{\Omega}) \cap C^2(\Omega)$ *such that*

$$Lu \leq -1 \quad \text{in } \Omega, \qquad |u| \leq C_0 \quad \text{in } \Omega \qquad (C_0 \text{ const}).$$

Then $t_{\Omega^c} < \infty$ *a.s. and, in fact,*

$$E_x t_{\Omega^c} \leq 2C_0 \qquad \text{for all} \quad x \in \Omega.$$

Proof. By Itô's formula,

$$E_x u(x(\lambda)) - u(x) = E_x \int_0^{\lambda} Lu(x(t)) \, dt \leq -E_x \lambda$$

for $\lambda = t_{\Omega^c} \wedge T$, $T > 0$. Hence

$$E_x \lambda \leq 2C_0.$$

Letting $T \uparrow \infty$ and noting that $\lambda \uparrow t_{\Omega^c}$, we obtain the asserted conclusion.

Corollary 2.2. *Let* (A_1) *hold, and let* Ω *be a domain contained in a strip* $-\infty < \beta \leq x_1 \leq \gamma < \infty$. *Assume also that* $a_{11}(x)\alpha^2 + b_1(x)\alpha \geq 1$ *for all*

$x \in \bar{\Omega}$. Then

$$E_x t_{\Omega^c} \leq C_1 < \infty \qquad \textit{for all} \quad x \in \Omega \qquad (C_1 \textit{ const}).$$

Consequently, the condition (A_2) *is satisfied, and* (2.1) *holds if f is a bounded function in $\bar{\Omega}$.*

Proof. The function

$$u(x) = -Ae^{\alpha x_1} \qquad (A > 0)$$

satisfies $Lu < -1$ in Ω, provided A is sufficiently large. Now use Lemma 2.1 to deduce that $E_x t_{\Omega^c} \leq C_1 < \infty$. Since $\partial\Omega \subset E \cap F$, the classes $\mathcal{A}_x$, $\mathcal{B}_x$ contain at least one element, namely, t_{Ω^c}. Finally, if $|f| \leq M$,

$$E_x \int_0^{\sigma \wedge \tau} e^{-\alpha t} |f(x(t))|\, dt \leq M E_x(\sigma \wedge \tau) \leq MC_1 < \infty.$$

Theorem 2.3. *Let* (A_1)–(A_3) *hold and suppose $(\hat{E}, \hat{F})$ is a saddle point of sets for the stochastic game G. Then the value function $V(x)$ satisfies the following properties:*

$$V(x) \leq \psi_1(x) \qquad \textit{if} \quad x \in E \setminus \hat{F}, \tag{2.2}$$

$$V(x) \geq \psi_2(x) \qquad \textit{if} \quad x \in F, \tag{2.3}$$

$$V(x) = \psi_1(x) \qquad \textit{if} \quad x \in \hat{E} \setminus \hat{F}, \tag{2.4}$$

$$V(x) = \psi_2(x) \qquad \textit{if} \quad x \in \hat{F}, \tag{2.5}$$

$$V(x) \leq E_x\left\{\int_0^\lambda e^{-\alpha t} f(x(t))\, dt + e^{-\alpha\lambda} V(x(\lambda))\right\} \qquad \textit{if} \quad \lambda \in \mathcal{C}_x, \quad \lambda \leq t_{\hat{F}}, \tag{2.6}$$

$$V(x) \geq E_x\left\{\int_0^\mu e^{-\alpha t} f(x(t))\, dt + e^{-\alpha\mu} V(x(\mu))\right\} \qquad \textit{if} \quad \mu \in \mathcal{C}_x, \quad \mu \leq t_{\hat{E}}. \tag{2.7}$$

Proof. From the definition of $V(x)$ we have

$$J_x(t_{\hat{E}}, \tau) \leq V(x) \leq J_x(\sigma, t_{\hat{F}}) \qquad \text{for any } \sigma \in \mathcal{A}_x, \quad \tau \in \mathcal{B}_x. \tag{2.8}$$

If $x \in E$, then $\sigma \equiv 0$ belongs to $\mathcal{A}_x$. If, further, $x \notin \hat{F}$, then $t_{\hat{F}} > 0$ a.s. Hence

$$J_x(\sigma, t_{\hat{F}}) = \psi_1(x).$$

The second inequality in (2.8) now yields the assertion (2.2).

Next, let $x \in F$. Then $\tau \equiv 0$ belongs to $\mathcal{B}_x$. Since $0 \leq t_{\hat{E}}$ a.s.,

$$J_x(t_{\hat{E}}, \tau) = \psi_2(x).$$

Using the first inequality in (2.8), we obtain (2.3).

To prove (2.4), notice that if $x \in \hat{E} \setminus \hat{F}$, then

$$t_{\hat{E}} = 0 < t_{\hat{F}} \quad \text{a.s.}$$

Hence

$$V(x) = J_x(t_{\hat{E}}, t_{\hat{F}}) = \psi_1(x).$$

Next, if $x \in \hat{F}$, then $t_{\hat{F}} = 0 \leq t_{\hat{E}}$ a.s., so that

$$V(x) = J_x(t_{\hat{E}}, t_{\hat{F}}) = \psi_2(x),$$

that is, (2.5) holds.

We proceed to prove (2.6). Set $\tau_0 = t_{\hat{F}}$. Then

$$\begin{aligned} V(x) = \inf_{\sigma \in \mathcal{Q}_x} J_x(\sigma, t_{\hat{F}}) &\leq \inf_{\sigma \in \mathcal{Q}_x,\, \sigma \geq \lambda} E_x\left\{ \int_0^{\sigma \wedge \tau_0} e^{-\alpha t} f(x(t))\, dt \right. \\ &\quad \left. + e^{-\alpha\sigma}\psi_1(x(\sigma))\chi_{\sigma<\tau_0} + e^{-\alpha\tau_0}\psi_2(x(\tau_0))\chi_{\tau_0\leq\sigma} \right\} \\ &= \inf_{\sigma \in \mathcal{Q}_x,\, \sigma \geq \lambda} E_x E_x\left\{ \int_0^{\sigma \wedge \tau_0} e^{-\alpha t} f(x(t)) + e^{-\alpha\sigma}\psi_1(x(\sigma))\chi_{\sigma<\tau_0} \right. \\ &\quad \left. + e^{-\alpha\tau_0}\psi_2(x(\tau_0))\chi_{\tau_0\leq\sigma} \right\} / \mathcal{F}_\lambda \end{aligned}$$

where $\mathcal{F}_\lambda$ is the σ-field generated by the process $x(t)$ for $t \leq \lambda$. By the strong Markov property, the right-hand side is equal to

$$\begin{aligned} &\inf_{\sigma \in \mathcal{Q}_x} E_x\left\{ \int_0^{\lambda} e^{-\alpha t} f(x(t))\, dt + E_{x(\lambda)}\left\{ \int_0^{\sigma \wedge \tau_0} e^{-\alpha t} f(x(t))\, dt \right.\right. \\ &\quad \left.\left. + e^{-\alpha\sigma}\psi_1(x(\sigma))\chi_{\sigma<\tau_0} + e^{-\alpha\tau_0}\psi_2(x(\tau_0))\chi_{\tau_0\leq\sigma} \right\}\right\}. \end{aligned}$$

Here we have used the fact that $\tau_0 \geq \lambda$. Since

$$\inf_{\sigma \in \mathcal{Q}_x} E_{x(\lambda)}\{\cdots\} = V(x(\lambda)) \qquad \text{for any} \quad \lambda = \lambda(\omega)$$

(where $\{\cdots\}$ stands for the content of the inner braces of the previous expression), the assertion (2.6) follows. The proof of (2.7) is similar.

Notice that the inequalities (2.2), (2.3) imply that

$$\psi_2(x) \leq \psi_1(x) \qquad \text{if} \quad x \in (E \cap F) \setminus \hat{F}. \tag{2.9}$$

Thus, for the existence of a saddle point of sets $(\hat{E}, \hat{F})$, it is necessary that (2.9) hold.

We shall now prove a converse of Theorem 2.2.

Theorem 2.4. *Let* (A_1)–(A_3) *hold. Suppose there exist closed sets* $\hat{E} \subset E$ *and*

$\hat{F} \subset F$ and a Borel measurable function $V(x)$ defined on $\bar{\Omega}$ such that

$$t_{\hat{E}} \in \mathcal{A}_x, \qquad t_{\hat{F}} \in \mathcal{B}_x, \tag{2.10}$$

(2.2)–(2.7) hold, and

$$\psi_1(x) = \psi_2(x) \qquad \text{if} \quad x \in \hat{E} \cap \hat{F}. \tag{2.11}$$

Then $(\hat{E}, \hat{F})$ is a saddle point of sets for the stochastic game G, and $V(x)$ is the value of the game.

Remark 1. If Ω is a bounded domain and (σ_{ij}) is nondegenerate in $\bar{\Omega}$, then, by Corollary 2.2, $t_{\Omega^c} < \infty$ a.s. Hence, if $\partial\Omega \in \hat{E}$, $\partial\Omega \in \hat{F}$, then the assumption (2.10) is satisfied.

Remark 2. The condition (2.11) means that the game is "fair." Indeed, from the form of $J_x(\sigma, \tau)$ we see that the player P_2 has a "slight" advantage, for he controls ψ_2 on the set $\tau = \sigma$. The condition (2.11) abolishes this advantage on the set $\hat{E} \cap \hat{F}$; in the complement of $\hat{E} \cap \hat{F}$ this advantage is irrelevant.

Proof. What we have to show is that

$$J_x(t_{\hat{E}}, \tau) \leq V(x) \qquad \text{if} \quad \tau \in \mathcal{B}_x, \tag{2.12}$$

$$J_x(\sigma, t_{\hat{F}}) \geq V(x) \qquad \text{if} \quad \sigma \in \mathcal{A}_x. \tag{2.13}$$

From the formula

$$J_x(t_{\hat{E}}, \tau) = E_x\left\{ \int_0^{t_{\hat{E}} \wedge \tau} e^{-\alpha t} f(x(t))\, dt \right.$$

$$\left. + e^{-\alpha t_{\hat{E}}} \psi_1(x(t_{\hat{E}}))\chi_{t_{\hat{E}} < \tau} + e^{-\alpha\tau}\psi_2(x(\tau))\chi_{\tau \leq t_{\hat{E}}} \right\},$$

it is clear that if we can prove the inequalities

$$\psi_1(x(t_{\hat{E}})) \leq V(x(t_{\hat{E}})), \tag{2.14}$$

$$\psi_2(x(\tau)) \leq V(x(\tau)), \tag{2.15}$$

then

$$J_x(t_{\hat{E}}, \tau) \leq E_x\left\{ \int_0^{t_{\hat{E}} \wedge \tau} e^{-\alpha t} f(x(t))\, dt + e^{-\alpha(t_{\hat{E}} \wedge \tau)} V(x(t_{\hat{E}} \wedge \tau)) \right\}.$$

Hence, by (2.7) with $\mu = t_{\hat{E}} \wedge \tau$,

$$J_x(t_{\hat{E}}, \tau) \leq V(x),$$

that is, (2.12) holds.

To prove (2.14) notice that $x(t_{\hat{E}}) \in \hat{E}$. Therefore, if $x(t_{\hat{E}}) \notin \hat{F}$, then, by

(2.4),

$$\psi_1(x(t_E)) = V(x(t_E)).$$

If, on the other hand, $x(t_E) \in \hat{F}$, then by (2.5), (2.11)

$$\psi_1(x(t_E)) = \psi_2(x(t_E)) = V(x(t_E)),$$

Thus, (2.14) is proved.

To prove (2.15), notice that $x(\tau) \in F$. Hence (2.15) is a consequence of (2.3). We have thus completed the proof of (2.12). The proof of (2.13) is similar.

3. Elliptic variational inequalities in bounded domains

The notation $W^{m,p}(\Omega)$, $W_0^{1,p}(\Omega)$ introduced in Sections 10.2, 10.3 will now be used. We shall denote by (,) the scalar product in $L^2(\Omega)$. Consider a partial differential operator

$$Lu = \frac{1}{2} \sum_{i,j=1}^{n} a_{ij} \frac{\partial^2 u}{\partial x_i \, \partial x_j} + \sum_{i=1}^{n} b_i \frac{\partial u}{\partial x_i} .$$

We shall need the following assumptions:

(B_1) Ω is a bounded domain with C^2 boundary $\partial\Omega$. The functions $a_{ij}(x)$, $b_i(x)$ are bounded and measurable in Ω, the a_{ij} are uniformly continuous in $\bar{\Omega}$, and

$$\tfrac{1}{2} \sum_{i,j=1}^{n} a_{ij}(x)\xi_i\xi_j \geq \beta|\xi|^2 \qquad \text{if } x \in \bar{\Omega}, \quad \xi \in R^n \qquad (\beta > 0).$$

(B_2) The functions $\psi_1(x)$ and $\psi_2(x)$ are continuous in $\bar{\Omega}$ and belong to $W^{2,p}(\Omega)$ for some $p > n$, and

$$\psi_1(x) \geq \psi_2(x) \qquad \text{in } \Omega, \tag{3.1}$$

$$\psi_1(x) = \psi_2(x) \qquad \text{on } \partial\Omega. \tag{3.2}$$

The function $f(x)$ belongs to $W^{0,p}(\Omega)$.

Let

$$\tilde{f} = f + (L\psi_2 - \alpha\psi_2), \qquad \psi = \psi_1 - \psi_2, \qquad \alpha \text{ nonnegative constant.} \tag{3.3}$$

Notice, by (3.1), (3.2), that $\psi \geq 0$ on Ω and $\psi = 0$ on $\partial\Omega$.

Consider the problem of finding a function u satisfying

$$u \in W^{2,2}(\Omega) \cap W_0^{1,2}(\Omega), \qquad 0 \leq u \leq \psi \ \text{ a.e.} \quad \text{in } \Omega,$$

$$\int_\Omega (Lu - \alpha u + \tilde{f})(v - u)\,dx \leq 0, \qquad \text{for any } v \in L^2(\Omega), \tag{3.4}$$

$$0 \leq v \leq \psi \ \text{ a.e.} \quad \text{in } \Omega.$$

This problem is called an *elliptic variational inequality.*

Theorem 3.1. *Let* (B_1), (B_2) *hold. Then there exists a continuous solution in* $W^{2,p}(\Omega)$ *of the variational inequality* (3.4). *If* a_{ij}, b_i *are Lipschitz continuous in* $\bar{\Omega}$, *then the solution is unique.*

Proof. Let $A = -L + \alpha$. For any $\epsilon > 0$, consider the problem

$$Au + \frac{1}{\epsilon}(u-\psi)^+ - \frac{1}{\epsilon}u^- = \tilde{f} \quad \text{a.e. in } \Omega, \qquad u \in W^{2,2}(\Omega) \cap W_0^{1,2}(\Omega), \tag{3.5}$$

or, equivalently,

$$A_\epsilon u = \frac{1}{\epsilon}K(x,u) + \tilde{f} \quad \text{a.e.} \quad \text{in } \Omega, \qquad u \in W^{2,2}(\Omega) \cap W_0^{1,2}(\Omega) \tag{3.6}$$

where

$$A_\epsilon u = Au + \frac{1}{\epsilon}u \qquad \text{and} \quad K(x,u) = u - (u-\psi)^+ + u^-.$$

It is clear that $K(x, u)$ is measurable in $(x, u) \in \bar{\Omega} \times R^1$, and

$$0 \leqslant K(x,u) \leqslant \psi(x). \tag{3.7}$$

Since the coefficient of u in A_ϵ is > 0, the maximum principle can be applied. Consequently (by Theorems 10.3.1, 10.3.2), for any $g \in L^p(\Omega)$, there exists a unique solution of the Dirichlet problem

$$A_\epsilon v = g \quad \text{a.e.} \quad \text{in } \Omega, \qquad v \in W^{2,2}(\Omega) \cap W_0^{1,2}(\Omega), \tag{3.8}$$

and

$$|v|_{2,p}^{\Omega} \leqslant C|g|_{0,p}^{\Omega} \qquad (C \text{ const}).$$

Using this result and the Schauder's fixed point theorem (see Problem 1), one can derive the existence of a solution u_ϵ of (3.6) which belongs to $W^{2,p}(\Omega)$.

Denote the solution v of (3.8) by $R_\epsilon g$. The maximum principle implies (see Problem 2) that

$$\text{if} \quad f \geqslant g \quad \text{a.e.} \quad \text{on } \Omega, \qquad \text{then} \quad R_\epsilon f \geqslant R_\epsilon g \quad \text{on } \Omega. \tag{3.9}$$

Recalling (3.7) we then have

$$u_\epsilon = R_\epsilon\left[\tilde{f} + \frac{1}{\epsilon}K(x,u_\epsilon)\right] \leqslant R_\epsilon\left[\tilde{f} + \frac{1}{\epsilon}\psi\right] = R_\epsilon(\tilde{f} + A\psi) + \psi,$$

where the relation

$$\frac{1}{\epsilon}R_\epsilon\psi = R_\epsilon A\psi + \psi \qquad (\psi \in W^{2,2}(\Omega) \cap W_0^{1,2}(\Omega))$$

has been used. It follows that

$$\frac{1}{\epsilon}(u_\epsilon - \psi) \leq \frac{1}{\epsilon} R_\epsilon(\tilde{f} + A\psi).$$

Hence

$$0 \leq \frac{1}{\epsilon}(u_\epsilon - \psi)^+ \leq \frac{1}{\epsilon} |R_\epsilon(\tilde{f} + A\psi)|.$$

Since, by Theorem 10.3.2,

$$\frac{1}{\epsilon} |R_\epsilon g|_{0,p}^{\Omega} \leq C |g|_{0,p}^{\Omega} \qquad \text{for any} \quad g \in L^p(\Omega), \tag{3.10}$$

where C is a constant independent of ϵ, we conclude that

$$\left| \frac{1}{\epsilon}(u_\epsilon - \psi)^+ \right|_{0,p}^{\Omega} \leq C \qquad (C \text{ independent of } \epsilon). \tag{3.11}$$

Next,

$$u_\epsilon = R_\epsilon\left[\tilde{f} + \frac{1}{\epsilon} K(x, u_\epsilon)\right] \geq R_\epsilon \tilde{f}$$

by (3.7), (3.9). Hence

$$0 \leq \frac{1}{\epsilon} u_\epsilon^- \leq \frac{1}{\epsilon} |R_\epsilon \tilde{f}|.$$

Using (3.10), we obtain

$$\left| \frac{1}{\epsilon} u_\epsilon^- \right|_{0,p}^{\Omega} \leq C \qquad (C \text{ independent of } \epsilon). \tag{3.12}$$

From (3.11), (3.12) and (3.5) it follows that

$$|Au_\epsilon|_{0,p}^{\Omega} \leq C.$$

Hence (by Theorem 10.3.2),

$$|u_\epsilon|_{2,p}^{\Omega} \leq C \tag{3.13}$$

where C is a constant independent of ϵ. Since $p > n$, by the Sobolev inequalities (Section 10.2), u_ϵ can be taken to be continuously differentiable in $\bar{\Omega}$.

Since $p > n$, the Sobolev inequalities also imply that

$$|u_\epsilon(x)| + \sum_{i=1}^{n} \left| \frac{\partial u_\epsilon(x)}{\partial x_i} \right| \leq C,$$

$$\sum_{i=1}^{n} \left| \frac{\partial u_\epsilon(x)}{\partial x_i} - \frac{\partial u_\epsilon(\bar{x})}{\partial x_i} \right| \leq C|x - \bar{x}|^{\mu} \tag{3.14}$$

for all x, $\bar{x}$ in Ω, where C, μ are positive constants independent of ϵ. Hence, we can choose a sequence $\{\epsilon_m\}$, decreasing monotonically to 0, such that

$$\begin{aligned} u_{\epsilon_m} \to u, \qquad \frac{\partial u_{\epsilon_m}}{\partial x_i} \to \frac{\partial u}{\partial x_i} \quad \text{uniformly in } \Omega, \\ \frac{\partial^2 u_{\epsilon_m}}{\partial x_i \partial x_j} \to \frac{\partial^2 u}{\partial x_i \partial x_j} \quad \text{weakly in } L^p(\Omega). \end{aligned} \tag{3.15}$$

We shall next prove that

$$(u - \psi)^+ = 0 \quad \text{in } \Omega, \tag{3.16}$$

$$u^- = 0 \quad \text{in } \Omega. \tag{3.17}$$

To prove (3.16), notice that, for any $v \in L^2(\Omega)$,

$$\left((u_\epsilon - \psi)^+ - (v - \psi)^+, u_\epsilon - v\right) \geq 0.$$

Taking $\epsilon = \epsilon_m \to 0$ and using (3.11), we get

$$-\left((v - \psi)^+, u - v\right) \geq 0.$$

Substituting $v = u - kw$ ($w \in L^2(\Omega)$, $k > 0$), we obtain

$$-\left((u - \psi - kw)^+, w\right) \geq 0,$$

and taking $k \to 0$ we find that

$$-\left((u - \psi)^+, w\right) \geq 0.$$

Since w is arbitrary, (3.16) follows.

To prove (3.17), we begin with the inequality

$$-\left(u_\epsilon^- - v^-, u_\epsilon - v\right) \geq 0 \qquad (v \in L^2(\Omega)),$$

and then proceed as before, making use of (3.12).

The assertions (3.16), (3.17) are equivalent to $0 \leq u \leq \psi$ in Ω. Recall that u is also continuous (in fact, continuously differentiable) in $\overline{\Omega}$, and that it belongs to $W_0^{1,2}(\Omega)$ (so that $u = 0$ on $\partial\Omega$). We shall now verify the inequality in (3.4).

Let $v \in W_0^{1,2}(\Omega)$, $0 \leq v \leq \psi$ a.e. in Ω. Then $(v - \psi)^+ = 0$, $v^- = 0$ a.e. in Ω. Multiplying both sides of the equation in (3.5) by $v - u_\epsilon$ and integrating, we get

$$\begin{aligned} &(Au_\epsilon, v - u_\epsilon) - (\tilde{f}, v - u_\epsilon) \\ &\quad = \frac{1}{\epsilon}\left((v - \psi)^+ - (u_\epsilon - \psi)^+, v - u_\epsilon\right) + \left[-\frac{1}{\epsilon}\left(v^- u_\epsilon^-, v - u_\epsilon\right)\right]. \end{aligned}$$

Each of the two terms on the right is ≥ 0. Hence, taking $\epsilon = \epsilon_m \to 0$ and using (3.15), the inequality in (3.4) follows.

To complete the proof of Theorem 3.1 it remains to prove uniqueness. In the next section we show, under the additional condition (A_1), that u is the value of the game associated with (1.1), (1.3) when $E = F = \bar{\Omega}$. This of course implies uniqueness assertion of the theorem.

Corollary 3.2. *The solution u established in Theorem 3.1 satisfies:*

$$Lu - \alpha u + \tilde{f} \geq 0 \quad a.e. \quad \text{on the set where} \quad u > 0, \tag{3.18}$$

$$Lu - \alpha u + \tilde{f} \leq 0 \quad a.e. \quad \text{on the set where} \quad u < \psi, \tag{3.19}$$

$$Lu - \alpha u + \tilde{f} = 0 \quad a.e. \quad \text{on the set where} \quad 0 < u < \psi. \tag{3.20}$$

Proof. Let $B = \{x \in \Omega, u(x) > 0\}$. Since u is continuous, B is an open set. Let w be any nonnegative and bounded function with support in B, and let $v = u - \epsilon w$. If ϵ is positive and sufficiently small, then $0 \leq v \leq \psi$ in Ω. Hence

$$-\int (Lu - \alpha u + \tilde{f}) w \, dx \leq 0.$$

Since w is arbitrary, (3.18) follows. The proof (3.19) is similar. Finally, (3.20) is a consequence of (3.18), (3.19).

4. Existence of saddle points in bounded domains

Consider the elliptic variational inequality: Find a function u satisfying

$$u \in W^{2,2}(\Omega) \cap W^{1,2}(\Omega), \qquad u \leq \psi \quad \text{a.e.} \quad \text{on } E, \qquad u \geq 0 \quad \text{a.e.} \quad \text{on } F, \tag{4.1}$$

$$\int_\Omega (Lu - \alpha u + \tilde{f})(v - u) \, dx \leq 0 \qquad \text{for any} \quad v \in L^2(\Omega),$$

$$v \leq \psi \quad \text{a.e.} \quad \text{on } E, \qquad v \geq 0 \quad \text{a.e. on } F. \tag{4.2}$$

This is a generalization of the problem (3.4).

Suppose

$$\psi_1 \geq \psi_2 \qquad \text{in} \quad E \cap F, \tag{4.3}$$

$$\psi_1 = \psi_2 \qquad \text{on} \quad \partial\Omega. \tag{4.4}$$

Suppose also that there is a solution of (4.1), (4.2) satisfying

$$u \text{ is continuous in } \bar{\Omega}. \tag{4.5}$$

Let $V = u + \psi_2$ and define sets $\hat{E}$, $\hat{F}$ by

$$\hat{E} = \{x \in E;\ V(x) = \psi_1(x)\}, \qquad \hat{F} = \{x \in F;\ V(x) = \psi_2(x)\}. \tag{4.6}$$

These are closed sets containing $\partial\Omega$.

Theorem 4.1. *Let* (A_1), (B_1) *hold, and let* ψ_1, ψ_2, f *be continuous functions in* $\bar{\Omega}$. *Let* (4.3)–(4.6) *hold. Then* $V(x)$ *is the value function and* $(\hat{E}, \hat{F})$ *is a saddle point of sets for the stochastic game associated with* (1.1), (1.3).

Proof. First we verify (2.2)–(2.7) and (2.10). Observe that (2.10) follows from Corollary 2.2. Since $u \leq \psi$ on E, $u \geq 0$ on F, the inequalities (2.2), (2.3) follow immediately from the definition of V. The equations (2.4), (2.5) follow from the definition of $\hat{E}$, $\hat{F}$. We proceed to prove (2.6).

Notice that

$$\int_\Omega (LV - \alpha V + f)(v - V)\, dx \leq 0$$

for any $v \in L^2(\Omega)$, $v \leq \psi_1$ a.e. on E, $v \geq \psi_2$ a.e. on F. Arguing as in the proof of Corollary 3.2, we find that

$$LV - \alpha V + f \geq 0 \quad \text{a.e.} \quad \text{on} \quad \Omega \backslash \hat{F}, \tag{4.7}$$

$$LV - \alpha V + f \leq 0 \quad \text{a.e.} \quad \text{on} \quad \Omega \backslash \hat{E}. \tag{4.8}$$

Let $g(x, t)$ be any bounded measurable function and let τ, τ_0 be stopping times, $0 \leq \tau_0 \leq \tau < s_0$ where s_0 is a positive constant. Then

$$E_x \int_{\tau_0}^{\tau} g(x(t), t)\, dt = \int_0^\infty \int_{R^n} g(y, t)\Gamma(y, t; x, 0) E_x\big(\chi_{\tau_0, \tau}(t) | x(t) = y\big)\, dy\, dt \tag{4.9}$$

where $\chi_{\tau_0, \tau}(t) = 1$ if $\tau_0 < t < \tau$ and $\chi_{\tau_0, \tau}(t) = 0$ if either $t \leq \tau_0$ or $t \geq \tau$, and $\Gamma(y, t; x, s)$ is the fundamental solution of $L - \partial/\partial t$.

Indeed, the left-hand side of (4.9) is equal to

$$E_x \int_{\tau_0}^{\tau} g(x(t), t) E_x\big(\chi_{\tau_0, \tau}(t) | x(t)\big)\, dt.$$

Using Theorems 6.5.4, 6.4.7, we find that the last expression is equal to the right-hand side of (4.9).

Let B_ϵ be a closed ϵ-neighborhood of $\hat{F}$. Let V_m be the mollifier $J_{1/m}V$ of V, where m is a positive integer, $m > 1/\epsilon$. Since V_m is in C^2 in a neighborhood of $\Omega \setminus B_\epsilon$ we can apply Itô's formula to obtain

$$E_x e^{-\alpha\lambda} V_m(x(\lambda)) = E_x e^{-\alpha\lambda_0} V_m(x(\lambda_0)) + E_x \int_{\lambda_0}^{\lambda} e^{-\alpha t}(L - \alpha) V_m(x(t))\, dt \qquad (x \in \Omega \backslash B_\epsilon) \tag{4.10}$$

where λ is any bounded stopping time in C_x, $\lambda \leq t_{B_\epsilon}$, $\lambda_0 = \lambda \wedge s$, and $s > 0$. By (4.9), the second term on the right-hand side of (4.10) is equal to

$$\int_0^\infty \int_{R^n} e^{-\alpha t}(L - \alpha) V_m(y) \cdot \Gamma(y, t; x, 0) h(y, t)\, dy\, dt \tag{4.11}$$

where

$$h(y,t)=E_x(\chi_{\lambda_0,\lambda}(t)|x(t)=y).$$

Noting that $h(y,t)=0$ if $t<s$, or if $t>\tilde{s}$ (where $\tilde{s}$ is any positive number such that $\lambda<\tilde{s}$ a.e.), or if $y\in\Omega^c\cup B_\epsilon$, we find (upon recalling (6.4.12)) that $\Gamma(y,t;x,0)h(y,t)$ belongs to $L^2[R^n\times(0,\infty)]$. Since

$$(L-\alpha)V_m(y)\to(L-\alpha)V(y)\qquad\text{in}\quad L^2(A),$$

where A is any compact subset of Ω, we conclude that, as $m\to\infty$, the expression in (4.11) converges to

$$\int_0^\infty\int_{R^n}e^{-\alpha t}(L-\alpha)V(y)\cdot\Gamma(y,t;x,0)h(y,t)\,dy\,dt,$$

provided $\lambda\leqslant t_{B_\epsilon}$. By (4.7), the integrand (in the last integral) is

$$\geqslant -e^{-\alpha t}f(y)\Gamma(y,t;x,0)h(y,t).$$

Hence

$$\begin{aligned}\lim_{m\to\infty}E_x\int_{\lambda_0}^{\lambda}&e^{-\alpha t}(L-\alpha)V_m(x(t))\,dt\\&\geqslant-\int_0^\infty\int_{R^n}e^{-\alpha t}f(y)\Gamma(y,t;x,0)h(y,t)\,dy\\&=E_x\int_{\lambda_0}^{\lambda}e^{-\alpha t}f(x(t))\,dt.\end{aligned}$$

Taking $m\to\infty$ in (4.10) and using the fact that $V_m\to V$ uniformly in compact subsets of Ω, we get

$$E_xe^{-\alpha\lambda}V(x(\lambda))\geqslant E_xe^{-\alpha(\lambda\wedge s)}V(x(\lambda\wedge s))-E_x\int_{\lambda\wedge s}^{\lambda}e^{-\alpha t}f(x(t))\,dt.$$

Taking $s\downarrow 0$ we obtain the inequality

$$V(x)\leqslant E_x\left\{\int_0^\lambda e^{-\alpha t}f(x(t))\,dt+e^{-\alpha\lambda}V(x(\lambda))\right\}.\tag{4.12}$$

Here λ is any bounded stopping time in $\mathcal{C}_x$ satisfying $\lambda<t_{B_\epsilon}$. If λ is any stopping time in $\mathcal{C}_x$ satisfying $\lambda\leqslant t_{\hat{F}}$, then (4.12) holds for λ replaced by $\lambda\wedge T\wedge t_{B_\epsilon}$ where $0<T<\infty$ and ϵ is any positive number. By Corollary 2.2, $E_x\lambda<\infty$. Hence, if we take $T\uparrow\infty$, $\epsilon\downarrow 0$ and apply the Lebesque bounded convergence theorem, we arrive at the inequality (4.12) for any $\lambda\in\mathcal{C}_x$, $\lambda\leqslant t_{\hat{F}}$. We have thus completed the proof of (2.6). The proof of (2.7) is similar.

In order to complete the proof of Theorem 4.1, we merely have to apply Theorem 2.4.

Now let u be the solution of (3.4) constructed in Theorem 3.1 and define

$$V(x) = u(x) + \psi_2(x), \tag{4.13}$$

$$\hat{E} = \{x \in \bar{\Omega};\ V(x) = \psi_1(x)\}, \tag{4.14}$$

$$\hat{F} = \{x \in \bar{\Omega};\ V(x) = \psi_2(x)\}. \tag{4.15}$$

We can then state the following:

Theorem 4.2. *Let the conditions* (A_1), (B_1), (B_2) *hold. Then the stochastic game associated with* (1.1), (1.3), *when* $E = F = \bar{\Omega}$, *has value* $V(x)$ *and a saddle point of sets* $(\hat{E}, \hat{F})$ *given by* (4.13)–(4.15). *Further.* $V \in C^1(\bar{\Omega})$ *and*

$$\frac{\partial V}{\partial x_i} = \frac{\partial \psi_1}{\partial x_i} \quad \text{on} \quad \hat{E} \cap \Omega, \qquad \frac{\partial V}{\partial x_i} = \frac{\partial \psi_2}{\partial x_i} \quad \text{on} \quad \hat{F} \cap \Omega \qquad (1 \leq i \leq n). \tag{4.16}$$

Proof. Notice that since ψ_1, ψ_2 belong to $W^{2,p}(\Omega)$ where $p > n$, and since they are continuous in $\bar{\Omega}$, the Sobolev inequalities imply that they are continuously differentiable in $\bar{\Omega}$. Now, all the assertions of Theorem 4.2, except for (4.16), follow from Theorems 4.1 and 3.1. To prove (4.16), notice that

$$V - \psi_1 \leq 0 \quad \text{in } \Omega, \qquad V - \psi_1 = 0 \quad \text{on } \hat{E}.$$

This implies that grad $(V - \psi_1) = 0$ on $\hat{E} \cap \Omega$. Similarly, grad $(V - \psi_2) = 0$ on $\hat{F} \cap \Omega$.

Notice, by Corollary 3.2 (cf. also (4.7), (4.8)),

$$LV - \alpha V + f \geq 0 \quad \text{a.e.} \quad \text{on} \quad \Omega \backslash \hat{F}, \tag{4.17}$$

$$LV - \alpha V + f \leq 0 \quad \text{a.e.} \quad \text{on} \quad \Omega \backslash \hat{E}, \tag{4.18}$$

$$LV - \alpha V + f = 0 \quad \text{a.e.} \quad \text{on} \quad \Omega \setminus (\hat{E} \cup \hat{F}). \tag{4.19}$$

It follows that

$$(LV - \alpha V + f)(v - V) \leq 0 \quad \text{a.e.} \quad \text{if} \quad v \in L^2(\Omega), \quad \psi_2 \leq v \leq \psi_1 \quad \text{a.e.} \tag{4.20}$$

5. Elliptic estimates for increasing domains

In Sections 6, 7 we generalize the results of Section 3, 4 to unbounded domains Ω. In this section we establish some estimates that will be needed in Section 6 in order to study elliptic variational inequalities in unbounded domains.

Let Ω be an unbounded domain in R^n with C^2 boundary $\partial\Omega$. Suppose there exists a sequence of bounded domains Ω_m with C^2 boundary $\partial\Omega_m$ such that:

(i) $\Omega_m \subset \Omega_{m+1} \subset \Omega$;
(ii) $\Omega \cap \{x; |x| < m\} = \Omega_m \cap \{x; |x| < m\}$;
(iii) there exist positive constants δ_0, C^* such that for each $m = 1, 2, \ldots,$ and for each $y \in \partial\Omega_m$, the set $\partial\Omega_m \cap \{x; |x-y| < \delta_0\}$ can be represented in the form

$$x_i = \Phi(x_1, \ldots, x_{i-1}, x_{i+1}, \ldots, x_n)$$

for some i, $1 \leqslant i \leqslant n$, and

$$\sum \left| \frac{\partial \Phi}{\partial x_j} \right| + \sum \left| \frac{\partial^2 \Phi}{\partial x_j\, \partial x_k} \right| \leqslant C^*.$$

We shall then say that Ω is in class $\mathcal{C}^2$. If further

$$\left| \frac{\partial^2 \Phi(\bar{x})}{\partial x_j\, \partial x_k} - \frac{\partial^2 \Phi(\bar{\bar{x}})}{\partial x_j\, \partial x_k} \right| \leqslant C^* |\bar{x} - \bar{\bar{x}}|^\alpha \qquad (0 < \alpha \leqslant 1)$$

for any $\bar{x}$, $\bar{\bar{x}}$, then we say that Ω is in class $\mathcal{C}^{2+\alpha}$.

In particular, if $\Omega = R^n$ or if Ω is the complement of a closed bounded domain with C^2 ($C^{2+\alpha}$) boundary, then Ω is in $\mathcal{C}^2$ ($\mathcal{C}^{2+\alpha}$), for (i)–(iii) hold with $\Omega_m = \{x; |x| < m\}$ for all large m.

Let k be a nonnegative integer, let $1 < p < \infty$, and let μ be any nonnegative number. Given a domain G, we introduce the space $W^{k,p,\mu}(G)$ consisting of all (real-valued) functions $u(x)$ whose first k weak derivatives exist and belong to L^p on compact subsets of G, and for which the norm

$$|u|^G_{k,p,\mu} \equiv \left\{ \sum_{|\alpha| \leqslant k} \int_G |e^{-\mu|x|} D^\alpha u(x)|^p \, dx \right\}^{1/p}$$

is finite. When $\mu = 0$, we denote the space by $W^{k,p}(G)$ and the norm of u by $|u|^G_{k,p}$. We shall denote by $W_0^{k,p,\mu}(G)$ the completion of $C_0^\infty(G)$ in the norm $|\ |^G_{k,p,\mu}$. When $\mu = 0$, we denote this space by $W_0^{k,p}(G)$. Note that $W^{0,p}(G) = W_0^{0,p}(G) = L^p(G)$.

Consider a partial differential operator

$$Au \equiv -\frac{1}{2} \sum_{i,j=1}^n \frac{\partial}{\partial x_i}\left(a_{ij}(x) \frac{\partial u}{\partial x_j} \right) + \sum_{i=1}^n \tilde{b}_i(x) \frac{\partial u}{\partial x_i} + c(x)u + \alpha u \qquad (5.1)$$

where α is a positive constant and $\tilde{b}_i = \frac{1}{2}\sum_{j=1}^n \partial a_{ij}/\partial x_j - b_i$. We shall assume:

(A) The functions a_{ij}, $\partial a_{ij}/\partial x_j$, b_i, c are measurable in Ω, and

$$\sum_{i,j} |a_{ij}(x)| + \sum_{i,j} \left| \frac{\partial a_{ij}(x)}{\partial x_j} \right| + \sum_i |b_i(x)| + |c(x)| \leq K; \tag{5.2}$$

$$c(x) \geq 0 \qquad \text{in } \Omega; \tag{5.3}$$

$$\tfrac{1}{2} \sum_{i,j=1}^{n} a_{ij}(x)\xi_i\xi_j \geq \beta|\xi|^2 \qquad \text{for all} \quad x \in \Omega, \quad \xi \in R^n \tag{5.4}$$

for all x, y in $\bar{\Omega}$, where β and K are positive constants;

$$|a_{ij}(x) - a_{ij}(y)| \leq C^*(|x - y|), \qquad C^*(r) \downarrow 0 \qquad \text{if} \quad r \downarrow 0; \tag{5.5}$$

$$\sup_{x\in\Omega} |\tilde{b}_i(x)| < \frac{2\sqrt{\alpha\beta}}{\sqrt{n}} . \tag{5.6}$$

From now on we fix $p \geq 2$. Let $f \in L^p(\Omega_m)$ and consider the Dirichlet problem

$$Au = f \quad \text{a.e.} \quad \text{in } \Omega_m, \qquad u \in W^{2,p}(\Omega_m) \cap W_0^{1,2}(\Omega_m). \tag{5.7}$$

By Theorem 10.3.1, this problem has a unique solution $u = u_m$.

Lemma 5.1. *Let $\Omega \in \mathcal{C}^2$ and let* (A) *hold. Then there exist a sufficiently small positive constant μ_0 and a positive constant M, independent of m, such that*

$$|u|_{2,p,\mu}^{\Omega_m} \leq M\left(|f|_{0,p,\mu}^{\Omega_m} + |f|_{0,2,\mu}^{\Omega_m}\right) \tag{5.8}$$

for any $0 \leq \mu \leq \mu_0$.

The importance of this lemma lies in the fact that μ_0 and M are independent of m.

Before proving the lemma, we shall state another lemma regarding the Dirichlet problem

$$Au + \lambda u = f \quad \text{a.e.} \quad \text{in } \Omega_m, \qquad u \in W^{2,p}(\Omega_m) \cap W_0^{1,2}(\Omega_m) \tag{5.9}$$

where $\lambda \geq 0$. Denote the solution by $u_{m,\lambda}$, and write

$$u_{m,\lambda} \equiv (A + \lambda I)^{-1} f \equiv R_{m,\lambda} f.$$

Lemma 5.2. *Let $\Omega \in \mathcal{C}^2$ and let* (A) *hold. Then there exist positive constants Λ, M^* independent of m, λ, such that*

$$|R_{m,\lambda} f|_{0,p,\mu}^{\Omega_m} \leq \frac{M^*}{1+\lambda} |f|_{0,p,\mu}^{\Omega_m} \qquad \text{if} \quad \lambda \geq \Lambda \quad \text{and} \quad 0 \leq \mu \leq \mu_0, \tag{5.10}$$

where μ_0 is as in Lemma 5.1.

Proof of Lemma 5.1. Let $\epsilon = \delta_0/3\sqrt{n}$, and introduce a mesh in R^n made up of cubes with sides parallel to the coordinate axes and having length ϵ. Denote by $\Gamma_1, \ldots, \Gamma_{h_0}$ those cubes whose closure intersects $\partial\Omega_m$. Denote the center of Γ_j by y_j. Let Γ_j', Γ_j'' be cubes with center y_j and with sides parallel to the coordinate axes, having lengths 2ϵ and 3ϵ respectively. Then $\Gamma_1', \ldots, \Gamma_{h_0}'$ form an open covering of $\partial\Omega_m$. Further, for any $y \in \partial\Omega_m$ there is a cube Γ_j' such that $y \in \Gamma_j$ and dist $(y, \partial\Gamma_j') > \epsilon/2$.

Let ψ be a C^∞ function such that

$$\psi(x) = 1 \quad \text{if } |x_i| < \epsilon \quad \text{for all } i = 1, 2, \ldots, n,$$
$$\psi(x) = 0 \quad \text{if } |x_i| > \tfrac{5}{4}\epsilon \quad \text{for some } i,$$
$$0 \leq \psi(x) \leq 1 \quad \text{elsewhere,}$$

and set $\psi_j(x) = \psi(y_j + x)$. Then $\psi_j = 1$ in Γ_j' and $\psi_j = 0$ in a small neighborhood of $\partial\Gamma_j''$ and outside Γ_j''.

Denote by $\Omega_{m,\epsilon}$ the set of all points in Ω_m whose distance to $\partial\Omega_m$ is $\geq \epsilon/2$. We now introduce a mesh made up of cubes with sides parallel to the coordinate axes and having length $\epsilon_0 = \epsilon/8\sqrt{n}$. Let $\Delta_1, \ldots, \Delta_{h_1}$ be those cubes whose closure intersects $\overline{\Omega_{m,\epsilon}}$. Let Δ_j', Δ_j'' be the cubes with the same center z_j as Δ_j and with sides parallel to the coordinate axes having length $2\epsilon_0$ and $3\epsilon_0$ respectively. The cubes $\Delta_1', \ldots, \Delta_{h_1}'$ form an open covering of $\overline{\Omega_{m,\epsilon}}$, and the cubes $\Delta_1'', \ldots, \Delta_{h_1}''$ lie entirely in Ω_m.

Let χ be the C^∞ function

$$\chi(x) = \psi\left(\frac{\epsilon}{\epsilon_0} x\right),$$

and let $\chi_j(x) = \chi(z_j + x)$. Let

$$\phi_j = \frac{\psi_j}{\sum \psi_k + \sum \chi_k} \qquad \text{if } 1 \leq j \leq h_0,$$

$$\phi_{j+h_0} = \frac{\chi_j}{\sum \psi_k + \sum \chi_k} \qquad \text{if } 1 \leq j \leq h_1,$$

$$G_j = \Gamma_j'', \qquad G_j' = \Gamma_j' \qquad \text{if } 1 \leq j \leq h_0,$$

$$G_{j+h_0} = \Delta_j'', \qquad G_{j+h_0}' = \Delta_j' \qquad \text{if } 1 \leq j \leq h_1,$$

and let $h = h_0 + h_1$. Then $\{G_1, \ldots, G_h\}$ form an open covering of $\overline{\Omega}_m$, and $\{\phi_1, \ldots, \phi_h\}$ form a partition of unity subordinate to this covering, such that:

(a) $G_1, \ldots, G_{h_0}$ intersect $\partial\Omega_m$, and $G_{h_0+1}, \ldots, G_h$ lie entirely in Ω_m;
(b) $\phi_k \in C_0^\infty(G_k)$;

(c) each $x \in \bar{\Omega}_m$ belongs to at most N_1 sets G_k, where N_1 is a positive integer independent of m;

(d) $\phi_k \geq 1/N_1$ on the set G_k', and the sets $\{G_1', \ldots, G_h'\}$ form an open covering of $\bar{\Omega}_m$;

(e) there is a constant N_2 independent of k, m, such that

$$|D^\alpha \phi_k| \leq N_2 \quad \text{if} \quad |\alpha| \leq 2, \quad x \in G_k, \quad 1 \leq k \leq h. \tag{5.11}$$

Let

$$G_{km} = G_k \cap \Omega_m.$$

Notice now that

$$Av = -\frac{1}{2} \sum_{i,j=1}^{n} a_{ij}(x) \frac{\partial^2 v}{\partial x_i \, \partial x_j} - \sum_{i=1}^{n} b_i(x) \frac{\partial v}{\partial x_i} + c(x)v + \alpha v$$

where $|b_i(x)| \leq K$, $|c| \leq K$. Since the a_{ij} satisfy (5.4), (5.5) and are bounded in Ω, it follows (by Theorem 10.3.1 and the remark at the end of Section 10.3) that for any $u_m \in W^{2,p}(\Omega_m) \cap W_0^{1,2}(\Omega_m)$

$$|u_m \phi_k|_{2,p}^{G_{km}} \leq c\{|A(u_m \phi_k)|_{0,p}^{G_{km}} + |u_m \phi_k|_{0,p}^{G_{km}}\} \tag{5.12}$$

where c is a constant independent of k, m. Here we use the condition (iii) in the definition of $\Omega \in \mathcal{C}^2$ and the fact that ϕ_k has compact support in G_k.

Note that

$$A(u_m \phi_k) = f\phi_k - \partial \sum_{i,j=1}^{n} a_{ij} \frac{\partial u_m}{\partial x_j} \frac{\partial \phi_k}{\partial x_i}$$

$$- u_m \left\{ \frac{1}{2} \sum_{i,j=1}^{n} \frac{\partial a_{ij}}{\partial x_i} \frac{\partial \phi_k}{\partial x_j} + \frac{1}{2} \sum_{i,j=1}^{n} a_{ij} \frac{\partial^2 \phi_k k}{\partial x_i \, \partial x_j} - \sum_{i=1}^{n} \tilde{b}_i \frac{\partial \phi_k}{\partial x_i} \right\}.$$

Hence, by (5.11),

$$|A(u_m \phi_k)|^p \leq C|f|^p + C|Du_m|^p + C|u_m|^p \quad \text{in } G_{km} \tag{5.13}$$

where Du is the gradient of u; the symbol C will be used to denote any one of various constants independent of k, m, f.

Taking the pth power of both sides of (5.12) and using the triangle inequality, we get

$$\int_{G_{km}} (|D^2 u_m| \phi_k)^p \, dx \leq C \int_{G_{km}} (|Du_m| \, |D\phi_k|)^p \, dx + C \int_{G_{km}} (|u_m| \, |D^2 \phi_k|)^p \, dx$$
$$+ Cc^p \int_{G_{km}} |A(u_m \phi_k)|^p \, dx + \int_{G_{km}} |u_m \phi_k|^p \, dx.$$

Multiplying both sides by $\exp(-p\mu|\zeta_k|)$, where ζ_k is the center of the cube G_k, and noting that

$$Ce^{-p\mu|x|} \leq e^{-p\mu|\zeta_k|} \leq Ce^{-p\mu|x|} \quad \text{if} \quad x \in G_k,$$

we obtain, after making use of (5.13) and (5.11),

$$\int_{G'_{km}} (e^{-\mu|x|}\,|D^2u_m|\phi_k)^p\,dx$$
$$\leq C\int_{G_{km}} (e^{-\mu|x|}\,|f|)^p\,dx + C\int_{G_{km}} (e^{-\mu|x|}\,|Du_m|)^p\,dx$$
$$+C\int_{G_{km}} (e^{-\mu|x|}\,|u_m|)^p\,dx; \tag{5.14}$$

here G'_{km} is the subset of G_{km} defined by

$$G'_{km} = G_{km} \cap G'_k = G'_k \cap \Omega_m.$$

Recalling the properties (c), (d) of ϕ_k, G_k, G'_k and summing the inequalities (5.14) for $k = 1, \ldots, h$, we obtain

$$\int_{\Omega_m} (e^{-\mu|x|}\,|D^2u_m|)^p\,dx \leq C\int_{\Omega_m} (e^{-\mu|x|}\,|f|)^p\,dx + C\int_{\Omega_m} (e^{-\mu|x|}\,|Du_m|)^p\,dx$$
$$+C\int_{\Omega_m} (e^{-\mu|x|}\,|u_m|)^p\,dx. \tag{5.15}$$

We next derive an estimate on the $W^{1,2}(\Omega_m)$ norm of u in terms of the $L^2(\Omega_m)$ norm of Au. Set $L^{p,\mu}(G) = W^{0,p,\mu}(G)$. The space $L^{2,\mu}(G)$ is a real Hilbert space with the scalar product

$$(u, v)_{\mu,G} = \int_G e^{-2\mu|x|}u(x)v(x)\,dx.$$

When $G = \Omega_m$, we write $(u, v)_{\mu,G} = (u, v)_{\mu,m}$. If $u \in W^{2,2}(\Omega_m) \cap W_0^{1,2}(\Omega_m)$, then

$$(Au, u)_{\mu,m} = \frac{1}{2}\sum_{i,j=1}^n \left[\left(a_{ij}\frac{\partial u}{\partial x_i}, \frac{\partial u}{\partial x_j}\right)_{\mu,m} - \mu\int_{\Omega_m} e^{-2\mu|x|}a_{ij}\frac{x_i}{|x|}\frac{\partial u}{\partial x_j}u\,dx\right]$$
$$+\sum_{i=1}^n \left(\tilde{b}_i\frac{\partial u}{\partial x_i}, u\right)_{\mu,m} + (cu + \alpha u, u)_{\mu,m}$$
$$\geq \beta\sum_{i=1}^n \int_{\Omega_m} e^{-2\mu|x|}\left(\frac{\partial u}{\partial x_i}\right)^2 dx$$
$$-\mu K\sum_{i=1}^n \int_{\Omega_m} e^{-2\mu|x|}\left(\frac{\partial u}{\partial x_i}\right)^2 dx - \mu K\int_{\Omega_m} e^{-2\mu|x|}u^2\,dx$$
$$-\left(\sup_{\substack{x\in\Omega\\ 1\leq i\leq n}} |\tilde{b}_i(x)|\right)\left\{\frac{\nu}{2}\sum_{i=1}^n \int_{\Omega_m} e^{-2\mu|x|}\left(\frac{\partial u}{\partial x_i}\right)^2 dx\right.$$
$$\left.+\frac{n}{2\nu}\int_{\Omega_m} e^{-2\mu|x|}u^2\,dx\right\} + \alpha\int_{\Omega_m} e^{-2\mu|x|}u^2\,dx$$
$$\geq \gamma\int_{\Omega_m} e^{-2\mu|x|}\left[\sum_{i=1}^n\left(\frac{\partial u}{\partial x_i}\right)^2 + u^2\right] dx$$

for some positive constant γ, if $\sup_{x\in\Omega}|\tilde{b}_i| < B$, $0 \leq \mu \leq \mu_0$, provided

$$\frac{Bn}{2(\alpha - \mu_0 K)} < \nu < \frac{2(\beta - \mu_0 K)}{B},$$

that is, provided

$$B^2 < 4(\alpha - \mu_0 K)(\beta - \mu_0 K)/n;$$

but this follows from (5.6) if μ_0 is sufficiently small. Notice that γ depends on α, β, n, μ_0, B, but is independent of μ, m. We have thus proved that

$$\int_{\Omega_m} e^{-2\mu|x|} uAu\, dx \geq \gamma\left(|u|_{1,2,\mu}^{\Omega_m}\right)^2. \tag{5.16}$$

In particular, when $u = u_m$,

$$\gamma\left(|u_m|_{1,2,\mu}^{\Omega_m}\right)^2 \leq (f, u_m)_{\mu,m} \leq |f|_{0,2,\mu}^{\Omega_m}|u_m|_{0,2,\mu}^{\Omega_m} \leq |f|_{0,2,\mu}^{\Omega_m}|u_m|_{1,2,\mu}^{\Omega_m}.$$

Consequently,

$$|u_m|_{1,2,\mu}^{\Omega_m} \leq \frac{1}{\gamma}|f|_{0,2,\mu}^{\Omega_m}. \tag{5.17}$$

In what follows we may assume that $p > 2$, for (5.8) is an immediate consequence of (5.15), (5.17) when $p = 2$.

We shall now use (5.15) in conjunction with (5.17) in order to derive the inequality (5.8). First we derive a variant of Sobolev's inequalities in Ω_m.

By Sobolev's inequalities (see Section 10.2)

$$\left[\int_{R^n}|w|^{q'}dx\right]^{1/q'} \leq C\left[\int_{R^n}|D^2w|^{r'}dx\right]^{a'/r'}\left[\int_{R^n}|w|^2dx\right]^{(1-a')/2}, \tag{5.18}$$

$$\left[\int_{R^n}|Dw|^{q}dx\right]^{1/q} \leq C\left[\int_{R^n}|D^2w|^{r}dx\right]^{a/r}\left[\int_{R^n}|w|^2dx\right]^{(1-a)/2} \tag{5.19}$$

for any $w \in C_0^2(R^n)$, where

$$\frac{1}{q'} = a'\left(\frac{1}{r'} - \frac{2}{n}\right) + \frac{1-a'}{2}, \qquad 0 < a' < 1, \tag{5.20}$$

$$\frac{1}{q} = \frac{1}{n} + a\left(\frac{1}{r} - \frac{2}{n}\right) + \frac{1-a}{2}, \qquad \frac{1}{2} < a < 1; \tag{5.21}$$

the constant C is independent of w.

Let $u \in C^2(\bar{\Omega}_m)$. Then we can extend it to a function w in $C_0^2(R^n)$ in such a way that

$$\sum_{j=0}^{i}\int_{R^n}|D^jw|^q dx \leq C\sum_{j=0}^{i}\int_{\Omega_m}|D^ju|^q dx$$

for $0 \leq i \leq 2$, $q \geq 1$, where C depends on q, but is independent of u, m.

Indeed, this follows from the proof of Problem 9, Chapter 10, provided we use the partition of unity $\{\phi_1, \ldots, \phi_{h_0}\}$ of a neighborhood of $\partial\Omega_m$ constructed above. Applying (5.18), (5.19), we conclude that

$$\left(\int_{\Omega_m} |u|^{q'}\,dx\right)^{1/q'} \leq C\left(\int_{\Omega_m} [|u| + |Du| + |D^2u|]^{r'}\,dx\right)^{a'/r'}\left(\int_{\Omega_m} |u|^2\,dx\right)^{(1-a')/2}, \tag{5.22}$$

$$\left(\int_{\Omega_m} |Du|^{q}\,dx\right)^{1/q} \leq C\left(\int_{\Omega_m} [|u| + |Du| + |D^2u|]^{r}\,dx\right)^{a/r}\left(\int_{\Omega_m} |u|^2\,dx\right)^{(1-a)/2} \tag{5.23}$$

for any $u \in C^2(\overline{\Omega}_m)$, where q', q are defined by (5.20), (5.21). Since $\partial\Omega_m$ is in C^2, the completion of $C^2(\overline{\Omega}_m)$ in $W^{2,p}(\Omega_m)$ $(p > 1)$ coincides with $W^{2,p}(\Omega_m)$ (see Friedman [2]). Consequently, the inequalities (5.22), (5.23) hold for all

$$u \in W^{2,r'}(\Omega_m) \cap W^{0,2}(\Omega_m) \qquad \text{and} \qquad u \in W^{2,r}(\Omega_m) \cap W^{0,2}(\Omega_m)$$

respectively.

We now substitute, in (5.22),

$$u = v \exp\left[-\mu(1 + |x|^2)^{1/2}\right]. \tag{5.24}$$

Since, as is easily seen,

$$|u| + |Du| + |D^2u| \leq C[|v| + |Dv| + |D^2v|] \exp\left[-\mu(1 + |x|^2)^{1/2}\right]$$

and since also

$$e^{-\mu(|x|+1)} \leq \exp\left[-\mu(1 + |x|^2)^{1/2}\right] \leq e^{-\mu|x|},$$

we obtain

$$|v|_{0,q',\mu}^{\Omega_m} \leq C\left(|v|_{2,r',\mu}^{\Omega_m}\right)^{a'}\left(|v|_{0,2,\mu}^{\Omega_m}\right)^{1-a'}. \tag{5.25}$$

Next we substitute u from (5.24) into (5.23). Noting that

$$|Du| \geq |Dv| \exp\left[-\mu(1 + |x|^2)^{1/2}\right] - \mu|v| \exp\left[-\mu(1 + |x|^2)^{1/2}\right],$$

we find that

$$|Dv|_{0,q,\mu}^{\Omega_m} \leq C\mu|v|_{0,q,\mu}^{\Omega_m} + C\left(|v|_{2,r,\mu}^{\Omega_m}\right)^{a}\left(|v|_{0,2,\mu}^{\Omega_m}\right)^{1-a}. \tag{5.26}$$

We shall now use the Sobolev type inequalities (5.25), (5.26) in order to derive the assertion (5.8) of Lemma 5.1 from (5.15), (5.17).

Notice that (5.15) holds not only for p, but also for any p_1 in the interval $2 \leq p_1 \leq p$. Taking $p_1 = 2$ and using (5.17), we get

$$|D^2u_m|_{0,1,\mu}^{\Omega_m} \leq C|f|_{0,2,\mu}^{\Omega_m}. \tag{5.27}$$

Applying (5.25) to $v = u_m$ with $r' = 2$, we then obtain, after using (5.17), (5.27),

$$|u_m|^{\Omega_m}_{0,q',\mu} \leqslant C|f|^{\Omega_m}_{0,2,\mu}, \tag{5.28}$$

provided

$$\frac{1}{q'} = a'\left(\frac{1}{2} - \frac{2}{n}\right) + \frac{1-a'}{2}, \qquad 0 < a' < 1,$$

that is, provided

$$q' > 2, \qquad \frac{1}{q'} > \frac{1}{2} - \frac{2}{n}. \tag{5.29}$$

Applying (5.26) with $r' = 2$, $v = u_m$ and using (5.17), (5.27), we obtain

$$|Du_m|^{\Omega_m}_{0,q,\mu} \leqslant C|f|^{\Omega_m}_{0,2,\mu} + C\mu|u_m|^{\Omega_m}_{0,q,\mu}, \tag{5.30}$$

provided

$$\frac{1}{q} = \frac{1}{n} + a\left(\frac{1}{2} - \frac{2}{n}\right) + \frac{1-a}{2}, \qquad \frac{1}{2} < a < 1,$$

that is, provided

$$q > 2, \qquad \frac{1}{q} > \frac{1}{2} - \frac{1}{n}. \tag{5.31}$$

If q satisfies (5.31), then $q' = q$ satisfies (5.29). Consequently, the second term on the right-hand side of (5.30) is bounded by $C|f|^{\Omega_m}_{0,2,\mu}$. Combining (5.30) with (5.28), we obtain

$$|u_m|^{\Omega_m}_{1,q,\mu} \leqslant C|f|^{\Omega_m}_{0,2,\mu} \tag{5.32}$$

for all q satisfying (5.31). Using (5.15) (with $p = q$), we then get

$$|u_m|^{\Omega_m}_{2,q_1,\mu} \leqslant |f|^{\Omega_m}_{0,2,\mu} + C|f|^{\Omega_m}_{0,q_1,\mu} \tag{5.33}$$

for any q_1 satisfying

$$2 < q_1 \leqslant p, \qquad \frac{1}{q_1} > \frac{1}{2} - \frac{1}{n}. \tag{5.34}$$

If

$$\frac{1}{p} > \frac{1}{2} - \frac{1}{n},$$

then we can take $q_1 = p$ in (5.33) and thus obtain the asserted inequality (5.8). Otherwise, we proceed to apply (5.25), (5.26) with

$$\frac{1}{r'} = \frac{1}{r} = \frac{1}{2} - \frac{1}{n} + \epsilon; \qquad \epsilon \text{ arbitrarily small}, \quad \epsilon > 0.$$

We conclude, by the same procedure as before, that

$$|u_m|_{2,q_2,\mu}^{\Omega_m} \le C|f|_{0,2,\mu}^{\Omega_m} + C|f|_{0,q_1,\mu}^{\Omega_m} + C|f|_{0,q_2,\mu}^{\Omega_m}$$

where

$$\frac{1}{q_1} = \frac{1}{2} - \frac{1}{n} + \epsilon, \qquad q_1 < q_2 \le p, \qquad \frac{1}{q_2} > \frac{1}{q_1} - \frac{1}{n}.$$

If the last inequality implies that $q_2 < p$, then we repeat the same argument again. After a finite number l of steps, we arrive at the inequality

$$|u_m|_{2,q_l,\mu}^{\Omega_m} \le C\sum_{i=0}^{l} |f|_{0,q_i,\mu}^{\Omega_m}$$

where

$$q_0 = 2, \qquad q_{i-1} < q_i < p \quad \text{if } 1 \le i \le l-1,$$

$$\frac{1}{q_{i+1}} > \frac{1}{q_i} - \frac{1}{n} \quad (0 \le i \le l-1), \qquad \text{and } q_{l-1} < q_l = p.$$

Since (see Problem 4)

$$|f|_{0,q_i,\mu}^{\Omega_m} \le |f|_{0,2,\mu}^{\Omega_m} + |f|_{0,p,\mu}^{\Omega_m}, \tag{5.35}$$

the assertion (5.8) follows.

Proof of Lemma 5.2. Using the notation of the previous proof, we shall now employ the inequalities (see Theorem 10.3.2 and the remark at the end of Section 10.3).

$$\lambda^p \int_{G'_{km}} |u\phi_k|^p\, dx \le c\int_{G_{km}} |(A+\lambda I)(u\phi_k)|^p\, dx, \tag{5.36}$$

$$\lambda^{p/2} \int_{G'_{km}} |D(u\phi_k)|^p\, dx \le c\int_{G_{km}} |(A+\lambda I)(u\phi_k)|^p\, dx, \tag{5.37}$$

which are valid for any $u \in W^{2,p}(\Omega_m) \cap W_0^{1,2}(\Omega_m)$, where $\lambda > \lambda_0 > 0$, and λ_0, c are positive constants independent of k, m; property (iii) in the definition of $\Omega \in \mathcal{C}^2$ is used here to deduce that λ_0 and c are independent of k, m. By (5.13) (with A replaced by $A + \lambda I$),

$$|(A+\lambda I)(u\phi_k)|^p \le C|(A+\lambda I)u|^p + C|Du|^p + C|u|^p \qquad \text{in } G_{km}. \tag{5.38}$$

Multiplying both sides of (5.36) by $\exp(-p\mu|\zeta_k|)$, where ζ_k is the center of G_k, we obtain, after summing over k and using (5.38) and the properties (c), (d) of the partition of unity $\{\phi_i\}$,

$$\lambda|u|_{0,p,\mu}^{\Omega_m} \le C|(A+\lambda I)u|_{0,p,\mu}^{\Omega_m} + C|u|_{1,p,\mu}^{\Omega_m}. \tag{5.39}$$

Since

$$\int_{G'_{km}} (|Du|\phi_k)^p\, dx \le C\int_{G'_{km}} |uD\phi_k|^p\, dx + C\int_{G'_{km}} |D(u\phi_k)|^p\, dx,$$

(5.37) implies that

$$\lambda^{p/2}\int_{G'_{km}} (|Du|\phi_k)^p\,dx$$
$$\leqslant C\int_{G_{km}} |(A+\lambda I)(u\phi_k)|^p\,dx + C\lambda^{p/2}\int_{G_{km}} |uD\phi_k|^p\,dx. \quad (5.40)$$

Proceeding in the same way that led to (5.39), we obtain from (5.40) the inequality

$$\lambda^{1/2}|Du|^{\Omega_m}_{0,p,\mu} \leqslant C|(A+\lambda I)u|^{\Omega_m}_{0,p,\mu} + C|u|^{\Omega_m}_{1,p,\mu} + C\lambda^{1/2}|u|^{\Omega_m}_{0,p,\mu}.$$

Combining this with (5.39) and taking λ sufficiently large, say $\lambda \geqslant \Lambda$, we get

$$\lambda|u|^{\Omega_m}_{0,p,\mu} + \lambda^{1/2}|Du|^{\Omega_m}_{0,p,\mu} \leqslant C|(A+\lambda I)u|^{\Omega_m}_{0,p,\mu}. \quad (5.41)$$

This inequality yields the assertion (5.10) of Lemma 5.2.

6. Elliptic variational inequalities

Let $\Omega \in \mathcal{C}^2$ and let (A) hold. Let ϕ_1, ϕ_2, f be functions defined in $\overline{\Omega}$ and satisfying:

(B) ϕ_1 and ϕ_2 belong to $W^{2,p,\mu}(\Omega) \cap W^{2,2,\mu}(\Omega)$ for some $0 \leqslant \mu \leqslant \mu_0$, $2 \leqslant p < \infty$, and

$$\phi_1 \geqslant \phi_2 \quad \text{a.e.}\quad \text{in } \Omega, \qquad \phi_1 \text{ and } \phi_2 \quad \text{belong to} \quad W_0^{1,2,\mu}(\Omega).$$

(C) $f \in L^{p,\mu}(\Omega) \cap L^{2,\mu}(\Omega)$.

The constant μ_0 is as in Lemma 5.1.

Introduce the set

$$K_\mu = \{ g \in L^{2,\mu}(\Omega);\ \phi_2 \leqslant g \leqslant \phi_1 \text{ a.e. in } \Omega \}$$

where $0 \leqslant \mu \leqslant \mu_0$. We consider the following elliptic variational inequality: Find a function u such that

$$u \in W^{2,2,\mu}(\Omega) \cap W_0^{1,2,\mu}(\Omega) \cap K_\mu, \quad (6.1)$$

$$\int_\Omega e^{-2\mu|x|}Au\cdot(v-u)\,dx \geqslant \int_\Omega e^{-2\mu|x|}f\cdot(v-u)\,dx \qquad \text{for any } v \in K_\mu. \quad (6.2)$$

Theorem 6.1. *Let $\Omega \in \mathcal{C}^2$ and let* (A)–(C) *hold. Then there exists a unique solution u of* (6.1), (6.2); *further, u belongs to $W^{2,p,\mu}(\Omega)$.*

We shall need the following lemma.

Lemma 6.2. *Let $\Omega \in \mathcal{C}^2$ and let* (A) *hold. If $w \in W^{2,2,\mu}(\Omega) \cap W_0^{1,2,\mu}(\Omega)$,* *where* $0 < \mu < \mu_0$, *then*

$$\int_\Omega e^{-2\mu|x|} wAw\, dx > \gamma\left(|w|^\Omega_{1,2,\mu}\right)^2 \tag{6.3}$$

where γ *is the constant appearing in* (5.16).

Proof. By the proof of (5.16),

$$\int_{\Omega_m} e^{-2\mu|x|} wAw\, dx > \gamma|w|^{\Omega_m}_{1,2,\mu} - C\int_{\partial\Omega_m} e^{-2\mu|x|}|Dw|\,|w|\, dS_x. \tag{6.4}$$

We shall use the partition of unity $\{\phi_1, \ldots, \phi_{h_0}\}$ of a δ' Ω_m-neighborhood of $\partial\Omega_m$ constructed in the proof of Lemma 5.1 ($\delta' = \delta_0/6\sqrt{n}$). In G_j we make a coordinate transformation $x \to y$ which takes $\partial\Omega_m$ into $y_n = 0$. Then $\phi_j w$, $D(\phi_j w)$ are transformed into $\tilde{w}$, $D_y\tilde{w}$. To these functions we apply the one-dimensional Sobolev inequality

$$|v(y_n)|^2 < C\int_0^c \left[(v(s))^2 + (v'(s))^2\right] ds.$$

Going back to the original coordinates, multiplying both sides by $\exp(-2\mu|\zeta_j|)$, where ζ_j = center of G_j, and using the properties (a)–(c) of the partition $\{\phi_1, \ldots, \phi_{h_0}\}$, we arrive at the inequality

$$\int_{\partial\Omega_m} e^{-2\mu|x|}\left(|w|^2 + |Dw|^2\right) dS_x < C\int_{\Omega_m^0} e^{-2\mu|x|}\left(|w|^2 + |Dw|^2 + |D^2w|^2\right) dx \tag{6.5}$$

where Ω_m^0 is the δ' Ω_m-neighborhood of $\partial\Omega_m$. Since $|Dw|\,|w| < (|Dw|^2 + |w|^2)/2$, the last term on the right-hand side of (6.4) is bounded by the right-hand side of (6.5) (with a different C). But the right-hand side of (6.5) tends to 0 if $m \to \infty$ (for $w \in W^{2,2,\mu}(\Omega)$). Hence, if we take $m \to \infty$ in (6.4), we obtain the assertion (6.3).

Proof of Theorem 6.1. To prove uniqueness, take $u = u_1$, $v = u_2$ in (6.2) and then $u = u_2$, $v = u_1$, and add the two inequalities. This results in

$$\int_\Omega e^{-2\mu|x|} Aw \cdot w\, dx < 0, \qquad \text{where} \quad w = u_1 - u_2.$$

In view of Lemma 6.2, $u_1 - u_2 = w = 0$.

We proceed to prove existence. Let $\zeta_m(x)$ be C^∞ functions in R^n such that $\zeta_m(x) = 1$ if $|x| < m - 2$, $\zeta_m(x) = 0$ if $|x| > m - 1$, $0 < \zeta_m(x) < 1$ if $m - 2 < |x| < m - 1$, and $|D^\alpha\zeta_m(x)| < C$ if $|\alpha| < 2$. Let $\phi_{im} = \zeta_m\phi_i$. Then

$$\phi_{im} \in W^{2,2}(\Omega_m) \cap W_0^{1,2}(\Omega_m), \tag{6.6}$$

$$|\phi_{im} - \phi_i|^\Omega_{2,2,\mu} \to 0 \qquad \text{if} \quad m \to \infty, \tag{6.7}$$

for $i = 1, 2$.

Let $\epsilon > 0$, and consider the Dirichlet problem

$$Au + \frac{1}{\epsilon}(u - \phi_{1m})^{+} - \frac{1}{\epsilon}(u - \phi_{2m})^{-} = f \qquad \text{in } \Omega_m, \tag{6.8}$$

$$u \in W^{2,2}(\Omega_m) \cap W_0^{1,2}(\Omega_m). \tag{6.9}$$

We can write (6.8) in the form

$$A_\epsilon u = \frac{1}{\epsilon} K(x, u) + f \tag{6.10}$$

where

$$A_\epsilon u = Au + \frac{1}{\epsilon} u \qquad \text{and} \qquad K(x, u) = u - (u - \phi_{1m})^{+} + (u - \phi_{2m})^{-}.$$

It is clear that $K(x, u)$ is a measurable function and

$$\phi_{2m}(x) \leq K(x, u) \leq \phi_{1m}(x). \tag{6.11}$$

Since the coefficient of u in $A_\epsilon u$ is positive, the maximum principle can be applied. Consequently, for any $g \in L^p(\Omega_m)$ there exists a unique solution of the Dirichlet problem

$$A_\epsilon v = g \quad \text{a.e.} \quad \text{in } \Omega_m, \qquad v \in W^{2,p}(\Omega_m) \cap W_0^{1,2}(\Omega_m) \tag{6.12}$$

and

$$|v|_{2,p}^{\Omega_m} \leq C^* |g|_{0,p}^{\Omega_m},$$

here C^* may depend on ϵ, m. Using this result and Schauder's fixed point theorem (cf. Problem 1), one can derive the existence of a solution u_ϵ of (6.8), (6.9) which belongs to $W^{2,p}(\Omega_m)$.

Denote the solution v of (6.12) by $R_\epsilon g$. It is easily seen that

$$\frac{1}{\epsilon} R_\epsilon \psi = -R_\epsilon A\psi + \psi \qquad \text{if} \quad \psi \in W^{2,2}(\Omega_m) \cap W_0^{1,2}(\Omega_m). \tag{6.13}$$

The maximum principle implies that

$$\text{if } f \geq g \text{ a.e. in } \Omega_m, \text{ then } R_\epsilon f \geq R_\epsilon g \text{ a.e. in } \Omega_m. \tag{6.14}$$

Recalling (6.11) and using (6.13) (we need here the relation (6.6) with $i = 1$), we get

$$u_\epsilon = R_\epsilon\left[f + \frac{1}{\epsilon} K(x, u_\epsilon)\right] \leq R_\epsilon\left(f + \frac{1}{\epsilon}\phi_{1m}\right) = R_\epsilon(f - A\phi_{1m}) + \phi_{1m}.$$

Hence

$$u_\epsilon - \phi_{1m} \leq R_\epsilon(f - A\phi_{1m}),$$

which implies that

$$\frac{1}{\epsilon}(u_\epsilon - \phi_{1m})^{+} \leq \frac{1}{\epsilon} |R_\epsilon(f - A\phi_{1m})|. \tag{6.15}$$

Similarly,

$$\frac{1}{\epsilon}(u_\epsilon - \phi_{2m})^- \leq \frac{1}{\epsilon}|R_\epsilon(f - A\phi_{2m})|. \tag{6.16}$$

Noting that R_ϵ is actually the operator $R_{m,1/\epsilon}$ appearing in Lemma 5.2, and using Lemma 5.2, we get

$$\left|\frac{1}{\epsilon}(R_\epsilon - A\phi_{im})\right|_{0,p,\mu}^{\Omega_m} \leq C \qquad \left(i = 1, 2; \epsilon \leq \frac{1}{\Lambda}\right) \tag{6.17}$$

where C is independent of ϵ, m. From (6.15), (6.16), we then conclude that

$$\left|\frac{1}{\epsilon}(u_\epsilon - \phi_{1m})^+\right|_{0,p,\mu}^{\Omega_m} \leq C, \qquad \left|\frac{1}{\epsilon}(u_\epsilon - \phi_{2m})^-\right|_{0,p,\mu}^{\Omega_m} \leq C. \tag{6.18}$$

The same proof (6.18) works also when $p = 2$, so that

$$\left|\frac{1}{\epsilon}(u_\epsilon - \phi_{1m})^+\right|_{0,2,\mu}^{\Omega_m} \leq C, \qquad \left|\frac{1}{\epsilon}(u_\epsilon - \phi_{2m})^-\right|_{0,2,\mu}^{\Omega_m} \leq C. \tag{6.19}$$

From (6.8) and (6.18), (6.19) we deduce that

$$|Au_\epsilon|_{0,p,\mu}^{\Omega_m} \leq C, \qquad |Au_\epsilon|_{0,2,\mu}^{\Omega_m} \leq C. \tag{6.20}$$

Hence, by Lemma 5.1 applied to the general value p and also to the special case $p = 2$,

$$|u_\epsilon|_{2,p,\mu}^{\Omega_m} \leq C, \qquad |u_\epsilon|_{2,2,\mu}^{\Omega_m} \leq C \tag{6.21}$$

where C is independent of ϵ, m.

We now take $\epsilon = 1/m$ and define

$$\tilde{u}_m(x) = \begin{cases} u_\epsilon(x) & \text{if } x \in \overline{\Omega}_m, \\ v_\epsilon(x) & \text{if } x \in \overline{\Omega}_m^c, \end{cases}$$

where v_ϵ is such that

$$\tilde{u}_m \in W^{2,p,\mu}(R^n) \cap W^{2,2,\mu}(R^n),$$

$\tilde{u}_m$ has compact support, and

$$|\tilde{u}_m|_{2,p,\mu}^{R^n} \leq C, \qquad |\tilde{u}_m|_{2,2,\mu}^{R^n} \leq C \tag{6.22}$$

where C is independent of m. The construction of v_ϵ can be performed by the method of Problem 9, Chapter 10.

By a compact imbedding theorem of Sobolev spaces (Theorem 10.2.6), there exists a subsequence $\{\tilde{u}_{m'}\}$, which is convergent to a function u in the norm of $W^{1,2}(K)$, for any compact subset K of R^n. Since $|\tilde{u}_{m'}|_{1,2,\mu}^{R^n} \leq C$, the same inequality holds for u. It follows that, as $m' \to \infty$,

$$|\tilde{u}_{m'} - u|_{1,2,\nu}^{\Omega} \to 0 \qquad \text{for any} \quad \nu > \mu.$$

We now extract from $\{\tilde{u}_{m'}\}$ a subsequence which is weakly convergent in

$W^{2,p,\mu}(\Omega) \cap W^{2,2,\mu}(\Omega)$. For simplicity, take this subsequence to be the sequence $\{\tilde{u}_m\}$. Then

$$\lim_{m\to\infty} |\tilde{u}_m - u|^{\Omega}_{1,2,\nu} = 0, \tag{6.23}$$

$$u \in W^{2,p,\mu}(\Omega) \cap W^{2,2,\mu}(\Omega) \cap W_0^{1,2,\mu}(\Omega). \tag{6.24}$$

(Since for any $\zeta \in C_0^\infty(R^n)$, $\{\zeta\tilde{u}_m\}$ is weakly convergent to ζu in $W^{1,2,\mu}(\Omega)$, ζu belongs to $W_0^{1,2,\mu}(\Omega)$. Hence also $u \in W_0^{1,2,\mu}(\Omega)$.)

We next show that

$$(u - \phi_1)^+ = 0 \qquad \text{in } \Omega, \tag{6.25}$$

$$(u - \phi_2)^- = 0 \qquad \text{in } \Omega. \tag{6.26}$$

To prove (6.25) notice that for any $v \in L^2(\Omega)$, with compact support, we have

$$\left((\tilde{u}_m - \phi_{1m})^+ - (v - \phi_{1m})^+, \tilde{u}_m - v\right)_{\nu,\Omega} \geq 0.$$

Letting $m\to\infty$ and using (6.19), (6.23) and the definition $\phi_{1m} = \zeta_m\phi_1$, we get

$$-\left((v - \phi_1)^+, u - v\right)_{\nu,\Omega} \geq 0.$$

By completion, this is true for any $v \in L^{2,\nu}(\Omega)$. Taking $v = u - kw$ ($w \in L^{2,\nu}(\Omega)$, $k > 0$), we obtain

$$\left((u - \phi_1 - kw)^+, w\right)_{\nu,\Omega} \geq 0.$$

Letting $k\to 0$, we find that

$$\left((u - \phi_1)^+, w\right)_{\nu,\Omega} \geq 0.$$

Since w is arbitrary, (6.25) follows. The proof of (6.26) is similar.

From (6.25), (6.26) we conclude that $u \in K_\mu$. Since (6.24) also holds, it remains to show that (6.2) is satisfied.

Let $v \in K_\mu$. Then

$$(\zeta_m v - \phi_{1m})^+ = (\zeta_m v - \phi_{2m})^- = 0 \quad \text{a.e.} \qquad \text{in } \Omega.$$

Multiplying both sides of (6.8) (with $\epsilon = 1/m$, $u = \tilde{u}_m$) by $(\zeta_m v - \tilde{u}_m) \cdot \exp(-2\nu|x|)$, where $\nu > \mu$, and integrating over Ω_m, we get

$$\int_{\Omega_m} e^{-2\nu|x|} A\tilde{u}_m \cdot (\zeta_m v - \tilde{u}_m)\, dx \geq \int_{\Omega_m} e^{-2\nu|x|} f \cdot (\zeta_m v - \tilde{u}_m)\, dx. \tag{6.27}$$

Since $\tilde{u}_m \to u$ weakly in $W^{2,2,\mu}(\Omega)$, $A\tilde{u}_m \to Au$ weakly in $L^{2,\mu}(\Omega)$. Using also (6.23), and taking $m\to\infty$ in (6.27), we get

$$\int_\Omega e^{-2\nu|x|} Au \cdot (v - u)\, dx \geq \int_\Omega e^{-2\nu|x|} f \cdot (v - u)\, dx. \tag{6.28}$$

Noting that both Au and $v - u$ belong to $L^{p,\mu}(\Omega)$, and letting $\nu \downarrow \mu$ in (6.28), we obtain the inequality (6.2).

Remark. From (6.2) we can deduce (cf. Corollary 3.2) that

$$Au - f \leq 0 \quad \text{if} \quad u > \phi_2, \tag{6.29}$$

$$Au - f \geq 0 \quad \text{if} \quad u < \phi_1. \tag{6.30}$$

Thus, (6.2) is equivalent to

$$Au \cdot (v - u) \geq f \cdot (v - u) \quad \text{a.e.} \quad \text{for any} \quad v \in K_\mu. \tag{6.31}$$

7. Existence of saddle points in unbounded domains

We shall need the following conditions:

(P) The functions ψ_1, ψ_2 belong to $W^{2,p,\mu}(\Omega) \cap W^{2,2,\mu}(\Omega)$ for some $p > n$ and to $L^\infty(\Omega)$, and

$$\psi_1 \geq \psi_2 \quad \text{in } \Omega, \qquad \psi_1 = \psi_2 \quad \text{on } \partial\Omega.$$

(Q) The function f belongs to $L^{p,\mu}(\Omega) \cap L^{2,\mu}(\Omega)$, and

$$E_x \int_0^\sigma e^{-\alpha t} |f(x(t))| \, dt < \infty \tag{7.1}$$

for all $x \in \Omega$, $\sigma \in \mathcal{C}_x$.

Let u be the solution of the elliptic variational inequality (6.1), (6.2) with $\phi_1 = \psi_1 - \psi_2$, $\phi_2 = 0$ and with f replaced by $\tilde{f} = f - A\psi_2$. Set

$$V(x) = u(x) + \psi_2(x). \tag{7.2}$$

Since $p > n$, u and V are continuously differentiable in $\overline{\Omega}$ (by Sobolev's inequality). Define

$$\hat{E} = \{x \in \overline{\Omega};\ V(x) = \psi_1(x)\}, \tag{7.3}$$

$$\hat{F} = \{x \in \overline{\Omega};\ V(x) = \psi_2(x)\}. \tag{7.4}$$

We shall need the condition:

$$P_x(t_{\hat{E}} < \infty) = 1, \qquad P_x(t_{\hat{F}} < \infty) = 1 \qquad \text{if} \quad x \in \Omega. \tag{7.5}$$

This condition is satisfied, of course, if $P_x(t_{\Omega^c} < \infty) = 1$.

Notice that if $E_x t_{\Omega^c} < \infty$ and f is bounded, then (7.1) holds.

Theorem 7.1. *Let $\Omega \in \mathcal{C}^2$ and suppose that* (A) *holds with $c \equiv 0$, and that* (A_1), (P), (Q), *and* (7.5) *hold. Then the stochastic game associated with* (1.1), (1.3), *and $E = F = \overline{\Omega}$ has value $V(x)$, and $(\hat{E}, \hat{F})$ form a saddle point of sets. Further,* (4.16) *and* (4.17)–(4.19) *hold.*

The proof is similar to the proof of Theorem 4.2 and Corollary 3.2. In verifying (4.12) for all $\lambda \leq t_{\hat{F}}$, we first take $\lambda \leq t_{B_\epsilon} \wedge t_{A_R} \wedge T$ where A_R

$= \{x;\ |x| > R\}$, B_ϵ is a closed ϵ-neighborhood of $\hat{F}$, and then take $R \uparrow \infty$, $T \uparrow \infty$, $\epsilon \downarrow 0$.

Remark. Let E, F be closed subsets of $\bar{\Omega}$ such that $\hat{E} \subset E$, $\hat{F} \subset F$, and such that

$$V < \psi_1 \quad \text{if} \quad x \in E \backslash \hat{F}, \qquad V > \psi_2 \quad \text{if} \quad x \in F.$$

Then V is also the value, and $(\hat{E}, \hat{F})$ a saddle point, of the stochastic game with σ, τ restricted by

$$\sigma \in \mathcal{C}_x, \quad x(\sigma) \in E; \qquad \tau \in \mathcal{C}_x, \quad x(\tau) \in F.$$

This follows from Theorem 2.4, since the properties (2.6), (2.7) are already satisfied.

8. The stopping time problem

Consider the stopping time problem associated with (1.1), (1.2) when σ varies over the set $\mathcal{Q}_x$ of all a.s. finite-valued stopping times σ with $\sigma < t_{\Omega^c}$.

We shall assume:

(P_0) $\psi_1 \in W^{2,p,\mu}(\Omega) \cap W^{2,2,\mu}(\Omega)$ for some $p > n$, and

$$\psi_1(x) > -C$$

for some positive constant C.

(Q_0) f belongs to $L^{p,\mu}(\Omega) \cap L^{2,\mu}(\Omega)$ and, for all $\sigma \in \mathcal{Q}_x$,

$$E \int_0^\sigma e^{-\alpha t} f(x(t))\, dt > -C \qquad \text{for some} \quad C > 0.$$

We now specialize the arguments of the previous sections. Thus, in the elliptic variational inequality (6.1), (6.2) we take

$$K_\mu = \{ g \in L^{2,\mu}(\Omega),\ g \leq 0 \text{ a.e. in } \Omega \}. \tag{8.1}$$

Denote by u the solution of (6.1), (6.2) when K_μ is given by (8.1) and f is replaced by $\tilde{f} = f - A\psi_1$, and set

$$V(x) = u(x) + \psi_1(x), \tag{8.2}$$

$$\hat{E} = \{ x \in \bar{\Omega};\ V(x) = \psi_1(x) \}. \tag{8.3}$$

We shall need the condition

$$P_x(t_{\hat{E}} < \infty) = 1 \qquad \text{for all} \quad x \in \Omega. \tag{8.4}$$

This condition is satisfied, of course, if $P_x(t_{\Omega^c} < \infty) = 1$.

Theorem 8.1. *Let $\Omega \in \mathcal{C}^2$ and suppose* (A) *holds with $c \equiv 0$, and* (A_1), (P_0), (Q_0), (8.4) *hold. Then the function $V(x)$ is the optimal cost and $\hat{E}$ is an optimal stopping set for the stopping time problem associated with* (1.1), (1.2). *Further, V and V_x are continuous in $\bar{\Omega}$, and*

$$\frac{\partial V}{\partial x_i} = \frac{\partial \psi_1}{\partial x_i} \quad \text{on} \quad \hat{E} \cap \Omega \qquad (1 \leqslant i \leqslant n). \tag{8.5}$$

The proof is left to the reader (Problems 7–10). If Ω is a bounded domain, then we can replace (A) (in Theorem 8.1) by (B_1), (B_2).

Notice that the variational inequality for u gives (cf. Corollary 3.2 and (4.17)–(4.19))

$$LV - \alpha V + f \geqslant 0 \quad \text{a.e.} \qquad \text{on } \Omega, \tag{8.6}$$

$$LV - \alpha V + f = 0 \quad \text{a.e.} \qquad \text{on } \Omega \backslash \hat{E}. \tag{8.7}$$

Part II. The Nonstationary Case

9. Characterization of saddle points

We shall extend the results of Part I to the case where the coefficients $b, \sigma, f, \psi_1, \psi_2$ depend on t. For simplicity we consider only the case analogous to $E = F = \bar{\Omega}$.

Consider a system of n stochastic differential equations

$$dx(t) = b(x(t), t)\, dt + \sigma(x(t), t)\, dw(t) \tag{9.1}$$

and an initial condition

$$x(s) = x, \tag{9.2}$$

where $s \geqslant 0$, $x \in R^n$. We shall assume:

(C_1) $\sigma(x, t)$ and $b(x, t)$ are continuous functions in $(x, t) \in R^n \times [0, \infty)$, and

$$|\sigma(x, t)| + |b(x, t)| \leqslant C(1 + |x|) \qquad (C \text{ const});$$

further, for any $R > 0$ there is a constant C_R such that

$$|\sigma(x, t) - \sigma(y, t)| + |b(x, t) - b(y, t)| \leqslant C_R |x - y|$$

for all $t \geqslant 0$, $x \in R^n$, $y \in R^n$, $|x| \leqslant R$, $|y| \leqslant R$.

Let Ω be a nonempty domain in R^n, and let

$$Q = \{(x, t);\ x \in \Omega,\ t > 0\},$$
$$Q^c = \{(x, t);\ x \in R^n \backslash \Omega,\ t > 0\}.$$

For any closed set A in the half-space $t > 0$, denote by $t_A = t_A^s$ the first time $t > s$ that $(x(t), t)$ hits A. Denote by $\mathfrak{D}_{x,s}$ the set of all a.s. finite-valued stopping times λ for the process $x(t)$ given by (9.1), (9.2) (the range of λ is in $[s, \infty)$) such that

$$\lambda \leq t_{Q^c} \quad \text{a.s.}$$

Let $f(x, t)$, $\psi_1(x, t)$, $\psi_2(x, t)$ be functions defined on $\overline{Q}$ and let α be a nonnegative number. To any pair of times σ, τ from $\mathfrak{D}_{x,s}$ we correspond the *payoff*

$$J_{x,s}(\sigma, \tau) = E_{x,s}\left\{ \int_s^{\sigma \wedge \tau} e^{-\alpha(t-s)} f(x(t), t)\, dt \right.$$
$$\left. + e^{-\alpha(\sigma - s)} \psi_1(x(\sigma), \sigma) \chi_{\sigma < \tau} + e^{-\alpha(\tau - s)} \psi_2(x(\tau), \tau) \chi_{\tau \leq \sigma} \right\}. \tag{9.3}$$

We call this scheme the *stochastic game* associated with (9.1)–(9.3), and we denote it by $G_{x,s}$. The set $G = \{G_{x,s};\ (x, s) \in \overline{Q}\}$ is called the stochastic game associated with (9.1)–(9.3) in $\overline{\Omega}$. If

$$\inf_{\sigma \in \mathfrak{D}_{x,s}} \sup_{\tau \in \mathfrak{D}_{x,s}} J_{x,s}(\sigma, \tau) = \sup_{\tau \in \mathfrak{D}_{x,s}} \inf_{\sigma \in \mathfrak{D}_{x,s}} J_{x,s}(\sigma, \tau) \tag{9.4}$$

then we say that $G_{x,s}$ has value $V(x, s)$, where $V(x, s)$ is the common number in (9.4). Suppose there exist closed sets $\hat{S}$, $\hat{T}$ in $\overline{Q}$ such that $t^s{}_{\hat{S}}$, $t^s{}_{\hat{T}}$ belong to $\mathfrak{D}_{x,s}$ for all (x, s) in $\overline{Q}$, and

$$J_{x,s}(t_{\hat{S}}, \tau) \leq J_{x,s}(t_{\hat{S}}, t_{\hat{T}}) \leq J_{x,s}(\sigma, t_{\hat{T}}) \tag{9.5}$$

for all $\sigma \in \mathfrak{D}_{x,s}$, $\tau \in \mathfrak{D}_{x,s}$, $(x, s) \in \overline{Q}$, then we say that $(t_{\hat{S}}, t_{\hat{T}})$ is a *saddle point* for G and that $(\hat{S}, \hat{T})$ forms a *saddle point of sets* for G.

We shall assume:

(C_2) $f(x, t)$, $\psi_1(x, t)$, $\psi_2(x, t)$ are continuous functions in $\overline{\Omega}$; ψ_1 and ψ_2 are bounded, and

$$E_{x,s} \int_s^{\lambda} e^{-\alpha t} |f(x(t), t)|\, dt < \infty \tag{9.6}$$

for all $(x, s) \in \overline{Q}$, $\lambda \in \mathfrak{D}_{x,s}$.

Theorem 9.1. *Let* (C_1), (C_2) *hold and suppose* $(\hat{S}, \hat{T})$ *is a saddle point of sets for the stochastic game* G. *Then the value* $V(x, s)$ *has the following* *properties*:

$$V(x, s) \leqslant \psi_1(x, s) \qquad \text{if} \quad (x, s) \in \overline{Q} \setminus \hat{S}, \tag{9.7}$$

$$V(x, s) \geqslant \psi_2(x, s) \qquad \text{if} \quad (x, s) \in \overline{Q}, \tag{9.8}$$

$$V(x, s) = \psi_1(x, s) \qquad \text{if} \quad (x, s) \in \hat{S} \setminus \hat{T}, \tag{9.9}$$

$$V(x, s) = \psi_2(x, s) \qquad \text{if} \quad (x, s) \in \hat{T}, \tag{9.10}$$

$$V(x, s) \leqslant E_{x,s}\left\{\int_s^{\lambda} e^{-\alpha(t-s)} f(x(t), t)\, dt + e^{-\alpha(\lambda - s)} V(x(\lambda), \lambda)\right\}$$

$$\text{if} \quad \lambda \in \mathfrak{D}_{x,s}, \quad \lambda \leqslant t_{\hat{T}}, \tag{9.11}$$

$$V(x, s) \geqslant E_{x,s}\left\{\int_s^{\mu} e^{-\alpha(t-s)} f(x(t), t)\, dt + e^{-\alpha(\mu - s)} V(x(\mu), \mu)\right\}$$

$$\text{if} \quad \mu \in \mathfrak{D}_{x,s}, \quad \mu \leqslant t_{\hat{S}}. \tag{9.12}$$

The proof is similar to the proof of Theorem 2.3, and it is left to the reader.

Theorem 9.2. *Let* (C_1), (C_2) *hold. Suppose* $V(x, s)$ *is a Borel measurable function in* $\overline{Q}$ *and* $\hat{S}$, $\hat{T}$ *are closed subsets of* $\overline{Q}$ *such that*

$$t_{\hat{S}},\ t_{\hat{T}} \quad \text{belong to } \mathfrak{D}_{x,s} \qquad (\text{for all } (x, s) \in \overline{Q}), \tag{9.13}$$

and suppose that (9.7)–(9.12) *hold and*

$$\psi_1 = \psi_2 \qquad \text{on} \quad \hat{S} \cap \hat{T}. \tag{9.14}$$

Then $V(x, s)$ *is the value of the stochastic game associated with* (9.1)–(9.3), *and* $(\hat{S}, \hat{T})$ *is a saddle point of sets.*

The proof is similar to the proof of Theorem 2.4, and it is left to the reader.

10. Parabolic variational inequalities

Let Ω be any unbounded domain in $\mathcal{C}^2$ and let $Q = \{(x, t);\ x \in \Omega,\ t > 0\}$. Consider a partial differential operator

$$A(t)u \equiv -\frac{1}{2}\sum_{i,j=1}^{n} \frac{\partial}{\partial x_i}\left(a_{ij}(x, t)\,\frac{\partial u}{\partial x_j}\right) + \sum_{i=1}^{n} \tilde{b}_i(x, t)\,\frac{\partial u}{\partial x_i} + c(x, t)u + \alpha \tag{10.1}$$

where α is a positive constant and $\tilde{b}_i = \frac{1}{2}\sum_{j=1}^n \partial a_{ij}/\partial x_j - b_i$. We shall assume (cf. the condition (A)):

(A*) (i) The functions a_{ij}, $\partial a_{ij}/\partial x_k$, b_i, c and their first t-derivatives are measurable functions in $\bar{Q}$ for all $1 \leqslant i, j, k \leqslant n$, bounded by a constant K; (ii) $c(x, t) \geqslant 0$ in $\bar{Q}$; (iii) there is a positive constant β such that

$$\frac{1}{2}\sum_{i,j=1}^{n} a_{ij}(x, t)\xi_i\xi_j \geqslant \beta|\xi|^2 \quad \text{if} \quad x \in \bar{\Omega}, \quad t \geqslant 0, \quad \xi \in R^n,$$

finally, (iv)

$$B \equiv \sup_{(x, t)\in Q} |\tilde{b}_i(x, t)| < \frac{2\sqrt{\alpha\beta}}{\sqrt{n}}.$$

For any $p \geqslant 2$, let $\hat{\mu}_p$ be a positive number sufficiently small so that

$$B^2 < 4(p-1)[\alpha - (p-1)\hat{\mu}_p K](\beta - \hat{\mu}_p K)/n.$$

In particular, $\hat{\mu}_2$ can be taken as μ_0 in Lemma 5.1.

Later on we shall define positive constants γ_p depending only on $\alpha, \beta, n, K, p, \hat{\mu}_p$. Let δ be any fixed number satisfying:

$$0 < \delta \leqslant \gamma_p.$$

Let $f(x, t)$, $\phi_1(x)$, $\phi_2(x)$ be functions defined in $\bar{Q}$ and satisfying:

(B*) $\phi_1(x)$ and $\phi_2(x)$ belong to $W^{2,p,\mu}(\Omega) \cap W^{2,2,\mu}(\Omega)$ for some $0 \leqslant \mu \leqslant \hat{\mu}_p$, and

$$\phi_1 \geqslant 0 \geqslant \phi_2 \quad \text{in } \bar{\Omega}, \qquad \phi_1, \phi_2 \quad \text{belong to} \quad W_0^{1,2,\mu}(\Omega).$$

(C*) $f(x, t)$ is a measurable function in $\bar{Q}$, and

$$e^{-\delta t}|f(\cdot, t)|_{0,p,\mu}^{\Omega} \leqslant C, \qquad e^{-\delta t}|f(\cdot, t)|_{0,2,\mu}^{\Omega} \in L^\infty(0, \infty) \cap L^1(0, \infty),$$

$$e^{-\delta t}\left|\frac{\partial}{\partial t} f(\cdot, t)\right|_{0,p,\mu}^{\Omega} \leqslant C, \qquad e^{-\delta t}\left|\frac{\partial}{\partial t} f(\cdot, t)\right|_{0,2,\mu}^{\Omega} \in L^1(0, \infty).$$

Here $\partial f(\cdot, t)/\partial t$ is taken as a strong derivative.

Introduce the set

$$K_\mu = \{g \in L^{2,\mu}(\Omega); \phi_2(x) \leqslant g(x) \leqslant \phi_1(x) \text{ a.e. in } \Omega\}.$$

We now consider a *parabolic variational inequality*: find a function $u(x, t)$ such that

$$\begin{aligned} &e^{-\delta t}u \in L^\infty(0, \infty; W^{2,2,\mu}(\Omega)) \cap L^\infty(0, \infty; W_0^{1,2,\mu}(\Omega)), \\ &e^{-\delta t}\frac{\partial u}{\partial t} \in L^\infty(0, \infty; L^{2,\mu}(\Omega)), \end{aligned} \tag{10.2}$$

and, for a.a. $t > 0$,

$$-\int_\Omega e^{-2\mu|x|} \frac{\partial u}{\partial t}(v-u)\,dx + \int_\Omega e^{-2\mu|x|} Au \cdot (v-u)\,dx$$
$$\geq \int_\Omega e^{-2\mu|x|} f(v-u)\,dx \qquad \text{for every} \quad v \in K_\mu, \tag{10.3}$$

$$\phi_2(x) \leq u(x,t) \leq \phi_1(x) \quad \text{a.e.} \quad \text{in } \Omega. \tag{10.4}$$

Theorem 10.1. *Let $\Omega \in \mathcal{C}^{2+\rho}$ for some $0 < \rho \leq 1$, and let* (A*), (B*), (C*) *hold. Then there exists a unique solution of* (10.2)–(10.4), *and*

$$e^{-\delta t}\left|\frac{\partial u}{\partial t}\right|_{0,p,\mu}^{\Omega} + e^{-\delta t}|u|_{2,p,\mu}^{\Omega} \leq C \qquad \textit{for a.a.} \quad t > 0. \tag{10.5}$$

In Section 11 we shall consider the case where ϕ_1, ϕ_2 depend also on t. We shall prove the existence and uniqueness of a solution that is not as "smooth" as the solution in Theorem 10.1.

Proof. To prove uniqueness, we suppose that u_1, u_2 are two solutions and let $w = u_1 - u_2$. Setting $u = u_1$, $v = u_2$ and then $u = u_2$, $v = u_1$ in (10.3) and adding, we obtain (cf. the proof of Theorem 6.1 and Problem 13)

$$\frac{1}{2}\frac{d}{dt}\int_\Omega e^{-2\mu|x|}|w|^2\,dx - 4\delta\int_\Omega e^{-2\mu|x|}|w|^2\,dx \geq 0 \qquad \text{for a.a.} \quad t > 0. \tag{10.6}$$

Hence, the function

$$\psi(t) = e^{-8\delta t}\int_\Omega e^{-2\mu|x|}|w|^2\,dx$$

satisfies $\dot{\psi}(t) \geq 0$. This implies that

$$e^{6\delta t}e^{-8\delta s}|w(s)|_{0,2,\mu}^{\Omega} \leq e^{-2\delta t}|w(t)|_{0,2,\mu}^{\Omega}$$

for all $0 \leq s < t$. Letting $t \to \infty$ we get $w(s) = 0$, that is, $u_1 = u_2$.

To prove existence, let $\epsilon > 0$ and consider the problem: find $u(x,t)$ satisfying:

$$u(\cdot,t) \in L^p(0,T;\, W^{2,p}(\Omega_m)) \cap L^\infty(0,T;\, W^{1,2}(\Omega_m)),$$
$$\frac{\partial u(\cdot,t)}{\partial t} \in L^p(0,T;\, L^p(\Omega_m)); \tag{10.7}$$

for a.a. $t \in (0,T)$,

$$-\frac{\partial u}{\partial t} + Au + \frac{1}{\epsilon}(u-\phi_{1m})^+ - \frac{1}{\epsilon}(u-\phi_{2m})^- = f \quad \text{a.e. in} \quad x \in \Omega_m, \tag{10.8}$$

and

$$u(x, T) = 0 \qquad \text{if} \quad x \in \Omega_m; \tag{10.9}$$

here $\phi_{im} = \zeta_m \phi_i$, as in the paragraph containing (6.6), (6.7).

The existence of a solution follows by using Theorem 10.4.2 (see Problem 14). We shall write this solution either as u or as $u_{T,\epsilon,m}$. In the sequel, various positive constants independent of T, ϵ, m will be denoted by the same symbol C.

We shall derive the following estimates:

$$e^{-\delta t}|u(t)|_{0,2,\mu}^{\Omega_m} \leq C \qquad \text{for all} \quad t \in (0, T), \tag{10.10}$$

$$\int_0^T e^{-2\delta t}\left[|u(t)|_{1,2,\mu}^{\Omega_m}\right]^2 dt \leq C, \tag{10.11}$$

$$e^{-\delta t}|u'(t)|_{0,2,\mu}^{\Omega_m} \leq C \qquad \text{for a.a.} \quad t \in (0, T), \tag{10.12}$$

$$\int_0^T e^{-2\delta t}\left[|u'(t)|_{1,2,\mu}^{\Omega_m}\right]^2 dt \leq C, \tag{10.13}$$

$$e^{-2\delta t}\left[|(u - \phi_{1m})^+|_{0,2,\mu}^{\Omega_m}\right]^2 \leq C\epsilon^2 \qquad \text{for a.a.} \quad t \in (0, T), \tag{10.14}$$

$$e^{-2\delta t}\left[|(u - \phi_{2m})^-|_{0,2,\mu}^{\Omega_m}\right] \leq C\epsilon^2 \qquad \text{for all} \quad t \in (0, T). \tag{10.15}$$

Proof of (10.10), (10.11). Denote the scalar product in $L^{2,\mu}(\Omega_m)$ by (,) and the norm by $|\ |$. Denote the norm of $W^{1,2,\mu}(\Omega_m)$ by $\|\ \|$. Notice that since $\phi_1 \geq 0 \geq \phi_2$,

$$(u - \phi_{1m})^+ u \geq 0, \qquad -(u - \phi_{2m})^- u \geq 0.$$

Hence, if we multiply (10.8) by $ue^{-2\mu|x|}$ and integrate over Ω_m, we obtain (see Problem 13)

$$\frac{1}{2}\frac{d}{dt}|u|^2 + (Au, u) \leq (f, u).$$

Integrating by parts in (Au, u) and arguing as in the derivation of (5.16), we find that if $\delta < \gamma/4$, γ as in (5.16) (γ is sufficiently small, depending on $\alpha, \beta, n, \mu_0, B$) then

$$-\frac{1}{2}\frac{d}{dt}|u|^2 + 4\delta\|u\|^2 \leq |(f, u)| \leq 2\delta|u|^2 + C|f|^2. \tag{10.16}$$

Hence

$$-\frac{d}{dt}|u|^2 + 4\delta\|u\|^2 \leq C|f|^2,$$

or

$$-\frac{d}{dt}\left(e^{-2\delta t}|u(t)|^2\right) + 2\delta e^{-2\delta t}\|u(t)\|^2 \leq Ce^{-2\delta t}|f(t)|^2. \tag{10.17}$$

Integrating and using (10.9), we get

$$e^{-2\delta t}|u(t)|^2 + 2\delta \int_t^T e^{-2\delta s}\|u(s)\|^2\, ds \leq C \int_t^T e^{-2\delta s}|f(s)|^2\, ds.$$

Making use of (C*), the inequalities (10.10), (10.11) follow.

Proof of (10.12), (10.13). Differentiating (10.8) formally with respect to t and taking the scalar product (in $L^{2,\mu}(\Omega_m)$) with $\partial u/\partial t$, we get

$$-\left(\frac{\partial^2 u}{\partial t^2}, \frac{\partial u}{\partial t}\right) + \left(A\frac{\partial u}{\partial t}, \frac{\partial u}{\partial t}\right) + \left(\frac{\partial A}{\partial t}u, \frac{\partial u}{\partial t}\right)$$
$$+ \frac{1}{\epsilon}\left(\frac{\partial(u-\phi_1)^+}{\partial t}, \frac{\partial u}{\partial t}\right) - \frac{1}{\epsilon}\left(\frac{\partial(u-\phi_2)^-}{\partial t}, \frac{\partial u}{\partial t}\right) = \left(\frac{\partial f}{\partial t}, \frac{\partial u}{\partial t}\right), \tag{10.18}$$

where $\partial A/\partial t$ is the operator obtained from A by differentiating all the coefficients of A once with respect to t.

Using the fact that

$$\{[u(x, t+h) - \phi_{1m}(x)]^+ - [u(x,t) - \phi_{2m}(x)]^+\}$$
$$\cdot \{u(x, t+h) - u(x,t)\} \geq 0,$$

we get the formal inequality

$$\left(\frac{\partial}{\partial t}(u-\phi_{1m})^+, \frac{\partial u}{\partial t}\right) \geq 0. \tag{10.19}$$

Similarly, we get the formal inequality

$$-\left(\frac{\partial}{\partial t}(u-\phi_{2m})^-, \frac{\partial u}{\partial t}\right) \geq 0. \tag{10.20}$$

Thus we find from (10.18) that

$$-\left(\frac{\partial^2 u}{\partial t^2}, \frac{\partial u}{\partial t}\right) + \left(A\frac{\partial u}{\partial t}, \frac{\partial u}{\partial t}\right) + \left(\frac{\partial A}{\partial t}u, \frac{\partial u}{\partial t}\right) \leq \left(\frac{\partial f}{\partial t}, \frac{\partial u}{\partial t}\right). \tag{10.21}$$

As in the proof of (5.6),

$$\left(A\frac{\partial u}{\partial t}, \frac{\partial u}{\partial t}\right) \geq 4\delta\left\|\frac{\partial u}{\partial t}\right\|^2.$$

We also have

$$\left|\left(\frac{\partial A}{\partial t}u, \frac{\partial u}{\partial t}\right)\right| \leq C\|u\|\left\|\frac{\partial u}{\partial t}\right\|;$$

in the terms

$$\left(\frac{\partial}{\partial x_i} \left(\frac{\partial a_{ij}}{\partial t} \frac{\partial u}{\partial x_j} \right), \frac{\partial u}{\partial t} \right)$$

occurring in $(\partial A/\partial t\, u, \partial u/\partial t)$ we perform integration by parts prior to estimating them. Putting these inequalities in (10.21), we get

$$-\frac{1}{2} \frac{d}{dt} |u'|^2 + 4\delta \|u'\|^2 \leq \left| \frac{\partial f}{\partial t} \right| |u'| + C\|u\| \, \|u'\|$$

$$\leq 2\delta \|u'\| + C\left(\|u\|^2 + \left| \frac{\partial f}{\partial t} \right|^2 \right) \quad (10.22)$$

where $u' = \partial u/\partial t$. The argument leading from (10.16) to (10.17) clearly leads from (10.22) to

$$-\frac{d}{dt}(e^{-2\delta t}|u'|^2) + 2\delta e^{-2\delta t}\|u'\|^2 \leq Ce^{-2\delta t}\left(\|u\|^2 + \left| \frac{\partial f}{\partial t} \right|^2 \right). \quad (10.23)$$

From (10.8), (10.9) we have (formally, since $u'(t)$ is not known to be continuous)

$$u'(T) = f(T).$$

Using (C*) we get

$$e^{-\delta T}|u'(T)| \leq C. \quad (10.24)$$

Integrating (10.23) and using (10.24), we find

$$e^{-2\delta t}|u'(t)|^2 + 2\delta \int_t^T e^{-2\delta s}\|u'(s)\|^2 \, ds$$

$$\leq C\int_t^T e^{-2\delta s}\left(\|u(s)\|^2 + \left| \frac{\partial f(s)}{\partial s} \right|^2 \right) ds + C. \quad (10.25)$$

Making use of (10.11) and of (C*), the estimates (10.12), (10.13) follow.

In the above derivation of (10.25) we have assumed that the derivatives $\partial^2 u/\partial t^2$, $\partial(u - \phi_{1m})^+/\partial t$, $\partial(u - \phi_{2m})^-/\partial t$ exist. In order to prove (10.25) rigorously, we proceed as follows:

Instead of differentiating (10.8) with respect to t, we take finite differences with respect to t, i.e., we write the parabolic equation for $(u(t+h) - u(t))/h$ $(h > 0)$. Then we take the scalar product of this equation with $(u(t+h) - u(t))/h$. Using the finite difference analog of (10.19), (10.20) we get (cf. (10.21))

$$-\left(\frac{\partial u_h}{\partial t}, u_h \right) + (Au_h, u_h) + (A_h \hat{u}, u_h) \leq (f_h, u_h)$$

where $\hat{u}(t) = u(t+h)$, $g_h(t) = (g(t+h) - g(t))/h$ and A_h is obtained from A by replacing each coefficient of A by its finite difference. Proceeding by the considerations following (10.21), we arrive at the inequality

$$-\frac{d}{dt}\left(e^{-2\delta t}|u_h(t)|^2\right) + 2\delta e^{-2\delta t}\|u_h(t)\|^2 \leq Ce^{-2\delta t}\left(\|u(t+h)\|^2 + |f_h(t)|^2\right).$$

Hence, by integration,

$$e^{-2\delta t}|u_h(t)|^2 - e^{-2\delta\tau}|u_h(\tau)|^2 + 2\delta\int_t^\tau e^{-2\delta s}\|u_h(s)\|^2\,ds$$
$$\leq C\int_t^\tau e^{-2\delta s}\left(\|u(s+h)\|^2 + |f_h(s)|^2\right)ds \qquad (\tau < T). \tag{10.26}$$

Taking $h \downarrow 0$, we get, for a.a. $0 < t < \tau < T$,

$$e^{-2\delta t}|u'(t)|^2 - e^{-2\delta\tau}|u'(\tau)|^2 + 2\int_t^\tau e^{-2\delta s}\|u'(s)\|^2\,ds$$
$$\leq C\int_t^\tau e^{-2\delta s}\left(\|u(s)\|^2 + |f'(s)|^2\right)ds. \tag{10.27}$$

If $D_x u$, $D_x^2 u$, $D_t u$ are continuous in $\bar{\Omega}_m \times [0, T]$, then, taking $\tau \uparrow T$ in (10.8), we conclude that $u_t(x, T) = f(x, T)$. We can then take $\tau \uparrow T$ in (10.27) and then use (10.24) in order to complete the proof of (10.25); thus (10.12), (10.13) hold in this case.

In the general case we approximate b_i, c, ϕ_{1m}, ϕ_{2m}, f by Hölder continuous functions (in $\bar{\Omega}_m \times [0, T]$) b_i^k, c^k, ϕ_{1m}^k, ϕ_{2m}^k, f^k with $f^k(x, T) = 0$ if $x \in \partial\Omega_m$. The approximation is in the norm $L^2(0, T; L^2(\Omega_m))$, and the condition (A*) holds for the approximating coefficients, with constants independent of k. The method used to prove the existence of a solution of (10.7)–(10.9) (see Problem 14) was based on Theorem 10.4.2. If instead we use the result stated in Remark 2 at the end of Section 10.1, then we obtain a solution u^k of (10.8), satisfying

$$u(x, T) = 0 \quad \text{if} \quad x \in \Omega_m, \qquad u(x, t) = 0 \quad \text{if} \quad x \in \partial\Omega_m, \qquad 0 < t \leq T$$

and $D_t u^k$, $D_x u^k$, $D_x^2 u^k$ are continuous in $\bar{\Omega}_m \times [0, T]$.

By the estimates of Theorem 10.4.2 and Theorem 10.2.6 we find that there is a subsequence of u^k, call it again u^k, which is convergent in $L^2(0, T; L^2(\Omega_m))$ to some function u. Applying the estimates of Theorem 10.4.3 with $p = 2$, we find that $\{u^k\}$ is weakly convergent in $L^2(0, T; W^{2,2}(\Omega_m))$, and $\{\partial u^k/\partial t\}$ is weakly convergent in $L^2(0, T; L^2(\Omega_m))$. It follows that u is a solution of (10.7)–(10.9). Finally, since (10.12), (10.13) hold for each u^k, they also hold for u (by Fatou's lemma).

Proof of (10.14), (10.15). One can show (see Problem 15) that

$$(Av, v^+) \geq 0 \qquad \text{if} \quad v \in W^{2,2,\mu}(\Omega_m) \cap W_0^{1,2,\mu}(\Omega_m). \tag{10.28}$$

Hence

$$\big(Au, (u - \phi_{1m})^+\big) = \big(A(u - \phi_{1m}), (u - \phi_{1m})^+\big) + \big(A\phi_{1m}, (u - \phi_{1m})^+\big)$$
$$\geq -C|(u - \phi_{1m})^+|.$$

We also have

$$(u - \phi_{1m})^+ (u - \phi_{2m})^- = 0.$$

Hence, taking the scalar product of (10.8) with $(u - \phi_{1m})^+$, we obtain

$$\frac{1}{\epsilon}|(u - \phi_{1m})^+|^2 \leq \big(f, (u - \phi_{1m})^+\big) + \big(u', (u - \phi_{1m})^+\big) + C|(u - \phi_{1m})^+|.$$

Using Schwarz's inequality and (10.12), the inequality (10.14) follows. The proof of (10.15) is obtained similarly, by taking the scalar product of (10.8) with $(u - \phi_{2m})^-$.

From (10.12), (10.14), (10.15), and (10.8) we deduce that

$$e^{-\delta t}|A(t)|_{0,2,\mu}^{\Omega_m} \leq C.$$

Hence, by Lemma 5.1,

$$e^{-\delta t}|u(t)|_{2,2,\mu}^{\Omega_m} \leq C. \tag{10.29}$$

We now extend $u = u_{T,\epsilon,m}$ into the half-space $t > 0$ so that the bounds (10.10)–(10.15) and (10.29) remain valid with T replaced by ∞ and Ω_m replaced by R^n. Denote these extended functions by $\tilde{u}_{T,\epsilon,m}$.

We next take a sequence $\tilde{u}_j = \tilde{u}_{T_j,\epsilon_j,m_j}$ with $T_j \uparrow \infty$, $\epsilon_j \downarrow 0$, $m_j \uparrow \infty$ which is convergent to a function u in the following sense:

$$e^{-\delta t}\tilde{u}_j \to e^{-\delta t}u \quad \text{weakly in} \quad L^2(0, \infty; W^{2,2,\mu}(R^n)).$$
$$e^{-\delta t}\tilde{u}_j \to e^{-\delta t}u \quad \text{in the weak star topology of} \quad L^\infty(0, \infty; L^{2,\mu}(R^n)); \tag{10.30}$$

$$e^{-\delta t}\frac{\partial \tilde{u}_j}{\partial t} \to e^{-\delta t}\frac{\partial u}{\partial t} \quad \text{weakly in} \quad L^2(0, \infty; W^{1,2,\mu}(R^n)),$$
$$e^{-\delta t}\frac{\partial \tilde{u}_j}{\partial t} \to e^{-\delta t}\frac{\partial u}{\partial t} \quad \text{in the weak star topology of} \quad L^\infty(0, \infty; L^{2,\mu}(R^n)). \tag{10.31}$$

It follows that u satisfies (10.2). By the compact imbedding theorem for Sobolev spaces (Theorem 10.2.6) we also have, for any $\nu > \mu$, $0 \leq t_1 < t_2 < \infty$,

$$\int_{t_1}^{t_2}\int_\Omega e^{-2\nu|x|}|\tilde{u}_j(x, t) - u(x, t)|^2\,dx\,dt \to 0 \quad \text{if} \quad j \to \infty. \tag{10.32}$$

Using (10.14), (10.15), and (10.32) one can prove (cf. Section 6) that, for any $v \in L^2(t_1, t_2; L^{2,\nu}(\Omega))$,

$$-\int_{t_1}^{t_2} \int_\Omega e^{-2\nu|x|}(v - \phi_1)^+ (u - v)\, dx\, dt \geq 0.$$

Taking $v = u - kw$, $k > 0$, $w \in L^2(t_1, t_2; L^{2,\nu}(\Omega))$ we get, after letting $k \to 0$,

$$\int_{t_1}^{t_2} \int_\Omega e^{-2\nu|x|}(u - \phi_1)^+ w\, dx\, dt \geq 0.$$

Since w is arbitrary,

$$(u - \phi_1)^+ = 0 \quad \text{a.e.} \tag{10.33}$$

Similarly,

$$(u - \phi_2)^- = 0 \quad \text{a.e.} \tag{10.34}$$

Now let $v \in K_\mu$. Multiply (10.8) (with $u = \tilde{u}_j$) by $(\zeta_m v - u) \exp(-2\nu|x|)$ $(\nu > \mu)$ and integrating with respect to $(x, t) \in \Omega \times (t_1, t_2)$; we find, after taking $j \to \infty$ and using (10.30)–(10.34) (cf. Section 6), that

$$-\int_{t_1}^{t_2} \int_\Omega e^{-2\nu|x|} \frac{\partial u}{\partial t} (v - u)\, dx\, dt + \int_{t_1}^{t_2} \int_\Omega e^{-2\nu|x|} A(t)u \cdot (v - u)\, dx\, dt$$
$$\geq \int_{t_1}^{t_2} \int_\Omega e^{-2\nu|x|} f(v - u)\, dx\, dt.$$

Dividing by $t_2 - t_1$ and letting $t_1 \to t_2$ we conclude that (10.3) holds for a.a. $t \geq 0$, with μ replaced by ν. Letting $\nu \downarrow \mu$ and recalling that $\partial u/\partial t$, $A(t)u$, and f belong to $L^{2,\mu}(\Omega)$ for a.a. $t \geq 0$, the inequality (10.3) follows.

In order to complete the proof of Theorem 10.1, it remains to derive the estimates (10.5).

Consider first the case where the coefficients of A are independent of t and $p = 2k$, k a positive integer. If we differentiate (10.8) formally with respect to t and multiply the resulting equation by

$$e^{-2k\mu|x|}\left(\frac{\partial u}{\partial t} \right)^{2k-1}$$

and integrate with respect to x, $x \in \Omega_m$, we get formally (cf. (10.21))

$$-\int_{\Omega_m} e^{-2k\mu|x|}\left(\frac{\partial u}{\partial t} \right)^{2k-1} \frac{\partial^2 u}{\partial t^2}\, dx + \int_{\Omega_m} e^{-2k\mu|x|} A\left(\frac{\partial u}{\partial t} \right) \cdot \left(\frac{\partial u}{\partial t} \right)^{2k-1} dx$$
$$\leq \int_\Omega e^{-2k\mu|x|} \frac{\partial f}{\partial t} \cdot \left(\frac{\partial u}{\partial t} \right)^{2k-1} dx$$
$$\leq Ce^{\delta t}\left[\int_{\Omega_m} e^{-2k\mu|x|}\left(\frac{\partial u}{\partial t} \right)^{2k} dx \right]^{(2k-1)/2k} \tag{10.35}$$

The proof of (5.16) shows that the second integral on the left-hand side of (10.35) is bounded below by

$$2\gamma' \int_{\Omega_m} e^{-2k\mu|x|} \left(\frac{\partial u}{\partial t} \right)^{2k-2} \left[\left| \nabla_x \left(\frac{\partial u}{\partial t} \right) \right|^2 + \left(\frac{\partial u}{\partial t} \right)^2 \right] dx$$

where γ' is a positive constant; here we use the fact that $0 \leq \mu \leq \mu_p$. We define γ_p such that $4k\gamma_p = \gamma'$, and take $0 < \delta < \gamma_p$. Thus, $\gamma' \geq 4k\delta$. It follows that the function

$$\Phi(t) = \int_{\Omega_m} e^{-2\mu|x|} \left(\frac{\partial u}{\partial t} \right)^{2k} dx$$

satisfies

$$-\frac{d}{dt}\Phi(t) + 8k\delta\Phi(t) \leq C[\Phi(t)]^{(2k-1)/2k} e^{\delta t}.$$

Hence

$$-\frac{d\Phi}{dt} + 4k\delta\Phi \leq Ce^{2k\delta t}. \tag{10.36}$$

Formally, $u'(T) = f(T)$. Hence

$$\Phi(T) = (|f(T)|_{0,2k,\mu})^{2k} \leq Ce^{2k\delta T}. \tag{10.37}$$

Integrating (10.36) and using (10.37), we conclude that $\Phi(t) \leq Ce^{2k\delta t}$, that is,

$$\int_{\Omega_m} e^{-2k\mu|x|} \left(\frac{\partial u}{\partial t} \right)^{2k} dx \leq Ce^{2k\delta}. \tag{10.38}$$

The rigorous justification of (10.18) can be accomplished by working with finite differences instead of taking the formal derivative of (10.3); cf. the proof of (10.12), (10.13).

Notice that in (10.38) u is the function $u_{T,\epsilon,m}$. Taking $T = T_j$, $\epsilon = \epsilon_j$, $m = m_j$, $j \uparrow \infty$, we arrive at the inequality

$$\int_{\Omega} e^{-2k\mu|x|} \left(\frac{\partial u}{\partial t} \right)^{2k} dx \leq Ce^{2k\delta t} \tag{10.39}$$

where $u(t)$ is now the solution of (10.2)–(10.4).

For fixed t, we can view $u(t)$ as the solution of the elliptic variational inequality

$$\int_{\Omega} e^{-2\mu|x|} Au \cdot (v - u)\, dx \geq \int_{\Omega} e^{-2\mu|x|} \tilde{f} \cdot (v - u)\, dx \tag{10.40}$$

for any $v \in K_\mu$, where

$$\tilde{f} = f + \partial u / \partial t.$$

By (10.39),

$$|\tilde{f}(t)|_{0,k,\mu} \leqslant Ce^{\delta t}.$$

Since (10.39) is valid also when $2k = 2$,

$$|\tilde{f}(t)|_{0,2,\mu} \leqslant Ce^{\delta t}.$$

Hence, by the proof of Theorem 6.1,

$$|u(t)|_{0,2k,\mu} \leqslant Ce^{\delta t}.$$

This completes the proof of (10.5) in case $p = 2k$.

If p is any positive number, we repeat the previous proof with $(\partial u/\partial t)^{2k-1}$ replaced by $|\partial u/\partial t|^{2p-2}\,\partial u/\partial t$.

Consider now the general case where the coefficients of A depend on t. If we start as in the special case where the coefficients of A are independent of t, then on the left-hand side of (10.35) there appears

$$\int_{\Omega_m} e^{-2k\mu|x|}\left(\frac{\partial A}{\partial t}u\right)\cdot\left(\frac{\partial u}{\partial t}\right)^{2k-1}dx.$$

Thus we have to handle the terms

$$I = -\int_{\Omega_m} e^{-2k\mu|x|}\frac{\partial a_{ij}}{\partial t}\frac{\partial^2 u}{\partial x_i\,\partial x_j}\left(\frac{\partial u}{\partial t}\right)^{2k-1}dx,$$

$$J = \int_{\Omega_m} e^{-2k\mu|x|}\frac{\partial b_i}{\partial t}\frac{\partial u}{\partial x_i}\left(\frac{\partial u}{\partial t}\right)^{2k-1}dx,$$

$$K = \int_{\Omega_m} e^{-2k\mu|x|}\frac{\partial c}{\partial t}u\left(\frac{\partial u}{\partial t}\right)^{2k-1}dx.$$

Consider the integral I. By integration by parts,

$$I = (2k-1)\int_{\Omega_m} e^{-2k\mu|x|}\frac{\partial a_{ij}}{\partial t}\frac{\partial u}{\partial x_j}\left(\frac{\partial u}{\partial t}\right)^{2k-2}\frac{\partial^2 u}{\partial x_i\partial t}\,dx$$

$$+\int_{\Omega_m}\frac{\partial}{\partial x_i}\left(e^{-2k\mu|x|}\frac{\partial a_{ij}}{\partial t}\right)\cdot\frac{\partial u}{\partial x_j}\left(\frac{\partial u}{\partial t}\right)^{2k-1}dx \equiv I_1 + I_2.$$

Next, we can write

$$|I_1| = \left|\int_{\Omega_m} e^{-2k\mu|x|}\frac{\partial a_{ij}}{\partial t}\frac{\partial u}{\partial x_j}\left(\frac{\partial u}{\partial t}\right)^{k-1}\cdot\frac{\partial^2 u}{\partial x_i\,\partial t}\left(\frac{\partial u}{\partial t}\right)^{k-1}dx\right|$$

$$\leqslant \frac{C}{\epsilon}\int_{\Omega_m} e^{-2k\mu|x|}\left|\frac{\partial u}{\partial x_j}\right|^2\left|\frac{\partial u}{\partial t}\right|^{2k-2}dx$$

$$+\epsilon\int_{\Omega_m} e^{-2k\mu|x|}\left|\nabla_x\left(\frac{\partial u}{\partial t}\right)\right|^2\left|\frac{\partial u}{\partial t}\right|^{2k-2}dx.$$

As for I_2, J, and K, they do not require any further treatment. Choosing ϵ sufficiently small, we then get the inequality

$$
\begin{aligned}
-\frac{d}{dt}\int_{\Omega_m} e^{-2k\mu|x|}\left|\frac{\partial u}{\partial t}\right|^{2k} dx &+ 2\gamma'\int_{\Omega_m} e^{-2k\mu|x|}\left|\frac{\partial u}{\partial t}\right|^{2k} dx \\
&\leq C\int_{\Omega_m} e^{-2k\mu|x|}(|u| + |\nabla_x u|)\left|\frac{\partial u}{\partial t}\right|^{2k-1} dx \\
&\quad + C\int_{\Omega_m} e^{-2k\mu|x|}|\nabla_x u|^2\left|\frac{\partial u}{\partial t}\right|^{2k-2} dx \\
&\quad + C\left\{\int_{\Omega_m} e^{-2k\mu|x|}\left|\frac{\partial u}{\partial t}\right|^{2k} dx\right\}^{(2k-1)/2k} e^{\delta t}.
\end{aligned}
$$

Using Hölder's inequality, we get

$$
\begin{aligned}
-\frac{d}{dt}\int_{\Omega_m} e^{-2k\mu|x|}\left|\frac{\partial u}{\partial t}\right|^{2k} dx &+ 2\gamma'\int_{\Omega_m} e^{-2k\mu|x|}\left|\frac{\partial u}{\partial t}\right|^{2k} dx \\
&\leq C\left\{\int_{\Omega_m} e^{-2k\mu|x|}(|u|^{2k} + |\nabla_x u|^{2k})\,dx\right\}^{1/2k} \\
&\quad \cdot \left\{\int_{\Omega_m} e^{-2k\mu|x|}\left|\frac{\partial u}{\partial t}\right|^{2k} dx\right\}^{(2k-1)/2k} \\
&\quad + C\left\{\int_{\Omega_m} e^{-2k\mu|x|}|\nabla_x u|^{2k}\,dx\right\}^{1/k} \cdot \left\{\int_{\Omega_m} e^{-2k\mu|x|}\left|\frac{\partial u}{\partial t}\right|^{2k} dx\right\}^{(2k-2)/2} \\
&\quad + C\left\{\int_{\Omega_m} e^{-2k\mu|x|}\left|\frac{\partial u}{\partial t}\right|^{2k} dx\right\}^{(2k-1)/2k} e^{\delta t}. \qquad (10.41)
\end{aligned}
$$

This inequality was proved for k a positive integer. However, if k is any positive number > 1, and if we differentiate (10.8) with respect to t and then multiply by

$$
e^{-2k\mu|x|}\left|\frac{\partial u}{\partial t}\right|^{2k-2}\left(\frac{\partial u}{\partial t}\right),
$$

then we again obtain (10.41).

Recalling (10.29), we deduce from Sobolev's inequality that

$$
|u|^{\Omega_m}_{1,2q,\mu} \leq C
$$

where $2q > 2$, $(2q)^{-1} \geq \frac{1}{2} - n^{-1}$. Hence, using (10.41) with $2k = 2q$, we get

$$
e^{-2q\delta t}\int_{\Omega_m} e^{-2q\mu|x|}\left|\frac{\partial u}{\partial t}\right|^{2q} dx \leq C.
$$

From the proof of Theorem 6.1 we deduce that

$$e^{-\delta t}|u(t)|^{\Omega_m}_{2,2q,\mu} \leq C.$$

By Sobolev's inequality

$$e^{-\delta t}|u(t)|^{\Omega_m}_{1,2q_1,\mu} \leq C$$

where $2q_1 > 2q$, $(2q_1)^{-1} \geq (2q)^{-1} - n^{-1}$. Now we can apply (10.41) with $2k = 2q_1$ and proceed as before. It is clear that after a finite number of steps we arrive at the inequality

$$e^{-\delta t}|u(t)|^{\Omega_m}_{2,p,\mu} \leq C.$$

Since also $e^{-\delta t}|u'(t)|_{0,p,\mu} \leq C$, the assertion (10.5) readily follows.

Remark. In Theorem 10.1 Ω is an unbounded domain in $\mathcal{C}^{2+\rho}$. The same theorem is clearly valid also if Ω is a bounded domain with $C^{2+\rho}$ boundary; in this case we take $\mu = 0$.

11. Parabolic variational inequalities (continued)

In this section we consider the case where ϕ_1, ϕ_2 are functions of t. Set

$$K_\mu(t) = \{ g \in L^{2,\mu}(\Omega);\ \phi_2(x, t) \leq g(x) \leq \phi_1(x, t) \text{ a.e. in } \Omega \}.$$

For simplicity we consider the parabolic variational inequality in a finite t-interval:

$$u \in L^\infty(0, T;\ W^{1,2,\mu}(\Omega)) \cap L^2(0, T;\ W^{2,2,\mu}(\Omega)),$$

$$\frac{\partial u}{\partial t} \in L^2(0, T;\ W^{0,2,\mu}(\Omega)); \tag{11.1}$$

for a.a. $t \in (0, T)$,

$$-\int_\Omega e^{-2\mu|x|}\frac{\partial u}{\partial t}(v - u)\,dx + \int_\Omega e^{-2\mu|x|}Au \cdot (v - u)\,dx$$

$$\geq \int_\Omega e^{-2\mu|x|}f(v - u)\,dx \qquad \text{for every} \quad v \in K_\mu(t); \tag{11.2}$$

$$\phi_2(x, t) \leq u(x, t) \leq \phi_1(x, t) \quad \text{a.e.} \quad \text{in} \quad (x, t) \in \Omega \times (0, T), \tag{11.3}$$

$$u(x, T) = 0 \quad \text{a.e.} \quad \text{in } \Omega. \tag{11.4}$$

The inequalities in (10.5) will be replaced by

$$u \in L^p(0, T;\ W^{2,p,\mu}(\Omega)), \qquad \frac{\partial u}{\partial t} \in L^p(0, T;\ W^{0,p,\mu}(\Omega)), \tag{11.5}$$

where $2 \leq p < \infty$.

The conditions (B*), (C*) will be replaced by:

(B**)

(i) ϕ_i, $D_x\phi_i$, $D_x^2\phi_i$, $D_t^i\phi$ belong to $L^p(0, T; W^{0,p,\mu}(\Omega)) \cap L^2(0, T; W^{0,2,\mu}(\Omega))$, for $i = 1, 2$;

(ii) $\partial^2\phi_i/\partial t^2$ belong to $L^2(0, T; W^{0,2,\mu}(\Omega))$, for $i = 1, 2$;

(iii) $\phi_2(x, t) \leq 0 \leq \phi_1(x, t)$ a.e., and, for a.a. $t \in (0, T)$, $\phi_1(x, t)$ and $\phi_2(x, t)$ belong to $W_0^{1,2,\mu}(\Omega)$.

(C**)

$$f \in L^p(0, T; W^{0,p,\mu}(\Omega)) \cap L^2(0, T; W^{0,2,\mu}(\Omega)),$$

$$\frac{\partial f}{\partial t} \in L^2(0, T; W^{0,2,\mu}(\Omega)).$$

Notice that (B**) and (C**) imply that

$$\phi_i \in C([0, T]; W^{0,p,\mu}(\Omega) \cap W^{0,2,\mu}(\Omega)) \qquad (i = 1, 2)$$

$$f \in C([0, T]; W^{0,2,\mu}).$$

Theorem 11.1. *Let $\Omega \in \mathcal{C}^{2+\rho}$ for some $0 < \rho \leq 1$. If (A*), (B**), (C**) hold, then there exists a unique solution of the parabolic variational inequality* (11.1)–(11.4), *and the solution satisfies* (11.5)

Proof. For simplicity we give the proof only in the special case where $A = -\Delta = -\sum \partial^2/\partial x_i^2$. Introduce functions $\phi_{im} = \zeta_m\phi_i$ as in (10.8). Consider the problem:

$$u(\cdot, t) \in L^p(0, T; W^{2,p}(\Omega_m)) \cap L^\infty(0, T; W_0^{1,2}(\Omega_m)), \tag{11.6}$$

$$\frac{\partial u(\cdot, t)}{\partial t} \in L^p(0, T; L^p(\Omega_m)) \qquad \text{for a.a.} \quad t \in (0, T), \tag{11.7}$$

$$-\frac{\partial u}{\partial t} - \Delta u + \frac{1}{\epsilon}(u - \phi_{1m})^+ - \frac{1}{\epsilon}(u - \phi_{2m})^- = f \quad \text{a.e.} \quad \text{in } \Omega_m, \tag{11.8}$$

$$u(x, T) = 0. \tag{11.9}$$

Set $\rho = \exp(-\mu|x|)$. Multiplying (11.8) by $-\rho^p|u|^{p-2}$ and integrating over Ω_m, we get

$$-\frac{1}{p}\frac{d}{dt}\int_{\Omega_m} \rho^p|u|^p\,dx - \int_{\Omega_m} \rho^p|u|^{p-2}u\,\Delta u\,dx$$

$$\leq \int_{\Omega_m} \rho^p\left[|u|^{p-1}(|f| + C) + C|u|^p + C|u|^{p-1}\left|\frac{\partial u}{\partial x}\right|\right]dx.$$

Since

$$-\int_{\Omega_m} \rho^p |u|^{p-2} u \,\Delta u \, dx$$

$$= \int_{\Omega_m} \left[(p-1)\rho^p |u|^{p-2} \left| \frac{\partial u}{\partial x} \right|^2 - p\mu\rho^p |u|^{p-2} u \sum_i \frac{x_i}{|x|} \frac{\partial u}{\partial x_i} \right] dx,$$

we get, for any $\gamma > 0$,

$$-\frac{1}{p}\frac{d}{dt}\int_{\Omega_m} \rho^p |u|^p \, dx + \int_{\Omega_m} \rho^p |u|^{p-2} \left| \frac{\partial u}{\partial x} \right|^2 \left((p-1) - \frac{p\mu\gamma}{2} \frac{C\gamma}{2} \right) dx$$

$$\leqslant \int_{\Omega_m} \rho^p (|f| + C)|u|^{p-1} \, dx + \int_{\Omega_m} |\rho|^p |u|^p \left(C + \frac{C}{2\gamma} + \frac{p\mu}{2\gamma} \right) dx$$

$$\leqslant \frac{1}{p}\int_{\Omega_m} \rho^p (|f| + C)^p \, dx + \int_{\Omega_m} \rho^p |u|^p \left(\frac{p-1}{p} + C + \frac{C}{2\gamma} + \frac{p\mu}{2\gamma} \right) dx. \tag{11.10}$$

Choosing γ so that

$$(p-1) - \frac{p\mu\gamma}{2} - \frac{C\gamma}{2} > 0$$

and setting

$$\Phi(t) = \int_{\Omega_m} \rho^p |u(x, t)|^p \, dx,$$

we get from (11.10) that

$$-\Phi'(t) \leqslant \delta\Phi(t) + C_1$$

where δ, C_1 are positive constants. It follows that $\phi(t) \leqslant$ const, i.e.,

$$\int_{\Omega_m} \rho^p |u(x, t)|^p \, dx \leqslant C. \tag{11.11}$$

Since the above analysis applies also for $p = 2$, we get

$$\int_{\Omega_m} \rho^2 |u(x, t)|^2 \, dx \leqslant C. \tag{11.12}$$

Taking $p = 2$ in (11.10) and integrating with respect to t, we get, after using (11.12),

$$\int_0^T \int_{\Omega_m} \rho^2 \left| \frac{\partial u}{\partial x} \right|^2 dx \, dt \leqslant C$$

for some positive constant C. Together with (11.12) this yields

$$\int_0^T \left(|u(t)|_{1,2,\mu}^{\Omega_m} \right)^2 dt \leqslant C. \tag{11.13}$$

Next we multiply (11.8) by $-\rho^p[(u - \phi_{1m})^+]^{p-1}$ and integrate with respect to x, t. Using the relations (with $\phi = \phi_{1m}$)

$$\int_0^T \int_{\Omega_m} \rho^p \frac{\partial u}{\partial t} [(u-\phi)^+]^{p-1}\, dx\, dt$$

$$= \int_0^T \int_{\Omega_m} \rho^p \frac{\partial (u-\phi)^+}{\partial t} [(u-\phi)^+]^{p-1}\, dx\, dt$$

$$+ \int_0^T \int_{\Omega_m} \rho^p \frac{\partial \phi}{\partial t} [(u-\phi)^+]^{p-1}\, dx\, dt$$

$$= -\frac{1}{p} \int_{\Omega_m} \rho^p [(u(x,t) - \phi(x,t))^+]^p dx$$

$$+ \int_0^T \int_{\Omega_m} \rho^p \frac{\partial \phi}{\partial t} [(u-\phi)^+]^{p-1}\, dx\, dt,$$

$$\int_0^T \int_{\Omega_m} \rho^p \Delta u \cdot [(u-\phi)^+]^{p-1}\, dx\, dt$$

$$= \int_0^T \int_{\Omega_m} \rho^p \Delta (u-\phi) \cdot [(u-\phi)^+]^{p-1}\, dx\, dt$$

$$+ \int_0^T \int_{\Omega_m} \rho^p \Delta\phi \cdot [(u-\phi)^+]^{p-1}\, dx\, dt,$$

$$= -\int_0^T \int_{\Omega_m} (p-1)\rho^p [(u-\phi)^+]^{p-2} \left| \frac{\partial}{\partial x} (u-\phi)^+ \right|^2 dx\, dt$$

$$+ p\mu \int_0^T \int_{\Omega_m} \rho^p [(u-\phi)^+]^{p-1} \sum \frac{x_i}{|x|} \frac{\partial}{\partial x_i} (u-\phi)^+\, dx\, dt$$

$$+ \int_0^T \int_{\Omega_m} \rho^p \Delta\phi \cdot [(u-\phi)^+]^{p-1}\, dx\, dt$$

$$\int_0^T \int_{\Omega_R} \rho^p k [(u-\phi)^+]^{p-1}\, dx\, dt$$

$$\leq \left[\int_0^T \int_{\Omega_m} \rho^p |k|^p\, dx\, dt \right]^{1/p} \left[\int_0^T \int_{\Omega_m} \rho^p [(u-\phi)^+]^p\, dx\, dt \right]^{(p-1)/p}$$

for $k = f$ and $k = \Delta\phi + \partial\phi/\partial t$, and noting that

$$\int_0^T \int_{\Omega_m} \rho^p [(u-\phi)^+]^{p-1} \frac{x_i}{|x|} \frac{\partial}{\partial x_i} (u-\phi)^+\, dx\, dt$$

$$\leq \left[\int_0^T \int_{\Omega_m} \rho^p [(u-\phi)^+]^p\, dx\, dt \right]^{(p-1)/p}$$

$$\cdot \left[\int_0^T \int_{\Omega_m} \rho^p \left| \frac{\partial}{\partial x_i} (u-\phi)^+ \right|^p dx\, dt \right]^{1/p},$$

we arrive at the inequality

$$\frac{1}{\epsilon}\int_0^T\int_{\Omega_m}\rho^p[(u-\phi)^+]^p dx\,dt$$
$$\leq C\left(1+\int_0^T\int_{\Omega_m}\rho^p\left|\frac{\partial u}{\partial x}\right|^p dx\,dt\right)^{1/p}\left[\int_0^T\int_{\Omega_m}\rho^p[(u-\phi)^+]^p dx\,dt\right]^{(p-1)/p};$$

here we have used (B**), (C**). It follows that

$$\int_0^T\left(\left|\frac{1}{\epsilon}(u-\phi)^+\right|_{0,p,\mu}^{\Omega_m}\right)^p dt \leq C + C\int_0^T\int_{\Omega_m}\rho^p\left|\frac{\partial u}{\partial x}\right|^p dx\,dt. \tag{11.14}$$

Similarly, if we multiply (11.8) by $-\rho^p((u-\phi_{2m})^-)^{p-1}$ and integrate, we get

$$\int_0^T\left(\left|\frac{1}{\epsilon}(u-\phi_{2m})^-\right|_{0,p,\mu}^{\Omega_m}\right)^p dt \leq C + C\int_0^T\int_{\Omega_m}\rho^p\left|\frac{\partial u}{\partial x}\right|^p dx\,dt. \tag{11.15}$$

From (11.8) and (11.14), (11.15) it follows that

$$\int_0^T\int_{\Omega_m}\rho^p\left|\frac{\partial u}{\partial t}+\Delta u\right|^p dx\,dt \leq C + C\int_0^T\int_{\Omega_m}\rho^p\left|\frac{\partial u}{\partial x}\right|^p dx\,dt. \tag{11.16}$$

By Theorem 10.4.3,

$$\int_0^T\int_{\Omega_m}\rho^p[|u|^p+|u_t|^p+|u_x|^p+|u_{xx}|^p]\,dx\,dt$$
$$\leq C\int_0^T\int_{\Omega_m}\rho^p|u_t+\Delta u|^p\,dx\,dt + C\int_0^T\int_{\Omega_m}\rho^p(|u|^p+|u_x|^p)\,dx\,dt \tag{11.17}$$

provided $\rho = 1$. The constant C may depend on m. We can show however that C can be chosen to be independent of m. Indeed, we take a partition of unity of $\overline{\Omega}_m$, say $\{\phi_i\}$, as in Section 5, and apply (11.17) (with $\rho = 1$) to $u\phi_i\exp(-\mu|x_i|)$, where x_i is a point in the support of ϕ_i. The constant C can be taken here to be independent of i, m. Summing the resulting inequalities over i, we end up with the inequality (11.17) (for $\rho = \exp(-\mu|x|)$) with a constant C that is independent of m.

We shall also need the inequality

$$\int_{\Omega_m}\rho^p|u_x|^p\,dx \leq \gamma\int_{\Omega_m}\rho^p|u_{xx}|^p\,dx + C(\gamma)\int_{\Omega_m}\rho^p|u|^p\,dx \tag{11.18}$$

for any $\gamma > 0$, where $C(\gamma)$ is a constant depending on γ but not on m. This is obtained by using the same partition of unity of $\overline{\Omega}_m$ as before, and Theorem 10.2.1 (which we apply to $u\phi_i\exp(-\mu|x_i|)$).

Estimating the right-hand side of (11.17) by using (1.16) and then (11.18), we obtain, after recalling (11.11),

$$\int_0^T \int_{\Omega_m} \rho^p(|u|^p + |u_t|^p + |u_x|^p + |u_{xx}|^p)\, dx\, dt \leq C \tag{11.19}$$

where C is a constant independent of ϵ, m.

If we use (11.19) in (11.14), (11.15), we get

$$\int_0^T \left(\left| \frac{1}{\epsilon}(u - \phi_{1m})^+ \right|_{0,p,\mu}^{\Omega_m} \right)^p dt \leq C, \tag{11.20}$$

$$\int_0^T \left(\left| \frac{1}{\epsilon}(u - \phi_{im})^- \right|_{0,p,\mu}^{\Omega_m} \right)^p dt \leq C. \tag{11.21}$$

We need one more estimate. Differentiate (11.8) with respect to t, multiply by $\rho^2(\partial u/\partial t)$ and integrate with respect to (x, t). Then,

$$\begin{aligned}\int_{\Omega_m} \rho^2|u_t(x,t)|^2\, dx - \int_{\Omega_m} \rho^2 |u_t(x,T)|^2\, dx + \int_t^T \int_{\Omega_m} \rho^2 |u_{xt}|^2\, dx\, dt \\ + \int_t^T \int_{\Omega_m} \frac{\rho^2}{\epsilon} \frac{\partial(u-\phi_{1m})^+}{\partial t} \frac{\partial u}{\partial t}\, dx\, dt \\ - \int_t^T \int_{\Omega_m} \frac{\rho^2}{\epsilon} \frac{\partial(u-\phi_{2m})^-}{\partial t} \frac{\partial u}{\partial t}\, dx\, dt = G\end{aligned} \tag{11.22}$$

where

$$|G| \leq C \int_t^T \int_{\Omega_m} \rho^2(|u_t|^2 + |f_t|^2)\, dx\, dt \leq C.$$

Now,

$$\begin{aligned}&\int_t^T \int_{\Omega_m} \frac{\rho^2}{\epsilon} \frac{\partial(u-\phi_{1m})^+}{\partial t} \frac{\partial u}{\partial t}\, dx\, dt \\ &= \int_t^T \int_{\Omega_m} \frac{\rho^2}{\epsilon} \frac{\partial(u-\phi_{1m})^+}{\partial t} \frac{\partial(u-\phi_{1m})}{\partial t}\, dx\, dt \\ &\quad + \int_t^T \int_{\Omega_m} \frac{\rho^2}{\epsilon} \frac{\partial(u-\phi_{1m})^+}{\partial t} \frac{\partial \phi_{1m}}{\partial t}\, dx\, dt \\ &\geq \int_t^T \int_{\Omega_m} \frac{\rho^2}{\epsilon} \frac{\partial(u-\phi_{1m})^+}{\partial t} \frac{\partial \phi_{1m}}{\partial t}\, dx\, dt \\ &= -\int_{\Omega_m} \frac{\rho^2}{\epsilon}(u-\phi_{1m})^+ \frac{\partial \phi_{1m}}{\partial t}\, dx \\ &\quad - \int_t^T \int_{\Omega_m} \frac{\rho^2}{\epsilon}(u-\phi_{1m})^+ \frac{\partial^2 \phi_{1m}}{\partial t^2}\, dx\, dt.\end{aligned}$$

Similarly

$$-\int_t^T \int_{\Omega_m} \frac{\rho^2}{\epsilon} \frac{\partial (u-\phi_{2m})^-}{\partial t} \frac{\partial u}{\partial t}\, dx\, dt$$
$$\geqslant \int_{\Omega_m} \frac{\rho^2}{\epsilon} (u-\phi_{2m})^- \frac{\partial \phi_{2m}}{\partial t}\, dx$$
$$+\int_t^T \int_{\Omega_m} \frac{\rho^2}{\epsilon} (u-\phi_{2m})^- \frac{\partial^2 \phi_{2m}}{\partial t^2}\, dx\, dt.$$

Since $u_t(x, T) = f(x, T)$, we also have

$$\int_{\Omega_m} \rho^2 |u_t(x, T)|^2\, dx \leqslant C.$$

Using these relations in (11.22) and using (11.20), (11.21), we get

$$\int_{\Omega_m} [\rho^2|u_t|^2]_{t=s}\, dx + \int_s^T \int_{\Omega_m} \rho^2 |u_{xt}|^2\, dx\, dt$$
$$-\int_{\Omega_m} \left[\frac{\rho^2}{\epsilon} (u-\phi_{1m})^+ \frac{\partial \phi_{1m}}{\partial t} \right]_{t=s} dx$$
$$+\int_{\Omega_m} \left[\frac{\rho^2}{\epsilon} (u-\phi_{2m})^- \frac{\partial \phi_{2m}}{\partial t} \right]_{t=s} dx \leqslant C.$$

Integrating with respect to s, we find that

$$\int_0^T \int_{\Omega_m} \rho^2 |u_t|^2\, dx\, dt + \int_0^T \left[\int_s^T \int_{\Omega_m} \rho^2 |u_{xt}|^2\, dx\, dt \right] ds \leqslant C. \quad (11.23)$$

Now, for any $\eta > 0$, all the functions given in Theorem 11.1 can be extended to $-\eta \leqslant t < 0$ in such a way that the conditions (A*), (B**), (C**) remain valid in the interval $[-\eta, T]$ instead of $[0, T]$. We can therefore carry out all the previous analysis in the interval $[-\eta, T]$. In particular, (11.23) yields

$$\int_{-\eta}^T \left[\int_s^T \int_{\Omega_m} \rho^2 |u_{xt}|^2\, dx\, dt \right] ds \leqslant C.$$

Hence

$$\int_{-\eta}^0 \left[\int_0^T \int_{\Omega_m} \rho^2 |u_{xt}|^2\, dx\, dt \right] ds \leqslant C,$$

i.e.,

$$\int_0^T \int_{\Omega_m} \rho^2 |u_{xt}|^2\, dx\, dt \leqslant C \quad (11.24)$$

with a different constant C.

Denote the solution of (11.6)–(11.9) by u_m. Extend u_m into a function $\tilde{u}_m$ defined for all $(x, t) \in R^n \times [0, T]$ such that $\tilde{u}_m$ has compact support and

$$\int_0^T \int_{R^n} \rho^2 \left[|\tilde{u}_m|^q + \left| \frac{\partial}{\partial t} \tilde{u}_m \right|^q + \left| \frac{\partial}{\partial x} \tilde{u}_m \right|^q + \left| \frac{\partial^2}{\partial x^2} \tilde{u}_m \right|^q \right] dx\, dt \leq C$$

$$(q = p, 2),$$

$$\int_0^T \int_{R^n} \rho^2 \left| \frac{\partial^2}{\partial t\, \partial x} \tilde{u}_m \right|^2 dx\, dt \leq C, \qquad |\tilde{u}_m(t)|_{1,2,\mu}^{R^n} \leq C.$$

Using the compact imbedding theorem for Sobolev spaces we obtain a subsequence of $\tilde{u}_m$, which we again denote by $\tilde{u}_m$, that is convergent in $L^2(\Omega^* \times (0, T))$, for any bounded set Ω^*, to a function u, together with its first x-derivative. We may further assume that

$$\tilde{u}_m \to u \quad \text{in } L^p(0, T; W^{2,p,\mu}(R^n)) \cap L^2(0, T; W^{2,2,\mu}(R^n)) \quad \text{weakly,}$$

$$\tilde{u}_m \to u \quad \text{in } L^\infty(0, T; W^{1,2,\mu}(R^n)) \quad \text{in the weak star topology},$$

$$\frac{\partial \tilde{u}_m}{\partial t} \to \frac{\partial u}{\partial t} \quad \text{in } L^p(0, T; W^{0,p,\mu}(R^n)) \cap L^2(0, T; W^{0,2,\mu}(R^n)) \quad \text{weakly.}$$

We can now complete the existence proof of Theorem 11.1 by the same arguments used in the proof of Theorem 10.1, following (10.32). The proof of uniqueness is similar to the corresponding proof for Theorem 10.1.

Remark 1. The condition (iv) in (A*) is not needed in Theorem 11.1. Thus, α can be any nonnegative number.

Remark 2. Theorem 10.1 is concerned with a parabolic variational inequality for $0 < t < \infty$. One can similarly formulate a parabolic variational inequality in a bounded interval $0 < t < T$, by adding a terminal condition

$$u(x, T) = 0 \quad \text{a.e.} \quad \text{in } \Omega. \tag{11.25}$$

The existence of a unique solution with

$$u \in L^\infty(0, T; W^{2,2,\mu}(\Omega) \cap W^{2,p,\mu}(\Omega)) \cap L^\infty(0, T; W_0^{1,2,\mu}(\Omega))$$

$$\frac{\partial u}{\partial t} \in L^\infty(0, T; W^{0,2,\mu}(\Omega) \cap W^{0,p,\mu}(\Omega))$$

follows by specializing the proof of Theorem 10.1. Note however that in this case the condition (iv) of (A*) is not needed. Thus, α can be any nonnegative number. The homogeneous condition (11.25) can also be replaced by a nonhomogeneous condition $u(x, T) = h(x)$, with $h \in W^{2,p,\mu}(\Omega) \cap W^{2,2,\mu}(\Omega) \cap W_0^{1,2,\mu}(\Omega)$.

Remark 3. In Theorem 11.1 and Remarks 1, 2, Ω is an unbounded domain in $\mathcal{C}^{2+\rho}$. The same results are clearly valid if Ω is a bounded domain with boundary in $C^{2+\rho}$.

12. Existence of a saddle point

Consider the stochastic game associated with (9.1)–(9.3). We shall need the following conditions:

$$\psi_1(x, t), \psi_2(x, t) \quad \text{are bounded functions in} \quad \Omega \times [0, \infty) \tag{12.1}$$

$$\psi_1(x, t) - \psi_2(x, t) = \phi(x), \tag{12.2}$$

$$\phi(x) \geq 0 \qquad \text{in } \Omega, \tag{12.3}$$

$$\phi \in W^{2, p, \mu}(\Omega) \cap W^{2, 2, \mu}(\Omega) \cap W_0^{1, 2, \mu}(\Omega), \tag{12.4}$$

$$e^{-\delta t}\psi_2 \in L^\infty(0, \infty; W^{2, p, \mu}(\Omega) \cap W^{2, 2, \mu}(\Omega)), \tag{12.5}$$

$$e^{-\delta t}\frac{\partial \psi_2}{\partial t} \in L^\infty(0, \infty; L^{p, \mu}(\Omega) \cap L^{2, \mu}(\Omega)), \tag{12.6}$$

$$\text{the condition (C*) holds for} \quad \tilde{f} = f - A\psi_2 + \frac{\partial \psi_2}{\partial t}. \tag{12.7}$$

If $\Omega \in \mathcal{C}^{2+\rho}$ and (A*), (12.1)–(12.7) hold, then, by Theorem 10.1 (with $\phi_1 = \phi$, $\phi_2 = 0$, and f replaced by $\tilde{f}$), there exists a function solution u satisfying (10.2) and, for a.a. $t > 0$,

$$-\int_\Omega e^{-2\mu|x|}\frac{\partial u}{\partial t}(v - u)\,dx + \int_\Omega e^{-2\mu|x|}Au \cdot (v - u)\,dx$$

$$\geq \int_\Omega e^{-2\mu|x|}\left(f - A\psi_2 + \frac{\partial \psi_2}{\partial t}\right)dx \qquad \text{for every} \quad v \in K_\mu, \tag{12.8}$$

$$u(\cdot, t) \in K_\mu, \tag{12.9}$$

where

$$K_\mu = \{g \in L^{2, \mu}(\Omega);\ 0 \leq g(x) \leq \phi(x) \quad \text{a.e.}\}.$$

If $p > n$, then $u(x, t)$ is continuous in $(x, t) \in \overline{Q}$ and $u_x(x, t)$ is continuous in $x \in \overline{\Omega}$, for any $t \geq 0$. Set

$$V(x, t) = u(x, t) + \psi_2(x, t), \tag{12.10}$$

$$\hat{S} = \{(x, t) \in \overline{Q};\ V = \psi_1\}, \tag{12.11}$$

$$\hat{T} = \{(x, t) \in \overline{Q};\ V = \psi_2\}. \tag{12.12}$$

We shall need the condition

$$P_x(t_{\hat{S}}^x < \infty) = 1, \qquad P_x(t_{\hat{T}}^x < \infty) = 1. \tag{12.13}$$

This condition is satisfied if

$$P_x(t^s_{Q^c} < \infty) = 1,$$

which is the case, for instance, if Ω is a bounded domain, or if Ω is a domain contained in a strip $-\infty < \alpha_1 \leq x_1 \leq \alpha_2 < \infty$.

Theorem 12.1. *Let $\Omega \in \mathcal{C}^{2+\rho}$ for some $0 < \rho \leq 1$, and assume that* (A*) (*with $c \equiv 0$*) *and* (12.1)–(12.7) *hold with $p > n$. Assume also that* (C_1), (9.6), *and* (12.13) *hold. Then the stochastic game associated with* (9.1)–(9.3) *has the value $V(x, t)$, and $(\hat{S}, \hat{T})$ form a saddle point of sets. Further, $V(x, t)$ is continuous in $\overline{Q}$, $V_x(x, t)$ is continuous in $x \in \overline{\Omega}$ for any $t \geq 0$, and*

$$\frac{\partial V}{\partial x_i} = \frac{\partial \psi_1}{\partial x_i} \quad \text{on} \quad \hat{S} \cap Q, \qquad \frac{\partial V}{\partial x_i} = \frac{\partial \psi_2}{\partial x_i} \quad \text{on} \quad \hat{T} \cap Q \qquad (1 \leq i \leq n). \tag{12.14}$$

The proof follows from Theorems 10.1, 9.2 by extending the argument used in the proof of Theorem 4.1. The details are left to the reader.

The variational inequality (12.8) leads to the following inequalities for V (cf. (4.17)–(4.20))

$$\left(\frac{\partial V}{\partial t} + LV - \alpha V + f\right)(v - V) \leq 0 \quad \text{a.e.} \qquad \text{for any } v, \quad \psi_1 \leq v \leq \psi_2, \tag{12.15}$$

or

$$\frac{\partial V}{\partial t} + LV - \alpha V + f \geq 0 \quad \text{a.e.} \qquad \text{on} \quad Q \setminus \hat{T}, \tag{12.16}$$

$$\frac{\partial V}{\partial t} + LV - \alpha V + f \leq 0 \quad \text{a.e.} \qquad \text{on} \quad Q \setminus \hat{S}, \tag{12.17}$$

$$\frac{\partial V}{\partial t} + LV - \alpha V + f = 0 \quad \text{a.e.} \qquad \text{on} \quad Q \setminus (\hat{S} \cup \hat{T}). \tag{12.18}$$

We consider now a stochastic game with *finite horizon T*. By this we mean that the stopping times σ, τ are restricted to vary in $\mathfrak{D}^T_{x,s}$: $\mathfrak{D}^T_{x,s}$ is the subset of $\mathfrak{D}_{x,s}$ consisting of all stopping times τ with $\tau \leq T$. The concepts of value and saddle point are defined in the obvious manner. Notice that

$$V(T, x) = \psi_2(x) \qquad \text{for all} \quad x \in \overline{\Omega}. \tag{12.19}$$

In the parabolic variational inequality for u we now take t only in $(0, T)$, and impose the terminal condition

$$u(x, T) = 0 \quad \text{a.e.} \qquad \text{on } \Omega. \tag{12.20}$$

The proof of Theorem 10.1 when specialized to this case again yields a solution u satisfying properties analogous to (10.2); see Remark 2 at the end of Section 11. The proofs of Theorems 9.1, 9.2 also extend, with trivial changes, to games with finite horizon. Here we define

$$\hat{S} = \{(x, t) \in \bar{\Omega} \times [0, T];\ V = \psi_1\} \cup \{(x, T);\ x \in \bar{\Omega}\}, \tag{12.21}$$

$$\hat{T} = \{(x, t) \in \bar{\Omega} \times [0, T];\ V = \psi_2\}. \tag{12.22}$$

By Remark 2 at the end of Section 11, the condition (iv) of (A*) is not needed in the finite horizon case. Thus *the discount coefficient can be any nonnegative number.*

Consider now the case where the restriction (12.2) is removed. Set $\phi_1(x, t) = \psi_1(x, t) - \psi_2(x, t)$ and assume

$$\text{the function } \phi_1 \text{ satisfies the conditions in (B**)}, \tag{12.23}$$

$$\psi_2 \in L^p(0, T;\ W^{2,p,\mu}(\Omega)), \quad \frac{\partial \psi_2}{\partial t} \in L^p(0, T;\ W^{0,p,\mu}(\Omega)), \tag{12.24}$$

$$\text{the condition (C**) holds for } \tilde{f} = f - A\psi_2 + \frac{\partial \psi_2}{\partial t}, \tag{12.25}$$

$$\psi_1, \psi_2, f \text{ are bounded functions.} \tag{12.26}$$

By Theorem 11.1 there exists a unique solution u of the parabolic variational inequality (11.1)–(11.4) with $\phi_2 \equiv 0$ and f replaced by $\tilde{f}$. Define V by (12.10). If $p > n$, then u and V are continuous in $(x, t) \in \bar{Q}$. Define $\hat{S}$, $\hat{T}$ by (12.21), (1.22). We then have:

Theorem 12.2. *Let* $\Omega \in \mathcal{C}^{2+\rho}$ *for some* $0 < \rho \leq 1$, *and assume that* (A*) (*with* $c \equiv 0$ *and without the condition* (iv)), (C_1) *and* (12.23)–(12.26) *with* $p > n$ *hold. Then the stochastic game with finite horizon T associated with* (9.1)–(9.3) *has the value* $V(x, t)$ *and a saddle point of sets* $(\hat{S}, \hat{T})$ *given by* (12.10) *and* (12.21), (1.22).

The proof follows by combining Theorem 11.1 (and Remark 1 following it) and an extension of Theorem 9.1 to the finite horizon case, using the argument appearing in the proof of Theorem 4.1. The details are left to the reader.

Remark. In Theorems 12.1, 12.2 the domain Ω is unbounded and in $\mathcal{C}^{2+\rho}$. The same theorems are clearly valid also if Ω is a bounded domain with boundary in $C^{2+\rho}$.

13. The stopping time problem

Consider the stopping time problem associated with (9.1), (9.2), and the cost

$$J_{x,s}(\sigma) = E_{x,s}\left\{ \int_s^\sigma e^{-\alpha(t-s)} f(x(t), t)\, dt + e^{-\alpha(\sigma-s)} \psi_1(x(\sigma), \sigma) \right\}. \tag{13.1}$$

This problem can be handled by specializing the proofs of the results for stochastic games. Thus, we specialize Theorem 10.1 to the case where

$$K_\mu = \{ g \in L^{2,\mu}(\Omega),\ g \leq 0 \quad \text{a.e.} \}, \tag{13.2}$$

and we specialize Theorem 9.2 to the case where (9.3) is replaced by (13.1). We shall just state the final results, leaving the details for the reader.

We shall need the following conditions:

$$\psi_1(x, t) \geq -C, \tag{13.3}$$

$$E\int_0^\sigma e^{-\alpha t}|f(x(t), t)|\, dt \geq -C \qquad \text{for any} \quad \sigma \in \mathfrak{D}_{x,s}, \tag{13.4}$$

where C is a constant;

$$e^{-\delta t}\psi_1 \in L^\infty(0, \infty;\ W^{2,p,\mu}(\Omega) \cap W^{2,2,\mu}(\Omega)), \tag{13.5}$$

$$e^{-\delta t}\frac{\partial \psi_1}{\partial t} \in L^\infty(0, \infty;\ W^{0,p,\mu}(\Omega) \cap W^{0,2,\mu}(\Omega)), \tag{13.6}$$

$$\text{the condition } (C^*) \text{ holds for } \quad \tilde{f} = f - A\psi_1 + \partial\psi_1/\partial t. \tag{13.7}$$

If (A^*) also holds, then there exists a unique solution of the parabolic variational inequality (10.2), (10.3) with K_μ defined by (13.2) and with $u(t) \in K_\mu$ for a.a. $t \geq 0$. Set

$$V(x, t) = u(x, t) + \psi_1(x, t). \tag{13.8}$$

If $p > n$, then u and V are continuous in $\overline{Q}$ and u_x, V_x are continuous in $x \in \overline{\Omega}$ for any $t \geq 0$. Set

$$\hat{S} = \{(x, t) \in \overline{Q};\ V = \psi_1\}. \tag{13.9}$$

We shall assume

$$P_{x,s}(t_{\hat{S}} < \infty) = 1 \qquad \text{for all} \quad (x, s) \in Q. \tag{13.10}$$

Then we have:

Theorem 13.1. *Let* $\Omega \in \mathcal{C}^{2+\rho}$ *for some* $0 < \rho \leq 1$, *and assume that* (A^*) (*with* $c \equiv 0$), (C_1), *and* (13.3)–(13.7), (13.10) *hold with* $p > n$. *Then* $V(x, t)$ *is the optimal cost and* $\hat{S}$ *is an optimal stopping set for the stopping time problem associated with* (9.1), (9.2), (13.1). *Further,* $V(x, t)$ *is continuous in* $(x, t) \in \overline{Q}$, $V_x(x, t)$ *is continuous in* $x \in \overline{\Omega}$ *for any* $t \geq 0$, *and*

$$\frac{\partial V}{\partial x_i} = \frac{\partial \psi_1}{\partial x_i} \qquad \text{on} \quad \hat{S} \cap Q \qquad (1 \leq i \leq n). \tag{13.11}$$

Theorem 13.1 extends to the case of finite horizon. In this case the condition (iv) in (A^*) is not needed, i.e., the discount coefficient α can be any nonnegative number. Further, the condition (13.10) becomes superfluous.

Notice finally that, under the conditions of Theorem 13.1,

$$\frac{\partial V}{\partial t} + LV - \alpha V + f \geq 0 \quad \text{a.e.} \qquad \text{in } Q, \tag{13.12}$$

$$\frac{\partial V}{\partial t} + LV - \alpha V + f = 0 \quad \text{a.e.} \qquad \text{in } Q\setminus\hat{S}. \tag{13.13}$$

PROBLEMS

1. Prove the existence of a solution of (3.6). [*Hint*: *Schauder's fixed point theorem* states: Let Y be a closed convex subset of a Banach space X and let T be an operator from Y into itself such that $TY = \{Ty; y \in Y\}$ is contained in a compact set. Then T has a fixed point, i.e., there is a point $y_0 \in Y$ such that $Ty_0 = y_0$. Take $w = Tu$ if $A_\epsilon w = \epsilon^{-1}K(x, u) + \tilde{f}$ in Ω, $w \in W^{2,2}(\Omega) \cap W_0^{1,2}(\Omega)$, $X = L^p(\Omega)$, $Y = \{g \in X, |g|_{0,p}^{\Omega} \leq M\}$.]
2. Prove (3.9). [*Hint*: Approximate f, g by smooth f_m, g_m. By the maximum principle, $R_\epsilon f_m \geq R_\epsilon g_m$.]
3. Generalize Theorems 2.3, 2.4, 3.1, 4.1, 4.2 to the case where $\alpha = \alpha(x)$ i.e. the function $e^{-\alpha t}$ in (1.3) is replaced by $\exp[-\int_0^t \alpha(x(s))\, ds]$.
4. Prove (5.35). [*Hint*: By Hölder's inequality $(\int|f|^q)^{1/q} \leq (\int|f|^2)^{(p-q)/(p-2)}(\int|f|^p)^{(q-2)/(p-2)}$.]
5. Let $Lu = \sum a_{ij}u_{x_ix_j} + \sum b_i u_{x_i} + cu$ be a uniformly elliptic operator in a closed ball N, with Hölder continuous coefficients (exponent α), and let $c \leq 0$ in N. Let $Lu = f \in W^{0,p}(N)$, $p > n/\alpha$, and $f < 0$ a.e. in N, and let $u \in W^{2,p}(N)$, $u \geq 0$ on ∂N. Prove that $u > 0$ in N. [*Hint*: Suppose $u \in W^{2,p}$ in a neighborhood of N and let $u_m = J_{1/m}u$ be a mollifier of u. Let $f_m = Lu_m$. Show that $|f_m - f|_{0,p}^N \to 0$. Next,

$$u_m = -\int_{\partial N} \frac{\partial G}{\partial \nu} u_m - \int_N G f_m \qquad (G = \text{Green's function in } N),$$

and $G > 0$ in N, $\partial G/\partial\nu \leq 0$. Take $m \to \infty$.]
6. Let the conditions of Theorem 7.1 hold and suppose $\psi_1 > \psi_2$ in Ω, $E = \Omega\setminus G_1$, $F = \Omega\setminus G_2$ where G_1, G_2 are open bounded subsets with closure in Ω, and

$$L\psi_1 - \alpha\psi_1 + f < 0 \quad \text{in } G_1, \qquad L\psi_2 - \alpha\psi_2 + f > 0 \quad \text{in } G_2.$$

Prove that the V and $(\hat{E}, \hat{F})$ in Theorem 7.1 are the value and saddle point of sets for the stochastic game associated with (1.1), (1.3) and the sets E, F. [*Hint*: By the remark at the end of Section 7, it suffices to prove that $V < \psi_1$ in G_1, $V > \psi_2$ in G_2. Let $w = \psi_1 - V$. If $x^0 \in G_1 \cap \hat{F}$, then $V(x^0) = \psi_2(x^0) < \psi_1(x^0)$. If $x^0 \in G_1 \setminus \hat{F}$, then $V(x) > \psi_2(x)$ in a ball N about x^0. Deduce $Lw - \alpha w < 0$ in N, $w \geq 0$ in N and use Problem 5.]
7. State and prove an analog of Theorem 2.3 for the stopping time problem (1.1), (1.2).

8. Do the same for Theorem 2.4.
9. Do the same for Theorems 3.1, 6.1. [*Hint*: In proving the existence of a solution u_ϵ of (3.6), notice that (3.7) does not hold for the new K; $K = u - u^+$. Since K is Lipschitz, $K(u) = k(u)u$ where $k(u)$ is a function of u, $|k(u)| \leq 2$. Define $w = Tu$ (cf. Problem 1) if $A_\epsilon w = \epsilon^{-1}k(u)w + \tilde{f}$ in Ω, $w \in W^{2,2}(\Omega) \cap W_0^{1,2}(\Omega)$.]
10. Prove Theorem 8.1.
11. Prove Theorem 9.1.
12. Prove Theorem 9.2.
13. If u, $\partial u/\partial t$ belong to $L^2(0, T; L^{2,\mu}(\Omega_m))$, then

$$\frac{d}{dt}|u(t)|^2 = 2(u(t), u_t(t)) \qquad \text{for a.a.} \quad t \in (0, T),$$

where $(\ , \ )$ denotes the scalar product in $L^{2,\mu}(\Omega_m)$. [*Hint*: It suffices to prove

$$|u(t)|^2 - 2\int_t^T (u(s), u_s(s))\, ds = \text{const.}$$

Approximate $u(t)$ by mollifiers.]
14. Prove that there exists a solution of (10.7)–(10.9). [*Hint*: Cf. Problem 1.]
15. Prove (10.28). [*Hint*: It suffices to take $v \in C_0^\infty(\Omega)$. For fixed $x_2, \ldots, x_n$, if I is the set $\{x_1;\, v(x) > 0\}$ then

$$\int_{-\infty}^{\infty} b\, \frac{\partial}{\partial x_1}\left(a\, \frac{\partial v}{\partial x_1}\right) v^+\, dx_1 = \int_I b\, \frac{\partial}{\partial x_1}\left(a\, \frac{\partial v}{\partial x_1}\right) v\, dx_1$$
$$= -\int_I a\, \frac{\partial v}{\partial x_1}\, \frac{\partial}{\partial x_1}(bv)\, dx_1.]$$

16. Let G be a bounded domain. If w is a uniformly Lipschitz continuous function in G, then w belongs to $W^{1,p}(G)$, for any $p > 0$. [*Hint*: Show that the weak derivatives of w exist and are bounded.]
17. Let $w_1, \ldots, w_m$ belong to $W^{1,p}(G)$, where G is a bounded domain and $1 \leq p < \infty$, and let $g(x_1, \ldots, x_m)$ be uniformly Lipschitz continuous in R^m. Prove that $g(w_1, \ldots, w_m)$ belongs to $W^{1,p}(G)$; thus, in particular, $\max(w_1, \ldots, w_m)$, $|w_1|$, w_1^+, w_1^- belong to $W^{1,p}(G)$.
18. Prove Theorem 12.1.
19. Prove Theorem 12.2.
20. Prove (12.15)–(12.18).
21. Let the conditions of Theorem 13.1 hold and let $\hat{\Omega} = Q \setminus \hat{S}$, $\partial\hat{\Omega}$ = boundary of $\hat{\Omega}$. We already know that

$$\frac{\partial V}{\partial t} + LV - \alpha V + f = 0 \quad \text{in } \hat{\Omega}, \tag{13.14}$$

$$V = \psi_1, \qquad \text{grad}_x V = \text{grad}_x \psi_1 \quad \text{on } \partial\hat{\Omega}. \tag{13.15}$$

The system (13.14), (13.15) constitutes a *free boundary problem*. Thus, one wants to solve (13.14) in a domain whose boundary is not known; however, on the unknown boundary $\partial\hat{\Omega}$ there are *two* prescribed conditions. The conditions (13.15) are called the "smooth fit" conditions. Note that we are interested only in solutions of (13.14), (13.15) for which $V \leqslant \psi_1$ in $\hat{\Omega}$.

Now specialize to $n = 1$, $L = \frac{1}{2}\,\partial^2/\partial x^2$, $\alpha = 0$, $f = 0$, and suppose that the continuation domain $\hat{\Omega}$ consists of all points (x, t) with $s_1(t) < x < s_2(t)$, where $s_1(t)$, $s_2(t)$ are continuously differentiable. Suppose also that the third derivatives of V are continuous in $\hat{\Omega}$. Prove that $w = \partial(V - \psi_1)/\partial t$ satisfies:

$$\begin{aligned}
&\frac{\partial w}{\partial t} + \frac{1}{2}\frac{\partial^2 w}{\partial x^2} = -\frac{\partial H(x,t)}{\partial t} && \text{in } \hat{\Omega},\\
&w(s_i(t), t) = 0 \qquad \text{for} \quad t > 0, \quad i = 1, 2, && (13.16)\\
&\frac{\partial w}{\partial x}(s_i(t), t) = 2H(s_i(t), t)\dot{s}_i(t) \qquad \text{for} \quad t > 0, \quad i = 1, 2,
\end{aligned}$$

where $H(x, t) = \partial\psi_1(x, t)/\partial t + \frac{1}{2}\,\partial^2\psi_1(x, t)/\partial x^2$. If $w > 0$ and $H \equiv -1$, then the system (13.16) is a *Stefan problem* and represents the standard model of water at temperature $w(x, t)$, occupying the interval $(s_1(t), s_2(t))$ at time t; this interval is surrounded by ice at zero temperature. If w takes also negative values, then we can think of it as a model with supercooled water.

22. Consider the special case of (13.14), (13.15) where $n = 1$, $\alpha = 0$, $f = 0$, $\psi_1 = \cos x$, $\Omega = R^1$. Show that $V(x, t) \equiv -1$ and find the domain of continuation.

23. Suppose that in Theorem 3.1 ψ_1, ψ_2 belong to $C^2(\bar{\Omega})$ and $f \in L^\infty(\Omega)$. Prove that $Au \in L^\infty(\Omega)$. [*Hint*: Let $\beta_\epsilon(x, u) = -\epsilon^{-1}(u - \psi(x))^+ - \epsilon^{-1}u^-$. At a point x_0 where $\beta_\epsilon(x, u_\epsilon(x))$ takes a positive maximum, $u - \psi$ takes a positive maximum, so that $A(u - \psi) \geqslant 0$. Hence $\beta_\epsilon(x, u(x)) \leqslant (\tilde{f} - A\psi) \times (x_0) \leqslant C$. Similarly, $\beta_\epsilon(x, u(x)) \geqslant -C$.]

24. Let ψ_1, ψ_2, f be as in the previous problem. Replace (3.5) by

$$Au + \beta_\epsilon(x, u) = \tilde{f}$$

where $\beta_\epsilon(x, u)$ is any C^2 function satisfying:

$$\begin{aligned}
&\beta_\epsilon(x, 0) = 0,\\
&\beta_\epsilon(x, u) \to -\infty \qquad \text{if} \quad u < 0, \quad \epsilon \to 0,\\
&\beta_\epsilon(x, u) \to +\infty \qquad \text{if} \quad u > \psi(x), \quad \epsilon \to 0,\\
&\frac{\partial}{\partial u}\beta_\epsilon(x, u) \geqslant 0.
\end{aligned}$$

Deduce that $|\beta_\epsilon(x, u_\epsilon(x))| \leqslant C$ and show that $u_\epsilon \to u$ uniformly as $\epsilon \to 0$, where u is the solution of (3.4).

25. Extend the result of the preceding problem to the setting of Theorem 6.1.

26. Extend the result of the preceding problem to the setting of Theorem 10.1, and deduce, when the coefficients of A are independent of t, that

$$e^{-\delta t}\left|\frac{\partial u}{\partial t}\right|_{0,\infty,\mu}^{\Omega}+e^{-\delta t}|Au|_{0,\infty,\mu}^{\Omega}\leqslant C \qquad \text{for a.a.} \quad t>0.$$

[*Hint*: Take the $(2k)$th root in (10.39), $k\to\infty$.]

17

Stochastic Differential Games

1. Auxiliary results

In this chapter we shall deal with stochastic differential equations in which control variables occur in the drift coefficient. The control variables are to be chosen so as to yield optimal results for a given set of cost functions or payoffs. We shall also introduce schemes whereby, in addition to control variables, optimal stopping is allowed.

In order to solve these problems we shall need some results on parabolic equations not stated earlier in this book.

The following notation will be used: Ω is a domain in R^n with boundary $\partial\Omega$,

$$Q_T = \{(x, t); x \in \Omega, 0 < t < T\},$$
$$S_T = \{(x, t); x \in \partial\Omega, 0 < t < T\},$$
$$\Omega_T = \{(x, T); x \in \Omega\},$$
$$\Gamma_T = S_T \cup \Omega_T.$$

We shall assume in this section that Ω is a bounded domain with C^2 boundary $\partial\Omega$. A function Φ defined on Γ_T is said to belong to $C^{2,1}(\Gamma_T)$ if (i) in terms of local C^2 representation of $\partial\Omega$ (say $x_i = \psi(x_1, \dots, x_{i-1}, x_{i+1}, \dots, x_n)$), the functions Φ, $\partial\Phi/\partial t$, $\partial\Phi/\partial x_j$, $\partial^2\Phi/\partial x_j \partial x_k$ (for all $j \neq i$, $k \neq i$) are uniformly continuous on S_T; (ii) $\Phi(T, x)$, $D_x\phi(T, x)$, $D_x^2\Phi(T, x)$ are uniformly continuous in Ω, and (iii) Φ is continuous on $\partial\Omega$.

If $\partial\Omega$ belong to $C^{2+\alpha}$ $(0 < \alpha \leq 1)$, then we say that Φ belongs to $C^{2+\alpha, 1+\alpha}(\Gamma_T)$ if $\Phi \in C^{2,1}(\Gamma_T)$ and, in addition, the functions occurring in (i), (ii) are uniformly Hölder continuous (exponent α) in (x, t).

We now take a fixed finite number of local representations of $\partial\Omega$ that cover $\partial\Omega$. If $\Phi \in C^{2,1}(\Gamma_T)$, then we denote by $\|\Phi\|_{2,1}^{\Gamma_T}$ an upper bound on all the derivatives in (i), (ii) occurring in the given fixed local representations of $\partial\Omega$. Denote by $H_\alpha(\Phi)$ an upper bound on the Hölder coefficients of all the functions in (i), (ii). If $\Phi \in C^{2+\alpha,\,1+\alpha}(\Gamma_T)$, we define

$$\|\Phi\|^{\Gamma_T}_{2+\alpha,\,1+\alpha} = \|\Phi\|^{\Gamma_T}_{2,\,1} + H_\alpha(\Phi).$$

Definition. A function $u(x, t)$ is said to belong to $W^{2,\,1}_p(Q_T)$ if its weak derivatives

$$D_x u, \quad D_t u, \quad D_x^2(u)$$

belong to $L^p(Q_T)$. We introduce the norm

$$\|u\|_{W^{2,1}_p(Q_T)} = \|u\|_{L^p(Q_T)} + \sum_{j=1}^{n} \left\| \frac{\partial u}{\partial x_j} \right\|_{L^p(Q_T)} + \left\| \frac{\partial u}{\partial t} \right\|_{L^p(Q_T)}$$
$$+ \sum_{j,k=1}^{n} \left\| \frac{\partial^2 u}{\partial x_j\, \partial x_k} \right\|_{L^p(Q_T)}.$$

Consider a partial differential equation

$$\frac{\partial u}{\partial t} + \frac{1}{2} \sum_{i,j=1}^{n} a_{ij}(x, t) \frac{\partial^2 u}{\partial x_i\, \partial x_j} + \sum_{i=1}^{n} b_i(x, t) \frac{\partial u}{\partial x_i} + c(x, t)u = f(x, t) \quad \text{in } Q_T, \tag{1.1}$$

with the initial and boundary conditions given by

$$u = \Phi \qquad \text{on } \Gamma_T. \tag{1.2}$$

We shall need the following conditions:

(A_1) For all $(x, t) \in \bar{Q}_T$ and for all $\xi \in R^n$,

$$\nu_0|\xi|^2 \le \sum_{i,j=1}^{n} a_{ij}(x, t)\xi_i\xi_j \le \nu_1|\xi|^2, \tag{1.3}$$

where ν_0, ν_1 are positive constants.

(A_2) For all (x, t) and $(\bar{x}, \bar{t})$ in $\bar{Q}_T$,

$$\sum |a_{ij}(x, t) - a_{ij}(\bar{x}, \bar{t})| \le \nu(|x - \bar{x}| + |t - \bar{t}|), \qquad \nu(r) \downarrow 0 \quad \text{if } r \downarrow 0. \tag{1.4}$$

(A_3) The derivatives $\partial a_{ij}(x, t)/\partial x_k$ are uniformly continuous in Q_T; let ν_3 be a constant such that

$$\sum_{i,j,k} \left| \frac{\partial a_{ij}(x, t)}{\partial x_k} \right| \le \nu_3 \qquad \text{for all} \quad (x, t) \in Q_T. \tag{1.5}$$

(A_4) The derivatives $\partial a_{ij}(x, t)/\partial t$ are continuous in Q_T; denote by ν_4 a constant for which

$$\sum_{i,j} \left| \frac{\partial a_{ij}(x, t)}{\partial t} \right| \le \nu_4 \qquad \text{for all} \quad (x, t) \in Q_T. \tag{1.6}$$

(B) The functions $b_i(x, t)$, $c(x, t)$ are measurable in Q_T, and

$$\sum |b_i(x, t)| \leq \nu_2, \qquad |c(x, t)| \leq \nu_3 \qquad \text{for all} \quad (x, t) \in \bar{Q}_T. \tag{1.7}$$

We shall need an extension of Theorems 10.4.2, 10.4.3 to the case of the nonhomogeneous boundary condition (1.2). We state it only for $p > n$.

Lemma 1.1. *Let $\partial\Omega \in C^2$, $\Phi \in C^{2,1}(\Gamma_T)$ and let* (A_1), (A_2), (B) *hold. Then, for any $p > n$, $f \in L^p(Q_T)$, there exists a unique solution u in $W_p^{2,1}(Q_T)$ of* (1.1), (1.2), *and*

$$\|u\|_{W_p^{2,1}(Q_T)} \leq C\|\Phi\|_{2,1}^{\Gamma_T} + \|f\|_{L^p(Q_T)} \tag{1.8}$$

where C is a constant depending only on ν_0, ν_1, ν, ν_2, and Q_T.

Notice, by the Sobolev inequality, that $u(x, t)$ is continuous in $\bar{Q}_T$. The condition (1.2) is understood in the classical sense.

Lemma 1.1 can be proved by constructing a function Ψ with continuous derivatives $D_t\Psi$, $D_x\Psi$, $D_x^2\Psi$ in $\bar{Q}_T$ such that $\Psi = \Phi$ on Γ_T, and then applying Theorems 10.4.2, 10.4.3, to $u - \Psi$. Solonnikov [1] has proved a stronger result than Lemma 1.1, requiring less differentiability of Φ.

For any set Q' in the (x, t)-space, write

$$|v|_{\alpha, Q'} = \text{l.u.b.} \frac{|v(x, t) - v(\bar{x}, \bar{t})|}{|x - \bar{x}|^\alpha + |t - \bar{t}|^{\alpha/2}}$$

where the l.u.b. is taken with respect to $(x, t) \in Q'$, $(\bar{x}, \bar{t}) \in Q'$, $(x, t) \neq (x, \bar{t})$.

Lemma 1.2. *Let $\partial\Omega \in C^2$, $\Phi \in C^{2,1}(\Gamma_T)$, and let* (A_1)–(A_3) *and* (B) *hold. Then there exists an α, $0 < \alpha < 1$, such that, for any $f \in L^\infty(Q_T)$, the solution u of* (1.1), (1.2) *satisfies*

$$|D_x u|_{\alpha, Q'} \leq C \tag{1.9}$$

for any set Q' whose closure is contained in Q_T. Here C is a constant depending only on ν_0, ν_1, ν, ν_2, ν_3, Q_T, Q', and $\text{l.u.b.}_{\Gamma_T}|\Phi|$.

This result is due to Ladyzhenskaja *et al.* [1; Chapter 6] in case u is a classical solution of (1.1), (1.2). The proof in the general case follows by approximation; see Friedman [4].

Lemma 1.3. *Let $0 < \alpha < 1$. Suppose $\partial\Omega \in C^{2+\alpha}$, $\Phi \in C^{2+\alpha, 1+\alpha}(\Gamma_T)$, and let* (A_1)–(A_4) *and* (B) *hold. Then, for any $f \in L^\infty(Q_T)$, the solution of* (1.1), (1.2) *satisfies*

$$\underset{Q_T}{\text{l.u.b.}}\, |u| + |D_x u|_{\alpha, Q_T} \leq C\left(\|\Phi\|_{2+\alpha, 1+\alpha}^{Q_T} + \underset{Q_T}{\text{l.u.b.}}\, |f|\right), \tag{1.10}$$

where C is a constant depending only on ν_0, ν_1, ν_2, ν_3, ν_4, Q_T.

This lemma is due to Friedman [1] in case $\Phi = 0$ and b_i, c, f are continuous. The case $\Phi \not\equiv 0$ can be reduced to the case $\Phi \equiv 0$ by considering $u - \Phi$. The case where b_i, c, f are not continuous follows by approximation.

We shall deal later on with nonlinear parabolic systems of the form

$$\frac{\partial u^k}{\partial t} + \frac{1}{2} \sum_{i,j=1}^{n} a_{ij}(x, t) \frac{\partial^2 u^k}{\partial x_i \, \partial x_j} + e_k(x, t, D_x u^1, \ldots, D_x u^N) = 0 \qquad \text{in } Q_T,$$

$$u_k = \Phi^k \qquad \text{on } \Gamma_T,$$

where $k = 1, \ldots, N$, and the e_k are nonlinear functions in the variables $D_x u^i$. The solution $u = (u^1, \ldots, u^N)$ is taken in the sense that each u^k is in $W_p^{2,1}(Q_T)$, is continuous in $\bar{Q}_T$, and $D_x u^k$ is continuous in Q_T.

We need the following conditions:

(C) (i) $f_i(x, t, y_1, \ldots, y_N)$ $(1 \leq i \leq n)$ and $h_i(x, t, y_1, \ldots, y_N)$ $(1 \leq i \leq N)$ are continuous functions in $(x, t, y_1, \ldots, y_N) \in R^n \times [0, T] \times Y_1 \times \cdots \times Y_N$, where $Y_1, \ldots, Y_N$ are compact subsets in some euclidean spaces $R^{k_1}, \ldots, R^{k_N}$ respectively;

(ii) $\partial\Omega$ is in C^2 and $g_i(x, t)$ $(1 \leq i \leq N)$ are functions belonging to $C^{2,1}(\Gamma_T)$.

Consider the functions

$$H_k(x, t, y_1, \ldots, y_N, p_k) = f(x, t, y_1, \ldots, y_N) \cdot p_k + h_k(x, t, y_1, \ldots, y_N) \tag{1.11}$$

where $f = (f_1, \ldots, f_n)$ and p_k is a variable point in R^n. We shall need the following *generalized minimax condition*:

(D) There exist functions $y_1^*(x, t, p), \ldots, y_N^*(x, t, p)$, where $p = (p_1, \ldots, p_N)$, such that:

(i) the $y_j^*(x, t, p)$ are measurable in $(x, t) \in \bar{Q}_T$ for every p, and continuous in p for every $(x, t) \in \bar{Q}_T$ with modulus of continuity independent of (x, t);

(ii) for all $(x, t) \in \bar{Q}_T$ and for all p,

$$y_j^*(x, t, p) \in Y_j \qquad (1 \leq j \leq N);$$

(iii) for all $(x, t) \in \bar{Q}_T$ and for all p,

$$\begin{aligned} &\min_{y_k \in Y_k} H_k(x, t, y_1^*(x, t, p), \ldots, y_{k-1}^*(x, t, p), \\ &\qquad y_k, y_{k+1}^*(x, t, p), \ldots, y_N^*(x, t, p), p_k) \\ &= H_k(x, t, y_1^*(x, t, p), \ldots, y_N^*(x, t, p), p_k) \qquad (1 \leq k \leq N). \end{aligned} \tag{1.12}$$

Theorem 1.4. *Let* (A_1)–(A_3), (C), *and* (D) *hold. Then there exists a solution*

$\phi^* = (\phi_1^*, \ldots, \phi_N^*)$ *of the nonlinear parabolic system*

$$\frac{\partial \phi_k}{\partial t} + \tfrac{1}{2} \sum_{i,j=1}^{n} a_{ij}(x, t) \frac{\partial^2 \phi_k}{\partial x_i \, \partial x_j} + f(x, t, y^*(x, t, D_x\phi)) \cdot D_x\phi_k$$

$$+ h_k(x, t, y^*(x, t, D_x\phi)) = 0 \qquad \text{in } Q_T, \qquad 1 \leq k \leq N, \tag{1.13}$$

$$\phi_k = g_k \qquad \text{on } \Gamma_T, \qquad 1 \leq k \leq N. \tag{1.14}$$

More precisely, ϕ^* *is continuous in* $\overline{Q}_T$ *and satisfies* (1.14), $D_x\phi^*$ *is a bounded function in* Q_T, *uniformly Hölder continuous (with some exponent* α*) in compact subsets of* Q_T, *the weak derivatives* $\partial^2\phi_k^*/\partial x_i \, \partial x_j$, $\partial\phi_k^*/\partial t$ *belong to* $L^r(Q_T)$ *for any* $r > 1$, *and* (1.13) *holds almost everywhere.*

The proof is given in Friedman [4]; it is based on Theorem 7.1 of Ladyzhenskaja *et al.* [1] and on Lemmas 1.1, 1.2.

We shall need the following *lemma of Filippov*:

Lemma 1.5. *Let* $g(t, u)$ *be a function with values in* R^n, *defined for* $t \in [a, b]$, $u \in U$ *where* U *is a compact set in* R^k. *Assume that* $g(t, u)$ *is continuous in* $(t, u) \in [a, b] \times U$. *Let* $\psi(t)$ *be a measurable function in* $[a, b]$ *such that*

$$\psi(t) \in g(t, U) \equiv \{ g(t, u); u \in U \}$$

for a.a. $t \in [a, b]$. *Then there exists a measurable function* $u(t)$ *such that* $u(t) \in U$ *and* $\psi(t) = g(t, u(t))$ *for a.a.* $t \in [a, b]$.

For the proof, see Filippov [1] or Friedman [3].

Corollary 1.6. *Let* $g(t, u)$ *be as in Lemma* 1.5 *with* $n = 1$. *Then there exist measurable functions* $u_1(t)$, $u_2(t)$ *with values in* U *such that*

$$\max_{u \in U} g(t, u) = g(t, u_1(t)), \qquad \min_{u \in U} g(t, u) = g(t, u_2(t))$$

for a.a. $t \in [a, b]$.

Indeed, apply Lemma 1.5 with $\psi(t) = \max_{u \in U} g(t, u)$ and with $\psi(t) = \min_{u \in U} g(t, u)$.

2. *N*-person stochastic differential games with perfect observation

We maintain the notation of Section 1, and take Ω to be a bounded domain in R^n. Consider a system of n stochastic differential equations

$$d\xi(t) = f(\xi(t), t, y_1, \ldots, y_N) \, dt + \sigma(\xi(t), t) \, dw(t) \tag{2.1}$$

for $s \leq t \leq T$, with initial condition

$$\xi(s) = x \qquad (x \in \Omega, 0 \leq s < T). \tag{2.2}$$

Here $y_1, \ldots, y_N$ are parameters to be chosen later on.

Denote by τ the first time t such that $(\xi(t), t)$ leaves Q_T and $t > s$.

Let $Y_1, \ldots, Y_N$ be compact sets in some euclidean spaces $R^{k_1}, \ldots, R^{k_N}$ respectively (as in condition (C)). We shall call Y_i the *control set* for the *player* y_i. A measurable function $y_i(x, t)$ defined on $R^n \times [0, T)$ with values in the control set Y_i is called a *control function* or a *pure strategy* for the player y_i. When each player y_i chooses a pure strategy $y_i(x, t)$, then (2.1) takes the form

$$d\xi(t) = f(\xi(t), t, y_1(\xi(t), t), \ldots, y_N(\xi(t), t))\, dt + \sigma(\xi(t), t)\, dw(t). \tag{2.3}$$

In addition to (2.1), (2.2) we are given *cost functionals*

$$J_i(y_1, \ldots, y_N) = E_{x,s}\left\{\int_s^\tau h_i(\xi, t, y_1, \ldots, y_N)\, dt + g_i(\xi(\tau), \tau)\right\} \tag{2.4}$$

for $1 \leq i \leq N$. If there exists a unique solution of (2.3), (2.2), then we can compute the costs $J_i(y_1, \ldots, y_N)$.

The above setting of the players choosing pure strategies represents a model of a game of perfect observation. In this model, the players know the position $\xi(t)$, at any time t. Furthermore, they make use of their knowledge of the present position only, i.e., they do not choose strategies based on the past observations (i.e., based on knowledge of $\xi(\lambda)$, $s \leq \lambda < t$).

The vector $y = (y_1, \ldots, y_N)$ will be called a pure strategy if each component is a pure strategy.

Suppose now that the functions f, σ are Lipschitz continuous in x. If the pure strategy $y(x, t)$ is also Lipschitz continuous in x, then there exists a unique solution of (2.3), (2.2). The costs $J_k(y)$ can then be computed. We shall presently derive a useful expression for the $J_k(y)$.

Set

$$a_{ij} = \sum_{k=1}^n \sigma_{ik}\sigma_{jk}$$

and let ψ_k be the solution of the parabolic initial–boundary value problem:

$$\frac{\partial \psi_k}{\partial t} + \frac{1}{2}\sum_{i,j=1}^n a_{ij}(x, t)\frac{\partial^2 \psi_k}{\partial x_i\, \partial x_j} + \sum_{i=1}^n f_i(x, t, y_1(x, t), \ldots, y_N(x, t))\frac{\partial \psi_k}{\partial x_i}$$

$$+ h_k(x, t, y_1(x, t), \ldots, y_N(x, t)) = 0 \qquad \text{in } Q_T, \tag{2.5}$$

$$\psi_k = g_k \qquad \text{on } \Gamma_T. \tag{2.6}$$

If there is a smooth solution of (2.5), (2.6), then by applying Itô's formula applied to $\psi_k(\xi(t), T - t)$ we find that

$$J_k(y) = \psi_k(x, s) \qquad (1 \leq k \leq N). \tag{2.7}$$

Now let the conditions (A_1), (A_2), (C) hold. Then, by Lemma 1.1, for *any* pure strategy $y(x, t)$ there exists a unique solution of (2.5), (2.6). The functional $J_k(y)$ defined by (2.7) will henceforth be called the *cost functional* corresponding to the pure strategy $y(x, t)$. This definition is a generalization of the definition (2.4).

Definitions. The system (2.1), (2.2), (2.4) is called an *N-person stochastic differential game with perfect observation*. Consider the following scheme: Each player chooses a pure strategy, and then the costs J_k are computed (from (2.7)). We refer to this scheme as a *game of perfect observation played by pure strategies*, or, briefly, a *Markovian game*.

Definition. A pure strategy

$$y^*(x, t) = (y_1^*(x, t), \ldots, y_N^*(x, t))$$

is called a *Nash equilibrium point in pure strategies* of the differential game if

$$J_k(y_1^*, \ldots, y_{k-1}^*, y_k, y_{k+1}^*, \ldots, y_N^*) > J_k(y_1^*, \ldots, y_{k-1}^*, y_k^*, y_{k+1}^*, \ldots, y_N^*) \tag{2.8}$$

for any pure strategy y_k, $1 < k < N$.

An equilibrium point is a "reasonable" solution for noncooperative game of N players. If $N = 2$ and $J_1 + J_2 = 0$, we say that we have a *zero sum two-person game*; the equilibrium point is then called a *saddle point in pure strategies.*

Theorem 2.1. *Let the conditions* (A_1)–(A_3), (C), *and* (D) *hold, and write*

$$y_j^*(x, t) = y_j^*(x, t, D_x\phi_1^*(x, t), \ldots, D_x\phi_N^*(x, t)) \tag{2.9}$$

where $\phi^*(x, t)$ *is the solution asserted in Theorem* 1.4. *Then*

$$y^*(x, t) = (y_1^*(x, t), \ldots, y_N^*(x, t)) \tag{2.10}$$

is a Nash equilibrium point in pure strategies of the differential game associated with (2.1), (2.2), (2.4).

Proof. Let $y_k(x, t)$ be any pure strategy for the player y_k. Denote by ϕ_k the unique solution of

$$\begin{aligned}\frac{\partial \phi_k}{\partial t} &+ \tfrac{1}{2} \sum_{i,j=1}^{n} a_{ij}(x, t) \frac{\partial^2 \phi_k}{\partial x_i\, \partial x_j} \\ &+ f(x, t, y_1^*(x, t), \ldots, y_{k-1}^*(x, t), y_k(x, t), y_{k+1}^*(x, t), \ldots, y_N^*(x, t)) \cdot D_x\phi_k \\ &+ h_k(x, t, y_1^*(x, t), \ldots, y_{k-1}^*(x, t), y_k(x, t), y_{k+1}^*(x, t), \ldots, y_N^*(x, t)) = 0 \\ &\qquad \text{in } Q_T,\end{aligned} \tag{2.11}$$

$$\phi_k = g_k \quad \text{on } \Gamma_T. \tag{2.12}$$

Using (1.12) we find that the function ϕ_k^* satisfies

$$\frac{\partial \phi_k^*}{\partial t} + \frac{1}{2} \sum_{i,j=1}^{n} a_{ij}(x, t) \frac{\partial^2 \phi_k^*}{\partial x_i \, \partial x_j}$$
$$+ f(x, t, y_1^*(x, t), \ldots, y_{k-1}^*(x, t), y_k(x, t), y_{k+1}^*(x, t), \ldots, y_N^*(x, t)) \cdot D_x\phi_k^*$$
$$+ h_k(x, t, y_1^*(x, t), \ldots, y_{k-1}^*(x, t), y_k(x, t), y_{k+1}^*(x, t), \ldots, y_N^*(x, t)) \geq 0$$

a.e. in Q_T. Setting

$$b(x, t) = f(x, t, y_1^*(x, t), \ldots, y_{k-1}^*(x, t), y_k(x, t), y_{k+1}^*(x, t), \ldots, y_N^*(x, t))$$

we see that $w = \phi_k^* - \phi_k$ satisfies:

$$\frac{\partial w}{\partial t} + \frac{1}{2} \sum_{i,j=1}^{n} a_{ij}(x, t) \frac{\partial^2 w}{\partial x_i \, \partial x_j} + b(x, t) \cdot D_x w \geq 0 \quad a.e. \qquad \text{in } Q_T,$$
$$w = 0 \qquad \text{on } \Gamma_T. \tag{2.13}$$

By the maximum principle (see Problem 1), $w \leq 0$ in Q_T. This gives (2.8).

For a zero-sum two-person game we can prove Theorem 2.1 under a condition weaker than (D), called the *minimax condition*:

(D′) For any $(x, t) \in Q_T$ and for any $p \in R^n$,

$$\min_{y_1 \in Y_1} \max_{y_2 \in Y_2} H_1(x, t, y_1, y_2, p_1) = \max_{y_2 \in Y_2} \min_{y_1 \in Y_1} H_1(x, t, y_1, y_2, p). \tag{2.14}$$

Note, by Corollary 1.6, that there exist measurable functions $y_1 = y_1^*(x, t, p_1)$, $y_2 = y_2^*(x, t, p_1)$ with values in Y_1 and Y_2, respectively, such that

$$\max_{y_2 \in Y_2} H_1(x, t, y_1^*(x, t, p_1), y_2, p_1) = \min_{y_1 \in Y_1} \max_{y_2 \in Y_2} H_1(x, t, y_1, y_2, p_1), \tag{2.15}$$
$$\min_{y_1 \in Y_1} H_1(x, t, y_1, y_2^*(x, t, p_1), p_1) = \max_{y_2 \in Y_2} \min_{y_1 \in Y_1} H_1(x, t, y_1, y_2, p_1). \tag{2.16}$$

From this we infer the condition (1.12). However we cannot infer, in general, that the functions $y_j^*(x, t, p_1)$ are continuous in p_1.

The function

$$H(x, t, p) = \min_{y_1 \in Y_1} \max_{y_2 \in Y_2} H_1(x, t, y_1, y_2, p) \tag{2.17}$$

is called the *Hamiltonian function* of the (zero-sum two-person) differential game.

Lemma 2.2. *Let $N = 2$, $J_2 = -J_1$ and assume that* (A_1)–(A_3) *and* (C) *hold. Then there exists a solution ϕ^* of the parabolic equation*

$$\frac{\partial \phi}{\partial t} + \frac{1}{2} \sum_{i,j=1}^{n} a_{ij}(x, t) \frac{\partial^2 \phi}{\partial x_i \, \partial x_j} + H(x, t, D_x\phi) = 0 \qquad \text{in } Q_T \tag{2.18}$$

with the initial-boundary conditions

$$\phi = g_1 \qquad \text{on } \Gamma_T. \tag{2.19}$$

The solution is taken in the same sense as in Theorem 1.4. In fact, the proof of the lemma is just slightly different from the proof of Theorem 1.4 when specialized to $N = 1$.

Theorem 2.3. *Let $N = 2$, $J_2 = -J_1$ and assume that* (A_1)–(A_3), (C), *and* (D′) *hold. Let $y_1^*(x, t)$, $y_2^*(x, t)$ be any control functions satisfying*

$$\max_{y_2 \in Y_2} H_1(x, t, y_1^*(x, t), y_2, D_x\phi^*(x, t)) = \min_{y_1 \in Y_1} \max_{y_2 \in Y_2} H_1(x, t, y_1, y_2, D_x^*\phi(x, t)),$$

$$\min_{y_1 \in Y_1} H_1(x, t, y_1, y_2^*(x, t), D_x\phi^*(x, t)) = \max_{y_2 \in Y_2} \min_{y_1 \in Y_1} H_1(x, t, y_1, y_2, D_x^*\phi(x, t)),$$

where ϕ^* is a solution of (2.18), (2.19) *(asserted in Lemma* 2.2*). Then $(y_1^*(x, t), y_2^*(x, t))$ is a saddle point in pure strategies of the differential game associated with* (2.1), (2.2), (2.4) *when $k = 1$, $N = 2$, $J_2 = -J_1$.*

Notice that the existence of $y_1^*(x, t)$, $y_2^*(x, t)$ follows from Corollary 1.6.

Proof. Let y_1 choose the strategy $y_1^*(x, t)$, and let y_2 choose any strategy $y_2(x, t)$. Denote by ψ the solution of

$$\frac{\partial\psi}{\partial t} + \frac{1}{2} \sum_{i,j=1}^{n} a_{ij}(x, t) \frac{\partial^2\psi}{\partial x_i\, \partial x_j} + f(x, t, y_1^*(x, t), y_2(x, t)) \cdot D_x\psi$$

$$+ h_1(x, t, y_1^*(x, t), y_2(x, t)) = 0 \qquad \text{in } Q_T,$$

$$\psi = g_1 \qquad \text{on } \Gamma_T.$$

Since $\phi^* = \psi$ on Γ_T and

$$\frac{\partial\phi^*}{\partial t} + \frac{1}{2} \sum_{i,j=1}^{n} a_{ij}(x, t) \frac{\partial^2\phi^*}{\partial x_i\, \partial x_j} + f(x, t, y_1^*(x, t), y_2(x, t)) \cdot D_x\phi^*$$

$$+ h_1(x, t, y_1^*(x, t), y_2(x, t)) \leqslant 0$$

almost everywhere in Q_T, we conclude (see Problem 1) that $\phi^* \geqslant \psi$ in Q_T. This gives

$$J_1(y_1^*, y_2^*) \geqslant J_1(y_1^*, y_2).$$

Similarly, one proves that

$$J_1(y_1^*, y_2^*) \leqslant J(y_1, y_2^*).$$

3. Stochastic differential games with stopping time

Consider a stochastic system of n equations

$$d\xi = f(\xi, t, y, z)\, dt + \sigma(\xi, t)\, dw, \tag{3.1}$$

$$\xi(s) = x \tag{3.2}$$

in an interval $[0, T_0]$, where $0 \leq s < T_0$. We are going to introduce a concept of a differential game with Markov stopping times S, T (with range in $[s, T_0]$) associated with (3.1), (3.2) and with a *payoff*

$$P_{x,s}(y, S; z, T) = E_{x,s}\left\{\int_s^{S\wedge T} e^{-\alpha(t-s)}h(\xi(t), t, y, z)\,dt + e^{-\alpha(S-s)}g_1(\xi(S), S)\chi_{S\leq T} + e^{-\alpha(T-s)}g_2(\xi(T), T)\chi_{T<S}\right\}; \tag{3.3}$$

α is a nonnegative number, called the *discount coefficient*.

The model will be a generalization of the stochastic game (without controls y, z) introduced in Chapter 16, and a generalization of the zero sum two-person stochastic differential game (without stopping times S, T) introduced in Section 2.

Set $\sigma = (\sigma_{ij})$, $f = (f_1, \ldots, f_n)$ and

$$a_{ij} = \sum \sigma_{ik}\sigma_{jk}.$$

We shall need the following conditions:

$$\sum a_{ij}(x, t)\xi_i\xi_j \geq \nu|\xi|^2, \qquad \nu > 0, \tag{3.4}$$

$$a_{ij}, \quad \frac{\partial a_{ij}}{\partial x_l}, \quad \frac{\partial^2 a_{ij}}{\partial x_l\,\partial t} \quad \text{are continuous and bounded in } (x, t) \in R^n \times [0, T_0], \tag{3.5}$$

$$f, \quad \frac{\partial f}{\partial x_l}, \quad \frac{\partial f}{\partial y}, \quad \frac{\partial f}{\partial z} \quad \text{are continuous and bounded in } (x, t) \in R^n \times [0, T_0], \tag{3.6}$$

$$h, \ h_t, \ f, \ f_t, \ g_1, \ \frac{\partial g_1}{\partial x_l}, \ \frac{\partial^2 g_1}{\partial x_l\,\partial t}, \ g_2 \quad \text{are continuous and bounded in } (x, t) \in R^n \times [0, T_0]. \tag{3.7}$$

Let Y, Z be compact sets in some euclidean spaces.

Recall that a control function for y is a measurable function $y(x, t)$ with values in Y. Similarly, a control function for z is a measurable function $z(x, t)$ with values in Z.

Suppose (3.4) holds, (a_{ij}) is continuous and bounded, and $f(x, t, y(x, t), z(x, t))$ is measurable and bounded. According to Stroock and Varadhan [1] one can construct a Markov process which is unique in law, and which satisfies (3.1) for suitably constructed Brownian motion. Thus, for any pair of control functions $y = y(x, t)$, $z = z(x, t)$ there is a unique solution of (3.1), (3.2) in the sense of Stroock and Varadhan.

On the other hand, if we restrict ourselves to control functions that are Lipschitz continuous in x, then there exists a unique solution of (3.1), (3.2) in the usual sense (provided we also assume that $f(x, t, y, z)$ is Lipschitz continuous in x, y, z).

Remark. For simplicity we shall assume that (3.6) holds. Then we can take the solution of (3.1), (3.2) in either sense. However the results of this section remain valid when the condition (3.6) is omitted, provided the solution of (3.1), (3.2) is taken in the Stroock-Varadhan sense.

The stopping times S, T are taken with respect to the σ-algebra $\mathfrak{F}_t$ generated by $\xi(\lambda)$, $s \leqslant \lambda \leqslant t$. We assume that

$$s \leqslant S, T \leqslant T_0.$$

The player y tries to choose $y(x, t)$, S so as to maximize the payoff, and the player z tries to choose $z(x, t)$, T so as to minimize the payoff.

The system made up of (3.1), (3.2), (3.3) and Y, Z will be referred to as a *stochastic differential game with stopping time.*

Definition. A pair $\{(y^*(x, t), S^*), (z^*(x, t), T^*)\}$ is called a *saddle point* if, for all $0 \leqslant s < T_0$, $x \in R^n$,

$$P_{x,s}(y, S; z^*, T^*) \leqslant P_{x,s}(y^*, S^*; z^*, T^*) \leqslant P_{x,s}(y^*, S^*; z, T) \quad (3.8)$$

for all controls y, z and stopping times S, T. the number $V(x, s) = P_{x,s}(y^*, S^*; z^*, T^*)$ is called the *value* of the game.

We shall now assume the *minimax condition*:

$$\max_{y\in Y}\min_{z\in Z}\{h(x, t, y, z) + p \cdot f(x, t, y, z)\}$$
$$= \min_{z\in Z}\max_{y\in Y}\{h(x, t, y, z) + p \cdot f(x, t, y, z)\} \equiv H(x, t, p). \quad (3.9)$$

Define

$$Lu = \frac{1}{2}\sum_{i,j=1}^{n} a_{ij}\frac{\partial^2 u}{\partial x_i\, \partial x_j} - \alpha u. \quad (3.10)$$

Notice that (3.7) implies the following condition:

(F) The function H(x, t, p) is continuous in $(x, t, p) \in R^n \times [0, T_0] \times R^n$, Lipschitz continues in (x, p), and a.e.,

$$|H(x, t, p)| \leqslant C(1 + |p|),$$
$$|H_t(x, t, p)| \leqslant C(1 + |p|),$$
$$|H_p(x, t, p)| \leqslant C$$

where C is a constant.

We shall write $W^{i,q,\mu}(R^n) = W^{i,q,\mu}$, and assume:

(G). (i) The functions $g_i(x, t)$ $(i = 1, 2)$ are measurable functions in $(x, t) \in R^n \times [0, T_0]$ and

$$g_i \in L^p(0, T_0; W^{2,p,\mu}) \cap L^2(0, T_0; W^{2,2,\mu}) \cap L^\infty(0, T_0; W^{1,2,\mu}),$$
$$\frac{\partial g_i}{\partial t} \in L^p(0, T_0; W^{0,p,\mu}) \cap L^2(0, T_0; W^{0,2,\mu})$$

for some number $p > n$;

(ii) $g_2 \geqslant g_1$ a.e.;

(iii) $\partial^2(g_2 - g_1)/\partial t^2 \in L^2(0, T_0; W^{0,2,\mu})$.

Consider the nonlinear parabolic variational inequality;

$$V \in L^p(0, T_0; W^{2,p,\mu}) \cap L^2(0, T_0; W^{2,2,\mu}) \cap L^\infty(0, T_0; W^{1,2,\mu}),$$
$$\frac{\partial V}{\partial t} \in L^p(0, T_0; W^{0,p,\mu}) \cap L^2(0, T_0; W^{0,2,\mu}); \tag{3.11}$$

for a.a. $t \in [0, T_0]$,

$$\left(\frac{\partial V}{\partial t} + LV + H(x, t, V_x)\right)(v - V) \leqslant 0 \quad \text{a.e.} \quad \text{for every} \quad v \in W^{0,2,\mu},$$
$$g_1 \leqslant v \leqslant g_2 \text{ a.e.}; \tag{3.12}$$

$$g_1 \leqslant V \leqslant g_2 \quad \text{a.e.}; \tag{3.13}$$

$$V(T_0, x) = g_1(x, T_0) \quad \text{a.e.} \tag{3.14}$$

This variational inequality has a unique notation. In fact, if $g_2 \geqslant 0 \equiv g_1$, then the proof is similar to the proof of Theorem 11.1. Thus, we first solve

$$\frac{\partial u}{\partial t} + Lu - \frac{1}{\epsilon}(u - g_2)^+ + \frac{1}{\epsilon}u^- + H(x, t, u_x) = 0 \quad \text{a.e.}$$
$$\text{in} \quad \Omega_m \times (0, T_0),\ u(t) \in W_0^{1,2}(\Omega_m), \qquad u(T_0) = 0, \tag{3.15}$$

where $\Omega_m = \{x; |x| < m\}$. By Problem 2, there exists a solution $u = u_m$ of (3.15). Next we obtain estimates as in Section 15.11, with just minor changes in the formulas. With these estimates at hand, we can then complete the proof of existence of a solution of (3.11)–(3.14). The proof of uniqueness is also similar to the proof in case $H(x, t, u_x)$ is linear in u_x.

If the condition $g_1(x) \equiv 0$ is not satisfied, then we first perform a transformation $u = V - g_1$ and then solve for u as before. Here we need the conditions that $\partial g_1/\partial x_l$, $\partial^2 g_1/\partial x_l\, \partial t$ are continuous and bounded.

Notice, by Sobolev's inequality, that $V(x, t)$ is a continuous function in $(x, t) \in R^n \times (0, T_0)$. Notice also that

$$\frac{\partial V}{\partial t} + LV + H(x, t, V_x) \geqslant 0 \qquad \text{if} \quad V > g_1 \tag{3.16}$$

$$\frac{\partial V}{\partial t} + LV + H(x, t, V_x) \leqslant 0 \qquad \text{if} \quad V < g_2. \tag{3.17}$$

Let $y^*(x, t)$, $z^*(x, t)$ be any control functions that render the $\max_y$ and $\min_z$ in (3.9) for $p = D_x V(x, t)$. By Corollary 1.6, such functions exist.

Define sets

$$G_1 = \{(x, t);\ V(x, t) = g_1(x, t)\}, \qquad G_2 = \{(x, t);\ V(x, t) = g_2(x, t)\}.$$

Denote by S^* and T^* the first hitting times of the sets G_1 and G_2 respectively. We shall show that $\{(y^*, S^*), (z^*, T^*)\}$ is a saddle point.

Before we do that we wish to point out that if $y^*(x, t)$, $z^*(x, t)$ are only known to be measurable functions (and not Lipschitz in x), then we must take the meaning of (3.1), (3.2) in the sense of Stroock and Varadhan [1]. If however they are Lipschitz in x, we can take (3.1), (3.2) in the usual sense, and restrict all control functions to be Lipschitz continuous in x.

Theorem 3.1. *Let the conditions* (3.4)–(3.7), (3.9), *and* (F), (G) *hold. Then* $\{(y^*, S^*), (z^*, T^*)\}$ *is a saddle point of the stochastic differential game with stopping time* (3.1)–(3.3).

Proof. We shall prove the second inequality in (3.8). Let $z(x, t)$ be a control function for z and T a stopping time for the process $\tilde{x}$ determined by (3.1), (3.2) when $z = z(x, t)$, $y = y^*(x, t)$. Denote by x^* the process determined by (3.1), (3.2) when $z = z^*(x, t)$, $y = y^*(x, t)$. If we apply Itô's formula to the function

$$V(x, t)\exp(-\alpha(t - s))$$

and the process x, formally, then we get

$$\begin{aligned} &E_{x,s}\{V(\tilde{x}(S^* \wedge T), S^* \wedge T)\exp[-\alpha(S^* \wedge T - s)]\} \\ &\quad = V(x, s) + E_{x,s}\left\{\int_s^{S^*\wedge T} \exp[-\alpha(t - s)]\left[\frac{\partial V}{\partial t} + V_x \cdot f(x, t, y^*, z)\right.\right. \\ &\qquad \left.\left. + LV\right](\tilde{x}(t), t)\, dt\right\}. \end{aligned} \tag{3.18}$$

Notice that if $t < S^* \wedge T \leqslant S^*$, then $V(\tilde{x}(t), t) > g_1(\tilde{x}(t), t)$; hence, by (3.16),

$$\left[\frac{\partial V}{\partial t} + LV + H(x, t, V_x)\right](\tilde{x}(t), t) \geqslant 0.$$

From the definition of y^* we then have

$$\left[\frac{\partial V}{\partial t} + LV + h(x, t, y^*(x, t), z) + V_x \cdot f(x, t, y^*(x, t), z)\right](\tilde{x}(t), t) \geqslant 0 \tag{3.19}$$

for every $z \in Z$, with equality when $z = z^*(x, t)$.

Taking $z = z(\tilde{x}(t), t)$ in (3.19) and using the resulting inequality in (3.18), we get

$$E_{x,s}\{V(\tilde{x}(S^* \wedge T), S^* \wedge T)[\exp[-\alpha(S^* \wedge T - s)]]\}$$
$$\geq V(x, s) - E_{x,s}\left\{\int_s^{S^* \wedge T} \{\exp[-\alpha(t-s)]\right.$$
$$\left.\times h(\tilde{x}(t), t, y^*(\tilde{x}(t), t), z(\tilde{x}(t), t))\}\, dt\right\}$$

with equality when $z = z^*(x, t)$, $\tilde{x}(t) = x^*(t)$.

Noticing that

$$V(\tilde{x}(S^* \wedge T), S^* \wedge T) \leq \chi_{S^* < T} g_1(\tilde{x}(S^*), S^*) + \chi_{T < S^*} g_2(\tilde{x}(T), T)$$

with equality when $z = z^*(x, t)$, $T = T^*$, the second inequality in (3.8) follows.

In order to justify (3.18) rigorously, we use mollifiers as in the proof of Theorem 16.4.1.

Remark 1. In Theorem 3.1 the differential game takes place in whole x-space R^n. The same methods apply as well in case the space variable x is restricted to a domain Ω. If Ω is a bounded domain, one requires that $\partial\Omega$ is in $C^{2+\rho}$ $(0 < \rho \leq 1)$, whereas if Ω is an unbounded domain, one requires that $\Omega \in \mathcal{C}^{2+\rho}$.

Remark 2. If $g_2(x, t) - g_1(x, t) = g(x)$, then we can use the method of proof of Theorem 10.1 instead of Theorem 11.1. We then find that

$$V \in L^\infty(0, T_0; W^{2,p,\mu}), \qquad \frac{\partial V}{\partial t} \in L^\infty(0, T_0; W^{0,p,\mu}). \tag{3.20}$$

Remark 3. Theorem 3.1 extends to the case where there is only one player. Thus, the variable y does not appear in (3.1), and (3.3) is replaced by the cost functional

$$P_{x,s}(z, T) = E_{x,s}\left\{\int_s^T e^{-\alpha(t-s)} h(\xi(t), t, z)\, dt + e^{-\alpha(T-s)} g_2(\xi(T), T)\right\}$$

where T is any stopping time with range in $[s, T_0]$. In this case, the value V satisfies (3.20).

4. Stochastic differential games with partial observation

We shall consider a zero-sum two-player stochastic differential game with partial observation. The dynamical system is given by n stochastic differential equations

$$d\xi = f(\xi, t, y, z)\, dt + \sigma(\xi, t)\, dw \qquad (s < t < T) \tag{4.1}$$

with initial condition

$$\xi(s) = x. \tag{4.2}$$

As in Section 2, the control sets Y, Z are compact subsets of some euclidean spaces R^p and R^q, respectively. A payoff is given by

$$P(y, z) = E_{x,s}\left\{\int_s^\tau h(\xi, t, y, z)\, dt + g(\xi(\tau), \tau)\right\} \tag{4.3}$$

where τ is the exit time of $(\xi(t), t)$ $(t > s)$ from Q_T. The player y wants to maximize the payoff, while the player z wants to minimize it.

If y and z make perfect observations, and if they use only pure strategies, then the existence of a saddle point follows by Theorem 2.3. Suppose now that y and z, at time t, can only observe a quantity $\eta(t)$, and suppose, further, that the manner by which $\eta(t)$ is related to $\xi(t)$ is known to have the form

$$d\eta = \tilde{f}(\xi, \eta, t, y, z)\, dt + \sigma(\xi, \eta, t)\, d\tilde{w},$$

where $\tilde{w}$ is a Brownian motion independent of w. We then consider the pair $\zeta = (\eta, \xi)$ as defining a diffusion process, governed by stochastic differential equations. With respect to this system, the players y and z observe a certain number of components of ζ, namely the components of η. The above setting is thus equivalent (with a different notation) to the following one:

The dynamics of the game is given by (4.1), and the players y, z observe just the first l components $\xi_1, \ldots, \xi_l$ of $\xi = (\xi_1, \ldots, \xi_n)$.

Set

$$\hat{\xi} = (\xi_1, \ldots, \xi_l), \qquad \hat{\hat{\xi}} = (\xi_{l+1}, \ldots, \xi_n),$$

so that $\xi = (\hat{\xi}, \hat{\hat{\xi}})$. We define a *pure strategy* for y as a measurable function $y = y(\hat{\xi}, t)$ from $R^l \times [s, T]$ into Y, and a *pure strategy* for z as a measurable function $z = z(\hat{\xi}, t)$ from $R^l \times [s, T]$ into Z.

As in Section 2, under some assumptions on f, σ the payoff (4.3) corresponding to the solution of (4.1), (4.2) with Lipschitz continuous $y = y(\hat{\xi}, t)$, $z = z(\hat{\xi}, t)$ can be given as follows: If

$$\frac{\partial\psi}{\partial t} + \frac{1}{2}\sum_{i,j=1}^n a_{ij}(x, t)\frac{\partial^2\psi}{\partial x_i\, \partial x_j} + f(x, t, y(\hat{x}, t), z(\hat{x}, t))\cdot D_x\psi$$

$$+ h(x, t, y(\hat{x}, t), z(\hat{x}, t)) = 0 \qquad \text{in } Q_T, \tag{4.4}$$

$$\psi = g \qquad \text{on } \Gamma_T, \tag{4.5}$$

then

$$P(y, z) = \psi(x, s). \tag{4.6}$$

Here $x = (x_1, \ldots, x_n)$, $\hat{x} = (x_1, \ldots, x_l)$.

From now on we *define* the payoff by (4.6), for any (measurable) pure strategies $y = y(\hat{x}, t)$, $z = z(\hat{x}, t)$.

We can define the concept of "a saddle point in pure strategies" as in Section 2. However, there is no simple connection between such saddle points and solutions of parabolic equations of the type (2.18). This makes it much more difficult to try to prove the existence of a saddle point in pure strategies. There is also an intuitive reason why one should not expect, in general, the existence of a saddle point in pure strategies: In the lack of perfect observation, it seems likely that each player should make use of all the past history of the game, not just the present state.

We shall now develop a model based on the partial observation of the whole past.

Let m be any positive integer, and let $\delta = (T - s)/m$. Denote by I_j the interval $t_{j-1} < t \leq t_j$, where $t_j = s + j\delta$. Denote by Y_j (Z_j) the set of all measurable functions $y_j(\hat{x}, t)$ $(z_j(\hat{x}, t))$ from $R^n \times I_j$ into $Y(Z)$. An *upper δ-strategy* Γ^δ for y is a vector

$$\Gamma^\delta = (\Gamma^{\delta, 1}, \ldots, \Gamma^{\delta, m}),$$

where $\Gamma^{\delta, j}$ is a map from

$$Z_1 \times Y_1 \times \cdots \times Z_{j-1} \times Y_{j-1} \times Z_j$$

into Y_j. A *lower δ-strategy* Δ_δ for z is a vector

$$\Delta_\delta = (\Delta_{\delta, 1}, \ldots, \Delta_{\delta, m}),$$

where $\Delta_{\delta, 1}$ is an element of Z_1, and $\Delta_{\delta, j}$ $(j \geq 2)$ is a map from

$$Z_1 \times Y_1 \times \cdots \times Z_{j-1} \times Y_{j-1}$$

into Z_j.

We shall assume:

(C′) $f(x, t, y, z)$ and $h(x, t, y, z)$ are continuous functions in $R^n \times [s, T] \times Y \times Z$, $\partial\Omega \in C^{2+\alpha}$ for some $\alpha \in (0, 1]$, and $g \in C^{2+\alpha, 1+\alpha}(\Gamma_T)$.

Any pair $(\Delta_\delta, \Gamma^\delta)$ defines a unique pair of pure strategies $(y^\delta(\hat{x}, t), z_\delta(\hat{x}, t))$, called the *outcome* of $(\Delta_\delta, \Gamma^\delta)$. If (A_1), (A_2), and (C′) hold, then there is a unique solution ψ^δ of (4.4), (4.5), when $y = y^\delta(\hat{x}, t)$, $z = z_\delta(\hat{x}, t)$ and a payoff

$$P(y^\delta, z_\delta) = \psi^\delta(x, s).$$

We denote this payoff also by $P[\Delta_\delta, \Gamma^\delta]$, or

$$P[\Delta_{\delta, 1}, \Gamma^{\delta, 1}, \ldots, \Delta_{\delta, m}, \Gamma^{\delta, m}].$$

The above scheme of corresponding a payoff $P[\Delta_\delta, \Gamma^\delta]$ to each pair $(\Delta_\delta, \Gamma^\delta)$ is called an *upper δ-game*, and is denoted by G^δ. The *upper δ-value* V^δ of this upper δ-game, is defined by

$$V^\delta = \inf_{\Delta_{\delta, 1}} \sup_{\Gamma^{\delta, 1}} \cdots \inf_{\Delta_{\delta, m}} \sup_{\Gamma^{\delta, m}} P[\Delta_{\delta, 1}, \Gamma^{\delta, 1}, \ldots, \Delta_{\delta, m}, \Gamma^{\delta, m}]. \tag{4.7}$$

Similarly, we define lower δ-game G_δ and lower δ-value V_δ. Here, y uses lower δ-strategies Γ_δ and z uses upper δ-strategies Δ^δ. Thus

$$V_\delta = \sup_{\Gamma_{\delta,1}} \inf_{\Delta^{\delta,1}} \cdots \sup_{\Gamma_{\delta,m}} \inf_{\Delta^{\delta,m}} P[\Gamma_{\delta,1}, \Delta^{\delta,1}, \ldots, \Gamma_{\delta,m}, \Delta^{\delta,m}].$$

One can show that

$$V^\delta = \inf_{z_1 \in Z_1} \sup_{y_1 \in Y_1} \cdots \inf_{z_m \in Z_m} \sup_{y_m \in Y_m} P(z_1, y_1, \ldots, z_m, y_m) \tag{4.8}$$

where $P(z_1, y_1, \ldots, z_m, y_m)$ stands for $P(y, z)$ when $y = y_j$, $z = z_j$ if $t \in I_j$. A similar formula holds for V_δ. Using these formulas one can easily verify

$$V^\delta \geqslant V_\delta, \tag{4.9}$$

$$V^\delta \geqslant V^{\delta'}, \quad V_\delta \leqslant V_{\delta'} \quad \text{if } \delta = \frac{T-s}{m}, \quad \delta' = \frac{T-s}{m'}, \quad m \text{ divides } m'. \tag{4.10}$$

The pair of sequences

$$G = (\{G^\delta\}, \{G_\delta\}) \qquad \left(\delta = \frac{T-s}{m}, \quad m = 1, 2, \ldots\right)$$

is called the *stochastic differential game with partial observation* associated with (4.1)–(4.3). If

$$V^+ = \lim_{\delta \to 0} V^\delta, \qquad V^- = \lim_{\delta \to 0} V_\delta$$

exist, we call them the *upper value* and the *lower value* of the game. If $V^+ = V^-$, then we say that the game has *value* V, where $V = V^+ = V^-$.

A sequence $\Gamma = \{\Gamma_\delta\}$ is called a *strategy* for y. Similarly, a sequence $\Delta = \{\Delta_\delta\}$ is called a strategy for z. Each pair $(\Delta_\delta, \Gamma_\delta)$ determines an outcome (y_δ, z_δ) and the corresponding solution ψ_δ of (4.4), (4.5). Suppose there exists a subsequence $\{\delta'\}$ of $\{\delta\}$ such that, as $\delta' \to 0$,

$$y_{\delta'}(\hat{x}, t) \to \bar{y}(\hat{x}, t) \quad \text{weakly in} \quad L^1(R^p \times (s, T)), \tag{4.11}$$

$$z_{\delta'}(\hat{x}, t) \to z(\hat{x}, t) \quad \text{weakly in} \quad L^1(R^q \times (s, T)), \tag{4.12}$$

$$\psi_{\delta'}(x, t) \to \bar{\psi}(x, t) \quad \text{for each} \quad (x, t) \in \bar{Q}_T, \tag{4.13}$$

where $\bar{y}(\hat{x}, t) \in Y$, $\bar{z}(\hat{x}, t) \in Z$ almost everywhere, and $\bar{\psi}$ is the solution of (4.4), (4.5) corresponding to $y = \bar{y}$, $z = \bar{z}$. Then we say that $(\bar{y}, \bar{z})$, or $(\bar{y}, \bar{z}, \bar{\psi})$ is an *outcome* of (Δ, Γ). The set of all numbers $\bar{\psi}(x, s)$, when $(\bar{y}, \bar{z}, \bar{\psi})$ varies over the set of all outcomes of (Δ, Γ), is called the *payoff set* of (Δ, Γ), and is denoted by $P[\Delta, \Gamma]$.

Given two sets of real numbers, A and B, we write $A \leqslant B$ if $a \leqslant b$ for all $a \in A$, $b \in B$. We write $A \leqslant B$ also in case A is empty or B is empty. Suppose the value V exists, and let Δ^*, Γ^* be strategies such that

$$P[\Delta^*, \Gamma] \leqslant P[\Delta^*, \Gamma^*] = \{V\} \leqslant P[\Delta, \Gamma^*]$$

for all strategies Δ, Γ. Then we call (Δ^*, Γ^*) a *saddle point*.

To every pure strategy $\tilde{y}(\hat{x}, t)$ we can correspond a *constant strategy* $\tilde{\Gamma}$ as follows:

$$\tilde{\Gamma} = \{\tilde{\Gamma}_\delta\}, \qquad \tilde{\Gamma}_\delta = (\tilde{\Gamma}_{\delta,1}, \ldots, \tilde{\Gamma}_{\delta,m}),$$

where $\hat{\Gamma}_{\delta,j}$ maps the whole space $Z_1 \times Y_1 \times \cdots \times Z_{j-1} \times Y_{j-1}$ into the point $\tilde{y}_j(\hat{x}, t)$, the restriction of $\tilde{y}(\hat{x}, t)$ to I_j. Using this correspondence, one can show that the saddle point in pure strategies established in Section 2 for a zero-sum two-person game, gives a saddle point in constant strategies—in the context of the present section.

We shall need the following condition:

(E) The controls y, z appear "separately" in f, h, i.e.,

$$f(x, t, y, z) = f^1(x, t, y) + f^2(x, t, z),$$

$$h(x, t, y, z) = h^1(x, t, y) + h^2(x, t, z).$$

Theorem 4.1. *Let the conditions* (A_1)–(A_4), (C'), *and* (E) *hold. Then the differential game with partial observation associated with* (4.1)–(4.3) *has value.*

Proof. It is sufficient to show that, for any $\epsilon > 0$, there is a $\delta_0 = \delta_0(\epsilon)$ such that

$$V^\delta - V_\delta \leqslant 3\epsilon \qquad \text{if} \quad \delta < \delta_0. \tag{4.14}$$

Indeed, it then follows that $V^{\delta'} - V_{\delta'} \leqslant 3\epsilon$ if $\delta' < \delta_0$. If $\delta = (T - s)/m$, $\delta' = (T - s)/m'$, $\delta'' = (T - s)/mm'$ then, by (4.9), (4.10),

$$V^\delta \geqslant V^{\delta''} \geqslant V_{\delta''} \geqslant V_{\delta'}.$$

Hence

$$V^{\delta'} - V^\delta \leqslant V^{\delta'} - V_{\delta'} < 3\epsilon.$$

Similarly $V^\delta - V^{\delta'} \leqslant \epsilon$. It follows that

$$|V^\delta - V^{\delta'}| < 3\epsilon \qquad \text{if} \quad \delta < \delta_0, \quad \delta' < \delta_0.$$

This implies that $V^+ = \lim_{\delta\to 0} V^\delta$ exists. Similarly one proves that $V^- = \lim_{\delta\to 0} V_\delta$ exists. Finally, from (4.14) it also follows that $V^+ = V^-$, so that the value exists.

In order to prove (4.14) we need the following lemma:

Lemma 4.2. *For any δ there exist an upper δ-strategy $\tilde{\Gamma}^\delta = (\tilde{\Gamma}^{\delta,1}, \ldots, \tilde{\Gamma}^{\delta,m})$ and an upper δ-strategy $\tilde{\Delta}^\delta = (\tilde{\Delta}^{\delta,1}, \ldots, \tilde{\Delta}^{\delta,m})$ such that that*

$$V^\delta \leqslant P[\Delta_\delta, \tilde{\Gamma}^\delta] + \epsilon \qquad \textit{for all} \quad \Delta_\delta, \tag{4.15}$$

$$V_\delta \geqslant P[\Gamma_\delta, \tilde{\Delta}^\delta] - \epsilon \qquad \textit{for all} \quad \Gamma_\delta. \tag{4.16}$$

To prove (4.15), one constructs the components of $\tilde{\Gamma}^{\delta}$: $\tilde{\Gamma}^{\delta,m}$, $\tilde{\Gamma}^{\delta,m-1}$, ... step by step, using the formula (4.8). The proof can be found in Friedman [3].

Fix an element $\bar{z}$ in Z_1, and consider the following game of G^{δ}: z chooses $z_1 = \bar{z}$. Then y chooses $y_1 = \Gamma^{\delta,1}z_1$ on I_1. In general, setting $z_k^{\tau}(\hat{x}, t) = z_k(\hat{x}, t + \delta)$ in I_{k-1}, we take

$$z_j(\hat{x}, t) = \tilde{\Delta}^{\delta,j-1}(y_1, z_2^{\tau}, \ldots, y_{j-2}, z_{j-1}^{\tau}, y_{j-1})(\hat{x}, t - \delta) \qquad \text{for} \quad t \in I_j,$$
$$y_j(\hat{x}, t) = \tilde{\Gamma}^{\delta,j}(z_1, y_1, \ldots, z_{j-1}, y_{j-1}, z_j)(\hat{x}, t) \qquad \text{for} \quad t \in I_j. \tag{4.17}$$

Denote by $y^{\delta}(\hat{x}, t)$, $z_{\delta}(\hat{x}, t)$ the control functions thus defined. Let $\hat{\Delta}_{\delta}$ be the constant lower δ-strategy for $z_{\delta}(\hat{x}, t)$. Applying (4.15) with $\Delta_{\delta} = \hat{\Delta}^{\delta}$, we get

$$V^{\delta} \leqslant P(y^{\delta}, z_{\delta}) + \epsilon. \tag{4.18}$$

We shall compare the above upper δ-game with the following lower δ-game. First y chooses on I_1 the restriction $y_1(\hat{x}, t)$ of $y^{\delta}(\hat{x}, t)$. Then z chooses $\zeta_1 = \tilde{\Delta}^{\delta,1}y_1$ for $t \in I_1$. In general,

$$\begin{aligned} &y \text{ chooses, for } t \in I_j, \text{ the restriction } \quad y_j(\hat{x}, t) \text{ of } y^{\delta}(\hat{x}, t), \\ &z \text{ chooses } \quad \zeta_j(\hat{x}, t) = \tilde{\Delta}^{\delta,j}(y_1, \zeta_1, \ldots, y_{j-1}, \zeta_{j-1}, y_j)(\hat{x}, t) \quad \text{for } t \in I_j. \end{aligned} \tag{4.19}$$

Denote the control of z thus obtained by $z^{\delta}(\hat{x}, t)$. Let $\hat{\Gamma}_{\delta}$ be the constant lower δ-strategy for $y^{\delta}(\hat{x}, t)$. Applying (4.16) with $\Gamma_{\delta} = \hat{\Gamma}_{\delta}$, we get

$$V_{\delta} \geqslant P(y^{\delta}, z^{\delta}) - \epsilon. \tag{4.20}$$

One can easily verify that $\zeta_j(\hat{x}, t) = z_{j+1}^{\tau}(\hat{x}, t)$ for $t \in I_j$, where z_j is the restriction of z_{δ} to I_j. Consequently,

$$z_{\delta}(\hat{x}, t) = z^{\delta}(\hat{x}, t - \delta) \qquad \text{for} \quad s + \delta \leqslant t \leqslant T. \tag{4.21}$$

We shall need the following lemma:

Lemma 4.3. *Let the assumptions of Theorem 4.1 hold. Let $y_{\lambda}(\hat{x}, t)$, $z_{\lambda}(\hat{x}, t)$ be control functions for y and z, respectively, for each λ from a sequence $\{\lambda_m\}$, $\lambda_m \downarrow 0$ if $m \uparrow \infty$. Let $\tilde{z}_{\lambda}(\hat{x}, t)$ be a control function for z satisfying $\tilde{z}_{\lambda}(\hat{x}, t) = z_{\lambda}(\hat{x}, t - \lambda)$ for $s + \lambda \leqslant t \leqslant T$, $\lambda \in \{\lambda_m\}$. Denote by ψ_{λ} and $\tilde{\psi}_{\lambda}$ the solutions of (4.4), (4.5) corresponding to $(y_{\lambda}, z_{\lambda})$ and $(y_{\lambda}, \tilde{z}_{\lambda})$, respectively. Then there exists a function $\alpha(\lambda)$, independent of the particular controls y_{λ}, z_{λ}, $\tilde{z}_{\lambda}$, such that $\alpha(\lambda_m) \to 0$ if $m \to \infty$ and*

$$\max_{(x,t) \in \bar{Q}_T} |\tilde{\psi}_{\lambda}(x, t) - \psi_{\lambda}(x, t)| \leqslant \alpha(\lambda), \qquad \lambda \in \{\lambda_m\}. \tag{4.22}$$

If the lemma is valid then, by combining (4.18) with (4.20) and using (4.21) and the lemma, we obtain the assertion (4.14). Thus, in order to complete the proof of Theorem 4.1, it remains to prove the lemma.

Proof of Lemma 4.3. Let $G(x, t; y, \tau)$ $(t < \tau)$ be Green's function in Q_T for the parabolic operator

$$\frac{\partial u}{\partial t} + \frac{1}{2} \sum_{i,j=1}^{n} a_{ij}(x, t) \frac{\partial^2}{\partial x_i \, \partial x_j}.$$

In the subsequent estimates, it is convenient to introduce a Banach space $X = L^r(\Omega)$, $r > n$, and the linear unbounded operator

$$A(t) = \frac{1}{2} \sum_{i,j=1}^{n} a_{ij}(x, t) \frac{\partial^2}{\partial x_i \, \partial x_j}.$$

The domain of $A(t)$ is $D_A = W^{2,r}(\Omega) \cap W_0^{1,2}(\Omega)$ and its range is in X. We also introduce the operator

$$U(t, \tau) = G(\,\cdot\,, t; \,\cdot\,, \tau),$$

which is a bounded operator in X; it is called the *fundamental solution* of $\partial/\partial t + A(t)$.

We shall denote by $\| \; \|$ both the norm in X and the norm of bounded linear operators in X.

We may assume that the resolvent of $A(t)$ exists for all λ with $\operatorname{Re} \lambda \geq 0$, for otherwise we first perform a transformation $\psi \to e^{\beta t}\psi$, where β is a suitable constant. But then, by Friedman [2], for $s \leq t < \sigma \leq T$,

$$\|A^{\theta}(t)U(t, \sigma)\| \leq \frac{C}{(\sigma - t)^{\theta}} \qquad (0 \leq \theta < 1). \tag{4.23}$$

Next, using the identity

$$U(t, \sigma + \lambda)x - U(t, \sigma)x = U(t, \sigma + \lambda) \int_{\sigma}^{\sigma+\lambda} A(\xi) U(\xi, \sigma)x \, d\xi$$

for $x \in D_A$ (see Friedman [2, p. 250]) and estimates on U given in Friedman [2, Section 2.14], we find that for $s \leq t < \sigma < \sigma + \lambda \leq T$

$$\|A^{\theta}(t)[U(t, \sigma + \lambda) - U(t, \sigma)]\| \leq C \frac{\lambda^{\rho - \theta}}{(\sigma - t)^{\rho'}} \qquad (0 < \theta < \rho < \rho' < 1); \tag{4.24}$$

here and in what follows, various different constants are denoted by the same symbol C.

Set $\phi_\lambda = \tilde{\psi}_\lambda - \psi_\lambda$. Then, with $\phi_\lambda(t) = \phi_\lambda(\cdot, t)$,

$$\begin{aligned} \frac{d\phi_\lambda}{dt} + A(t)\phi_\lambda = &- [f^1(x, t, y_\lambda(\hat{x}, t)) \cdot D_x\phi_\lambda] \\ &- [f^2(x, t, \tilde{z}_\lambda(\hat{x}, t)) \cdot D_x\tilde{\psi}_\lambda - f^2(x, t, z_\lambda(\hat{x}, t)) \cdot D_x\psi_\lambda] \\ &- [h^2(x, t, \tilde{z}_\lambda(\hat{x}, t)) - h^2(x, t, z_\lambda(\hat{x}, t))] \\ \equiv &- B_1 - B_2 - B_3. \end{aligned} \tag{4.25}$$

We shall write $B_i(t) = B_i(\cdot, t)$.

Suppose $y_\lambda(\hat{x}, t)$, $z_\lambda(\hat{x}, t)$, $\tilde{z}_\lambda(\hat{x}, t)$, f, and h are all continuously differentiable. By Lemma 1.3, $D_x\psi_\lambda$, $D_x\tilde{\psi}_\lambda$ are then uniformly Hölder continuous in Q_T. Hence, by Friedman [2, p. 109],

$$-\phi_\lambda(t) = \int_t^T U(t, \sigma)B_1(\sigma)\, d\sigma + \int_t^T U(t, \sigma)B_2(\sigma)\, d\sigma + \int_t^T U(t, \sigma)B_3(\sigma)\, d\sigma$$
$$\equiv \Phi_1 + \Phi_2 + \Phi_3.$$

We shall estimate the Φ_i. First, for any $0 < \theta < 1$,

$$\|A^\theta(t)\Phi_1(t)\| \leq C\int_t^T \|A^\theta(t)U(t, \sigma)\|\, \|D_x\phi_\lambda(\sigma)\|\, d\sigma$$
$$\leq C\int_t^T \frac{\|D_x\phi_\lambda(\sigma)\|}{(\sigma - t)^\theta}\, d\sigma. \tag{4.26}$$

Since (see Friedman [2, p. 179])

$$\|D_x\phi_\lambda(\sigma)\| \leq C\|A^\theta(\sigma)\phi_\lambda(\sigma)\| \qquad \text{if} \quad \tfrac{1}{2} < \theta < 1, \tag{4.27}$$

we get

$$\|A^\theta(t)\Phi_1(t)\| \leq C\int_t^T \frac{\|A^\theta(\sigma)\phi_\lambda(\sigma)\|}{(\sigma - t)^\theta}\, d\sigma \qquad \text{if} \quad \tfrac{1}{2} < \theta < 1. \tag{4.28}$$

By Lemma 1.3,

$$|D_x\psi_\lambda|_{\alpha, Q_T} \leq C, \qquad |D_x\tilde{\psi}_\lambda|_{\alpha, Q_T} \leq C, \tag{4.29}$$

$$\underset{Q_T}{\text{l.u.b.}}\, |D_x\psi_\lambda(x, t)| \leq C, \qquad \underset{Q_T}{\text{l.u.b.}}\, |D_x\tilde{\psi}_\lambda(x, t)| \leq C. \tag{4.30}$$

To estimate Φ_3, write

$$\Phi_3 = \left[\int_{t+\lambda}^T U(t, \sigma)h^2(\cdot, \sigma, z_\lambda(\cdot, \sigma - \lambda))\, d\sigma - \int_t^{T-\lambda} U(t, \sigma)h^2(\cdot, \sigma, z_\lambda(\cdot, \sigma))\, d\sigma\right]$$
$$+ \int_t^{t+\lambda} U(t, \sigma)h^2(\cdot, \sigma, \tilde{z}_\lambda(\cdot, \sigma))\, d\sigma - \int_{T-\lambda}^T U(t, \sigma)h^2(\cdot, \sigma, z_\lambda(\cdot, \sigma))\, d\sigma$$
$$\equiv I_1 + I_2 - I_3. \tag{4.31}$$

We can write

$$I_1 = \int_t^{T-\lambda} [U(t, \sigma + \lambda) - U(t, \sigma)]h^2(\cdot, \sigma + \lambda, z_\lambda(\cdot, \sigma))\, d\sigma$$
$$+ \int_t^{T-\lambda} U(t, \sigma + \lambda)[h^2(\cdot, \sigma + \lambda, z_\lambda(\cdot, \sigma)) - h^2(\cdot, \sigma, z_\lambda(\cdot, \sigma))]\, d\sigma$$
$$\equiv I_{11} + I_{12}.$$

By (4.24),

$$\|A^{\theta}(t)I_{11}\| \leq C\int_t^{T-\lambda} \frac{\lambda^{\rho-\theta}}{(\sigma-t)^{\rho'}}\|h^2(\cdot,\sigma+\lambda,z_\lambda(\cdot,\sigma))\|\,d\sigma \leq C\lambda^{\rho-\theta}$$

if $\theta < \rho < \rho' < 1$. We also have

$$\|A^{\theta}(t)I_{12}\| \leq C\epsilon(\lambda), \qquad \epsilon(\lambda)\to 0 \quad \text{if} \quad \lambda\to 0,$$

where $\epsilon(\lambda)$ depends on the modulus of continuity of $h^2(x, t, z)$ with respect to t. Hence,

$$\|A^{\theta}(t)I_1\| \leq C\lambda^{\rho-\theta} + C\epsilon(\lambda). \tag{4.32}$$

Next,

$$\|A^{\theta}(t)I_2\| \leq \int_t^{t+\lambda} \frac{C}{(\sigma-t)^{\theta}}\,d\sigma \leq C\lambda^{1-\theta}.$$

Similarly $\|A^{\theta}(t)I_3\| \leq C\lambda^{1-\theta}$. We conclude that

$$\|A^{\theta}(t)\Phi_3\| \leq C\lambda^{\rho-\theta} + C\epsilon(\lambda) \qquad \text{for any} \quad 0<\theta<\rho<1. \tag{4.33}$$

Next,

$$\begin{aligned}\Phi_2 &= \int_t^T U(t,\sigma)[f^2(x,\sigma,\tilde{z}_\lambda(\hat{x},\sigma))\cdot D_x\tilde{\psi}_\lambda(x,\sigma) - f^2(x,\sigma,z_\lambda(\hat{x},\sigma))\\ &\quad \cdot D_x\tilde{\psi}_\lambda(x,\sigma)]\,d\sigma + \int_t^T U(t,\sigma)[f^2(x,\sigma,z_\lambda(x,\sigma))\cdot D_x\phi_\lambda(x,\sigma)]\,d\sigma\\ &\equiv \Phi_{21}+\Phi_{22}.\end{aligned} \tag{4.34}$$

As for Φ_{22}, we have, by (4.27),

$$\|A^{\theta}(t)\Phi_{22}\| \leq \int_y^T \frac{C}{(\sigma-t)^{\theta}}\|D_x\phi_\lambda(\sigma)\|\,d\sigma \leq C\int_t^T \frac{\|A^{\theta}(\sigma)\phi_\lambda(\sigma)\|}{(\sigma-t)^{\theta}}\,d\sigma$$

if $\frac{1}{2} < \theta < 1$. As for $A^{\theta}(t)\Phi_{21}$, it can be estimated in the same manner as $A^{\theta}(t)\Phi_3$. Here we make use of (4.29), (4.30). The inequality we get is

$$\|A^{\theta}(t)\Phi_{31}\| \leq C\lambda^{\rho-\theta} + C\epsilon(\lambda) + C\lambda^{\alpha}.$$

We conclude that

$$\|A^{\theta}(t)\Phi_2\| \leq C\lambda^{\rho-\theta} + C\epsilon(\lambda) + C\lambda^{\alpha} + C\int_t^T \frac{\|A^{\theta}(\sigma)\phi_\lambda(\sigma)\|}{(\sigma-t)^{\theta}}\,d\sigma. \tag{4.35}$$

Combining this with (4.33), (4.28), and (4.25), and setting

$$\gamma_\lambda(t) = \|A^{\theta}(t)\phi_\lambda(t)\|, \qquad \beta(\lambda) = \min\{\epsilon(\lambda), \lambda^{\rho-\theta}, \lambda^{\alpha}\},$$

we get

$$\gamma_\lambda(t) \leq C\beta(\lambda) + C\int_t^T \frac{\gamma_\lambda(\sigma)}{(\sigma-t)^{\theta}}\,d\sigma.$$

By iteration we find that

$$\gamma_\lambda(t) \leqslant \beta^*(\lambda), \qquad \beta^*(\lambda) \to 0 \quad \text{if} \quad \lambda \to 0,$$

i.e.,

$$\|A^\theta(t)\phi_\lambda(t)\| \leqslant \beta^*(\lambda). \tag{4.36}$$

In deriving (4.36) we have assumed that y_λ, z_λ, $\tilde{z}_\lambda$, f, and h are continuously differentiable. However, the function $\beta^*(\lambda)$ occurring in (4.36) depends only on the constants which enter into the conditions (A_1)–(A_4) and on bounds and moduli of continuity of f, h. Hence, by approximating y_λ, z_λ, $\tilde{z}_\lambda$, f, and h by smooth functions and applying (4.36) to each of the corresponding ϕ_λ, we conclude that (4.36) holds in general.

Since (by Friedman [2], for instance)

$$|\phi_\lambda(x, t)| \leqslant C\|A^\theta(t)\phi_\lambda(\cdot, t)\| \qquad \text{if} \quad r > n, \quad \tfrac{1}{2} < \theta < 1,$$

the assertion of the lemma follows from (4.36).

We shall now prove the existence of a saddle point under the additional condition:

(F) $f(x, t, y, z)$ and $h(x, t, y, z)$ are linear functions of y, z, i.e.,

$$f(x, t, y, z) = f^0(x, t) + F^1(x, t)y + F^2(x, t)z,$$
$$h(x, t, y, z) = h^0(x, t) + h^1(x, t) \cdot y + h^2(x, t) \cdot z,$$

and Y, Z are convex sets.

Theorem 4.4. *Let the conditions of Theorem* 4.1 *hold and let* (F) *hold. Then there exists a saddle point for the game of partial observation associated with* (4.1)–(4.3).

We shall need the following lemma:

Lemma 4.5. *Let the assumptions of Theorem* 4.4 *hold. Then, for any strategies* Δ, Γ, *the payoff set* $P[\Delta, \Gamma]$ *is nonempty.*

Proof. Suppose first that $g \equiv 0$.

Let $y_m(\hat{x}, t)$, $z_m(\hat{x}, t)$ be pure strategies and let $\psi_m(x, t)$ be the corresponding solution of (4.4), (4.5). Since the sets of all pure strategies for y and z are bounded convex and weakly closed in $L^1(R^p \times [s, T])$ and $L^1(R^q \times [s, T])$, respectively, it follows that

$$y_{m'} \to \bar{y} \qquad \text{weakly in} \quad L^1(R^p \times [s, T]),$$
$$z_{m'} \to \bar{z} \qquad \text{weakly in} \quad L^1(R^q \times [s, T])$$

for some subsequence $\{m'\}$, where $\bar{y}$, $\bar{z}$ are pure strategies.

By Lemma 1.3, $|\psi_m(x, t)| \le C$ and

$$|\psi_m(x, t) - \psi_m(x', t')| + |D_x\psi_m(x, t) - D_{x'}\psi_m(x', t')| \le C(|x - x'|^\alpha - |t - t'|^{\alpha/2}).$$

Hence, by the Ascoli–Arzela lemma, there is a subsequence of $\{m'\}$, which we again denote by $\{m'\}$, such that

$$\psi_{m'}(x, t) \to \bar{\psi}(x, t), \qquad D_x\psi_{m'}(x, t) \to D_x\bar{\psi}(x, t)$$

uniformly in Q_T, for some function $\bar{\psi}$. We can write

$$- \psi_m(x, t) = \int_t^T U(t, \sigma)[f^0(x, \sigma) + F^1(x, \sigma)y_m(\hat{x}, \sigma) + F^2(x, \sigma)z_m(\hat{x}, \sigma)] \cdot D_x\psi_m(x, \sigma)\, d\sigma + \int_t^T U(t, \sigma)[h^0(x, \sigma) + h^1(x, \sigma) \cdot y_m(\hat{x}, \sigma) + h^2(x, \sigma) \cdot z_m(\hat{x}, \sigma)]\, d\sigma.$$

Taking $m = m' \to 0$, we find that

$$- \bar{\psi}(x, t) = \int_t^T U(t, \sigma)[f^0(x, \sigma) + F^1(x, \sigma)\bar{y}(\hat{x}, \sigma) + F^2(x, \sigma)\bar{z}(\hat{x}, \sigma)] \cdot D_x\bar{\psi}(x, \sigma)\, d\sigma + \int_t^T U(t, \sigma)[h^0(x, \sigma) + h^1(x, \sigma) \cdot \bar{y}(\hat{x}, \sigma) + h^2(x, \sigma) \cdot \bar{z}(\hat{x}, \sigma)]\, d\sigma. \tag{4.37}$$

Now, the solution of (4.4), (4.5) (when $g = 0$) corresponding to $\bar{y}$, $\bar{z}$ also satisfies the integral equation (4.37). Further, from the estimates (4.23), (4.27) we can deduce that there is at most one solution $\bar{\psi}$ of (4.37). It follows that $\bar{\psi}$ is the solution of (4.4), (4.5) corresponding to $\bar{y}$, $\bar{z}$. Thus the set $P[\Delta, \Gamma]$ is nonempty, and contains the point $\bar{\psi}(x, s)$.

So far we have assumed that $g \equiv 0$. If $g \not\equiv 0$, then we apply the preceding proof to $\psi_m - \hat{g}$, where $\hat{g}$ is a smooth extension of g into $\bar{Q}_T$.

Proof of Theorem 4.4. By a variant of Lemma 4.2, for any δ there exists a lower σ-strategy Δ_δ^* such that

$$V^\delta > P[\Delta_\delta^*, \Gamma^\delta] - \delta \qquad \text{for any } \Gamma^\delta. \tag{4.38}$$

Similarly, there exists a lower δ-strategy Γ_δ^* such that

$$V_\delta < P[\Gamma_\delta^*, \Delta^\delta] + \delta \qquad \text{for any } \Delta^\delta. \tag{4.39}$$

Set

$$\Delta^* = \{\Delta_\delta^*\}, \qquad \Gamma^* = \{\Gamma_\delta^*\}.$$

We shall prove that (Δ^*, Γ^*) is a saddle point.

Let $\Gamma = \{\Gamma_\delta\}$ be any strategy for y. Denote by (y_δ, z_δ) the outcome of $(\Delta_\delta^*, \Gamma_\delta)$. Since Γ_δ may be viewed as an upper δ-strategy, (4.38) gives

$$P(y_\delta, z_\delta) < V^\delta + \delta. \tag{4.40}$$

Let $\{\delta'\}$ be any subsequence of $\{\delta\}$ such that

$$\begin{aligned} &y_{\delta'} \to \bar{y} \qquad \text{weakly in} \quad L^1(R^p \times [s, T]),\\ &z_{\delta'} \to \bar{z} \qquad \text{weakly in} \quad L^1(R^q \times [s, T]),\\ &\psi_{\delta'}(x, s) \to \bar{\psi}(x, s), \end{aligned} \tag{4.41}$$

where $\psi_{\delta'}$ and $\bar{\psi}$ are the solutions of (4.4), (4.5) corresponding to $(y_{\delta'}, z_{\delta'})$ and $(\bar{y}, \bar{z})$ respectively. Notice that the last relation in (4.41) gives

$$P(y_{\delta'}, z_{\delta'}) \to P(\bar{y}, \bar{z}).$$

Hence, by (4.40) and Theorem 4.1,

$$P(\bar{y}, \bar{z}) \leq V.$$

We have thus proved that $P[\Delta^*, \Gamma] \leq V$ for any Γ.

Similarly one shows that $P[\Delta, \Gamma^*] \geq V$ for any Δ. Since, by Lemma 4.5, the payoff set $P[\Delta^*, \Gamma^*]$ is nonempty, it follows that $P[\Delta^*, \Gamma^*] = \{V\}$. This completes the proof of Theorem 4.4.

PROBLEMS

1. If w is a solution of (2.13), then $w \leq 0$ in Q_T. [*Hint*: Approximate a_{ij}, b_i by smooth a_{ij}^k, b_i^k. Apply the maximum principle to the corresponding $w = w^k$, and take $k \to \infty$.]
2. Prove that there exists a solution of (3.15). [*Hint*: Write the parabolic equation as

$$\frac{\partial u}{\partial t} + Lu + F(x, t, u, u_x) = 0.$$

Write

$$\begin{aligned} F(x, t, u, p) = \sum [F(x, t, u, \hat{p}_i) - F(x, t, u, \hat{p}_{i-1})]\\ + [F(x, t, u, 0) - F(x, t, 0, 0)] + F(x, t, 0, 0) \end{aligned}$$

where $\hat{p}_i = (p_1, \ldots, p_{i-1}, p_i, 0, \ldots, 0)$. Then,

$$F(x, t, u, p) = \sum b_i^u(x, t) \frac{\partial u}{\partial x_i} + c^u(x, t)u + e(x, t)$$

where $|b_i^u(x, t)| \leq K$, $|c^u(x, t)| \leq K$, K independent of u. Define $w = Tu$

where

$$\frac{\partial w}{\partial t} + Lw + \sum b_i^u(x, t) \frac{\partial w}{\partial x_i} + c^u(x, t)w + e(x, t) = 0 \quad \text{in} \quad \Omega_m \times (0, T_0),$$

$$w(t) \in W_0^{1,2}(\Omega_m), \qquad w(T_0) = 0.$$

Apply Theorems 10.4.3, 10.4.4 to deduce that T maps a set $\{u; |u|_{W_p^{2,1}(Q_T)} < M\}$ into itself. Apply Lemma 1.3 to deduce compactness, and use Schauder's fixed point theorem.

3. Complete the proof that (3.11)–(3.14) has a unique solution.

4. Prove (4.8).

5. Prove (4.9).

6. Prove (4.10).

7. Let the conditions of Theorem 4.1 hold. Denote the value of the game (4.1)–(4.3) by $V(x, s)$. Prove that $V(x, s)$ is a continuous function in (x, s).

Bibliographical Remarks

Chapter 10. The material of Sections 1–3 is based on Friedman [1, 2]. The Schauder estimates and the L^p estimates for general elliptic equations have been proved by Agmon *et al.* [1]. The Sobolev inequality in the general form of Theorem 10.1 is due to Nirenberg [1] and Gagliardo [1, 2].

Chapter 11. The results of this chapter are due to Friedman [8]. A special case of Theorem 4.1 is due to Bonami *et al.* [1]. Problems 2–9 are based on Pinsky [1].

Chapter 12. Stability theorems have been proved by Khasminskii [1, 2], Kushner [1], Kozin and Prodromou [1]. Various concepts of stability are given by Khasminskii [2] and in Kushner [1]. The concept adopted here is that used by Pinsky [2]. The results of Section 1 are due to Pinsky [2]. The results of Section 2 are due to Friedman and Pinsky [2]. Theorem 3.1 is due to Khasminskii [1]; the present proof and Theorems 3.2, 3.3 are due to Pinsky [2]. The method of descent was established by Pinsky [2]. The results of Sections 5, 6, under some stricter conditions, were proved by Friedman and Pinsky [2]; in their present form they are due to Pinsky [2]. Section 7 is based on Friedman and Pinsky [1].

Chapter 13. The results of this chapter are essentially due to Friedman and Pinsky [3]. In that article, however, the condition (2.7) is replaced by a more restrictive condition; the present improvement is due to Pinsky [2]; it is based on the results of Section 1, on the method of descent and on Problems 6–9, which were communicated to us by Pinsky. Freidlin [1] studied the Dirichlet problem for degenerate elliptic equations. A typical result from his paper is described in Problem 5.

The Dirichlet problem for degenerate elliptic equations was also studied by probabilistic methods by Stroock and Varadhan [2]. For nonprobabilistic methods, see Kohn and Nirenberg [1] and the references given there. These nonprobabilistic methods require some "coercivity" and, therefore, they usually assume that the coefficient $c(x)$ of u satisfies $c(x) \leq -\alpha < 0$, where α is sufficiently large.

Chapter 14. The estimates of Theorem 2.2 and of Theorem 3.1 (in the special case $\Gamma = R^n$, $\Delta = \emptyset$) are due to Ventcel and Freidlin [1]. They also give in this paper the results of Sections 1, 4. The results of Sections 5, 6 are due to Varadhan [1, 2] in case $b(x) \equiv 0$, and to Friedman [9] in the general case. The exit problem (of Section 7) and Problem 18 are based on Ventcel and Freidlin [1]. Section 8 is due to Friedman [9]. The results of Sections 10, 11 are based on Friedman [5]. Ventcel [1, 2] has stated similar results. Lemma 10.2 is taken from Courant and Hilbert [1]. Other results on the behavior of the principal eigenvalue, as $\epsilon \to 0$, were obtained by nonprobabilistic methods, by Devinatz *et al.* [1].

Chapter 15. The results of this chapter are due to Friedman [10]. Some ideas of S. Itô [2] are used in Section 1.

Chapter 16. Stopping time problems were studied by Chernoff [1], Samuelson [1], McKean [1], Grigelionis and Shirayev [1], Dynkin and Yushkevich [1]. Bensoussan and Lions [1] were the first to consider the stopping time problem by variational inequalities. They worked only with L^2 estimates for parabolic variational inequalities. Friedman [6, 7] considered stochastic games and also proved the L^p regularity theorems for elliptic and parabolic variational inequalities in bounded or unbounded domains. The existence of a saddle point for the stationary case in a bounded domain was first proved by Krylov [1] (without variational inequalities). A stochastic game was also considered by Gusein-Zade [1].

For general regularity theorems for variational inequalities in bounded domains, see Lewy and Stampacchie [1], and Brezis [1]. Brezis [2] also considered variational inequalities with the convex set varying in t.

The results of Section 11 are due to Bensoussan and Friedman [1].

Van Moerbeke [1–3] has studied the nonstationary stopping time problem in case of one-dimensional Brownian motion, by reducing it to a problem of the Stefan type (cf. Problem 21). He studied the optimal cost function and obtained various results on the shape of the free boundary. Kotlow [1] considered one-dimensional Brownian motion with absorption and with time-independent cost function. He shows that the free boundary is monotone. Results of Lewy and Stampacchia [1] and of Kinderlehrer [1] yield information on the smoothness of the boundary of the continuation domain in the stationary stopping time problem for two-dimensional Brownian motion in a convex bounded domain: if $f \equiv 0$ and, roughly, $\Delta\psi_1 < 0$, then the free boundary is a smooth curve; if ψ_1 is analytic, so is the free boundary.

More recently Friedman and Kinderlehrer [1] have obtained results on the shape and smoothness of the free boundary for a class of parabolic variational inequalities in n-dimensions arising from the Stefan problem. The results of van Moerbeke [3] have recently been rederived and extended by Friedman [11] by methods of variational inequalities. Jensen [1] has recently derived a scheme to approximate the free boundary by polygonal curves.

Chapter 17. The results of Sections 2, 4 are due to Friedman [4]. The result of Theorem 2.3 in the case of one player was first established by Fleming [1]. Section 3 is due to Bensoussan and Friedman [1]. They also considered the case where $a_{ij} = \epsilon\delta_{ij}$, $\epsilon \downarrow 0$ and proved that the value $V(x, s)$ converges to a limit as $\epsilon \downarrow 0$.

A comprehensive review of the theory of stochastic control is given in Fleming [2]. For more recent results see Fleming [3], Davis and Varaiya [1], Duncan and Varaiya [1].

References

S. AGMON

[1] *Lectures on Elliptic Boundary Value Problems*. Van Nostrand Reinhold, Princeton, New Jersey, 1965.

S. AGMON, A. DOUGLIS, and L. NIRENBERG

[1] Estimates near the boundary for elliptic partial differential equations satisfying the general boundary condition (I), *Comm. Pure Appl. Math.* 12 (1959), 623–727.

D. G. ARONSON

[1] The fundamental solution of a linear parabolic equation containing a small parameter, *Illinois J. Math.* 3 (1959), 580–619.

[2] Non-negative solutions of linear parabolic equations, *Ann. Scuola Norm. Sup. Pisa* 22 (1968), 607–694.

A. BENSOUSSAN and S. FRIEDMAN

[1] Nonlinear variational inequalities and differential games with stopping times, *J. Functional Analysis* 16 (1974), 305–352.

A. BENSOUSSAN and J. L. LIONS

[1] Problèmes de temps d'arrét optimal et inéquations variationelles parabolic, *Applicable Analy* 3 (1973), 267–295.

A. BONAMI, N. KAROUI, B. ROYNETTE, and H. REINHARD

[1] Processus de diffusion associé à un opérateur elliptique dégénéré, *Ann. Inst. H. Poincaré Sect. B* 7 (1971), 31–80.

H. BREZIS

[1] Problèmes unilateraux, *J. Math. Pures Appl.* 51 (1972), 1–168.

[2] Un problème d'évolution avec contraintes unilatérales dépendant du temps, *C. R. Acad. Sci. Paris Sér. A-B* 274 (1972), 310–112.

H. BREZIS and G. STAMPACCHIA

[1] Sur la régularité de la solution d'inéquations elliptiques, *Bull Soc. Math. France* 96 (1968), 153–180.

H. CHERNOFF

[1] Optimal stochastic control, *Sankhyā Ser. A* 30 (1968), 221–252.

E. A. CODDINGTON and N. LEVINSON

[1] *Theory of Ordinary Differential Equations*. McGraw-Hill, New York, 1955.

R. COURANT and D. HILBERT

[1] *Methods of Mathematical Physics*, Vol. 2. Wiley (Interscience), New York, 1962.

M. H. A. DAVIS and P. VARAIYA

[1] Dynamic programming conditions for partially observable stochastic systems, *SIAM J. Control* 11 (1973), 226–261.

A. DEVINATZ, R. ELLIS, and A. FRIEDMAN
[1] The asymptotic behavior of the first real eigenvalue of second order elliptic operators with a small parameter in the highest derivatives, II. *Indiana Univ. Math. J.* **23** (1974), 991–1011.
T. DUNCAN and P. VARAIYA
[1] On the stochastic solutions of a stochastic control problem. *SIAM J. Control* **9** (1971), 354–371.
E. B. DYNKIN and A. A. YUSHKEVICH
[1] *Markov Processes*. Plenum, New York, 1969.
E. B. FABES and N. M. RIVIÈRE
[1] *Lp*-estimates near the boundary for solutions of the Dirichlet problem, *Ann. Scuola Norm. Sup. Pisa* **24** (1970), 491–553.
I. P. FILIPPOV
[1] On certain questions in the theory of optimal control, *SIAM J. Control* **1** (1962), 76–84. [Translated from *Vestnik Moscow Univ. Ser. Mat. Mech. Astr.* **2** (1959), 25–32.]
W. H. FLEMING
[1] Some Markovian optimization problems, *J. Math. Mech.* **12** (1963), 131–140.
[2] Optimal continuous-parameter stochastic control. *SIAM Rev.* **11** (1969), 470–509.
[3] Stochastic control for small noise intensities, *SIAM J. Control* **9** (1971), 473–517.
M. I. FREIDLIN
[1] On the smoothness of solutions of degenerate elliptic equations, *Math. USSR-Izv.* **2** (1968), 1337–1359 [*Izv. Akad. Nauk SSSR Ser. Mat.* 32 (1968), 1391–1413].
A. FRIEDMAN
[1] *Partial Differential Equations of Parabolic Type*. Prentice-Hall, Englewood Cliffs, New Jersey, 1964.
[2] *Partial Differential Equations*. Holt, New York, 1969.
[3] *Differential Games*. Wiley (Interscience), New York, 1971.
[4] Stochastic differential games, *J. Differential Equations* **11** (1972), 79–108.
[5] The asymptotic behavior of the first real eigenvalue of a second order elliptic operator with small parameter in the highest derivatives, *Indiana Univ. Math. J.* **22** (1973), 1005–1015.
[6] Stochastic games and variational inequalities, *Arch. Rational Mech. Anal.* **51** (1973), 321–346.
[7] Regularity theorems for variational inequalities in unbounded domains and applications to stopping time problems, *Arch. Rational Mech. Anal.* **52** (1973), 134–160.
[8] Non-attainability of a set by a diffusion process, *Trans. Amer. Math. Soc.* **197** (1974), 245–271.
[9] Small random perturbations of dynamical systems and application to parabolic equations. *Indiana Univ. Math. J.* **24** (1974), 533–553; Erratum, *ibid.* **25** (1975).
[10] Fundamental solutions for degenerate parabolic equations, *Acta Math.* **133** (1974), 171–217.
[11] Parabolic variational inequalities in one space dimension and smoothness of the free boundary, *J. Functional Analysis* **18** (1975), 151–176.
A. FRIEDMAN and D. KINDERLEHRER
[1] A one phase Stefan problem, *Indiana Univ. Math. J.* **25** (1975).
A. FRIEDMAN and M. A. PINSKY
[1] Asymptotic behavior of solutions of linear stochastic differential systems, *Trans. Amer. Math. Soc.* **181** (1973), 1–22.
[2] Asymptotic stability and spiraling properties of solutions of stochastic equations, *Trans. Amer. Math. Soc.* **186** (1973), 331–358.
[3] Dirichlet problem for degenerate elliptic equations, *Trans. Amer. Math. Soc.* **186** (1973), 359–383.

E. Gagliardo
[1] Propietà di alcune classi di funzioni in più variabili, *Ricerche Mat.* 7 (1958), 102–137.
[2] Ulteriori proprietà di alcune classi di funzioni in più variabili, *Richerche Mat.* 8 (1959), 24–51.

B. I. Grigelionis and A. N. Shirayev
[1] On Stefan's problem and optimal stopping rules for Markov processes, *Theor. Probability Appl.* 11 (1966), 541–558 [*Teor. Verojatnost. i Primenen.* 11 (1966), 612–631].

S. M. Gusein-Zade
[1] A certain game connected wtih a Wiener process, *Theor. Probability Appl.* 14 (1969), 701–704 [*Teor. Verojatnost. i Primenen.* 14 (1969), 732–735].

S. Itô
[1] The fundamental solution of the parabolic equation in a differentiable manifold, *Osaka J. Math.* 5 (1953). 75–92.
[2] Fundamental solutions of parabolic differential equations and boundary value problems, *Japan. J. Math.* 27 (1957), 55–102.

R. Jensen
[1] Finite difference approximation to the free boundary of a parabolic variational inequality, *to appear.*

R. Z. Khasminskii
[1] Necessary and sufficient conditions for the asymptotic stability of linear stochastic systems, *Theor. Probability Appl.* 12 (1967), 144–147 [*Teor. Verojatnost. i Primenen.* 12 (1967), 167–172].
[2] *Stability of Systems of Differential Equations under Random Perturbations of their Parameters*. Izdat. Nauka, Moscow, 1969.

D. Kinderlehrer
[1] The coincidence set of solutions of certain variational inequalities, *Arch. Rational Mech. Anal.* 40 (1971), 231–250.

J. J. Kohn and L. Nirenberg
[1] Degenerate elliptic-parabolic equations of second order, *Comm. Pure Appl. Math.* 20 (1967), 797–872.

D. B. Kotlow
[1] A free boundary problem associated with the optimal stopping problem for diffusion processes, *Trans. Amer. Math. Soc.* 184 (1973), 457–478.

F. Kozin and S. Prodromou
[1] Necessary and sufficient conditions for almost sure sample stability of linear Itô equations, *SIAM J. Appl. Math.* 21 (1971), 413–424.

M. A. Krasnoselskii
[1] *Positive Solutions of Operator Equations*. Groningen, Nordhoff, 1964.

N. V. Krylov
[1] Control of Markov processes and W-spaces, *Math. USSR-Izv.* 5 (1971), 233–266 [*Izv. Akad. Nauk SSSR Ser. Mat.* 35 (1971), 224–255].

H. Kushner
[1] *Stochastic Stability and Control*. Academic Press, New York, 1967.

D. A. Ladyzhenskaja, V. A. Solonnikov, and N. N. Uraltseva
[1] *Quasilinear Equations of Parabolic Type* (Amer. Math Soc. Transl. Vol. 23). Amer. Math. Soc., Providence, Rhode Island, 1968.

H. Lewy and G. Stampacchia
[1] On the regularity of the solution of a variational inequality, *Comm. Pure Appl. Math.* 22 (1969), 152–188.

H. P. McKean
[1] Appendix: A free boundary problem for the heat equation arising from a problem in mathematical economics, *Industrial Management Rev.* 6 (1965), 32–39.

L. NIRENBERG

[1] On elliptic partial differential equations, *Ann. Scuola Norm. Sup. Pisa* **13** (1959), 1–48.

M. A. PINSKY

[1] A note on degenerate diffusion processes, *Theor. Probability Appl.* **14** (1969), 502–506 [*Teor. Verojatnost. i Primenen.* **14** (1969), 522–527].

[2] Stochastic stability and the Dirichlet problem, *Comm. Pure Appl. Math.* **27** (1974), 311–350.

M. H. PROTTER and H. F. WEINBERGER

[1] On the spectrum of general second order operators, *Bull. Amer. Math. Soc.* **72** (1966), 251–255.

P. A. SAMUELSON

[1] Rational theory of warrant pricing, *Industrial Management Rev.* **6** (1965), 13–31.

V. A. SOLONNIKOV

[1] A priori estimates for equations of second order of parabolic type, *Trudy Mat. Inst. Steklov* **10** (1964), 133–212 [Amer. Math. Soc. Transl. Vol. 65 (1967), 51–137].

D. W. STROOCK and S. R. S. VARADHAN

[1] Diffusion processes with continuous coefficients, Part I, *Comm. Pure Appl. Math.* **22** (1969), 345–400; Part II, *ibid.* 479–530.

[2] On degenerate elliptic-parabolic operators of second order and their associated diffusions, *Comm. Pure Appl. Math.* **25** (1972), 651–713.

P. VAN MOERBEKE

[1] Optimal stopping problems (Conference on Stochastic Differential Equations, Edmonton (Alberta) 1972), *Rocky Mountains J. Math.* **4** (1974), 539–577.

[2] Optimal stopping and free boundary problems, *Arch. Rational Mech. Anal.* to appear.

[3] An optimal stopping problem for linear reward, *Acta Math.* **132** (1974), 1–41.

S. R. S. VARADHAN

[1] On the behavior of the fundamental solution of the heat equation with variable coefficients, *Comm. Pure Appl. Math.* **20** (1967), 431–455.

[2] Diffusion processes in a small time interval, *Comm. Pure Appl. Math.* **20** (1967), 659–685.

A. D. VENTCEL

[1] On the asymptotic behavior of the greatest eigenvalue of a second order elliptic differential operator with a small parameter in the higher derivatives, *Soviet Math. Dokl.* **13** (1972), 13–17 [*Dokl. Akad. Nauk SSSR* **202** (1972), no. 1].

[2] On the asymptotics of eigenvalues of matrices with elements of order $\exp[-V_{ij}/(2\epsilon^2)]$, *Soviet Math. Dokl.* **13** (1972), 65–68 [*Dokl. Akad. Nauk SSSR* **202** (1972), no. 2].

A. D. VENTCEL and M. I. FREIDLIN

[1] On small random perturbations of dynamical systems, *Russian Math. Surveys* **25** (1970), 1–56 [*Uspehi Mat. Nauk.* **25** (1970), 3–55].

Index–Volume 1

Index–Volume 2

A CATALOG OF SELECTED

DOVER BOOKS

IN SCIENCE AND MATHEMATICS

Astronomy

CHARIOTS FOR APOLLO: The NASA History of Manned Lunar Spacecraft to 1969, Courtney G. Brooks, James M. Grimwood, and Loyd S. Swenson, Jr. This illustrated history by a trio of experts is the definitive reference on the Apollo spacecraft and lunar modules. It traces the vehicles' design, development, and operation in space. More than 100 photographs and illustrations. 576pp. 6 3/4 x 9 1/4. 0-486-46756-2

EXPLORING THE MOON THROUGH BINOCULARS AND SMALL TELESCOPES, Ernest H. Cherrington, Jr. Informative, profusely illustrated guide to locating and identifying craters, rills, seas, mountains, other lunar features. Newly revised and updated with special section of new photos. Over 100 photos and diagrams. 240pp. 8 1/4 x 11. 0-486-24491-1

WHERE NO MAN HAS GONE BEFORE: A History of NASA's Apollo Lunar Expeditions, William David Compton. Introduction by Paul Dickson. This official NASA history traces behind-the-scenes conflicts and cooperation between scientists and engineers. The first half concerns preparations for the Moon landings, and the second half documents the flights that followed Apollo 11. 1989 edition. 432pp. 7 x 10. 0-486-47888-2

APOLLO EXPEDITIONS TO THE MOON: The NASA History, Edited by Edgar M. Cortright. Official NASA publication marks the 40th anniversary of the first lunar landing and features essays by project participants recalling engineering and administrative challenges. Accessible, jargon-free accounts, highlighted by numerous illustrations. 336pp. 8 3/8 x 10 7/8. 0-486-47175-6

ON MARS: Exploration of the Red Planet, 1958-1978--The NASA History, Edward Clinton Ezell and Linda Neuman Ezell. NASA's official history chronicles the start of our explorations of our planetary neighbor. It recounts cooperation among government, industry, and academia, and it features dozens of photos from Viking cameras. 560pp. 6 3/4 x 9 1/4. 0-486-46757-0

ARISTARCHUS OF SAMOS: The Ancient Copernicus, Sir Thomas Heath. Heath's history of astronomy ranges from Homer and Hesiod to Aristarchus and includes quotes from numerous thinkers, compilers, and scholasticists from Thales and Anaximander through Pythagoras, Plato, Aristotle, and Heraclides. 34 figures. 448pp. 5 3/8 x 8 1/2. 0-486-43886-4

AN INTRODUCTION TO CELESTIAL MECHANICS, Forest Ray Moulton. Classic text still unsurpassed in presentation of fundamental principles. Covers rectilinear motion, central forces, problems of two and three bodies, much more. Includes over 200 problems, some with answers. 437pp. 5 3/8 x 8 1/2. 0-486-64687-4

BEYOND THE ATMOSPHERE: Early Years of Space Science, Homer E. Newell. This exciting survey is the work of a top NASA administrator who chronicles technological advances, the relationship of space science to general science, and the space program's social, political, and economic contexts. 528pp. 6 3/4 x 9 1/4. 0-486-47464-X

STAR LORE: Myths, Legends, and Facts, William Tyler Olcott. Captivating retellings of the origins and histories of ancient star groups include Pegasus, Ursa Major, Pleiades, signs of the zodiac, and other constellations. "Classic." – *Sky & Telescope.* 58 illustrations. 544pp. 5 3/8 x 8 1/2. 0-486-43581-4

A COMPLETE MANUAL OF AMATEUR ASTRONOMY: Tools and Techniques for Astronomical Observations, P. Clay Sherrod with Thomas L. Koed. Concise, highly readable book discusses the selection, set-up, and maintenance of a telescope; amateur studies of the sun; lunar topography and occultations; and more. 124 figures. 26 halftones. 37 tables. 335pp. 6 1/2 x 9 1/4. 0-486-42820-6

Chemistry

MOLECULAR COLLISION THEORY, M. S. Child. This high-level monograph offers an analytical treatment of classical scattering by a central force, quantum scattering by a central force, elastic scattering phase shifts, and semi-classical elastic scattering. 1974 edition. 310pp. 5 3/8 x 8 1/2. 0-486-69437-2

HANDBOOK OF COMPUTATIONAL QUANTUM CHEMISTRY, David B. Cook. This comprehensive text provides upper-level undergraduates and graduate students with an accessible introduction to the implementation of quantum ideas in molecular modeling, exploring practical applications alongside theoretical explanations. 1998 edition. 832pp. 5 3/8 x 8 1/2. 0-486-44307-8

RADIOACTIVE SUBSTANCES, Marie Curie. The celebrated scientist's thesis, which directly preceded her 1903 Nobel Prize, discusses establishing atomic character of radioactivity; extraction from pitchblende of polonium and radium; isolation of pure radium chloride; more. 96pp. 5 3/8 x 8 1/2. 0-486-42550-9

CHEMICAL MAGIC, Leonard A. Ford. Classic guide provides intriguing entertainment while elucidating sound scientific principles, with more than 100 unusual stunts: cold fire, dust explosions, a nylon rope trick, a disappearing beaker, much more. 128pp. 5 3/8 x 8 1/2. 0-486-67628-5

ALCHEMY, E. J. Holmyard. Classic study by noted authority covers 2,000 years of alchemical history: religious, mystical overtones; apparatus; signs, symbols, and secret terms; advent of scientific method, much more. Illustrated. 320pp. 5 3/8 x 8 1/2. 0-486-26298-7

CHEMICAL KINETICS AND REACTION DYNAMICS, Paul L. Houston. This text teaches the principles underlying modern chemical kinetics in a clear, direct fashion, using several examples to enhance basic understanding. Solutions to selected problems. 2001 edition. 352pp. 8 3/8 x 11. 0-486-45334-0

PROBLEMS AND SOLUTIONS IN QUANTUM CHEMISTRY AND PHYSICS, Charles S. Johnson and Lee G. Pedersen. Unusually varied problems, with detailed solutions, cover of quantum mechanics, wave mechanics, angular momentum, molecular spectroscopy, scattering theory, more. 280 problems, plus 139 supplementary exercises. 430pp. 6 1/2 x 9 1/4. 0-486-65236-X

ELEMENTS OF CHEMISTRY, Antoine Lavoisier. Monumental classic by the founder of modern chemistry features first explicit statement of law of conservation of matter in chemical change, and more. Facsimile reprint of original (1790) Kerr translation. 539pp. 5 3/8 x 8 1/2. 0-486-64624-6

MAGNETISM AND TRANSITION METAL COMPLEXES, F. E. Mabbs and D. J. Machin. A detailed view of the calculation methods involved in the magnetic properties of transition metal complexes, this volume offers sufficient background for original work in the field. 1973 edition. 240pp. 5 3/8 x 8 1/2. 0-486-46284-6

GENERAL CHEMISTRY, Linus Pauling. Revised third edition of classic first-year text by Nobel laureate. Atomic and molecular structure, quantum mechanics, statistical mechanics, thermodynamics correlated with descriptive chemistry. Problems. 992pp. 5 3/8 x 8 1/2. 0-486-65622-5

ELECTROLYTE SOLUTIONS: Second Revised Edition, R. A. Robinson and R. H. Stokes. Classic text deals primarily with measurement, interpretation of conductance, chemical potential, and diffusion in electrolyte solutions. Detailed theoretical interpretations, plus extensive tables of thermodynamic and transport properties. 1970 edition. 590pp. 5 3/8 x 8 1/2. 0-486-42225-9

Engineering

FUNDAMENTALS OF ASTRODYNAMICS, Roger R. Bate, Donald D. Mueller, and Jerry E. White. Teaching text developed by U.S. Air Force Academy develops the basic two-body and n-body equations of motion; orbit determination; classical orbital elements, coordinate transformations; differential correction; more. 1971 edition. 455pp. 5 3/8 x 8 1/2. 0-486-60061-0

INTRODUCTION TO CONTINUUM MECHANICS FOR ENGINEERS: Revised Edition, Ray M. Bowen. This self-contained text introduces classical continuum models within a modern framework. Its numerous exercises illustrate the governing principles, linearizations, and other approximations that constitute classical continuum models. 2007 edition. 320pp. 6 1/8 x 9 1/4. 0-486-47460-7

ENGINEERING MECHANICS FOR STRUCTURES, Louis L. Bucciarelli. This text explores the mechanics of solids and statics as well as the strength of materials and elasticity theory. Its many design exercises encourage creative initiative and systems thinking. 2009 edition. 320pp. 6 1/8 x 9 1/4. 0-486-46855-0

FEEDBACK CONTROL THEORY, John C. Doyle, Bruce A. Francis and Allen R. Tannenbaum. This excellent introduction to feedback control system design offers a theoretical approach that captures the essential issues and can be applied to a wide range of practical problems. 1992 edition. 224pp. 6 1/2 x 9 1/4. 0-486-46933-6

THE FORCES OF MATTER, Michael Faraday. These lectures by a famous inventor offer an easy-to-understand introduction to the interactions of the universe's physical forces. Six essays explore gravitation, cohesion, chemical affinity, heat, magnetism, and electricity. 1993 edition. 96pp. 5 3/8 x 8 1/2. 0-486-47482-8

DYNAMICS, Lawrence E. Goodman and William H. Warner. Beginning engineering text introduces calculus of vectors, particle motion, dynamics of particle systems and plane rigid bodies, technical applications in plane motions, and more. Exercises and answers in every chapter. 619pp. 5 3/8 x 8 1/2. 0-486-42006-X

ADAPTIVE FILTERING PREDICTION AND CONTROL, Graham C. Goodwin and Kwai Sang Sin. This unified survey focuses on linear discrete-time systems and explores natural extensions to nonlinear systems. It emphasizes discrete-time systems, summarizing theoretical and practical aspects of a large class of adaptive algorithms. 1984 edition. 560pp. 6 1/2 x 9 1/4. 0-486-46932-8

INDUCTANCE CALCULATIONS, Frederick W. Grover. This authoritative reference enables the design of virtually every type of inductor. It features a single simple formula for each type of inductor, together with tables containing essential numerical factors. 1946 edition. 304pp. 5 3/8 x 8 1/2. 0-486-47440-2

THERMODYNAMICS: Foundations and Applications, Elias P. Gyftopoulos and Gian Paolo Beretta. Designed by two MIT professors, this authoritative text discusses basic concepts and applications in detail, emphasizing generality, definitions, and logical consistency. More than 300 solved problems cover realistic energy systems and processes. 800pp. 6 1/8 x 9 1/4. 0-486-43932-1

THE FINITE ELEMENT METHOD: Linear Static and Dynamic Finite Element Analysis, Thomas J. R. Hughes. Text for students without in-depth mathematical training, this text includes a comprehensive presentation and analysis of algorithms of time-dependent phenomena plus beam, plate, and shell theories. Solution guide available upon request. 672pp. 6 1/2 x 9 1/4. 0-486-41181-8